ENCYCLOPEDIA OF MATHEMATICS AND ITS APPLICATIONS

EDITED BY G.-C. ROTA

Algorithmic algebraic number theory

ENCYCLOPEDIA OF MATHEMATICS AND ITS APPLICATIONS

6 H. Minc *Permanents*
18 H. O. Fattorini *The Cauchy problem*
19 G. G. Lorentz, K. Jetter, and S. D. Riemenschneider *Birkhoff interpolation*
22 J. R. Bastida *Field extensions and Galois theory*
23 J. R. Cannon *The one-dimensional heat equation*
25 A. Salomaa *Computation and automata*
27 N. H. Bingham, C. M. Goldie, and J. L. Teugels *Regular variation*
28 P. P. Petrushev and V. A. Popov *Rational approximation of real functions*
29 N. White (ed.) *Combinatorial geometries*
30 M. Pohst and H. Zassenhaus *Algorithmic algebraic number theory*
31 J. Aczel and J. Dhombres *Functional equations containing several variables*
32 M. Kuczma, B. Chozewski, and R. Ger *Iterative functional equations*
33 R. V. Ambartzumian *Factorization calculus and geometric probability*
34 G. Gripenberg, S.-O. Londen, and O. Staffans *Volterra integral and functional equations*
35 G. Gasper and M. Rahman *Basic hypergeometric series*
36 E. Torgersen *Comparison of statistical experiments*
37 A. Neumaier *Interval methods for systems of equations*
38 N. Korneichuk *Exact constants in approximation theory*
39 R. A. Brualdi and H. J. Ryser *Combinatorial matrix theory*
40 N. White (ed.) *Matroid applications*
41 S. Sakai *Operator algebras in dynamical systems*
42 W. Hodges *Model theory*
43 H. Stahl and V. Totik *General orthogonal polynomials*
44 R. Schneider *Convex bodies*
45 G. Da Prato and J. Zabczyk *Stochastic equations in infinite dimensions*
46 A. Bjorner, M. Las Vergnas, B. Sturmfels, N. White, and G. Ziegler *Oriented matroids*
47 E. A. Edgar and L. Sucheston *Stopping times and directed processes*
48 C. Sims *Computation with finitely presented groups*
49 T. Palmer *Banach algebras and the general theory of *-algebras I*
50 F. Borceux *Handbook of categorical algebra I*
51 F. Borceux *Handbook of categorical algebra II*
52 F. Borceux *Handbook of categorical algebra III*
54 A. Katok and B. Hasselblatt *Introduction to the modern theory of dynamical systems*
55 V. N. Sachkov *Combinatorial methods in discrete mathematics*
56 V. N. Sachkov *Probabilistic methods in combinatorial analysis*
57 P. M. Cohn *Skew fields*
58 Richard J. Gardner *Geometric tomography*
59 George A. Baker, Jr., and Peter Graves-Morris *Padé approximants*
60 Jan Krajíček *Bounded arithmetic, propositional logic, and complexity theory*
61 H. Groemer *Geometric applications of Fourier series and spherical harmonics*
62 H. O. Fattorini *Infinite dimensional optimization and control theory*
63 A. C. Thompson *Minkowski geometry*
64 R. B. Bapat and T. E. S. Raghavan *Nonnegative matrices and applications*
66 D. Cvetković, P. Rowlinson and S. Simić *Eigenspaces of graphs*

ENCYCLOPEDIA OF MATHEMATICS AND ITS APPLICATIONS

Algorithmic algebraic number theory

M. POHST
University of Düsseldorf

H. ZASSENHAUS
Late, Ohio State University

PUBLISHED BY THE PRESS SYNDICATE OF THE UNIVERSITY OF CAMBRIDGE
The Pitt Building, Trumpington Street, Cambridge CB2 1RP, United Kingdom

CAMBRIDGE UNIVERSITY PRESS
The Edinburgh Building, Cambridge CB2 2RU, United Kingdom
40 West 20th Street, New York, NY 10011-4211, USA
10 Stamford Road, Oakleigh, Melbourne 3166, Australia

© Cambridge University Press 1989

This book is in copyright. Subject to statutory exception
and to the provisions of relevant collective licensing agreements,
no reproduction of any part may take place without
the written permission of Cambridge University Press.

First published 1989
Reprinted 1990, 1993
First paperback edition 1997

Typeset in Times 10/13 pt

A catalogue record for this book is available from the British Library

Library of Congress Cataloguing in Publication data

Pohst, M.
Algorithmic algebraic number theory/M. Pohst and H. Zassenhaus.
p. cm. Bibliography: p.
Includes index.
ISBN 0 521 33060 2
1. Algebraic number theory. 2. Algorithms. I. Zassenhaus, Hans. II. Title
QA247.P58 1989
512′.74—dc19 88-2960 CIP

ISBN 0 521 33060 2 hardback
ISBN 0 521 59669 6 paperback

Transferred to digital printing 2002

CONTENTS

PREFACE

This book is a first step in a new direction: to modify existing theory from a constructive point of view and to stimulate the readers to make their own computational experiments. We are thoroughly convinced that their observations will help to build a new basis from which to venture into new theory on algebraic numbers. History shows that in the long run, number theory always followed the cyclic movement from theory to construction to experiment to conjecture to theory.

Consequently, this book is addressed to all lovers of number theory. On the one hand, it gives a comprehensive introduction to (constructive) algebraic number theory and is therefore especially suited as a textbook for a course on that subject. On the other hand, many parts go far beyond an introduction and make the user familiar with recent research in the field. For experimental number theoreticians we developed new methods and obtained new results (e.g., in the tables at the end of the book) of great importance for them. Both computer scientists interested in higher arithmetic and in the basic makeup of digital computers, and amateurs and teachers liking algebraic number theory will find the book of value.

Many parts of the book have been tested in courses independently by both authors. However, the outcome is not presented in the form of lectures, but, rather, in the form of developed methods and problems to be solved. Algorithms occur frequently throughout the presentation. Though we do not give a thorough definition of an algorithm (but just a rough explanation in 1.1), the underlying idea is that a definite output is obtained from prescribed input data by certain arithmetical rules in a finite number of computational steps. Clearly, an upper bound for the number of those computational steps depending on the input data should be desirable in each case. However, the bounds obtainable for many well-known, frequently used algorithms are completely unrealistic. Hence, we usually do without a complexity analysis.

(The derivation of rough estimates is a good exercise for the reader interested in that topic, however.) This approach is justified by the fact that the algorithms under consideration yield good to excellent results for number fields of small degree and not too large discriminants. In those cases O-estimates are not very helpful in general. Rather, our intention is to make the readers conscious of weak performances of (parts of) algorithms and to strengthen their ability to improve them. From our experiences those weak links in the chain of operations can be detected often only by numerical computation. Hence, we highly recommend the interaction of developing algorithms, observing their performance in practical application, followed by improving them.

Moreover, new algorithms are used to replace older proofs of theorems by means of using their output to show the existence of certain mathematical objects, such as the shortest vector in a lattice, or of a polynomial in the elementary symmetric functions representing an arbitrary symmetric function (principal theorem on symmetric functions). Any such algorithm – respectively, its performance for specified data – yields new observations, giving rise to new conjectures and thus to an improvement of the theory. That is one of the major goals of this book since many of the available numerical invariants of algebraic number fields were already obtained without the use of modern electronic computers. So there is still very little known about algebraic number fields other than abelian extensions of the rational number field.

The contents of the book are divided into six chapters. The first chapter serves as a kind of an introduction. Some basic material (e.g. the Euclidean algorithm, quadratic extensions, Gaussian integers) is to stimulate the readers and to make them curious for more systematic theory. The second chapter gives a self-contained account of Galois theory and elementary prerequisites (e.g. a good knowledge of finite field theory). The reader is introduced to E. Galois' idea of studying the algebraic relations between the roots of a given algebraic equation and thus to recognition of the algebraic background generated by the solutions. Eventually, a method of determining the Galois group of an equation is developed.

The third chapter contains an independent introduction to those parts of the geometry of numbers which will be used in later chapters. Most of Minkowski's classical theorems are presented, as well as some recent reduction methods. The fourth chapter discusses the problem of embedding an equation order into its maximal order, thereby establishing the arithmetical background of a given equation. An algorithm for the computation of an integral basis of an algebraic number field is included. A local account (using valuation theory and the theory of algebraically ordered fields) of the Hilbert–Dedekind–Krull ideal theory is part of the exposition.

The last two chapters deal with the main difference between arithmetics of the rational numbers and of the higher algebraic number fields. Chapter 5 gives a logarithm free proof of Dirichlet's famous unit theorem. It is followed by developing several methods (some new ones) for the computation of the roots of unity and of a full system of fundamental units of an order. In chapter 6 the maximal order of an algebraic number field is studied as a Dedekind ring. We then present efficient methods for the computation of the class number and the class group of an algebraic number field. Primarily they are based on a normal presentation of an ideal by two elements and a fast method for solving norm equations, both of them developed only recently. As an Appendix we present several tables with numerical data concerning the calculation of Galois groups, integral bases, unit groups and class groups.

Chapters 1–4 are essentially self-contained, using only formal results but no conceptual theory of other chapters. The last two chapters rely on the knowledge of parts of chapters 3 and 4; chapter 6 also on parts of chapter 5. Throughout this book, we only assume that the readers have a proper basic knowledge of algebra. Should they not be familiar with some topic supposed to be known they will certainly find it in the book on algebra by S. Lang to which we refer quite frequently in the early chapters. We have also provided a bibliography for each chapter at the end of the book.

We hope to succeed in encouraging some of our readers to engage in enlightened experimentation with numbers and obtain deeper insights into their structure.

M. Pohst
H. Zassenhaus 1987

Preface to paperback edition

Since the first edition of this book in 1989 algorithmic algebraic number theory has developed rapidly. In order to keep the changes to a minimum I have mainly corrected the many typos and errors which have been found. The new developments are sketched in a new chapter with numerous references.

Unfortunately, my coauthor Hans Zassenhaus, one of the pioneers of computational algebraic number theory, passed away on 21 November, 1991. The mathematical community lost one of its outstanding members and a great person.

M. Pohst

ACKNOWLEDGEMENTS

We are much indebted to many students and colleagues for valuable suggestions, criticisms and incentives to do better. In particular we wish to acknowledge the help of D. Shanks, John McKay and J. Buchmann.

We wish to acknowledge the generous support of the production of the manuscript and the research which went into it which we received from the Department of the Ohio State University, the Mathematical Institute of the University of Düsseldorf, the National Science and Engineering Council of Canada and the Centre de Recherches Mathématiques, Université de Montréal.

Our thanks go out to the continued support and interest by the editors of Cambridge University Press, M. Gilchrist and D. Tranah, as well as to the production staff of the Encyclopedia of Mathematics and its Applications.

We would also like to acknowledge the help in proof reading we received from U. Schröter, M. Slawik, J. von Schmettow and essentially by U. Halbritter, and we thank the many secretaries who typed parts of the manuscript.

We were constantly encouraged in completing the work by the support and understanding of our wives Christel and Lieselotte.

My thanks go to Katherine Roegner and to the members of the KANT group, who helped me in locating most of the errors of the first edition. I also thank Cambridge University Press for their kind support during the preparation of the paperback edition.

SYMBOLS USED IN THE TEXT

Symbols used throughout the book are listed in connection with the mathematical terms with which they are associated in the text.

Arithmetic

δ_{ij} is Kronecker's symbol; it is one for $i = j$, zero otherwise;
sign(x) is one for $x > 0$, minus one for $x < 0$, 0 for $x = 0$, $|x| = \text{sign}(x)x$;
$\lfloor x \rfloor$ denotes the largest integer less than or equal to x;
$\lceil x \rceil$ denotes the smallest integer greater than or equal to x;
$\{x\}$ denotes the integer closest to x, for $x + \frac{1}{2} \in \mathbb{Z}$ it is either $x + \frac{1}{2}$ or $x - \frac{1}{2}$;
$a \mid b$ means that there is an element c satisfying $b = ac$;
$a \nmid b$ means that there is no element c satisfying $b = ac$;
$p^k \| b$ means that $p^k \mid b$ and $p^{k+1} \nmid b$;
gcd denotes the greatest common divisor;
lcm denotes the least common multiple;
glb denotes the greatest lower bound;
$a \equiv b \bmod c$ means that $c \mid (a - b)$;
$a = Q(a, b)b + R(a, b)$ denotes division with remainder in a Euclidean ring;
$\gcd(a, b) = X(a, b)a + Y(a, b)b$ denotes a presentation of the *gcd* in a Euclidean ring;
$\text{Re}(a), \text{Im}(a)$ real, respectively imaginary part of $a \in \mathbb{C}$;
$\max\{a_1, \ldots, a_k\}$ denotes $a \in \mathbb{R}$ satisfying $a \in \{a_1, \ldots, a_k\}$ and $a \geqslant a_i$ $(1 \leqslant i \leqslant k)$.

Functions and mappings

Γ, φ, μ denote the Gamma function, Euler φ-function, Möbius μ-function, respectively;
id $= \mathbf{1}$ identity mapping, $\mathbf{1}$ also the identity permutation;
ker denotes the kernel of a homomorphic mapping;
D_t derivation with respect to the variable t;

N, Tr norm, respectively trace;
ind index in finite fields.

Groups

$\mathfrak{S}_n$ symmetric group on n letters;
$\mathfrak{A}_n$ subgroup of $\mathfrak{S}_n$ consisting of all even permutations;
$\mathfrak{V}_4$ the Klein Four Group;
Hol the holomorph of a group;
$\rtimes$ denotes the semidirect product of two groups;
$\wr$ denotes the wreath product;
D_{2n} dihedral group on n letters;
ord(x) order of the group element x.

Matrices

I_n denotes the $n \times n$ unit matrix;
diag$(a_1, \ldots, a_n)$ denotes the $n \times n$ matrix $(a_{ij})_{1 \leqslant i,j \leqslant n}$ with $a_{ii} = a_i$ $(1 \leqslant i \leqslant n)$ and $a_{ij} = 0$ for $i \neq j$;
det(M) determinant of the matrix M;
$H(M)$ Hermite normal form of the matrix M;
$GL(r, \mathbb{Z})$, $SL(r, \mathbb{Z})$ general linear group, special linear group of degree r.

Orders

$\mathfrak{D}(\Lambda/R)$ discriminant ideal of the R-order Λ;
$\mathfrak{D}_0(\Lambda/R)$ reduced discriminant ideal;
AR(Λ) arithmetic radical;
$\mathfrak{E}_i(\Lambda/R)$ elementary ideals ($i \in \mathbb{Z}^{\geqslant 0}$);
$\mathfrak{N}(\Lambda/R)$ exponent ideal;
$(\Lambda :_R \tilde{\Lambda})$ index ideal;
$(\Lambda :_R 0)$ order ideal;
Tor(Λ/R) set of all x of Λ for which there exists a non-zero divisor $\lambda \in R$ such that $\lambda x = 0$.

Polynomials

$d(f)$ discriminant of the polynomial f;
deg(f) degree of f;
$l(f)$ denotes the coefficient of the term of f of highest degree;
Res(f, g) resultant of the polynomials f, g;
$P_{\xi/\mathbb{Q}}$ principal polynomial of ξ over $\mathbb{Q}$;
$M_{\xi/\mathbb{Q}}$ minimal polynomial of ξ over $\mathbb{Q}$.

Rings and fields

$C(R)$ center of the ring R;
$\mathfrak{Q}(R)$ quotient ring of R;
$J(R)$ Jacobson radical;
$\mathrm{NR}(R)$ nilradical;
PID principal ideal domain;
$[A/B]=\{x\in R \mid xB\subseteq A\}$ for subsets A, B of the ring R;
$\simeq_R$ R-isomorphic;
$\dot{+}$ inner direct sum (see 1. (3.8));
$\oplus$ direct sum;
$\otimes_R$ tensor product over R;
$F^\times = F\backslash\{0\}$ for fields F.

Algebraic number fields F

$o_F = \mathrm{Cl}(\mathbb{Z}, F)$ ring of algebraic integers of F;
d_F discriminant of F (respectively of o_F);
$\beta^{(j)}$ jth conjugate of $\beta\in F$;
$U(R)$ unit group of the order R of F, $U_F := U(o_F)$;
$TU(R)$ torsion subgroup (elements of finite order) of $U(R)$;
$\mathrm{Reg}(U(R))$ regulator of $U(R)$, $\mathrm{Reg}_F = \mathrm{Reg}(U_F)$;
I_R semigroup of R-fractional ideals of the order R of F;
H_R group of principal R-fractional ideals;
h_R class number of R, $h_F := h_{o_F}$;
Cl_F class group of F;
$\alpha\sim\beta \Leftrightarrow \alpha/\beta\in U(R)$;
$R_A := \{\beta\in o_F \mid \beta A\subseteq A\}$ for subsets A of o_F.

Special sets of numbers

$\mathbb{N}, \mathbb{P}, \mathbb{Z}$ natural numbers, prime numbers, rational integers, respectively;
$\mathbb{Q}, \mathbb{R}, \mathbb{C}$ rational numbers, real numbers, complex numbers, respectively;
$\mathbb{F}_q$ finite field of $q=p^n$ elements ($p\in\mathbb{P}, n\in\mathbb{N}$); $\mathbb{F}_p = \mathbb{Z}/p\mathbb{Z} = \mathbb{Z}/p$;
$\mathbb{Z}^{\leqslant m}$ all $x\in\mathbb{Z}$ subject to $x\leqslant m$ (analogously $\mathbb{Z}^{\geqslant m}, \mathbb{Z}^{<m}, \mathbb{Z}^{>m}$);
$[a, b]$ interval of real numbers x satisfying $a\leqslant x\leqslant b$.

Other

$|S| = \#S$ denotes the number of elements of the set S;
$\| \;\; \|$ denotes the norm of a vector;
$\langle S\rangle$ denotes the generation of a subgroup, subring, etc. by the elements of the set S;
$\lim\limits_{t\to 1} = \lim\limits_{\substack{t\to 1\\ t>1}}$;
$\square$ marks the end of a proof.

We refer to a formula or theorem of the same chapter by its number $(m.n)$, where m denotes the number of the corresponding section and n the number within that section. Formulae of different chapters are referred to by also listing the number of that chapter, for example 2 (11.12).

1

Basics of constructive algebraic number theory

1.1. Introduction

Algebraic numbers are defined as complex numbers x satisfying an *algebraic equation* of the form

$$a_0x^n + a_1x^{n-1} + \cdots + a_n = 0 \; (n \in \mathbb{N}; a_i \in \mathbb{Z} \; (0 \leqslant i \leqslant n), a_0 \neq 0). \qquad (1.1)$$

We are not satisfied merely with the existence of algebraic numbers such as, for instance,

the natural numbers $1, 2, 3, \ldots (\mathbb{N})$,
rational integers $0, \pm 1, \pm 2, \ldots (\mathbb{Z})$,
rational numbers $0, \pm 1, \pm \frac{1}{2}, \pm 2, \pm \frac{1}{3}, \pm \frac{2}{3}, \pm 3, \ldots (\mathbb{Q})$,
surds $r^{1/n}$ $(r \in \mathbb{Q}, n \in \mathbb{N}, n \geqslant 2)$;

we also inspect the means of constructing them. For this purpose we employ *algorithms.*

We shall not endeavour to give a definition of algorithms in terms of mathematical logics. Our algorithms consist of a stated input, a stated output and a finite number of well-defined steps. The input and output will usually be rational integers and quantities (such as fractions, algebraic integers, integral matrices) derived from rational integers by stated rules. The steps are numbered from 1 to n $(n \in \mathbb{N})$. They consist of statements which use the known data (from the input, or already calculated) to obtain new data by unique mathematical rules. The steps are usually carried out one after the other. But there can also be a jump from step i to step k $(k \neq i + 1)$ depending on the value of some data. The mathematical rules for the computation of new data and for the decision, whether a jump occurs, are fixed throughout the whole algorithm. During the execution of the algorithm a certain step i $(1 \leqslant i \leqslant n)$ may be carried out several times, say $N(i)$ times, depending on the input data. However, all numbers $N(i)$ $(1 \leqslant i \leqslant n)$ must be finite.

The rational integers occuring in the algorithm usually signify what they stand for. Sometimes they denote algebraic numbers, sometimes they signify a decision of questions such as: is this real algebraic number r positive, zero, or negative?

$$\operatorname{sign}(r) = \begin{cases} 1 \text{ if } r > 0 \text{ (i.e. } r \text{ is positive)} \\ 0 \text{ if } r = 0 \\ -1 \text{ if } r < 0 \text{ (i.e. } r \text{ is negative).} \end{cases} \tag{1.2}$$

As an introductory but very instructive example for an algorithm, we present *Euclid's algorithm for rational integers.* For a better understanding of algorithms it is useful to present the underlying ideas in advance. The theory of Euclid's algorithm in $\mathbb{Z}$ is easily explained. For two natural numbers a, b with $a \geqslant b$ there exists a third natural number c subject to

$$bc \leqslant a, \; b(c+1) > a. \tag{1.3}$$

Hence, there exists a non-negative integer d such that

$$a = bc + d \quad \text{and} \quad 0 \leqslant d < b. \tag{1.4}$$

The process (1.4) is called *division with remainder* or, simply, *long division.* The numbers c, d are uniquely determined by a, b. The same procedure for $a, b \in \mathbb{Z}$ instead of a, $b \in \mathbb{N}$ is a little more complicated. One possibility of generalizing (1.3), (1.4) for arbitrary integers a, b, $b \neq 0$, is to stipulate

$$a = bc + d \quad \text{with} \quad 0 \leqslant d < |b|, \quad \operatorname{sign}(c) = \operatorname{sign}(a)\operatorname{sign}(b) \quad \text{for} \quad c, d \in \mathbb{Z}. \tag{1.5}$$

For a number theorist, however, the following generalization is more 'natural' and more useful:

$$a = bc + d \quad \text{and} \quad 0 \leqslant |d| \leqslant \left|\frac{b}{2}\right| \quad (c, d \in \mathbb{Z}). \tag{1.6}$$

It makes the remainder as small as possible in absolute value. On the other hand, IBM chose still another solution for its 360 computer series. Their way of long division for a, b yields c, $d \in \mathbb{Z}$ subject to

$$a = bc + d, \quad 0 \leqslant |d| < |b|, \quad \operatorname{sign}(b) = \operatorname{sign}(d). \tag{1.7}$$

Let us finally remark that there are other computers, such as the CDC Cyber 70 series, without any hardware instruction for integer division.

We just pointed out that long division for rational integers is not unique. However, all possibilities have the following properties in common: for given $a, b \in \mathbb{Z}, b \neq 0$, they determine a *quotient* $c = Q(a, b)$ and a *remainder* $d = R(a, b)$ satisfying

$$a = Q(a,b)b + R(a,b) \quad \text{and} \quad 0 \leqslant |R(a,b)| < |b|. \tag{1.8}$$

If we repeat long division in case $R(a, b) \neq 0$ with b, $R(a, b)$ in place of a, b we obtain a remainder $R(b, R(a, b))$ which is smaller in absolute value than

$R(a, b)$. Going on we get a sequence of remainders becoming smaller and smaller in absolute value. Hence, this process must come to a halt after finitely many steps, the last remainder being zero.

For example, let $a_1 = a$, $a_2 = b \neq 0$, then there are integers $k \in \mathbb{N}$, $a_i \in \mathbb{Z}$ $(1 \leqslant i \leqslant k+2)$, $q_i \in \mathbb{Z}$ $(2 \leqslant i \leqslant k+1)$, subject to

$$a_i = q_{i+1} a_{i+1} + a_{i+2} \tag{1.9}$$

and

$$0 < |a_{i+1}| < |a_i| \quad (2 \leqslant i < k+1), \quad a_{k+2} = 0. \tag{1.10}$$

It is clear that we can specify the a_i $(2 \leqslant i \leqslant k+1)$ analogous to (1.5) or (1.6) or (1.7). For each case, however, we obtain

$$|a_{k+1}| = \gcd(a_1, a_2), \tag{1.11}$$

i.e. $|a_{k+1}|$ is the greatest common divisor of a_1 and a_2 $(\gcd(a_1, a_2))$. Namely, each divisor e of a_1 and a_2 divides a_3 because of (1.9). Inductively we conclude $e | a_{k+1}$. On the other hand, a_{k+1} is a divisor of a_k because of (1.9) and $a_{k+2} = 0$. Applying (1.9) repeatedly it follows that a_{k+1} also divides $a_{k-1}, a_{k-2}, \ldots, a_2, a_1$. Both results together prove $|a_{k+1}| = \gcd(a_1, a_2)$.

Sometimes it does not suffice to compute the greatest common divisor c of two integers a, b but one also needs to determine integers $x = x(a, b)$, $y = y(a, b)$ for its presentation

$$c = xa + yb \tag{1.12}$$

by a and b. (Such a presentation exists since $\mathbb{Z}$ is a principal ideal ring.) This task can be easily incorporated into (1.9), (1.10). At the beginning we have $a_1 = x_1 a_1 + y_1 a_2$, $a_2 = x_2 a_1 + y_2 a_2$ for $x_1 = y_2 = 1$, $x_2 = y_1 = 0$. We show that there exists a presentation $a_i = x_i a_1 + y_i a_2$ at each level i. Namely, after the computation of a_{i+2} from a_{i+1}, a_i according to (1.9) we obtain

$$a_{i+2} = a_i - q_{i+1} a_{i+1} = (x_i - q_{i+1} x_{i+1}) a_1 + (y_i - q_{i+1} y_{i+1}) a_2 \quad \text{for } 1 \leqslant i \leqslant k-1.$$

Therefore the x_i, y_i satisfy the recurrence relations

$$x_{i+2} = x_i - q_{i+1} x_{i+1}, \quad y_{i+2} = y_i - q_{i+1} y_{i+1} \quad (1 \leqslant i \leqslant k-1). \tag{1.13}$$

With these explanations it is easy to write down the following algorithm. We note that $\gcd(0, 0) = 0$ by definition.

Euclid's algorithm with presentation of the gcd (1.14)

Input. $a, b \in \mathbb{Z}$.

Output. c, $x = x(a, b)$, $y = y(a, b) \in \mathbb{Z}$, $c \geqslant 0$ satisfying
$\gcd(a, b) = c = xa + yb$.

Step 1. (Initialization).
For $b = 0$ set $x \leftarrow \operatorname{sign}(a)$, $c \leftarrow xa$, $y \leftarrow 0$ and terminate. Else set $m \leftarrow a$, $n \leftarrow b$, $\tilde{x} \leftarrow 1$, $\tilde{y} \leftarrow 0$, $x \leftarrow 0$, $y \leftarrow 1$.

Step 2. (Long division).
Compute $Q(m,n)$, $R(m,n)$ according to (1.8).
Step 3. (Done?).
For $R(m,n)=0$ set $c \leftarrow \operatorname{sign}(n)\cdot n$, $x \leftarrow \operatorname{sign}(n)\cdot x$, $y \leftarrow \operatorname{sign}(n)\cdot y$ and terminate. Else set $\tilde{\tilde{x}} \leftarrow \tilde{x} - Q(m,n)x$, $\tilde{\tilde{y}} \leftarrow \tilde{y} - Q(m,n)y$ and then $m \leftarrow n$, $n \leftarrow R(m,n)$, $\tilde{x} \leftarrow x$, $\tilde{y} \leftarrow y$, $x \leftarrow \tilde{\tilde{x}}$, $y \leftarrow \tilde{\tilde{y}}$ and go to 2.

For a detailed discussion of Euclid's algorithm we refer to Knuth [1].

We note that the greatest common divisor of two rational integers a, b is unique by definition if it exists. As already stated algorithm (1.14) comes to halt since each decreasing sequence of non-negative integers becomes constant after finitely many terms. *Thus the algorithm actually proves the existence of the gcd.* This observation is one of the guiding principles of our exposition.

Exercises

1. Show by using (1.14): $a = bc$ and $\gcd(a,b)=1 \Rightarrow a \mid c$ for any a, b, $c \in \mathbb{Z}$.
2. Show that any rational number r can be presented in precisely one way in the 'reduced form' $r = a/b (a \in \mathbb{Z}, b \in \mathbb{Z}^{>0}, \gcd(a,b)=1)$. Write an algorithm which transforms $r = \tilde{a}/\tilde{b} (\tilde{a}, \tilde{b} \in \mathbb{Z}, \tilde{b} \neq 0)$ into its reduced form.
3. Let $n \in \mathbb{Z}$ and $a_i \in \mathbb{Z}$ $(1 \leqslant i \leqslant n+1)$. Show that the greatest common divisor of $a_1, \ldots, a_{n+1}$ satisfies the recursive law
$$\gcd(a_1, \ldots, a_{n+1}) = \gcd(\gcd(a_1, \ldots, a_n), a_{n+1}), \tag{1.15}$$
where $\gcd(0, \ldots, 0) = 0$ by definition. Use (1.14) to write an algorithm which computes $\gcd(a_1, \ldots, a_{n+1})$.
4. Let $n \in \mathbb{N}$ and $a_i \in \mathbb{Z}$ $(1 \leqslant i \leqslant n+1)$. Prove the existence of a relation
$$\gcd(a_1, \ldots, a_{n+1}) = \sum_{i=1}^{n+1} x_i a_i \quad (x_i \in \mathbb{Z}, 1 \leqslant i \leqslant n+1). \tag{1.16}$$
Use (1.14) to write an algorithm for determining $x_i = x_i(a_1, \ldots, a_{n+1})$ $(1 \leqslant i \leqslant n+1)$ satisfying (1.16).

1.2. The main task of constructive algebra

Diophantine analysis begins with the problem of solving (1.1) by a rational integer. Constructively speaking: find an algorithm deciding, whether (1.1) has a rational integral solution and if the answer is 'yes' to exhibit a solution $x \in \mathbb{Z}$. For the calculation of x we now interpret the left-hand side of (1.1) as the 'value' of the polynomial
$$f(t) = a_0 t^n + a_1 t^{n-1} + \cdots + a_n \in \mathbb{Z}[t] \tag{2.1}$$
for a fixed ring element (in that case x). Then (1.1) is shortly written as
$$f(x) = 0 \tag{2.2}$$
and x is called *a zero of* $f(t)$ in that case. For $a \in \mathbb{Z}$ we compute $f(a)$ in n steps

by Horner's method. At step j we assume that we have already calculated $S_j = \sum_{i=0}^{j} a_i t^{j-i}$, and for $j < n$ we proceed to S_{j+1} via $S_{j+1} = S_j \cdot a + a_{j+1}$.

Horner's algorithm (2.3)

Input. Coefficients $a_0, \ldots, a_n$ of an nth degree polynomial $f(t) = \sum_{i=0}^{n} a_i t^{n-i}$ $(n \geqslant 1)$ and a number a.
Output. $S = f(a)$.
Step 1. (Initialization). Set $S \leftarrow a_0, j \leftarrow 0$.
Step 2. (Computation of S_j). Set $j \leftarrow j + 1$, $S \leftarrow S \cdot a + a_j$.
Step 3. (Done?). In case $j = n$ terminate, else go to 2.

We note that the algorithm remains valid, if $a, a_0, a_1, \ldots, a_n$ are elements of a commutative ring.

In case $f(t) \in \mathbb{Z}[t]$ Horner's algorithm is very useful in determining all zeros of $f(t)$ in $\mathbb{Z}$. As we saw it suffices to compute $f(a)$ for those $a \in \mathbb{Z}$ dividing a_n, i.e. we must check $f(\pm a)$ for $a | a_n$, $a > 0$. A fast computation of $f(\pm a)$ uses the even and odd part of $f(t)$:

$$f(t) = f_e(t) + f_o(t)$$

defined by

$$f_e(t) = \sum_{\substack{i=0 \\ n-i \equiv 0 \bmod 2}}^{n} a_i t^{n-i}, \quad f_o(t) = \sum_{\substack{i=0 \\ n-i \equiv 1 \bmod 2}}^{n} a_i t^{n-i} \tag{2.4}$$

We recommend it as an excercise to write an algorithm analogous to (2.3) for the computation of $f(\pm a)$ using $f_e(a)$ and $f_o(a)$.

Knowing Horner's method an algorithm for the computation of a solution x of (1.1) (if it exists) is immediate.

Algorithm Diophant (2.5)

Input. Coefficients $a_0, \ldots, a_n$ of $f(t) = \sum_{i=0}^{n} a_i t^{n-i} \in \mathbb{Z}[t]$.
Output. $\varnothing$ if $f(t)$ has no zero in $\mathbb{Z}$, or $a \in \mathbb{Z}$ for which $f(a) = 0$.
Step 1. (Initialization). For $a_n = 0$ set $a \leftarrow 0$ and terminate. Else set $a \leftarrow 1$.
Step 2. (Zero found?). Compute $Q(a_n, a)$, $R(a_n, a)$. For $R(a_n, a) \neq 0$ go to 3. Else calculate $f(\pm a)$, $f(\pm Q(a_n, a))$ by (2.3) and (2.4). If $f(x) = 0$ for $x \in \{\pm a, \pm Q(a_n, a)\}$ set $a \leftarrow x$ and terminate.
Step 3. (Increase a). Set $a \leftarrow a + 1$. For $a^2 \leqslant |a_n|$ go to 2, else print $\varnothing$ and terminate.

(Obviously (2.5) is useful only if $|a_n|$ is small.)

In case (1.1) has no rational integral solution the question is, whether one can find a solution in a larger number system. The number system $\mathbb{Z}$ formed by the rational integers uses two basic operations: addition and multiplication which are connected by certain basic rules of operation (associativity, commutativity, distributivity, invertibility of addition), which, taken together, establish the axioms of a *commutative ring*. We observe that the commutative ring $\mathbb{Z}$ is *unital*, i.e. it contains a neutral element n with respect to multiplication:

$$nx = x = xn \text{ for all } x \in \mathbb{Z} \text{ is satisfied for } n = 1 \neq 0.$$

Not every ring is unital, for example the even integers form the ring $2\mathbb{Z}$ which is contained in $\mathbb{Z}$ as a subring (the operations in $2\mathbb{Z}$ are obtained by restricting the operations of $\mathbb{Z}$ to $2\mathbb{Z}$), but $2\mathbb{Z}$ has no neutral element of multiplication. Since number theory is just swarming with such non-unital rings (viz. the proper ideals of orders) we shall retain the distinction made between commutative unital rings and commutative rings contrary to the usage in S. Lang's algebra book.

However, every ring R (not necessarily commutative) can be embedded into the unital ring $R \oplus \mathbb{Z}$ formed by the couples $x \oplus \lambda$ ($x \in R, \lambda \in \mathbb{Z}$) with the following rules of operation:

$$\left.\begin{aligned}
&x \oplus \lambda = y \oplus \mu \Leftrightarrow x = y, \lambda = \mu;\\
&x \oplus \lambda = y \Leftrightarrow x = y, \lambda = 0;\\
&x \oplus \lambda = \mu \Leftrightarrow x = 0, \lambda = \mu;\\
&(x \oplus \lambda) + (y \oplus \mu) = (x + y) \oplus (\lambda + \mu);\\
&(x \oplus \lambda)(y \oplus \mu) = (xy + \lambda y + \mu x) \oplus \lambda\mu \qquad \text{for all } x, y \in R, \lambda, \mu \in \mathbb{Z}.
\end{aligned}\right\} \quad (2.6)$$

The *permanence principle* as established by Peacock and his British contemporaries implies that any number system should satisfy the axioms of a commutative ring.

In general terms the task of constructive algebra assumes the following form. Let a commutative ring R be given in such a way that for any two of its elements a, b

(i) there is a clearcut answer whether a is equal to b ($a = b$) or whether a, b are distinct ($a \neq b$);
(ii) there are elements $a + b$, $a - b$, ab of R explicitly known (viz. sum, difference, product of a, b) such that the axioms of a commutative ring are satisfied.

Then we say that the *commutative ring R is given constructively*. For example, the rational integer ring $\mathbb{Z}$ as introduced in customary high-school mathematics is constructively given.

Now let us consider an algebraic equation (1.1) with coefficients $a_0, a_1, \ldots, a_n$ in a commutative ring R. Can it be solved in a commutative overring Λ of R, i.e. in a commutative ring Λ containing R such that the operations of Λ restrict to the given operations on the elements of R?

Moreover, can we give Λ constructively in terms of R and exhibit a solution x of (1.1) in Λ? For example, the linear diophantine equation (1.1) for $n = 1$ has a solution in $\mathbb{Z}$ only if a_0 divides a_1. However, there is always a unique solution in terms of the rational number

$$x = \frac{-a_1}{a_0}. \tag{2.7}$$

As a matter of fact, the history of mathematics shows that the desire to solve (1.1) in case $n = 1$ has led to the creation of the rational number system $\mathbb{Q}$ as quotient ring of $\mathbb{Z}$: $\mathbb{Q} = \mathfrak{Q}(\mathbb{Z})$. (In the sequel we denote the *quotient ring* of a commutative ring R – relative to the semigroup formed by the non-zero divisors of R – by $\mathfrak{Q}(R)$.) Again the fractional calculus as taught by well-trained math teachers presents a constructive solution of the task.

Exercises

1. Develop an algorithm to determine all solutions of (1.1) in $\mathbb{Z}$.
2. Examine whether the embedding (2.6) of R into $R \oplus \mathbb{Z}$ is constructive.

1.3. On the construction of overmodules and overrings

The computation of overrings Λ of a given ring R involves the discussion of the additive structure of Λ relative to the multiplicative action of R as well as the multiplicative structure of Λ.

1. Modules

Speaking purely in terms of addition the operation $+$ on Λ satisfies the module axioms (commutativity, associativity and invertibility of addition) which are the additive version of the axioms of a commutative (abelian) group. We refer to the basic definitions and concepts of module theory as contained in chapter 3 of S. Lang, *Algebra*, pp. 74–93. Of course the admission of non-unital rings R leads to a slightly more general concept of (*left*) *R-modules M*:

M is an additive abelian group together with a left multiplication of R and M resulting in a mapping $(R, M) \to M$: $(a, x) \mapsto ax$ satisfying $a(x + y) = ax + ay$, $(a + b)x = ax + bx$, $(ab)x = a(bx)$ for all $a, b \in R$, $x, y \in M$. (3.1)

As an example of the more general definition let us mention the *null-modules* for which every product is defined to be zero.

If R is unital and if

$$1x = x \quad \text{for all } x \in M, \tag{3.2}$$

then M is said to be a *(left) unital R-module.* The last, more special concept is the only one considered by S. Lang. We stipulate that in case R is unital the (left) R-modules considered are unital, too, if not otherwise stated. This convention leaves open the possibility of considering also R-modules over non-unital rings in which case there will be no restrictions.

Regarding direct sums, the direct sum of a family $\{M_i\}_{i\in I}$ of R-modules (I an index set) can be defined categorically as on p. 79 of S. Lang. From the constructive viewpoint it is more satisfactory to define the direct sum

$$M = \prod_{i\in I} M_i \tag{3.3}$$

of the R-module family $\{M_i\}_{i\in I}$ as the set of 'vectors' $(x_i)_{i\in I}$ with 'components' x_i in M_i such that $x_i = 0$ for almost all i, and the rules of vector calculus

$$a(x_i)_{i\in I} = (ax_i)_{i\in I}, \tag{3.4a}$$

$$(x_i)_{i\in I} + (y_i)_{i\in I} = (x_i + y_i)_{i\in I} \tag{3.4b}$$

are used to define the two basic operations.

If the given index set happens to be finite, say, I consists of $1, 2, \ldots, n$ ($n \in \mathbb{N}$), then we also write

$$M = M_1 \oplus M_2 \oplus \cdots \oplus M_n. \tag{3.5}$$

On the other hand, if one discovers that a given R-module M contains two R-submodules M_1, M_2 such that

$$M = M_1 + M_2, \quad M_1 \cap M_2 = \{0\} \tag{3.6}$$

then there is the R-module isomorphism

$$\alpha\colon M_1 \oplus M_2 \to M\colon m_1 \oplus m_2 \mapsto m_1 + m_2, \tag{3.7}$$

but of course it is imprecise and confusing to identify M with $M_1 \oplus M_2$. Therefore we introduce a new symbol $\dot{+}$ to describe the discovered relation (3.6) briefly as

$$M = M_1 \dot{+} M_2. \tag{3.8}$$

For example, the module $M = \mathbb{Z}/2\mathbb{Z} \oplus \mathbb{Z}/2\mathbb{Z}$ is a $\mathbb{Z}$-module of four elements with operation table

	n	b	c	d
n	n	b	c	d
b	b	n	d	c
c	c	d	n	b
d	d	c	b	n

for the elements $n = 0 \oplus 0$ (neutral element).
$b = 1 \oplus 0$
$c = 0 \oplus 1$
$d = 1 \oplus 1$

Using addition we have a module with n as zero element, using multiplication as operation we have the *Klein Four Group* with n as unit element. Adopting module language the only submodule of M other than M and $\{n\}$ and the two direct summands $M_1=\{n,b\}$, $M_2=\{n,c\}$ is the submodule $M_3=\{n,d\}$. It is easy to see that $M=M_1+M_3$, $M_1\cap M_3=\{n\}$ so that $M=M_1\dot{+}M_3$. However, $M_1\oplus M_3$ is a construct which is isomorphic to M but it should not be confused with M.

2. Rings over commutative rings

For the rest of this section R always denotes a commutative ring. An *R-ring* is defined as a ring Λ which is also an R-module such that scalar multiplication and ring multiplication commute:

$$\lambda(ab)=(\lambda a)b=a(\lambda b) \quad \text{for all } \lambda\in R,\ a,b\in\Lambda. \tag{3.9}$$

Examples are the commutative overrings of R, where (3.9) is trivially satisfied.

For constructive purposes the R-rings Λ with a *basis* (over R) are of importance. They are defined as unital R-rings which – considered as R-modules – have an R-basis B. In other words, the module theoretic structure underlying the ring Λ is a *free R-module.* $|B|$ is called the *rank* of the free R-module Λ. The ring structure of Λ is derived from the basis multiplication table:

$$bb'=\sum_{b''\in B}\gamma_{b,b',b''}b''\ (b,b'\in B,\gamma_{b,b',b''}\in R), \tag{3.10a}$$

$$\gamma_{b,b',b''}=0 \text{ for all but a finite number of } b''\in B, \text{ if } b,b'\in B \text{ are fixed.} \tag{3.10b}$$

The associativity of ring multiplication implies the associativity relations

$$\sum_{d\in B}\gamma_{a,b,d}\gamma_{d,c,e}=\sum_{d\in B}\gamma_{a,d,e}\gamma_{b,c,d} \quad (a,b,c,e\in B). \tag{3.10c}$$

Conversely, if a set of elements $\gamma_{b,b',b''}(b,b',b''\in B)$ of R is known for some subset B of Λ satisfying (3.10b) and (3.10c), then it can be interpreted as the set of multiplication constants of the R-ring RB formed by the formal linear combinations

$$\sum_{b\in B}\lambda(b)b, \tag{3.10d}$$

where $\lambda:B\to R$ is a restricted mapping of B into R (i.e. $\lambda(b)=0$ for all but a finite number of $b\in B$) subject to the rules of operations of a free R-module and the multiplication rule

$$\left(\sum_{b\in B}\lambda(b)b\right)\left(\sum_{b'\in B}\mu(b')b'\right)=\sum_{b''\in B}\left(\sum_{b,b'\in B}\lambda(b)\mu(b')\gamma_{b,b',b''}\right)b'', \tag{3.10e}$$

which is implied by (3.10a) and linearity.

Thus a very powerful tool for the construction of R-rings is obtained. However, it must be used with care, since the number of associativity relations to be satisfied is equal to the cube of the number of basis elements so that it becomes impractical to verify them by direct means already for quite small R-dimensions.

It is clear that the commutativity of an R-ring with an R-basis is implied by the commutativity of the multiplication of basis elements, i.e. by the commutativity rules

$$\gamma_{b',b'',b} = \gamma_{b'',b',b} \tag{3.10f}$$

for the multiplication constants.

We shall very frequently consider unital commutative rings Λ containing a given unital commutative ring R such that $1_R = 1_\Lambda$. In such cases we simply speak of a *unital overring* Λ *of* R (but see exercise 7 for an example where $1_R \neq 1_\Lambda$).

3. Polynomial rings

For example, we interpret the polynomial ring $R[t]$ in one variable t over a unital commutative ring R as the R-ring with basis

$$t^0 = 1, t^1 = t, t^2, \ldots,$$

and basis multiplication law

$$t^i t^k = t^{i+k} \quad (i, k \in \mathbb{Z}^{\geqslant 0}),$$

which corresponds to the multiplication constants

$$\gamma_{i,k,m} = \delta_{i+k,m} \quad (i, k, m \in \mathbb{Z}^{\geqslant 0}).$$

(Here $\delta_{x,y}$ is the well-known *Kronecker symbol* which is 1 for $x = y$ and otherwise 0.)

Elementary properties of polynomials are explained in S. Lang, chapter V.3. We denote the images of elements of R under the canonical injection

$$\iota: R \to R[t]: a \mapsto at^0$$

as the *constant polynomials*. For a polynomial $A \in R[t]$ we denote its *degree* by $\deg(A)$ and – in case $A \neq 0$ – its *leading coefficient* by $l(A)$, i.e. $l(A)$ is the coefficient of $t^{\deg(A)}$, hence $l(A) \neq 0$. In case of $l(A) = 1$ the polynomial A is called *monic*.

4. Long division of polynomials

Though the quotient of two polynomials $A, B \in R[t]$ may not exist in $R[t]$, not even if both A and B are distinct from zero, there is a *division with remainder* in $R[t]$. It assumes its simplest form in case $B \neq 0$ and $l(B) = 1$. In that case the following equation holds:

$$A = Q(A, B)B + R(A, B) \tag{3.11a}$$

in $R[t]$ with $Q(A, B)$ a polynomial of degree $\deg(A) - \deg(B)$ in case $\deg(A) \geqslant \deg(B)$, otherwise $Q(A, B) = 0$, and with $R(A, B)$ a polynomial of degree less than $\deg(B)$. (This includes the case $R(A, B) = 0$ since then $\deg(R(A, B)) = -\infty$ by definition.) $R(A, B)$ being zero is tantamount to the divisibility of A by B:

$$B \mid A \Leftrightarrow R(A, B) = 0. \qquad (3.11b)$$

This division with remainder is essentially taught in high-school algebra (see exercise 2).

If $B \neq 0$ but B is not monic, then (3.11a) need not be valid since the leading coefficient of B does not necessarily divide the leading coefficient of A or of intermediary remainders. In that case we have a *modified division with remainder*, called *pseudo-division* for short:

$$L(A, B)A = Q(A, B)B + R(A, B), \qquad (3.12a)$$

where

$$L(A, B) = \begin{cases} 1 \text{ for } \deg(A) < \deg(B) \\ l(B)^{\deg(A) - \deg(B) + 1} \text{ otherwise} \end{cases} \qquad (3.12b)$$

and

$$\deg(R(A, B)) < \deg(B) \quad (\text{or } R(A, B) = 0). \qquad (3.12c)$$

Moreover, the quotient $Q(A, B)$ satisfies the following conditions: if $\deg(L(A, B)A) \geqslant \deg(B)$, then $\deg(Q(A, B)) = \deg(L(A, B)A) - \deg(B)$, but if $\deg(L(A, B)A) < \deg(B)$, then $Q(A, B) = 0$. For a detailed discussion of pseudo-division we refer to Knuth's book [1]. We just note that the reason for adopting the $(\deg(A) - \deg(B) + 1)$th power of $l(B)$ lies in the number of degree reductions by 1 which can be necessary to obtain $\deg(R(A, B)) < \deg(B)$. We observe that pseudo-division assumes the previous form (3.11a) in case B is monic (see exercise 2).

Long division or division with remainder of polynomials is used similarly as long division of rational integers in order to bring about a pseudo-Euclidean division algorithm which similarly as in section 1 leads to an equation

$$F(A, B) = X(A, B)A + Y(A, B)B, \qquad (3.13)$$

where $F(A, B)$ is a well-determined polynomial which always divides the product of A and some other leading coefficients as well as the product of B and some other leading coefficients, but $F(A, B)$ does not necessarily divide A or B themselves. Only in case R is a field, we can be sure that $F(A, B)$ is a common divisor of A, B and then (3.13) tells us that $F(A, B)$ is even a greatest common divisor.

5. Specializations

Let R be a unital commutative ring. The R-homomorphisms ϕ (i.e., $\phi(r) = r$ for all $r \in R$) of $R[t]$ into R or into a unital overring Λ of R are said to be *specializations* of $R[t]$. They are uniquely determined by the image x of t

since for any ring homomorphism,

$$\phi: R[t] \rightarrow \Lambda \text{ subject to } \phi(t) = x, \phi(r) = r \text{ for all } r \in R, \tag{3.14a}$$

we have

$$\phi\Big(\sum_{\text{finite}} r_i t^i\Big) = \sum_{\text{finite}} r_i x^i. \tag{3.14b}$$

See Horner's algorithm (2.3) for the evaluation of a polynomial f at x. It provides the 'value' $f(x)$ of $f \in R[t]$ at x. If f specializes to 0 at x, but f itself is not zero, then x is said to be a *root* or *zero* of f in Λ. The equation

$$f(x) = 0 \tag{3.15a}$$

is tantamount to the statement

$$R(f, t - x)(x) = 0 \quad \text{(to be computed in } \Lambda[t]) \tag{3.15b}$$

for any x of Λ, which implies

$$f(x) = R(f, t - x)(x). \tag{3.15c}$$

Taking for granted a knowledge of elementary properties of ideals and kernels (as expounded in S. Lang, *Algebra*, for example), the kernel of the specialization $\Lambda[t] \rightarrow \Lambda : t \mapsto x$ is the principal ideal of $\Lambda[t]$ generated by $t - x$. It consists of 0 and of all polynomials with root x. Its factor ring is represented by Λ.

The factorring of $R[t]$ modulo the principal ideal $fR[t]$ generated by a non-constant monic polynomial f has an R-basis formed by the $\deg(f)$ residue classes

$$1/f, t/f, \ldots, (t/f)^{\deg(f)-1}, \tag{3.16}$$

where we denote by A/f the residue class of $R[t]$ modulo the principal ideal $fR[t]$ which is represented by $A \in R[t]$. The polynomials of degree less than $\deg(f)$ may be taken as a representative system for $R[t]/fR[t]$ in case of f being monic and non-constant. We will use this construction in later sections frequently.

Exercises

1. Write an algorithm for the multiplication of two polynomials of $\mathbb{Z}[t]$.
2. Write an algorithm for the pseudo-division of polynomials over a unital commutative ring R. Specialize it to the case where the divisor polynomial is monic.
3. Using exercise 2, transfer algorithm (1.14) to the case of polynomial rings with pseudo-division, i.e. develop an algorithm for determining $F(A, B)$, $X(A, B)$, $Y(A, B)$ of (3.13) for any two polynomials $A, B \neq 0$ of $R[t]$, R a unital commutative ring. Test the algorithm for $A(t) = 2t^4 + 5t^3 + 6t^2 + 8t + 10$, $B(t) = 4t^2 + 6t + 3$ and $R = \mathbb{Z}, \mathbb{Z}/2\mathbb{Z}, \mathbb{Z}/8\mathbb{Z}, \mathbb{Z}/12\mathbb{Z}$.

4. Prove that polynomial rings are functorial in the following sense: Let R, Λ be unital commutative rings and $\phi: R \to \Lambda$ a ring homomorphism. Then ϕ extends uniquely to a homomorphism $\bar{\phi}: R[t] \to \Lambda[t]$ such that $\bar{\phi}|_R = \phi$, $\bar{\phi}(t) = t$. Moreover, for any non-constant polynomial $f(t) \in R[t]$ there is a unique homomorphism $\phi_f: R[t]/fR[t] \to \Lambda[t]/\bar{\phi}(f)\Lambda[t]$ satisfying $\phi_f(t/f) = t/\bar{\phi}(f)$.
5. Let Λ be a unital overring of R with R-basis B. Let M be a Λ-module with Λ-basis B'. Prove that $BB' := \{bb' \mid b \in B, b' \in B'\}$ is an R-basis of M considered as an R-module. (This is the so-called *degree theorem*.)
6. Let R be a unital commutative ring.
   (a) Let $R[t_i]$ be the polynomial ring in t_i over R and $R[t_i][t_j]$ the polynomial ring in t_j over $R[t_i]$ $(i,j = 1,2,\dots,n,\ i \neq j; n \in \mathbb{N})$. Show that there is precisely one (standard) isomorphism of $R[t_i][t_j]$ on $R[t_j][t_i]$ fixing R elementwise and mapping t_i on t_i, t_j on t_j. (Write $R[t_i, t_j]$ or $R[t_j, t_i]$.)
   (b) Similarly for $R[t_i][t_j, t_k]$ and $R[t_i, t_j][t_k]$. Write $R[t_i, t_j, t_k]$.
   (c) Interpret the meaning of $R[t_1, t_2, \dots, t_n] = R\,[t_{\pi 1}, t_{\pi 2}, \dots, t_{\pi n}]$ where π is any permutation of $1, 2, \dots, n$. (This is the polynomial ring in n variables $t_1, t_2, \dots, t_n$ over R.)
7. Let R be a unital commutative ring. Construct a unital commutative overring Λ of R such that $1_\Lambda \neq 1_R$. Show that it contains a divisor of zero.

1.4. The ring of an equation

We now consider equation (1.1) over a unital commutative ring R (instead of $\mathbb{Z}$). It implies the equation

$$(a_0 x)^n + a_1 (a_0 x)^{n-1} + \cdots + a_n a_0^{n-1} = 0 \tag{4.1a}$$

upon multiplication by a_0^{n-1}. In case a_0 is not a zero divisor, any solution y of the equation

$$y^n + b_1 y^{n-1} + \cdots + b_n = 0 \tag{4.1b}$$

in R with coefficients

$$b_i = a_i a_0^{i-1} \quad (1 \leqslant i \leqslant n) \tag{4.1c}$$

leads to the solution

$$x = \frac{y}{a_0} \tag{4.1d}$$

in $\mathfrak{Q}(R)$.

Equation (4.1b) is of the same type as (1.1) but with leading coefficient 1, it is said to be a *monic equation*. In case of $R = \mathbb{Z}$ any rational solution is of the form

$$y = \frac{a}{b} \quad (a \in \mathbb{Z}, b \in \mathbb{Z}^{>0}, \gcd(a, b) = 1), \tag{4.2a}$$

which leads to the equation

$$a^n + b_1 a^{n-1} b + \cdots + b_n b^n = 0, \tag{4.2b}$$

implying $b|a^n$ and hence because of (4.2a) $b|a$, i.e. $b = 1$ and $y \in \mathbb{Z}$. It suffices to look for the integral solutions y of (4.1b) which – as we know – are among the finitely many divisors of b_n. They yield all rational solutions of (1.1) in the form (4.1d).

Thus we are led to narrow down the discussion of algebraic equations (1.1) to the monic case. Hence, in the sequel we assume

$$a_0 = 1. \tag{4.3}$$

In section 3 we saw that any solution of (1.1) in any commutative overring Λ of R is a root of the corresponding monic polynomial

$$f(t) = t^n + a_1 t^{n-1} + \cdots + a_n. \tag{4.4}$$

A commutative overring Λ of R in which f has a root may be called a *solution ring* of (1.1). An example for a solution ring is the ring of residue classes of $R[t]$ modulo the principal ideal $fR[t]$ defined at the end of the previous section. The ring $R[t]/fR[t]$ has the following three properties:

(a) It is a unital overring of R.
(b) It is generated by R and the root $x = t/f$ of f. (Indeed, the elements $1, x, \ldots, x^{n-1}$ form an R-basis of $R[t]/fR[t]$.)
(c) For any solution ring Λ of (1.1) and any root y of f in Λ there is a ring homomorphism

$$\phi\colon R[t]/fR[t] \to \Lambda\colon x \mapsto y.$$

A ring with these three properties may be said to be the *ring of the equation* (1.1) (in the sense of category theory) from which all other solution rings are derived by means of homomorphisms. Obviously, it is uniquely determined up to R-isomorphisms. Exercise 4 of section 3 shows the functorial property of the construction. In particular, we always view the construction of the equation ring $R[t]/fR[t]$ on the algebraic background provided by the equation ring $\mathfrak{Q}(R)[t]/f\mathfrak{Q}(R)[t]$ of f over the quotient ring $\mathfrak{Q}(R)$ of R. We see that the natural embedding of R in $\mathfrak{Q}(R)$ is extended to an embedding of $R[t]$ into $\mathfrak{Q}(R)[t]$ which is compatible with residue class formation modulo f so that the residue class x represented by t provides a basis over R, $\mathfrak{Q}(R)$, respectively, consisting of the first $\deg(f)$ powers of x: $1, x, \ldots, x^{\deg(f)-1}$.

Exercises

1. Let R be a unital commutative ring and

$$f(t) = a_0 t^n + a_1 t^{n-1} + \cdots + a_n \in R[t]$$

be non-constant. Show that the natural homomorphism $R \to R[t]/fR[t]: \lambda \mapsto \lambda/f$ is a monomorphism, if and only if the coefficients of f satisfy the following condition: If $a_0 z = 0$ for $z \in R$, then also $a_i z = 0$ $(1 \leqslant i \leqslant n)$. Give an example of a ring R and a polynomial $f \in R[t]$ where these conditions are not satisfied.

1.5. The Gaussian integer ring $\mathbb{Z}[i]$

1. Introduction

C.F. Gauss defined the integers bearing his name as the complex numbers of the form

$$a + bi \quad (a, b \in \mathbb{Z}). \tag{5.1a}$$

It is easily seen that they form a unital commutative entire ring by using the most elementary properties of complex numbers. The additive structure is also easily analyzed, the Gaussian integers form a free $\mathbb{Z}$-module of rank 2 with basis elements 1, i. Hence, the multiplicative structure can be obtained from the multiplication table

$$\begin{array}{c|cc} \cdot & 1 & i \\ \hline 1 & 1 & i \\ i & i & -1 \end{array}. \tag{5.1b}$$

However, if we don't want to use our knowledge of complex numbers, we can use the results of the preceding section. We have already learnt the 'principle behind the thing': The Gaussian integer ring is the equation ring over $\mathbb{Z}$ of

$$x^2 + 1 = 0, \tag{5.2a}$$

with basis elements

$$1 := 1/(t^2 + 1), \quad i := t/(t^2 + 1) \tag{5.2b}$$

of $\mathbb{Z}[t]/(t^2 + 1)\mathbb{Z}[t]$. This information should suffice to derive the elementary properties of the Gaussian integer ring intrinsically. (As a matter of fact C.F. Gauss found the sign determination of certain Gaussian sums which are contained in a ring of complex numbers – similarly constructed as the Gaussian integers – to be a problem with which he could not deal intrinsically but only by means of complex analysis.)

It will appear that the study of the algebraic background and the arithmetic of the Gaussian integer ring provides an instructive model for the study of equation rings, in particular of those corresponding to monic quadratic equations

$$x^2 + a_1 x + a_2 = 0 \tag{5.3}$$

over $\mathbb{Z}$ (or over unital commutative rings of characteristic $\neq 2$).

2. Algebraic background of the Gaussian integers

We embed the Gaussian integer ring into the ring $R := \mathbb{Q}[t]/(t^2+1)\mathbb{Q}[t]$ with basis 1, i over $\mathbb{Q}$ and multiplication table (5.1b). We observe that the factorization

$$t^2 + 1 = (t - i)(t + i) \tag{5.4}$$

holds, which implies that $-i$ is another solution of (5.2a). It is clear that also 1, $-i$ form a basis of R over $\mathbb{Q}$. In other words, the mapping

$$\alpha\colon R \to R\colon a + bi \mapsto a - bi \quad (a, b \in \mathbb{Q}) \tag{5.5}$$

is bijective and preserves both addition and multiplication, it is an *automorphism* of the ring R. It is obvious that α is just the restriction of complex conjugation to $\mathbb{Q} + \mathbb{Q}i$. But from the constructive viewpoint it is important to verify this directly. Because of $\alpha^2 = id_R$ (the *identity mapping* of R) and $\alpha \neq id_R$ the automorphism α is of order 2, it is *involutory*. It is clear that the elements of R which are *invariant under* α (*fixed by* α) are the elements of $\mathbb{Q}$. This implies that any algebraic expression using only arguments of the form $\xi, \alpha(\xi)$ $(\xi \in R)$ in a symmetric way with rational coefficients must be rational.

For example, for $\xi = a + bi \in R$

$$\mathrm{Tr}(\xi) = \xi + \alpha(\xi) = 2a \quad (\text{'}\textit{trace of } \xi\text{'}), \tag{5.6a}$$

$$\mathrm{N}(\xi) = \xi \cdot \alpha(\xi) = a^2 + b^2 \quad (\text{'}\textit{norm of } \xi\text{'}) \tag{5.6b}$$

are both rational. (They correspond to twice the real part and the square of the absolute value of ξ viewed as a complex number.) Trace and norm have the following properties which are easily verified.

$$\left.\begin{array}{ll} \text{The trace is linear:} & \mathrm{Tr}(\xi+\eta) = \mathrm{Tr}(\xi) + \mathrm{Tr}(\eta) \quad \text{for all } \xi, \eta \in R, \\ & \mathrm{Tr}(a\xi) = a\,\mathrm{Tr}(\xi) \quad \text{for all } \xi \in R, a \in \mathbb{Q}. \end{array}\right\} \tag{5.7a}$$

The norm is multiplicative: $\mathrm{N}(\xi\eta) = \mathrm{N}(\xi)\mathrm{N}(\eta)$ for all $\xi, \eta \in R$; (5.7b)
and homogeneous: $\mathrm{N}(a) = a^2$ and $\mathrm{N}(a\xi) = a^2\mathrm{N}(\xi)$ for all $\xi \in R, a \in \mathbb{Q}$.

For all $\xi \in R$: $\mathrm{N}(\xi) \geqslant 0$ and $\mathrm{N}(\xi) = 0 \Leftrightarrow \xi = 0$. (5.7c)

Actually, the equation ring R is a field. That follows from the fact that $(t^2+1)\mathbb{Q}[t]$ is a maximal ideal in $\mathbb{Q}[t]$ because of the irreducibility of t^2+1. But we can also obtain this result intrinsically by noting that for $\xi \in R$, $\xi \neq 0$, we obtain

$$\xi^{-1} = \mathrm{N}(\xi)^{-1}\alpha(\xi) \in R. \tag{5.8}$$

In the sequel we write $\mathbb{Q}(i)$ instead of R indicating that it is the quotient ring of the subring generated by $\mathbb{Q}$ and i. Similarly, we write $\mathbb{Z}[i]$ for the Gaussian integers $\mathbb{Z} + \mathbb{Z}i$. Hence, $\mathbb{Q}(i) = \mathfrak{Q}(\mathbb{Z}[i])$.

Because of (5.6a), (5.6b) any element $\xi \in \mathbb{Q}(i)$ satisfies the quadratic equation

$$\xi^2 - \mathrm{Tr}(\xi)\xi + \mathrm{N}(\xi) = 0 \tag{5.9a}$$

over $\mathbb{Q}$, the corresponding monic polynomial

$$P_{\xi/\mathbb{Q}}(t) = t^2 - \mathrm{Tr}(\xi)t + \mathrm{N}(\xi) \in \mathbb{Q}[t] \tag{5.9b}$$

is said to be the *characteristic polynomial* of ξ over $\mathbb{Q}$. Over $\mathbb{Q}(i)$ it factors into linear factors ('splits') as follows:

$$P_{\xi/\mathbb{Q}}(t) = (t - \xi)(t - \alpha(\xi)). \tag{5.9c}$$

If $\xi \in \mathbb{Q}(i)$ is the root of any monic polynomial $f(t) \in \mathbb{Q}[t]$, then the application of the automorphism α to the equation $f(\xi) = 0$ implies $f(\alpha(\xi)) = 0$. Because of $f = (t - \xi)Q(f, t - \xi)$ we obtain $Q(f, t - \xi)(\alpha(\xi)) = 0$ in case of $\xi \neq \alpha(\xi)$. Hence, in that case f is divisible by $P_{\xi/\mathbb{Q}}$. In other words, for $\xi \in \mathbb{Q}(i) \backslash \mathbb{Q}$ the polynomial $P_{\xi/\mathbb{Q}}$ is the monic polynomial of lowest degree over $\mathbb{Q}$ which has ξ as a root. It is therefore said to be the *minimal polynomial of ξ over* $\mathbb{Q}$ and the equation (5.9a) is called the *minimal equation of ξ over* $\mathbb{Q}$. We denote the minimal polynomial by $M_{\xi/\mathbb{Q}}$:

$$M_{\xi/\mathbb{Q}} = P_{\xi/\mathbb{Q}} \quad \text{for } \xi \in \mathbb{Q}(i) \backslash \mathbb{Q}. \tag{5.10}$$

If, however, ξ is rational, then its minimal polynomial is $M_{\xi/\mathbb{Q}}(t) = t - \xi$, and we have $P_{\xi/\mathbb{Q}} = M_{\xi/\mathbb{Q}}^2$. In that case the minimal equation is simply $\xi - \xi = 0$.

We conclude this discussion by showing that the identity and α are the only automorphisms of $\mathbb{Q}(i)$.

Proposition (5.11)
The automorphism group of $\mathbb{Q}(i)$ consists of $\mathrm{id}_{\mathbb{Q}(i)}$ and α.

Proof
Any automorphism β of $\mathbb{Q}(i)$ must satisfy $\beta(1) = 1$, hence $\beta(n) = n$ for all $n \in \mathbb{N}$. Furthermore we conclude $\beta(-n) = -n$ and finally $\beta(p/q) = p/q$ for all $p, q \in \mathbb{Z}$, $q \neq 0$. Applying β to the equation $i^2 = -1$ we obtain $\beta(i)^2 = -1$ and therefore $\beta(i) = \pm i$. Thus either $\beta = \mathrm{id}_{\mathbb{Q}(i)}$ or $\beta = \alpha$. □

We note that proposition (5.11) does not merely follow from the 2-dimensionality of $\mathbb{Q}(i)$ over $\mathbb{Q}$, as exercise 1 shows.

3. Algebraic background of the 8th roots of unity

The algebraic background of the Gaussian integers shows a group theoretic feature which is revealed in even greater detail in the equation ring of the polynomial $t^4 + 1 \in \mathbb{Z}[t]$. It has four basis elements

$$1 := 1/(t^4 + 1), \quad \zeta_8 = t/(t^4 + 1), \quad i := t^2/(t^4 + 1), \quad i\zeta_8 = t^3/(t^4 + 1) \tag{5.12a}$$

over $\mathbb{Z}$ with multiplication table

$\cdot$	1	ζ_8	i	$i\zeta_8$
1	1	ζ_8	i	$i\zeta_8$
ζ_8	ζ_8	i	$i\zeta_8$	-1
i	i	$i\zeta_8$	-1	$-\zeta_8$
$i\zeta_8$	$i\zeta_8$	-1	$-\zeta_8$	$-i$

(5.12b)

We observe that $\zeta_8^4 = -1, \zeta_8^8 = 1$, hence ζ_8 is a primitive 8th root of unity. The quotient ring $\mathfrak{Q}(\mathbb{Z}[\zeta_8]) = \mathbb{Q}(\zeta_8)$ has the same four basis elements.

It is easily seen that the $\mathbb{Q}$-linear mappings

$$\left.\begin{array}{l} id_R\colon \zeta_8^j \mapsto \zeta_8^j, \quad \rho\colon \zeta_8^j \mapsto \zeta_8^{3j}, \\ \sigma\colon \zeta_8^j \mapsto \zeta_8^{5j}, \quad \tau\colon \zeta_8^j \mapsto \zeta_8^{7j}, \quad (0 \leqslant j \leqslant 7) \end{array}\right\} \tag{5.13a}$$

form automorphisms with the multiplication behavior of Klein's Four Group:

(5.13b)

$\cdot$	id_R	ρ	σ	τ
id_R	id_R	ρ	σ	τ
ρ	ρ	id_R	τ	σ
σ	σ	τ	id_R	ρ
τ	τ	σ	ρ	id_R

The only elements of $R = \mathbb{Q}(\zeta_8)$ which are fixed by all four automorphisms are the rational numbers.

The only elements of R which are fixed by σ are the $\mathbb{Q}$-linear combinations of $1, i$. They form the subfield $\mathbb{Q}(i)$ isomorphic to the quotient ring of the Gaussian integers. As a matter of fact we may look at $\mathbb{Q}(\zeta_8)$ as a $\mathbb{Q}(i)$-ring with basis $1, \zeta_8$ and multiplication table

$\cdot$	1	ζ_8
1	1	ζ_8
ζ_8	ζ_8	i

(5.14a)

Thus we can form the *relative norm* of $\xi = x + \zeta_8 y$ ($x, y \in \mathbb{Q}(i)$):

$$\mathrm{N}_{\mathbb{Q}(\zeta_8)/\mathbb{Q}(i)}(\xi) := \xi\sigma(\xi) = x^2 - iy^2 \in \mathbb{Q}(i). \tag{5.14b}$$

The relative norm is multiplicative and homogeneous of degree 2:

$$\mathrm{N}_{\mathbb{Q}(\zeta_8)/\mathbb{Q}(i)}(\xi\eta) = \mathrm{N}_{\mathbb{Q}(\zeta_8)/\mathbb{Q}(i)}(\xi)\cdot\mathrm{N}_{\mathbb{Q}(\zeta_8)/\mathbb{Q}(i)}(\eta), \tag{5.14c}$$

$$\mathrm{N}_{\mathbb{Q}(\zeta_8)/\mathbb{Q}(i)}(a\xi) = a^2\mathrm{N}_{\mathbb{Q}(\zeta_8)/\mathbb{Q}(i)}(\xi) \quad \text{for all } \xi, \eta \in \mathbb{Q}(\zeta_8), a \in \mathbb{Q}(i). \tag{5.14d}$$

The relative norm vanishes only trivially. Namely, $\mathrm{N}_{\mathbb{Q}(\zeta_8)/\mathbb{Q}(i)}(\xi) = 0$ implies $\xi = 0$ or $(x/y)^2 = i$. In the latter case we obtain $x/y = a + ib \in \mathbb{Q}(i)$ and $(a^2 - b^2) + i2ab = i$. But then $a = \pm b$ ($a, b \in \mathbb{Q}$!) yielding $\pm 2a^2 = 1$ which is impossible over $\mathbb{Q}$.

As above we conclude that every non-zero element of $\mathbb{Q}(\zeta_8)$ has the inverse

$$\xi^{-1} = (\mathrm{N}_{\mathbb{Q}(\zeta_8)/\mathbb{Q}(i)}(\xi))^{-1}\sigma(\xi). \tag{5.15}$$

Therefore $\mathbb{Q}(\zeta_8)$ is a field which we knew already from t^4+1 being irreducible in $\mathbb{Q}[t]$.

Next we consider the automorphism group of $\mathbb{Q}(\zeta_8)$. Any automorphism ω of $\mathbb{Q}(\zeta_8)$ fixes $\mathbb{Q}$. Application of ω to $i^2+1=0$ yields $\omega(i)=\pm i$ and to $\zeta_8^2-i=0$ yields $\omega(\zeta_8)=\zeta_8^j$ $(j\in\{1,3,5,7\})$. Hence, id_R, ρ,σ,τ are the only automorphisms of $R=\mathbb{Q}(\zeta_8)$. We determine the fixed field for each automorphism. For $\xi=a_0+a_1\zeta_8+a_2i+a_3i\zeta_8$ $(a_j\in\mathbb{Q}, 0\leqslant j\leqslant 3)$ the condition $\rho(\xi)=\xi$ is equivalent to $a_1=a_3, a_2=0$. Therefore ξ must be of the form $\xi=a_0+a_1(i+1)\zeta_8$. We observe that $((i+1)\zeta_8)^2=-2$ and so we may write $(i+1)\zeta_8=(-2)^{\frac{1}{2}}$. Hence, the fixed field of ρ is $\mathbb{Q}((-2)^{\frac{1}{2}})$. Similarly, we see that the fixed field of τ is $\mathbb{Q}(2^{\frac{1}{2}})$ (see exercise 2).

These considerations lead to the *Hasse diagram* for the subfields of $\mathbb{Q}(\zeta_8)$ which is in 1–1-correspondence with the diagram of subgroups of the Klein-Four Group:

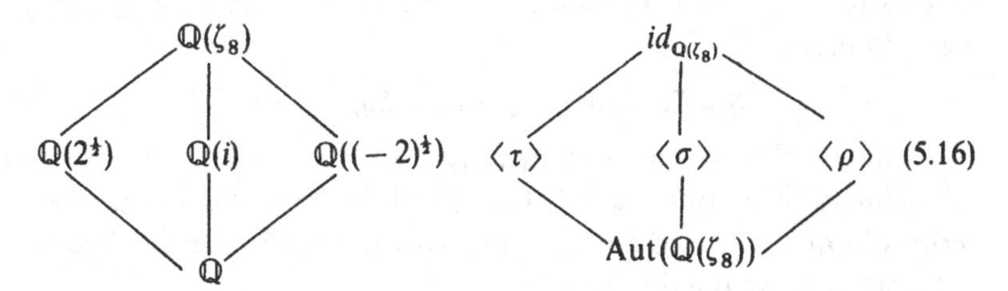

(5.16)

In both diagrams the connection of two items means inclusion. In the field diagram the larger field is marked higher; for the groups it is obviously the opposite. The connection of both diagrams is explained by the principle of *Galois correspondence*: to each subgroup in the subgroup diagram there corresponds that field of the Hasse diagram which is fixed under this subgroup; and to every subfield F of the Hasse diagram there corresponds that subgroup of $\mathrm{Aut}(\mathbb{Q}(\zeta_8))$ which is the relative automorphism group of $\mathbb{Q}(\zeta_8)$ over F.

Can there be subfields F of $\mathbb{Q}(\zeta_8)$ which are not contained in the Hasse diagram above? By exercise 5 of section 3 the field $\mathbb{Q}(\zeta_8)$ would have two basis elements over F, e.g. $1, \zeta_8$. Hence, there would be a monic irreducible polynomial $f(t)=t^2+at+b\in F[t]$ satisfying $f(\zeta_8)=0$. This implies $\gcd(f(t), t^4+1)=f(t)$ in $F[t]$ and therefore $f(t)=(t-\zeta_8)(t-\zeta_8^j)$ $(j\in\{3,5,7\})$. But then F is already one of the subfields $\mathbb{Q}(i)$, $\mathbb{Q}(2^{\frac{1}{2}})$, $\mathbb{Q}((-2)^{\frac{1}{2}})$ of the Hasse diagram.

The constructive realization of the Galois correspondence is obtained by means of a *normal basis* of $\mathbb{Q}(\zeta_8)$ over $\mathbb{Q}$, i.e. of a basis consisting of the images ('*conjugates*') of one element of $\mathbb{Q}(\zeta_8)$ under $\mathrm{Aut}(\mathbb{Q}(\zeta_8))$. The conjugates of ζ_8 itself are $\zeta_8=id_R(\zeta_8)$, $\zeta_8^3=i\zeta_8=\rho(\zeta_8)$, $\zeta_8^5=-\zeta_8=\sigma(\zeta_8)$, $\zeta_8^7=-i\zeta_8=\tau(\zeta_8)$. Obviously they do not form a $\mathbb{Q}$-basis. But the conjugates of

$\zeta = 1 + \zeta_8 + \zeta_8^2 = 1 + i + \zeta_8$ are

$$\left.\begin{array}{ll} \zeta_{id} := \zeta = 1 + \zeta_8 + \zeta_8^2, & \zeta_\rho := \rho(\zeta) = 1 - i + i\zeta_8, \\ \zeta_\sigma := \sigma(\zeta) = 1 + i - \zeta_8, & \zeta_\tau := \tau(\zeta) = 1 - i - i\zeta_8, \end{array}\right\} \tag{5.17a}$$

and they form a normal $\mathbb{Q}$-basis.

Suppose now, a subgroup S of $G = \mathrm{Aut}(\mathbb{Q}(\zeta_8))$ is given and we want to find out which elements ξ of $\mathbb{Q}(\zeta_8)$ are fixed by S; ξ is necessarily of the form

$$\xi = \sum_{g \in G} \lambda(g)\xi_g \quad (\lambda(g) \in \mathbb{Q}), \tag{5.17b}$$

and we wish to solve $\omega(\xi) = \xi$ for every $\omega \in S$. But for a normal basis we have $\omega(\zeta_g) = \zeta_{\omega g}$ for all $g \in G$ and we obtain

$$\omega(\xi) = \sum_{g \in G} \lambda(g)\xi_{\omega g}. \tag{5.17c}$$

To satisfy $\omega(\xi) = \xi$ the mapping λ must be constant on each S-right coset of G. This remark gave C.F. Gauss the idea to form the sums $\sum_{\omega \in S} \xi_{\omega g_j}$ over the right cosets

$$S = Sg_1 = Sid_{\mathbb{Q}(\zeta_8)}, Sg_2, \ldots, Sg_r \quad (r := (G:S)) \tag{5.17d}$$

of G modulo S represented by $g_1, \ldots, g_r$. They are called the *Gauss periods* belonging to the normal basis ζ_g ($g \in G$) and the subgroup S of G. The Gauss periods form a standard basis for the fixed field of S over the field $\mathbb{Q}$.

In our case we obtain

$$\left.\begin{array}{l} 2 + 2i, 2 - 2i \quad \text{for } \mathbb{Q}(i), \\ 2 + (-2)^{\frac{1}{2}}, 2 - (-2)^{\frac{1}{2}} \quad \text{for } \mathbb{Q}((-2)^{\frac{1}{2}}) \\ 2 + 2^{\frac{1}{2}}, 2 - 2^{\frac{1}{2}} \quad \text{for } \mathbb{Q}(2^{\frac{1}{2}}), \\ 1 + i + \zeta_8, 1 - i + \zeta_8^3, 1 + i + \zeta_8^5, 1 - i + \zeta_8^7 \quad \text{for } \mathbb{Q}(\zeta_8) \text{ (see exercise 3).} \end{array}\right\} \tag{5.17e}$$

4. Arithmetic of the Gaussian integers

The concept of divisibility in $\mathbb{Z}[i]$ will be familiar to the reader. For $\xi, \eta \in \mathbb{Z}[i]$, $\xi \neq 0$, we say

$$\xi \textit{ divides } \eta \text{ (`}\xi | \eta\text{')} :\Leftrightarrow \exists \gamma \in \mathbb{Z}[i] : \eta = \xi\gamma. \tag{5.18}$$

To verify the simple rules of divisibility is left as exercise 4. More important is the equivalence relation on $\mathbb{Z}[i] \setminus \{0\}$:

$$\xi \text{ is } \textit{equivalent} \text{ to } \eta \text{ (`}\xi \sim \eta\text{')} :\Leftrightarrow \xi | \eta \wedge \eta | \xi. \tag{5.19}$$

We note that in entire rings R we also say ξ is *associate* to η if ξ/η and η/ξ are in R. It is clear that for $\xi \sim \eta$ the quotient ξ/η is a unit of R. We shall denote the unit group of R by $U(R)$.

Proposition (5.20)

$U(\mathbb{Z}[i]) = \{\pm 1, \pm i\} = \{i^m \mid m = 0, 1, 2, 3\}$.

Proof
If ζ is in $U(R)$, then it has a multiplicative inverse, say $\tilde{\zeta}$. But then $1 = \mathrm{N}(1) = \mathrm{N}(\zeta\tilde{\zeta}) = \mathrm{N}(\zeta)\mathrm{N}(\tilde{\zeta})$ and $\mathrm{N}(\zeta) = 1$ because of (5.7c). On the other hand, $\zeta = a + ib$ $(a, b \in \mathbb{Z})$ and $\mathrm{N}(\zeta) = 1$ is equivalent to $a^2 + b^2 = 1$ with $a = \pm 1$, $b = 0$ and $a = 0$, $b = \pm 1$ as only solutions. That ± 1, $\pm i$ are in $U(\mathbb{Z}[i])$ is obvious. □

Is there some kind of long division in $\mathbb{Z}[i]$? We shall prove that the Gaussian integers are a *Euclidean ring* R, i.e. a unital commutative entire ring with a *height function* $E: R\backslash\{0\} \to \mathbb{N}$ such that for $a, b \in R$, $b \neq 0$, there always exist $c, d \in R$ subject to $a = bc + d$ and either $d = 0$ or $E(d) < E(b)$. It is clear that this assumption suffices to establish a Euclidean algorithm in R which determines $\gcd(a, b)$ in at most $E(b)$ steps.

An appropriate function E for $\mathbb{Z}[i]$ is the norm function.

Proposition (5.21)
$\mathbb{Z}[i]$ is a Euclidean ring using the norm as height function.

Proof
Let α, $\beta \in \mathbb{Z}[i]$, $\beta \neq 0$. We must determine γ, $\delta \in \mathbb{Z}[i]$, viz. $\gamma = \mathrm{Q}(\alpha, \beta)$, $\delta = R(\alpha, \beta)$, such that

$$\alpha = \gamma\beta + \delta, \mathrm{N}(\delta) < \mathrm{N}(\beta). \tag{5.21a}$$

(This also covers the case $\delta = 0$!) (5.21a) is obviously equivalent to

$$\alpha = \gamma\beta + \delta, \mathrm{N}\left(\frac{\alpha}{\beta} - \gamma\right) = \mathrm{N}\left(\frac{\delta}{\beta}\right) < \mathrm{N}\left(\frac{\beta}{\beta}\right) = 1. \tag{5.21b}$$

But $\alpha/\beta = r + is$ $(r, s \in \mathbb{Q})$, $\gamma = c + id$ $(c, d \in \mathbb{Z})$ and (5.21b) becomes

$$\alpha = \gamma\beta + \delta, \quad (r - c)^2 + (s - d)^2 < 1. \tag{5.21c}$$

But this is always solvable via $c = \{r\}$, $d = \{s\}$ ($\{r\}$, $\{s\}$ denoting the closest integers to r, s). Since (5.21a)–(5.21c) are equivalent the proposition is proved. □

It is easy to see that a Euclidean ring R is a *principal ideal ring* (two generators of an ideal can be substituted by their gcd and because of the height function this can happen only a finite number of times, otherwise, the ideal would be $R = 1R$ itself). Principal ideal rings are *factorial* or *unique factorization rings* (see chapter 2, section 4 in S. Lang's book). Hence, every element of $\mathbb{Z}[i]$ is uniquely presentable as the product of a unit and finitely many *prime elements* of $\mathbb{Z}[i]$. For the prime elements of $\mathbb{Z}[i]$ see exercise 6 and – in a more general context chapter 6 section 2.

5. The maximal order property of $\mathbb{Z}[i]$

Any $\xi \in \mathbb{Q}(i)$ is presentable as

$$\xi = \frac{\alpha}{\beta} \quad (\alpha, \beta \in \mathbb{Z}[i], \beta \neq 0, \gcd(\alpha, \beta) = 1). \tag{5.22}$$

This presentation is unique up to equivalence. Analogously as in the beginning of section 4 it is shown that any root of a monic polynomial of $\mathbb{Z}[t]$ in $\mathbb{Q}(i)$ is already in $\mathbb{Z}[i]$ (see exercise 5).

In particular the elements of $\mathbb{Z}[i]$ are characterized among the elements of $\mathbb{Q}(i)$ by the property that they occur as roots of monic polynomials of $\mathbb{Z}[t]$ in $\mathbb{Q}(i)$. Such numbers are called *algebraic integers.*

Proposition (5.23)

$\mathbb{Z}[i]$ consists exactly of the algebraic integers of $\mathbb{Q}(i) = \mathfrak{Q}(\mathbb{Z}[i])$.

The Gaussian integers form a unital subring R of $\mathbb{Q}(i)$ with the two properties

$$\mathfrak{Q}(R) = \mathbb{Q}(i), \tag{5.24a}$$

$$R \text{ is a finitely generated } \mathbb{Z}\text{-module.} \tag{5.24b}$$

Such rings are called *orders.* We shall characterize the orders of $\mathbb{Q}(i)$. Obviously the distinct subrings

$$\mathbb{Z}[ki] := \mathbb{Z} \dotplus k\mathbb{Z}i \quad (k \in \mathbb{N}) \tag{5.25}$$

satisfy (5.24a), (5.24b). We want to show that they are the only orders of $\mathbb{Q}(i)$.

Any order R of $\mathbb{Z}[i]$ must contain $\mathbb{Z}$ and some element $\alpha = a + bi \in \mathbb{Z}[i]$ with $b \neq 0$. Let $\alpha_0 \in R$ such that its imaginary part is as small as possible in absolute value, say $\alpha_0 = a_0 + b_0 i$, $b_0 > 0$ wlog. Then R also contains the element $b_0 i$. For arbitrary $\alpha \in R$, $\alpha = a + bi$, also $\alpha - a - \{b/b_0\} b_0 i$ is in R implying that b is a rational integral multiple of b_0 because of the minimality of b_0 in R. Hence, $R = \mathbb{Z}[b_0 i]$.

In general the additive group R^+ of R is a finitely generated abelian group with 0 as its only element of finite order (*torsion element*). By the basis theorem for abelian groups there is a finite $\mathbb{Z}$-basis of R, say $b_1, \ldots, b_n$. It also serves as $\mathbb{Q}$-basis of $\mathfrak{Q}(R)$, hence $n = 2$ in our case.

There must hold equations

$$b_\nu b_i = \sum_{j=1}^{n} m_{ij}^{(\nu)} b_j \quad (m_{ij}^{(\nu)} \in \mathbb{Z}, 1 \leqslant i, \nu \leqslant n). \tag{5.26}$$

In our case, b_ν satisfies the *Hamilton–Cayley equation* $(b_\nu^2 + a_1 b_\nu + a_2) b_i = 0$ $(i = 1, 2)$ with coefficients $a_1 = -m_{11}^{(\nu)} - m_{22}^{(\nu)}$, $a_2 = m_{11}^{(\nu)} m_{22}^{(\nu)} - m_{12}^{(\nu)} m_{21}^{(\nu)}$ in $\mathbb{Z}$. Therefore $b_\nu^2 + a_1 b_\nu + a_2 = 0$ and b_ν is an algebraic integer of $\mathbb{Q}(i)$ $(\nu = 1, 2)$. Hence, R is contained in $\mathbb{Z}[i]$ and therefore has the form (5.25).

We note that the Gaussian integer ring is maximal among the orders of $\mathbb{Q}(i)$. It is therefore called the *maximal order of* $\mathbb{Q}(i)$.

Exercises

1. Let F be a field of characteristic zero. The *Study numbers* are the algebra with basis elements 1, ε and multiplication table

.	1	ε
1	1	ε
ε	ε	0

over F.

Show that it is unital, commutative and associative and exhibits the infinitely many automorphisms of the Study numbers over the field of reference F.

2. Show that the fixed field of the automorphism τ of $\mathbb{Q}(\zeta_8)$ is $\mathbb{Q}(2^{\frac{1}{2}})$ where $2^{\frac{1}{2}} = (i-1)\zeta_8$.
3. Compute the Gauss periods (5.17e).
4. Verify the following simple properties of divisibility in $\mathbb{Z}[i]$:

(i) $\xi | \xi$,
(ii) $\xi | \eta$ and $\eta | \zeta \Rightarrow \xi | \zeta$,
(iii) $\xi | \eta \Rightarrow \xi | \eta\zeta$,
(iv) $\xi | \eta$ and $\varepsilon | \zeta \Rightarrow \xi\varepsilon | \eta\zeta$,
(v) $\xi | \eta$ and $\xi | \zeta \Rightarrow \xi | (\eta + \zeta) (\xi, \zeta, \eta, \varepsilon \in \mathbb{Z}[i])$.

5. Prove proposition (5.23). Which properties of $\mathbb{Z}[i]$ are essential for the proof?
6. (a) Let π be a prime element of $\mathbb{Z}[i]$. Show that there exists a rational prime number p which is divisible by π. Any prime number $p \in \mathbb{P}$ is divisible by at most two prime elements of $\mathbb{Z}[i]$.
(b) Let $p \in \mathbb{P}$ be an odd prime number. Show by group theoretical arguments that -1 is a square in $\mathbb{Z}/p\mathbb{Z}$ if and only if $p \equiv 1 \bmod 4$.
(c) For prime numbers $p \in \mathbb{P}$ show that the following decomposition law holds:

$$p \sim \begin{cases} \pi^2 & \text{for } p = 2 \\ \pi & \text{for } p \equiv 3 \bmod 4, \\ \pi\pi' & \text{for } p \equiv 1 \bmod 4 \end{cases}$$

where π is a prime element of $\mathbb{Z}[i]$ depending on p and $\pi' = \alpha(\pi)$.
(d) Determine the decomposition of the first ten prime numbers into prime elements of $\mathbb{Z}[i]$.

1.6. Factorial monoids and divisor cascades

The rational integer ring, the Gaussian integer ring, and polynomial rings in one variable over a field are examples of principal entire rings. We observe that the property of unique factorization into irreducible elements of a principal entire ring R essentially depends on the monoid $M = R^{\times}$ formed by the non-zero elements of R under multiplication.

Clearly, the notions of units, equivalence of two elements (notation: $a \sim b$, see (5.19)), irreducible elements, divisibility $(a|b)$, and unique factorization into irreducible elements can be transfered to arbitrary commutative monoids M; exercises 1 and 2 show that the usual properties of divisibility hold there, too. A commutative monoid M is said to be *factorial* if every non-zero element of M has a unique factorization into irreducible elements. Two elements a, b of a factorial monoid M are called *associate* if $a|b$ and $b|a$ (i.e. $a \sim b$). (We note that we have to assume that the set of irreducible elements of M does not contain associate elements, otherwise all factorizations into irreducible elements are unique up to equivalence only.)

Using the same arguments as S. Lang in his book, chapter II, section 4, we conclude that a (commutative) monoid M is factorial if and only if

every properly ascending chain $a_1 M \subset a_2 M \subset \cdots (a_i \in M,\ i \in \mathbb{N})$ *is finite,* (6.1a)

and

for every irreducible element $p \in M$ *which divides* ab $(a, b \in M)$ *either* $p|a$ *or* $p|b$. (6.1b)

In the sequel let M be a factorial monoid. Arithmetic in M can be done via the unique presentation of non-zero elements $a \in M$ in the form

$$a = u \prod_{i=1}^{r} p_i \quad (u \in U(M), r \in \mathbb{Z}^{\geq 0}, p_i \in M \text{ irreducible } (1 \leq i \leq r)). \tag{6.2}$$

(See also exercises 1, 2.)

However, it is usually difficult to obtain such a factorization, even in the case $M = \mathbb{N}$. We shall therefore develop a kind of substitute for (6.2). Let S be a finite subset of M. A (finite!) subset $B(S)$ of M is called a *basis* of S if any $a \in S$ has a unique presentation

$$a = u \prod_{b \in B(S)} b^{m_b} \quad (m_b \in \mathbb{Z}^{\geq 0}, u \in U(M)). \tag{6.3}$$

For example, all irreducible elements of M dividing at least one $a \in S$ form a basis $B(S)$. But in general we can obtain a basis which consists of less elements and is easier to determine than by factorizing all $a \in S$ into irreducible elements. As an example let $M = \mathbb{N}$ and $S = \{14700, 5040\}$, then $B(S) = \{12, 35\}$ will do as we shall see below.

The instrument for computing 'nice' bases are *divisor cascades*, by which we derive from S a set $\delta(S)$ which is closed under division and gcd-formation.

Definition (6.4)

Let S be a non-empty finite subset of a factorial monoid M and $0 \notin S$. Then a **divisor cascade** *$\delta(S)$ of S is a smallest subset of M with the properties:*

(i) $S \subseteq \delta(S)$, $1 \in \delta(S)$;
(ii) *If $a, b \in \delta(S)$, then $\delta(S)$ also contains an element c, $c \sim \gcd(a, b)$.*
(iii) *If a, $b \in \delta(S)$ and $a|b$, then $\delta(S)$ contains an element $c \sim b/a$.*

It is clear that a basis $B(S)$ of S is obtained from those elements of $\delta(S)$ which are not in $U(M)$ and are not divisible by any other $y \in \delta(S)$, $y \sim b$, $y \in U(M)$. Any $x \in \delta(S)$ then has a presentation (6.3).

Example (6.5)
Let $M = \mathbb{N}$ and $S = \{14\,700, 5040\}$. In accordance with (6.4) we compute $\gcd(14\,700, 5040) = 420$, $14\,700/420 = 35$, $5040/420 = 12$, $14\,700/12 = 1225$, $5040/35 = 144$ yielding $\delta(S) = \{14\,700, 5040, 420, 35, 12, 1225, 144, 1\}$. The following diagram needs no further explanation:

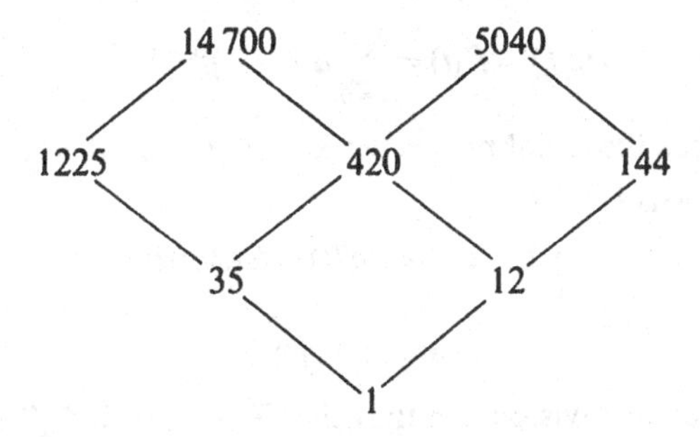

Indeed, we have $B(S) = \{12, 35\} = B(\delta(S))$.

Hence, a basis for a finite subset of M can be computed just by forming quotients and gcds but without any factorization into irreducible elements. In the sequel we assume that we can calculate gcds. In praxis M would have to be a subset of a Euclidean ring for this purpose. We note that for computing $\delta(S)$ associate elements of S should be eliminated.

Algorithm for computation of a divisor cascade (6.6)

Input. Let $S = \{s_1, \ldots, s_r\}$ $(r \in \mathbb{N})$ a subset of a factorial monoid not containing 0.

Output: $\delta(S)$.

Step 1. (Eliminate associate elements and units.) Set $k \leftarrow 1$, $t_k \leftarrow 1$, $\delta(S) \leftarrow \{t_1\}$. For $i = 1, \ldots, r$: check whether s_i is associated to one of the elements of $\delta(S)$; if this is not the case set $k \leftarrow k + 1$, $t_k \leftarrow s_i$, $\delta(S) \leftarrow \delta(S) \cup \{t_k\}$. If $\delta(S) \leqslant 2$, go to 8.

Step 2. (Initialization). Set $A \leftarrow \{(i, j) \mid 2 \leqslant i < j \leqslant k\}$, $B \leftarrow A$.

Step 3. (Choose elements for division). For $A = \emptyset$ go to 7. Else choose $(i, j) \in A$ and set $A \leftarrow A \setminus \{(i, j)\}$.

Step 4. $(s_j | s_i?)$. For $s_j \nmid s_i$ go to 6, else set $s \leftarrow s_i / s_j$.

Step 5. (s associated to an element of $\delta(S)$?). If s is associated to an element of $\delta(S)$ go to 3. Else set $k \leftarrow k + 1$, $t_k \leftarrow s$, $\delta(S) \leftarrow \delta(S) \cup \{t_k\}$, $A \leftarrow A \cup \{(i, k) \mid 2 \leqslant i < k\}$, $B \leftarrow B \cup \{(i, k) \mid 2 \leqslant i < k\}$ and go to 3.

Step 6. $(s_i|s_j?)$. For $s_i\nmid s_j$ go to 3. Else set $s \leftarrow s_j/s_i$ and go to 5.
Step 7. (Choose elements for gcd). For $B \neq \emptyset$ go to 8. Else choose $(i,j)\in B$ and set $B \leftarrow B\backslash\{(i,j)\}$, $s \leftarrow \gcd(s_i, s_j)$ and go to 5.
Step 8. (Done). Set $\delta(S) \leftarrow \delta(S) \cup S$ and terminate.

We leave it as excercise 3 to the reader to extend (6.6) such that also $B(\delta(S))$ is in the output.

If the monoid M consists of polynomials over a factorial ring R then we have additionally the option to factor out squares. For $f(t) = \sum_{i=0}^{n} a_i t^{n-i} \in R[t]$ the *derivation of* f is defined as

$$D_t(f) = f'(t) = \sum_{i=0}^{n-1} a_i(n-i)t^{n-1-i}. \tag{6.7}$$

The derivation has the usual properties known from calculus. Especially, for $f(t) = g^2(t)h(t)$ in $R[t]$ we obtain

$$f'(t) = 2g(t)h(t)g'(t) + h'(t)g^2(t), \tag{6.8a}$$

hence

$$g \,|\, \gcd(f, f') \tag{6.8b}$$

in the sense of pseudo-division. On the other hand, if $\gcd(f, f')$ is a polynomial of positive degree in $R[t]$, say $g(t)$, then $g^2 | f$ (see exercise 4). Therefore, we can use the derivation of a polynomial to exhibit square factors.

With respect to divisor cascading we can therefore add a fourth condition to (6.4), namely:

for $x \in M$, $x \notin U(M)$, *and* $x^2|a$, $a \in \delta(S)$, *then* $\delta(S)$ *contains an element* b *such that* $x|b$, $x^2 \nmid b$. (6.9)

Finally, let M be a subset of a *perfect* field F with $\chi(F) = p > 0$. (In case $\chi(F) = p > 0$ this means $F^p = F$, for $\chi(F) = 0$ it implies no restriction.) In that case we can also extract pth roots from polynomials. Namely,

if f *is a non-constant polynomial with derivative zero, then* $f(t) = g(t^p)$. (6.10)

It is clear how to change (6.6) to obtain $\delta(S)$ including (6.9), (6.10). We just give an illustration.

Example (6.11)
Let $M = \mathbb{Z}/2\mathbb{Z}[t]$, $S = \{s_1\}$, $s_1 = t^6 + t^5 + t^3 + t^2 (= t^2(t+1)^2(t^2+t+1))$. Then the computation of $\delta(S)$ goes as follows:

$$s_2 = s_1' = t^4 + t^2,$$
$$s_3 = t^2 + t \quad (s_3(t^2) = s_2(t)),$$
$$s_4 = t^4 + t + 1 = \frac{s_1}{s_2}$$

implying

$$\delta(S) = \{s_1, s_2, s_3, s_4, 1\}.$$

Exercises

1. Let M be a commutative monoid and $a, b, c, d \in M$. Prove:

   (i) $a|b$ and $a \sim c$, $b \sim d$ implies $c|d$.

   (ii) $a|b$ and $c|d$ implies $ac|bd$.

2. Let M be a factorial monoid and X a non-empty subset of M. Then $x \in M$ is called a *common divisor* of X, if x divides all elements of X. An element $y \in M$ is called *a common multiple* of X, if every $a \in X$ divides y. A *greatest common divisor* (*gcd*) of X is a common divisor which is divisible by all other common divisors of X. A *least common multiple* (*lcm*) of X is a common multiple which divides all other common multiples of X. Prove:

(i) If $x, \tilde{x}$ are two greatest common divisors of X, then they are associate. (The same for least common multiples.) Hence, gcd (X), lcm (X) (if it exists) are unique up to equivalence.

(ii) Let P be a set of non equivalent irreducible elements of M such that $x \in X$ has a presentation

$$x = u_x \prod_{p \in P} p^{\nu(x,p)} \quad (u_x \in U(M), \nu(x,p) \in \mathbb{Z}^{\geq 0}).$$

Then

$$\gcd(X) \sim \prod_{p \in P} p^{\min\{\nu(x,p) | x \in X\}}, \operatorname{lcm}(X) \sim \prod_{p \in P} p^{\max\{\nu(x,p) | x \in X\}} \text{ (if it exists).}$$

3. Modify algorithm (6.6) such that the output also contains a basis $B(S)$ with a minimal number of elements.
4. Let R be a factorial ring and $f(t) \in R[t]$. Prove: If $h(t) := gcd(f, f')$ is non-constant, then $(h(t))^2 | f(t)$ in the sense of pseudo-division.
5. Compute $\delta(S)$ and $B(S)$ for

   (i) $S = \{17\,640, 3366, 3534, 1254, 7158\}$ in $\mathbb{N}$,

   (ii) $S = \{t^5 + t^4 + 5t^3 + 2t^2, t^4 + 7t^3 + 15t^2 + 9t, t^4 + 6t^3 + 11t^2 + 6t\}$ in $\mathbb{Z}[t]$, $\mathbb{Z}/2\mathbb{Z}[t]$, $\mathbb{Z}/3\mathbb{Z}[t]$.

2

The group of an equation

2.1. Splitting rings

1. Introduction

E. Galois pronounced the following theorem:

Soit une équation donnée, dont $a, b, c, \ldots$ sont les m racines. Il y aura toujours un groupe de permutations des lettres $a, b, c, \ldots$ qui jouira de la propriété suivante: 1° que toute fonction des racines, invariables par les substitutions de ce groupe, soit rationellement connue; 2° réciproquement, que toute fonction des racines, détermin-able rationellement, soit invariable par ces substitutions.

in his Mémoire sur les conditions de résolubilité des équations par radicaux (1–16–1831) [4].

E. Galois studied systematically the algebraic relations between the roots of an algebraic equation

$$\left.\begin{array}{l} a_0x^n + a_1x^{n-1} + \cdots + a_n = 0 \\ (n \in \mathbb{Z}^{>0};\ a_i \in R, 0 \leqslant i \leqslant n,\ a_0 \neq 0) \end{array}\right\} \tag{1.1}$$

over unital commutative rings R which most of the time he considered to be subfields of the complex number field for his purpose. Not always, however, since Galois also established finite fields (Galois fields) on the basis of Gauss' earlier work.

As was known previously already they are tied together by the factorization

$$f(t) = a_0 \prod_{i=1}^{n} (t - x_i) \tag{1.2}$$

of the polynomial

$$f(t) = a_0t^n + a_1t^{n-1} + \cdots + a_n \tag{1.3}$$

over a field (*domain of rationality* = 'déterminable rationellement') into the product of the leading coefficient a_0 and the linear factors $t-x_1$, $t-x_2,\ldots,t-x_n$ where the complex numbers $x_1, x_2,\ldots,x_n$ are the roots of f.

We analyze the same problem under the more general assumption that R is any unital commutative ring and nothing else is known in advance. We have seen already many instances where the existence of a factorization like (1.2) does not take place in R. The question arises, whether there are unital overrings Λ of R in which a factorization of f (given by (1.3)) like (1.2) is possible. Any overring of this kind may be said to be a *splitting ring* of f (relative to R). It contains as a subring the ring generated by R and $x_1, x_2,\ldots,x_n$ which is also a splitting ring and which we call a *minimal splitting ring* of f over R.

We deal here only with the case that f is monic:

$$a_0 = 1. \tag{1.4}$$

(However see exercise 1 for the more general case that $a_0 \neq 1$.) Now the equation

$$f(t) = t^n + a_1 t^{n-1} + \cdots + a_n = \prod_{i=1}^{n} (t - x_i) \tag{1.5a}$$

is tantamount to the n Vieta equations

$$\left.\begin{array}{l}
\sigma_1(x_1, x_2,\ldots,x_n) := x_1 + x_2 + \cdots + x_n = -a_1, \\
\sigma_2(x_1, x_2,\ldots,x_n) := \sum\limits_{1\leqslant i<j\leqslant n} x_i x_j = a_2, \\
\cdots\cdots\cdots\cdots\cdots\cdots\cdots\cdots\cdots\cdots\cdots\cdots \\
\sigma_h(x_1, x_2,\ldots,x_n) := \sum\limits_{1\leqslant i_1<i_2<\cdots<i_h\leqslant n} x_{i_1}\cdots x_{i_h} = (-1)^h a_h, \\
\cdots\cdots\cdots\cdots\cdots\cdots\cdots\cdots\cdots\cdots\cdots\cdots \\
\sigma_n(x_1, x_2,\ldots,x_n) := x_1 x_2 \cdots x_n = (-1)^n a_n,
\end{array}\right\} \tag{1.5b}$$

where the polynomial expressions $\sigma_1, \sigma_2,\ldots,\sigma_n$ are said to be the n *basic symmetric functions* of the arguments $x_1,\ldots,x_n$.

2. Universal splitting rings

Theorem (1.6)

There is a **universal splitting ring** $S(f/R)$ *of f relative to R which is obtained by using an R-basis of the $n!$ elements*

$$\xi_1^{i_1}\xi_2^{i_2}\cdots\xi_n^{i_n} \quad (0 \leqslant i_j < j;\ 1 \leqslant j \leqslant n) \tag{1.6a}$$

(referred to as **standard basis** *in the sequel) such that*

$$f(t) = \prod_{j=1}^{n} (t - \xi_j) \tag{1.6b}$$

holds in $S(f/R)[t]$. For every homomorphism

$$\phi: R \to R' \tag{1.6c}$$

of R into a unital commutative ring R' with $\phi(1_R) = 1_{R'}$ and for every splitting ring Λ of the monic polynomial

$$\begin{aligned}\phi f &:= t^n + \phi(a_1)t^{n-1} + \cdots + \phi(a_n)\\ &= \prod_{j=1}^{n} (t - x_j) \quad (x_1, x_2, \ldots, x_n \in \Lambda)\end{aligned} \tag{1.6d}$$

over R' there is an extension

$$\Phi: S(f/R) \to \Lambda \tag{1.6e}$$

of ϕ such that

$$\Phi(\xi_j) = x_j \quad (1 \leqslant j \leqslant n). \tag{1.6f}$$

Proof

If $n = 1$ then $f(t) = t + a_1$ and $R = S(f/R)$ with $\xi_1 = -a_1$ is a universal splitting ring of f with the desired property.

Now let $n > 1$ and assume the construction works for monic polynomials of degree $n - 1$ over unital commutative rings. Let

$$\xi_n := t/f \quad \text{in} \quad R[t]/fR[t].$$

Note that the n elements

$$1, \xi_n, \ldots, \xi_n^{n-1}$$

form an R-basis of $R[t]/fR[t]$. Set

$$S(f/R) = S(Q(f, t - \xi_n)/(R[t]/fR[t])). \tag{1.7}$$

By assumption there are $(n-1)!$ basis elements

$$\xi_1^{i_1} \cdots \xi_{n-1}^{i_{n-1}} \quad (0 \leqslant i_j < j; 1 \leqslant j \leqslant n-1)$$

of $S(f/R)$ over $R[t]/fR[t]$ such that

$$Q(f, t - \xi_n) = \prod_{j=1}^{n-1} (t - \xi_j)$$

holds in $S(f/R)$. Hence (1.6b) is established. Moreover, the $n!$ elements (1.6a) form an R-basis in accordance with exercise 5 of chapter 1, section 3.

Now let ϕ and Λ as in the theorem. It follows that ϕ extends to a homomorphism

$$\begin{aligned}\phi_1 &: R[t]/fR[t] \to \Lambda,\\ \phi_1|_R &= \phi,\\ \phi_1(\xi_n) &= x_n,\end{aligned}$$

as was shown at the end of chapter 1, section 4. For the polynomial ϕf there are two factorizations

$$\begin{aligned}\phi f &= (t - x_n)Q(\phi f, t - x_n)\\ &= (t - x_n)\prod_{j=1}^{n-1} (t - x_j).\end{aligned}$$

But the polynomial $t - x_n$ with leading coefficient 1 is not a zero divisor of $\Lambda[t]$. Hence

$$Q(\phi f, t - x_n) = \prod_{j=1}^{n-1} (t - x_j).$$

By induction hypothesis ϕ_1 extends to a homomorphism

$$\Phi: S(Q(f, t - \xi_n)/(R[t]/fR[t])) \to \Lambda,$$

satisfying

$$\Phi|_{R[t]/fR[t]} = \phi_1,$$
$$\Phi(\xi_j) = x_j \quad (1 \leqslant j < n).$$

But because of (1.7) it follows that Φ satisfies the conditions required by the second assertion of theorem (1.6). □

3. Permutation automorphisms

If R, f are constructively given then so is $S(f/R)$. As a consequence of theorem (1.6) for every permutation π of the indices $1, 2, \ldots, n$ there is a homomorphism

$$\bar{\pi}: S(f/R) \to S(f/R), \tag{1.8a}$$

satisfying

$$\bar{\pi}|_R = id_R, \tag{1.8b}$$
$$\bar{\pi}(\xi_i) = \xi_{\pi(i)} \quad (i = 1, 2, \ldots, n), \tag{1.8c}$$

where we denote by $\pi(i)$ the effect of π on the index i. Note that $\bar{\pi}$ is uniquely determined by (1.8b), (1.8c). For any two permutations π, ρ we have

$$\pi\rho(i) = \pi(\rho(i)),$$
$$\bar{\pi}\bar{\rho}(\xi_i) = \bar{\pi}(\bar{\rho}(\xi_i)) = \bar{\pi}(\xi_{\rho(i)}) = \xi_{\pi\rho(i)},$$
$$\overline{\pi\rho} = \bar{\pi}\bar{\rho},$$

and, in particular,

$$\bar{1} = id_{S(f/R)}$$

for the identity permutation **1**. Thus it follows that

$$\overline{\pi\rho} = \bar{\pi}\bar{\rho},$$

so that the $\bar{\pi}$ form a group of R-automorphisms of $S(f/R)$. It may be called the *group of the permutation automorphisms* of the universal splitting ring of f over R.

Since $1, \xi_2, \xi_3, \ldots, \xi_n$ are among the standard basis of $S(f/R)$ and

$$\xi_1 = -a_1 - \xi_2 - \cdots - \xi_n, \tag{1.9a}$$

it follows from linear independency that the roots $\xi_1, \ldots, \xi_n$ are distinct unless

$$a_1 = 0, \quad n = 2, \quad 2 = 0 \tag{1.9b}$$

in which case $\xi_1 = \xi_2$. In any other case we may view the permutation automorphisms as extensions of the $n!$ root permutations. In other words the mapping of π on $\bar{\pi}$ is a (standard) monomorphism

$$\mu: \mathfrak{S}_n \to \mathrm{Aut}(S(f/R)/R)$$

of the symmetric permutation group $\mathfrak{S}_n$ of n letters formed by all permutations of $1, 2, \ldots, n$ into the group $\mathrm{Aut}(S(f/R)/R)$ of all automorphisms of the universal splitting ring of f over R which leave R elementwise fixed. The image group may be denoted as $\bar{\mathfrak{S}}_n(f/R)$.

There can be other automorphisms of $S(f/R)$ but the permutation automorphisms are of primary interest in Galois theory.

4. Quadratic monic equations

For example, for quadratic equations

$$x^2 + a_1 x + a_2 = 0, \tag{1.10a}$$

we have

$$f(t) = t^2 + a_1 t + a_2 \quad (a_1, a_2 \in R), \tag{1.10b}$$

and

$$S(f/R) = R[t]/fR[t] = R[t]/f \tag{1.10c}$$

as the proof of theorem (1.6) has already shown.

Indeed, setting

$$\left.\begin{array}{l} \xi_2 = t/f, \\ \xi_1 = -a_1 - \xi_2, \end{array}\right\} \tag{1.10d}$$

we find that $1, \xi_2$ is the R-basis of both $R[t]/f$ and $S(f/R)$ with basic multiplication table:

$$\begin{array}{c|cc} \cdot & 1 & \xi_2 \\ \hline 1 & 1 & \xi_2 \\ \xi_2 & \xi_2 & -a_1\xi_2 - a_2 \end{array} \tag{1.10e}$$

and that ξ_1, ξ_2 satisfy (1.6) for $n = 2$.

The permutation automorphisms are the identity mapping $\bar{1} = id_{S(f/R)}$ of $S(f, R)$, and the automorphism

$$\alpha: S(f/R) \to S(f/R): \lambda + \mu\xi_2 \mapsto \lambda + \mu\xi_1 = (\lambda - a_1\mu) - \mu\xi_2 \ (\lambda, \mu \in R). \tag{1.10f}$$

Of course α fixes every element of R. Conversely, every element of $S(f/R)$ which is fixed by α belongs to R when the R-ideals

$$[0/2] = \{\mu \in R \mid 2\mu = 0\}, \quad [0/a_1] = \{\mu \in R \mid \mu a_1 = 0\}$$

have zero intersection. The subring of $S(f/R)$ formed by the fixed elements of α is

$$S(f/R)^\alpha = R + ([0/2] \cap [0/a_1])\xi_2. \tag{1.10g}$$

Among the fixed elements of $\mathfrak{S}_n(f/R)$ we have the trace

$$\mathrm{Tr}(\xi) = \xi + \alpha(\xi) = 2\lambda - a_1\mu \tag{1.10h}$$

and the norm

$$\mathrm{N}(\xi) = \xi\alpha(\xi) = \lambda^2 - a_1\lambda\mu + \mu^2 a_2 \tag{1.10i}$$

of the element

$$\xi = \lambda + \mu\xi_2 \quad (\lambda, \mu \in R) \tag{1.10j}$$

of $S(f/R)$. As we saw already in the case of the Gaussian integers the trace is additive and the norm is multiplicative. Trace and norm are related by

$$\mathrm{Tr}(\xi^2) = \mathrm{Tr}(\xi)^2 - 2\mathrm{N}(\xi). \tag{1.10k}$$

The norm satisfies the additive relation

$$\mathrm{N}(\xi + \eta) + \mathrm{N}(\xi - \eta) = 2\mathrm{N}(\xi) + 2\mathrm{N}(\eta). \tag{1.10l}$$

Trace and norm are important because they provide a natural monic equation, the *characteristic equation*

$$\xi^2 - \mathrm{Tr}(\xi)\xi + \mathrm{N}(\xi) = 0 \tag{1.11a}$$

for every element ξ of $S(f/R)$ over R. The quadratic polynomial

$$P_{\xi/R}(t) = t^2 - \mathrm{Tr}(\xi)t + \mathrm{N}(\xi) \tag{1.11b}$$

is the *characteristic polynomial of* ξ *over* R. The characteristic polynomial factorizes in $S(f/R)$ into the linear factors $t - \xi$, $t - \alpha(\xi)$:

$$P_{\xi/R}(t) = (t - \xi)(t - \alpha(\xi)). \tag{1.11c}$$

Let us consider the special case $R = \mathbb{Q}$. By the usual high-school method we obtain the corresponding *pure quadratic equation*

$$y^2 = \frac{d}{4} \tag{1.12a}$$

for

$$y = x + (a_1/2) \tag{1.12b}$$

and

$$d = d(f) = a_1^2 - 4a_2, \tag{1.12c}$$

where d is called the *discriminant* of the polynomial f because it discriminates between solvability: d is a rational square (number); and insolvability: d is a rational non-square (number) of (1.10a) in $\mathbb{Q}$. If (1.10a) is solvable, then the solutions are

$$x_1 = -\frac{a_1}{2} + \frac{d^{\frac{1}{2}}}{2}, \quad x_2 = -\frac{a_1}{2} - \frac{d^{\frac{1}{2}}}{2}, \tag{1.12d}$$

where $d^{\frac{1}{2}}$ is a root of the pure quadratic equation

$$z^2 - d = 0. \tag{1.12e}$$

Analyzing the splitting ring $S(f/\mathbb{Q})$ we discover that

$$f(t) = (t - x_1)(t - x_2), \tag{1.13a}$$

$$0 = f(\xi_2) = \xi_2^2 + a_1\xi_2 + a_2 = (\xi_2 - x_1)(\xi_2 - x_2), \tag{1.13b}$$

so that there are the zero divisors $\xi_2 - x_1, \xi_2 - x_2$. As a matter of fact there are the two ideals

$$S_i = (\xi_2 - x_i)S(f/\mathbb{Q}) \quad (i = 1, 2) \tag{1.13c}$$

of $S(f/\mathbb{Q})$ which annihilate each other and which are α-conjugate:

$$S_1S_2 = 0, \tag{1.13d}$$

$$S_2 = \alpha(S_1), \quad S_1 = \alpha(S_2). \tag{1.13e}$$

Hence,

$$0 \subset S_i \subset S(f/\mathbb{Q}), \tag{1.13f}$$

$$\dim_{\mathbb{Q}} S_i = 1, \tag{1.13g}$$

$$S_i = \mathbb{Q}(\xi_2 - x_i) \quad (i = 1, 2). \tag{1.13h}$$

Because of the basis property of $1, \xi_2$ we find that

$$S_1 = S_2 \Leftrightarrow x_1 = x_2 \Leftrightarrow d = 0. \tag{1.13i}$$

In this case, when the two roots x_1, x_2 of the quadratic equation coincide, we call it *inseparable*. As (1.13i) shows this is precisely the case for $d = 0$. As a matter of fact over any unital ring of reference R we see that

$$d(f) = a_1^2 - 4a_2 = (\xi_1 + \xi_2)^2 - 4\xi_1\xi_2 = (\xi_1 - \xi_2)^2. \tag{1.13j}$$

Hence by theorem (1.6)

$$d(f) = (x_1 - x_2)^2 \tag{1.13k}$$

in any splitting ring of f over R permitting the factorization (1.2). In general ring theory it can very well happen that $d(f) = 0$ but that the roots x_1, x_2 are distinct in any splitting ring. For example, let $R = (\mathbb{Z}/2\mathbb{Z})[t]/t^2(\mathbb{Z}/2\mathbb{Z})[t]$ with $(\mathbb{Z}/2\mathbb{Z})[t]$-basis $1, \varepsilon$ subject to $\varepsilon^2 = 0$ and let $f(t) = t^2 + \varepsilon t$. From this reason we call the monic quadratic polynomial $f(t) = t^2 + a_1t + a_2$ *separable* if $d(f)$ is not a zero divisor; otherwise we say it is *inseparable*. Now it follows from theorem (1.6) that the roots x_1, x_2 of a separable polynomial are distinct in any splitting ring of f. In the case of inseparability one can find a splitting ring with roots $x_1 = x_2$ in case R is an entire ring. But if R is not an entire ring it can be either way.

Continuing the discussion of above for the case of separability of the rationally solvable monic quadratic equation $f = 0$ we find now that

$$S_1 \neq S_2, \tag{1.13l}$$

$$S(f/R) = S_1 \dot{+} S_2, \tag{1.13m}$$

$$1 = e_1 + e_2, \tag{1.13n}$$

with uniquely determined S_i-components e_i of the unit element $1 = 1_{S(f/\mathbb{Q})}$ such that

$$S_i = 1S_i = (e_1 + e_2)S_i = e_1S_i + e_2S_i = e_iS_i, \tag{1.13o}$$

$$e_1e_2 = 0 = e_2e_1, \tag{1.13p}$$

$$e_i = e_i1 = e_i(e_1 + e_2) = e_i^2. \tag{1.13q}$$

Direct computation tells us that

$$e_1 = (-a_1 - 2x_1)^{-1}(\xi_2 - x_1), \quad e_2 = (-a_1 - 2x_1)^{-1}(\xi_1 - x_1), \tag{1.13r}$$

but the important information is contained in (1.13d), (1.13m–p) telling us that $S(f/\mathbb{Q})$ may be viewed as a 2-vector space over $\mathbb{Q}$ with componentwise multiplication upon writing $\xi \in S(f/\mathbb{Q})$ as

$$\xi = \phi_1(\xi)e_1 + \phi_2(\xi)e_2 \quad (\phi_i(\xi) \in \mathbb{Q}, i = 1, 2),$$

and representing ξ by the $\mathbb{Q}$-vector

$$\phi(\xi) = (\phi_1(\xi), \phi_2(\xi)) \quad (\xi \in S(f/\mathbb{Q})).$$

Thus essentially the solvable separable case provides a universal splitting ring that is algebraically 'decomposed' into the direct sum of two isomorphic copies of $\mathbb{Q}$. We note that there are two factorizations of f in $S(f/\mathbb{Q})$

$$\begin{aligned} f(t) &= (t - x_1e_1 - x_2e_2)(t - x_2e_1 - x_1e_2) \\ &= (t - x_1)(t - x_2). \end{aligned} \tag{1.13s}$$

Which of the two is the standard one (1.6b)?

In case f is inseparable we can even find infinitely many distinct factorizations. What are they? Correspondingly we find infinitely many distinct automorphisms of $S(f/\mathbb{Q})$ over $\mathbb{Q}$.

Finally we turn to the case that $d(f)$ is non-square. In this case there is no factorization

$$f = gh$$

into non-constant factors g, h of $\mathbb{Q}[t]$ possible, the polynomial f is irreducible over $\mathbb{Q}$. The universal splitting ring $S(f/\mathbb{Q})$ contains no divisor of zero. It is a field, called the *splitting field* of f. In this splitting field, however, there holds the factorization

$$f(t) = (t - \xi_1)(t - \xi_2).$$

Also just as above (1.12d) holds so that we obtain

$$\xi_1 = -\frac{a_1}{2} + \frac{d^{\frac{1}{2}}}{2}, \quad \xi_2 = -\frac{a_1}{2} - \frac{d^{\frac{1}{2}}}{2}, \tag{1.14}$$

when the square root of the discriminant is defined as one of the two solutions of the pure quadratic equation (1.12e) for z in $S(f/\mathbb{Q})$. We observe that in this case any $\mathbb{Q}$-homomorphism of $S(f/\mathbb{Q})$ into a splitting ring of f over $\mathbb{Q}$ is a monomorphism.

On the other hand, in case f is reducible (= non-irreducible) over $\mathbb{Q}$ then $\mathbb{Q}$ itself is a splitting ring of f, of a $\mathbb{Q}$-dimension less than the $\mathbb{Q}$-dimension of $S(f/\mathbb{Q})$.

Exercises

1. Let f be an arbitrary non-constant polynomial in t over the unital commutative ring R, say

$$f(t) = a_0 t^n + a_1 t^{n-1} + \cdots + a_n$$
$$(n \in \mathbb{Z}^{>0};\ a_i \in R, 0 \leqslant i \leqslant n;\ a_0 \neq 0).$$

Then show that the natural homomorphism of R into

$$R[t, t_1, \ldots, t_n] \Big/ \left(f - \prod_{i=1}^{n} (t - t_i) \right) R[t, t_1, \ldots, t_n]$$

($t, t_1, \ldots, t_n$ are $n+1$ polynomial variables) which maps λ on $\lambda/(f - \prod_{i=1}^{n}(t - t_i))$ for λ of R is a monomorphism if and only if the coefficients $a_0, a_1, \ldots, a_n$ satisfy the same condition as in exercise 1 of chapter 1, section 4:

If $a_0 z = 0$ $(z \in R)$ then $a_i z = 0$ $(0 < i \leqslant n)$.

2. Let a monic polynomial f of degree $n > 0$ over $\mathbb{Z}$ and a polynomial P of $\mathbb{Z}[t_1, t_2, \ldots, t_n]$ be given. Write an algorithm in order to represent $P(\xi_1, \xi_2, \ldots, \xi_n)$ as a linear combination of the standard basis of $S(f/\mathbb{Z})$.
3. (a) Compute the universal splitting ring of $f(t) = t^3 - 7t - 7$ over $\mathbb{Z}$.
   (b) Show that $\mathbb{Z}[\rho]$ is a splitting ring for f, if ρ denotes a zero of f (in $\mathbb{R}$). (Hint: $3\rho^2 - 5\rho - 14$ is another zero.)
   (c) Determine an epimorphism of $S(f/\mathbb{Z})$ onto $\mathbb{Z}[\rho]$.
   (Compare proposition (4.6).)
4. Find all factorizations of the monic quadratic polynomial

$$f(t) = t^2 + a_1 t + a_2 \in \mathbb{Q}[t]$$

in $S(f/\mathbb{Q})$.

2.2. The fixed subring of the permutation automorphisms

Firstly, we remark that the R-dimension $n!$ grows stronger than exponentially with n since by Stirling's formula

$$n! \sim n^n e^{-n} (2\pi n)^{\frac{1}{2}} \tag{2.1}$$

so that direct computations in $S(f/R)$ become impractical for large values of n. The value of the concept of splitting rings derives from its constructive nature and its use as scaffolding for more efficient constructions.

1. Tschirnhausen transformations

Secondly, our discussion of quadratic equations (1.10a) over $\mathbb{Q}$ amounts to the statement that $S(f/\mathbb{Q})$ is isomorphic to $S(g/\mathbb{Q})$ where

$$g(t) = t^2 - d(f)$$

is a pure quadratic polynomial.

More generally speaking a *Tschirnhausen transformation* [6] of (1.1) with $a_0 = 1$ is defined as the transition from the generator

$$\xi = t/f$$

of the equation ring $R[t]/fR[t]$ to another generator

$$\eta = h(\xi) \tag{2.2a}$$

of $R[t]/fR[t]$ with

$$h \in R[t], \tag{2.2b}$$

$$R[t]/fR[t] = R[\xi] = R[\eta]. \tag{2.2c}$$

As will be shown soon the minimal polynomial m of η is again a monic polynomial of degree n over R and thus

$$R[t]/fR[t] \simeq_R R[t]/mR[t] \tag{2.2d}$$

where the R-isomorphism maps η on t/m.

It follows readily that

$$S(f/R) \simeq_R S(m/R), \tag{2.2e}$$

so that, in essence, we deal with the same problem as before, but the use of m rather than f may confer both conceptual and computational advantages. Note that there is a polynomial j of $R[t]$ such that $\xi = j(\eta)$ so that the transition from η to ξ also is a Tschirnhausen transformation. Prior to E. Galois' revolutionary treatment of algebraic equation theory, for several centuries the algebraists tried to apply a Tschirnhausen transformation to (1.1) over an extension E of $\mathbb{Q}$ obtained by the adjunction of some radicals such that m factors into a product of pure monic polynomials of the form

$$t^{\mu} - \beta \tag{2.2f}$$

with β in E. This is possible for $n = 1, 2, 3, 4$. Ruffini and Abel showed that it cannot be done in general if $n > 4$. A few years later E. Galois revealed the group-theoretic reason for that impossibility and gave a satisfactory criterion as to when it can be done. The simplest Tschirnhausen transformations are obtained by translations:

$$\eta = \xi + a \quad (a \in R). \tag{2.3}$$

If n has an inverse in R then a translation by a_1/n is used to produce a polynomial

$$m = t^n + b_2 t^{n-2} + \cdots,$$

with second highest coefficient 0. This technique is well known from high-school algebra.

Thus our treatment of quadratic equations works over any field in which $1 + 1 \neq 0$. But if R is a field in which

$$1 + 1 = 0$$

(i.e. a field of characteristic 2) then pure equations are inseparable so that there can be no Tschirnhausen transformation of (1.10a) into a pure equation in case

$$a_1 \neq 0.$$

However, the dilatation

$$\eta = a_1^{-1} \xi \tag{2.4}$$

leads to the simplified form (*Artin–Schreier normal form*)

$$\eta^2 + \eta + b_2 = 0, \tag{2.5a}$$

with

$$b_2 = a_1^{-2} a_2. \tag{2.5b}$$

Just like the pure monic polynomial it depends only on one parameter, viz. b_2. If we apply a translation now, say,

$$\zeta = \eta + a, \tag{2.5c}$$

then the second coefficient stays 1, but the last coefficient is modified by the summand $a^2 + a$:

$$\zeta^2 + \zeta + b_2 + a^2 + a = 0. \tag{2.5d}$$

For example over $R = \mathbb{Z}/2\mathbb{Z}$ we have the pure normal form

$$x^2 = 0, \tag{2.6a}$$

and the following Artin–Schreier normal forms

$$x^2 + x + 1 = 0, \tag{2.6b}$$

which is irreducible, and

$$x^2 + x = 0, \tag{2.6c}$$

which is reducible.

For equations of 5th degree defined over a field R of zero characteristic there is the *Bring–Jerrard normal form* [7]

$$x^5 - x + a = 0 \tag{2.7}$$

which can be obtained upon adjunction of specified radicals of degree < 5 for any 5th degree equation even if it cannot be solved by radicals. D. Hilbert analyzed the question how many coefficients $a_1, a_2, \ldots$ can be transformed into 0 for given degree n [6]. But in general the question which normal forms can be devised for equations of degree $n > 6$ after 'trivial' adjunctions (any adjunction of degree $< n$) and a suitable Tschirnhausen transformation and how to construct them, is still unsolved. E. Galois's approach to the task of solving an algebraic equation is radically different, as we shall see.

2. Normal forms of monic quadratic equations over $\mathbb{Z}$

Thirdly, we remark that our treatment of monic quadratic equations yields the following normal forms for $R = \mathbb{Z}$ after suitable translations:

I. Inseparable equations:

$$x^2 = 0. \tag{2.8a}$$

II. Separable equations:

$$\left.\begin{aligned} &\text{(a) } x^2 - \frac{d}{4} = 0 \quad \text{if } d \equiv 0 \bmod 4, \\ &\text{(b) } x^2 - x + \frac{1-d}{4} \quad \text{if } d \equiv 1 \bmod 4. \end{aligned}\right\} \tag{2.8b}$$

We observe that $d = d(f)$ is invariant under translation and that the corresponding factorizations are

I. $$f = t^2 \tag{2.9a}$$

II. $$\left.\begin{aligned} &\text{(a) } f = \left(t - \frac{d^{\frac{1}{2}}}{2}\right)\left(t + \frac{d^{\frac{1}{2}}}{2}\right) \\ &\text{(b) } f = (t - (1 + d^{\frac{1}{2}})/2)(t - (1 - d^{\frac{1}{2}})/2). \end{aligned}\right\} \tag{2.9b}$$

Reducibility occurs if and only if d is a square integer. It turns out that the discriminant is the perfect invariant in the case of quadratic equations over $\mathbb{Z}$.

3. Idempotents

Fourthly, in discussing quadratic equation rings we discovered the important role played by elements e of a (commutative) ring R satisfying the conditions

$$e^2 = e \neq 0. \tag{2.10}$$

They are said to be the *idempotents* of R. The *Peirce right-decomposition* holds:

$$x = ex + (x - ex), \tag{2.11a}$$

with first component

$$x_1 = ex = (ee)x = e(ex), \tag{2.11b}$$

characterized by the property

$$ex_1 = x_1, \tag{2.11c}$$

and second component

$$x_2 = x - ex, \tag{2.11d}$$

characterized by

$$ex_2 = 0, \tag{2.11e}$$

since indeed

$$e(x - ex) = ex - eex = ex - ex = 0.$$

Namely, if

$$x = x_1 + x_2,$$

with summands x_1, x_2 satisfying (2.11c), (2.11e), then we have

$$ex = ex_1 + ex_2 = ex_1 = x_1.$$

Now all the first components form the right-ideal

$$R_1 = eR \tag{2.11f}$$

and the second components form the right-ideal

$$R_2 = \{x - ex | x \in R\} \tag{2.11g}$$

so that

$$R_1 \cap R_2 = 0, \quad R_1 + R_2 = R.$$

Similarly R is represented as the direct sum of the two left-ideals

$$R'_1 = Re, \quad R'_2 = \{x - xe | x \in R\}, \tag{2.11h}$$

derived from the Peirce left-decomposition

$$x = xe + (x - xe). \tag{2.11i}$$

In case the idempotent e is a *central idempotent*, i.e. an idempotent commuting with all elements of R we obtain the *Peirce decomposition* (2.11a), (2.11i) where $xe = ex$ and the presentation

$$R = R_1 \dot{+} R_2 \tag{2.11j}$$

of R as the direct sum of the two ideals

$$R_1 = eR = Re, \quad R_2 = \{x - xe | x \in R\} = \{x - ex | x \in R\}.$$

Which conclusions can we draw regarding the structure of R?

Because of the directness of (2.11j) it follows that

$$R_1 R_2 + R_2 R_1 \subseteq R_1 \cap R_2 = 0, \quad R_1 R_2 = R_2 R_1 = 0. \tag{2.11k}$$

Thus, for every element x of R there is a unique presentation in the form

$$x = x_1 + x_2 \quad (x_i \in R_i, i = 1, 2). \tag{2.11l}$$

Writing

$$x = (x_1, x_2), \tag{2.11m}$$

we have componentwise addition:

$$(x_1, x_2) + (y_1, y_2) = (x_1 + y_1, x_2 + y_2). \tag{2.11n}$$

Moreover, we have also componentwise multiplication:

$$(x_1, x_2)(y_1, y_2) = (x_1 y_1, x_2 y_2). \tag{2.11o}$$

Conversely, if R_1, R_2 are two rings and we deal with the elements (x_1, x_2) of $R = R_1 \oplus R_2$ according to the multiplication rule (2.11o), then we obtain again a ring R into which R_1, R_2 are canonically embedded as ideals yielding R as their direct sum. This ring is said to be the *algebraic sum* of the rings R_1, R_2. This type of sum formations of rings is commutative and associative in the same sense as it is in module theory.

If R is unital then we have

$$1_R = 1_{R_1} + 1_{R_2}$$

where of course

$$1_{R_1} 1_{R_2} = 1_{R_2} 1_{R_1} = 0,$$
$$1_{R_i}^2 = 1_{R_i} \quad (i = 1, 2).$$

Two idempotents e_1, e_2 of a ring are said to be *orthogonal* if

$$e_1 e_2 = 0 = e_2 e_1.$$

The sum of two orthogonal idempotents is an idempotent:

$$(e_1 + e_2)^2 = e_1^2 + e_1 e_2 + e_2 e_1 + e_2^2 = e_1 + e_2, \quad 0 \neq e_1 = e_1(e_1 + e_2)$$

implying $e_1 + e_2 \neq 0$.

Conversely, if for two idempotents e, e_1 one has

$$e e_1 = e_1 e = e_1 \neq e,$$

then

$$e - e_1 =: e_2$$

also is an idempotent, the two idempotents e_1, e_2 are orthogonal and their sum is e.

An idempotent is said to be *primitive*, if it is not the sum of two orthogonal idempotents. Commuting distinct primitive idempotents e_1, e_2 are orthogonal. This is because of

$$e_1 e_2 = e_2 e_1, \quad e_1 = (e_1 - e_1 e_2) + e_1 e_2, \quad e_2 = (e_2 - e_1 e_2) + e_1 e_2. \tag{2.12a}$$

If $e_1 e_2 \neq 0$ then the primitivity of e_1, e_2 requires that $e_1 - e_1 e_2 = 0 = e_2 - e_1 e_2$, hence $e_1 = e_2$ contrary to assumption. Therefore $e_1 e_2 = e_2 e_1 = 0$. If $e_1, e_2, \ldots, e_s$ is a maximal set of commuting pairwise distinct primitive idempotents then any finite sum of some of them, but without repetition, is also an idempotent and the $2^s - 1$ idempotents e thus obtained are the only ones commuting with each e_i $(1 \leqslant i \leqslant s)$.

If the ring R is unital and if 1 is the sum of primitive idempotents of the center $C(R)$ of R (*primitive central idempotents*)

$$1 = e_1 + e_2 + \cdots + e_s,$$

then the $2^s - 1$ non-repetitive sums over the e_i are the only central idempotents of the ring R. Otherwise there are infinitely many idempotents; as a matter of fact there is a sequence of central idempotents $e_1, e_2, \ldots$ for which

$$e_i e_{i+1} = e_i = e_{i+1} e_i \neq e_{i+1} \quad (i = 1, 2, \ldots), \tag{2.12b}$$

and hence there are infinitely many (pairwise) orthogonal idempotents, viz.:

$$e_1 - e_2, e_2 - e_3, \ldots. \tag{2.12c}$$

Thus the behavior of central idempotents under multiplication, complementation and addition of orthogonal idempotents reflects the behaviour of the ideals which they generate under intersection (multiplication), complementation and direct addition in as much as for any two central idempotents e, e' of R we have one of the following five alternatives:

$$e = e'e = e' \colon eR = e'R = eReR, \tag{2.13a}$$

$$ee' = e \neq e' \colon e'R = eR \dotplus (e' - e)R \supset eR, \tag{2.13b}$$

$$ee' = e' \neq e \colon eR = e'R \dotplus (e - e')R \supset e'R, \tag{2.13c}$$

$$\left.\begin{aligned} &ee' \neq e, ee' \neq e', ee' \neq 0 \colon 0 \subset ee'R = eR \cap e'R = eRe'R, \\ &ee'R \subset eR = ee'R \dotplus (e - ee')R, \\ &ee'R \subset e'R = ee'R \dotplus (e' - ee')R, \\ &eR + e'R = (e - ee')R \dotplus ee'R \dotplus (e' - ee')R, \end{aligned}\right\} \tag{2.13d}$$

$$\left.\begin{aligned} &ee' = 0 \colon eR \cap e'R = 0, \\ &(e + e')R = eR \dotplus e'R. \end{aligned}\right\} \tag{2.13e}$$

Note, if the ideal $[0/2] := \{x \in R \mid 2x = 0\}$ is 0, then $e + e'$ is idempotent if and only if e, e' are orthogonal. More generally speaking, any finite set X of distinct commuting idempotents $\{e_1, e_2, \ldots, e_s\}$ generates under the operations:

join ee' to the set if $0 \neq ee'$, $e \neq ee'$, $e' \neq ee'$ and if ee' does not yet belong to the set; (2.14a)

join $e - e'$ to the set if $ee' = e' \neq e$ and if it did not yet belong to it; (2.14b)

join $e + e'$ to the set if $ee' = 0$ and if $e + e'$ did not yet belong to it; (2.14c)

another finite set $\hat{X}$ of at most 7^{s-1} idempotents containing X as a subset and closed under (2.14a–c) (proof by induction over s). Thus we see the importance of idempotents for structural investigations. Let us give two examples.

4. Idempotents and factorization of polynomials

Suppose we find an idempotent $e \neq 1/f$ of the equation ring $\Lambda = R[t]/fR[t]$ of a monic non-constant polynomial f over the unital commutative ring R. Which advantage can we get for the factorization of f? By assumption we have the decomposition

$$1_\Lambda = e_1 + e_2 \tag{2.15a}$$

of 1_Λ into the sum of the orthogonal idempotents

$$e_1 = e, \quad e_2 = 1_\Lambda - e. \tag{2.15b}$$

Hence,

$$e_1 = g_1/f, \quad e_2 = g_2/f \ (g_1, g_2 \in R[t], \deg(g_i) < \deg(f), i = 1, 2), \tag{2.15c}$$

$$1_{R[t]} = g_1 + g_2, \tag{2.15d}$$

$$f \mid g_1 g_2, \quad f \mid g_1(1 - g_1), \quad f \nmid g_1. \tag{2.15e}$$

Conversely, if we succeed in finding any polynomial $g_1 \in R[t]$ satisfying (2.15e) then e_1 as defined by (2.15c) becomes an idempotent and g_1 is uniquely determined by e_1, replacing g_1 by $R(g_1, f)$ if necessary. For example, if there holds a factorization

$$f = f_1 f_2 \tag{2.15f}$$

of f into the product of two monic non-constant polynomials f_1, f_2, such that we also know a presentation

$$1 = x_1 f_1 + x_2 f_2 \tag{2.15g}$$

of the unit element as linear combination of f_1, f_2 over $R[t]$, then

$$e_i = x_i f_i / f \quad (i = 1, 2) \tag{2.15h}$$

form a pair of orthogonal idempotents of $R[t]/fR[t]$ which add up to 1.

5. Idempotents and Chinese remainder theorem

Another example is taken from elementary number theory. It was dealt with in a similar fashion by C.F. Gauss in the *Disquisitiones Arithmeticae* [5].

Let N be a natural number > 1. Suppose we discover an idempotent residue class e of $R = \mathbb{Z}/N$ which is not $1/N$. What does it mean in terms of a factorization of N? As above we have (2.15a), (2.15b). Hence

$$e_1 = g_1/N, \quad e_2 = g_2/N, \tag{2.16a}$$

$$1 = g_1 + g_2 \quad (g_i \in \mathbb{Z}; i = 1, 2), \tag{2.16b}$$

$$N \mid g_1 g_2, \tag{2.16c}$$

$$N \mid g_1(1 - g_1), \quad N \nmid g_1; \tag{2.16d}$$

and conversely. However, now we can form

$$f_1 = \gcd(N, g_1), \quad f_2 = \gcd(N, g_2),$$

and find that

$$N \mid f_1 f_2 \tag{2.16e}$$

$$g_1 = x_1 f_1, \quad g_2 = x_2 f_2, \tag{2.16f}$$

$$1 = x_1 f_1 + x_2 f_2, \tag{2.16g}$$

$$\gcd(f_1, f_2) = 1, \quad 1 < f_i < N \quad (i = 1, 2). \tag{2.16h}$$

Let

$$N = f_1 x, \quad f_2 \mid f_1 x$$

and

$$\gcd(f_2, f_1) = 1$$

implying

$$f_2 \mid x, \quad f_1 f_2 \mid N,$$

hence by (2.16e)

$$N = f_1 f_2. \tag{2.16i}$$

Conversely, let N be the product of two mutually prime natural numbers f_1, f_2 both less than N so that (2.16h), (2.16i) holds. Then we compute rational integers x_1, x_2 satisfying (2.16g) and find that

$$e_1 = g_1/N,$$

with g_1 as defined by (2.16f) is a non trivial idempotent of $\mathbb{Z}/N$.

In this way a 1–1-correspondence between the non-trivial idempotents of $\mathbb{Z}/N$ and those divisors of N which are not 1 and coprime to their complement in N arises. Now we find that

$$\mathbb{Z}/N = e_1 \mathbb{Z}/N \dotplus e_2 \mathbb{Z}/N. \tag{2.16j}$$

But by construction

$$e_1 \mathbb{Z}/N \simeq \mathbb{Z}/f_2, \quad e_2 \mathbb{Z}/N \simeq \mathbb{Z}/f_1,$$

in other words

$$\mathbb{Z}/N \simeq \mathbb{Z}/f_1 \dotplus \mathbb{Z}/f_2. \tag{2.16k}$$

A more precise expression of (2.16k) is the *Chinese Remainder Theorem* for $\mathbb{Z}$: If the natural number N is factorized into the product of two mutually prime natural numbers f_1, f_2, then for any two rational integers m_1, m_2 there is a rational integer x satisfying the congruences

$$x \equiv m_i \bmod f_i \quad (i = 1, 2),$$

and x is uniquely determined modulo N. This is a specialization of the general *Chinese Remainder Theorem for unital (commutative) rings*:

Let R be a unital ring and $\mathfrak{a}_1, \ldots, \mathfrak{a}_n$ be ideals such that $\mathfrak{a}_i + \mathfrak{a}_j = R$ $(1 \leqslant i < j \leqslant n)$. Given elements $x_1, \ldots, x_n \in R$, there exists $x \in R$ such that $x \equiv x_i \bmod \mathfrak{a}_i$ $(1 \leqslant i \leqslant n)$. (2.17)

(A proof is given in S. Lang's book [10].)

The construction given above enables us to present x in the simple form

$$x = m_2 x_1 f_1 + m_1 x_2 f_2$$

as Gauss points out, who was the first to use idempotents though he did not want to give this concept a name. An important application is the *multimodular calculus* which enables us to deal with very large integers by considering them modulo sufficiently many mutually prime (relatively small) rational integers.

6. The fixed subring of the permutation automorphisms

Returning to the discussion of the universal splitting ring $S(f/R)$ of a monic polynomial (2.3) over the unital commutative ring R we ask the question: Which elements of $S(f/R)$ are invariant under all permutation automorphisms? Certainly the elements of R.

Theorem. *The subring* (2.18)

$$S(f/R)^{\bar{\mathfrak{S}}_n} = \{\kappa \in S(f/R) \mid \forall\, \pi \in \mathfrak{S}_n : \bar{\pi}(\kappa) = \kappa\} \tag{2.18a}$$

of $S(f/R)$ formed by the elements which are invariant under all permutation automorphisms equals

$$S(f/R)^{\bar{\mathfrak{S}}_n} = R + [0/2]^{\bar{\mathfrak{S}}_n}, \tag{2.18b}$$

where

$$[0/2] = \{\kappa \in S(f/R) \mid 2\kappa = 0\} (= ([0/2] \cap R) S(f/R)). \tag{2.18c}$$

In particular, if R is an entire ring of characteristic $\neq 2$ then

$$S(f/R)^{\bar{\mathfrak{S}}_n} = R. \tag{2.18d}$$

Proof

Trivially,

$$S(f/R) = S(f/R)^{\bar{\mathfrak{S}}_1} = R \text{ in case } n = 1.$$

The case $n = 2$ was discussed already in section 1. Let $n > 2$ and apply induction over n. Clearly, the right-hand side of (2.18b) is contained in the left-hand side. Conversely, let

$$y \in S(f/R)^{\bar{\mathfrak{S}}_n}.$$

By construction $S(f/R) = S(Q(f, t - \xi_n)/R[\xi_n])$ and $\bar{\mathfrak{S}}_{n-1}(Q(f, t - \xi_n)/R[\xi_n])$ consists precisely of those permutation automorphisms of $S(f/R)$ which leave ξ_n invariant. By hypothesis,

$$S(f/R)^{\bar{\mathfrak{S}}_n} \subset R[\xi_n] + [0/2]^{\bar{\mathfrak{S}}_{n-1}};$$

using theorem (1.6) we find that

$$S(f/R)^{\bar{\mathfrak{S}}_n} \subset R[\xi_{n-1}] + [0/2]^{\bar{\mathfrak{S}}'_{n-1}},$$

where $\bar{\mathfrak{S}}'_{n-1}$ consists of those permutation automorphisms of $S(f/R)$ which leave ξ_{n-1} invariant. Hence we obtain for y of $S(f,R)^{\bar{\mathfrak{S}}_n}$:

$$y = y_0 + \sum_{j=0}^{n-1} \lambda_j \xi_n^j, \tag{2.19a}$$

$$y = \bar{\pi}(y_0) + \sum_{j=0}^{n-1} \lambda_j \xi_{n-1}^j \quad (\lambda_0, \lambda_1, \ldots, \lambda_{n-1} \in R, \quad y_0 \in [0/2]^{\bar{\mathfrak{S}}'_{n-1}}), \tag{2.19b}$$

where π transposes n and $n-1$ and fixes all $i < n-1$. We have

$$Q(f, t - \xi_n) = t^{n-1} + (a_1 + \xi_n)t^{n-2} + \cdots,$$
$$0 = \xi_{n-1}^{n-1} + (a_1 + \xi_n)\xi_{n-1}^{n-2} + \cdots,$$
$$\xi_{n-1}^{n-1} = -(a_1 + \xi_n)\xi_{n-1}^{n-2} + \cdots,$$

hence the basis element $\xi_{n-1}^{n-2}\xi_n$ has the coefficient $-\lambda_{n-1} \bmod [0/2]$. Therefore,

$$\lambda_{n-1} \equiv 0 \bmod [0/2]$$

and upon comparison of (2.19a), (2.19b) we find that

$$\lambda_i \equiv 0 \bmod [0/2] \quad (0 < i < n-1).$$

Hence

$$y_0 = \bar{\pi}(y_0) \bmod [0/2],$$

proving the theorem. □

Exercises

1. Given a unital ring R which is the direct sum of two non-zero ideals A, B:

$$R = A \dot{+} B.$$

Show that the corresponding presentation

$$1_R = e_1 + e_2$$
$$(e_1 \in A, e_2 \in B)$$

defines uniquely two orthogonal idempotents e_1, e_2 both belonging to the center $C(R)$ of R such that

$$1_A = e_1, \quad 1_B = e_2,$$
$$A = e_1 R = Re_1, \quad B = e_2 R = Re_2.$$

Conversely, if $e_1, e_2, \ldots, e_s$ are mutually orthogonal central idempotents of R for which $e_1 + e_2 + \cdots + e_s = 1_R$, then there holds the decomposition $A = \dot{\sum}_{i=1}^{s} e_i A$ of A into the direct sum of non zero ideals $e_i A$ with e_i as unit element.

2. Compute all idempotents of

(i) $\mathbb{Z}[t]/P\mathbb{Z}[t]$ for $P(t) = t^4 - 3t^2 + 2$;
(ii) $\mathbb{Z}/N\mathbb{Z}$ for $N = 32, 48, 59, 60$.

2.3. Symmetric polynomials

1. The theorem on symmetric functions

As an application of theorem (2.18) we study the polynomial ring $R_n := R_0[t_1, t_2, \dots, t_n]$ in n independent variables $t_1, t_2, \dots, t_n$ over a unital commutative ring R_0. Let

$$g(t) = t^n - t_1 t^{n-1} + t_2 t^{n-2} + \cdots + (-1)^n t_n \in R_n[t] \tag{3.1a}$$

be the 'generic' monic polynomial of degree n over R_0. Then $S(g/R_n) = R_n[\xi_1, \dots, \xi_n]$ is a polynomial ring in the variables $\xi_1, \dots, \xi_n$ over R_0:

$$S(g/R_n) = R_0[\xi_1, \dots, \xi_n], \tag{3.1b}$$

where

$$g(t) = \prod_{i=1}^{n} (t - \xi_i) \tag{3.1c}$$

in $S(g/R_n)$. The polynomials in $\xi_1, \dots, \xi_n$ over R_0 which are fixed by the $n!$ permutation automorphisms of $S(g/R_n)$ are said to be the *symmetric polynomials* in $\xi_1, \dots, \xi_n$ over R_0. According to theorem (2.18) they belong to R_n for $[0/2] = 0$. In the sequel we shall show that the restriction $[0/2] = 0$ is superfluous.

One of the symmetric polynomials is

$$d(g) := \prod_{1 \leqslant i < k \leqslant n} (\xi_i - \xi_k)^2 = (-1)^{\binom{n}{2}} \prod_{i \neq k} (\xi_i - \xi_k). \tag{3.2}$$

As a matter of fact setting $R_0 = \mathbb{Z}$ we obtain an identity of the form

$$\prod_{1 \leqslant i < k \leqslant n} (\xi_i - \xi_k)^2 = P_n(t_1, \dots, t_n), \tag{3.3a}$$

where

$$P_n \in \mathbb{Z}[t_1, \dots, t_n] \tag{3.3b}$$

and

$$t_i = \sum_{1 \leqslant j_1 < j_2 < \cdots < j_i \leqslant n} \xi_{j_1} \xi_{j_2} \cdots \xi_{j_i} \quad (1 \leqslant i \leqslant n) \tag{3.3c}$$

are the basic symmetric functions of $\xi_1, \dots, \xi_n$.

This identity remains in force upon application of the standard homomorphism

$$\phi: \mathbb{Z} \to R_0: m \mapsto m 1_{R_0}.$$

It follows that $d(g)$ defined by (3.2) belongs to R_n for any R_0 and has coefficients in the *prime ring* $\phi(\mathbb{Z})$ of R_0. We observe that each factor $\xi_i - \xi_k$ is a non-zero divisor of $R_0[\xi_1, \dots, \xi_n]$ (consider the highest ξ_i-term of P in any equation $(\xi_i - \xi_k)P = 0$ and show that it is zero!). Hence, the product $d(g)$ also is a non-zero divisor of $R_0[\xi_1, \dots, \xi_n]$ and – *a fortiori* – of R_n.

Furthermore, for any splitting ring Λ of any monic polynomial

$$f(t) = t^n + a_1 t^{n-1} + \cdots + a_n \in R_0[t], \tag{3.4a}$$

the expression

$$d(f) = \prod_{1 \leq i < k \leq n} (x_i - x_k)^2, \tag{3.4b}$$

derived from the factorization

$$f(t) = \prod_{i=1}^{n} (t - x_i) \tag{3.4c}$$

of f over Λ turns out to be a polynomial expression in $a_1, \ldots, a_n$ over R_0. In fact,

$$d(f) = P_n(-a_1, a_2, \ldots, (-1)^n a_n), \tag{3.4d}$$

where P_n is as in (3.3a). We call the expression $d(f)$ the *discriminant* of the monic polynomial $f \in R_0[t]$. It belongs to R_0. If it is not a zero divisor, then f is said to be *separable*, otherwise *inseparable*.

As had already been remarked for $n = 2$ the roots occurring in any factorization (3.4c) of a separable monic polynomial are distinct. The converse need not be true unless we restrict our considerations to entire rings.

Correspondingly, for two monic polynomials, say f as in (3.4a) and

$$h(t) = t^m + b_1 t^{m-1} + \cdots + b_m \quad (m \in \mathbb{Z}^{>0}; b_i \in R_0, 1 \leqslant i \leqslant m), \tag{3.5a}$$

we find that in any common splitting ring of f and

$$h(t) = \prod_{i=1}^{m} (t - y_i), \tag{3.5b}$$

the expression

$$\operatorname{Res}(f, h) := \prod_{i=1}^{n} \prod_{k=1}^{m} (y_k - x_i) = (-1)^{mn} \operatorname{Res}(h, f) \tag{3.5c}$$

is a polynomial in $a_1, \ldots, a_n, b_1, \ldots, b_m$ with rational integral coefficients. It is said to be the *resultant* of f and h. The discriminant of the product fh is given by the formula

$$d(fh) = d(f)d(h) \operatorname{Res}(f, h)^2 \tag{3.5d}$$

which implies that the product is separable, if and only if each factor is separable and the resultant of f, h is a non-zero divisor of R_0.

Our discussion of quadratic equations in section 1 showed that for a separable monic quadratic polynomial f over R the elements of R are the only elements of $S(f/R)$ which are invariant under the permutation automorphisms. Now the same inductive procedure which was applied in the proof of (2.18) shows

Corollary of theorem (2.18) (3.6)

If the polynomial f is separable under the premises of (2.18), *then only the elements of R are invariant under all permutation automorphisms.*

(Actually we only need to demand that $d(f)$ is 2-regular: if $\xi \in R$, $\xi \neq 0$, $2\xi = 0$, then $d(f)\xi \neq 0$.)
As an application we have the

Theorem on symmetric functions (3.7)
Let R be a unital commutative ring. Then any polynomial in n variables $\xi_1, \ldots, \xi_n$ over R which remains unchanged for any permutation of the variables is representable as a polynomial in the basic symmetric functions

$$\sigma_i(\xi_1, \ldots, \xi_n) := \sum_{1 \leqslant j_1 < \cdots < j_i \leqslant n} \xi_{j_1} \xi_{j_2} \cdots \xi_{j_i}$$

over R.

The subtle specialization argument developed here in order to overcome the possibility that $[0/2] \neq 0$ is shown to be unnecessary by the following algorithm which produces a constructive proof.

Algorithm for the theorem on symmetric functions (3.8)

Input. A symmetric polynomial $P(\xi_1, \ldots, \xi_n)$ and the basic symmetric functions $\sigma_1, \ldots, \sigma_n (n \geqslant 2)$.
Output. A polynomial Q of n variables satisfying $P(\xi_1, \ldots, \xi_n) = Q(\sigma_1, \ldots, \sigma_n)$.
Step 1. (Initialization). Set $F_n(\xi_1, \ldots, \xi_n) \leftarrow P(\xi_1, \ldots, \xi_n)$, $Q_{n,0} \leftarrow 0$, $\sigma_j^{(k)} \leftarrow \sigma_j(\xi_1, \ldots, \xi_k, 0, \ldots, 0)\ (1 \leqslant j \leqslant n, 1 \leqslant k \leqslant n)$, $k \leftarrow n$, $i_n \leftarrow 0$.
Step 2. (Decrease k). Set $k \leftarrow k-1$, $F_k(\xi_1, \ldots, \xi_k) \leftarrow F_{k+1}(\xi_1, \ldots, \xi_k, 0)$, $Q_{k,0} \leftarrow 0$, $i_k \leftarrow 0$.
Step 3. ($k = 1$?). For $k > 1$ go to 2, else set
$Q_1(\sigma_1^{(1)}) \leftarrow F_1(\sigma_1^{(1)})$.
Step 4. (Increase k). Set $k \leftarrow k+1$. For $k > n$ terminate. Else set
$F_k(\xi_1, \ldots, \xi_k) \leftarrow F_k(\xi_1, \ldots, \xi_k) - Q_{k-1}(\sigma_1^{(k)}, \ldots, \sigma_{k-1}^{(k)})$,
$Q_{k,i_k} \leftarrow Q_{k-1}(\sigma_1^{(k)}, \ldots, \sigma_{k-1}^{(k)})$.
Step 5. ($F_k = 0$?). For $F_k(\xi_1, \ldots, \xi_k) \neq 0$ go to 7.
Step 6. (Compute Q_k). Set $Q_k(\sigma_1^{(k)}, \ldots, \sigma_k^{(k)}) \leftarrow \sum_{j=0}^{i_k} Q_{k,j}(\sigma_1^{(k)}, \ldots, \sigma_{k-1}^{(k)}) \cdot (\sigma_k^{(k)})^j$ and go to 4.
Step 7. (Divide F_k by $\sigma_k^{(k)}$). Set $\mu \leftarrow \max\{m \in \mathbb{N} \mid (\sigma_k^{(k)})^m \mid F_k(\xi_1, \ldots, \xi_k)\}$, $Q_{k,\nu} \leftarrow 0\ (i_k + 1 \leqslant \nu \leqslant i_k + \mu)$, $i_k \leftarrow i_k + \mu$. Then set $F_k(\xi_1, \ldots, \xi_k) \leftarrow F_k(\xi_1, \ldots, \xi_k)/(\sigma_k^{(k)})^\mu$. For $F_k(\xi_1, \ldots, \xi_k)$ non-constant go to 2.
Step 8. (Update Q_{k,i_k}). Set $Q_{k,i_k} \leftarrow F_k$ and go to 6.

Remarks (3.9)
For a better understanding of the algorithm we explain the roles of the polynomials F_k, Q_k in greater detail. At the beginning we have

$F_k(\xi_1,\ldots,\xi_k) = P(\xi_1,\ldots,\xi_k, 0,\ldots,0)$. Then F_k is gradually modified via basic symmetric functions until we finally obtain

$$F_k(\xi_1,\ldots,\xi_k) = \sum_{j=0}^{i_k} Q_{k,j}(\sigma_1^{(k)},\ldots,\sigma_{k-1}^{(k)})\cdot(\sigma_k^{(k)})^j$$
$$= Q_k(\sigma_1^{(k)},\ldots,\sigma_k^{(k)}).$$

Example (3.10)
Let us compute the representation of $P(\xi_1,\xi_2,\xi_3) := (\xi_1-\xi_2)^2(\xi_1-\xi_3)^2(\xi_2-\xi_3)^2$ by the symmetric functions $\sigma_1,\sigma_2,\sigma_3$. The algorithm produces the polynomials $F_3 = P$, $Q_{3,0} = 0$ and the restricted symmetric functions $\sigma_j^{(k)}$ for $1 \leq j, k \leq 3$ in step 1. In step 2 we obtain $F_2(\xi_1,\xi_2) = (\xi_1-\xi_2)^2(\xi_1\xi_2)^2$, $Q_{2,0} = 0$, $i_2 = 0$ and then again $F_1 = 0$, $Q_{1,0} = 0$, $i_1 = 0$. Then step 3 yields $Q_1 = 0$. We proceed to step 4: F_2 and $Q_{2,0}$ remain unchanged. Then steps 7, 2, 3, 4, 5, 7, 8 and 6 result in $Q_2 = (\sigma_1^{(2)})^2(\sigma_2^{(2)})^2 - 4(\sigma_2^{(2)})^3 = F_2(\xi_1, \xi_2)$, i.e. we have approximated P correctly for ξ_1, ξ_2. Going on we finally obtain

$$P(\xi_1,\xi_2,\xi_3) = \sigma_1^2\sigma_2^2 - 4\sigma_2^3 + \sigma_3(18\sigma_1\sigma_2 - 4\sigma_1^3 - 27\sigma_3).$$

This result expresses the discriminant of a monic third degree polynomial by its coefficients.

We conclude these considerations on symmetric polynomials with a correspondence between the basic symmetric polynomials and power sums:

Definition (3.11)
Let $\xi_1,\ldots,\xi_n$ be as in (3.7), *then $S_k := \sum_{i=1}^n \xi_i^k$ is called the kth* **power sum** *of $\xi_1,\ldots,\xi_n$ $(k \in \mathbb{Z}^{\geq 0})$.*

Theorem (3.12)
Between the basic symmetric functions and the power sums we have the so-called 'Newton's relations'

$$\sum_{i=0}^{k-1}(-1)^i\sigma_i S_{k-i} + (-1)^k k\sigma_k = 0 \quad (\sigma_0 := 1, 0 \leq k \leq n), \qquad (3.12a)$$

$$\sum_{i=0}^{n}(-1)^i\sigma_i S_{k-i} = 0 \quad (\sigma_0 := 1, k \geq n). \qquad (3.12b)$$

Proof
Because of (3.1a–c), (3.3a–c) we have

$$0 = \sum_{j=0}^{n}(-1)^j\sigma_j\xi_i^{n-j} \quad (1 \leq i \leq n),$$

and therefore

$$0 = \sum_{j=0}^{n}(-1)^j\sigma_j\xi_i^{m-j} \quad (1 \leq i \leq n; m \geq n).$$

A summation over i yields

$$0=\sum_{j=0}^{n}(-1)^j\sigma_j\sum_{i=1}^{n}\xi_i^{m-j}=\sum_{j=0}^{n}(-1)^j\sigma_jS_{m-j},$$

hence (3.12b), respectively (3.12a) in case $k=n$. The rest of the proof is done by induction over n for fixed k. The initialization $n=k$ was already shown. Now let us assume that (3.12a) is correct for $n-1$. We set

$$F(\xi_1,\dots,\xi_n):=\sum_{i=0}^{k-1}(-1)^i\sigma_iS_{k-i}+(-1)^kk\sigma_k.$$

This is a symmetric expression in $\xi_1,\dots,\xi_n$ of degree $\leqslant k<n$. Furthermore,

$$F(\xi_1,\dots,\xi_{n-1},0)=\sum_{i=0}^{k-1}(-1)^i\sigma_i^{(n-1)}S_{k-i}^{(n-1)}+(-1)^kk\sigma_k^{(n-1)}=0$$

because of the induction assumption. Therefore $F(\xi_1,\dots,\xi_n)$ is divisible by ξ_n. Because of the symmetry it is also divisible by σ_n. Because of $\deg(F)<n$ this implies $F(\xi_1,\dots,\xi_n)=0$, and thus the induction from $n-1$ to n. □

Newton's relations allow us to compute the power sums from the basic symmetric functions and vice versa. For example,

$$\begin{aligned}S_1&=\sigma_1 &\qquad&& \sigma_1&=S_1\\ S_2&=\sigma_1^2-2\sigma_2 &\quad\text{and}&& \sigma_2&=\tfrac{1}{2}(S_1^2-S_2)\\ S_3&=\sigma_1^3-3\sigma_1\sigma_2+3\sigma_3 &&& \sigma_3&=\tfrac{1}{6}S_1^3-\tfrac{1}{2}S_1S_2+\tfrac{1}{3}S_3\end{aligned}$$

(if 2, 3 are invertible in R). We remark that it is sometimes useful also to consider power sums for $k\in\mathbb{Z}^{<0}$ in the corresponding quotient ring.

2. The principal equation

In order to define the *principal equation*

$$P_{\xi/R}(\xi)=0 \tag{3.13a}$$

of an element ξ of the equation ring $R[t]/fR[t]$ of a monic equation

$$f(x)=0 \tag{3.13b}$$

of degree n over R we proceed as follows. Firstly, we consider the generic monic polynomial (3.1a) of degree n over the polynomial ring $R_n=R_0[t_1,\dots,t_n]$ with $R_0=\mathbb{Z}[y_1,\dots,y_n]$. We form the 'generic element'

$$y=\sum_{i=0}^{n-1}y_{i+1}\xi_n^i \tag{3.14a}$$

contained in $S(g/R_n)=R_0[\xi_1,\dots,\xi_n]$ for

$$g(t)=\prod_{i=1}^{n}(t-\xi_i). \tag{3.14b}$$

Realizing that the powers of the n-cycle $\zeta=(1,2,\dots,n)\in\mathfrak{S}_n$, say $1,\zeta,\dots,\zeta^{n-1}$ form a representative set of $\mathfrak{S}_n$ modulo $\mathfrak{S}_{n-1}$, the stabilizer of n, we see that the coefficients of

$$P_{y/R_n} = \prod_{i=0}^{n-1} (t - \bar{\zeta}^i(y)) \tag{3.15a}$$

are fixed by the permutation automorphisms of $S(g/R_n)$. Hence,

$$P_{y/R_n} \in R_n[t] \tag{3.15b}$$

is a monic polynomial of degree n over R_n satisfying the equation

$$P_{y/R_n}(y) = 0. \tag{3.15c}$$

Now we apply the homomorphism

$$\phi: S(g/R_n) \to S(f/R)$$

mapping

$$\lambda \mapsto \lambda 1_R \quad (\lambda \in \mathbb{Z}), \quad t_i \mapsto (-1)^i a_i \quad (1 \leqslant i \leqslant n), \quad y_i \mapsto \alpha_i \quad (\alpha_i \in R, 1 \leqslant i \leqslant n)$$

in order to obtain the principal polynomial

$$P_{\xi/R} := \phi P_{y/R_n} \tag{3.16a}$$

for

$$\xi := \sum_{i=0}^{n-1} \alpha_{i+1} \phi(\xi_n)^i, \tag{3.16b}$$

so that there holds the principal equation

$$P_{\xi/R}(\xi) = 0. \tag{3.16c}$$

In particular, we obtain

$$P_{\phi(\xi_n)/R} = f, \tag{3.17a}$$

$$P_{a/R} = (t - a)^n \quad (a \in R). \tag{3.17b}$$

The principal polynomial is of the form

$$P_{\xi/R}(t) = t^n - \mathrm{Tr}(\xi) t^{n-1} + \cdots + (-1)^n N(\xi), \tag{3.18}$$

where the trace mapping $\mathrm{Tr}: R[t]/fR[t] \to R$ is linear and the norm mapping $N: R[t]/fR[t] \to R$ is multiplicative and homogeneous of degree n:

$$\left.\begin{array}{l} \mathrm{Tr}(\xi + \eta) = \mathrm{Tr}(\xi) + \mathrm{Tr}(\eta), \quad \mathrm{Tr}(a\xi) = a\mathrm{Tr}(\xi), \quad \mathrm{Tr}(1) = n, \\ N(\xi\eta) = N(\xi)N(\eta), \quad N(a\xi) = a^n N(\xi), \quad N(1) = 1 \end{array}\right\} \tag{3.19}$$

for all

$$\xi, \eta \in R[t]/fR[t], \quad a \in R.$$

The intermediate coefficients of the principal polynomial are also of interest, but have been studied relatively little.

We observe that by construction

$$P_{\xi/R}(t) = \prod_{i=1}^{n} (t - \bar{\zeta}^i(\xi)), \tag{3.20}$$

so that for a nilpotent element ξ of $R[t]/fR[t]$ also $\bar{\zeta}^i(\xi)$ is nilpotent

$(1 \leqslant i \leqslant n)$, and hence the coefficients of $P_{\xi/R}$ are nilpotent elements of R. If R contains no nilpotent element other than 0, then $P_{\xi/R} = t^n$, $\xi^n = 0$.

We note that

$$d(f) = \prod_{1 \leqslant i < k \leqslant n} (\xi_i - \xi_k)^2 = \det((\xi_i^{j-1})_{1 \leqslant i,j \leqslant n}), \tag{3.21a}$$

according to the well-known rule for the evaluation of Vandermonde's determinant. Therefore

$$\det((\xi_i^{j-1}))^2 = \det((\xi_j^{i-1}))\det((\xi_i^{j-1})) = \det(\mathrm{Tr}(\xi_n^{i+k-2})) \tag{3.21b}$$

implies that $d(f)$ is equal to the discriminant of the trace bilinear form

$$Q(y_1, \ldots, y_n; z_1, \ldots, z_n) = \sum_{i=1}^{n} \sum_{j=1}^{n} \mathrm{Tr}(\xi_n^{i-1}\xi_n^{j-1}) y_i z_j. \tag{3.21c}$$

It follows that *for non-zero divisor discriminants the trace bilinear form is non degenerate.* In other words, the equations

$$\mathrm{Tr}(\xi\eta) = 0 \tag{3.22a}$$

for all $\eta \in R[t]/fR[t]$ imply that

$$\xi = 0. \tag{3.22b}$$

An important application is made in the case that R has no nilpotent element other than zero, e.g. if R is an entire ring. In such a case the separability of f makes it impossible that $R[t]/fR[t]$ contains nilpotent elements other than zero. This was shown above and follows again from the fact that the nilpotency of ξ implies the nilpotency of $\xi\eta$ for $\eta \in R[t]/fR[t]$, i.e. $\mathrm{Tr}(\xi\eta) = 0$ and therefore $\xi = 0$.

Proposition (3.23)
If the unital commutative ring R contains no nilpotent elements other than zero, then for any separable monic polynomial $f \in R[t]$ also the equation ring $R[t]/fR[t]$ contains no nilpotent elements other than zero.

We have seen already that each monic divisor of a separable polynomial is separable. Therefore a repeated application of (3.23) yields

Proposition (3.24)
If the unital commutative ring R contains no nilpotent elements other than 0, then also the universal splitting ring $S(f/R)$ contains no nilpotent elements $\neq 0$.

We extend the notion of principal equation, norm and trace to arbitrary overrings Λ of R with a finite R-basis $\omega_1, \ldots, \omega_n$. For any element ξ of Λ there holds an equation

$$\xi(\omega_1,\ldots,\omega_n)=(\omega_1,\ldots,\omega_n)M(\xi), \tag{3.25a}$$

with a matrix $M(\xi)=(m_{ik})\in R^{n\times n}$, meaning that

$$\xi\omega_k=\sum_{i=1}^{n} m_{ik}\omega_i \quad (1\leqslant k\leqslant n). \tag{3.25b}$$

The mapping

$$M\colon \Lambda\to R^{n\times n}\colon \xi\mapsto M(\xi) \tag{3.25c}$$

is said to be the *regular representation* of Λ over R relative to the basis $\omega_1,\ldots,\omega_n$. We verify directly the R-homomorphism properties of M:

$$M(\xi+\eta)=M(\xi)+M(\eta),\quad M(\lambda\xi)=\lambda M(\xi),\quad M(\xi\eta)=M(\xi)M(\eta)$$
$$\text{for all } \xi,\eta\in\Lambda,\ \lambda\in R. \tag{3.25d}$$

The *characteristic polynomial* of the matrix $M(\xi)$

$$P_{M(\xi)}=P_\xi=\det(tI_n-M(\xi)) \tag{3.26}$$

is defined as the *principal polynomial* of ξ over R. It is independent of the choice of the R-basis $\omega_1,\ldots,\omega_n$ of Λ since the transition to another basis changes M to an equivalent representation of the form

$$U^{-1}MU \tag{3.27a}$$

with invertible transition matrix U of degree n over R (see chapter 3 (2.4)). Thus

$$\begin{aligned} P_{U^{-1}M(\xi)U} &= \det(tI_n-U^{-1}M(\xi)U)\\ &= \det(U^{-1}(tI_n-M(\xi))U)\\ &= \det(U^{-1})\det(tI_n-M(\xi))\det U\\ &= P_{M(\xi)} \quad \text{for all } \xi\in\Lambda. \end{aligned} \tag{3.27b}$$

By the 'generic argument' (that means by using generic elements and generic polynomials and specialization) we show that the principal equation always holds:

$$P_\xi(\xi)=0. \tag{3.28}$$

In particular, if $\Lambda=R[t]/fR[t]$ for some monic non-constant polynomial $f\in R[t]$, then the principal polynomial defined as characteristic polynomial of the regular representation coincides with the principal polynomial defined above. (Generically speaking there can be only one monic equation of degree n for ξ over R!) Correspondingly, trace and norm are defined by

$$\mathrm{Tr}_{\Lambda/R}(\xi)=\mathrm{Tr}(\xi)=\mathrm{Tr}(M(\xi))=\sum_{i=1}^{n} m_{ii}, \tag{3.29a}$$

$$N_{\Lambda/R}(\xi)=N(\xi)=\det(M(\xi))=\det((m_{ij})), \tag{3.29b}$$

so that the rules (3.19) remain in force.

3. The resultant

Let $\bar{R} = \mathbb{Z}[\bar{a}_0, \ldots, \bar{a}_n, \bar{b}_0, \ldots, \bar{b}_m]$ be the polynomial ring in $n + m + 2$ variables over $\mathbb{Z}$, let

$$\bar{A}(t) = \bar{a}_0 t^n + \bar{a}_1 t^{n-1} + \cdots + \bar{a}_n, \quad \bar{B}(t) = \bar{b}_0 t^m + \bar{b}_1 t^{m-1} + \cdots + \bar{b}_m \quad (3.30a)$$

be the 'generic' polynomials of degree n, m of $\bar{R}[t]$, let

$$\mathfrak{Q}(\bar{R}) = \mathbb{Q}(\bar{a}_0, \ldots, \bar{a}_n, \bar{b}_0, \ldots, \bar{b}_m) \quad (3.30b)$$

be the quotient field of $\bar{R}$, and let

$$\bar{A}(t) = \bar{a}_0 \prod_{i=1}^{n} (t - \bar{x}_i), \quad \bar{B}(t) = \bar{b}_0 \prod_{k=1}^{m} (t - \bar{y}_k) \quad (3.30c)$$

be the factorizations of $\bar{A}, \bar{B}$ in the extension algebra $S(\bar{B}/S(\bar{A}/\mathfrak{Q}(\bar{R})))$ of $\mathfrak{Q}(\bar{R})$ with $n!m!$ basis elements

$$\prod_{i=1}^{n} \prod_{j=1}^{m} \bar{x}_i^{\mu_i} \bar{y}_j^{\nu_j} \quad (0 \leqslant \mu_i < i, 0 \leqslant \nu_j < j). \quad (3.30d)$$

Then

$$\bar{\Lambda} := \mathbb{Z}[\bar{a}_0, \bar{x}_1, \ldots, \bar{x}_n, \bar{b}_0, \bar{y}_1, \ldots, \bar{y}_m] \quad (3.30e)$$

is a minimal splitting ring of $\bar{A}\bar{B}$ over $\bar{R}$ and the $n + m + 2$ generators $\bar{a}_0, \bar{x}_1, \ldots, \bar{x}_n, \bar{b}_0, \bar{y}_1, \ldots, \bar{y}_m$ are algebraically independent over $\mathbb{Z}$. Therefore $\bar{\Lambda}$ is a polynomial ring in $\bar{a}_0, \bar{x}_1, \ldots, \bar{x}_n, \bar{b}_0, \bar{y}_1, \ldots, \bar{y}_m$ over $\mathbb{Z}$ with

$$S(\bar{B}/S(\bar{A}/\mathfrak{Q}(\bar{R}))) = \mathbb{Q}(\bar{a}_0, \bar{x}_1, \ldots, \bar{x}_n, \bar{b}_0, \bar{y}_1, \ldots, \bar{y}_m) \quad (3.30f)$$

as quotient field. There are the $n!m!$ permutation automorphisms of $\mathfrak{Q}(\bar{\Lambda})$ over $\mathfrak{Q}(\bar{R})$ which permute the $\bar{x}_i$ among themselves as well as the $\bar{y}_j$ among themselves. They form the group

$$\mathrm{Aut}\,(\mathfrak{Q}(\bar{\Lambda})/\mathfrak{Q}(\bar{R})) \simeq \mathfrak{S}_n \times \mathfrak{S}_m, \quad (3.30g)$$

and the fixed subring of this group if $\mathfrak{Q}(\bar{R})$. It follows that the expression

$$\mathrm{Res}\,(\bar{A}, \bar{B}) := \bar{a}_0^m \bar{b}_0^n \prod_{i=1}^{n} \prod_{j=1}^{m} (\bar{y}_j - \bar{x}_i) \quad (3.31)$$

belongs to $\bar{R}$ and is irreducible in $\bar{a}_0, \bar{a}_1, \ldots, \bar{a}_n, \bar{b}_0, \bar{b}_1, \ldots, \bar{b}_m$ because of the transitivity of the action of $\mathfrak{S}_n \times \mathfrak{S}_m$ on the $\bar{y}_j - \bar{x}_i$ and the fact that the product of those differences is in $\bar{R}$ only after multiplication by suitable powers of $\bar{a}_0, \bar{b}_0$.

For any unital commutative ring R and for any two polynomials

$$A(t) = a_0 t^n + a_1 t^{n-1} + \cdots + a_n, \quad B(t) = b_0 t^m + b_1 t^{m-1} + \cdots + b_m \in R[t] \quad (3.32a)$$

the specialization

$$\bar{R} \to R \text{ via } 1_{\mathbb{Z}} \mapsto 1_R, \quad \bar{a}_i \mapsto a_i \quad (0 \leqslant i \leqslant n), \quad \bar{b}_j \mapsto b_j \quad (0 \leqslant j \leqslant m) \quad (3.32b)$$

leads to a polynomial expression $\mathrm{Res}\,(A, B) = \mathrm{Res}_{n,m}(A, B)$ in $a_0, \ldots, a_n,$

$b_0, \ldots, b_m$ with coefficients contained in the subring of R generated by 1_R (the prime ring of R), which is said to be *resultant* of A, B.

For any overring Λ of R which is a splitting ring of both A, B in accordance with

$$A(t) = a_0 \prod_{i=1}^{n} (t - x_i), \quad B(t) = b_0 \prod_{j=1}^{m} (t - y_j) \tag{3.33a}$$

the specialization (3.32b) can be uniquely extended to the specialization

$$\bar{\Lambda} \to \Lambda \text{ via } 1_Z \mapsto 1_\Lambda; \ \bar{a}_0 \mapsto a_0, \bar{x}_i \mapsto x_i \ (1 \leqslant i \leqslant n); \ \bar{b}_0 \mapsto b_0, \bar{y}_j \mapsto y_j \ (1 \leqslant j \leqslant m), \tag{3.33b}$$

such that $\operatorname{Res}(\bar{A}, \bar{B})$ specializes to

$$\operatorname{Res}(A, B) = a_0^m b_0^n \prod_{i=1}^{n} \prod_{j=1}^{m} (y_j - x_i). \tag{3.33c}$$

Hence, the new definition of $\operatorname{Res}(A, B)$ coincides with the one formerly given in case A and B are both monic. We note that the definition of $\operatorname{Res}(A, B)$ depends somewhat on the 'formal' degrees n, m which may not be the actual degrees in case a_0 or b_0, or both, vanish. However, this dependence is not too serious since the result will be about the same, if for at least one of the two polynomials A, B the formal degree agrees with the actual degree. This is because of the formula

$$\operatorname{Res}(\bar{A}, \bar{B}) = (-1)^{mn} \bar{a}_0^m \prod_{i=1}^{n} \bar{B}(\bar{x}_i) = \bar{b}_0^n \prod_{k=1}^{m} \bar{A}(\bar{y}_k), \tag{3.34a}$$

which by specialization turns into

$$\operatorname{Res}(A, B) = (-1)^{mn} a_0^m \prod_{i=1}^{n} B(x_i) = b_0^n \prod_{j=1}^{m} A(y_j) \tag{3.34b}$$

in any splitting ring of A, B over R in which (3.33a) holds. Hence, in case of $a_0 = a_1 = \cdots = a_{k-1} = 0$ $(k \leqslant n)$ we have

$$b_0^k \operatorname{Res}_{n-k,m}(A, B) = \operatorname{Res}_{n,m}(A, B). \tag{3.34c}$$

Analogously, for $b_0 = b_1 = \cdots = b_{l-1} = 0$ $(l \leqslant m)$ we have

$$(-1)^{nl} a_0^l \operatorname{Res}_{n,m-l}(A, B) = \operatorname{Res}_{n,m}(A, B). \tag{3.34d}$$

Also the order in which A, B are given presents no serious problem since we obtain again by specialization

$$\operatorname{Res}(B, A) = (-1)^{nm} \operatorname{Res}(A, B). \tag{3.34e}$$

Similarly we have for scalar multiplication:

$$\operatorname{Res}(\lambda A, B) = \lambda^m \operatorname{Res}(A, B), \quad \operatorname{Res}(A, \lambda B) = \lambda^n \operatorname{Res}(A, B) \quad \text{for all } \lambda \in R. \tag{3.34f}$$

The resultant is of course a very formidable expression, in fact, as derived from (3.31), (3.33c) it is a homogeneous polynomial of degree m in the variables

$x_1,\dots,x_n$ and homogeneous of degree n in the $y_1,\dots,y_m$. If we give the 'weight' i to a_i and the 'weight' j to b_j, then the resultant is homogeneous of degree n in the weighted variables $b_1,\dots,b_m$ and homogeneous of degree m in the weighted variables $a_1,a_2,\dots,a_n$.

It is not advisable to compute the resultant term by term. However, for $n \geqslant m$ any long division like

$$A = QB + R_0 \quad (Q, R_0 \in R[t], R_0 \text{ of formal degree } \tilde{m} < m) \tag{3.35a}$$

leads via

$$\operatorname{Res}(A,B) = b_0^{n-\tilde{m}} \operatorname{Res}(R_0, B) \tag{3.35b}$$

(derived from (3.34b), (3.34c), (3.35a)) to an easy transition from $\operatorname{Res}(A,B)$ to $\operatorname{Res}(R_0,B)$. Thus long division as explained in chapter 1, section 3 leads to a quite easy computation of $\operatorname{Res}(A,B)$, if we keep in mind that our formula remains intact for constant polynomials yielding

$$\left.\begin{aligned} \operatorname{Res}(a_0,B) &= a_0^m = \operatorname{Res}_{0,m}(a_0,B),\\ \operatorname{Res}(A,b_0) &= b_0^n = \operatorname{Res}_{n,0}(A,b_0),\\ \operatorname{Res}(a_0,b_0) &= 1 = \operatorname{Res}_{0,0}(a_0,b_0). \end{aligned}\right\} \tag{3.36}$$

Traditionally, the resultant is interpreted as the determinant connected with the homogeneous linear system for the unknowns $\xi_0,\xi_1,\dots,\xi_{m-1},\eta_0,\eta_1,\dots,\eta_{n-1}$ with the property that the polynomials

$$X(t) = \xi_0 + \xi_1 t + \cdots + \xi_{m-1}t^{m-1}, \quad Y(t) = \eta_0 + \eta_1 t + \cdots + \eta_{n-1}t^{n-1} \tag{3.37a}$$

provide a solution of the polynomial equation

$$XA + YB = 0. \tag{3.37b}$$

If the resultant is zero, then there is a non-trivial solution to (3.37b) but the reverse is true for entire rings only.

More important is the remark that generically speaking the resultant $\operatorname{Res}(\bar{A},\bar{B})$ is not zero (for example, specialize $\bar{A}$ to t^n, $\bar{B}$ to $(t-1)^m$). Hence, there is a unique solution of

$$\bar{X}\bar{A} + \bar{Y}\bar{B} = 1, \tag{3.38a}$$

in terms of polynomials $\bar{X},\bar{Y} \in \mathfrak{Q}(\bar{R})[t]$ of degree $m-1$, $n-1$, respectively, such that the coefficients have $\operatorname{Res}(\bar{A},\bar{B})$ as common denominator relative to $\bar{R}$. Hence, there is a unique presentation

$$X_1(\bar{A},\bar{B})\bar{A} + Y_1(\bar{A},\bar{B})\bar{B} = \operatorname{Res}(\bar{A},\bar{B}), \tag{3.38b}$$

in terms of polynomials $X_1, Y_1 \in \bar{R}[t]$ of degrees $m-1, n-1$, respectively. As a consequence of the irreducibility of $\operatorname{Res}(\bar{A},\bar{B})$ it follows that any presentation

$$X_1\bar{A} + Y_2\bar{B} = \bar{R}_1 \quad (X_2, Y_2 \in \bar{R}[t], \deg(X_2) < m, \deg(Y_2) < n, \bar{R}_1 \in \bar{R}) \tag{3.39a}$$

is a multiple of (3.38b) so that.

$$X_2 = \lambda X_1, \quad Y_2 = \lambda Y_1, \quad \bar{R}_1 = \lambda \operatorname{Res}(\bar{A}, \bar{B}) \quad \text{for some } \lambda \in \bar{R}. \tag{3.39b}$$

Thus, for example, there hold equations (compare chapter 1 (3.13))

$$\left.\begin{aligned} X(\bar{A}, \bar{B}) &= X_1(\bar{A}, \bar{B}) L(\bar{A}, \bar{B}), \\ Y(\bar{A}, \bar{B}) &= Y_1(\bar{A}, \bar{B}) L(\bar{A}, \bar{B}), \\ F(\bar{A}, \bar{B}) &= \operatorname{Res}(\bar{A}, \bar{B}) L(\bar{A}, \bar{B}), \end{aligned}\right\} \tag{3.39c}$$

which upon specialization assume the form

$$\left.\begin{aligned} &X(A, B) = X_1(A, B) L(A, B), \\ &Y(A, B) = Y_1(A, B) L(A, B), \quad F(A, B) = \operatorname{Res}(A, B) L(A, B), \end{aligned}\right\} \tag{3.39d}$$

$$\operatorname{Res}(A, B) = X_1(A, B) A + Y_1(A, B) B. \tag{3.39e}$$

Examples show that $F(A, B)$ and $\operatorname{Res}(A, B)$ need not concide.

Example (3.40)
Let $A_1(t) = B(t) = t^5 - t + 1$, $A_2(t) = A(t) = 5t^4 - 1$. Then $25 A_1(t) = 5t A_2(t) + A_3(t)$ for $A_3(t) = -20t + 25$, and $\operatorname{Res}(A, B) = \operatorname{Res}(A_2, A_1) = 25^{-4} \operatorname{Res}(A_2, 25 A_1) = 5^{-4} \operatorname{Res}(A_2, A_3)$. In the next step we obtain

$$(-20)^4 A_2(t) = (-40\,000t^3 - 50\,000t^2 - 62\,500t - 79\,125) A_3 + A_4$$

for $A_4 = 1\,797\,125$ and therefore

$$\begin{aligned} \operatorname{Res}(A_2, A_3) &= (-20)^{-4} \operatorname{Res}((-20)^4 A_2, A_3) \\ &= (-20)^{-4} (-20)^4 \operatorname{Res}(1\,797\,125, A_3) \\ &= 1\,797\,125 = F(A, B). \end{aligned}$$

On the other hand, $\operatorname{Res}(A, B) = 2869$ and thus

$$\begin{aligned} &L(A, B) = 625, \quad X_1(A, B) = -320t^4 - 400t^3 - 500t^2 - 625t + 256, \\ &Y_1(A, B) = -1600t^3 - 2000t^2 - 2500t - 3125. \end{aligned}$$

If R is a factorial ring and the non-zero polynomials $A, B \in R[t]$ have no common divisor, then there are polynomials X, $Y \in R[t]$ with $\deg(X) < \deg(B)$, $\deg(Y) < \deg(A)$ such that

$$XA + YB = 1. \tag{3.41}$$

As the example $A(t) = t^2 + 1$, $B(t) = t^2 + 4$ with $\operatorname{Res}(A, B) = 9$ shows, any divisor of $\operatorname{Res}(A, B)$ has a presentation in the form $XA + YB$ in that case. See also exercise 4.

We conclude our discussion of resultants by presenting an algorithm for their computation.

Algorithm for computing resultants (3.42)

Input. $A, B \in R[t]$, $\deg(A) \geqslant \deg(B) > 0$, R a unital commutative entire ring (or $R = R_0[\bar{a}_0, \ldots, \bar{a}_n, \bar{b}_0, \ldots, \bar{b}_m]$).

Output. $\mathrm{Res}(A, B) \in R$ and polynomials $X_1, Y_1 \in R[t]$ satisfying (3.39e), respectively (3.38b).

Step 1. (Initialization). Set $\mathrm{Res}(A, B) \leftarrow 1$, $F \leftarrow A$, $G \leftarrow B$, $n \leftarrow \deg(A)$, $m \leftarrow \deg(B)$; $A_i \leftarrow t^{m-i}$, $\tilde{A}_i \leftarrow 0$ $(1 \leqslant i \leqslant m)$; $B_j \leftarrow 0$, $\tilde{B}_j \leftarrow t^{n-j}$ $(1 \leqslant j \leqslant n)$.

Step 2. (Pseudo-division). Set $b_0 \leftarrow l(G)$ and compute via 1 (3.12) polynomials $Q(F, G)(t) = \sum_{i=0}^{n-m} q_i t^{n-m-i}$ and $R(F, G)(t)$. Set $s \leftarrow n - \deg(R(F, G))$, $\mathrm{Res}(A, B) \leftarrow \mathrm{Res}(A, B)(b_0^{n-m+1})^{-m} b_0^s(-1)^{mn}$, $A_i(t) \leftarrow b_0^{n-m+1} A_i(t) - \sum_{j=1}^{n-m+i} q_{j-i} B_j(t)$, $\tilde{A}_i(t) \leftarrow b_0^{n-m+1} \tilde{A}_i(t) - \sum_{j=i}^{n-m+i} q_{j-1} \tilde{B}_j(t) (1 \leqslant i \leqslant m)$. In case $R(F, G)$ is constant go to 4, else to 3.

Step 3. (Interchange F, G). Set $F \leftarrow G$, $G \leftarrow R(F, G)$, $n \leftarrow m$, $m \leftarrow \deg(R(F, G))$; $C_i \leftarrow B_{i+s}$, $\tilde{C}_i \leftarrow \tilde{B}_{i+s}$ $(1 \leqslant i \leqslant m)$; $B_i \leftarrow A_i$, $\tilde{B}_i \leftarrow \tilde{A}_i$ $(1 \leqslant i \leqslant n)$; $A_i \leftarrow C_i$, $\tilde{A}_i \leftarrow \tilde{C}_i$ $(1 \leqslant i \leqslant m)$ and go to 2.

Step 4. (Termination). For $R(F, G) = 0$ set $\mathrm{Res}(A, B) \leftarrow 0$, $X_1 \leftarrow A_m$, $Y_1 \leftarrow \tilde{A}_m$; otherwise set $\mathrm{Res}(A, B) \leftarrow \mathrm{Res}(A, B)(R(F, G))^m$, $X_1 \leftarrow A_m(R(F, G))^{m-1}$, $Y_1 \leftarrow \tilde{A}_m(R(F, G))^{m-1}$. Then terminate.

Remarks (3.43)

(i) The polynomials X_1, Y_1 obviously satisfy $\deg(X_1) \leqslant m - 1$, $\deg(Y_1) \leqslant n - 1$. The example $A(t) = t^2 + 1$, $B(t) = t^2 + 4$, however, shows that equality need not hold. Namely, in that case $9 = \mathrm{Res}(A, B) = -3(t^2 + 1) + 3(t^2 + 4)$.

(ii) The change of $\mathrm{Res}(A, B)$ in step 2 can usually not be carried out in R but only in $\mathfrak{Q}(R)$. This can be avoided by storing the multipliers b_0 and their exponents in each step separately and computing $\mathrm{Res}(A, B)$ only at the end of the algorithm, since we know that $\mathrm{Res}(A, B) \in R$. If zero divisors occur, then it is advisable to carry out the computations of $\mathrm{Res}(A, B)$ in $R_0[\bar{a}_0, \ldots, \bar{a}_n, \bar{b}_0, \ldots, \bar{b}_m]$ and then to determine $\mathrm{Res}(A, B)$ at the end via a specialization procedure.

Example (3.44)

Compute $\mathrm{Res}(t^3 + 1, 2t^2 - 2)$ in $\mathbb{Z}[t]$. The single steps of the algorithm produce:

1. $F(t) = t^3 + 1$, $n = 3$, $G(t) = 2t^2 - 2$, $m = 2$; $A_1 = t$, $A_2 = 1$, $\tilde{A}_1 = \tilde{A}_2 = 0$; $\tilde{B}_1 = t^2$, $\tilde{B}_2 = t$, $\tilde{B}_3 = 1$, $B_1 = B_2 = B_3 = 0$; $\mathrm{Res}(A, B) = 1$.
2. $2^2(t^3 + 1) = 2t(2t^2 - 2) + 4t + 4$, hence $Q(F, G) = 2t$, $R(F, G) = 4t + 4$, yielding $s = 2$, $\mathrm{Res}(A, B) = (2^2)^{-2} 2^2 = 2^{-2}$, $A_1 = 4t$, $A_2 = 4$, $\tilde{A}_1 = -2t^2$, $\tilde{A}_2 = -2t$.

3. $F(t)=2t^2-2$, $n=2$, $G(t)=4t+4$, $m=1$; $A_1=0$, $\tilde{A}_1=1$, $B_1=4t$, $B_2=4$, $\tilde{B}_1=-2t^2$, $\tilde{B}_2=-2t$.
2. $4^2(2t^2-2)=(8t-8)(4t+4)+0$, hence $Q(F,G)=8t-8$, $R(F,G)=0$, yielding $s=2$, $\operatorname{Res}(A,B)=2^{-2}(4^2)^{-1}4^2=2^{-2}$, $A_1=-32t+32$, $\tilde{A}_1=16t^2-16t+16$.
4. $\operatorname{Res}(A,B)=0$, $X_1(t)=32(-t+1)$, $Y_1(t)=16(t^2-t+1)$.

4. Discriminant computation

Let

$$\bar{R}=\mathbb{Z}[\bar{a}_0,\bar{a}_1,\ldots,\bar{a}_n] \tag{3.45a}$$

be the polynomial ring in $\bar{a}_0, \bar{a}_1,\ldots,\bar{a}_n$; let

$$\bar{A}(t)=\bar{a}_0t^n+\bar{a}_1t^{n-1}+\cdots+\bar{a}_n \tag{3.45b}$$

be the 'generic' polynomial of degree n of $\bar{R}[t]$; let

$$\mathfrak{Q}(\bar{R})=\mathbb{Q}(\bar{a}_0,\bar{a}_1,\ldots,\bar{a}_n) \tag{3.45c}$$

be the quotient field of $\bar{R}$, and let

$$\bar{A}(t)=\bar{a}_0\prod_{i=1}^{n}(t-\bar{x}_i) \tag{3.45d}$$

be the factorization of $\bar{A}$ in $S(\bar{A}/\mathfrak{Q}(\bar{R}))$. We see as before that $S(\bar{A}/\mathfrak{Q}(\bar{R}))$ is an extension of degree $n!$ of $\mathfrak{Q}(\bar{R})$ with the permutation automorphisms of $\bar{x}_1,\ldots,\bar{x}_n$ as automorphism group over $\mathfrak{Q}(\bar{R})$ fixing simultaneously only the elements of $\mathfrak{Q}(\bar{R})$. We also see that $\mathbb{Z}[\bar{a}_0,\bar{x}_1,\ldots,\bar{x}_n]$ is a minimal splitting ring of $\bar{A}$ over $\bar{R}$ and that the generators $\bar{a}_0$, $\bar{x}_1,\ldots,\bar{x}_n$ are algebraically independent such that the quotient field is

$$S(\bar{A}/\mathfrak{Q}(\bar{R}))=\mathfrak{Q}(\mathbb{Z}[\bar{a}_0,\bar{x}_1,\ldots,\bar{x}_n])=\mathbb{Q}(\bar{a}_0,\bar{x}_1,\ldots,\bar{x}_n).$$

It follows that the expression

$$d(\bar{A})=\bar{a}_0^n\prod_{1\leqslant i<k\leqslant n}(\bar{x}_i-\bar{x}_k)^2 \tag{3.45e}$$

belongs to $\bar{R}$.

For any polynomial

$$A(t)=a_0t^n+a_1t^{n-1}+\cdots+a_n\in R[t], \tag{3.46a}$$

we obtain, by the specialization

$$\bar{R}\to R \quad \text{via} \quad 1_{\bar{R}}\mapsto 1_R,\ \ \bar{a}_i\mapsto a_i \ \ (0\leqslant i\leqslant n) \tag{3.46b}$$

applied to $d(\bar{A})$, the *discriminant* $d(A)$ of A. For any splitting ring Λ of A over R we have a factorization

$$A(t)=a_0\prod_{i=1}^{n}(t-x_i) \quad (x_i\in\Lambda,\ \ 1\leqslant i\leqslant n), \tag{3.46c}$$

such that the specialization

$$\bar{R}\to\Lambda \quad \text{via} \quad 1_{\bar{R}}\mapsto 1_{\Lambda},\quad \bar{a}_0\mapsto a_0,\quad \bar{x}_i\mapsto x_i \quad (1\leqslant i\leqslant n) \tag{3.46d}$$

extends (3.46b), leading to the equation

$$d(A)=a_0^n \prod_{1\leqslant i<k\leqslant n}(x_i-x_k)^2, \tag{3.46e}$$

which shows that the new definition coincides with the former one in case A is monic.

Upon differentiation of (3.45d) we obtain

$$\bar{A}'(t)=\bar{a}_0\sum_{i=1}^{n}\prod_{\substack{j=1\\ j\neq i}}^{n}(t-\bar{x}_j), \tag{3.47a}$$

hence

$$A'(\bar{x}_i)=\bar{a}_0\prod_{\substack{j=1\\ j\neq i}}^{n}(\bar{x}_i-\bar{x}_j) \tag{3.47b}$$

and

$$\begin{aligned}\operatorname{Res}(\bar{A},\bar{A}')&=\bar{a}_0^{n-1}\prod_{i=1}^{n}\bar{A}'(\bar{x}_i)=\bar{a}_0^{2n-1}\prod_{i=1}^{n}\prod_{\substack{j=1\\ j\neq i}}^{n}(\bar{x}_i-\bar{x}_j)\\ &=(-1)^{\binom{n}{2}}\bar{a}_0^{n-1}d(\bar{A}).\end{aligned}$$

Again, by specialization we obtain

$$\operatorname{Res}(A,A')=(-1)^{\binom{n}{2}}a_0^{n-1}d(A). \tag{3.48}$$

Thus discriminants can be computed quite easily as resultants of the polynomial and its derivative.

For example, in (3.40) we found for $R=\mathbb{Z}$, $A(t)=t^5-t+1$ that $d(A)=2869$.

In case $A\in R[t]$ is monic we set $x=t/A$ in $R[t]/AR[t]$. Then the left-hand side of (3.48) appears as the norm from $R[t]/AR[t]$ to R yielding

$$\mathrm{N}(A'(x))=(-1)^{\binom{n}{2}}d(A), \tag{3.49}$$

an often useful relation.

Exercises

1. Carry out example (3.10) in detail.
2. Write

   $P(\xi_1,\xi_2,\xi_3):=2(\xi_1^3+\xi_2^3+\xi_3^3)-3(\xi_1^2\xi_2+\xi_1^2\xi_3+\xi_1\xi_2^2+\xi_1\xi_3^2+\xi_2^2\xi_3+\xi_2\xi_3^2)$

   as a polynomial in the basic symmetric functions of ξ_1,ξ_2,ξ_3.
3. Compute $\operatorname{Res}(A,B)$ for $A(t)=t^3+2t^2+3t+5$, $B(t)=6t^2+8t+9$ over $\mathbb{Z}$.
4. Let $A,B\in\mathbb{Z}[t]$. Give a necessary and sufficient condition for $\operatorname{Res}(A,B)=\gcd(A,B)$.
5. Specialize algorithm (3.42) to the case $B=A'$.

6. Compute the discriminant of $t^6 + t^3 + 3$ over $\mathbb{Z}$ and over $\mathbb{Z}/3$.

7. (a) Compute the discriminant $d(t^n - a_n)$.
   (b) Prove $d(t^n + at + b) = (n-1)^{n-1}(-a)^n + (-1)^{\binom{n}{2}} n^n b^{n-1}$.
   (c) Compute $d(t^4 + a_2 t^3 + a_3 t + a_4)$.

8. Let Λ be *a* commutative unital overring of the ring R with R-basis $b_1, \ldots, b_n$. Let Ψ be a Λ-ring with Λ-basis $c_1, \ldots, c_m$ and corresponding regular representation

$$M_1 : \Psi \to \Lambda^{m \times m} : \xi \mapsto M_1(\xi) = (\alpha_{ik}(\xi))$$

Also let $M_2 : \Lambda \to R^{n \times n}$ be the regular representation of Λ with respect to the R-basis $b_1, \ldots, b_n$.

   (a) Show that

$$M_3 : \Psi \to R^{mn \times mn} : \xi \mapsto M_3(\xi) = (M_2(\alpha_{ik}(\xi)))$$

   is the regular representation of Ψ with respect to the R-basis formed by the products $b_i c_k (1 \leqslant i \leqslant n, 1 \leqslant k \leqslant m)$ in some order.
   (b) Show that

$$\mathrm{Tr}_{\Psi/R}(\xi) = \mathrm{Tr}_{\Lambda/R}(\mathrm{Tr}_{\Psi/\Lambda}(\xi)), \mathrm{N}_{\Psi/R}(\xi) = \mathrm{N}_{\Lambda/R}(\mathrm{N}_{\Psi/\Lambda}(\xi))$$

   for all ξ of Ψ.

2.4. Indecomposable splitting rings

Let us study the algebraic decomposition of universal splitting rings. For simplicity's sake we shall assume that the coefficient ring is indecomposable, e.g. any entire ring will do.

Theorem (4.1)
The universal splitting ring $S(f/R)$ of a monic polynomial (1.3) *over an indecomposable unital commutative ring R is the algebraic sum of a finite number of indecomposable component ideals*

$$S(f/R) = \sum_{i=1}^{s} e_i S(f/R) \tag{4.1a}$$

generated by the primitive idempotents

$$e_1, \ldots, e_s \tag{4.1b}$$

of $S(f/R)$.

Proof
Let e be an idempotent of $S(f/R)$ and let $e_1 = e, e_2, \ldots, e_\mu$ be the distinct $\mathfrak{S}_n$-conjugates (= images) of e. They generate under multiplication, complementation and addition of orthogonal idempotents a finite system of

idempotents which is invariant under the application of $\bar{\mathfrak{S}}_n$ and which is generated by summing up without repetition certain orthogonal idempotents $e'_1, \ldots, e'_{\mu'}$ of the system. By construction also $\{e'_1, \ldots, e'_{\mu'}\}$ is invariant under the application of $\bar{\mathfrak{S}}_n$. We number the e'_i in such a way that $\{e'_1, e'_2, \ldots, e'_{\mu''}\}$ is a $\bar{\mathfrak{S}}_n$-conjugacy class where μ'' is a certain index for which $1 \leqslant \mu'' \leqslant \mu'$. Hence, the sum $e' = e'_1 + e'_2 + \cdots + e'_{\mu''}$ is an idempotent of $S(f/R)$ which is invariant under the action of $\bar{\mathfrak{S}}_n$. We shall prove

$$e' = 1_R. \tag{4.2}$$

For this purpose we show that $e' \in R$ implying (4.2) because of the assumed indecomposability of R. For $n=2$ we obtain $f(t) = t^2 + a_1 t + a_2 \in R[t]$, $e' = \lambda + \mu\xi_2$ $(\lambda, \mu \in R)$, $\mu \in [0/2]$ by theorem (2.18), $2\mu = 0$;

$$\begin{aligned}
\xi_1 &= -a_1 - \xi_2, \\
e' &= \lambda + \mu\xi_1 = \lambda + \mu(-a_1 - \xi_2) \\
&= \lambda - \mu a_1 - \mu\xi_2 \\
&= \lambda + \mu\xi_2, \\
\mu a_1 &= 0, \\
e'^2 &= \lambda^2 + 2\lambda\mu\xi_2 + \mu^2\xi_2^2 = \lambda^2 + 0 + \mu^2(-a_1\xi_2 - a_2) \\
&= \lambda^2 - \mu^2 a_2 = e' = \lambda + \mu\xi_2, \quad \mu = 0, \quad e' = \lambda \in R.
\end{aligned}$$

For $n > 2$ we apply induction as in the proof of theorem (2.18) so that $e' \in R[\xi_{n-1}] \cap R[\xi_n]$,

$$e' = \lambda_0 + \lambda_1\xi_{n-1} + \cdots + \lambda_{n-1}\xi_{n-1}^{n-1} \quad (\lambda_i \in R, 0 \leqslant i \leqslant n-1), \tag{4.3a}$$

$$e' = \lambda_0 + \lambda_1\xi_n + \cdots + \lambda_{n-1}\xi_n^{n-1}. \tag{4.3b}$$

Furthermore,

$$0 = \frac{f}{t - \xi_n}(\xi_{n-1}) = \xi_{n-1}^{n-1} + (a_1 + \xi_n)\xi_{n-1}^{n-2} + \cdots,$$

and upon comparison of (4.3a), (4.3b) the coefficient of $\xi_n\xi_{n-1}^{n-2}$ must vanish: $\lambda_{n-1} = 0$. Hence, again by comparison of (4.3a), (4.3b): $\lambda_1 = \cdots = \lambda_{n-2} = 0$ and $e' \in R$.

As stated above, this implies $e' = 1_R$. But 1_R cannot be orthogonal to any idempotent, and we conclude $\mu' = \mu''$, i.e. all the e'_i are conjugate under $\bar{\mathfrak{S}}_n$.

Next we want to show that the idempotent e annihilates no element of R other than 0. At first, we show this for e'_1. Indeed, if $e'_1 a = 0$ for an element $a \in R$, then it follows $0 = \bar{\pi}(e'_1 a) = \bar{\pi}(e'_1)a$ for all $\bar{\pi} \in \bar{\mathfrak{S}}_n$ implying $0 = \sum_{i=1}^{\mu'} e'_i a = 1_R a = a$. But e itself is a sum of some of the e'_i, say $e = \sum_{i=1}^{\kappa} e'_i$, where $1 \leqslant \kappa \leqslant \mu'$. Hence, any equation $ea = 0$ $(a \in R)$ implies that $e'_1 a = (e'_1 e)a = e'_1(ea) = 0$ and therefore $a = 0$.

Any system of orthogonal idempotents $\hat{e}_1, \ldots, \hat{e}_m$ of $S(f/R)$ is linearly independent over R since any R-relation $\sum_{i=1}^{m} \lambda_i \hat{e}_i = 0$ $(\lambda_i \in R, \; 1 \leqslant i \leqslant m)$

implies upon multiplication by $\hat{e}_j$ that $\lambda_j\hat{e}_j = 0$, hence $\lambda_j = 0$, as we saw above. We conclude from this remark that $m \leqslant n!$. Otherwise there would be $n! + 1$ orthogonal idempotents $\hat{e}_1, \ldots, \hat{e}_{n!+1}$ in $S(f/R)$ which we can represent as R-linear combinations of the standard R-basis B given in theorem (1.6), say $\hat{e}_j = \sum_{b\in B} \lambda_{jb} b$ ($\lambda_{jb} \in R$) so that the rectangular matrix $L := (\lambda_{jb})_{(n!+1)\times n!}$ would be of rank $n! + 1$ according to the assumed linear independence of the $\hat{e}_j$. But the rank of L is clearly $\leqslant n!$, and therefore there can be at most $n!$ orthogonal idempotents in $S(f/R)$.

It follows that there are only finitely many primitive idempotents, say $e_1, \ldots, e_s$ in $S(f/R)$. They are mutually orthogonal implying (4.1a). □

Corollary (4.4)

Under the premises of (4.1) *the primitive idempotents* $e_1, \ldots, e_s$ *of* $S(f/R)$ *are conjugate under the permutation automorphisms of* $S(f/R)$, *and the same applies to the component rings* $e_iS(f/R)$ $(1 \leqslant i \leqslant s)$.

Proof

We saw in the proof of (4.1) that the permutation automorphisms permute the primitive idempotents and that there is only one orbit under the action of $\mathfrak{S}_n$. □

Corollary (4.5)

Under the premises of (4.1) *there are monomorphisms* $\mu_i : R \to e_iS(f/R) : a \mapsto e_ia$ $(1 \leqslant i \leqslant s)$, *and the* i*th component ring* $e_iS(f/R)$ *is a minimal splitting ring of* $\bar{\mu}_i(f)$ *over* e_iR. (*Here* $\bar{\mu}_i$ *denotes the canonical extension of* μ_i *to* $R[t]$.)

Proof

From the proof of (4.1) we know that any relation $e_ia = 0$ for $a \in R$ implies $a = 0$ $(1 \leqslant i \leqslant s)$. Therefore μ_i is injective. The factorization (1.6b) implies upon multiplication with e_i that $e_if(t) = \prod_{j=1}^{n}(t - e_i\xi_j)$, hence $e_iS(f/R)$ is a splitting ring of e_if over e_iR.

Since $e_iS(f/R)$ is generated by the $n!$ elements $e_i\xi_1^{i_1}e_i\xi_2^{i_2}\cdots e_i\xi_n^{i_n}$ $(0 \leqslant i_j < j;$ $1 \leqslant j \leqslant n)$ it is clear that $e_iS(f/R)$ is a minimal splitting ring of e_if over e_iR. □

Proposition (4.6)

Let R, $S(f/R)$, $e_1, \ldots, e_s$ *as in* (4.1). *If* Λ *is an indecomposable minimal splitting ring of* f *over* R, *say with factorization* (1.2) *for* $a_0 = 1$, *then there is the epimorphism* $\varepsilon_i : e_iS(f/R) \to \Lambda$ *satisfying* $\varepsilon_i(e_ia) = a$ *for all* $a \in R$ *and* $\varepsilon_i(e_i\xi_k) = x_{\pi(k)}$ $(1 \leqslant k \leqslant n)$ *for some permutation* $\pi = \pi_i \in \mathfrak{S}_n$.

Proof

By theorem (1.6) there is an R-epimorphism $\varepsilon : S(f/R) \to \Lambda$ mapping ξ_k onto x_k $(1 \leqslant k \leqslant n)$. Under this epimorphism, the equation $1 = \sum_{i=1}^{s} e_i$ becomes

$1_\Lambda = \sum_{i=1}^{s} \varepsilon(e_i)$, where $\varepsilon(e_i)\,\varepsilon(e_j) = \varepsilon(e_ie_j) = \delta_{ij}\varepsilon(e_i)$ $(1 \leqslant i,j \leqslant s)$. Because of the indecomposability of Λ there is an index i' such that $\varepsilon(e_{i'}) = 1_\Lambda$, and $\varepsilon(e_j) = 0$ $(1 \leqslant j \leqslant s, j \neq i')$. But $e_1,\ldots,e_s$ are $\mathfrak{S}_n$-conjugate because of (4.4). Hence there is $\bar\pi = \bar\pi_i \in \bar{\mathfrak{S}}_n$ satisfying $\bar\pi(e_i) = e_{i'}$ for each i $(1 \leqslant i \leqslant s)$, and $\varepsilon_i := \varepsilon\bar\pi|_{e_iS(f/R)}$ does the job. □

Proposition (4.7)
Let R, $S(f/R)$, $e_1,\ldots,e_s$ as in (4.1).

(i) *The elements of e_iR are invariant under the $\bar{\mathfrak{S}}_n$ stabilizer $G_i := \{\bar\pi \in \bar{\mathfrak{S}}_n \mid \bar\pi(e_i) = e_i\}$ of e_i $(1 \leqslant i \leqslant s)$. They are the only invariant elements of $e_iS(f/R)$ in case $d(f)$ is 2-regular,* i.e. *$2d(f)x = 0$ implies $d(f)x = 0$ for all $x \in S(f/R)$.*
(ii) *The restriction of the $\bar{\mathfrak{S}}_n$-stabilizer G_i of e_i to $e_iS(f/R)$ is faithful in case $d(f) \neq 0$ $(1 \leqslant i \leqslant s)$.*

Proof
(i) Let $\bar\pi \in G_i$. Obviously, $\bar\pi(ae_i) = a\bar\pi(e_i) = ae_i$ for all $a \in R$. In case $d(f)$ is 2-regular we know already that each element of $S(f/R)$ which is fixed by $\bar{\mathfrak{S}}_n$ belongs to R. Now let $y = e_iy$ be an element of $e_iS(f/R)$ which is fixed by G_i. There holds a coset-decomposition $\bar{\mathfrak{S}}_n = \bigcup_{j=1}^{s} g_jG_i$ with g_j depending on e_i such that $g_1 = id$ and $\{g_1(e_i),\ldots,g_s(e_i)\} = \{e_1,\ldots,e_s\}$. Hence, for each element $g \in \bar{\mathfrak{S}}_n$ there hold equations $gg_jG_i = g_{\rho(g,j)}G_i$, where $\rho: \bar{\mathfrak{S}}_n \to \mathfrak{S}_s$ is a transitive permutation representation of $\bar{\mathfrak{S}}_n$ based on the left-multiplication action on the left-cosets g_jG_i. Hence, $z = \sum_{j=1}^{s} g_j(y) = y +$ other components is invariant under all $g \in \bar{\mathfrak{S}}_n$ and therefore $z \in R$ by (2.18) implying $y \in e_iR$.

(ii) For $0 \neq d(f) = \prod_{1 \leqslant j<k \leqslant n}(\xi_j - \xi_k)^2$ we obtain upon multiplication by e_i: $0 \neq e_id(f) = \prod_{1 \leqslant j<k \leqslant n}(e_i\xi_j - e_i\xi_k)^2$, hence $e_i\xi_j \neq e_i\xi_k$. For any $\bar\pi \in G_i$, $\bar\pi \neq id$, there is an index j such that $\pi(j) \neq j$, such that $\pi(j) \neq j$, hence $\xi_j \neq \bar\pi(\xi_j)$, $\bar\pi(e_i\xi_j) = e_i\bar\pi(\xi_j) \neq e_i\xi_j$. Therefore the restriction of G_i to $e_iS(f/R)$ is faithful. □

We have studied only the restriction of the permutation automorphisms to the minimal splitting rings $e_iS(f/R)$ of e_if over e_iR.

In general, there are other automorphisms too. It may not even be possible to single out the restriction of the permutation automorphisms from among them, purely by looking at the structure of $e_iS(f/R)$. This situation changes, if R is a field.

Theorem (4.8)
Let R, $S(f,R)$, e_i $(1 \leqslant i \leqslant s)$ as in (4.1) *and μ_i, $\bar\mu_i$ as in* (4.5). *If R is a field and f is separable, then $e_iS(f,R)$ is a minimal splitting extension of $\bar\mu_i(f)$ over $\mu_i(R)$. Its degree is equal to the order of G_i. Every automorphism of $e_iS(f/R)$ over e_iR is obtained by restricting some element of G_i to $e_iS(f/R)$.*

Proof

If R is a field then the R-dimension of each of the s ideals $e_i S(f/R)$ is the same number d by (4.4). By (4.1a)

$$sd = \dim_R(S(f/R)) = n!.$$

According to (3.24) there is no nilpotent element $\neq 0$ in $S(f/R)$. If there is a zero divisor $\xi \neq 0$ of $e_i S(f/R)$ then it may be chosen in such a way that it generates an R-subalgebra M of $e_i S(f/R)$ of minimal R-dimension μ. There is a non-zero element η of $e_i S(f/R)$ for which $\xi\eta = 0$, hence $M\eta = 0$, hence every non-zero element ζ of M is a zero divisor of $e_i S(f/R)$, and each of them generates an F-subalgebra of the same dimension μ. Hence, $\zeta M = M$, ζM is a field, it has a unit element e, e is an idempotent of $e_i S(f/R)$ distinct from e_i, contrary to (4.1). It follows that $e_i S(f/R)$ is a field.

Using (4.5) it follows that $e_i S(f/R)$ is a minimal splitting extension of $\bar{\mu}_i(f)$ over $\mu_i(R)$. The restriction of G_i to $e_i S(f/R)$ is a group of automorphisms of order d because of $(\bar{\mathfrak{S}}_i : G_i) = s$.

Now we show that there cannot be more than d automorphisms of $e_i S(f/R)$ over $e_i R$ by an induction argument. If all roots $e_i \xi_j$ belong to $e_i R$ then we have $e_i S(f/R) = e_i R$, $d = 1$, and we are done. Choose $e_i \xi_j$ in such a way that $e_i R[e_i \xi_j] = F$ is an extension of $e_i R$ of degree $d_1 > 1$. It follows that the minimal polynomial m of $e_i \xi_j$ over $e_i R$ is irreducible of degree d_1. Denoting by G_{ij} the stabilizer of $e_i \xi_j$ in G_i and setting

$$G_i = \bigcup_{k=1}^{(G_i:G_{ij})} g_k G_{ik},$$

it follows that the G_i-orbit of $e_i \xi_j$ consists of the $(G_i : G_{ij})$ elements

$$g_k e_i \xi_j \quad (1 \leqslant k \leqslant (G_i : G_{ij})).$$

The equation $m(e_i \xi_j) = 0$ implies the equation $m(g_k e_i \xi_j) = 0$. Since the polynomial m has at least d_1 distinct roots it follows that

$$d_1 \geqslant (G_i : G_{ij}), \quad |G_{ij}| \geqslant d/d_1,$$

hence by induction the degree of $e_i S(f/R)$ over F is at least $|G_{ij}|$. On the other hand, it is d/d_1 by the degree theorem (exercise 5 in chapter 1, section 3). Hence, $|G_{ij}| = d/d_1$, $(G_i : G_{ij}) = d_1$,

$$G_{ij}|_{e_i S(f/R)} = \operatorname{Aut}(e_i S(f/R)/F),$$
$$G_i|_{e_i S(f/R)} = \operatorname{Aut}(e_i S(f/R)/e_i R).$$

Thus (4.8) is demonstrated. □

The last remark shows that for every chain or finite extensions

$$F = F_0 \subseteq F_1 \subseteq F_2 \subseteq \cdots \subseteq F_r = E, \tag{4.9a}$$

and

$$F_{i+1} = F_i(\eta_{i+1}) \quad (0 \leqslant i < r) \tag{4.9b}$$

of the field F there cannot be more than $[F_{i+1}:F_i]$ distinct automorphisms of F_{i+1} over F_i $(0 \leqslant i < r)$ and upon using the degree theorem (exercise 5 in chapter 1, section 3)

$$[E:F] = \prod_{i=0}^{r-1} [F_{i+1}:F_i] \tag{4.9c}$$

so that the maximal number of automorphisms of E over F is equal to the degree. But every finite extension E of F is obtainable as in (4.9a), (4.9b). Indeed, for $E = F$ this is trivial. Otherwise E contains an element η_1 not contained in F and hence $F_1 = F(\eta_1)$ has degree > 1 over F. Continuing this process yields (4.9a), (4.9b). Thus it is shown that any finite extension E of F has at most $[E:F]$ automorphisms over F.

Definition (4.10)
The finite extension E of F is said to be a **normal extension** *of F if the number of automorphisms of E over F is equal to the degree* $[E:F]$.

For example any minimal splitting extension of a separable polynomial over F is a normal extension.

Conversely, we show that every normal extension E of F can be presented (in many ways) as minimal splitting extension of a suitable monic polynomial over F. Firstly let us point out that the fixed field of $\mathrm{Aut}(E/F)$ is F. Otherwise it would be a subfield E_0 properly containing F, hence $[E_0:F] > 1$ and by the degree theorem $[E:E_0] < [E:F]$. But there cannot be more than $[E:E_0]$ distinct automorphisms of E over E_0. Hence $E_0 = F$. Secondly we show that every element x of E satisfies a separable minimal equation over F. Indeed, let the monic non-constant polynomial m_x of $F[t]$ be the minimal polynomial of x, i.e.

$$m_x(x) = 0. \tag{4.11a}$$

Applying the elements α of $G = \mathrm{Aut}(E/F)$ to the minimal equation we find that

$$m_x(\alpha(x)) = 0. \tag{4.11b}$$

Upon forming the polynomial

$$f(t) = \prod_{\alpha \in G}{}' (t - \alpha(x)), \tag{4.11c}$$

where the accent at the product symbol indicates that only distinct factors occur, we discover that f is invariant under $\mathrm{Aut}(E/F)$, which means the coefficients of f belong to the fixed field F. But (4.11a–c) imply that f divides m_x. Hence, $f = m_x$, m_x is separable, the minimal equation is a separable equation for x. Finally, we point out that E can be finitely generated over F, for example any F-basis will do.

Let $E = F(x_1, \ldots, x_r)$, then E is a minimal splitting field of $m_{x_1} m_{x_2} \cdots m_{x_r}$. If some of the factors are equal, omit repetitions, the remaining polynomial will be a separable monic polynomial of $F[t]$ with E as minimal splitting field.

Normal extensions form the fundament of algebraic number theory. It is their automorphism group which plays the key role in the discussion of structure. To have this discovered and clearly established forms the immortal contribution of E. Galois to the theory.

Exercises

1. Let E/F be a normal extension with Galois group G. Prove for $\xi \in E$:

$$\mathrm{Tr}(\xi) = \sum_{\sigma \in G} \sigma(\xi), \quad \mathrm{N}(\xi) = \prod_{\sigma \in G} \sigma(\xi).$$

2. (a) Form the universal splitting ring $S(f/\mathbb{Z})$ for $f(t) = (t^2 + 1)(t^2 + 4)$ and show that its primitive idempotents generate components with zero divisors. How many are there?
   (b) What happens over $\mathbb{Q}$ in place of $\mathbb{Z}$?

2.5. Finite fields

Historically speaking C.F. Gauss [5] was the first mathematician to make extensive computations in finite fields other than $\mathbb{Z}/p\mathbb{Z} = \mathbb{Z}/p$. The projected last chapter of the *Disquisitiones arithmeticae* (chapter 8) was devoted to their thorough study. Gauss introduced them as equation rings of irreducible polynomials over $\mathbb{Z}/p$. Thus he used new domains of rationality possibly independent of their origins in number theory, but he refused to give the concept a name. E. Galois introduced the finite field concept in his classical mémoire: Sur la Théorie des nombres (*Bulletin des sciences mathématiques de Férussac*, XIII, §218, June 1830) as follows [4]:

Soit une pareille équation ou congruence, $Fx = 0$, *et* p le module. Supposons d'abord, pour plus de simplicité, que la congruence en question n'admette aucun facteur commensurable, c'est-à-dire qu'on ne puisse pas trouver 3 fonctions ϕx, ψx, χx telles que

$$\phi x \cdot \psi x = Fx + p\chi x.$$

Dans ce cas, la congruence n'admettra donc aucune racine entière, ni même aucune racine incommensurable du degré inférieur. Il faut donc regarder les racines de cette congruence comme des espèces de symboles imaginaires, puis qu'elles ne satisfont pas aux questions des nombres entiers, symboles dont l'emploi dans le calcul sera souvent aussi utile que celui de l'imaginaire $(-1)^{\frac{1}{2}}$ dans l'analyse ordinaire.

From this reason we speak about finite fields also as *Galois fields.* In the sequel we discuss finite fields from a constructive viewpoint. For a detailed discussion of Galois fields we refer the reader to the book by Lidl and Niederreiter [11].

1. Gauss' routine for primitive roots

In order to find a *primitive root*, i.e. a generator of the multiplicative group $\mathbb{F}^\times$ of a finite field $\mathbb{F} = \mathbb{F}_q$ of q elements we make use of Gauss' algorithm [5]. We enumerate the non-zero elements of $\mathbb{F}$ in some suitable way, say

$$x_1, x_2, \ldots, x_{q-1}, \tag{5.1}$$

where it is not demanded to have a big list or table ready for use, we merely need to know a method to find another element $x_{k+1} \in \mathbb{F}^\times$ in case $x_1, \ldots, x_k$ already are known as distinct elements of $\mathbb{F}^\times$ and $k < q-1$. We shall call an algorithm which produces all elements of $\mathbb{F}^\times$ without repetition a *potential list* of $\mathbb{F}^\times$.

For each element x_i we need to determine its order $\operatorname{ord}(x_i)$ in $\mathbb{F}^\times$. For this purpose we use the Fermat theorem yielding

$$x_i^{q-1} = 1 \tag{5.2}$$

and a prime factorization

$$q - 1 = \prod_{i=1}^{s} p_i^{\mu_i} \quad (\mu_i \in \mathbb{N}) \tag{5.3a}$$

of $q-1$. Suppose we know already that

$$x_i^m = 1 \tag{5.3b}$$

for some divisor

$$m = \prod_{i=1}^{s} p_i^{\nu_i} \quad (\nu_i \in \mathbb{Z}, 0 \leqslant \nu_i \leqslant \mu_i) \tag{5.3c}$$

of $q-1$, then we test in sequence the equations

$$x_i^{m/p_i} = 1 \tag{5.3d}$$

for each prime number p_i $(1 \leqslant i \leqslant s)$ for which $\nu_i > 0$. If (5.3d) holds for some index i, we replace m by m/p_i and start all over again. If none of the equations (5.3d) holds, then we know that

$$m = \operatorname{ord}(x_i). \tag{5.3e}$$

Next we form

$$m_1 = \operatorname{ord}(x_1),\ m_2 = lcm(m_1, \operatorname{ord}(x_2)), \ldots, m_k = lcm(m_{k-1}, \operatorname{ord}(x_k)). \tag{5.4}$$

We note that $\operatorname{ord}(xy) = \operatorname{ord}(x)\operatorname{ord}(y)$ in case $\gcd(\operatorname{ord}(x), \operatorname{ord}(y)) = 1$.

Proposition (5.5)
There is an index k, $k \leqslant q-1$, for which

$$m_k = q - 1, \tag{5.5a}$$

and a primitive root of $\mathbb{F}^\times$ can be constructed as a power product of $x_1, \ldots, x_k$.

Proof
Since we know already that there exists a primitive element of $\mathbb{F}^\times$, say x, that element is among those of (5.1), hence (5.5a) is established. But of course (5.5a) will usually arise much earlier than the occurence of x. This is because the numbers $m_1, m_2, \ldots$ form an ascending divisor chain terminating with $q-1$.

For the second statement of (5.5) we merely have to construct a power product $x^i y^j$ for any two elements x, y of $\mathbb{F}^\times$ such that $\operatorname{ord}(x^i y^j) = \operatorname{lcm}(\operatorname{ord}(x), \operatorname{ord}(y))$. Let B be a basis of the divisor cascade of $S = \{\operatorname{ord}(x), \operatorname{ord}(y)\}$. Then we obtain

$$\operatorname{ord}(x) = \prod_{b \in B} b^{\mu_x}, \quad \operatorname{ord}(y) = \prod_{b \in B} b^{\mu_y}$$

and therefore $\operatorname{lcm}(\operatorname{ord}(x), \operatorname{ord}(y)) = \prod_{b \in B} b^{\max\{\mu_x, \mu_y\}}$. Hence, for

$$i := \prod_{\substack{b \in B \\ \mu_x \geqslant \mu_y}} b^{\mu_x}, \quad j := \prod_{\substack{b \in B \\ \mu_y \leqslant \mu_x}} b^{\mu_y},$$

the elements x^i, y^j have orders

$$\operatorname{ord}(x^i) = \prod_{\substack{b \in B \\ \mu_x \geqslant \mu_y}} b^{\mu_x}, \quad \operatorname{ord}(y^j) = \prod_{\substack{b \in B \\ \mu_y > \mu_x}} b^{\mu_y}.$$

(Obviously, for $i \mid \operatorname{ord}(x)$ we have $\operatorname{ord}(x^i) = \operatorname{ord}(x)/i$.) Therefore the product $x^i y^j$ has the desired order since $\operatorname{ord}(x^i)$, $\operatorname{ord}(y^j)$ are now coprime. □

Remarks
(i) It is clear that the method of the proof also applies to elements of a finite abelian group.

(ii) For an efficient performance of Gauss' algorithm we remark that for each new element x_i we first test whether $x_i^{m_{i-1}} = 1$ already. In that case x_i is of no help in increasing m_{i-1} and we proceed to x_{i+1} with $m_i = m_{i-1}$. Especially, $x_i = 1$ and $x_i = -1$ (for $2 \mid m_{i-1}$) can be omitted.

(iii) It should be pointed out once again that the algorithm of Gauss can also be used to establish the existence of a primitive root. The field property of $\mathbb{F}$ is invoked at the point when we obtained a power product of $x_1, \ldots, x_k$ of order $m_k = q-1$, since there cannot be more than $q-1$ solutions of the equation $x^{q-1} = 1$ in $\mathbb{F}$. On the other hand, if one is not sure whether a finite unital commutative ring R is a field, then Gauss' algorithm forms the basis for many tests. The main difficulty in that case is to establish equations $x^e = 1$ $(e \in \mathbb{N})$ for any $x \neq 0$ of R. This method is used, for example, to decide whether a given large number x is a prime number. It turns out to be easier to search for a primitive element than for a zero divisor.

(iv) For $\mathbb{F} = \mathbb{Z}/2$ the unit element is the primitive root. For $\mathbb{F} = \mathbb{Z}/p$ $(p > 2)$ Gauss used the sequence $x_1 = -1$, $x_2 = 2$, $x_3 = 3$, $x_4 = 5, \ldots$. Experience shows that this procedure yields a primitive root in about $\log p$ steps, though with the present knowledge of analytic number theory no comparable estimate is demonstrable.

(v) If $\mathbb{F}_{p^n}$ is given as $\mathbb{F}_p[t]/f\mathbb{F}_p[t]$, where f is a monic irreducible polynomial of degree $n > 1$ of $\mathbb{F}_p[t]$, then one may arrange

$$x_i = \phi_i(t/f) \quad (\phi_i \in \mathbb{F}_p[t], \deg(\phi_i) < n),$$

such that linear polynomials ϕ_i come first, then quadratic polynomials, etc. Of course the ϕ_i should be monic and irreducible.

Primitive roots are useful in order to simplify calculations in a fixed finite field $\mathbb{F}_q$. Let ω be a primitive root of $\mathbb{F}_q$. Then the powers ω^i $(1 \leqslant i \leqslant q-1)$ present the elements of $\mathbb{F}_q^\times$ canonically. The exponents which appear in this presentation are said to be the *indices* of the corresponding elements. We use them for the purpose of tagging the elements of $\mathbb{F}_q^\times$ by means of an *index table.*

For example, the residue class 2/5 is a primitive root of $\mathbb{F}_5$. Its four distinct powers are

$$1/5 = (2/5)^0, \quad 2/5 = (2/5)^1, \quad -1/5 = (2/5)^2, \quad -2/5 = (2/5)^{-1}.$$

The corresponding index table is

x	1	2	-1	-2
ind (x)	0	1	2	-1

,

with the integers $1, 2, -1, -2$ representing the elements x of $\mathbb{F}_5^\times$ in the first row.

Regarding multiplication and exponentiation indices behave like logarithms according to the rules

$$\begin{aligned} \operatorname{ind}(xy) &\equiv \operatorname{ind}(x) + \operatorname{ind}(y) \mod (q-1), \\ \operatorname{ind}(x^m) &\equiv m \operatorname{ind}(x) \qquad\quad \mod (q-1), \end{aligned}$$

i.e.

$$\begin{aligned} \operatorname{ind}(xy) &= R(\operatorname{ind}(x) + \operatorname{ind}(y), q-1), \\ \operatorname{ind}(x^m) &= R(m \operatorname{ind}(x), q-1) \quad (x, y \in \mathbb{F}_q^\times, m \in \mathbb{Z}). \end{aligned}$$

Indices can be used in place of the elements of $\mathbb{F}_q$ upon tagging the zero element by some additional symbol, say ∞: $\operatorname{ind}(0) = \infty$. Then both rows of the index table become in case of $q = 5$:

x	1	2	-1	-2	0
ind (x)	0	1	2	-1	∞

.

In any case

$$\mathrm{ind}:\mathbb{F}_q \to \{\infty\} \cup \{x \in \mathbb{Z} \mid -q < 2x \leqslant q\}$$

is a bijection.

For the purpose of doing additions as well we use a second table of 'additive indices' connecting the values ind (x) and ind $(1+x)$ for $x \in \mathbb{F}_q$. Again, for $q = 5$ we obtain

ind (x)	∞	0	1	2	-1
ind $(1+x)$	0	1	-1	∞	2

By the use of $x + y = x(1 + (y/x))$ we get

$$\begin{aligned}\mathrm{ind}(x+y) &= R\left(\mathrm{ind}(x) + \mathrm{ind}\left(1 + \frac{y}{x}\right), q-1\right) \\ &= R(\mathrm{ind}(x) + \mathrm{ind}(1 + \mathrm{ind}^{-1}(R(\mathrm{ind}(y) - \mathrm{ind}(x), q-1))), q-1).\end{aligned} \tag{5.6}$$

2. The construction of finite fields

Let p be a rational prime number and $n \in \mathbb{N}$. The Galois field $\mathbb{F}_{p^n}$ is usually defined as a minimal splitting field of $t^{p^n} - t$ over $\mathbb{Z}/p$. But that construction is too time consuming. Of course it suffices to know an irreducible monic polynomial f of degree n of $\mathbb{Z}/p[t] = \mathbb{F}_p[t]$ since

$$\mathbb{F}_{p^n} \cong \mathbb{F}_p[t]/f\mathbb{F}_p[t]. \tag{5.7a}$$

In greater generality we deal with the construction of

$$\mathbb{F}_{q^m} \cong \mathbb{F}_q[t]/f\mathbb{F}_q[t], \tag{5.7b}$$

where $q = p^n$ and $f \in \mathbb{F}_q[t]$ is monic and irreducible.

As in many constructive problems there are two ways competing for consideration, one more aleatoric (from alea = dice; one believes that dice throwing produces the numbers 1–6 randomly), the other one more deterministic.

The *aleatoric method* is based on the observation that the number $A_{m,q}$ of monic irreducible polynomials of degree m over $\mathbb{F}_q$ is roughly $1/m$ of the total number q^m of monic polynomials of degree m, as we shall show below. In other words, if one chooses randomly a monic polynomial

$$f(t) = t^m + a_1 t^{m-1} + \cdots + a_m \in \mathbb{F}_q[t], \tag{5.8}$$

the probability that f is irreducible is about $1/m$. One believes that picking out m elements $a_1, \ldots, a_m \in \mathbb{F}_q$ independently can be done randomly so that an irreducible f can be produced 'completely by chance'. The fundamental problems arising from the construction of random sequences are very well discussed in the paper by Niederreiter [12].

In order to produce a formula for the number $A_{m,q}$ of monic irreducible polynomials of degree m over $\mathbb{F}_q$ we need some preparations in the form of propositions and theorems.

Proposition (5.9)
Let R be a unital commutative ring and $r, s \in \mathbb{N}$. Then we have: $(t^s - 1) | (t^r - 1)$ in $R[t] \Leftrightarrow s | r$ in $\mathbb{N}$. (For $R = \mathbb{Z}$ the specialization $t \mapsto m \in \mathbb{Z}^{\geq 2}$ yields: $(m^s - 1) | (m^r - 1) \Leftrightarrow s | r$.)

Proof
Let $r = Q(r, s)s + R(r, s)$ such that $0 \leq R(r, s) < s$. Then we obtain

$$\begin{aligned} t^r - 1 &= t^{Q(r,s)s} t^{R(r,s)} - t^{R(r,s)} + t^{R(r,s)} - 1 \\ &= t^{R(r,s)}((t^s)^{Q(r,s)} - 1) + t^{R(r,s)} - 1, \end{aligned}$$

and because of

$$(t^s)^{Q(r,s)} - 1 = (t^s - 1) \sum_{i=0}^{Q(r,s)-1} (t^s)^i,$$

the equivalence

$$R(r, s) = 0 \Leftrightarrow (t^s - 1) | (t^r - 1)$$

holds. □

Theorem (5.10)
Let $\mathbb{F}_q = \mathbb{F}_{p^n}$ and $m \in \mathbb{N}$. Then $t^{q^m} - t \in \mathbb{F}_q[t]$ is the product of all monic irreducible polynomials $g \in \mathbb{F}_q[t]$ for which $\deg(g) | m$.

Proof
The proof is carried out in three steps.

(i) Let $g \in \mathbb{F}_q[t]$ be monic irreducible with $\deg(g) | m$. We shall show $g(t) | (t^{q^m} - t)$. The case $g(t) = t$ is trivial. We therefore assume that $g(t) \neq t$ and $x \neq 0$ is a zero of g. Then $\mathbb{F} := \mathbb{F}_q(x)$ is a field extension of $\mathbb{F}_q$ of degree $\deg(g)$ and g is the minimal polynomial of x. Since $\mathbb{F}^\times$ is cyclic we must have $x^{q^{\deg(g)} - 1} = 1$ and therefore $g(t) | (t^{q^{\deg(g)} - 1} - 1)$. Then (5.9) implies $(q^{\deg(g)} - 1) | (q^m - 1)$ and also $g(t) | (t^{q^{\deg(g)} - 1} - 1) | (t^{q^m - 1} - 1)$.

(ii) Let $g(t) \in \mathbb{F}_q[t]$ be monic, irreducible and dividing $t^{q^m} - t$. We show: $\deg(g) | m$. Again the case $g(t) = t$ is trivial. We assume that $g(t) \neq t$ and that $x \neq 0$ is a zero of $g(t)$. Let y be a primitive root of $\mathbb{F} := \mathbb{F}_q(x)$, i.e. $\operatorname{ord}(y) = q^{\deg(g)} - 1$ in $\mathbb{F}^\times$.

There is a presentation

$$y = \sum_{i=1}^{\deg(g)} y_i x^{i-1} \quad (y_i \in \mathbb{F}_q, 1 \leq i \leq \deg(g)),$$

from which we obtain

$$y^{q^m} = y^{p^{nm}} = \left(\sum_{i=1}^{\deg(g)} y_i x^{i-1}\right)^{p^{nm}} = \sum_{i=1}^{\deg(g)} y_i^{p^{nm}} x^{(i-1)p^{nm}} = \sum_{i=1}^{\deg(g)} y_i x^{i-1} = y.$$

Hence, $q^m - 1$ is an exponent of y which is divisible by $\operatorname{ord}(y)$. Again with (5.9) we conclude $(q^{\deg(g)} - 1)|(q^m - 1) \Rightarrow \deg(g)|m$.

(iii) Let $g(t) \in \mathbb{F}_q[t]$ be monic, irreducible and dividing $t^{q^m} - t$. We show $(g(t))^2 \nmid (t^{q^m} - t)$. Otherwise we had $\gcd(t^{q^m} - t, (t^{q^m} - t)') \neq 1$. But because of $(t^{q^m} - t)' = q^m t^{q^m} - 1 = -1$ in $\mathbb{F}_q[t]$, obviously the gcd is 1. □

Corollary (5.11)

A monic polynomial $f \in \mathbb{F}_q[t]$ is irreducible, if and only if $\gcd(f(t), t^{q^j} - t) = 1$ *for all j $(1 \leqslant j \leqslant \deg(f)/2)$.*

Corollary (5.11) can be used to test polynomials $f \in \mathbb{F}_q[t]$ for irreducibility. For larger values of $q = p^n$ and $\deg(f)$, however, we need improved methods for the computation of the corresponding gcd's. We first compute

$$R(t^p, f(t)) =: \phi_0(t), \tag{5.12a}$$

and then successively

$$\begin{array}{c} \phi_1(t) := R(\phi_0(\phi_0(t)), f(t)) \equiv t^{p^2} \bmod f \\ \hline \phi_k(t) := R(\phi_{k-1}(\phi_{k-1}(t)), f(t)) \equiv t^{p^{2^k}} \bmod f \end{array} \tag{5.12b}$$

up to a bound k determined by the inequalities

$$2^k \leqslant n \frac{\deg f}{2} < 2^{k+1}. \tag{5.12c}$$

Using the 2-adic representation

$$j = \sum_{i=0}^{k} b_i 2^i \quad (b_i \in \{0, 1\}, 0 \leqslant i \leqslant k), \tag{5.12d}$$

with $b_i = 1$ exactly for the indices $i_1 < i_2 < \ldots < i_l$, we obtain the congruence

$$t^{p^j} \equiv \phi_{i_1}(\phi_{i_2}(\ldots(\phi_{i_l}(t)))) \bmod f, \tag{5.12e}$$

permitting us to determine

$$R(t^{p^j}, f) = R(\phi_{i_1}(\phi_{i_2}(\ldots(\phi_{i_l}))), f) \tag{5.12f}$$

in a modest number of steps. Finally, we get

$$R(t^{p^j} - t, f) = R(t^{p^j}, f) - t \tag{5.12g}$$

and

$$\gcd(t^{p^j} - t, f) = \gcd(R(t^{p^j} - t, f), f). \tag{5.12h}$$

The formula for the numbers $A_{m,q}$ will come out as a consequence of (5.10). However, for the proof we need a well-known function from elementary number theory which will be useful also in later sections.

Definition (5.13)

Let the function μ be defined on $\mathbb{N}$ as follows: for $x=1$ set $\mu(x)=1$. If x contains a square factor set $\mu(x)=0$. Finally, if x is a product of r distinct primes set $\mu(x)=(-1)^r$; μ is called the **Möbius μ-function**.

Its useful properties are contained in

Proposition (5.14)

(a) *For $n\in\mathbb{N}$ we have* $\sum_{d|n}\mu(d)=\begin{cases}1 \text{ for } n=1\\ 0 \text{ else}\end{cases}$.

(b) *Let $f,g:\mathbb{N}\to R$ (R a unital commutative ring), then* $(\forall n\in\mathbb{N}: f(n)=\sum_{d|n}g(d))\Leftrightarrow(\forall n\in\mathbb{N}: g(n)=\sum_{d|n}\mu(d)f(n/d))$. *(This is well-known as the* **Möbius inversion formula.**)

Proof

(a) The case $n=1$ is obvious. For $n>1$, $n\in\mathbb{N}$, let $n=p_1^{m_1}\cdot\ldots\cdot p_r^{m_r}$ ($m_i\in\mathbb{N}$, $1\leqslant i\leqslant r$) the prime factorization of n. Only those divisors of n yield a summand different from zero for which the exponent of p_i is either 1 or 0 ($1\leqslant i\leqslant r$). There are exactly $\binom{r}{k}$ divisors of n for which k exponents are 1, the others 0. Therefore we have

$$\sum_{d|n}\mu(d)=\sum_{k=0}^{r}(-1)^k\binom{r}{k}=\sum_{k=0}^{r}\binom{r}{k}1^{r-k}(-1)^k=(1-1)^k=0.$$

(b) "$\Rightarrow$" $\sum_{d|n}\mu(d)f\left(\frac{n}{d}\right)=\sum_{d|n}\mu(d)\sum_{e|\frac{n}{d}}g(e)=\sum_{d|n}\mu(d)\sum_{ed|n}g(e)$

$$=\sum_{e|n}g(e)\sum_{d|\frac{n}{e}}\mu(d)=g(n) \quad \text{by} \quad (a).$$

"$\Leftarrow$" $\sum_{d|n}g(d)=\sum_{d|n}\sum_{e|d}\mu(e)f\left(\frac{d}{e}\right)=\sum_{e\tilde{e}f=n}\mu(e)f(\tilde{e})$

$$=\sum_{\tilde{e}|n}f(\tilde{e})\sum_{e|\frac{n}{\tilde{e}}}\mu(e)=f(n) \quad \text{by} \quad (a).$$ □

Finally, we compute the $A_{m,q}$.

Theorem (5.15)

Let $A_{m,q}$ denote the number of monic irreducible polynomials of degree m of $\mathbb{F}_q[t]$. Then we have:

(a) $q^m=\sum_{d|m}dA_{d,q}$,

(b) $A_{m,q} = \frac{1}{m} \sum_{d|m} \mu(d) q^{m/d}$,

(c) $A_{m,q} \geqslant 1$,

(d) $A_{m,q} \sim q^m/m$ *for* $m \to \infty$.

Proof

(a) This is a consequence of (5.10) by considering the degrees of the involved polynomials.

(b) Let $f(m) := q^m$, $g(m) := mA_{m,q}$ for all $m \in \mathbb{N}$. From (a) we obtain $f(m) = \sum_{d|m} g(d)$ and the Möbius inversion formula yields

$$g(m) = mA_{m,q} = \sum_{d|m} \mu(d) f\left(\frac{m}{d}\right) = \sum_{d|m} \mu(d) q^{m/d}$$

which proves (b).

(c) According to (b) we get

$$A_{m,q} \geqslant \frac{1}{m}\left(q^m - \sum_{1<d|m} q^{m/d}\right) \geqslant \frac{1}{m}\left(q^m - \sum_{d=0}^{m-1} q^d\right)$$

$$= \frac{1}{m}\left(q^m - \frac{q^m - 1}{q - 1}\right)$$

$$\geqslant \frac{1}{m}(q^m - (q^m - 1)) = \frac{1}{m} > 0.$$

Then $A_{m,q} \in \mathbb{Z}$ yields (c). □

(d) Obvious.

Examples (5.16)

(a) If m is a prime number, then $A_{m,q} = (1/m)(q^m - q)$.

(b) For $q = 2$ we find

m	1	2	3	4	5	6
$A_{m,q}$	2	1	2	3	6	9
corresp. $f(t)$	$t, t+1$	t^2+t+1	t^3+t+1	t^4+t+1		
			t^3+t^2+1	t^4+t^3+1		
				$t^4+t^3+t^2+t+1.$		

The determination of the irreducible polynomials of degree 5, 6 is left as exercise 4.

Deterministic methods of constructing $\mathbb{F}_{q^m}$ seek to construct specific monic irreducible polynomials of degree m over $\mathbb{F}_q$. Let

$$m = \prod_{i=1}^{s} p_i^{v_i} \quad (s \in \mathbb{Z}^{>0};\ p_1, \ldots, p_s \text{ distinct prime numbers};\ v_i \in \mathbb{Z}^{>0},\ 1 \leqslant i \leqslant s), \tag{5.17}$$

$$\tilde{m} := \gcd(m, q-1). \tag{5.18}$$

Then the first step is the construction of an extension E over $\mathbb{F}_q$ of degree $\tilde{m}$. It is based on the knowledge of a primitive root ω of $\mathbb{F}_q$.

Proposition. (5.19)

Let ω be a primitive root of $\mathbb{F}_q$. Then $E := \mathbb{F}_q[t]/(t^{\tilde{m}} - \omega)\,\mathbb{F}_q[t]$ is an extension of $\mathbb{F}_q$ of degree $\tilde{m}$.

Proof

We note that $\mathbb{F}_q$ contains a primitive $\tilde{m}$th root of unity, say $\zeta_{\tilde{m}}$. Hence, any extension of $\mathbb{F}_q$ containing one root of $t^{\tilde{m}} - \omega$ does contain all roots of that polynomial. We conclude that every irreducible factor of $t^{\tilde{m}} - \omega$ in $\mathbb{F}_q[t]$ has the same degree, say k. Let $g(t) \in \mathbb{F}_q[t]$ be such an irreducible factor. Its constant term is of the form $\omega^{k/\tilde{m}}\zeta_{\tilde{m}}^{\nu}$ for a suitable integer ν. On the other hand it must be in $\mathbb{F}_q$ and therefore be representable in the form $\omega^{\mu}\zeta_{\tilde{m}}^{\nu}$ for an appropriate exponent μ. Hence, $\omega = \omega^{\mu\tilde{m}/k}$ and because of $k \mid \tilde{m} \mid (q-1)$ we obtain $\tilde{m} = k$. □

In the second step an extension E of degree p of the finite field $\mathbb{F}$ of characteristic p is constructed.

For this purpose we study the *Weierstrass mapping* for rings R of characteristic p, denoted *by* $\wp$ because of the surprising connection between elliptic function fields of characteristic 0 and of characteristic p which was discovered in this century. It is defined by

$$\wp_p = \wp : R \to R : x \mapsto x^p - x, \tag{5.20a}$$

and is an R^+-endomorphism because of

$$\wp(x+y) = (x+y)^p - (x+y) = x^p - x + y^p - y = \wp(x) + \wp(y) \tag{5.20b}$$

for all $x, y \in R$. If R is unital, the *prime ring* of R $(= \{m \cdot 1_R \mid m \in \mathbb{Z}\})$ belongs to the kernel of $\wp$. If R is also an entire ring, then the elements of the prime ring are the only elements of R which lie in $\ker(\wp)$. Hence, in case of R being a finite field $\mathbb{F}$ of $q = p^n$ elements we find that $\wp(\mathbb{F})$ is a submodule of index p in $\mathbb{F}^+$:

$$\#\wp(\mathbb{F}) = p^{n-1}. \tag{5.20c}$$

At least one of the n elements $\omega, \omega^2, \ldots, \omega^n$ (ω a primitive root of $\mathbb{F}$) is not in $\wp(\mathbb{F})$. For $\xi \in \mathbb{F}$ with $\xi \notin \wp(\mathbb{F})$ the extension

$$E := \mathbb{F}[t]/(t^p - t - \xi) \tag{5.20d}$$

is of degree p over $\mathbb{F}$. To prove this we make use of the theorem of Artin–Schreier:

Theorem (5.21)
Let F be a field of prime characteristic p and let E be a normal extension of F of degree p. Then there exists an element $\eta \in E$ such that

$$E = F(\eta) \tag{5.21a}$$

$$\xi := \eta^p - \eta \in F, \tag{5.21b}$$

$$\xi \notin \wp(F). \tag{5.21c}$$

Conversely, given $\xi \in F$, the polynomial $f(t) = t^p - t - \xi \in \mathbb{F}[t]$ either splits over F or is irreducible. In the latter case a root η of f generates a cyclic extension $E = F(\eta)$ over F of degree p.

Proof
The proof is done in several steps.

(i) We show that $\mathrm{Tr}_{E/F}(\beta) = \mathrm{Tr}(\beta) = 0$ for $\beta \in E$, if and only if there is an element $\alpha \in E$ such that $\beta = \alpha - \sigma(\alpha)$, where σ generates the cyclic group $\mathrm{Aut}(E/F)$ of order p. (This is a special case of the additive version of Hilbert's Theorem 90.) For each β of the form $\beta = \alpha - \sigma(\alpha)$ the trace vanishes:

$$\mathrm{Tr}(\beta) = \sum_{i=1}^{p} \sigma^i(\beta) = \sum_{i=1}^{p} (\sigma^i(\alpha) - \sigma^{i+1}(\alpha)) = \sigma(\alpha) - \sigma^{p+1}(\alpha) = 0.$$

Conversely, we assume $\mathrm{Tr}(\beta) = 0$ for some $\beta \in E$. Since the trace bilinear form is non degenerate ((3.21a–c), (3.22a, b)), there exists $\gamma \in E$ with $\mathrm{Tr}(\gamma) \neq 0$. Define α by

$$\alpha = \frac{1}{\mathrm{Tr}(\gamma)}(\beta\sigma(\gamma) + (\beta + \sigma(\beta))\sigma^2(\gamma) + \cdots + (\beta + \sigma(\beta) + \cdots + \sigma^{p-2}(\beta))\sigma^{p-1}(\gamma))$$

$$= \frac{1}{\mathrm{Tr}(\gamma)} \sum_{i=0}^{p-2} \left(\sum_{j=i+1}^{p-1} \sigma^j(\gamma) \right) \sigma^i(\beta)$$

to obtain $\mathrm{Tr}(\gamma)(\alpha - \sigma(\alpha)) = \beta \, \mathrm{Tr}(\gamma)$.

(ii) We show the existence of $\eta \in E$ subject to (5.21a–c). Since $\mathrm{Tr}(-1) = 0$ because of characteristic p, there exists $\eta \in E$ such that $\sigma(\eta) - \eta = 1$ by (*i*). This implies $\sigma^i(\eta) = \eta + i$ $(1 \leqslant i \leqslant p)$, i.e. η has p distinct conjugates. Therefore, we obtain $[F(\eta):F] \geqslant p$, hence $E = F(\eta)$. Furthermore,

$$\sigma(\eta^p - \eta) = \sigma(\eta)^p - \sigma(\eta) = (\eta + 1)^p - (\eta + 1) = \eta^p - \eta,$$

i.e. the element $\xi := \eta^p - \eta$ is fixed under σ (and therefore under $\mathrm{Aut}(E/F)$) and must be in F. Finally, to prove (5.21c) we note that in case of $\xi \in \wp(F)$, say $\xi = \wp(\zeta)$, we would have that η and ζ would differ only by an element of $\ker(\wp) \subset F$ which is of course impossible.

(iii) Let $f(t) = t^p - t - \xi \in F[t]$ and η be a root of f. Then also $\eta + i$

$(1 \leqslant i \leqslant p)$ is a root of f. Hence, the roots of f are all distinct and if one of them belongs to F they all do. For the rest of the proof we assume that f has no root in F. We show that f is irreducible in that case. For any $g \in F[t]$, $g | f$, $\deg(g) \geqslant 1$ we consider the coefficient of g of the power $t^{\deg(g)-1}$. It is a sum of terms $-(\eta + i)$ for precisely $\deg(g)$ different integers i $(1 \leqslant i \leqslant p)$, hence it is of the form $-\deg(g)\eta + k$ for some integer k $(1 \leqslant k \leqslant p)$ and does not belong to F, proving the irreducibility of f. All roots of f lie in $E := F(\eta)$, and they are all distinct. Therefore E/F is a normal extension of degree p. Its (cyclic) Galois group is generated by the automorphism $\sigma: \eta \mapsto \eta + 1$.

□

In the third and last step we construct an extension E of $\mathbb{F} = \mathbb{F}_q$ of degree $\tilde{p}$, where $\tilde{p} \neq p$ is a prime number not dividing $q - 1$ (especially $\tilde{p}$ is not 2). Let d be the order of $q \bmod \tilde{p}$ so that

$$p^{nd} \equiv 1 \bmod \tilde{p}, \quad p^{nj} \not\equiv 1 \bmod \tilde{p} \quad \text{for } 1 \leqslant j < d. \tag{5.22a}$$

Then there is an exponent $\nu \geqslant 1$ such that

$$p^{nd} \equiv 1 \bmod \tilde{p}^{\nu}, \quad p^{nd} \not\equiv 1 \bmod \tilde{p}^{\nu+1}. \tag{5.22b}$$

Using induction and the methods of the first two steps we construct an extension E_1 of $\mathbb{F}$ of degree d over $\mathbb{F}$. It contains a primitive $\tilde{p}^{\nu}$th root of unity, say ζ_1. By construction

$$E_1 := E_1[t]/(t^{\tilde{p}} - \zeta_1)E_1[t] = E_1(\zeta) \quad (\zeta = t/(t^{\tilde{p}} - \zeta_1) = \zeta_1^{1/\tilde{p}}) \tag{5.22c}$$

is an extension of degree $\tilde{p}$ of E_1. We know from elementary number theory the existence of an integer j of order $\phi(\tilde{p}^{\nu})$ modulo $\tilde{p}^{\nu+1}$ (ϕ is Euler's function, see section 12 and chapter 5 (4.5)). Now it follows from cyclotomic theory (section 12) that the polynomial

$$f_{\tilde{p},\nu}(t) := \prod_{i=1}^{\tilde{p}} \left(t - \sum_{k=1}^{\phi(\tilde{p}^{\nu})} \zeta^{j^k \tilde{p} + i} \right) \tag{5.22d}$$

is irreducible of degree $\tilde{p}$ over $\mathbb{F}$. (See also exercise 8.)

3. Factorization of polynomials

Finally we investigate the task of factorizing a polynomial f of $\mathbb{F}_q[t]$ into a product of (monic) irreducible factors.

Without loss of generality we can assume that f is monic. Using divisor cascading (which includes formal differentiation and pth root formation here) we find a factorization

$$f = \prod_{i=1}^{\deg(f)} f_i^i, \tag{5.23}$$

with separable mutually prime factors $f_1, f_2, \ldots, f_{\deg(f)}$ (many of them being 1). There remains the task to factorize a given separable monic non-constant polynomial f.

If $f(0) = 0$ then we find a factorization $f(t) = g(t)t$ with $g \in \mathbb{F}_q[t]$, $g(0) \neq 0$.

There remains the task of factorizing a separable monic non-constant polynomial f of $\mathbb{F}_q[t]$ with $f(0) \neq 0$. Following Št. Schwarz ([13], [14], [15]) we make use of the formula (established in (5.10))

$$t^{q^d - 1} - 1 = \prod_{i=1}^{b_{q,d}} g_i(t), \tag{5.24}$$

where $g_1, \ldots, g_{b_{q,d}}$ are the irreducible monic polynomials $\neq t$ of degree dividing d $(d \in \mathbb{N})$. For $m = 1, 2, \ldots$ we determine a factorization

$$f = \prod_{i=1}^{\deg(f)} h_i \tag{5.25}$$

into the product of mutually prime monic polynomials h_i, h_i being the product of all irreducible monic polynomials of degree i dividing f. The h_i are computed in the following algorithm.

Equal-degree factorization of f (5.26)

Input. $f \in \mathbb{F}_q[t]$ monic, separable, non-constant with $f(0) \neq 0$.

Output. $h_i \in \mathbb{F}_q[t]$ $(1 \leqslant i \leqslant \deg(f))$ such that h_i is the product of all irreducible monic polynomials of degree i over $\mathbb{F}_q$ dividing f.

Step 1. (Initialization). Set $h_i \leftarrow 1 (1 \leqslant i \leqslant \deg(f))$, $h \leftarrow f$, $i \leftarrow 1$.

Step 2. (Divisors of degree i). Set $h_i \leftarrow \gcd(h, t^{q^i - 1} - 1)$, $h \leftarrow h/h_i$. For $h = 1$ terminate, else go to 3.

Step 3. (Increase i). Set $i \leftarrow i + 1$. For $2i \leqslant \deg(h)$ go to 2, else set $h_{\deg(h)} \leftarrow h$ and terminate.

It remains the task to factorize f in case of f being the product of distinct monic irreducible polynomials all of degree d and different from t. For $d = \deg(f)$ the polynomial f is obviously irreducible and we are done. Hence, we assume $\deg(f) > d$.

If q is odd and $d = 1$, then we suggest a probabilistic method based on the observation that the squares of elements of $\mathbb{F}_q^\times$ form the cyclic subgroup $(\mathbb{F}_q^\times)^2$ of order $(q-1)/2$. They are the solutions of the equation

$$x^{(q-1)/2} - 1 = 0 \tag{5.27}$$

in $\mathbb{F}_q$. Hence, the expectation to find an element of $\mathbb{F}_q^\times$ to be a square is $\frac{1}{2}$. We note that $\gcd(f, t^{(q-1)/2} - 1)$ is 1, if every root of f is a non-square; it is f, if every root of f is a square. Otherwise it yields a proper non-constant divisor of f. Therefore the probability of not finding such a divisor of f is only

$$\alpha_{q,\deg(f)} := \frac{2\binom{(q-1)/2}{\deg(f)}}{\binom{q-1}{\deg(f)}}. \tag{5.28}$$

If we do not obtain a proper factor of f this way, then we try the same procedure for $f(t+a)$, obtained from f by Tschirnhausen translations, where $a \in \mathbb{F}_q^\times$ is randomly chosen. From a prospective factorization

$$f(t+a) = g(t)h(t) \ (g, h \in \mathbb{F}_q[t] \text{ monic non-constant}) \tag{5.29a}$$

we find the corresponding proper factorization of $f(t)$ via

$$f(t) = g(t-a)h(t-a). \tag{5.29b}$$

This method is bound to break up f, if enough elements a are tried out. The expectation of no break in k tests is at most

$$(\alpha_{q,\deg(f)})^k. \tag{5.30}$$

If q is odd and $d > 1$, then we construct $\mathbb{F}_{q^d}$ as above so that f will become a product of linear factors over $\mathbb{F}_{q^d}$. Thereafter we apply the same method as above. Once we obtain a factorization $f = gh$ of f into monic non-constant polynomials g, h over $\mathbb{F}_{q^d}$, we apply the *Frobenius automorphism*

$$\pi: \mathbb{F}_{q^d} \to \mathbb{F}_{q^d}: x \mapsto x^q \tag{5.31}$$

(to the coefficients) on both sides. Certainly $\pi(f) = f$. For $\pi(g) = g$ also g is in $\mathbb{F}_q[t]$ and we have succeeded in breaking up f over $\mathbb{F}_q$. Then we can apply the same procedure repeatedly. For $\pi(g) \neq g$ we obtain d factorizations $f = \pi^i(g)\pi^i(h)$ $(1 \leq i \leq d)$. A basis of the divisor cascade of $\{g, \pi(g), \ldots, \pi^{d-1}(g), h, \pi(h), \ldots, \pi^{d-1}(h)\}$ is a system B of monic non-constant polynomials of $\mathbb{F}_{q^d}[t]$ the product of which is f:

$$f = \prod_{b \in B} b. \tag{5.32}$$

Moreover, B is invariant under π. Again, if $\pi(b) = b$ for some $b \in B$, then that $b \in \mathbb{F}_q[t]$ is a proper divisor of f. If $\pi(b) \neq b$ and B does not only consist of $\langle \pi \rangle$-conjugates of b, then there is a proper divisor $\tilde{d}$ of d such that

$$1 < \tilde{d} | d; \quad \pi^{\tilde{d}}(b) = b; \quad b, \pi(b), \ldots, \pi^{\tilde{d}-1}(b) \text{ are distinct,} \tag{5.33a}$$

$$g = \prod_{i=1}^{\tilde{d}} \pi^i(b) \in \mathbb{F}_q[t], \tag{5.33b}$$

and g is a proper divisor of f over $\mathbb{F}_q$. There remains the case

$$1 < \tilde{d} | d, \; \pi^{\tilde{d}}(b) = b, \tag{5.34a}$$

and B consists of the $\tilde{d}$ distinct polynomials b, $\pi(b), \ldots, \pi^{\tilde{d}-1}(b)$ with $f = \prod_{i=1}^{\tilde{d}} \pi^i(b)$. In this case b is not irreducible over $\mathbb{F}_{q^d}$. We apply the method described above to get a partial factorization $b = b'b''$ with monic non-constant polynomials b', b'' over $\mathbb{F}_{q^d}$. Again, upon forming a basis B_1 of the divisor cascade of $\{b', \pi(b'), \ldots, \pi^d(b'), b'', \pi(b''), \ldots, \pi^d(b'')\}$ we reduce the task as above to the case that $B_1 = \{b_1, \pi(b_1), \ldots, \pi^{d_1-1}(b_1)\}$, where

$$1 < d_1 | d, \; d_1 > \tilde{d}. \tag{5.34b}$$

Continuing in this way we arrive at a break of f after finitely many steps. If q is even, say $q = 2^n$, then we set $d_1 = \begin{cases} d & \text{for } 2|nd \\ 2d & \text{else} \end{cases}$ and form

$$g := g_a = \gcd(f, (t+a)^{(q^{d_1}-1)/3} - 1) \tag{5.35}$$

for randomly chosen elements $a = a_1, a_2, \ldots$ of $\mathbb{F}_{q^{d_1}}$, beginning with $a_1 = 0$. Trying sufficiently many elements a we obtain a proper factor of f in $\mathbb{F}_{q^{d_1}}$ by means of a non-constant polynomial g_a of degree $\deg(g_a) < \deg(f)$. The probability that k attempts fail is only

$$\left(\frac{\binom{\frac{1}{3}(q^{d_1}-1)}{\deg f} + \binom{\frac{2}{3}(q^{d_1}-1)}{\deg f}}{\binom{q^{d_1}-1}{\deg f}} \right)^k \tag{5.36}$$

(See D. Cantor-H. Zassenhaus [2].)
Once a break is achieved we apply the Frobenius automorphism

$$\pi : \mathbb{F}_{q^{d_1}} \to \mathbb{F}_{q^{d_1}} : x \mapsto x^q, \tag{5.37}$$

as above until a break of f over $\mathbb{F}_q$ results.

In case of q and $\deg(f)$ being small it is easier to factorize f with the *Berlekamp method*, once f is squarefree. In that case f has a factorization

$$f = \prod_{i=1}^{r} f_i, \tag{5.38}$$

with monic non-constant irreducible distinct polynomials $f_i \in \mathbb{F}_q[t]$, $1 \leqslant i \leqslant r$. This factorization is closely related to those polynomials $g \in \mathbb{F}_q[t]$ which satisfy

$$g(t)^q \equiv g(t) \bmod f(t) \quad \text{and} \quad \deg(g) < \deg(f). \tag{5.39}$$

Namely, if (5.39) holds, then f divides

$$g(t)^q - g(t) = \prod_{x\in\mathbb{F}_q} (g(t) - x),$$

and for every f_i of (5.38) there exists a unique element $x = x_i \in \mathbb{F}_q$ such that $f_i | (g(t) - x_i)$. On the other hand, let $x_1, \ldots, x_r \in \mathbb{F}_q$, then by (2.17) there is a unique polynomial $g \in \mathbb{F}_q[t]$, $\deg(g) < \deg(f)$ such that $g(t) \equiv x_i \bmod f_i(t)$ $(1 \leqslant i \leqslant r)$, and we obtain successively

$$g(t)^q - x_i \equiv g(t)^q - x_i^q \equiv (g(t) - x_i)^q \equiv 0 \bmod f_i(t),$$

hence

$$g(t)^q - g(t) \equiv 0 \bmod f_i(t) \quad (1 \leqslant i \leqslant r),$$

and thus (5.39). We note that there are q^r possibilities of choosing $x_1, \ldots, x_r \in \mathbb{F}_q$ and therefore q^r possibilities for g. Hence, the polynomials of $\mathbb{F}_q[t]$ satisfying (5.39) form an $\mathbb{F}_q$-vector space of dimension r. In order to apply the methods of linear algebra we transform (5.39) into a system of linear equations. Let us assume

$$f(t) = \sum_{i=0}^{m} a_i t^{m-i} \quad (a_0 = 1), \quad g(t) := \sum_{i=0}^{m-1} g_i t^i, \quad x = t/f.$$

Then we can reduce powers $x^k (k \geqslant m)$, and we determine a matrix $Q = (q_{ij}) \in \mathbb{F}_q^{m\times m}$ such that

$$x^{kq} = \sum_{\nu=0}^{m-1} q_{k,\nu} x^\nu \quad (0 \leqslant k \leqslant m-1). \tag{5.40a}$$

Then the condition $g(x)^q = g(x)$ yields

$$\begin{aligned} \sum_{i=0}^{m-1} g_i x^i &= g(x) = g(x)^q = g(x^q) \\ &= \sum_{i=0}^{m-1} g_i \sum_{\nu=0}^{m-1} q_{i,\nu} x^\nu \\ &= \sum_{\nu=0}^{m-1} x^\nu \left(\sum_{i=0}^{m-1} g_i q_{i,\nu} \right). \end{aligned} \tag{5.40b}$$

Setting $\mathbf{g}^t = (g_0, g_1, \ldots, g_{m-1}) \in \mathbb{F}_q^{1\times m}$ we obtain

$$\mathbf{g}^t(Q - I_m) = \mathbf{0}. \tag{5.40c}$$

Therefore all possible q^r solutions g can be obtained from solving (5.40c). This especially implies that

$$m - \operatorname{rank}(Q) = r. \tag{5.40d}$$

By $\mathbf{v}_1^t, \ldots, \mathbf{v}_r^t$ we denote a basis of the solution space of (5.40c), and by $g_i(t)$ the polynomial corresponding to $\mathbf{v}_i^t$ $(1 \leqslant i \leqslant r)$. Because the first row of Q is $(1\,0 \cdots 0)$ we assume $\mathbf{v}_1^t = (1\,0 \cdots 0)$ also. The rank r of Q is easily calculated.

In case $r=1$ we know that f is irreducible. For $r \geqslant 2$ there is a proper irreducible divisor f_i of f which divides $g_2(t)-u$ for suitable $u \in \mathbb{F}_q$. Hence, on computing

$$\gcd(g_2(t)-u, f(t)) \quad \text{for } u \in \mathbb{F}_q, \tag{5.40e}$$

we obtain a factorization of f into proper factors which need not be irreducible, however. If we do not obtain r factors f from (5.40e), then we repeat this process for g_3 instead of g_2 and replace f by the factors already found. We show that this indeed leads to a factorization of f into its irreducible factors. For distinct irreducible divisors f_i, f_j $(1 \leqslant i < j \leqslant r)$ of f there exists a polynomial $g \in \mathbb{F}_q[t]$ satisfying (5.39) such that $g(t) \equiv x_i \bmod f_i(t)$, $g(t) \equiv x_j \bmod f_j(t)$ and $x_i, x_j \in \mathbb{F}_q$, $x_i \neq x_j$. Since $g_1, \ldots, g_r$ form a basis of all polynomials g satisfying (5.39) there must be at least one g_k among them $(1 < k \leqslant r)$ which is congruent to distinct elements x_i, x_j of $\mathbb{F}_q$ modulo f_i, f_j. Hence $\gcd(f(t), g_k(t)-x_i)$, $\gcd(f(t), g_k(t)-x_j)$ yield proper factors h_{ki}, h_{kj} of f such that

$$f_i \mid h_{ki},\ f_i \nmid h_{kj},\ f_j \mid h_{kj},\ f_j \nmid h_{ki}. \tag{5.40f}$$

These ideas are the basis of the following algorithm for factorizing square-free polynomials of $\mathbb{F}_q[t]$.

Berlekamp's algorithm (5.41)

Input. $f \in \mathbb{F}_q[t]$ monic, square-free, non-constant of degree $\deg(f)=m$.
Output. Monic irreducible polynomials $f_i \in \mathbb{F}_q[t]$ $(1 \leqslant i \leqslant r)$ such that

$$f = \prod_{i=1}^{r} f_i.$$

Step 1. (Initialization). Compute $Q=(q_{ij})_{0 \leqslant i,j \leqslant m-1} \in \mathbb{F}_q^{m \times m}$ according to (5.40a).

Step 2. (Solution space of (5.40c)). Compute basis $\mathbf{v}_1^t = (1\ 0 \cdots 0)$, $\mathbf{v}_2^t, \ldots, \mathbf{v}_r^t$ of the solution space of (5.40c). In case of $r=1$ set $f_1 \leftarrow f$ and terminate. Else set $g_i(t) \leftarrow \sum_{j=0}^{m-1} v_{ij} t^j$ for $\mathbf{v}_i^t = (v_{i0}, \ldots, v_{im-1})$ $(1 \leqslant i \leqslant r)$.

Step 3. (Set number of determined factors). Set $v \leftarrow 1$, $\rho \leftarrow 2$, $q_1 \leftarrow f$.

Step 4. (Factorization). For $i=1, \ldots, v$ compute $\gcd(q_i, g_\rho(t)-u)$ for all $u \in \mathbb{F}_q$. This yields a factorization of q_i into r_i proper factors $q_{i,1}, \ldots, q_{i,r_i}$ over $\mathbb{F}_q$.

Step 5. (Done?). For $\sum_{i=1}^{v} r_i = r$ set $\{f_1, \ldots, f_r\} \leftarrow \{q_{1,1}, \ldots, q_{1,r_1}, \ldots, q_{v,1}, \ldots, q_{v,r_v}\}$ and terminate. Else set $\tilde{v} \leftarrow \sum_{i=1}^{v} r_i$, $\{q_1, \ldots, q_{\tilde{v}}\} \leftarrow \{q_{1,1}, \ldots, q_{1,r_1}, \ldots, q_{v,1}, \ldots, q_{v,r_v}\}$, $v \leftarrow \tilde{v}$, $\rho \leftarrow \rho + 1$ and go to 4.

For an estimate of the number of computational steps and a non-trivial example we refer to Knuth's book [8].

Exercises

1. Let p be a rational prime number. Develop an algorithm for computing

   (i) x^{-1} modulo p for given $x \in \mathbb{Z}$, $p \nmid x$;
   (ii) a primitive root of $\mathbb{Z}/p\mathbb{Z}$;
   (iii) an index table containing $\operatorname{ind}(x)$ and $\operatorname{ind}(1+x)$ for all $x \in \mathbb{Z}/p\mathbb{Z}$.

2. Let $f \in \mathbb{F}_p[t]$ be irreducible, monic and of degree $m > 1$. Develop an algorithm for computing

   (i) a primitive root of $\mathbb{F}_p[t]/f\mathbb{F}_p[t]$;
   (ii) an index table for $\mathbb{F}_p[t]/f\mathbb{F}_p[t]$.

3. Compute all monic irreducible polynomials of degree 5, 6 over $\mathbb{F}_2$.
4. Factorize $t^4 + 1 \in \mathbb{C}[t]$ over $\mathbb{Z}[i]$, $\mathbb{Z}[2^{\frac{1}{2}}]$, $\mathbb{Z}[(-2)^{\frac{1}{2}}]$ and derive that $t^4 + 1$ factors in two quadratic polynomials over $\mathbb{Z}/p\mathbb{Z}$, p any prime number. (Hence, it is in general not possible to obtain a factorization over $\mathbb{Z}$ from factorizations over $\mathbb{Z}/p\mathbb{Z}$ without additional considerations.)
5. Factorize into irreducible polynomials

   (i) $t^8 + t^7 + t + 1$ over $\mathbb{F}_2$,
   (ii) $t^{11} + t^{10} + t^7 + t^6 + t^5 + t^4 + t^3 + t + 1$ over $\mathbb{F}_2$,
   (iii) $t^8 + t^6 - 3t^4 - 3t^3 + 8t^2 + 2t - 5$ over $\mathbb{F}_5$.

6. An irreducible polynomial $f \in \mathbb{F}_q[t]$ is called *primitive*, if a root α of $f(t)$ generates $\mathbb{F}_q(\alpha)^\times$.

   (i) Prove that $t^4 + t + 1$, $t^4 + t^3 + 1 \in \mathbb{F}_2[t]$ are primitive polynomials over $\mathbb{F}_2$ while $t^4 + t^3 + t^2 + t + 1$ is not primitive.
   (ii) Determine all primitive monic polynomials of degree 5 over $\mathbb{F}_2$ and of degree 3 over $\mathbb{F}_3$.
   (iii) Give a sufficient condition for a monic irreducible polynomial of $\mathbb{F}_q[t]$ to be primitive.

7. Let $\mathbb{F} = \mathbb{F}_3$. Construct an extension of degree $m = 5$ over $\mathbb{F}$. (Hint: $E := \mathbb{F}((1+\beta)^{\frac{1}{2}})$, $\beta^2 + 1 = 0$, is an extension of degree 4 over $\mathbb{F}$, and $\beta(2 + \beta + (1+\beta)^{\frac{1}{2}})$ an element of order 5 in $E^\times$.)
8. (Application of idempotents) Let $F = \mathbb{F}_q$ and $n \in \mathbb{N}$. For any set $\mathfrak{B}$ of polynomials of $F[t_1, \ldots, t_n]$ we define the solution set $\mathfrak{C}(\mathfrak{P}/F) := \{\mathbf{x} \in F^n \mid P(x_1, \ldots, x_n) = 0 \, \forall P \in \mathfrak{P}\}$. Perform the following tasks:

(a) For any subset S of F^n all $P \in F[t_1, \ldots, t_n]$ satisfying $P(x_1, \ldots, x_n) = 0$ for all $\mathbf{x} \in S$ form an ideal $I(S/F)$ of $F[t_1, \ldots, t_n]$ such that the factorring of $F[t_1, \ldots, t_n]$ modulo $I(S/F)$ is isomorphic to the algebraic sum of $|S|$ fields isomorphic to F and that $\mathfrak{C}(I(S/F)/F) = S$.
(b) $I(S/F) \supseteq I(S'/F) \Leftrightarrow S \subseteq S'$.
(c) $I(F^n/F)$ is generated by the n polynomials $t_i^q - t_i$ $(1 \leqslant i \leqslant n)$.
(d) For any $P \in F[t_1, \ldots, t_n]$ there is precisely one polynomial $\bar{P} \in F[t_1, \ldots, t_n]$ of degree

less than q in each variable t_i $(1 \leqslant i \leqslant n)$ such that $P \equiv \bar{P} \bmod I(F^n/F)$. Construct $\bar{P}$.

(e) $P, Q \in F[t_1, \ldots, t_n]$ have the same value for each argument $(x_1, \ldots, x_n) \in$ if and only if $\bar{P} = \bar{Q}$.

(f) $\bar{\bar{P}} = \bar{P}$, $\overline{P + Q} = \bar{P} + \bar{Q}$, $\overline{PQ} = \overline{\bar{P}\bar{Q}}$ for all $P, Q \in F[t_1, \ldots, t_n]$.

(g) For each set $\mathfrak{P}$ of polynomials of $F[t_1, \ldots, t_n]$ there is precisely one standard polynomial $\sigma(\mathfrak{P}) = \sigma(\mathfrak{P}/F)$ characterized as a polynomial of $F[t_1, \ldots, t_n]$ with the properties: $\overline{\sigma(\mathfrak{P})^2} = \sigma(\mathfrak{P})$, $\mathfrak{C}(\mathfrak{P}/F) = \mathfrak{C}(\{\sigma(\mathfrak{P})\}/F)$.

(h) If $\mathfrak{P} \subseteq F[t_1, \ldots, t_n]$ is finite then $\sigma(\mathfrak{P})$ can be constructed in accordance with the formula $\sigma(\mathfrak{P}) = 1 - \overline{\prod_{P \in \mathfrak{P}}(1 - \overline{P^{q-1}})}$.

(i) $1 - \sigma(\mathfrak{P} \cup \mathfrak{P}'/F) = \overline{(1 - \sigma(\mathfrak{P}/F))(1 - \overline{\sigma(\mathfrak{P}'/F)})}$, $\mathfrak{C}(\mathfrak{P}/F) \supseteq \mathfrak{C}(\mathfrak{P}'/F) \Leftrightarrow \overline{\sigma(\mathfrak{P})\sigma(\mathfrak{P}')} = \sigma(\mathfrak{P})$, $\mathfrak{C}(\mathfrak{P}/F) = \varnothing \Leftrightarrow \sigma(\mathfrak{B}/F) = 1$, $\mathfrak{C}(\mathfrak{P}/F) = F^n \Leftrightarrow \sigma(\mathfrak{P}/F) = 0$.

2.6. The main theorem of Galois theory

We assume familiarity with the basic concepts and results on tensor products as expounded in S. Lang, *Algebra*, chapter 16, pp. 408–27. Though the results of this section are taught in most algebra courses (see the book by Bastida [1], for example) we give here a different – and as we believe very interesting – proof based on the ideas developed in the first four sections of this chapter.

Following the precedent set by E. Artin we formulate the *Main Theorem of Galois Theory* as follows:

Theorem (6.1)

Let G be a finite subgroup of the automorphism group of a field E and let

$$F := E^G = \{x \in E \mid \forall \sigma \in G : \sigma(x) = x\} \tag{6.2a}$$

be the fixed subfield of E under the action of G. Then E is a normal extension of F and G is the full automorphism group of E over F:

$$G = \mathrm{Aut}(E/F). \tag{6.2b}$$

Proof

The extension E of F is algebraic because every element x of E is root of the separable polynomial

$$f(t) = \prod_{\sigma \in G}{}' (t - \sigma(x)),$$

where the accent at the product sign indicates that only distinct G-conjugates of x are admitted among the linear factors. Indeed for every element τ of G we have

$$\tau f(t) = \prod_{\sigma \in G}{}' \tau(t - \sigma(x)) = \prod_{\sigma \in G}{}' (t - \tau\sigma(x)) = f(t),$$

so that $f(t)$ is a monic non-constant polynomial of $F[t]$. In fact f is the minimal polynomial of x over F.

Firstly, let E be a finite extension of F of degree $[E:F]$. Hence E can be generated from F by the adjunction of finitely many distinct elements $x_1, x_2, \ldots, x_s$ of E (e.g. an F-basis of E will do). Since the application of G to the element x_i produces a finite number of conjugates which generate again E over F we can assume without loss of generality that the set $\{x_1, x_2, \ldots, x_s\}$ is invariant under the application of G. As above it follows that

$$g(t) := \prod_{i=1}^{s} (t - x_1) \in F[t]. \tag{6.3}$$

Hence E is a minimal splitting field of the separable monic non-constant polynomial g over F.

There are the F-isomorphisms

$$\phi: E \to E \otimes_F E : a \mapsto a \otimes 1,$$
$$\psi: E \to E \otimes_F E : a \mapsto 1 \otimes a,$$

of E into the commutative F-algebra

$$A = E \otimes_F E$$

of dimension $[E:F]^2$ over F. We observe that A is a unital overring of dimension $[E:F]$ of the subfield $\phi(E)$. A is generated over $\phi(E)$ by the elements

$$1 \otimes x_i \quad (1 \leqslant i \leqslant s),$$

subject to the equations

$$\phi(g)(1 \otimes x_i) = 0,$$

in consequence of (6.3). Hence the minimal polynomial m_i of $1 \otimes x_i$ over $\phi(E)$ is a divisor of the separable monic polynomial $\phi(g)$ of $\phi(E)[t]$. Since $\phi(g)$ is factorized into the product of the distinct linear factors $t - \phi(x_j)$ $(1 \leqslant j \leqslant s)$, also $m_i = m_{1 \otimes x_i}$ can be factorized into a product of distinct linear factors over $\phi(E)$. Hence, $\phi(E)[t]/m_i$ is the algebraic sum of subfields of the form $\phi(E)e$ where e is a primitive idempotent. Moreover, there is the $\phi(E)$-isomorphism

$$\psi_i : \phi(E)[t]/m_i \to \phi(E)[1 \otimes x_i]$$

with

$$\psi_i(t/m_i) = 1 \otimes x_i.$$

Hence $\phi(E)[1 \otimes x_i]$ is the algebraic sum of subfields of the form $\phi(E)\psi_i(e)$ where $\psi_i(e)$ is an idempotent of A. It follows that also A itself is the algebraic sum of subfields of the form $\phi(E)e_i$:

$$A = \sum_{i=1}^{n} \phi(E)e_i \tag{6.4}$$

where $e_1, e_2, \ldots, e_n$ are the primitive idempotents of A. Hence,

$$\dim_F A = [E:F]n = [E:F]^2, \quad n = [E:F]. \tag{6.5}$$

There is the monomorphism

$$\chi: G \to \operatorname{Aut}(A/E \otimes_F F),$$

defined by

$$\chi(\sigma)(a \otimes b) = a \otimes \sigma(b) \quad (\sigma \in G; a, b \in E)$$

of G into the automorphism group of A over $\phi(E)$. Since the fixed subfield of $\bar{G} := \chi(G)$ operating on $\psi(E)$ is $\psi(F) = F \otimes F$ it follows that the fixed subring of $\bar{G}$ acting on A is the field

$$\phi(E) = E \otimes_F F.$$

Moreover, the primitive idempotents of A are permuted under the action of $\bar{G}$. For any $\bar{G}$-orbit the sum over the elements of the $\bar{G}$-orbit belongs to the fixed subring of $\bar{G}$. Hence it is an idempotent of the field $\phi(E)$. Hence it is

$$1_A = \sum_{i=1}^{n} e_i. \tag{6.6}$$

Thus $\bar{G}$ permutes the e_i's transitively (compare the proof of (4.1)) and

$$n \mid (G:1).$$

On the other hand we have seen already that

$$(G:1) \leqslant [E:F] = n,$$

and that equality happens only if E is normal over F and (6.2b) holds. Hence, we obtain

$$(G:1) = n \tag{6.7}$$

and (6.2b).

Now we drop the condition that $[E:F]$ is finite. Let us choose for every element $\sigma \neq id_E$ of G an element y_σ of E for which $y_\sigma \neq \sigma(y_\sigma)$, and we form the G-conjugates of the y_σ's. Thus we obtain a finite set $\{x_1, \ldots, x_s\}$ of elements of E which is invariant under the action of G such that for every element $\sigma \neq id_E$ of G there is at least one x_i such that $\sigma(x_i) \neq x_i$. Now let E_1 be the subfield of E obtained by the adjunction of $x_1, \ldots, x_s$ to F. It is invariant under the action of G with F as fixed subfield. Hence as above E_1 is a normal extension of F with

$$G|_{E_1} = \operatorname{Aut}(E_1/F),$$
$$[E_1:F] = (G:1).$$

For any element x of E we form the finite set X obtained by joining $\{x_1, \ldots, x_s\}$ with the G-conjugates of x and the extension E_2 of F obtained by the adjunction of X to E_1. It is G-invariant with F as fixed subfield and it contains E_1. Again as above we find that

$$[E_2:F]=(G:1),$$
$$x\in E_1,$$
$$E=E_1.$$

□

Corollary (6.8)
For any normal extension E of the field F there holds the **Galois correspondence** *between the subgroups of the full automorphism group*

$$G=\mathrm{Aut}(E/F),$$

and the intermediate fields of F and E: if S is a subgroup of G then the fixed subfield

$$E^S=\{x\in E\,|\,\forall\sigma\in S:\sigma(x)=x\} \tag{6.8a}$$

of E relative to S is an intermediate field of F and E such that

$$S=\mathrm{Aut}(E/E^S).$$

If X is an intermediate field of F and E then it is the fixed subfield of E relative to Aut(E/X) *and E is normal over X.*

Corollary (6.9)
In the situation of corollary (6.8) *the intermediate field X of F and E is a normal extension of F, if and only if* Aut(E/X) *is a normal subgroup of G.*

Proof
Any $\sigma\in G=\mathrm{Aut}(E/F)$ maps X onto another intermediate field $\sigma(X)$ and E is also normal over $\sigma(X)$. Clearly, $\mathrm{Aut}(E/\sigma(X))=\sigma\,\mathrm{Aut}(E/X)\sigma^{-1}$ since we have $\sigma\tau\sigma^{-1}(\sigma(\xi))=\sigma\tau(\xi)=\sigma(\xi)$ for all $\tau\in\mathrm{Aut}(E/X)$, $\xi\in X$, and vice versa.

We show that all F-isomorphisms of X are obtained as restrictions of F-isomorphisms of E to X. Hence, X is normal over F, if and only if every F-isomorphism of X is an automorphism, and we just saw that the latter is tantamount to Aut(E/X) being a normal subgroup of G.

Two automorphisms σ,τ of G have the same restriction of X, if and only if $\sigma^{-1}\tau$ belongs to Aut(E/X). Thus the number of distinct restrictions of G to X equals

$$(G:\mathrm{Aut}(E/X))=[E:F]/[E:X]=[X:F].$$

On the other hand, every F-isomorphism of X can be extended to an F-isomorphism of the normal extension E of X. Thus there are precisely $[X:F]$ distinct F-isomorphisms of X. □

2.7. Minimal splitting fields

Reference is made to chapter 10, section 7 of S. Lang [10]. However, since mainly derivations of fields are treated by S. Lang, let us state some simple properties of derivations of rings without proof.

A mapping

$$D:R\to R$$

of a ring into itself satisfying the following laws of differential calculus

$$D(u+v)=D(u)+D(v),\quad D(uv)=uD(v)+D(u)v\quad(u,v\in R)\tag{7.1}$$

is said to be a *derivation* of the ring R.

For example, there is the derivation D_t of the polynomial ring $R[t]$ in t over the unital commutative ring R defined in chapter 1 (6.10). It generates the full derivation ring of $R[t]$ over R:

$$\mathrm{Der}_R R[t]=R[t]D_t.\tag{7.2}$$

There is the monomorphism

$$\mu:\mathrm{Der}(R)\to\mathrm{Der}(R[t]),\tag{7.3}$$

defined by

$$\mu(D)\left(\sum_{i=0}^{n}a_it^i\right)=D(a_0)\quad(n\in\mathbb{Z}^{\geqslant 0},\ a_0,a_1,\ldots,a_n\in R, D\in\mathrm{Der}(R)),\tag{7.4}$$

and the direct decomposition

$$\mathrm{Der}(R[t])=\mu(\mathrm{Der}(R))\dotplus R[t]D_t.\tag{7.5}$$

The set $\mathrm{Der}(R)$ of the derivations of R is closed under addition and under Lie multiplication: if D_1, $D_2\in\mathrm{Der}(R)$ then also D_1+D_2, $[D_1,D_2]=D_1D_2-D_2D_1$ are in $\mathrm{Der}(R)$. Thus $\mathrm{Der}(R)$ is a Lie ring. It is invariant under multiplication by central elements of R:

$$C(R)\,\mathrm{Der}(R)\subseteq\mathrm{Der}(R).\tag{7.6}$$

For any subset X of R the derivations of R annihilating X form a subring $\mathrm{Der}_X(R)$ of $\mathrm{Der}(R)$. It coincides with $\mathrm{Der}_{\langle X\rangle}(R)$ where $\langle X\rangle$ is the subring of R generated by X.

Any ring epimorphism

$$\eta:R\to\Lambda\tag{7.7a}$$

induces the homomorphism

$$\eta:\ \mathrm{Der}_{\ker(\eta)}(R)\to\mathrm{Der}(\Lambda):\ D\mapsto\eta\circ D\tag{7.7b}$$

via

$$\eta\circ D(\eta(x))=\eta(D(x))\quad(D\in\mathrm{Der}_{\ker(\eta)}(R),x\in R).$$

For any derivation D and central idempotent e of R we obtain

$$\begin{aligned} D(e) &= D(e^2) = D(e)e + eD(e) = 2eD(e) = e(2D(e)), \\ eD(e) &= e^2(2D(e)) = e(2D(e)) = D(e), \\ D(e) &= 2eD(e) = 2D(e), \\ &\left.\begin{aligned} D(e) &= 0, \\ D(ex) &= D(e)x + eD(x) = eD(x), \end{aligned}\right\} \end{aligned} \tag{7.8}$$

$$D(eR) \subseteq eR. \tag{7.9}$$

Hence

$$\operatorname{Der}\left(\bigoplus_{i=1}^{s} R_i\right) = \bigoplus_{i=1}^{s} \operatorname{Der}(R_i) \tag{7.10}$$

for the algebraic sum of s unital rings $R_1, R_2, \ldots, R_s$. Derivations are used to great advantage to clarify and to extend the concept of *separability*.

Proposition (7.11)

Let R be a unital commutative ring. Let f be a monic non-constant polynomial of R[t] of degree n. Then f is separable precisely if $\operatorname{Der}_R(R[t]/fR[t]) = 0$.

Proof

Indeed, for any unital commutative overring Λ with R-derivation D and element x we have

$$D(g(x)) = D_t(g)(x)D(x) \quad (g \in R[t]) \tag{7.12}$$

as a consequence of the basic rules of differential calculus. Thus $f(x) = 0$ implies

$$D_t(f)(x)D(x) = 0. \tag{7.13}$$

If f is separable then the discriminant $d(f)$ is a regular element of R (see (3.4d)). Hence if

$$R[t]/fR[t] = R[x] \quad (x = t/f), \tag{7.14}$$

then $d(f)$ is a regular element of $R[x]$ because of the existence of an R-basis. Moreover, we have seen already that

$$D_t(f)(x) \mid d(f).$$

Hence $D_t(f)(x)$ is a regular element of $R[x]$ and (7.13) implies that $D(x) = 0$.

Conversely, if $D_t(f)(x)$ is a regular element of $R[x]$, then it is also a regular element of $S(f/R)$ because of the existence of a basis of $S(f/R)$ over $R[x]$. Thus also $d(f)$, being the product of certain $\mathfrak{S}_n$-conjugates of $D_t(f)(x)$, is a regular element of $S(f/R)$.

On the other hand, if $D_t(f)(x)$ is a zero divisor of $R[x]$ then there is a non-zero element y of $R[x]$ for which $D_t(f)(x)y = 0$ and this yields the R-derivation D of $R[x]$ making $D(x) = y$ which is not zero. □

Let us remark also that for any separable monic non-constant polynomial f over a field R and for any unital overring $R[x]$ of R for which $f(x)=0$ any derivation D of R can be extended to a derivation $\bar{D}$ of $R[x]$. Indeed, using (7.3) we set

$$\bar{D}(g(x)) = \mu(D)(g)(x) + D_t(g)(x)\bar{D}(x) \quad (g \in R[t]), \tag{7.15}$$

where

$$\bar{D}(x) = -D_t(f)(x)^{-1}\mu(D)(f)(x). \tag{7.16}$$

The element x of the unital commutative overring Λ of the ring R is said to be *separable* over R if

$$\mathrm{Der}_R(R[x]) = 0.$$

A separable element always is algebraic because for any non-algebraic element x of Λ over R, the ring $R[x]$ generated by x and R is isomorphic to the polynomial ring in one variable so that D_x is a non-zero derivation over R. If R is a field if follows that there is the R-isomorphism

$$\phi\colon R[t]/m_xR[t] \to R[x]\colon \quad t/m_x \mapsto x.$$

As was shown above, an element of Λ is separable over R precisely if it is algebraic with separable minimal polynomial over R.

Definition (7.17)

We say that the unital commutative overring Λ is **separable** *over R if every element Λ is separable over R.*

While we use the separability concept only for field extensions here, its ring theoretical implication will be explored in section 9.

Proposition (7.18)

An extension Λ of the field R is separable over R if and only if it can be generated by separable elements of Λ.

Proof

We must show that any element z of the subfield $R[x,y]$ generated by two separable elements x, y of an extension Λ of R is separable. Since x is separable over R and hence m_x is separable it follows that x is separable over $R[z]$. Hence, any R-derivation D of $R[z]$ can be extended to an R-derivation $\tilde{D}$ of $R[z,x]$. Similarly y is separable over $R[z,x]$ and $\tilde{D}$ can be extended to an R-derivation $\bar{D}$ of $R[z,x,y] = R[x,y]$. But because of the separability of x we have

$$\bar{D}|_{R[x]} = 0,$$

and because of the separability of y over $R[x]$ it follows that $\bar{D}=0$. Hence $D=0$, z is separable over R. □

Theorem (7.19)

Let Λ be a unital commutative overring of the field R. Then the separable elements of Λ over R form a separable overring $\mathrm{Sep}(\Lambda/R)$ *of R. It is the* **maximal separable overring** of *R in Λ.*

Proof

Note that a polynomial $f(t)$ of $R[t]$ which is separable over an extension of R, also is separable over R, and conversely. Namely, the discriminant of f is up to sign just $\mathrm{Res}(f, D_t(f))$ (see (3.48)) which is computed in $R[t]$. Any divisor of a separable polynomial is separable and also the least common multiple of finitely many separable polynomials is separable.

If e is an idempotent of Λ then for any separable element x of Λ the projection ex of x on $e\Lambda$ is separable over eR. Hence ex is separable over R. This is because either $ex=0$ or $m_{ex/eR}(0)\neq 0$, $m_{ex/eR}=eg$, g separable in $R[t]$, $g(0)\neq 0$, $m_{ex/R}=tg$ is separable. If e, e' are two orthogonal idempotents of Λ then the sum of a separable element x of $e\Lambda$ and a separable element y of $e'\Lambda$ is separable. Indeed, $f=\mathrm{lcm}\,(m_{x/R}, m_{y/R})$ is separable and $f(x+y)=0$.

For any two separable elements x, y of Λ the subring $R[x, y]$ is of finite dimension over R. Let $e_1, e_2, \ldots, e_s$ be the primitive idempotents of $R[x, y]$.

Since e_1 is the only idempotent of $X:=e_1R[x, y]$ it follows that $e_1R[x]=e_1R[e_1x]$ is a finite extension of e_1R; e_1y is separable over $e_1R[x]$; $e_1R[x, y]$ is a finite extension of $e_1R[x]$, the extension $e_1R[x, y]$ of e_1R is separable. Similarly $e_iR[x, y]$ is a separable extension of e_iR. Hence $R[x, y]$ is separable over R. Therefore $\mathrm{Sep}(\Lambda/R)$ is a separable overring of R. □

Corollary (7.20)

If in theorem (7.19) *the field R is a separable extension of the field R_0 then we have*

$$\mathrm{Sep}(\Lambda/R)=\mathrm{Sep}(\Lambda/R_0).$$

Corollary (7.21)

$\mathrm{Der}_R(\mathrm{Sep}(\Lambda/R))=0.$

Corollary (7.22)

If Λ is of finite dimension over R then there are finitely many primitive idempotents $e_1, e_2, \ldots, e_s$ of Λ such that

$$e_i^2=e_i\neq 0 \quad (1\leqslant i\leqslant s),$$
$$e_ie_k=e_ke_i=0 \quad (1\leqslant i<k\leqslant s),$$
$$1=\sum_{i=1}^{s} e_i,$$

and

$$\Lambda = \sum_{i=1}^{s} e_i \Lambda,$$

$$\mathrm{Sep}(\Lambda/R) = \sum_{i=1}^{s} e_i \,\mathrm{Sep}(\Lambda/R),$$

and each of the algebraic components $e_i \,\mathrm{Sep}(\Lambda/R)$ *of* $\mathrm{Sep}(\Lambda/R)$ *is a separable extension of the field* $e_i R$.

Corollary (7.23)
If E is a separable extension of R then we have

$$\mathrm{Sep}(E \otimes_R \Lambda) = E \otimes_R \mathrm{Sep}(\Lambda/R)$$

in terms of the standard embedding.

Corollary (7.24)
For any two unital overrings Λ_1, Λ_2 *of R we have*

$$\mathrm{Sep}(\Lambda_1 \oplus_R \Lambda_2/R) = \mathrm{Sep}(\Lambda_1/R) \oplus \mathrm{Sep}(\Lambda_2/R).$$

Definition (7.25)
The extension Λ *of the field R is said to be* **purely inseparable** *if* Λ *is algebraic over R and if* $\mathrm{Sep}(\Lambda/R) = R$.

Trivially R itself is purely inseparable over R. For any algebraic extension Λ of R we find that Λ is purely inseparable over $\mathrm{Sep}(\Lambda/R)$.

Theorem (7.26)
Let R be a field. If there exists a non-trivial purely inseparable (field) extension Λ *of R, then R is an imperfect field of characteristic* $p \neq 0$ (i.e. R is not perfect, compare chapter 1, section 6) *and for every non-zero element x of* Λ *there is a power* p^μ $(\mu > 0)$ *such that* $\mathrm{ord}(x/R^\times) \doteq p^\mu$ (i.e. $x^{p^\mu} \in R, x^{p^{\mu-1}} \notin R$).

Conversely, if R is of prime characteristic $p > 0$ *and* Λ *is a non-trivial extension of R with the property that* $\mathrm{ord}(x/R^\times)$ *is a power of p for every non-zero element x* of Λ *then* Λ *is a non-trivial purely inseparable extension of R.*

Proof
Let Λ be a purely inseparable extension of R then every element x of Λ not in R has inseparable minimal polynomial m_x over R.

Since m_x is irreducible it follows that $\gcd(D_t(m_x), m_x) = 1$ in case $D_t(m_x) \neq 0$. Thus from the inseparability of m_x we conclude that

$$D_t(m_x) = 0,$$

hence the characteristic of R is a prime number $p>0$ and

$$m_x(t)=t^{p\lambda}+\sum_{i=1}^{\lambda}a_it^{p(\lambda-i)}$$
$$(\lambda\in\mathbb{Z}^{>0},a_i\in R,1\leqslant i\leqslant\lambda).$$

Hence,

$$m_{x^p}(t)=t^{\lambda}+\sum_{i=1}^{\lambda}a_it^{\lambda-i}.$$

But since every element of Λ which does not belong to R is inseparable over R it follows by repeated substitution of x^p for x that

$$m_x(t)=t^{p^\mu}+a_\mu \quad\text{and}\quad p^\mu=\operatorname{ord}(x/R^\times).$$

Since there is at least one element x of Λ not belonging to R it can be assumed without loss of generality that

$$m_x=t^p-a,\quad a\in R^\times,\quad \operatorname{ord}(x/R^\times)=p,\quad x\notin R^\times,$$

hence

$$a\in R^\times,\quad a\notin(R^\times)^p:=\{b^p|b\in R^\times\},\quad (R^\times)^p\subset R^\times,$$

R is imperfect.

Conversely, if $\chi(R)=p>0$ and Λ is a non-trivial extension of R with the property that

$$\operatorname{ord}(x/R^\times)=p^\mu\quad(\mu=\mu(x)\in\mathbb{Z}^{>0},0\neq x\in\Lambda),$$

then we have

$$a=x^{p^\mu}\in R^\times,\quad m_x=t^{p^\mu}-a,\quad D_t(m_x)=0$$

for all x of $\Lambda^\times$. Hence, $\operatorname{Sep}(\Lambda/R)=R,\Lambda$ is purely inseparable over R. □

Corollary (7.27)
The degree of a finite purely inseparable extension of an imperfect field is a power of the characteristic.

Corollary (7.28)
In any extension Λ of a field R of characteristic $p>0$ the elements of $\Lambda^\times$ with p-power order modulo $R^\times$ and 0 form a purely inseparable extension $R^{p^{-\infty}}\cap\Lambda$ of R, the **maximal purely inseparable extension** *of R in Λ.*

Exercises

1. Let Λ/R be a field extension of the field R of characteristic p. Show that

$$\bar{R}:=\operatorname{Sep}(\Lambda/R)(R^{p^{-\infty}}\cap\Lambda)$$

is the maximal algebraic extension of R in Λ and that there holds the R-module isomorphism

$$R\simeq\operatorname{Sep}(\Lambda/R)\otimes_R(R^{p^{-\infty}}\cap\Lambda).$$

2. Prove:
   If a minimal splitting field Λ of the monic non-constant polynomial f over the field F is not a normal extension, then

   1. $\chi(F) = p > 0$.
   2. F is imperfect.
   3. f is divisible by an inseparable irreducible polynomial over F.
   4. Λ is normal over $F^{p^{-\infty}} \cap \Lambda = \Lambda^{\mathrm{Aut}(\Lambda/F)}$.
   5. $\mathrm{Sep}(\Lambda/F)$ is separable over F such that

$$\mathrm{Aut}(\Lambda/F)\,\mathrm{Sep}(\Lambda/F) = \mathrm{Aut}(\mathrm{Sep}(\Lambda/F)/F) \simeq \mathrm{Aut}(\Lambda/F).$$

Conversely, if Λ is a finite extension of the field F with the property that the fixed subfield of $\mathrm{Aut}(\Lambda/F)$ is a proper extension of F then Λ is a minimal splitting field of some monic non-constant polynomial f over F, and f has property 3.

2.8. The Lagrange resolvent

E. Artin's interpretation of Galois theory as study of the action of a finite group G (viz. the automorphism group of E over F) on a normal extension E of a field F, was anticipated in the cyclic case by about 200 years by J. Lagrange.

He considered 'operators'

$$\mathfrak{x} = \sum_{h \in G} \lambda(h)h \neq 0 \quad (\lambda : G \to F), \tag{8.1}$$

which is to say he used elements of the group algebra FG of G over F with the property that

$$FG\mathfrak{x} \subseteq F\mathfrak{x} \tag{8.2}$$

For such operators we have

$$g\mathfrak{x} = \psi(g)\mathfrak{x} \quad (g \in G) \tag{8.3}$$

when the mapping

$$\psi : G \to F$$

has the properties of a proper representation of degree 1 over F:

$$\psi(g) \neq 0, \quad \psi(gh) = \psi(g)\psi(h) \quad (g, h \in G). \tag{8.4}$$

It follows that the ψ-image elements form the finite subgroup $\psi(G)$ of $F^\times$. As we saw in section 5 any such group is cyclic:

$$\psi(G) = \langle \zeta \rangle, \tag{8.5}$$

where ζ is an element of $F^\times$ of finite order n. The cyclic group $\psi(G)$ is isomorphic to the factor group $G/\ker(\psi)$. We observe that the characteristic

$\chi(F)$ of F either is 0 or it is a prime number which does not divide n:

$$\chi(F) \nmid n. \tag{8.6}$$

Namely, for $\chi(F) > 0$ any equation

$$\xi^{\chi(F)} = 1 \quad (\xi \in F^\times)$$

implies $\xi = 1$ because of $0 = \xi^p - 1 = (\xi - 1)^p$.

The construction of the operator $\mathfrak{x}$ of (8.1) follows from the equations

$$\begin{aligned} g\mathfrak{x} &= g \sum_{h \in G} \lambda(h)h = \sum_{h \in G} \lambda(h)gh = \sum_{h \in G} \lambda(g^{-1}h)h \\ &= \psi(g) \sum_{h \in G} \lambda(h)h = \sum_{h \in G} \psi(g)\lambda(h)h \end{aligned}$$

(by (8.3)), i.e.

$$\begin{aligned} \lambda(g^{-1}h) &= \psi(g)\lambda(h) \quad (g, h \in G), \\ \mathfrak{x} &= \lambda(1) \sum_{h \in G} \psi(h)^{-1}h. \end{aligned}$$

Hence $\mathfrak{x}$ is uniquely determined by ψ up to a scalar factor. Conversely, let

$$\psi : G \to F^\times \tag{8.7}$$

be a homomorphism of G into the multiplicative group of F, also called a *multiplicative character of G over F*. Then (8.5) holds with ζ being an element of $F^\times$ of finite non-zero order n and the element

$$\mathfrak{x} = \sum_{h \in G} \psi(h)^{-1}h \tag{8.8}$$

of FG has the property

$$\begin{aligned} g\mathfrak{x} &= \sum_{h \in G} \psi(h)^{-1}gh \\ &= \sum_{h \in G} \psi(g)\psi(h)^{-1}h \\ &= \psi(g)\mathfrak{x}. \end{aligned}$$

For every element ξ of the normal extension E of F and every $g \in G$ we have

$$\begin{aligned} g(\mathfrak{x}(\xi)) &= (g\mathfrak{x})(\xi) = (\psi(g)\mathfrak{x})(\xi) = \psi(g)(\mathfrak{x}(\xi)), \\ \psi(g)^n &= 1, \\ g(\mathfrak{x}(\xi)^n) &= g(\mathfrak{x}(\xi))^n = (\psi(g)\mathfrak{x}(\xi))^n = \psi(g)^n\mathfrak{x}(\xi)^n = \mathfrak{x}(\xi)^n. \end{aligned}$$

By the main theorem of Galois theory, respectively by (6.8), we obtain

$$\mathfrak{x}(\xi)^n = a \in F,$$

i.e. $\mathfrak{x}(\xi)$ is an nth *radical relative to F*. Conversely, if an element η of E satisfies the condition

$$\eta^n = a \in F,$$

then it is of the form

$$\eta = \mathfrak{x}(\xi)$$

for some ξ of E and some element $\mathfrak{x}$ of FG satisfying (8.1), (8.2). That is trivial for $\eta = 0$.

Let $\eta \neq 0$. There is a normal F-basis (see 1 (5.17) and section 12)

$$B = \{g(\omega) | g \in G\}$$

for some element ω of E, so that

$$\eta = \sum_{h \in G} \mu(h)h(\omega) \quad (\mu : G \to F),$$

$$F^\times \ni \eta^n = g(\eta^n) = g(\eta)^n, \quad g(\eta) = \psi(g)\eta \text{ (defining } \psi),$$

$$\psi(g)^n = 1, \quad \psi(g) \in \langle \zeta \rangle,$$

$$\psi(g_1 g_2)\eta = g_1 g_2(\eta) = g_1((\psi(g_2)\eta) = \psi(g_2) g_1 \eta = \psi(g_2)\psi(g_1)\eta,$$

$$\psi(g_1 g_2) = \psi(g_1)\psi(g_2),$$

$$g(\eta) = \sum_{h \in G} \mu(h) gh(\omega)$$

$$= \sum_{h \in G} \mu(g^{-1}h)h(\omega)$$

$$= \psi(g)\eta = \sum_{h \in G} \psi(g)\mu(h)h(\omega),$$

$$\mu(g^{-1}h) = \psi(g)\mu(h),$$

$$\mu(g^{-1}) = \psi(g)\mu(1), \quad \text{respectively } \mu(1) = \mu(g)/\psi(g^{-1}),$$

$$\eta = \mathfrak{x}(\mu(1)\omega) \quad \text{for} \quad \mathfrak{x} = \sum_{g \in G} \psi(g^{-1})g.$$

We summarize our account of Lagrange's work by

Theorem (8.9)

The application of the Lagrange operators to the elements of a normal extension E of a field F produces precisely all radical elements of E relative to F, provided F contains the pertinent roots of unity.

The question arises how the radical elements of an extension relative to the field of reference F are related to each other and how they are related to the automorphism group of E over F in case E is normal over F.

Theorem (8.10)

(a) *For any separable extension E of the field F and for every natural number n the elements η of $E^\times$ with the property*

$$\eta^n = a \in F^\times \tag{8.11}$$

form a subgroup $S_{E,F,n}$ of $E^\times$ containing $F^\times$ as a subgroup. The exponent $e_{E,F,n}$

of the factor group $S_{E/F,n} := S_{E,F,n}/F^{\times}$, *i.e. the smallest natural number e such that* $\eta^e \in F^{\times}$ *for all* $\eta \in S_{E,F,n}$, *is a divisor of n which is not divisible by the characteristic of F. Trivially we have*

$$S_{E,F,n} = S_{E,F,e}. \tag{8.12}$$

(b) *The intermediary field*

$$\phi_{E/F,n} = F(S_{E,F,n})$$

generated by the radical elements satisfying (8.11) *is an algebraic extension of F with the property that any representative system of* $S_{E,F,n}$ *mudulo* $F^{\times}$ *is an F-basis of* $\phi_{E/F,n}$.

Proof

(a) follows from the identity

$$(\eta_1\eta_2)^n = \eta_1^n\eta_2^n$$

for elements η_1, η_2 of E and from the remark that

$$\eta^{\chi(F)} \in F \quad (\eta \in E) \quad \text{for} \quad \chi(F) = p > 0$$

implies that $F(\eta)$ is purely inseparable over F, hence

$$F(\eta) \subseteq F \tag{8.13}$$

(compare (7.25) and (7.26)). (8.12) is trivial.

(b) An element η of E of prime order m modulo $F^{\times}$ satisfies the separable equation

$$\eta^m - a = 0, \tag{8.14}$$

with some non-zero element a of $F^{\times}$. Hence the minimal splitting field of $m_{\eta/F}$ is a normal extension E_2 of F containing at least one element $\eta' \neq \eta$ satisfying $\eta'^m - a = 0$, and $\eta' = \zeta\eta$, where ζ is a primitive mth root of unity of E_2. Now E_2 also is a normal extension of the subfield $E_1 = F(\zeta)$. For any F-automorphism $\alpha \neq id_{E_2}$ of E_2 we have $\alpha(\eta)^m = \alpha(\eta^m) = \alpha(a) = a$, $\alpha(\alpha(\eta)\eta^{-1})^m = 1$, $\alpha(\eta) = \zeta^k\eta$ $(0 \leqslant k < m)$, hence $E_1(\eta)$ already contains all E_2/E_1-conjugates of η, hence

$$E_2 = E_1(\eta), \quad \alpha^m(\eta) = \eta, \quad \alpha^m = id_{E_2},$$

α is of order m (because of $m \in \mathbb{P}$), hence the polynomial $t^m - a$ is irreducible over E_1. *A fortiori*, it is irreducible over F.

If $t^m - \tilde{a}$ is reducible over F, then $\tilde{\eta} \in E_1$. We observe that E_1 is a minimal splitting field of the separable polynomial $t^m - 1$, hence E_1 is normal over F.

The automorphism group of E_1 over F is Abelian because for any two automorphisms α_1, α_2 of E_1 over F we have

$$\alpha_i(\zeta^m) = \alpha_i(\zeta)^m = 1, \quad \alpha_i(\zeta) = \zeta^{\nu_i} \quad (\nu_i \in \mathbb{Z}, i = 1, 2),$$
$$\alpha_1\alpha_2(\zeta) = \alpha_1(\zeta^{\nu_2}) = (\zeta^{\nu_1})^{\nu_2} = \alpha_2\alpha_1(\zeta). \tag{8.15}$$

The subfield $F(\tilde{\eta})$ of the abelian extension $F(\zeta)$ is normal. It contains at least one element $\eta' \neq \tilde{\eta}$ satisfying $\eta'^m - a = 0$, hence $(\eta'\tilde{\eta}^{-1})^m = 1$, $\eta' = \zeta^\nu \tilde{\eta}$ for some $\nu \in \mathbb{Z}$, $m \nmid \nu$, hence $\zeta \in F(\tilde{\eta})$, $F(\tilde{\eta}) = F(\zeta)$. The group of the equation (8.14) is a regular permutation group of its roots (those are easy to understand terms of applied Galois theory explained in section 9), hence

$$1 < [F(\tilde{\eta}):F] | m,$$
$$[F(\tilde{\eta}):F] = m,$$

because m is a prime number. But on the other hand $F(\tilde{\eta}) = F(\zeta)$, $[F(\zeta):F] < m$ because ζ satisfies the equation

$$\zeta^{m-1} + \zeta^{m-2} + \cdots + 1 = 0 \tag{8.16}$$

of degree $m - 1$.

Since this is a contradiction we conclude that the polynomial $t^m - \tilde{a}$ is irreducible so that $1, \tilde{\eta}, \ldots, \tilde{\eta}^{m-1}$ form an F-basis of $F(\eta)$. Set $\tilde{\eta} = \eta$.

For any radical element η' of $F(\eta)$ of prime order m' modulo $F^\times$ we show as above that $F(\eta')$ is an intermediate extension of F of degree m'. Because of the degree theorem it follows that $m = m'$, $F(\eta) = F(\eta')$,

$$\eta' := \sum_{i=j}^{m-1} \lambda_i \eta^i \quad (j \in \mathbb{Z}^{>0}, j < m; \lambda_j, \ldots, \lambda_{m-1} \in F, \lambda_j \neq 0),$$

$\eta'' := \eta' \eta^{-j} = \sum_{i=j}^{m-1} \lambda_i \eta^{i-j}$ is a radical element of $F(\eta)$ satisfying $\eta''^m \in F^\times$.

We observe that $\mathrm{Tr}(\eta) = 0$ where Tr denotes the trace from $F(\eta)$ to F because η satisfies the irreducible equation (8.14) over F. Similarly, it follows that every radical element of $F(\eta)$ which does not belong to F has zero trace.

For example,

$$\mathrm{Tr}(\eta^i) = 0 \quad (0 < i < m).$$

Hence

$$\mathrm{Tr}(\eta'') = \sum_{i=j}^{m-1} \lambda_i \mathrm{Tr}(\eta^{i-j})$$
$$= \lambda_j \mathrm{Tr}(1) = m\lambda_j \neq 0,$$

so that

$$\eta'' \in F^\times,$$
$$\eta' \in F^\times \eta^j.$$

In other words the group formed by the non-zero radical elements of $F(\eta)$ coincides with the set theoretical union

$$\bigcup_{i=0}^{m-1} F^\times \eta^i.$$

Now let S be a subgroup of $S_{E,F,n}$ containing $F^\times$ as subgroup of finite index. We must show that any representative set X of S modulo F forms an F-basis of $F(S)$. That was done already in case $(S:F^\times)$ is a prime number. It is trivial if $S = F^\times$. Apply induction over $(S:F^\times)$.

Let $(S:F^\times)$ be a composite number. There is an element η of S of prime order m modulo $F^\times$. It generates an intermediate field $F(\eta)$ of $F(S)$ which has the F-basis $1, \eta, \ldots, \eta^{m-1}$. It was shown above that

$$S \cap F(\eta)^\times = \bigcup_{i=0}^{m-1} F^\times \eta^i.$$

Hence by an isomorphism theorem of group theory any representative set of S modulo $S \cap F(\eta)^\times$ also is a representative set of $SF(\eta)^\times$ modulo $F(\eta)^\times$ so that it forms an $F(\eta)$-basis of $F(S)$ by induction assumption. Hence any representative set of S modulo $F^\times$ forms an F-basis of $F(S)$ in accordance with the degree theorem.

If $S_{E,F,n}$ is infinite we apply the above analysis to any finite subgroup S of $S_{E,F,n}$ which contains $F^\times$ as a subgroup of finite index. □

Theorem (8.17)

Let E be a normal extension of F with automorphism group $G = \mathrm{Aut}(E/F)$ and let $e = e_{E,F,n}$ be the exponent of $S_{E,F,n}/F^\times = S_{E/F,n}$.

(a) *E contains a primitive* eth *root of unity, and $\phi_{E/F,n} = F(S_{E,F,n})$ is normal over F.*

(b) *If F contains a primitive* eth *root of unity, then*

$$\mathrm{Aut}(\phi_{E/F,n}/F) \simeq S_{E/F,n}, \tag{8.18}$$

and

$$\mathrm{Aut}(\phi_{E/F,n}/F) \simeq G/G^e DG, \tag{8.19}$$

where $G^e := \{g^e \mid g \in G\}$ and $DG = [G, G]$ denotes the commutator subgroup of G.

Proof

(a) If E is normal over F then we conclude from the separable equation

$$\eta^m - a = 0 \quad (a \in F^\times),$$

satisfied by any element η of $S_{E,F,n}$ of order m modulo $F^\times$, $m|n$, that the conjugates of η under the group

$$G = \mathrm{Aut}(E/F)$$

are of the form $\eta\zeta'$ with ζ' being an mth root of unity.

Since unit roots of E always satisfy separable equations over F it follows that $\phi_{E/F,n}$ is a normal extension of F and that E contains a primitive eth root of unity ζ.

(b) Moreover, if ζ belongs to F then the F-automorphisms σ of $\phi_{E/F,n}$ are uniquely determined by their action on the generators of $S_{E,F,n}$ and we have

$$\sigma(\eta) = \psi(\sigma)\eta,$$

$$\psi(\sigma)\in\mathrm{Hom}\,(S_{E/F,n},\langle\zeta\rangle),$$
$$\psi(\sigma)(\alpha)=1 \quad \text{if } \alpha\in F^{\times},$$

where ψ is a monomorphism of $\mathrm{Aut}\,(\phi_{E/F,n}/F)$ into $\mathrm{Hom}\,(S_{E/F,n},\langle\zeta\rangle)$. But it follows from the basis theorem for finite Abelian groups S of exponent e that

$$S\simeq\mathrm{Hom}\,(S,\langle\zeta\rangle),$$

where $\mathrm{Hom}\,(S,\langle\zeta\rangle)$ is the group of abelian characters of S. Indeed, S is the direct product of finitely many cyclic subgroups

$$S=\langle\sigma_1\rangle\times\langle\sigma_2\rangle\times\cdots\times\langle\sigma_r\rangle,$$

each of an order n_i dividing e.

Hence there are the characters

$$\chi_i\colon S\to\langle\zeta\rangle\colon \sigma_k\mapsto\zeta^{(e/n_i)\delta_{ik}}$$

of order:

$$\mathrm{ord}\langle\chi_i\rangle=n_i=\mathrm{ord}\langle\sigma_i\rangle,$$

and each element of $\mathrm{Hom}\,(S,\langle\zeta\rangle)$ is of the form

$$\chi\colon S\to\langle\zeta\rangle\,\mathrm{via}\,\sigma_i\mapsto\zeta^{(e/n_i)x_i}\ (x_i\in\mathbb{Z},1\leqslant i\leqslant r),$$

and the character χ is obtained as

$$\chi=\prod_{i=1}^{r}\chi_i^{x_i},$$

where the exponents x_i are uniquely determined modulo n_i. It follows that

$$\mathrm{Hom}\,(S,\langle\zeta\rangle)=\prod_{i=1}^{r}\langle\chi_i\rangle,$$

where $\prod$ denotes the direct product of the cyclic groups $\langle\chi_i\rangle$ and the order of the group $\langle\chi_i\rangle$ is equal to n_i. Thus there is the isomorphism of S on $\mathrm{Hom}\,(S,\langle\zeta\rangle)$ which maps σ_i on χ_i $(i=1,2,\ldots,r)$. According to (8.10b) the degree of $\phi_{E/F,n}$ over F is equal to the order of $S_{E/F,n}$. It is also equal to the order of $\mathrm{Aut}\,(\phi_{E/F,n}/F)$. Hence the monomorphism established above is in fact an isomorphism as stated in (8.18).

Since $\phi_{E/F,n}$ is an abelian extension of F with automorphism group of exponent e it follows that the restriction of $G=\mathrm{Aut}\,(E/F)$ to $\phi_{E/F,n}$ establishes an epimorphism of G on $\mathrm{Aut}\,(\phi_{E/F,n}/F)$ with G^eDG in its kernel. Hence the fixed subfield E' of G^eDG contains $\phi_{E/F,n}$. We want to show that $E'=\phi_{E/F,n}$. But by construction E' is abelian over F with automorphism group of exponent e. Using the basis theorem for abelian groups we obtain E' as the composite of cyclic extensions of F each of degree dividing e. By Lagrange each of those extensions is generated by the adjunction of an element of $S_{E,F,e}$ to F. Hence E' is contained in $\phi_{E/F,n}$. □

Corollary (8.20)
Given a field F containing a primitive nth root of unity. Then the finite abelian extension E of F with the property that the exponent e of $G = \mathrm{Aut}(E/F)$ *divides n are obtained as follows:*

Choose finitely many elements $a_1, a_2, \ldots, a_s$ *of* $F^\times$ *with the property that*

$$A = \langle a_1, a_2, \ldots, a_s, (F^\times)^n \rangle / (F^\times)^n$$
$$\simeq \prod_{i=1}^{s} \langle a_i, (F^\times)^n \rangle / (F^\times)^n.$$

Then

$$E = F(a_1^{1/n}, \ldots, a_s^{1/n})$$

is an abelian extension of F with

$$\mathrm{Aut}(E/F) \simeq A.$$

(Note that the ambiguity of the root symbols $a_i^{1/n}$ does not matter here because of the presence of a primitive nth root of unity in the field of reference.)

The extensions dealt with by Corollary (8.20) are also called *Kummer extensions.*

The Lagrange method of gathering the nth roots of elements of a base field which contains a primitive nth root of unity in normal extensions entirely bypasses the construction of abelian extensions with degree a power of the characteristic.

In such cases there is an additive counterpart to Lagrange's construction. It was discovered by Artin and Schreier in 1923.

We had already observed in section 5 that for any commutative ring R of prime characteristic p the mapping

$$\wp = \wp_p \colon R \to R \colon x \mapsto x^p - x$$

is an R^+-endomorphism of R. Its kernel is a subring of R.

In case R is unital then the kernel of $\wp$ contains the prime ring of R and if R is entire then $\ker(\wp)$ coincides with the prime ring.

Theorem (8.21)
Let E be an extension of the field F of characteristic p and let

$$\wp_E^{-1}(F) := \wp^{-1}(F) = \{x \in E \mid \wp(x) \in F\}. \tag{8.22}$$

(a) $\wp^{-1}(F)$ *generates a separable extension* $F(\wp^{-1}(F))$ *of F.*
(b) *Every element of* $F(\wp^{-1}(F))$ *which does not belong to F generates a cyclic extension of degree p of F.*
(c) *Conversely, every cyclic extension of degree p over F contained in E is generated by an element of* $\wp^{-1}(F)$.

(d) *If $F(\wp^{-1}(F))$ is of finite degree over F, then it is a normal extension of F and its group is an abelian (elementary abelian) group of exponent p.*

Proof

(a) and (b). Let $\xi \in E \backslash F$ with $\wp(\xi) \in F$. Then ξ satisfies an Artin–Schreier equation

$$\xi^p - \xi - a = 0 \tag{8.23}$$

for some a of F which does not belong to $\wp(F)$. The polynomial

$$f(t) = t^p - t - a \in F[t] \tag{8.24}$$

has the distinct roots (compare (5.21))

$$\xi, \xi + 1, \ldots, \xi + p - 1 \tag{8.25}$$

in E. Hence the extension $F(\xi)$ of F is normal of degree p over F, its automorphism group over F is generated by the F-automorphism

$$\sigma\colon F(\xi) \to F(\xi)\colon \xi \mapsto \xi + 1 \tag{8.26}$$

of prime order p.

(c) and (d). If E_1 is a normal extension of F of degree p over F and contained in E then there is an F-automorphism w of order p which generates $\mathrm{Aut}(E_1/F)$.

The equation

$$w^p = \mathbf{1}$$

where

$$\mathbf{1} = id_{E_1}$$

implies

$$0 = w^p - \mathbf{1} = (w - \mathbf{1})^p.$$

On the other hand

$$w \neq \mathbf{1}, \quad w - \mathbf{1} \neq 0,$$

hence there is an index $j' \in \mathbb{N}$ such that

$$1 \leqslant j' < p, \quad (w - \mathbf{1})^{j'} \neq 0, \quad (w - \mathbf{1})^{j'+1} = 0.$$

It follows that

$$(w - \mathbf{1})(w - \mathbf{1})^{j'}(E_1) = \{0\}, \quad \{0\} \subset (w - \mathbf{1})^{j'}(E_1) \subseteq E_1, \quad (w - \mathbf{1})^{j'}(E_1) = F,$$

hence there is an element

$$\xi \in (w - \mathbf{1})^{j'-1}(E_1),$$

for which

$$(w - \mathbf{1})(\xi) = 1, \quad w(\xi) = \xi + 1, \quad w^i(\xi) = \xi + i,$$

and the conjugates of ξ are

$$\xi, \xi + 1, \xi + 2, \ldots, \xi + p - 1,$$

$$\wp(\xi)=\xi(\xi+1)\cdots(\xi+p-1)=a\in F,$$
$$\xi^p-\xi-a=0,$$
$$a\notin\wp(F),$$
$$E_1=F(\xi).$$

The elements η of E_1/F satisfying $\wp(\eta)\in F$ may be called *Artin–Schreier generators* of E_1 (over F). Since all their conjugates are of the form $\eta+i$ ($i\in\mathbb{F}_p$), they clearly satisfy $(w-1)(\eta)\in\mathbb{F}_p$. In order to determine them we note that

$$w(\xi)=\xi+1,$$
$$w(\xi^i)=(\xi+1)^i,$$
$$(w-1)(\xi^i)=(\xi+1)^i-\xi^i=\sum_{j=0}^{i-1}\binom{i}{j}\xi^j=i\xi^{i-1}+\cdots,$$
$$(w-1)^{p-1}(\xi^{p-1})=(p-1)(p-2)\cdots 1=-1 \quad \text{(see also exercise 2)},$$
$$j'=p-1,$$
$$(w-1)^{p-1}(E_1)=F. \tag{8.27}$$

Also

$$\ker(w-1)=F, \tag{8.28}$$

hence the only elements ξ' of E_1 for which $(w-1)(\xi')\in F$, are the elements of $F\xi+F$, hence

$$\wp(E_1)\cap F=\mathbb{F}_p a+\wp(F). \tag{8.29}$$

For the completion of (*d*) we may use either *a* 'basis argument' or an 'abstract argument'. Since this alternative frequently occurs in the application of group theory to number theory let us do both this time.

Basis argument: Let $\xi_1,\ldots,\xi_j$ be a basis of $\wp^{-1}(F)$ modulo F. Then the automorphisms α of $\mathrm{Aut}(F(\wp^{-1}(F))/F)$ are characterized by their effect on $\xi_1,\ldots,\xi_j$: $\alpha(\xi_i)=\xi_i+\phi(\alpha)(\xi_i)$ ($\phi(\alpha)(\xi_i)\in\mathbb{F}_p$), where all vectors $\mathbf{x}\in\mathbb{F}_p^j$ occur in the form $\mathbf{x}=(\phi(\alpha)(\xi_1),\ldots,\phi(\alpha)(\xi_j))^t$, since otherwise there is some non-zero linear combination of $\xi_1,\ldots,\xi_j$ over $\mathbb{F}_p$ which is annihilated by all α in contradiction to the main theorem of Galois theory. Therefore, the automorphism group of $F(\wp^{-1}(F))$ over F is isomorphic to $\mathbb{F}_p^j$ which is an elementary Abelian additive group of order p^j.

Abstract argument: For each automorphism α of $F(\wp^{-1}(F))$ over F we have the homomorphism

$$\phi(\alpha)\colon F(\wp^{-1}(F))/F\to\mathbb{F}_p\colon \xi/F\mapsto\alpha(\xi)-\xi,$$

such that ϕ is a homomorphism of the multiplicative group $\mathrm{Aut}(F(\wp^{-1}(F))/F)$ into the additive group $\mathrm{Hom}(F(\wp^{-1}(F))/F,\mathbb{F}_p)$. The kernel of this mapping is 1. The image of $\mathrm{Aut}(F(\wp^{-1}(F))/F)$ is isomorphic to the factor group of $\wp^{-1}(F)$ over those elements of $\wp^{-1}(F)$ which are fixed by all automorphisms of $F(\wp^{-1}(F))$ over F (see exercise 7). By the main theorem of Galois theory those fixed elements form F. Hence ϕ defines an automorphism between $\mathrm{Aut}(F(\wp^{-1}(F))/F)$ and the character group of the elementary abelian additive group $F(\wp^{-1}(F))/F$. □

Exercises

1. Let $\mathbb{F}_q$ be a finite field of characteristic p with primitive root ρ of $\mathbb{F}_q^\times$. Let r be a prime number dividing $q-1$. Then show for each natural number m that the polynomial
$$t^{r^m} - \rho$$
is irreducible.
2. Let F_q be a finite field of characteristic p. Prove *Wilson's theorem* for $\mathbb{F}_q$: $(p-1)! = -1$.
3. Let F be a field of prime characteristic p, let
$$a \in F, \quad a \notin \wp(F),$$
$$E = F[t]/(t^p - t - a) = F[\xi] \quad (\xi = t/(t^p - t - a)),$$
then show that $\xi^{p-1} \notin \wp(E)$.
4. Let $\mathbb{F}_q$ be as in exercise 2. Construct recursively irreducible monic polynomials of degree $p, p^2, p^3, \ldots$ using exercise 3.
5. Let $\mathbb{F}_q$ be a finite field and let m, n be two coprime natural numbers. Show that
$$\mathbb{F}_{q^m} \otimes_{\mathbb{F}_q} \mathbb{F}_{q^n} \simeq \mathbb{F}_{q^{mn}}.$$
6. Let f, g be two monic non-constant polynomials over the unital commutative ring R. Denote by $f \times g$ the principal polynomial of $t/f \otimes_R t/g$ in $R[t]/f \otimes_R R[t]/g$.

   (a) Suppose
$$f = \prod_{i=1}^{m} (t - x_i), \quad g = \prod_{k=1}^{n} (t - y_k),$$
   then show
$$f \times g = \prod_{i=1}^{m} \prod_{k=1}^{n} (t - x_i y_k).$$

   (b) If R is a finite field and f, g are irreducible polynomials of mutually prime degrees m, n then show $f \times g$ is an irreducible polynomial of degree mn.

7. Prove $Tm(\phi) \simeq \wp^{-1}(F)/\wp^{-1}(F)^{\mathrm{Aut}(F(\wp^{-1}(F))/F)}$ in the context of the abstract argument at the end of the proof of (8.21). More generally, let G be a finite Abelian group with character group $\chi(G)$ isomorphic to G. Show that every subgroup S of $\chi(G)$ satisfies $S \simeq \chi(G/G_s)$ for $G_s := \{g \in G \mid \forall s \in S; s(g) = 1\}$.

2.9. The group of an equation

1. Introduction

Definition (9.1)

The group of the monic equation

$$x^n + a_1 x^{n-1} + \cdots + a_n = 0 \tag{9.2}$$

related to the monic polynomial

$$f(t) = t^n + a_1 t^{n-1} + \cdots + a_n \tag{9.3}$$

of positive degree n over the indecomposable unital commutative ring R is defined as the $\bar{\mathfrak{S}}_n$-stabilizer

$$\mathrm{Stab}(e_i/\bar{\mathfrak{S}}_n) = \{\pi \in \mathfrak{S}_n \mid \bar{\pi}(e_i) = e_i\} \tag{9.4}$$

of a primitive idempotent e_i of the universal splitting ring $S(f/R)$ of f over R.

We have seen in section 4 already that there are only finitely many primitive idempotents of $S(f/R)$, say $e_1, e_2, \ldots, e_s$, that they are $\bar{\mathfrak{S}}_n$-conjugates and that there holds the decomposition

$$S(f/R) = \sum_{i=1}^{\cdot s} e_i S(f/R) \tag{9.5}$$

of $S(f/R)$ into the algebraic sum of algebraically indecomposable R-rings $e_i S(f/R)$ $(1 \leqslant i \leqslant s)$. Hence the index of $\mathrm{Stab}(e_i/\bar{\mathfrak{S}}_n)$ in $\bar{\mathfrak{S}}_n$ is s and the s stabilizers $\mathrm{Stab}(e_i/\bar{\mathfrak{S}}_n)$ $(1 \leqslant i \leqslant s)$ are $\bar{\mathfrak{S}}_n$-conjugates.

It was shown that the projection mapping

$$P(e_i)\colon S(f/R) \to e_i S(f/R)\colon x \mapsto e_i x \tag{9.6}$$

amounts to an R-epimorphism of $S(f/R)$ on the ith component of (9.5) restricting to an isomorphism of R on $e_i R$. The universal splitting ring $S(f/R)$ is characterized as unital commutative overring of R with $n!$ basis elements that is generated by n roots $x_1, x_2, \ldots, x_n$ of f such that there holds the factorization

$$f(t) = \prod_{k=1}^{n} (t - x_k) \tag{9.7}$$

in $S(f/R)[t]$. Application of $P(e_i)$ yields the factorization

$$e_i f(t) = \prod_{k=1}^{n} (t - e_i x_k) \tag{9.8}$$

of the monic polynomial

$$e_i f(t) = t^n + e_i a_1 t^{n-1} + \cdots + e_i a_n \tag{9.9}$$

of $e_iR[t]$ so that $e_iS(f/R)$ is a minimal splitting ring of e_if over e_iR which is invariant under the stabilizer subgroup $\mathrm{Stab}(e_i/\bar{\mathfrak{S}}_n)$.

If f is separable then the roots $e_ix_1, \ldots, e_ix_n$ are distinct. This is because the discriminant

$$d(f) = (-1)^{\binom{n}{2}} \prod_{j \neq k} (x_j - x_k) \tag{9.10}$$

is a non-zero divisor of R, hence

$$d(e_if) = (-1)^{\binom{n}{2}} \prod_{j \neq k} (e_ix_j - e_ix_k)$$

is a non-zero divisor of e_iR which implies that

$$e_ix_j \neq e_ix_k \quad \text{for } j \neq k.$$

Hence in this case the restriction of $\mathrm{Stab}(e_i/\bar{\mathfrak{S}}_n)$ to $e_iS(f/R)$ is faithful. Moreover, the elements of R are the only elements fixed by $\bar{\mathfrak{S}}_n$ which implies that the elements of e_iR are the only fixed elements of $e_iS(f/R)$ under the restriction of $S(f/R)$ to $e_iS(f/R)$. Indeed, if

$$\forall \bar{\pi} \in \bar{\mathfrak{S}}_n : \bar{\pi}_j(x) = x$$

for some x of $e_iS(f/R)$, and if

$$\bar{\pi}_j(e_i) = e_j \quad (1 \leqslant j \leqslant s),$$

for some π_j of $\mathfrak{S}_n$, then we have

$$\bar{\mathfrak{S}}_n = \bigcup_{j=1}^{s} \bar{\pi}_j \mathrm{Stab}(e_i/\bar{\mathfrak{S}}_n),$$

and

$$y = \sum_{j=1}^{s} \bar{\pi}_j(x)$$

is an element $S(f/R)$ with $e_jS(f/R)$-component $\bar{\pi}_j(x)$ that is invariant under $\bar{\mathfrak{S}}_n$. As stated above $y \in R$, hence $e_iy = x \in e_iR$. Thus the group of a separable equation is presented as a permutation group of the n roots e_ix_k $(1 \leqslant k \leqslant n)$ of the separable polynomial e_if of $e_iR[t]$.

In case R is a field then $e_iS(f/R)$ is a minimal splitting field of e_if over the field e_iR and the group of the equation is the full automorphism group of the minimal splitting field over the field of reference according to the main theorem of Galois theory.

Let us deal for example with the group of a monic cubic equation

$$f(x) = 0,$$

where

$$f(t) = t^3 + a_1t^2 + a_2t + a_3 \in R[t].$$

Up to $\bar{\mathfrak{S}}_n$-conjugacy there are the following four possibilities for the stabilizer of e_i:

I. Even subgroups of $\mathfrak{S}_3$

no.	group	order	comment
1	1	1	$S(f/R)$ decomposes into the algebraic sum of six rings isomorphic to R
2	$\mathfrak{A}_3$	3	$S(f/R)$ decomposes into the algebraic sum of two isomorphic algebraically indecomposable rings

II. Odd subgroup of $\mathfrak{S}_3$

no.	group	order	comment
3	$\langle(12)\rangle$	2	$S(f/R)$ decomposes into the algebraic sum of three isomorphic algebraically indecomposable rings
4	$\mathfrak{S}_3$	6	$S(f/R)$ is algebraically indecomposable

We see that for $n = 3$ it is always true that subgroups of $\mathfrak{S}_3$ of the same order also are $\mathfrak{S}_3$-conjugate.

Let us assume now that R is a field and that the polynomial f is separable, i.e. $d(f)$ is a non-zero divisor of R and that we have an algebraically indecomposable minimal splitting ring X of f over R so that there is given a generation

$$X = R(y_1, y_2, y_3),$$

where

$$f(t) = \prod_{i=1}^{3} (t - y_i).$$

Then we can find out which of the four possibilities listed above applies to f upon trying to collect the linear factors $t - y_i$ into subproducts with coefficients in R. Using suitable numbering of y_1, y_2, y_3 we obtain three possibilities (the numbers in column 1 referring to no. of tables above):

1	$y_1, y_2, y_3 \in R$	f splits over R,
3	$y_1 \in R, f(t) = (t - y_1)\dfrac{f(t)}{t - y_1}$	f has a root in R, but does not split over R,
2,4	f is irreducible in $R[t]$	f has no root in R.

Quite the same result is obtained upon analysing the equation ring of f over R in regard to the three possibilities listed above.

In general there holds the

Proposition (9.11)

Let f be a monic separable polynomial of degree $n>0$ over the field R and let the group G of the equation $f(x)=0$ be a permutation group with s orbits, say

$$\{1,2,\ldots,n_1\},\ \{n_1+1,\ldots,n_1+n_2\},\ldots,\{n-n_s+1,\ldots,n\}$$

$$\left(s\in\mathbb{Z}^{>0}, n_i\in\mathbb{Z}^{>0}\quad (1\leqslant i\leqslant s);\quad \sum_{i=1}^{s} n_i = n\right),$$

then there holds a factorization

$$f=\prod_{i=1}^{s} f_i$$

of f into the product of s distinct monic irreducible polynomials f_i of degree n_i over R $(1\leqslant i\leqslant s)$.

Proof

Assuming $e_1,e_2,\ldots,e_s$ to be the primitive idempotents of $S(f/R)$ and setting $G=\mathrm{Stab}(e_1/\mathfrak{S}_n)$ we find that

$$e_1S(f/R)=\langle R,e_1x_1,\ldots,e_1x_n\rangle$$

and that the coefficients of the polynomials

$$e_1f_i(t):=\sum_{j=1}^{n_i}\quad (t-e_1x_{m_i+j}),$$

$$m_i=\sum_{j=1}^{i-1} n_j\quad (1\leqslant i\leqslant s),$$

stay fixed under the group G. Because of the separability of f each polynomial f_i is in $R[t]$. Because R is a field it follows that f_i is irreducible over R. Indeed, if there is a factorization

$$f_i=f_{i1}f_{i2}$$

of f_i into the product of two non-constant monic polynomials f_{i1}, f_{i2} of $R[t]$ then the equation $f_i(x_{m_i+1})=0$ implies that one of the two equations $f_{ij}(x_{m_i+1})=0$ holds $(j=1$ or $2)$, say $f_{i1}(x_{m_i+1})=0$. Application of G to this equation yields the equations

$$f_{i1}(x_{m_i+j})=0\quad (1\leqslant j\leqslant n_i).$$

But a polynomial of degree less than n_i cannot have n_i distinct roots. Hence f_i is irreducible over R. □

We see from the list given above that cases 2, 4 cannot be distinguished merely by factoring f over R. But we can distinguish those two cases by means of the following odd–even criterion.

Proposition (9.12)

The group of the monic separable equation $f(x)=0$ of degree $n>1$ over a unital commutative ring R with $[0/2]=0$ is even if and only if the discriminant of f is a square element of R.

Proof

The different expression of f

$$\delta(f):=\prod_{i<k}(x_i-x_k)$$

in $S(f/R)$, i.e. the product of all ordered root differences, has the property that

$$\delta(f)^2=d(f).$$

Moreover, it is invariant under all even permutations, whereas it changes its sign for all odd permutations:

$$\pi(\delta(f))=\operatorname{sign}(\pi)\delta(f).$$

Hence, if G is even then $0\neq\delta(f)\in R$ because of the separability of f, but if G is odd then $0\neq\delta(f)\notin R$ because of the absence of non-zero elements equalling their negative. □

Examples for each of the four cases of monic cubic equations over $\mathbb{Z}$ are given by

(1)	$f(t)=t^3-t$	$d(f)=4,$
(2)	$f(t)=t^3-7t-7$	$d(f)=49,$
(3)	$f(t)=t(t^2+1)$	$d(f)=-4,$
	$f(t)=t(t^2-2)$	$d(f)=32,$
(4)	$f(t)=t^3-t+1$	$d(f)=-23,$
	$f(t)=t^3+t+1$	$d(f)=-31,$
	$f(t)=t^3+3t^2-t-1$	$d(f)=148.$

2. A categorical characterization of the group of an equation

Lemma (9.13)

Let R be an algebraically indecomposable unital commutative ring, and let f be a monic polynomial of degree $n>0$ of $R[t]$. Let $e_1,\ldots,e_s$ be the primitive idempotents of $S(f/R)$.

(a) *For any epimorphism*

$$\varepsilon: S(f/R)\to\Lambda$$

of $S(f/R)$ on the algebraically indecomposable unital ring Λ there is precisely one index i such that

$$\varepsilon(e_i)=1_\Lambda, \tag{9.13a}$$

but

$$\sum_{\substack{j=1\\ j\neq i}}^{s} e_j S(f/R) \subseteq \ker(\varepsilon). \tag{9.13b}$$

(b) *Furthermore* $\varepsilon(R)$ *is algebraically indecomposable,* $\varepsilon(S(f/R)) = \varepsilon(e_i S(f/R))$ *is a minimal splitting ring of* εf *over* $\varepsilon(R)$.

(c) *The automorphisms of* $\varepsilon(S(f/R))$ *over* $\varepsilon(R)$ *which merely permute the images* $\varepsilon x_1, \varepsilon x_2, \ldots, \varepsilon x_n$ *form a finite subgroup* X *of* $\mathrm{Aut}(\varepsilon(S(f/R))/\varepsilon(R))$. *The elements* $\bar{\pi}$ *of* $\mathfrak{S}_n$ *for which* $\varepsilon x_{\pi(j)} = \varepsilon x_j$ $(1 \leqslant j \leqslant n)$ *form a subgroup* Y *of* $\mathrm{Stab}(e_i/\bar{\mathfrak{S}}_n)$ *such that there is the epimorphism*

$$\bar{\varepsilon}\colon Y \to X$$

satisfying

$$\bar{\varepsilon}(\bar{\pi})(x) = \varepsilon(\bar{\pi}(x)) \quad (x \in S(f/R)).$$

Proof

(a) Since $e_1, e_2, \ldots, e_s$ are orthogonal idempotents it follows that

$$\varepsilon(e_j)^2 = \varepsilon(e_j), \quad \varepsilon(e_j)\varepsilon(e_k) = \varepsilon(e_j e_k) = 0 \quad (j \neq k).$$

Because of the algebraic indecomposability of Λ there cannot be two orthogonal idempotents in Λ. Hence there is an index i $(1 \leqslant i \leqslant s)$ such that $\varepsilon(e_j) = 0$ $(1 \leqslant j \leqslant s, j \neq i)$ and (9.13b). But then

$$\varepsilon(e_i) = \varepsilon\left(\sum_{j=1}^{s} e_j\right) = \varepsilon(1_{S(f/R)}) = 1_\Lambda.$$

(b) The image ring $\varepsilon(R)$ is algebraically indecomposable because it is a subring of the algebraically indecomposable ring $\varepsilon(S(f/R))$ and therefore it cannot contain two orthogonal idempotents. Since $S(f/R) = \langle R, x_1, \ldots, x_n \rangle$ with (9.7) it follows that $\varepsilon(S(f/R)) = \langle \varepsilon(R), \varepsilon(x_1), \ldots, \varepsilon(x_n) \rangle$ and

$$\varepsilon f(t) = \prod_{j=1}^{n} (t - \varepsilon(x_j)) \tag{9.13c}$$

so that $\varepsilon(S(f/R))$ is a minimal splitting ring of εf over $\varepsilon(R)$.

(c) If $\pi \in \mathfrak{S}_n$ and $\varepsilon x_{\pi(j)} = \varepsilon x_j$ $(1 \leqslant j \leqslant n)$ so that there is an automorphism $\bar{\bar{\pi}}$ of $\varepsilon(S(f/R))$ satisfying $\bar{\bar{\pi}}\varepsilon(x) = \varepsilon(\bar{\pi}(x))$ $(x \in S(f/R))$ then of course the elements of $\varepsilon(R)$ stay fixed, so that $\bar{\bar{\pi}}$ belongs to X. Moreover $\bar{\pi} e_i = e_j$ for some index j satisfying $1 \leqslant j \leqslant s$. Hence

$$1_{\varepsilon(S(f/R))} = \bar{\bar{\pi}}(1_{\varepsilon(S(f/R))}) = \bar{\bar{\pi}}(\varepsilon e_i) = \varepsilon\bar{\pi}(e_i) = \varepsilon(e_j),$$

hence $j = i$, $\bar{\pi} \in \mathrm{Stab}(e_i/\bar{\mathfrak{S}}_n)$, $Y \subseteq \mathrm{Stab}(e_i/\bar{\mathfrak{S}}_n)$. Clearly, $\bar{\varepsilon} Y = X$. □

Corollary (9.14)

The equation $$Y = \mathrm{Stab}(e_i/\bar{\mathfrak{S}}_n)$$

holds if and only if the ideal $\ker(\varepsilon)\cap e_iS(f/R)$ *of* $e_iS(f/R)$ *is invariant under* $\mathrm{Stab}(e_i/\bar{\mathfrak{S}}_n)$.

3. Polynomials with multiple factors, lifting of idempotents

Let us deal with the group of an inseparable monic equation (9.2) over a field R.

One reason for inseparability can be the occurrence of multiple factors. Suppose there is the factorization

$$f=\prod_{i=1}^{m} f_i^i, \quad 1<m\leqslant n, f_m\neq 1 \tag{9.15}$$

of f into the power product of monic mutually prime polynomials $f_1, f_2, \ldots, f_m$ of $R[t]$ none of which contains a multiple factor.

Let $f(t)=\prod_{i=1}^{n}(t-x_i)$ be the standard factorization of f in the universal splitting ring $S(f/R)$. In this case the R-algebra $e_iS(f/R)$ is not a field because it contains the non-zero nilpotent element e_iy_k where

$$y_k=\prod_{j=1}^{m} f_j(x_k) \quad (1\leqslant k\leqslant n).$$

Indeed, since f is the minimal polynomial of x_k over R it follows that

$$y_k^{m-1}\neq 0.$$

On the other hand,

$$y_k^m=f(x_k)\prod_{j=1}^{m} f_j(x_k)^{m-j}=0, \quad (e_iy_k)^m=e_iy_k^m=0.$$

For each index k' there is a component $e_iy_{k'}^{m-1}\neq 0$. Since $\bar{\mathfrak{S}}_n$ permutes the e_i's transitively it follows that there is an index k for which

$$0\neq(e_iy_k)^{m-1}=e_iy_k^{m-1}, \quad (e_iy_k)^m=0.$$

On the other hand, lemma (9.13) shows for any minimal splitting field Λ of f over R the existence of an epimorphism

$$\varepsilon: e_iS(f/R)\to\Lambda.$$

What is its kernel? Certainly $\ker(\varepsilon)$ contains all the nilpotent elements of $e_iS(f/R)$. Any non-nilpotent element x of $e_iS(f/R)$ is not contained in $\ker(\varepsilon)$ because its minimal polynomial over the field e_iR is of the form $t^\mu g(t)$ with $\mu\geqslant 0$ and with monic non-constant polynomial g of $e_iR[t]$ satisfying $g(0)\neq 0$. If $\mu>0$ then there holds an equation

$$A(t)t^\mu+B(t)g(t)=e_i \quad (A(t), B(t)\in e_iR[t]),$$

implying the existence of the orthogonal idempotents

$$A(x)x^\mu, \quad B(x)g(x),$$

adding up to e_i, contradicting the primitivity of e_i. Hence the kernel of ε consists exactly of all nilpotent elements of $e_iS(f/R)$.

Definition (9.16)
An ideal of a ring consisting entirely of nilpotent elements is said to be a **nilideal.**

We recommend to the reader to show the following as an exercise:

The intersection of any set of nilideals is a nilideal. If A is a nilideal of the ring X and B/A is a nilideal of the factoring X/A then B is a nilideal of X. The sum of two nilideals is a nilideal. The ideal generated by any set of nilideals is a nilideal. The ideal generated by all nilideals of the ring X is the largest nilideal of X, said to be the *nilradical* $\mathrm{NR}(X)$ of X.

For any nilideal A of X we have

$$\mathrm{NR}(X/A) = \mathrm{NR}(X)/A.$$

Theorem (9.17)
For any field R and every monic non-constant polynomial $f(t)$ of $R[t]$ the R-factor algebra

$$e_iS(f/R)/\mathrm{NR}(e_iS(f/R))$$

of the indecomposable component $e_iS(f/R)$ of the universal splitting ring $S(f/R)$ over its nilradical is a minimal splitting field of f over R. Its automorphism group over R is a factor group of the group of the equation (9.1). *(See also exercise* 1.)

If a ring Λ is algebraically indecomposable then the same is true for the factor ring of Λ over any nilideal A. The reason is that any idempotent $\bar{e}$ of Λ/A can be 'lifted' to an idempotent e of Λ such that $\bar{e} = e/A$. Moreover, if there are two orthogonal idempotents $\bar{e}_1$, $\bar{e}_2$ of Λ/A then they can be simultaneously lifted to a pair of orthogonal idempotents e_1, e_2 of Λ such that $\bar{e}_j = e_j/A$ ($j = 1,2$). More generally, we have

Lemma (9.18)
Let Λ be a ring with a nilideal A and let $\bar{e}_1, \bar{e}_2, \ldots$ be a countable set of orthogonal idempotents of Λ/A then there are orthogonal idempotents e_1, $e_2, \ldots$ of Λ such that

$$\bar{e}_j = e_j/A.$$

Proof
Let $e_{10} \neq 0$ be an element of $\bar{e}_1$. By assumption the element

$$n_0 = e_{10}^2 - e_{10}$$

of A is nilpotent, say

$$n_0^r = 0$$

for some natural number r. We note that n_0 is a polynomial in e_{10} and therefore commutes with e_{10}. We form

$$\begin{aligned} e_{11} &= 3e_{10}^2 - 2e_{10}^3 \\ &= 3(e_{10} + n_0) - 2e_{10}(e_{10} + n_0) \\ &= 3e_{10} + 3n_0 - 2e_{10}^2 - 2e_{10}n_0 \\ &= e_{10} + n_0 - 2e_{10}n_0 \\ &\equiv e_{10} \bmod A, \end{aligned}$$

implying

$$\begin{aligned} n_1 = e_{11}^2 - e_{11} &= (e_{10} + n_0 - 2e_{10}n_0)^2 - (e_{10} + n_0 - 2e_{10}n_0) \\ &= e_{10}^2 + 2e_{10}n_0 - 4e_{10}^2n_0 + n_0^2 - 4e_{10}n_0^2 + 4e_{10}^2n_0^2 - e_{10} - n_0 + 2e_{10}n_0 \\ &= -3n_0^2 + 4n_0^3. \end{aligned}$$

We set

$$e_{1,j+1} = 3e_{1j}^2 - 2e_{1j}^3 \quad (j = 1, 2, 3, \ldots),$$

recursively, and obtain

$$\begin{aligned} e_{1j} &\equiv e_{10} \bmod (A \cap \langle e_{10} \rangle), \\ e_{1j}^2 - e_{1j} &\in n_0^{2^j}\Lambda. \end{aligned}$$

If $2^j \geq r$ then

$$\begin{aligned} e_{1j}^2 &= e_{1j}, \\ e_{1,j+1} &= e_{1j}. \end{aligned}$$

Thus we find an idempotent e_1 in $\bar{e}_1$ by an algorithm requiring at most $\lceil \log_2 r \rceil$ steps.

Suppose we have already constructed orthogonal idempotents $e_1, e_2, \ldots, e_k$ of Λ satisfying

$$e_i \in \bar{e}_i \quad (1 \leq i \leq k),$$

and suppose there is $\bar{e}_{k+1}$. We pick any element $e_{k+1,0}$ of $\bar{e}_{k+1}$ and form

$$e_{k+1,1} = e_{k+1,0} - ee_{k+1,0} - e_{k+1,0}e + ee_{k+1,0}e,$$

where

$$e = \sum_{i=1}^{k} e_i.$$

It follows that

$$\begin{aligned} e_{k+1,1} \in \bar{e}_{k+1}, \quad e_ie_{k+1,1} = e_{k+1,1}e_i = 0 \quad (1 \leq i \leq k), \\ e_{k+1,1}^2 - e_{k+1,1} = n_{k+1} \in A, \end{aligned}$$

where n_{k+1} *is nilpotent*. Now we form as above an idempotent e_{k+1} of $\langle e_{k+1,1} \rangle$ belonging to $\bar{e}_{k+1}$. It follows that

$$e_ie_{k+1} = e_{k+1}e_i = 0 \quad (1 \leq i \leq k).$$

Thus we construct a countable chain of orthogonal idempotents $e_1, e_2, \ldots$ such that $e_i \in \bar{e}_i$ for $i = 1, 2, \ldots$. □

Thus the lifting of idempotents is seen as an effective algorithm which will be used in many connections.

4. Purely inseparable equations

Another reason for inseparability occurs over base rings R with the property that for some prime number p every element of R has *p-power characteristic*:

We say R has characteristic p^ν ($\nu \in \mathbb{N}$) if

$$p^\nu R = 0, \quad p^{\nu-1} R \neq 0.$$

We say R has characteristic p^∞ if $p^\nu R \neq 0$ for all $\nu \in \mathbb{N}$.

Let Λ be a unital commutative overring of the ring R with p-power characteristic.

Definition (9.19)

The element x of Λ is said to be **purely inseparable** *over R if it satisfies an equation*

$$x^{p^\mu} - a = 0 \quad (a \in R, \mu \in \mathbb{Z}^{\geq 0})$$

over R.

Proposition (9.20)

All purely inseparable elements of Λ over R form a unital subring $R^{p^{-\infty}} \cap \Lambda$ of Λ containing R.

Proof

If x is in R then we have

$$0 = x^{p^0} - x = x - x$$

so that x is purely inseparable over R.

For any two purely inseparable elements x, y over R we have equations

$$x^{p^\mu} - a = 0, \quad y^{p^\nu} - b = 0, \quad p^{\mu'} a = 0, \quad p^{\nu'} b = 0 \quad (a, b \in R; \mu, \nu, \mu', \nu' \in \mathbb{Z}^{\geq 0}).$$

The identity

$$(x + y)^{p^\lambda} = \sum_{i=0}^{p^\lambda} \binom{p^\lambda}{i} x^i y^{p^\lambda - i}$$

is used for sufficiently large exponents λ to produce only terms

$$\binom{p^\lambda}{i} x^i y^{p^\lambda - i} \quad (0 \leq i \leq p^\lambda)$$

on the right which are already in R.

This is always true if i is divisible by $p^{\max(\mu,\nu)}$. If i is not divisible by $p^{\max(\mu,\nu)}$

then we use the fact that p^λ divides $i\binom{p^\lambda}{i}$ (see exercise 5). Assuming already that $\lambda > \max(\mu, \nu)$ we know that either x^i or $y^{p^\lambda - i}$ has a factor in R. Now we require even

$$\lambda > \max(\mu, \nu) + \max(\mu', \nu')$$

and conclude that

$$\binom{p^\lambda}{i} x^i y^{p^\lambda - i} = 0.$$

Hence $x + y$ is purely inseparable over R. The same is true for xy because of

$$(xy)^{p^{\max(\mu,\nu)}} = a^{\max(\mu,\nu) - \mu} b^{\max(\mu,\nu) - \nu}.$$

Thus proposition (9.20) is demonstrated. □

Corollary (9.21)

$$(R^{p^{-\infty}} \cap \Lambda)^{p^{-\infty}} \cap \Lambda = R^{p^{-\infty}} \cap \Lambda.$$

The overring Λ is purely inseparable over R if every element of Λ is purely inseparable over R.

Proposition (9.22)
If R is algebraically indecomposable then also any purely inseparable overring Λ of R is algebraically indecomposable.

Proof
This is because any idempotent e of Λ satisfies an equation

$$e^{p^\mu} - a = 0$$

with a in R so that

$$e = e^{p^\mu} = a \in R.$$ □

If R is a field then R is a field of prime characteristic p.

Definition (9.23)
The equation (9.2) *is said to be* **purely inseparable** *over the ring R if the universal splitting ring $S(f/R)$ is purely inseparable over R.*

From Proposition (9.22) it follows that the group of a purely inseparable equation of degree n over an algebraically indecomposable unital commutative ring R of characteristic p^ν ($\nu \in \mathbb{Z}^{>0}$ or $\nu = \infty$) is $\mathfrak{S}_n$. For example, any polynomial of the form

$$f(t) = \prod_{i=1}^{s} (t^{p^{\mu_i}} - a_i) \quad (\mu_i \in \mathbb{Z}^{\geqslant 0}, a_i \in R, 1 \leqslant i \leqslant s)$$

leads to a purely inseparable equation (9.2) over R. If R is a field of prime characteristic p then this is the only possibility.

A field R of characteristic p was defined to be *perfect* if every element of R is a pth power. For such fields R itself is the only purely inseparable extension of R. In general we form the purely inseparable extension $R^{p^{-\infty}}$ of R consisting of all symbols

$$a^{p^{-\lambda}} \quad (a\in R, \lambda\in\mathbb{Z}^{\geq 0}),$$

with operational rules given by

$$a^{p^{-\lambda}} = b^{p^{-\mu}} \Leftrightarrow a^{p^{\mu}} = b^{p^{\lambda}}$$

$$a^{p^{-\lambda}} + b^{p^{-\mu}} = (a^{p^{\mu}} + b^{p^{\lambda}})^{p^{-\lambda-\mu}}, \quad a^{p^{-\lambda}} b^{p^{-\mu}} = (a^{p^{\mu}} b^{p^{\lambda}})^{p^{-\lambda-\mu}},$$

with identification rule for the new symbols

$$a^{p^{-\lambda}} = b \Leftrightarrow b^{p^{\lambda}} = a$$

and pth root operations given by

$$(a^{p^{-\lambda}})^{p^{-1}} = a^{p^{-\lambda-1}} \quad (a, b\in R;\ \lambda, \mu\in\mathbb{Z}^{\geq 0})$$

as the smallest perfect extension of R which is embedded (up to isomorphy) as subfield into any perfect extension of R.

For any finite extension E of R we have the *maximal purely inseparable subfield*

$$E\cap R^{p^{-\infty}} = \{x\in E \mid x^{p^{\nu}}\in R\}$$

where p^{ν} is the greatest p-power dividing $[E:R]$. We also form the subfield

$$E^{p^{\nu}} := \{x^{p^{\nu}} y \mid x\in E, y\in R\}$$

as the smallest subfield X of E with the property that E is purely inseparable over X and X contains R. We observe that E is separable over $E\cap R^{p^{-\infty}}$ and that $E^{p^{\nu}}$ is separable over R, moreover

$$E\cap R^{p^{-\infty}} \cap E^{p^{\nu}} = R,$$

$$(E\cap R^{p^{-\infty}})E^{p^{\nu}} = E.$$

5. Transition to overrings

We generalize a remark of C. Jordan to

Proposition (9.24)

Upon transition from any algebraically indecomposable unital commutative ring R to an algebraically indecomposable unital commutative overring Λ the group of the equation (9.1) *changes to a subgroup.*

Proof

The decomposition (9.5) of the universal splitting ring $S(f/R)$ of f over R into the algebraic sum of indecomposable subrings gives rise to the decomposition

$$S(f/\Lambda) = S(f/R) \otimes_R \Lambda = \bigoplus_{i=1}^{s} e_i S(f/\Lambda) \tag{9.25}$$

of the universal splitting ring $S(f/\Lambda)$ of f over Λ into the algebraic sum of subrings $e_i S(f/\Lambda)$ that are conjugate under the permutation automorphism group

$$\bar{\mathfrak{S}}_{n,\Lambda} = \bar{\mathfrak{S}}_{n,R} \otimes_R \mathbf{1}_\Lambda,$$

and that can be refined into the algebraic sum of algebraically indecomposable subrings:

$$e_i S(f/\Lambda) = \bigoplus_{j=1}^{m} e_{ij} S(f/\Lambda),$$

where the e_{ij} are the primitive idempotents of $S(f/\Lambda)$. It follows that $\bar{\mathfrak{S}}_{n,\Lambda}$ restricts faithfully to $\bar{\mathfrak{S}}_{n,R}$ on $S(f/\Lambda)$:

$$\bar{\mathfrak{S}}_{n,\Lambda}|_{S(f/R)} = \bar{\mathfrak{S}}_{n,R}.$$

Hence $e_1, \ldots, e_s$ form an $\bar{\mathfrak{S}}_{n,\Lambda}$-orbit. The equations

$$e_{ij} e_i = e_{ij}, \quad e_{ij} e_k = 0 \quad (1 \leqslant j \leqslant m, i \neq k)$$

imply that the $\bar{\mathfrak{S}}_{n,\Lambda}$-stabilizer of e_{ij} fixes e_i so that

$$\mathrm{Stab}\,(e_{ij}/\bar{\mathfrak{S}}_{n,\Lambda})|_{S(f/R)} \subseteq \mathrm{Stab}\,(e_i/\bar{\mathfrak{S}}_{n,R}).$$

□

6. The principal equation of the universal splitting ring

In order to apply the criterion (9.24) we first study the principal equation

$$P_{\xi/R}(\xi) = 0$$

of an arbitrary element ξ of the universal splitting ring $S(f/R)$ of the monic polynomial (9.3) over R. The polynomial $P_{\xi/R}$ of $R[t]$ is defined as the characteristic polynomial of the regular representation

$$P\colon S(f/R) \to R^{n! \times n!}\colon \xi \mapsto (\alpha_{ik}(\xi)),$$

$$\xi b_i = \sum_{k=1}^{n!} \alpha_{ik}(\xi) b_k \quad (1 \leqslant i \leqslant n!),$$

where $b_1, \ldots, b_{n!}$ is an R-basis of $S(f/R)$, for example we may take the standard basis defined in section 1.

Proposition (9.26)

$$P_{\xi/R}(t) = \prod_{\pi \in \mathfrak{S}_{n!}} (t - \bar{\pi}(\xi)) \quad (\xi \in S(f/R)).$$

Proof
By the 'generic argument'! Let

$$\bar{\Lambda} = \mathbb{Z}[\bar{x}_1,\ldots,\bar{x}_n,\bar{y}_\nu \mid \nu=(\nu_1,\ldots,\nu_n), 0\leqslant \nu_i < i, 1\leqslant i\leqslant n](\bar{y}_\nu := \bar{y}_1^{\nu_1}\cdots\bar{y}_n^{\nu_n})$$

be a polynomial ring in $n+n!$ variables over $\mathbb{Z}$. Form

$$\bar{f}(t) = \prod_{i=1}^{n}(t-\bar{x}_i) = t^n + \bar{a}_1 t^{n-1} + \cdots + \bar{a}_n \quad (\bar{a}_i \in \mathbb{Z}[\bar{x}_1,\ldots,\bar{x}_n], 1\leqslant i\leqslant n).$$

Let

$$\bar{R} = \mathbb{Z}[\bar{a}_1,\ldots,\bar{a}_n,\bar{y}_\nu \mid \ 0\leqslant \nu_i < i, 1\leqslant i\leqslant n]$$

be the subring of $\bar{\Lambda}$ which is again a polynomial ring in $\bar{a}_1,\ldots,\bar{a}_n,\bar{y}_\nu$ $(0\leqslant \nu_i < i, 1\leqslant i\leqslant n)$ over $\mathbb{Z}$ so that

$$\bar{\Lambda} = S(\bar{f}/\bar{R}).$$

Let

$$\bar{\xi} = \sum_{0\leqslant \nu_i < i, 1\leqslant i\leqslant n} \bar{y}_\nu \bar{x}_1^{\nu_1}\cdots\bar{x}_n^{\nu_n}$$

be the 'generic' element of $\bar{\Lambda}$. By the specialization argument it follows that the monic polynomial $P_{\bar{\xi}/\bar{R}}(t)$ is irreducible of degree $n!$ over $\bar{R}$. On the other hand, the monic polynomial

$$\bar{f}(t) = \prod_{\pi\in\mathfrak{S}_{n!}} (t-\bar{\pi}(\bar{\xi}))$$

has all its coefficients fixed under each permutation automorphism. Since every non-zero element of $\bar{R}$ is of zero characteristic it follows that $\bar{f}$ belongs to $\bar{R}[t]$. Since $f(\xi)=0$ and f is monic of degree $n!$ it follows that

$$f(t) = P_{\xi/R}(t). \tag{9.27}$$

There is the homomorphism

$$\eta\colon\ \bar{\Lambda}\to S(f/R)\colon \ \begin{Bmatrix} \bar{x}_i \to x_i \\ \bar{a}_i \to a_i \\ \bar{\xi}\to\xi \end{Bmatrix} \ (1\leqslant i\leqslant n)$$

so that (9.27) yields (9.26).

7. Discriminant ideal and trace radical

Definition (9.28)
Let a unital commutative ring R and a unital overring S of R with trace bilinear form $\mathrm{Tr}\colon S\times S\to R$ *be given. If S has a finite R-basis X which is ordered then* $\partial_{|X|}(S) := \det((\mathrm{Tr}(xy))_{x,y\in X})$ *is called the discriminant of S over R as we already know. The discriminant generates an R-submodule of $\mathfrak{Q}(R)$ which is said to be the* **discriminant ideal** *of S over R.*

We saw already in section 4 that the discriminant of the monic non-constant polynomial f generates the discriminant ideal of the equation ring $R[t]/f$.

Hence f is separable over R if and only if the discriminant ideal contains no zero divisor.

Theorem (9.29)

Let R be a unital commutative ring. Let Λ be a unital commutative overring of R with finite R-basis $b_1, \ldots, b_m$. Let E be a Λ-ring with finite basis $c_1, \ldots, c_n$ over Λ. By the degree theorem the mn elements $\omega_{(i-1)n+k} := b_i c_k$ $(1 \leqslant i \leqslant m, 1 \leqslant k \leqslant n)$ form an R-basis of E. Let $d(\Lambda/R) = \det(\mathrm{Tr}_{\Lambda/R}(b_i b_j))$, $d(E/\Lambda) = \det(\mathrm{Tr}_{E/\Lambda}(c_h c_k))$, $d(E/R) = \det(\mathrm{Tr}_{E/R}(\omega_\mu \omega_\nu))$ be the **discriminants** *of Λ/R, E/Λ, E/R relative to the bases $b_1, \ldots, b_m$; $c_1, \ldots, c_n$; $\omega_1, \ldots, \omega_{mn}$ respectively. Then there holds the* **discriminant composition formula**

$$d(E/R) = d(\Lambda/R)^n \mathrm{N}_{\Lambda/R}(d(E/\Lambda)). \tag{9.29a}$$

Proof

The regular representation of Λ over R relative to the R-basis $b_1, \ldots, b_m$ is defined as the matrix representation $M: \Lambda \to R^{m \times m}: x \mapsto (\alpha_{ik}(x))$ subject to $x b_i = \sum_{k=1}^{m} \alpha_{ik}(x) b_k$ $(1 \leqslant i \leqslant m)$. The trace of E over R of the product of $\omega_{(i-1)n+h}$ and $\omega_{(j-1)n+k}$ is equal to

$$\mathrm{Tr}_{E/R}(b_i c_h b_j c_k) = \mathrm{Tr}_{\Lambda/R}(\mathrm{Tr}_{E/\Lambda}(b_i b_j c_h c_k)) = \mathrm{Tr}_{\Lambda/R}(b_i b_j \mathrm{Tr}_{E/\Lambda}(c_h c_k)).$$

Setting

$$\mathrm{Tr}_{E/\Lambda}(c_h c_k) = \gamma_{hk} \in \Lambda, \quad \mathrm{Tr}_{\Lambda/R}(b_i b_j) = \beta_{ij} \in R$$

we have

$$\begin{aligned}\mathrm{Tr}_{E/R}(\omega_{(i-1)n+k}\omega_{(\bar{\imath}-1)n+\bar{k}}) &= \mathrm{Tr}_{\Lambda/R}(b_i \gamma_{k\bar{k}} b_{\bar{\imath}}) \\ &= \mathrm{Tr}_{\Lambda/R}\left(\sum_{j=1}^{m} \alpha_{ij}(\gamma_{k\bar{k}}) b_j b_{\bar{\imath}}\right) = \sum_{j=1}^{m} \alpha_{ij}(\gamma_{k\bar{k}}) \mathrm{Tr}_{\Lambda/R}(b_j b_{\bar{\imath}}) \\ &= \sum_{j=1}^{m} \alpha_{ij}(\gamma_{k\bar{k}}) \beta_{j\bar{\imath}}.\end{aligned}$$

Suitable row and column interchanging of $(M(\gamma_{k\bar{k}}))_{1 \leqslant k, \bar{k} \leqslant n}$ yields

$$d(E/R) = \det((M(\gamma_{k\bar{k}}))(B \times I_n)) = \det(M(\gamma_{k\bar{k}})) \det(B \times I_n) \tag{9.29b}$$

with

$$B = (\mathrm{Tr}(b_i b_j))_{1 \leqslant i,j \leqslant m},$$

hence

$$\det(B \times I_n) = (\det(B))^n = d(\Lambda/R)^n. \tag{9.29c}$$

By the 'multilinear argument' we prove that

$$\det(M(\gamma_{k\bar{k}})) = \det(M(\det(\gamma_{k\bar{k}}))) \tag{9.29d}$$

and because of $\det(\gamma_{k\bar{k}}) = d(E/\Lambda)$, $\det(M(x)) = N_{\Lambda/R}(x)$ $(x \in \Lambda)$, we derive (9.29a) from (9.29b, c, d).

The multilinear argument is used in order to prove the identity

$$\det(M(\xi_{ik})) = \det(M(\det(\xi_{ik}))) \quad (\xi_{ik} \in \Lambda, 1 \leqslant i, k \leqslant \tilde{m}, \tilde{m} \in \mathbb{N}). \tag{9.29e}$$

Both, left- and right-hand side of (9.29e) define functions $f_i: \Lambda^{\tilde{m} \times \tilde{m}} \to R$ $(i = 1, 2)$ subject to the three conditions (9.29f–h):

$$f_i(X) = f_i(X_1) + f_i(X_2) \tag{9.29f}$$

in case the three matrices X, X_1, X_2 have all rows but the jth one in common and the jth row of X is the sum of the jth rows of X_1, X_2;

$$f_i(X) = \det(M(\lambda)) f_i(Y) \tag{9.29g}$$

in case the two matrices X, Y have all rows but the jth one in common and the jth row of X is λ times the jth row of Y with $\lambda \in \Lambda$;

$$f_i(I_{\tilde{m}}) = 1. \tag{9.29h}$$

By the usual rules of determinantal calculus we show that $f_i(X)$ is the sum of the f_i-values over the matrices arising from X after replacing in each row all but one entry by zero. Clearly, those summands which contain a zero column turn out to be zero. There remain $\tilde{m}!$ terms giving rise to the Laplace development formula

$$f_i(X) = \sum_{\pi \in \mathfrak{S}_{\tilde{m}}} \mathrm{N}_{\Lambda/R}(\operatorname{sign}(\pi) \xi_{1\pi(1)} \cdots \xi_{n\pi(n)}) \quad (X = (\xi_{ik}) \in \Lambda^{\tilde{m} \times \tilde{m}}), \tag{9.29i}$$

hence (9.29e). □

Proposition (9.30)

Let R be a unital commutative ring and Λ an R-ring with finite R-basis $b_1, \ldots, b_n$.

(a) $\Lambda^{\perp} := \{x \in \Lambda \mid \forall y \in \Lambda: \mathrm{Tr}_{\Lambda/R}(xy) = 0\}$ *is an ideal of Λ – the so-called* ***trace radical*** *– which contains $NR(\Lambda)$.*

(b) *If R is a field and $NR(\Lambda)$ is zero, then Λ is the algebraic sum of finitely many field extensions Λ_i $(1 \leqslant i \leqslant s, s \in \mathbb{N})$ of subfields $1_{\Lambda_i} R$ being isomorphic to R:*

$$\Lambda = \dot{\sum_{i=1}^{s}} \Lambda_i.$$

Proof

(a) If $\mathrm{Tr}(x_i y) = 0$ for $x_i \in \Lambda$ $(i = 1, 2)$ and all $y \in \Lambda$ then we have $\mathrm{Tr}((x_1 + x_2)y) = 0$, $\mathrm{Tr}((\lambda x_1)y) = \mathrm{Tr}(x_1(\lambda y)) = 0$ for all $\lambda \in R$, $y \in \Lambda$, hence $\Lambda^{\perp}$ is an ideal. For any nilpotent element x of Λ it follows from the 'generic argument' that the principal equation of the regular representation of Λ over R specialized to x is t^n implying $x^n = 0$, $\mathrm{Tr}_{\Lambda/R}(x) = 0$ and therefore $\mathrm{NR}(\Lambda) \subseteq \Lambda^{\perp}$.

(b) Since the rank of Λ over R is finite there are only finitely many primitive idempotents, say $e_1, \ldots, e_s$, of Λ. Therefore we have $\Lambda = \dot{\sum}_{i=1}^{s} \Lambda_i$ with $\Lambda_i = e_i \Lambda$

$(1 \leqslant i \leqslant s)$ and $\Lambda_0 = \{x - ex | x \in \Lambda, e = e_1 + \cdots + e_s\}$. We note that Λ_0 contains no idempotent. Hence, the minimal polynomial of every element $y \in \Lambda_0$ over R is of the form $m_y(t) = t^\mu f(t)$, where $\mu \in \mathbb{N}$ and $f(t) \in R[t]$ is monic and coprime to t^μ. But in case $f(t) \neq 1$ there would hold an equation $a(t)t^\mu + b(t)f(t) = 1$ $(a(t), b(t) \in R[t])$ yielding the idempotent $a(y)y^\mu$ in Λ_0 contrary to our construction. Hence, we have $f(t) = 1$ and Λ_0 is a nilideal, i.e. $\Lambda_0 = 0$ according to our assumption. It remains to show that each Λ_i $(1 \leqslant i \leqslant s)$ is a field. For any element $x \neq 0$ of Λ_i the minimal polynomial m_x of x over $e_i R$ is a monic irreducible polynomial of $e_i R[t]$ satisfying $m_x(0) \neq 0$, hence we obtain $-x^{-1} = a_\mu^{-1} \sum_{i=1}^{\mu} a_{i-1} x^{\mu - i}$ for $m_x(t) = t^\mu + a_1 t^{\mu - 1} + \cdots + a_\mu$, $a_0 := 1$, i.e. x is invertible in Λ_i and Λ_i is a field $(1 \leqslant i \leqslant s)$. □

Proposition (9.31)

Let R be a unital commutative ring satisfying $R = \mathfrak{Q}(R)$, i.e. every $x \in R \backslash U(R)$ is a zero divisor. Let Λ be an overring of R with finite R-basis $b_1, \ldots, b_n$.

(a) $\Lambda^\perp = 0 \Leftrightarrow d(\Lambda/R) \in U(R)$.

(b) *Let R be the algebraic sum of finitely many fields and Λ be commutative. Then Λ is separable over R, if and only if $d(\Lambda/R)$ is a unit of R.*

Proof

(a) The determinant $d(\Lambda/R)$ is a zero divisor of R, precisely if there is a non-zero vector $\xi = (\xi_1, \ldots, \xi_n)^t \in R^{n \times 1}$ satisfying $\xi^t X = \mathbf{0}$ for $X = (\mathrm{Tr}_{\Lambda/R}(b_i b_k))_{1 \leqslant i,k \leqslant n}$. Hence, in that case there is an element $x = \sum_{i=1}^{n} \xi_i b_i$ of $\Lambda \backslash \{0\}$ such that $\mathrm{Tr}_{\Lambda/R}(x b_k) = 0$ $(1 \leqslant k \leqslant n)$ and therefore $\mathrm{Tr}_{\Lambda/R}(xy) = 0$ for all $y \in \Lambda$. However, if $d(\Lambda/R)$ is a unit of R, the matrix X is invertible in $R^{n \times n}$ and any equation $\xi^t X = 0$ $(\xi \in R^{n \times 1})$ has the unique solution $\xi = \mathbf{0}$.

(b) We first prove this in case Λ is a field (which of course implies that R is a field, too).

Any element x of Λ satisfies a minimal equation $m_x(x) = 0$ over the subfield R of Λ, where $m_x(t)$ is a monic irreducible polynomial, $R[x] \simeq R[t]/m_x$, the elements $1, x, \ldots, x^{\mu - 1}$ with $\mu = \deg(m_x)$ form an R-basis of $R[x]$ with discriminant $d(m_x) = d(R[x]/R)$. Supposing Λ can be obtained from R by a chain of separable adjunctions, then it follows from theorem (9.29) that $d(\Lambda/R) \neq 0$.

On the other hand, if Λ cannot be generated in this way, then the characteristic of Λ is a prime number p and there is an element x of Λ for which $x \notin R$, $x^p = \alpha \in R$, hence $m_x(t) = t^p - \alpha$, $d(m_x) = 0$, and therefore $d(\Lambda/R) = 0$ by theorem (9.29).

Before we turn to the general case let us investigate the behavior of the discriminant of a finite R-basis $b_1, \ldots, b_n$ of the R-ring Λ in case that R is a unital commutative ring and that the R-basis is the disjoint union of two non-empty subsets, say $b_1, \ldots, b_m$ and $b_{m+1}, \ldots, b_n$ $(m \in \mathbb{N}$ subject to $1 \leqslant m < n)$, such that $b_i b_k = b_k b_i = 0$ $(1 \leqslant i \leqslant m < k \leqslant n)$. In other words

$$\Lambda = \Lambda_1 \dot{+} \Lambda_2 \quad \text{for} \quad \Lambda_1 = \dot{\sum_{i=1}^{m}} Rb_i, \quad \Lambda_2 = \dot{\sum_{i=m+1}^{n}} Rb_i.$$

In this case the regular representation P of Λ with respect to the given R-basis decomposes into

$$P(x) = \begin{pmatrix} P_1(x) & 0 \\ 0 & P_2(x) \end{pmatrix}, \quad P_j(x) = (\alpha_{jik}(x)) \quad (j = 1, 2) \quad \text{for}$$

$$xb_i = \sum_{h=1}^{m} \alpha_{1ih} b_h, \quad xb_{m+j} = \sum_{k=1}^{n-m} \alpha_{2jk} b_{m+k} \quad (1 \leqslant i \leqslant m, 1 \leqslant j \leqslant n - m)$$

so that

$$\begin{aligned} \mathrm{Tr}_{\Lambda/R}(x) &= \mathrm{Tr}(P_1(x)) + \mathrm{Tr}(P_2(x)), \\ \mathrm{Tr}_{\Lambda/R}(b_i b_k) &= \mathrm{Tr}_{\Lambda/R}(b_k b_i) = 0 \quad (1 \leqslant i \leqslant m < k \leqslant n), \\ \mathrm{Tr}_{\Lambda/R}(b_i b_k) &= \mathrm{Tr}_{\Lambda_1/R}(b_i b_k) \quad (i, k = 1, 2, \ldots, m), \\ \mathrm{Tr}_{\Lambda/R}(b_{m+i} b_{m+k}) &= \mathrm{Tr}_{\Lambda_2/R}(b_{m+i} b_{m+k}) \quad (i, k = 1, 2, \ldots, n - m), \end{aligned}$$

hence

$$d((\Lambda_1 \dot{+} \Lambda_2)/R) = d(\Lambda_1/R) d(\Lambda_2/R).$$

Now suppose both $d(\Lambda_1/R)$ and $d(\Lambda_2/R)$ are non-zero divisors of R then the same is true for $d((\Lambda_1 \dot{+} \Lambda_2)/R)$.

Let us also study the behavior of an algebraic sum $\Lambda = \Lambda_1 \oplus \Lambda_2 \oplus \cdots \oplus \Lambda_s$ of finitely many separable finite extensions Λ_i of a field $1_{\Lambda_i} R$ isomorphic to the field R. The separability of Λ_i over $1_{\Lambda_i} R$ means that the minimal polynomial $m_{x_i}(t)$ of any element x_i of Λ_i over $1_{\Lambda_i} R$ is a monic separable irreducible polynomial. The minimal polynomial $m_x(t)$ of $x = x_1 \oplus x_2 \oplus \cdots \oplus x_s$ is the least common multiple of the polynomials $m_{x_i}(t)$ of $R[t]$ which is the product of the distinct ones among them and therefore a separable polynomial. Hence, $R[x] \simeq R[t]/m_x(t)$ and the only R-derivation of $R[x]$ is 0. Thus it follows that Λ is separable over R. In this way we establish (9.31) (b) in the event that R is a field and Λ is an algebraic sum of finitely many finite extensions of R.

Now let R be a field and Λ a commutative R-ring with finite R-basis such that $\Lambda^{\perp} = 0$. Then Λ is a unital overring of R because of $\Lambda = \dot{\sum}_{i=1}^{n} Rb_i$ and by the previous results we see at once that Λ is separable over R.

Conversely, let R be a field and let Λ be a separable commutative R-ring with finite R-basis $b_1, \ldots, b_n$. Then for any nilpotent element x of Λ we find that $R[x]$ has the R-basis $1, x, \ldots, x^{\mu-1}$, where $\mu \in \mathbb{N}$ and $x^{\mu} = 0$, $x^{\mu-1} \neq 0$. If μ is greater than one then there is the non-zero derivation $D: R[x] \to R[x]$, $D(f(x)) = D_t(f)(x)x$ $(f(t) \in R[t])$ of $R[x]$ over R contrary to assumption. Hence we must have $\mu = 1$, there is no non-zero nilpotent element in Λ. Therefore we can apply (9.31) (a).

By assumption each Λ_i is separable over $1_{\Lambda_i} R$. It was shown above that $d(\Lambda_i / 1_{\Lambda_i} R) \neq 0$ $(1 \leqslant i \leqslant s)$ implying $d(\Lambda/R) \neq 0$.

If R is the algebraic sum of finitely many fields R_i $(1 \leqslant i \leqslant s)$ then we set $\Lambda_i = 1_{R_i}\Lambda$ $(1 \leqslant i \leqslant s)$ with algebraic decomposition $\Lambda = \dot{\sum}_{i=1}^{s} \Lambda_i$ of Λ into ideals Λ_i which have the basis $1_{R_i}b_1, \ldots, 1_{R_i}b_n$ over R_i. Applying the previous results we prove (9.31) (b) in the most general case. □

The subsequent two corollaries are now immediate.

Corollary (9.32)

Let R, Λ, E as in (9.29). Then E is separable over R if and only if E is separable over Λ and Λ is separable over R.

Corollary (9.33)

The universal splitting ring $S(f/R)$ of a separable monic polynomial f over R is separable over R.

Proposition (9.34)

The discriminant ideal of the universal splitting ring $S(f/R)$ of a monic polynomial of degree $n > 1$ over R is generated by $d(f)^{n!/2}$.

Proof

In the 'generic case' we have $\bar{R} = \mathbb{Z}[\bar{a}_1, \ldots, \bar{a}_n]$, where $\bar{a}_1, \ldots, \bar{a}_n$ are polynomial variables, and $\bar{f}(t) = t^n + \bar{a}_1 t^{n-1} + \cdots + \bar{a}_n = \prod_{i=1}^{n}(t - \bar{x}_i)$, $S(f/R) = \mathbb{Z}[\bar{x}_1, \ldots, \bar{x}_n]$. The discriminant $d(\bar{f}) = (-1)^{n(n-1)/2}\prod_{i \neq k}(\bar{x}_i - \bar{x}_k)$ is an irreducible polynomial in $\bar{a}_1, \ldots, \bar{a}_n$. This is because any non-unit divisor of $d(\bar{f})$ in $\bar{R}$ also is a divisor of $d(\bar{f})$ in the polynomial ring $S(\bar{f}/\bar{R})$ in the variables $\bar{x}_1, \ldots, \bar{x}_n$ over $\bar{R}$. Hence it must have some divisor $\bar{x}_i - \bar{x}_k$ $(i \neq k)$. Because of $\mathfrak{S}_n$-invariance it is either $d(\bar{f})$ or it is the different expression

$$\delta(\bar{f}) = \prod_{1 \leqslant i < k \leqslant n} (\bar{x}_i - \bar{x}_k)$$

up to a factor ± 1. But $\delta(\bar{f})$ does not belong to $\bar{R}$ because for $n > 1$ the group $\mathfrak{S}_n$ contains odd permutations π and for each of those

$$\bar{\pi}\delta(\bar{f}) = -\delta(\bar{f}) \neq \delta(\bar{f}).$$

Because of proposition (9.31) the discriminant ideal of $S(\bar{f}/\bar{R})$ over $\bar{R}$ is generated by some non-zero element d' of $\bar{R}$. If there is an irreducible polynomial $P \neq \pm d(\bar{f})$ dividing d' then it follows upon specialization that the discriminant ideal of $S(\bar{f}/P\bar{R})/(\bar{R}/P\bar{R})$ is 0, but the discriminant of $\bar{f}/P\bar{R}$ is the non-zero divisor $d(\bar{f})/P\bar{R}$, a contradiction. Hence d' is a power of $d(\bar{f})$ up to ± 1. Comparing degrees in the $\bar{x}_j$'s in the formula

$$\pm d' = \det(\operatorname{Tr}(\bar{x}_1^{\mu_1+\nu_1} \cdots \bar{x}_n^{\mu_n+\nu_n}) \quad (0 \leqslant \mu_i < i, \quad 0 \leqslant \nu_i < i, 1 \leqslant i \leqslant n)$$

yields $d' = \pm d(\bar{f})^{n!/2}$. Now we obtain proposition (9.34) upon specialization. □

8. The van der Waerden criterion

Proposition (9.35)

Let R be a unital commutative ring, let f be a monic non-constant polynomial of $R[t]$ with discriminant $d(f)$ not contained in the prime ideal $\mathfrak{p}$ of R. Then the group of the equation

$$f/\mathfrak{p}(x) = 0$$

over $\mathfrak{Q}(R/\mathfrak{p})$ is $\mathfrak{S}_n$-conjugate to a subgroup of the group of the equation (9.1) *over* $\mathfrak{Q}(R)$.

Proof

Let $e_1, e_2, \ldots, e_s$ be the primitive idempotents of $S(f/\mathfrak{Q}(R))$. We observe that for any element x of $S(f/R)$ we find that

$$e_j(\mathrm{Tr}_{S(f/\mathfrak{Q}(R))/\mathfrak{Q}(R)}(e_i x)) = \sum_{\pi \in \mathfrak{S}_n} e_j \bar{\pi}(e_i x)$$

$$= \sum_{\pi \in \mathfrak{S}_n} e_j \bar{\pi}(e_i)\bar{\pi}(x) \in e_j S(f/R) \cap e_j \mathfrak{Q}(R) = e_j R,$$

$$\mathrm{Tr}_{S(f/\mathfrak{Q}(R))/\mathfrak{Q}(R)}(e_i x) \in \mathfrak{Q}(R) \bigcap \sum_{j=1}^{s} e_j R = R.$$

On the other hand let X be some R-basis of $S(f/R)$ (e.g. the standard R-basis). Then we have equations

$$e_i = \sum_{y \in X} \alpha_{iy} y \quad (\alpha_{iy} \in \mathfrak{Q}(R)),$$

$$\mathrm{Tr}_{S(f/\mathfrak{Q}(R))/\mathfrak{Q}(R)}(e_i x) = \sum_{y \in X} \alpha_{iy} \mathrm{Tr}(yx) \quad (x \in X).$$

Since according to proposition (9.34)

$$\det(\mathrm{Tr}(yx)) = d(f)^{n!/2},$$

it follows that

$$d(f)^{n!/2} \alpha_{iy} \in R, \tag{9.36}$$

$$d(f)^{n!/2} e_i \in S(f/\mathfrak{Q}(R)). \tag{9.37}$$

The unital commutative overring R^* of R that is generated by R and by all components α_{iy} $(y \in X)$ is a subring of $\mathfrak{Q}(R)$ with the property that $e_1, \ldots, e_s$ are the primitive idempotents of $S(f/R^*)$. Hence the group of the equation (9.1) over R^* is the same as it is over $\mathfrak{Q}(R)$. We shall prove that the subset

$$\mathfrak{p}^* := \{x \in R^* \mid \exists \mu \in \mathbb{Z}^{>0} : d(f)^{\mu} x \in \mathfrak{p}\}$$

of R^* is a prime ideal of R^* intersecting R in $\mathfrak{p}$.

Indeed, the ideal property of $\mathfrak{p}^*$ is trivial, since by (9.36) for every element y of R^* there is a power $d(f)^{\nu}$ $(\nu \in \mathbb{Z}^{>0})$ such that

$$d(f)^{\nu} y \in R.$$

Clearly $\mathfrak{p} \subseteq \mathfrak{p}^*$. If $x \in \mathfrak{p}^* \cap R$, then there is a positive number μ for which

$d(f)^\mu x \in \mathfrak{p}$. Since by assumption $d(f) \notin \mathfrak{p}$ it follows that $x \in \mathfrak{p}$ from the prime ideal property of $\mathfrak{p}$. Hence

$$\mathfrak{p}^* \cap R = \mathfrak{p}. \tag{9.38}$$

If

$$y \in R^*, \quad y \notin \mathfrak{p}^*,$$
$$z \in R^*, \quad yz \in \mathfrak{p}^*,$$

then there is a natural number ν for which

$$d(f)^\nu y \in R,$$
$$d(f)^\nu z \in R, \quad d(f)^\nu yz \in \mathfrak{p},$$
$$d(f)^\nu y d(f)^\nu z \in \mathfrak{p}.$$

By assumption

$$d(f)^\nu y \notin \mathfrak{p},$$

hence

$$d(f)^\nu z \in \mathfrak{p}$$

because of the prime ideal property of $\mathfrak{p}$, hence $z \in \mathfrak{p}^*$. It follows that $\mathfrak{p}^*$ is a prime ideal of R^*.

We form

$$N = (d(f)/\mathfrak{p})^{-1}$$

in $\mathfrak{Q}(R/\mathfrak{p})$. Now the elements

$$e_i^* = N^{n!/2} \sum_{y \in X} (d(f)^{n!/2} \alpha_{iy}/\mathfrak{p}) y$$

of $S(f/\mathfrak{Q}(R/\mathfrak{p}))$ are orthogonal central idempotents conjugate under $\mathfrak{S}_n$. They are also contained in $S((f/\mathfrak{p}^*)/(R^*/\mathfrak{p}^*))$. Hence the group of the equation

$$f/\mathfrak{p}\mathfrak{Q}(R/\mathfrak{p})(x) = 0$$

is contained in an $\mathfrak{S}_n$-conjugate of the group of the equation

$$f/\mathfrak{p}^*(x) = 0, \tag{9.39}$$

and the group of the equation (9.39) is contained in an $\mathfrak{S}_n$-conjugate of the group of the equation (9.1) over R^*. □

The van der Waerden criterion is chiefly applied in case the residue class ring $R/\mathfrak{p}$ is finite. In that case it is a finite field.

Proposition (9.40)

The group of a monic separable equation (9.1) *of degree* $n > 0$ *over a finite field* R *of* q *elements is a cyclic subgroup of* $\mathfrak{S}_n$ *generated by a permutation of type*

$$n = \sum_{i=1}^{s'} n_i$$

when $n_1, n_2, \ldots, n_{s'}$ *are the degrees of the irreducible factors of* f *over* R.

Proof
Let $e_1,\ldots,e_s$ be the primitive idempotents of $S(f/R)$. Because of the separability of f it follows that $e_1S(f/R)$ is a finite extension of e_1R and that the group of the equation (9.1) is isomorphic to the automorphism group of $e_1S(f/R)$ over e_1R which is cyclic. In fact, it is generated by the Frobenius automorphism raising every element of $e_1S(f/R)$ to its qth power. Hence there is an element π of $\mathfrak{S}_n$ such that $\bar{\pi}(e_1)=e_1$,

$$\bar{\pi}(x)=x^q \quad (x\in e_1S(f/R)). \tag{9.41a}$$

Let

$$f=\prod_{i=1}^{s'} f_i \tag{9.41b}$$

be the factorization of f into a product of distinct monic irreducible polynomials of $R[t]$.

Since in $e_1S(f/R)$ there holds the factorization

$$e_1f(t)=\prod_{k=1}^{n}(t-e_1x_k) \tag{9.41c}$$

it follows after suitable renumbering of $e_1x_1,\ldots,e_1x_n$ that

$$e_1f_1(t)=\prod_{k=1}^{n_1}(t-e_1x_k),$$

$$e_1f_2(t)=\prod_{k=n_1+1}^{n_1+n_2}(t-e_1x_k),$$

$$\vdots$$

$$e_1f_{s'}(t)=\prod_{k=n-n_{s'}+1}^{n}(t-e_1x_k),$$

$$\bar{\pi}^je_1x_k=e_1x_k^j=e_1x_{k+j} \quad \left(0\leqslant j<n_i, k=\left(\sum_{g<i}n_g\right)+1\right),$$

hence $\bar{\pi}$ permutes the root projections as a product of cycles of length $n_1,n_2,\ldots,n_{s'}$.

The permutation automorphism $\bar{\pi}$ permutes $x_1,\ldots,x_n$ since we have

$$\bar{\pi}(e_1x_k)=\bar{\pi}(e_1)\bar{\pi}(x_k)=e_1x_{\pi(k)}. \qquad \square$$

9. Application of the van der Waerden criterion. A theorem of Čebotarev

The van der Waerden criterion is a powerful tool for determining the group of a monic separable equation (9.1) over $R=\mathbb{Z}$.

In order to apply the test we choose prime numbers p and factorize f modulo p into a product of irreducible factors in accordance with the methods of section 5. If multiple factors occur then $p|d(f)$ and we cannot apply the criterion. Otherwise $p \nmid d(f)$ and we obtain a congruence factorization

$$f \equiv \prod_{i=1}^{s'} f_i \mod p\mathbb{Z}[t],$$

where $f_1, \ldots, f_{s'}$ are monic polynomials over $\mathbb{Z}$ that are distinct and irreducible mod $p\mathbb{Z}[t]$. In each case we find that the group of the equation (9.1) interpreted as subgroup of $\mathfrak{S}_n$ contains a permutation with cycle decomposition of type

$$n = \deg f_1 + \deg f_2 + \cdots + \deg f_{s'}.$$

This knowledge is cumulative in the sense that the number of possibilities for the group of the equation decreases as one finds new patterns of cycle decompositions. The great number theorist and mathematician N. Čebotarev showed in 1927 that the knowledge we gain by repeated congruence factorizations of f also is cumulative in a deeper way [3]:

Denoting by $A(n_1, \ldots, n_{s'}, x)$ the number of times that the congruence factorization modulo a prime number $p \leqslant x$ yields the partition

$$n = n_1 + n_2 + \cdots + n_{s'} \quad (s' \in \mathbb{Z}^{>0}; n_i \in \mathbb{Z}^{>0}, 1 \leqslant i \leqslant s') \tag{9.42}$$

($\pi(x)$ the total number of prime numbers $\leqslant x$) then $A(n_1, \ldots, n_{s'}, x)/\pi(x)$ determines a frequency that converges for $x \to \infty$ to the frequency of permutations of the cycle decomposition type (9.42) in the group of the equation (9.1).

A proof of this theorem which is capable of even sharper formulation shall not be attempted here. Error estimates exist, but are not yet sharp enough to compare with actual number theoretic testing.

Example (9.43)

We consider a cubic equation

$$f(x) = 0 \quad \text{for } f(t) = t^3 + a_1 t^2 + a_2 t + a_3$$

over $\mathbb{Z}$ with $d(f) \neq 0$. We have four possibilities for the group of the equation which were listed already at the end of the first subsection. The group specific frequency distribution is as follows:

no.	group G	order	frequency × cycle decomposition type
1	1	1	1×1
2	$\mathfrak{A}_3$	3	$\frac{1}{3} \times 1 + \frac{2}{3} \times 3$
3	$C_2 = \langle (12) \rangle$	2	$\frac{1}{2} \times 1 + \frac{1}{2} \times 2$
4	$\mathfrak{S}_3$	6	$\frac{1}{6} \times 1 + \frac{1}{2} \times 2 + \frac{1}{3} \times 3$

Here we denote the partition

$$n = \sum_{i=1}^{n} a_i i \quad (a_i \in \mathbb{Z}^{\geq 0})$$

briefly as

$$\prod_{\substack{i=2 \\ a_i > 0}}^{n} i^{a_i}$$

with the exception of the partition

$$n = 1 + 1 + \cdots + 1$$

which is indicated as **1**; also we write the frequency in front so that $\frac{1}{2} \times 2$ means that half of the group consists of 2-cycles. For example for $f(t) = t^3 - t - 1$ we obtain the factorization:

p	congruence factorization mod p	partition type
2	irreducible	3
3	irreducible	3
5	$(t-2)(t^2+2t-2)$	2
7	$(t+2)(t^2-2t+3)$	2
11	$(t+5)(t^2-5t+2)$	2
13	irreducible	3
17	$(t-5)(t^2+5t+7)$	2
19	$(t-6)(t^2+6t-3)$	2
23	$(t-3)(t-10)^2$	$-(23 = -d(f))$
29	irreducible	3
31	irreducible	3
37	$(t-10)(t^2+13t-17)$	2
41	irreducible	3
43	$(t-10)(t^2+10t+13)$	2
47	irreducible	3
53	$(t+16)(t^2-16t-10)$	2
59	$(t-4)(t-13)(t+17)$	1

If we know $d(f)$ (i.e. that $d(f)$ is a non-square) then already the first test shows $G = \mathfrak{S}_3$. Not knowing $d(f)$ we see already after three steps that $G = \mathfrak{S}_3$. The frequency up to 59 is distributed 7:8:1 in favor of 3, 2, 1, excluding $d(f) = -23$.

For $n = 4, 5$ the simple criteria mentioned above suffice to guess the unknown group G after a few trials, in accordance with the following table:

I. d(f) square, f has no linear factor (9.44)

degree n	group G	order $\|G\|$	cycle distribution	comments	isomorphic to
4	$\langle(12)(34)\rangle$	2	$1\times 1+1\times 2^2$	2+2	C_2
	$\mathfrak{B}_4$	4	$1\times 1+3\times 2^2$	3-impr.	$C_2\times C_2$
	$\mathfrak{A}_4$	12	$1\times 1+3\times 2^2+8\times 3$	2-trans.	
5	$\langle(12345)\rangle$	5	$1\times 1+4\times 5$	prim.	C_5
	$\langle(12345),(25)(34)\rangle$	10	$1\times 1+4\times 5+5\times 2^2$	prim.	D_{10}
	$\mathfrak{A}_5$	60	$1\times 1+15\times 2^2+20\times 3$ $+24\times 5$	3-tr.	$PSL(2,5)$

II. d(f) non-square, f has no linear factor (9.45)

degree n	group G	order $\|G\|$	cycle distribution	comments	isomorphic to
4	$\langle(1234)\rangle$	4	$1\times 1+1\times 2^2+2\times 4$	3-impr. 2+2	C_4
	$\langle(1234),(24)\rangle$	8	$1\times 1+2\times 2+3\times 2^2$ $+2\times 4$	impr. 2+2	D_8
	$\mathfrak{S}_4$	24	$1\times 1+3\times 2^2+6\times 2$ $+8\times 3+6\times 4$	4-trans.	
5	$\langle(12),(345)\rangle$	6	$1\times 1+1\times 2+2\times 3$ $+2\times 2.3$	2+3	C_6
	$\langle(12345),(2354)\rangle$	20	$1\times 1+5\times 2^2$ $+10\times 4\times 5$	2-trans.	$\mathrm{Hol}(C_5)$
	$\mathfrak{S}_5$	120	$1\times 1+15\times 2^2+20\times 3$ $+24\times 5+10\times 2$ $+20\times 2.3$ $+30\times 4$	5-trans.	$PGL(2,5)$

Here the partitions are denoted in the same way as in (9.43). But the symbol $m\times\prod_{\substack{i>1\\ a_i>0}} i^{a_i}$ now indicates that the permutation group G contains precisely m permutations of cycle decomposition type employing a_i cycles of length i. The frequency is $m/|G|$. The comment 'k-trans.' = k-transitive means the group G is k-fold transitive, but not $(k+1)$-transitive. In this case the order is divisible by $n(n-1)\cdots(n-k+1)$. The comment 'prim.' = primitive indicates that the permutation group G is primitive, but not 2-transitive. The comment 'impr.' = imprimitive indicates that the permutation group G is transitive, but imprimitive. In this case the permuted letters $1,2,\ldots,n$ can be grouped into blocks which are permuted as a whole by the elements of G.

Each grouping corresponds to a partition of n given below the symbol impr. If the number of groupings is $k > 1$ then we write 'k-impr'.

If the permutation group is intransitive then the distribution of the permuted letters into orbits corresponds to a partition of n which is given in the usual way: $n = \sum_i a_i \cdot i$ $(a_i \in \mathbb{Z}^{\geqslant 0}, 1 \leqslant i \leqslant n)$.

For the computation of cycle distributions we refer to [8].

Exercises

1. The group G is said to be the *semidirect product of two subgroups A, B*:

$$G = A \rtimes B = B \ltimes A,$$

if A is a normal subgroup of G (denoted by $A \lhd G$) and B is a representative subgroup of G modulo A:

$$AB = BA = G,$$
$$A \cap B = 1.$$

(a) Show that there is the homomorphism

$$\beta: B \to \operatorname{Aut}(A)$$

of B into the automorphism group $\operatorname{Aut}(A)$ of A such that

$$\beta(b)(a) = b^{-1}ab$$

and that the elements of G are uniquely presented in the form

$$g = ab \quad (a \in A, b \in B) \tag{9.46a}$$

with multiplication rule

$$gg' = a\beta(b^{-1})(a')bb' \quad (a, a' \in A; b, b' \in B). \tag{9.46b}$$

(b) Show the converse of (a): if β is a homomorphism of the group B into the group $\operatorname{Aut}(A)$ then the element (9.46a) with multiplication rule (9.46b) and equality rule:

$$ab = a'b' \Leftrightarrow a = a', \quad b = b' \quad (a, a' \in A; b, b' \in B)$$

form a group G of the form (a).

(c) Show that the semidirect product $A \rtimes B$ is direct if and only if $\beta(B) = 1$.

(d) If $B = \operatorname{Aut}(A)$ then $\operatorname{Hol}(A) := A \rtimes B$ is said to be the *holomorph* of A. Show that $\operatorname{Hol}(C_n)$ is isomorphic to the permutation group formed by the permutations

$$\begin{pmatrix} z \\ az + b \end{pmatrix} \quad (a \in U(\mathbb{Z}/n), b \in \mathbb{Z}/n)$$

of the elements of $\mathbb{Z}/n$.

2. The wreath product $G \wr H$ of a subgroup G of $\mathfrak{S}_n$ and the group H is the group

$$(\theta_1 H \times \theta_2 H \times \cdots \times \theta_n H) \rtimes G,$$

obtained as the semidirect product of the direct product of n isomorphic copies of H (θ_i isomorphisms of H on $\theta_i H$ $(1 \leqslant i \leqslant n)$) and G with transformation action defined by $\pi(\theta_i h)\pi^{-1} = \theta_{\pi(i)}h$ $(\pi \in G, h \in H, 1 \leqslant i \leqslant n)$.

(a) Show that $|G \wr H| = |H|^n |G|$.

(b) Let $f = g^m$, g a monic non-constant polynomial of $R[t]$, H the group of the equation $g(x) = 0$, then the group of the equation $f(x) = 0$ is $\mathfrak{S}_m \wr H$.

(c) Let R be a field of prime characteristic p, let $f = g(t^{p^\mu})$, g a monic separable polynomial of $R[t]$, H the group of the equation $g(x) = 0$ over R, then the group of the equation $f(x) = 0$ over R is $\mathfrak{S}_{p^\mu} \wr H$.

3. Let R be a unital commutative ring, f a monic non-constant polynomial of $R[t]$.

(a) The idempotent e of $S(f/R)$ is said to be $\bar{\mathfrak{S}}_n$-orthogonal if the $\bar{\mathfrak{S}}_n$-orbit of e consists of orthogonal idempotents. For example any idempotent of R is $\bar{\mathfrak{S}}_n$-orthogonal with itself being the only member of its $\bar{\mathfrak{S}}_n$-orbit. Prove: If e, e' are two $\bar{\mathfrak{S}}_n$-orthogonal idempotents of $S(f/R)$ with $ee' \neq 0$ then ee' also is a $\bar{\mathfrak{S}}_n$-orthogonal idempotent of $S(f/R)$.

(b) The idempotent e of $S(f/R)$ is said to be $\bar{\mathfrak{S}}_n$-extreme if for any idempotent e' of $S(f/R)$ satisfying $ee' = e'$ we have $\mathrm{Stab}(e/\bar{\mathfrak{S}}_n) = \mathrm{Stab}(e'/\bar{\mathfrak{S}}_n)$. Show that every idempotent of $S(f/R)$ is the sum of finitely many $\bar{\mathfrak{S}}_n$-extreme idempotents.

(c) There are finitely many orthogonal idempotents $E_1, E_2, \ldots, E_r$ of R such that

(1) $E_1 + E_2 + \cdots + E_r = 1$.

(2) There is an $\bar{\mathfrak{S}}_n$-extreme idempotent e_i of $S(f/R)$ satisfying $e_i E_i = e_i$ $(1 \leqslant i \leqslant r)$.

(3) For any $\bar{\mathfrak{S}}_n$-extreme idempotent e'_i of $S(f/R)$ satisfying $e'_i E_i = e'_i$ the subgroup $\mathrm{Stab}(e'_i/\bar{\mathfrak{S}}_n)$ is $\bar{\mathfrak{S}}_n$-conjugate to $\mathrm{Stab}(e_i/\bar{\mathfrak{S}}_n)$. If $e_i e'_i \neq 0$ then $\mathrm{Stab}(e_i/\bar{\mathfrak{S}}_n) = \mathrm{Stab}(e'_i/\bar{\mathfrak{S}}_n)$.

(4) The $\bar{\mathfrak{S}}_n$-extreme idempotent e_i can be chosen in such a way that E_i is the sum over the $\bar{\mathfrak{S}}_n$-orbit of e_i. Show that the $\bar{\mathfrak{S}}_n$-orbit of e_i is uniquely determined by E_i.

(d) Generalize the theorems on groups of an equation of section 9 to arbitrary unital commutative rings.

4. (a) Let M_j be two R-modules and $B_j: M_j \otimes_R M_j \to R$ be R-bilinear forms on M_j $(j = 1, 2)$. Then there is the R-bilinear form

$$B = B_1 \times B_2 : M_1 \otimes_R M_2 \to R:$$
$$(u_1 \otimes u_2) \times (v_1 \otimes v_2) \mapsto B_1(u_1 \otimes v_1) B_2(u_2 \otimes v_2) \quad (u_i, v_i \in M_i, i = 1, 2)$$

of $M = M_1 \otimes_R M_2$. Show that the ideals $\partial_{i,B/R}(M)$ generated by the determinants $\det(B(u_j \otimes v_k))$ $(u_j, v_k \in M; j, k = 1, \ldots, i)$ satisfy

$$\partial_{i,B/R}(M) = \sum_{j=0}^{i} \partial_{j,B_1/R}(M_1) \partial_{i-j,B_2/R}(M_2) \quad (\partial_{0,B_j/R}(M_j) = R).$$

(b) Let R be an algebraically indecomposable unital commutative ring. Let f be a monic non-constant polynomial of $R[t]$. Let $e_1, e_2, \ldots, e_s$ be the primitive idempotents of $S(f/R)$. Then show we have

$$\partial_{n!/s,R}(e_i S(f/R)) = \partial_{n!/s,R}(e_1 S(f/R) \quad (1 \leqslant i \leqslant s),$$
$$\partial_{n!/s,R}(e_1 S(f/R))^s = d(f)^{n!/2} R.$$

5. (a) Show the validity of the ván der Waerden criterion for all ideals of R with the property that $d(f)/\mathfrak{p}$ is a non-zero divisor of $R/\mathfrak{p}$.
   (b) Let A be a unital commutative ring with nilradical $\mathrm{NR}(A)$ such that the factoring $A/\mathrm{NR}(A)$ is finite of p-power characteristic (p a prime number). Then show every automorphism of $A/\mathrm{NR}(A)$ raising every element x of $A/\mathrm{NR}(A)$ to x^{p^α} ($\alpha \in \mathbb{Z}^{\geqslant 0}$ fixed) can be lifted to an automorphism of A.

6. Let F be a field, f a monic non-constant polynomial in t over F. Show that every idempotent of $S(f/R)$ is generated from the idempotents of $R[x_1]$ upon application of $\bar{\mathfrak{S}}_n$, multiplication of non orthogonal idempotents and complementation of idempotents e of relative to idempotents e' by $e' - e$ in case $ee' = e = e'e \neq e'$.

2.10. How to determine the group of a separable equation over a field

1. Root finder

The direct approach to the determination of the group of the equation

$$f(x) = 0, \tag{10.1a}$$

with

$$f(t) = t^n + a_1 t^{n-1} + \cdots + a_n \in F[t], \quad n > 0, \tag{10.1b}$$
$$d(f) \neq 0 \tag{10.1c}$$

over the field F proceeds by constructing a primitive idempotent e of the universal splitting ring $S(f/F)$ and forming the $\bar{\mathfrak{S}}_n$-stabilizer of e. It is obvious that any method which must deal with an overring of F with $n!$ basis elements is bound to be slow. Still, we use it to obtain

Proposition (10.2)

(a) *Given the constructive field F of zero characteristic and given a construction for a primitive idempotent e of the universal splitting ring $S(f/F)$ of the separable equation* (10.1a–c) *then we obtain a construction of the roots of f in F.*

(b) *Conversely, let F be a constructive field of known characteristic with the following* **root finder:** *For any monic polynomial g of $F[t]$ it can be decided in finitely many computational steps whether g has a root in F. If the decision is 'yes' then a root of g in F can be constructed.*

   Then we obtain a construction of a primitive idempotent of $S(f/F)$ for any monic polynomial f over F.

Proof

(a) Suppose we have constructed e then the e-projection of $S(f/F)$ is the finite extension $eS(f/F)$ of eF generated by the roots $ex_1,\ldots,ex_n$ of ef where $x_1,\ldots,x_n$ are the standard generators of $S(f/F)$ over F so that

$$ef(t) = \prod_{i=1}^{n} (t - ex_i)$$

in $eS(f/F)$. The permutation group $G = \{\pi \in \mathfrak{S}_n | \bar{\pi}(e) = e\}$ is the group of the equation (10.1a) over F. If it has no orbit of length 1 then f has no root in F.

Now let $\{i_1\}, \{i_2\},\ldots,\{i_r\}$ $(r \in \mathbb{Z}^{>0}; 1 \leqslant i_1 < i_2 < \cdots < i_r \leqslant n)$ be the orbits of length 1 of G then precisely the elements $ex_{i_1}, ex_{i_2},\ldots, ex_{i_r}$ are the roots of $ef(t)$ in eF.

Thus we know already that there must hold equations of the form $ex_{i_j} = \alpha_j e$ $(1 \leqslant j \leqslant r)$ where $\alpha_1, \alpha_2,\ldots,\alpha_r$ are the roots of f in F. Only we don't know yet which particular elements of F they are.

The principal polynomial of ex_{i_j} over eF is obtained in the form

$$P_{ex_{i_j}}(t) = t^{n!} + \sum_{i=1}^{m} b_i t^{n!-i}$$
$$(m \in \mathbb{Z}^{>0}; \quad 0 < m \leqslant n!, \quad b_i \in eF \quad (1 \leqslant i \leqslant m), \quad b_m \neq 0).$$

We know already that it factorizes into

$$P_{ex_{i_j}}(t) = t^{n!-m} g_j(t),$$

where m is the degree of $eS(f/F)$ over eF because $eS(f/F)$ annihilates the complementary component $(1-e)S(f/F)$. We find g_j upon division of $P_{ex_{i_j}}$ by $t^{n!-m}$. Since the characteristic of F is zero we obtain $t - \alpha_j$ as that basis element of the divisor cascade of g_j, $D_t(g_j)$ which divides g_j. We find the negative $-\alpha_j$ of the root α_j of f as the constant term of $t - \alpha_j$.

(b) Conversely, let us assume that F is a constructive field of known characteristic with root finder. Let f be any monic non-constant polynomial in t over F. Then we construct a primitive idempotent of $S(f/F)$ as follows: Initially set $e = 1_F$. Determine the group of permutations

$$G = \{\pi \in \mathfrak{S}_n | \bar{\pi}(e) = e\}.$$

If $G = \mathbf{1}$ then e is primitive.

If $G \neq \mathbf{1}$ then form a representative list $L(G, \mathfrak{S}_n)$ of the $\mathfrak{S}_n$-conjugacy classes of proper subgroups of G.

If e is not primitive then there must be an idempotent e' of $S(f/F)$ that is stabilized precisely by one of the members of $L(G, \mathfrak{S}_n)$ such that $e'e = e'$. We test the members H of $L(G, \mathfrak{S}_n)$ in the order given by the list. Form the F-subalgebra

$$eS(f/R)^H = \{x \in eS(f/F) | \forall \pi \in H : \bar{\pi}(x) = x\}.$$

It contains Fe.

If $eS(f/R)^H = Fe$ then there is no idempotent e' precisely stabilized by H such that $e'e = e$.

Now assume that we find an element x of $eS(f/F)^H$ which is not contained in Fe. Then the principal polynomial of x is of the form

$$P_x(t) = t^{n!-m}g(t),$$

where

$$m = \dim_F eS(f/F) = \deg(g),$$

and g is a monic polynomial of $F[t]$ that is not of the form $(t-\alpha)^m$.

We test whether g has a root in F. If the answer is 'no' then there is no idempotent e' of $S(f/F)$ which is stabilized by H such that $e'e = e$. In this case we proceed to test another member of $L(G, \mathfrak{S}_n)$ as candidate for H.

If there is no other member of $L(G, \mathfrak{S}_n)$ to test then we conclude that e is primitive and G is the group of the equation. But if the answer to the test of g is 'yes' then construct a root α of g in F so that

$$\begin{gathered} g(\alpha) = 0, \\ g(t) = (t-\alpha)^\mu g_1(t) \quad (\mu \in \mathbb{Z}^{>0}, g_1(t) \in F[t]), \\ g_1(\alpha) \neq 0, \\ A(t)(t-\alpha)^\mu + B(t)g_1(t) = 1 \quad (A(t), B(t) \in F[t]), \\ e' = B(x)g_1(x) = e'e' = e'e \neq 0. \end{gathered}$$

Now replace e by e' and go on as above. After a finite number of tests we find e itself to be primitive, and G will be the group of the equation $f(x) = 0$. □

2. Root finder and polynomial factorization over fields

Proposition (10.3)
Let F be a constructive field with root finder. Then there is an algorithm to factorize any given non-constant monic polynomial f in t over F into a product of irreducible monic factors $f_1, \ldots, f_s$ over F.

Proof
Apply induction over the degree n of f. If $n = 1$ then f is irreducible. Let $n > 1$. The proof of proposition (10.2) resulted in the construction of a primitive idempotent e of $S(f/F)$. Applying $\bar{\mathfrak{S}}_n$ to e we find all primitive idempotents of $S(f/F)$ as the distinct $\bar{\mathfrak{S}}_n$-conjugates of e.

Let $x_1, \ldots, x_n$ be the standard root generators of $S(f/F)$ where n is the degree of f. The subalgebra $F[x_1]$ is isomorphic to the equation ring $F[t]/f$ with x_1 as standard generator. For any non-empty sum over finitely many primitive idempotents of $S(f/F)$ we test whether it belongs to $F[x_1]$. In this way we find all idempotents of $F[x_1]$; in particular the primitive idempotents of $F[x_1]$ are characterized as those sums over distinct primitive idempotents of $S(f/F)$ for which no proper non-empty subsum belongs to $F[x_1]$. Say

$$e_i = g_i(x_1) \quad (g_i \in F[t], \deg g_i < n, 1 \leqslant i \leqslant \sigma, \sigma \in \mathbb{Z}^{>0})$$

are the primitive idempotents of $F[x_1]$, then

$$h_i = f/\gcd(f, g_i) \quad (1 \leqslant i \leqslant \sigma)$$

are mutually prime monic polynomials in t over F for which

$$f = \prod_{i=1}^{\sigma} h_i,$$

and we know that there exist equations

$$h_i = f_i^{v_i} \quad (v_i \in \mathbb{Z}^{>0}, 1 \leqslant i \leqslant \sigma)$$

with f_i an irreducible monic polynomial in t over F. If $\sigma > 1$ then we obtain the full factorization of f over F by application of the induction assumption to h_i $(1 \leqslant i \leqslant \sigma)$. Now let $\sigma = 1$.

If $D_t h_1 \neq 0$ then we find that

$$f_1 = h_1/\gcd(h_1, D_t(h_1)).$$

Now let $D_t(h_1) = 0$. Hence the characteristic of F is a prime number p which divides n. It is determined by prime factorization of n and testing the equation $q 1_F = 0$ for each prime divisor q of n.

Applying the root finder to the coefficients of h_1 we test the equation

$$h_1 = j^p \quad (j \in F[t]).$$

If there is a solution then we obtain f_1 by application of the induction hypothesis to j. Otherwise we have

$$p \mid v_1,$$
$$f_1(t) = j_1(t^p),$$
$$h_1(t) = k_1(t^p),$$
$$k_1(t) = j_1(t)^{v_1},$$

where j_1 is a monic irreducible polynomial in t over F. Applying the induction hypothesis to k_1 we construct j_1. Thus $f_1(t) = j_1(t^p)$ also is constructed. □

3. Separable extensions of a constructive field with root finder

Proposition (10.4)

If F is a constructive field with a root finder, then for every monic irreducible separable polynomial f in t over F also the finite extension $E = F[t]/f$ is a constructive field with a root finder.

Proof

It is clear that E is a constructive extension of F. If $\deg(f)=1$ then $E=F$ has a root finder. Apply induction over $\deg(f)$. Let $\deg(f)>1$.

Now let g be a monic non-constant polynomial in t over E. We want to decide whether g has a root in E, and if it has one we want to construct it. This is trivial if g is linear. Apply induction over $\deg(g)$. Assume that $\deg(g)>1$.

The norm of g from E to F is a monic polynomial h of degree $\deg(f)\deg(g)$ in t over F and g divides h in $E[t]$. Let $h=\prod_{i=1}^{s} h_i$ be a factorization of h into a product of irreducible monic polynomials in t over F. If $d=\gcd(h_i, g)$ is a monic proper non-constant divisor of g then we test d and g/d according to the induction hypothesis. If one of them has a root in E then also g has one and we can construct it. If neither d nor g/d have a root in E then also g has no root in E. In either case we are done.

It remains to discuss the case that for each h_i either $\gcd(h_i, g)=1$ or g divides h_i. Hence, after suitable numbering, $h_1=h_2=\cdots=h_j$, $h_1\neq h_i$ $(1<i\leqslant s)$, $g|h_1$, $1<\deg(h_1)$. If g has a root α in E then $(t-\alpha)|g$, $N_{E/F}(t-\alpha)|h$, $\deg(N_{E/F}(t-\alpha))=\deg(f)$, $\deg(h_1)\leqslant\deg(f)$ and h_1 must be separable. Assume that h_1 is separable of degree $\leqslant\deg(f)$. Now if $\deg(h_1)<\deg(f)$ then we have a root finder for the finite extension $E_1=F[t]/h_i$ in accordance with the first induction hypothesis. Now E_1 is isomorphic over F to a proper subfield of E. Hence we can find a proper monic irreducible divisor f_1 of f in $E_1[t]$ such that

$$E_2=E_1[t]/f_1=E_1[\xi],\quad \xi=t/f_1,\quad f_1(\xi)=0,$$
$$f(\xi)=0,\quad \deg(h_1)\deg(f_1)=\deg(f).$$

Suppose this test is confirmed. Then E_2 must be isomorphic to E and by the second induction hypothesis we decide whether the polynomial g in t over E_2 has a root in E_2; and if it has one then we construct it. It remains to discuss the case that $\deg(h_1)=\deg(f)$. We know that fh_1 is separable. We construct a primitive idempotent e of $S(h_1/F)$. Let $x_1,\ldots,x_m$ be the standard root generators of $S(fh_1/F)$ over F $(m=\deg(f)+\deg(h_1))$. If g has a root in E then there must be roots x_i, x_j such that $f(ex_i)=0$ and F-isomorphisms

$$\phi: F\to eF[ex_i],\quad \phi(\lambda)=\lambda e\quad (\lambda\in F),$$
$$\phi(t/f)=ex_i,\quad g(ex_j)=0,\quad ex_j=\phi(\alpha)$$

for some α of E. Hence $g(\alpha)=0$. Otherwise the decision is negative. □

4. Primitivity test

Definition (10.5)

The extension E of the field F is said to be **primitive** *if there is no field between*

E and F, i.e. *for for every intermediate extension Ψ of F in E either $\Psi = E$ or $\Psi = F$; F is the trivial primitive extension of F.*

If E is a non-trivial primitive extension of F then the extension E of F is generated over F by any element ξ of E that does not belong to F:

$$E = F(\xi). \tag{10.5a}$$

Definition (10.6)
We say that the element ξ of the extension E of the field F is a **primitive element** *of E over F if* (10.5a) *applies.*

Not every extension E with a primitive element over F also is a primitive extension of F. For example for $F = \mathbb{Q}$, $E = \mathbb{Q}[t]/(t^4 + 1)$ we have $E = F(\xi)$ ($\xi = t/(t^4 + 1)$), but there are 3 intermediate subfields. Or, if $E = F(\xi)$ is the rational function field in one variable ξ over F then $F(\xi^2)$ is a proper intermediate extension $\neq F$ so that E is imprimitive. It follows that a primitive extension E of F must be finite. If it is not separable then F is of prime characteristic p and $E = F(\xi)$, $\xi^p \in F$, $\xi \notin F$.

Proposition (10.7)
Let F be a constructive field with a root finder and let f be a monic separable irreducible polynomial of $F[t]$. Then the extension

$$E = F[t]/f = F(\xi), \quad \xi = t/f$$

of F is primitive if and only if the factorization

$$f(t) = \prod_{i=1}^{s} f_i(t) \quad (f_1(t) = t - \xi)$$

of f into the product of distinct monic irreducible polynomials of $E[t]$ has the property that the coefficients of any divisor of f

$$d(t) = f_1(t) f_{i_2}(t) \cdots f_{i_r}(t), \tag{10.7a}$$

subject to the conditions

$$1 = i_1 < i_2 < \cdots < i_r \leqslant s, \quad 1 \leqslant r < s, \quad \sum_{j=1}^{r} i_j | \deg(f) \tag{10.7b}$$

generate E over F.

Proof
If there is a proper intermediate extension Ψ such that

$$F \subset \Psi \subset E$$

then we have $E = \Psi(\xi)$ so that the minimal polynomial d of ξ over Ψ is of the form (10.7a) subject to (10.7b), and conversely. □

We remark that the primitivity of the separable extension $E = F[t]/f$ is equivalent to the statement that the group G of the equation $f(x)=0$ over F considered as permutation group of the roots of the irreducible separable polynomial f is transitive and that there is no proper subgroup of G properly containing the G-stabilizer of any one of the $\deg(f)$ roots of f. This remark made already by E. Galois became the motive for defining the concept of primitive permutation groups as outlined above. In case of feasible polynomial factorization algorithms we have seen how to reduce the solution of a separable equation to a sequence of separable irreducible equations each with primitive group.

The properties of primitive permutation groups of finitely many letters have been researched extensively, but we are still far away from a full classification. Only for solvable primitive permutation groups a classification along the lines of C. Jordan's *Traité des Substitutions* and B. Huppert's thesis is feasible. For the convenience of the reader we present a table of the primitive permutation groups of six and seven letters:

(10.8)

no.	number of permuted letters	notation of group	order	comments
Even permutation groups				
1	6	$\mathfrak{A}_6$	360	4-trans.
2	6	$PSL(2,5)$	60	2-trans. $\simeq \mathfrak{A}_5$
3	7	$\mathfrak{A}_7$	2520	5-trans.
4	7	$PSL(3,2)$	168	2-trans.
5	7	$\mathrm{Hol}(G_7) \cap \mathfrak{A}_7$	21	
6	7	C_7	7	
Odd permutation groups				
7	6	$\mathfrak{S}_6$	720	6-trans.
8	6	$PGL(2,5)$	120	3-trans. $\simeq \mathfrak{S}_5$
9	7	$\mathfrak{S}_7$	5040	7-trans.
10	7	$\mathrm{Hol}(C_7)$	42	2-trans.
11	7	D_{14}	14	

Note: $\mathrm{Hol}(G)$, the holomorph of the group G, is the semidirect product of G (as normal subgroup) with its automorphism group.

5. Permutation representations

Before we go on with the analysis of the group verification task let us survey briefly the basic concepts on permutation representations in this subsection. Reference is made to H. Zassenhaus [18] and H. Wielandt [17].

Definition (10.9)

For any non-empty set S the bijections of S form the symmetric permutation group $\mathfrak{S}_S$. We use as composition law of π, $\pi' \in \mathfrak{S}_S$ the left–right application $\pi\pi'(s) = \pi(\pi'(s))$ $(s \in S)$. A homomorphism

$$\Gamma : G \to \mathfrak{S}_S \tag{10.9a}$$

of a group G into $\mathfrak{S}_S$ is said to be a **permutation representation** *of G over S (or of degree $|S|$ given by the cardinality of S).*

Thus there are just the conditions

$$\Gamma(gh) = \Gamma(g)\Gamma(h) \quad (g, h \in G), \tag{10.9b}$$

$$\Gamma(1_G) = id_S \tag{10.9c}$$

to be met by Γ. Two permutation representations, say (10.9a) and

$$\Gamma' : G \to \mathfrak{S}_{S'} \tag{10.9d}$$

are said to be *equivalent*:

$$\Gamma \sim \Gamma' \tag{10.9e}$$

if there is a 1–1-correspondence

$$\kappa : S \to S', \tag{10.9f}$$

such that

$$\Gamma'(x) = \kappa\Gamma(x)\kappa^{-1} \quad (x \in G). \tag{10.9g}$$

In other words equivalence of permutation representations amounts to a mere 'change of tag' of the permuted objects. It is reflexive, symmetric and transitive. The *sum of two permutation representations*, say (10.9a) and (10.9d), is defined as the permutation representation

$$\Gamma + \Gamma' : G \to \mathfrak{S}_{S \dot\cup S'} : (\Gamma + \Gamma')(g) = \Gamma(g) + \Gamma'(g), \tag{10.10a}$$

where the sums of two permutations

$$\pi \in \mathfrak{S}_S, \quad \pi' \in \mathfrak{S}_{S'} \tag{10.10b}$$

is defined as the permutation

$$(\pi + \pi')(a) = \begin{cases} \pi(a) & \text{if } a \in S \\ \pi'(a) & \text{if } a \in S' \end{cases} \tag{10.10c}$$

of the disjoint union $S \dot\cup S'$ of the two sets S, S'. This addition of permutation representation satisfies the substitutional law:

$$\Gamma \sim \Delta, \Gamma' \sim \Delta' \Rightarrow \Gamma + \Gamma' \sim \Delta + \Delta'. \tag{10.10d}$$

It is also commutative and associative:

$$\Gamma + \Gamma' \sim \Gamma' + \Gamma, \quad (\Gamma + \Gamma') + \Gamma'' \sim \Gamma + (\Gamma' + \Gamma''). \tag{10.10e}$$

Moreover, it satisfies the cancellation law

$$\Gamma + \Delta \sim \Gamma' + \Delta \Rightarrow \Gamma \sim \Gamma', \tag{10.10f}$$

so that the equivalence classes of permutation representations of G form a *half-module.* Each permutation representation (10.9a) establishes the equivalence relation

$$x \underset{G}{\sim} y \quad (x, y \in S) \tag{10.10g}$$

on S that is defined by

$$y \in Gx \tag{10.10h}$$

i.e.

$$\exists g \in G\colon g(x) = y.$$

Note that for simplicity's sake we have used $g(x)$ instead of $\Gamma(g)(x)$ with operational rules:

$$gg'(x) = g(g'(x)) \quad (g, g' \in G;\ x \in S).$$

We show that $\underset{G}{\sim}$ is indeed an equivalence relation.

Reflexivity: $x = 1_G x \Rightarrow x \underset{G}{\sim} x$,

Symmetry: $x \underset{G}{\sim} y \Rightarrow \exists g \in G\colon y = g(x) \Leftrightarrow \exists g \in G\colon x = g^{-1}(y) \Rightarrow y \underset{G}{\sim} x$,

Transitivity: $x \underset{G}{\sim} y,\ y \underset{G}{\sim} z \Rightarrow \exists g, g' \in G\colon y = g(x) \quad \text{and} \quad z = g'(y)$

$$\Rightarrow \exists g, g' \in G\colon z = g'(g(x)) = (g'g)(x) \Rightarrow x \underset{G}{\sim} z.$$

The equivalence classes of $\underset{G}{\sim}$ on S are said to be the *orbits* of G acting on S via Γ or, simply, the G-*orbits.*

They provide a partition of S. If there is just one orbit then Γ is said to be *transitive* otherwise Γ is said to be *intransitive.* Each intransitive permutation representation (10.9a) with finitely many orbits, say $S_1, S_2, \ldots, S_r$, is equivalent to the sum of the transitive representations

$$\Gamma_i\colon G \to \mathfrak{S}_{S_i}\colon \Gamma_i(g)(x) = g(x) \quad (g \in G;\ x \in S, 1 \leqslant i \leqslant r), \tag{10.10i}$$

$$\Gamma \sim \Gamma_1 + \Gamma_2 + \cdots + \Gamma_r. \tag{10.10j}$$

Any transitive permutation representation is *indecomposable* in the sense that it is not equivalent to the sum of two permutation representations. The equivalence classes of permutation representations with a finite number of orbits form a half-module with the equivalence classes of transitive permutation representations as basis.

For a given permutation representation (10.9a) and an element x of S the elements g of G fixing x according to

$$g(x) = x \tag{10.11a}$$

form the *G-stabilizer* of x, a subgroup of G variously denoted as

$$G_x = \mathrm{Stab}(x/G) = \{g \in G \mid g(x) = x\}. \tag{10.11b}$$

The G-stabilizers of G-equivalent elements, say x and $y = g(x)$ (g some element of G) are G-conjugate:

$$G_y = gG_xg^{-1}.$$

An example of a transitive permutation representation of a group G is provided by the left-action of G on the set $(G/U)_R$ of the right-cosets xU ($x \in G$) of G modulo a given subgroup U:

$$\Gamma_{G,U}: G \to \mathfrak{S}_{(G/U)_R}: \Gamma_{G,U}(g)(xU) = g(x)U. \tag{10.11c}$$

The G-stabilizer of the right-coset U is U. The G-stabilizer of the right coset gU is the G-conjugate subgroup gUg^{-1} ($g \in G$).

If (10.9a) is transitive then there is the 1–1 correspondence

$$\pi: S \to (G/G_x)_R: y \mapsto \{g \in G \mid g(x) = y\}$$

for any fixed element x of S establishing equivalence with Γ_{G,G_x}:

$$\Gamma \sim \Gamma_{G,G_x} \quad (x \in G).$$

In other words, we obtain a full set of representations of the indecomposable equivalence classes of permutation representations of G by forming a representative set of the G-conjugacy classes of subgroups U of G and forming the associated transitive permutations $\Gamma_{G,U}$.

If $U = 1$ then we speak of the *regular permutation representation* $\Gamma_{G,1}$ which was derived earlier directly from the multiplication table of G. It is *faithful*, i.e. its kernel is 1_G so that G is isomorphic to the image $\Gamma_{G,1}(G)$.

If $U = G$ then we speak of the *trivial permutation representation* which maps every element of G on the identity permutation of a set formed by only one object. Its kernel is G and it is faithful only if $G = 1$. The permutation representation (10.9a) is said to be *primitive* if it is transitive and if for every subset X of S consisting of more than one, but less than all elements of S there is an element g of G for which

$$\varnothing \subset g(X) \cap X \subset X.$$

If it is not primitive then it is said to be *imprimitive*. In that case either Γ is intransitive or Γ is transitive and there is a partition of S into *blocks of imprimitivity* consisting of more than one but less than all elements of S such that the application of any element of G merely permutes the blocks. Such a partition is said to be a *system of imprimitivity* of Γ.

The transitive permutation representation $\Gamma_{G,U}$ is primitive precisely if either $U = G$ or U is a maximal subgroup of G. It is imprimitive precisely if there is a subgroup V of G intermediate to U and G:

$$U \subset V \subset G.$$

In this case the right-cosets of G modulo U belonging to the same right-coset of G modulo V form subsets of $(G/U)_R$ establishing a system of imprimitivity of $\Gamma_{G,U}$. In fact, every system of imprimitivity of $\Gamma_{G,U}$ is obtained in this way.

6. Indicator method

Let us develop another method of determining the group of an equation which is more efficient than the direct approach.

The group of an equation is uniquely determined as an abstract group. But if the group of an equation is viewed as a permutation group of the n root generators of the universal splitting ring then it is only unique up to conjugacy under $\mathfrak{S}_n$ so that it is more correct to speak of '*a* group of the equation'.

Keeping this remark in mind let f be a monic separable polynomial of degree $n > 0$ over $\mathbb{Q}$. The application of the van der Waerden criterion yields a list of cycle decomposition patterns which must occur in any group G of the equation (10.1a). Then we can establish a finite list L of non $\mathfrak{S}_n$-conjugate subgroups of $\mathfrak{S}_n$ such that G is $\mathfrak{S}_n$-conjugate to precisely one of the members of L. Among the members of L there is surely $\mathfrak{S}_n$. But if $d(f)$ is a square number then $\mathfrak{S}_n$ ceases to be a candidate, actually $\mathfrak{A}_n$ becomes the top candidate which must contain every other member of L.

Assume now we have found already that G is contained, up to $\mathfrak{S}_n$-conjugacy in the member H of L. Without loss of generality we may assume that

$$G \subseteq H.$$

Let $L(H)$ be the subset of L formed by all those members of L which are $\mathfrak{S}_n$-conjugate to a maximal subgroup of H, but not $\mathfrak{S}_n$-conjugate to a non-maximal subgroup of H. If $L(H)$ is empty then we have $G = H$. Now let $L(H)$ not be empty and replace every member of $L(H)$ by an $\mathfrak{S}_n$-conjugate contained in H so as to yield a new list $L'(H)$ with the following properties:

(a) Every member of $L'(H)$ is a maximal subgroup of H.
(b) No member of $L'(H)$ is conjugate to a non-maximal subgroup of H.
(c) Every maximal subgroup of H is $\mathfrak{S}_n$-conjugate to precisely one member of $L'(H)$.

If $G \subset H$ then G is $\mathfrak{S}_n$-conjugate to a subgroup of some member J of $L'(H)$. Without loss of generality we may assume that

$$G \subseteq J \in L'(H). \tag{10.12a}$$

Now we must test whether (10.12a) applies or whether

$$G = H. \tag{10.12b}$$

If (10.12a) applies then we must be able to tell which member J of $L'(H)$ contains G and proceed with the same method replacing H by J until we reach the stage where (10.12b) holds. This will always happen after a finite number of applications of the verification test.

In order to carry out the test we need a *generator system of the invariants of the permutation group* H *relative to a given subgroup* J *of* H. Let

$$f(t) = t^n + a_1 t^{n-1} + \cdots + a_n \tag{10.13a}$$

be the 'generic' monic polynomial of degree n in t over the ring

$$R = \mathbb{Z}[a_1, \ldots, a_n]$$

of polynomials in the n variables $a_1, \ldots, a_n$ over $\mathbb{Z}$.

Then the universal splitting ring of f over R is the polynomial ring

$$\Lambda = S(f/R) = \mathbb{Z}[x_1, \ldots, x_n]$$

in the standard root generators of $S(f/R)$ over R as polynomial variables over $\mathbb{Z}$. It gives rise to the factorization

$$f(t) = \prod_{i=1}^{n} (t - x_i). \tag{10.13b}$$

There holds the decomposition

$$\Lambda = \dot{\sum}_{i=0}^{\infty} M_i \tag{10.13c}$$

over Λ into the direct sum of the modules M_i formed by the homogeneous polynomials of degree i in $x_1, \ldots, x_n$ over $\mathbb{Z}$. Thus M_i has the $\mathbb{Z}$-basis formed by the monomials

$$x_1^{\nu_1} \cdots x_n^{\nu_n} \quad \left(\nu_j \in \mathbb{Z}^{\geqslant 0}, 1 \leqslant j \leqslant n, \sum_{j=1}^{n} \nu_j = i \right). \tag{10.13d}$$

The group $\mathfrak{S}_n$ of the permutation automorphisms of

$$\begin{aligned} E &= \mathbb{Q}(x_1, \ldots, x_n) = \mathfrak{Q}(\Lambda) \\ &= S(f/\mathbb{Q}(a_1, \ldots, a_n)) \end{aligned}$$

has the fixed subfield

$$F = \mathbb{Q}(a_1, \ldots, a_n)$$

with F-basis

$$\prod_{j=1}^{n} x_j^{\nu_j} \quad (0 \leqslant \nu_j < j, 1 \leqslant j \leqslant n) \tag{10.13e}$$

of E. The subgroup $\bar{H}$ formed by the permutation automorphisms $\bar{\pi}$ with π in H has the fixed subfield E^H. Hence

$$F = E^{\mathfrak{S}_n} \subset E^H \subset E = E^1.$$

According to the theorem on symmetric functions the monomials

$$a_1^{\mu_1} \cdots a_n^{\mu_n} \quad \left(\mu_j \in \mathbb{Z}^{\geqslant 0}, 1 \leqslant j \leqslant n; \sum_{j=1}^{n} \mu_j j = i \right) \tag{10.13f}$$

in $a_1,\ldots,a_n$ form a $\mathbb{Z}$-basis of M_i over R. A $\mathbb{Z}$-basis of Λ is obtained by multiplication of (10.13e) and (10.13f). What happens upon transition from $\mathfrak{S}_n$, 1, to H, J?

The answer is given by

Theorem (10.14)

(a) *For any subgroup J of $\mathfrak{S}_n$ the ring $S(f/R)^J$ of the fixed elements of Λ under the permutation automorphisms π of J splits into its homogeneous components*

$$S(f/R)^J = \dot{\sum}_{i=0}^{\infty} M_{iJ} \tag{10.15a}$$

for

$$M_{iJ} = S(f/R)^J \cap M_i.$$

(b) *For any subgroup H of $\mathfrak{S}_n$ containing J we construct a finite J–H-indicator set $X_{J,H}$ of homogeneous polynomials in $x_1,\ldots,x_n$ over $\mathbb{Z}$ with the following properties:*

I. $$X_{J,H} \subseteq S(f/R)^J. \tag{10.15b}$$

II. *There is a natural number $v_{J,H}$ such that for any element P of $S(f/R)^J$ there is a denominator $v \in \mathbb{Z}^{>0}$ with the properties that each prime divisor of v divides $v_{J,H}$ and that vP belongs to the ring generated by $X_{J,H}$ and $S(f/R)^H$.*

Proof

(a) For any non-zero element y of $S(f/R)^J$ there is a unique representation of the form

$$y = \sum_{i=0}^{m} y_i \quad (m \in \mathbb{Z}^{>0};\ y_i \in M_i, 0 \leqslant i \leqslant m;\ y_m \neq 0). \tag{10.16}$$

For any element π of J we have

$$y = \bar{\pi}(y) = \sum_{i=0}^{m} \bar{\pi}(y_i), \quad \bar{\pi}(y_i) \in M_i \quad (0 \leqslant i \leqslant m).$$

Due to the uniqueness of (10.15a)

$$\bar{\pi}(y_i) = y_i \quad (\pi \in J), \quad y_i \in S(f/R)^J \quad (0 \leqslant i \leqslant m).$$

(b) Any element y of $S(f/R)^J$ is of the form

$$y = \sum_{\substack{0 \leqslant \mu_j < j \\ 1 \leqslant j \leqslant n}} \lambda_{\mu_1\mu_2\cdots\mu_n} x_1^{\mu_1} x_2^{\mu_2} \cdots x^{\mu_n} \quad (\lambda_{\mu_1\mu_2\cdots\mu_n} \in R).$$

We apply the endomorphism

$$\hat{J} = \sum_{\pi \in J} \bar{\pi}$$

of the module Λ in order to produce the submodule of $S(f/R)^J \cap M_i$ generated by all elements

$$\hat{J}y = \sum_{\substack{0 \leqslant \mu_j < j \\ 1 \leqslant j \leqslant n}} \lambda_{\mu_1 \cdots \mu_n} \hat{J}(x_1^{\mu_1} \cdots x_n^{\mu_n})$$

$$\left(\lambda_{\mu_1 \cdots \mu_n} \in R, \quad i = \sum_{j=1}^{n} \mu_j\right).$$

It contains all elements

$$\hat{J}y = |J|y \quad (y \in S(f/R)^J \cap M_i).$$

Hence for any element P of $S(f/R)^J$ there is a denominator $\nu \in \mathbb{Z}^{>0}$ with the properties that each prime divisor of ν divides $|J|$ and that νP belongs to the ring generated by R and by the finite set formed by the elements

$$\hat{J}(x_1^{\mu_1} \cdots x_n^{\mu_n}) \quad (0 \leqslant \mu_j < j, 1 \leqslant j \leqslant n). \tag{10.17a}$$

In the interest of economy we select a minimal subset $X_{J,H,i}$ of the set (10.17a) subject to the degree conditions

$$\sum_{j=1}^{n} \mu_j = i, \tag{10.17b}$$

and the condition that the module generated by $X_{J,H,i}$ and by the module

$$\sum_{j=1}^{i} (S(f/R)^H \cap M_j)(S(f/R)^J \cap M_{i-j})$$

is of finite index in $S(f/R)^J \cap M_i$ for

$$i = 1, 2, \ldots, \sum_{j=1}^{n} (j-1) = \binom{n}{2} = n(n-1)/2.$$

We determine $\nu_{J,H}$ as the product of all prime numbers dividing $|H|$ or any one of the $\binom{n}{2}$ indices. We order $X_{J,H} = \bigcup_{i=1}^{\binom{n}{2}} X_{J,H,i}$ so that members of higher degree are later than members of lower degree. □

As an application of theorem (10.14) to the verification task we have

Proposition (10.18)

Let F be a field. Let

$$f(t) = t^n + a_1 t^{n-1} + \cdots + a_n$$

be a monic separable polynomial in t over F. Let e be an idempotent of $S(f/R)$ and

$$\operatorname{Stab}(e/\bar{\mathfrak{S}}_n) = H,$$

where H is a subgroup of $\mathfrak{S}_n$ and distinct $\bar{\mathfrak{S}}_n$-conjugates of e are orthogonal. Let J be a member of the list $L(H)$ of certain maximal subgroups of H. Let $x_1, x_2, \ldots, x_n$ be the standard root generators of $S(f/F)$ so that

$$f(t) = \prod_{i=1}^{n} (t - x_i)$$

in $S(f/F)$. Let ϕ be the homomorphism of $\bar{\Lambda}$ in $S(f/F)$ that maps $\bar{x}_i$ on x_i.

We assume that the characteristic of F does not divide $\nu_{J,H}$. Then there is a group of the equation $f=0$ contained in J precisely if there is an element

$$y \in \phi(X_{J,H}),$$

such that

$$ey \in Fe, \tag{10.19}$$

but the principal polynomial of ey of $eS(f/F)$ over Fe has a root in Fe.

Proof

Let us observe that under the assumption of proposition (10.18)

$$S(f/F)^J = \{x \in S(f/F) \mid \forall \pi \in J : \bar{\pi}(x) = x\} = F\phi(\bar{\Lambda})^J.$$

Indeed, using once again the module endomorphism

$$\hat{J} = \sum_{\pi \in J} \bar{\pi}$$

of $\bar{\Lambda}$ we derive the direct decomposition

$$|J|\bar{\Lambda} = \hat{J}\bar{\Lambda} \dotplus (|J|1 - \hat{J})\bar{\Lambda},$$

corresponding to the presentation

$$S(f/F) = \phi(\bar{\Lambda})F = |J|S(f/F) = \phi(|J|\bar{\Lambda})F = \hat{J}S(f/F) \dotplus (|J|1 - \hat{J})S(f/F),$$

where

$$\hat{J}S(f/F) = S(f/F)^J = \{x \in S(f/F) \mid Jx = x\},$$
$$(|J|1 - \hat{J})S(f/F) = \{x \in S(f/F) \mid \hat{J}x = 0\},$$
$$\hat{J}S(f/F) \cap (|J|1 - \hat{J})S(f/F) = \{0\},$$

so that

$$\phi(\bar{\Lambda}^J)F = \hat{J}S(f/F) = S(f/F)^J, \tag{10.20a}$$
$$e\phi(\bar{\Lambda}^J)F = eS(f/F)^J = (eS(f/F))^J. \tag{10.20b}$$

If J contains a group of the equation $f=0$ then there is an idempotent e' of $S(f/F)$ for which

$$e'e = e', \quad \mathrm{Stab}(e'/\mathfrak{S}_n) = J,$$

and hence the $\bar{\mathfrak{S}}_n$-conjugates of e' are orthogonal. If $e\phi(X_{J,H}) \subseteq Fe$, then

$$eS(f/F)^J = \langle eS(f/F)^H, e\phi(X_{J,H}) \rangle \subseteq Fe,$$

contradicting the fact that e' is contained in $(eS(f/F))^J$. Hence there is an element y of $\phi(X_{J,H})$ so that ey does not belong to eF. But

$$e'y = e'ey \in (e'S(f(F))^J = Fe',$$

hence the principal polynomial of ey of $eS(f/F)$ over Fe has a root in Fe.

Conversely, let

$$y \in \phi(X_{J,H}),$$

such that

$$ey \notin Fe,$$

but the principal polynomial of ey of $eS(f/F)$ over Fe has a root λe with λ in F. Since f is separable it follows that $S(f/F)$ contains no nilpotent element over F. Hence there is an idempotent e' of $eS(f/F)$ such that $e'y = \lambda e'$, but the principal equation of $(e-e')y$ of $(ee')S(f/F)$ over $(e-e')F$ has not the root $\lambda(e-e')$. Since $ey \notin Fe$, it follows that $ey \neq \lambda e$, $e \neq e'$. Since $ey \in (eS(f/F))^J$ it follows that

$$e' \in (eS(f/F))^J, \quad \mathrm{Stab}(e'/\bar{\mathfrak{S}}_n) = J,$$

a group of the equation $f = 0$ is contained in J.

We present an *indicator table of the primitive permutation groups of* 6 *and* 7 *letters.*

(10.21)

no.	number of permuted letters	notation of groups H	J	indicator set $X_{J,H}$
1	6	$\mathfrak{S}_6$	$\mathfrak{A}_6$	$a_1, \ldots, a_6, d(f)$
2	6	$\mathfrak{A}_6$	$PSL(2,5)$	$\hat{J}(x_1x_2x_3)$
3	7	$\mathfrak{S}_7$	$\mathfrak{A}_7$	$a_1, \ldots, a_7, d(f)$
4	7	$\mathfrak{A}_7$	$PSL(3,2)$	$\sum_{i=1}^{7} x_i x_{i+1} x_{i+3}$
5	7	$PSL(3,2)$	$\mathrm{Hol}(C_7) \cap \mathfrak{A}_7$	$\hat{J}(x_1x_2)$
6	7	$\mathrm{Hol}(C_7) \cap \mathfrak{A}_7$	C_7	$\hat{J}(x_1x_2), \hat{J}(x_1x_4)$ $\hat{J}(x_7x_3), \hat{J}(x_7x_6)$
7	6	$\mathfrak{S}_6$	$\mathfrak{S}_6$	$a_1, \ldots, a_6$
8	6	$\mathfrak{S}_6$	$PGL(2,5)$	$\hat{J}(x_1x_2x_3x_4)$, $\hat{J}(x_1x_2x_3x_5)$
9	7	$\mathfrak{S}_7$	$\mathfrak{S}_7$	$a_1, \ldots, a_7$
10	7	$\mathfrak{S}_7$	$\mathrm{Hol}(C_7)$	$\hat{J}(x_1x_2x_3)$, $\hat{J}(x_1x_2x_4)$, $J(x_1x_2x_5)$, $\hat{J}(x_1x_2x_6)$
11	7	$\mathrm{Hol}(C_7)$	D_{14}	$\hat{J}(x_7x_1)/2$, $\hat{J}(x_7x_2)/2$

Note that the denominators always are one. Here we view $PSL(3,2)$ as the automorphism group of the seven Steiner triples $(i, i+1, i+3)$ $(1 \leqslant i \leqslant 7)$ in seven letters $i = 1, 2, \ldots, 7$ (identifying 1 and 8, 2 and 9, 3 and 10). It is a doubly transitive permutation group of $1, 2, \ldots, 7$ containing the 7-Sylow subgroup $C_7 = ((1\,234\,567))$ and its normalizer $\mathrm{Hol}(C_7) \cap \mathfrak{A}_7 = \langle C_7, (241)(365) \rangle$.

Let us remark here that the work involved in testing the condition (10.19) can be shortened considerably upon choosing the basis $X_{J,H}$ subject to the additional condition

$$\hat{H} X_{J,H} = 0. \tag{10.22}$$

It is realized upon application of $|H|\hat{J} - \hat{H}$ in place of $\hat{J}$ in the proof of theorem (10.14). With this new choice of $X_{J,H}$ it follows that (10.19) is satisfied precisely if the root of the principal polynomial of ey of $e(f/F)$ over Fe is contained in Fe, but not zero.

7. Root estimates over $\mathbb{C}$

In order to simplify the group verification task over the complex number field $\mathbb{C}$ we apply some simple estimates of the roots of a monic polynomial

$$f(t) = t^n + a_1 t^{n-1} + \cdots + a_n = \prod_{i=1}^{n} (t - \xi_i) \tag{10.23a}$$

with coefficients $1, a_1, \ldots, a_n$ and roots $\xi_1, \ldots, \xi_n$ in $\mathbb{C}$. From the Vieta formulae

$$(-1)^i a_i = \sum_{1 \leqslant j_1 < j_2 < \cdots < j_i \leqslant n} \xi_{j_1} \xi_{j_2} \cdots \xi_{j_i} \tag{10.23b}$$

we derive the estimates

$$|a_i| \leqslant \sum_{1 \leqslant j_1 < j_2 < \cdots < j_i \leqslant n} |\xi_{j_1}| \cdots |\xi_{j_i}| \leqslant \binom{n}{i} \max_{1 \leqslant j \leqslant n} |\xi_j|^i \quad (1 \leqslant i \leqslant n), \tag{10.23c}$$

$$\max_{1 \leqslant j \leqslant n} |\xi_j| \geqslant \phi(f) := \max_{1 \leqslant i \leqslant n} \left| \left(a_i \Big/ \binom{n}{i} \right)^{1/i} \right| \geqslant |a_n^{1/n}| = \left| \left(\prod_{i=1}^{n} \xi_i \right)^{1/n} \right| \geqslant \min_{1 \leqslant j \leqslant n} |\xi_j|, \tag{10.23d}$$

where the non-negative number $\phi(f)$ defined by (10.23d) is a kind of average root magnitude. It vanishes if and only if all roots are zero. It is easily computable in terms of the coefficients of f. In order to find an upper estimate of the maximal root magnitude of the equation

$$\xi_i^n + a_1 \xi_i^{n-1} + \cdots + a_n = 0,$$

we derive from it the estimate

$$|\xi_i^n| = |a_1 \xi_1^{n-1} + \cdots + a_n| \leqslant \sum_{j=1}^{n} |a_j| |\xi_i|^{n-j} \leqslant \sum_{j=1}^{n} \binom{n}{j} \phi(f)^j |\xi_i|^{n-j}$$

$$= (\phi(f) + |\xi_i|)^n - |\xi_i|^n, \quad 2^{1/n}|\xi_i| \leqslant \phi(f) + |\xi_i|,$$

$$\max_{1 \leqslant i \leqslant n} |\xi_i| \leqslant \phi(f)/(2^{1/n} - 1), \tag{10.23e}$$

where $2^{1/n}$ is the real positive solution of $\xi^n = 2$.

In case $a_n \neq 0$ we obtain by application of (10.23c) to the polynomial

$$\prod_{i=1}^{n} (t - \xi_i^{-1}) = t^n + a_n^{-1} a_{n-1} t^{n-1} + \cdots + a_n^{-1},$$

the estimate

$$\min_{1 \leqslant i \leqslant n} |\xi_i| \geqslant (2^{1/n} - 1) / \max_{1 \leqslant i \leqslant n} \left| \left(a_{n-i} \Big/ \binom{n}{i} a_n \right)^{1/n} \right|. \tag{10.23f}$$

It assumes the elegant form

$$\min_{1 \leqslant i \leqslant n} |\xi_i| \geqslant (2^{1/n} - 1)/\phi(f) \tag{10.23g}$$

in case $|a_n| = 1$.

Though the estimates are sharp for certain polynomials like $(t + \alpha)^n$, $t^n + \frac{1}{2}((t + \alpha)^n - t^n)$, in general better estimates can be found with little effort.

As a matter of fact most pocket calculators have built-in algorithms for obtaining approximate solutions of $f(x) = 0$. We remark that

$$\phi(f(t)) = \phi((t \pm \phi(f))^n), \tag{10.23h}$$

in view of the definition of $\phi(f)$. There holds the majorization estimate:

$$\phi(t^n + a_1 t^{n-1} + \cdots + a_n) \leqslant \phi(t^n + b_1 t^{n-1} + \cdots + b_n) \quad \text{if} \quad |a_i| \leqslant |b_i| \quad (1 \leqslant i \leqslant n). \tag{10.23i}$$

Let us note that two polynomials

$$f(t) = t^n + \sum_{i=1}^{n} a_i t^{n-i}, \quad g(t) = t^m + \sum_{i=1}^{m} b_i t^{m-i},$$

which are majorized by polynomials f^*, g^* with only non-negative coefficients the product fg is majorized by f^*g^*. From this remark it follows that

$$\phi(fg) \leqslant \max(\phi(f), \phi(g)). \tag{10.23j}$$

Upon skilful use of g and use of (10.23j) we can usually improve the estimate (10.23e).

8. Group verification over $\mathbb{Q}$

Suppose we have a monic separable polynomial

$$f(t) = t^n + a_1 t^{n-1} + \cdots + a_n$$

with coefficients in $\mathbb{Z}$ of which it is known already that a group of the equation $f(x) = 0$ is contained in the subgroup H of $\mathfrak{S}_n$. Let J be a member of the list $L(H)$ referred to earlier. We would like to know whether a group of the equation $f(x) = 0$ is already contained in J.

For this purpose we like to employ finite field methods. In other words we form a minimal splitting extension $\mathbb{F}_{p^\mu}$ of the polynomial f/p over $\mathbb{F}_p$ where f/p is the monic polynomial in t over $\mathbb{F}_p$ that is obtained upon replacement of the coefficients of f by their residue class modulo p.

In order to avoid repeated roots we assume that

$$p \nmid d(f), \tag{10.24a}$$

hence there holds a factorization

$$f(t) = \prod_{i=1}^{n} (t - \xi_i) \tag{10.24b}$$

of f in n distinct linear factors of $\mathbb{F}_{p^\mu}[t]$.

The automorphism group of $\mathbb{F}_{p^\mu}$ over $\mathbb{F}_p$ is generated by the Frobenius automorphism of $\mathbb{F}_{p^\mu}$ which raises each element of $\mathbb{F}_{p^\mu}$ to its pth power. This automorphism generates the permutation ρ of $\mathfrak{S}_n$ defined by setting

$$\xi_i^p = \xi_{\rho(i)}. \tag{10.24c}$$

In order that there is any chance at all that J contains a group of the equation $f(x) = 0$ there must be an $\mathfrak{S}_n$-conjugate of ρ contained in J, as follows from the proof of the van der Waerden criterion. We assume without loss of generality that

$$\rho \in J. \tag{10.24d}$$

Now let $S(f/\mathbb{C})$ be the universal splitting ring of f over the complex number field $\mathbb{C}$. It is generated by the standard root generators $x_1, \ldots, x_n$ so that

$$f(t) = \prod_{i=1}^{n} (t - x_i). \tag{10.24e}$$

It contains as subrings $S(f/\mathbb{Q})$, $S(f/R_1)$, $S(f/\mathbb{Z})$, where

$$R_1 = \langle \mathbb{Z}, d(f)^{-1} \rangle.$$

By assumption there is an idempotent e of $S(f/R_1)$ such that

$$\operatorname{Stab}(e/\bar{\mathfrak{S}}_n) = \bar{H}, \tag{10.24f}$$

and that the distinct $\bar{\mathfrak{S}}_n$-conjugates of e are orthogonal. There is an epimorphism

$$\tau : S(f/R_1) \to \mathbb{F}_{p^\mu} \quad \text{satisfying} \quad \tau(x_i) = \xi_i \quad (1 \leqslant i \leqslant n). \tag{10.24g}$$

In fact there is a factorization

$$f(t) = \prod_{i=1}^{n} (t - \eta_i) \tag{10.24h}$$

of f in $\mathbb{C}[t]$ such that $\mathbb{Q}(\eta_1, \ldots, \eta_n)$ is a minimal splitting field of f over $\mathbb{Q}$

and there is the epimorphism

$$\left.\begin{array}{l}\varepsilon: S(f/R_1)\to R_1[\eta_1,\ldots,\eta_n],\\ \varepsilon(x_i)=\eta_i \quad (1\leqslant i\leqslant n),\\ \varepsilon(e)=1,\end{array}\right\} \tag{10.24i}$$

and the epimorphism

$$\left.\begin{array}{l}\psi: R_1[\eta_1,\ldots,\eta_n]\to \mathbb{F}_{p^{\mu}},\\ \psi(\eta_i)=\xi_i \quad (1\leqslant i\leqslant n),\end{array}\right\} \tag{10.24j}$$

so that

$$\tau=\psi\circ\varepsilon. \tag{10.24k}$$

Now we make the further assumption that

$$p\nmid \nu_{J,H}. \tag{10.24l}$$

Then there is the epimorphism $\Phi:\bar{\Lambda}\to S(f/\mathbb{Z})$ such that

$$e(R_1[\eta_1,\ldots,\eta_n])^J=\langle eR_1, e\Phi(X_{J,H})\rangle.$$

Each member y of $X_{J,H}$ is of the form

$$y=\hat{J}(x_1^{\mu_1}x_2^{\mu_2}\cdots x_n^{\mu_n}),$$

hence the roots of the principal polynomial of $\Phi(y)$ of $S(f/\mathbb{C})$ over $\mathbb{C}$ are of the form

$$\sum_{\pi\in J}\eta_{\pi(1)}^{\mu_1}\eta_{\pi(2)}^{\mu_2}\cdots\eta_{\pi(n)}^{\mu_n},$$

and the same is true up to the factor e for the roots of the principal polynomial of $e\Phi(y)$ of $eS(f/\mathbb{C})$ over $e\mathbb{C}$. It follows that the absolute value is bounded by

$$|J|\left(\max_{1\leqslant k\leqslant n}|\eta_k|\right)^{\sum_{i=1}^n\mu_i}$$

On the other hand, we have $\Phi(y)\in\mathbb{F}_p$ because of the assumption (10.24d). We represent the elements of $\mathbb{F}_p$ by the least remainder $R(a,p)$ of the rational integer a mod p. Suppose now that $\varepsilon\Phi y$ is a rational integer. Then we have the rough estimate

$$|\tau\Phi(y)|\leqslant|J|\left(\frac{\phi(f)}{2^{1/n}-1}\right)^{\binom{n}{2}}=:M_1 \tag{10.25a}$$

which can be greatly improved upon by taking into account the special choice of y from $X_{J,H}$ and more precise knowledge of the magnitude of the roots of f in $\mathbb{C}$.

If J contains the group of the equation $f=0$ then one of the $\bar{H}$-conjugates of y must be a rational integer. On the other hand, if J does not contain the group of the equation or any $\bar{H}$-conjugate of it then it follows from the fact that Φ is an H-module isomorphism that no $\bar{H}$-conjugate of $\Phi(y)$ is a rational integer. Setting

$$H = \bigcup_{i=1}^{(H:J)} \pi_i J$$

it follows that for any rational integer b the elements $\varepsilon\bar{\pi}_i\Phi(y) - b$ $(1 \leqslant i \leqslant (H:J))$ are the algebraic conjugates of $\varepsilon\Phi(y) - b$ and that their product is a rational integer which is non-zero in case J does not contain the group of the equation nor any H-conjugate of it.

Similar to (10.25a) we also have the estimate

$$|\varepsilon\bar{\pi}_i\Phi(y)| \leqslant M_1. \tag{10.25b}$$

Hence, if

$$|b| \leqslant M_1, \tag{10.25c}$$

then we have

$$|\varepsilon\bar{\pi}_i\Phi(y) - b| \leqslant 2M_1,$$

$$\left|\prod_{i=1}^{(H:J)} (\varepsilon\bar{\pi}_i\Phi(y) - b)\right| \leqslant (2M_1)^{(H:J)}.$$

Suppose now that

$$\tau\Phi(y) \equiv \tilde{b} \bmod p, \quad \tilde{b} = R(b, p), \quad |b| \leqslant M_1, \tag{10.25d}$$

then we find that

$$\tau\left(\prod_{i=1}^{(H:J)} (\varepsilon\bar{\pi}_i\Phi(y) - \tilde{b})\right) \equiv 0 \bmod p, \quad p \,\Big|\, \prod_{i=1}^{(H:J)} (\varepsilon\bar{\pi}_i\Phi(y) - \tilde{b}).$$

Hence, if J does not contain the group of the equation nor any H-conjugate of it and if (10.25c), (10.25d) hold then we have the inequality

$$p \leqslant (2M_1)^{(H:J)}. \tag{10.25e}$$

As an example let us study *W. Trinks' equation* [14]

$$x^7 - 7x + 3 = 0.$$

The monic polynomial

$$f(t) = t^7 - 7t + 3 = \prod_{i=1}^{7} (t - \eta_i)$$

is irreducible over $\mathbb{F}_2$ since f is not divisible by the irreducible polynomials t, $t+1$, t^2+t+1, t^3+t+1, t^3+t^2+1 of $\mathbb{F}_2[t]$ of degree $\leqslant 3$. Hence f is irreducible over $\mathbb{F}_2$. The discriminant of the trinomial f is

$$d(f) = 6^6 \cdot 7^7 - 7^7 3^6 = 3^8 \cdot 7^8 = (3^4 \cdot 7^4)^2,$$

hence the group of f is even and contains a 7-cycle. The derivative

$$D_t(f) = 7t^6 - 7$$

has two real zeros viz. 1 and -1. Since $f(1) = -3$, $f(-1) = 9$, $f(+\infty) = +\infty$, $f(-\infty) = -\infty$ it follows that f has precisely three real roots by Sturm's rule, hence f splits over the real number field into the product of three linear factors say $t - \eta_1$, $t - \eta_3$, $t - \eta_7$, and 2 quadratic factors, say $(t - \eta_2)(t - \eta_4)$,

$(t-\eta_5)(t-\eta_6)$ with η_2, η_5 complex non-real and η_4 complex conjugate to η_2, η_6 complex conjugate to η_5. The group of f over $\mathbb{R}$ is generated by the double transposition (24) (56). Hence the group G of f over $\mathbb{Q}$ contains a double transposition. Consulting the table of primitive groups of 7 letters we discover that either $G=\mathfrak{A}_7$ or $G=PSL(3,2)$. And in the latter case G is generated by the double transposition (24) (56) and the 7-cycle $\rho=(1\,234\,567)$. In order to test the latter possibility we set

$$J=PSL(3,2),\quad H=\mathfrak{A}_7,\quad X_{J,H}=\{y\},\quad y=\hat{J}(x_1x_2x_4).$$

We must test whether one of the six complex numbers

$$\zeta=\sum_{i=1}^{7}\eta_{\pi(i)}\eta_{\pi(i+1)}\eta_{\pi(i+3)}\quad(\eta_8=\eta_1,\ \eta_9=\eta_2,\ \eta_{10}=\eta_3;\ \pi \text{ permutes } 1,3,7)$$

is a rational integer. In our case we don't have to seek for larger prime numbers. We simply take good approximations of the complex numbers $\eta_1,\ldots,\eta_7$ and test whether one of the six ζ's approximates a rational integer a well enough such that the product of the 15 H-conjugates:

$$\zeta^{(j)}-a=\left(\sum_{i=1}^{7}\eta_i\eta_{i+h}\eta_{i+k}\right)-a\quad(0<h<k\leqslant 7;\ \zeta^{(1)}=\zeta)$$

is <1 so that $\zeta-a=0$ is obtained. We use the estimate

$$|\zeta^{(j)}+a|\leqslant|a|+7bcd.$$

where b, c, d are the maximum, the next highest and the second highest among the $|\eta_i|$. In our case we find that indeed

$$\prod_{j=1}^{15}|(\zeta^{(j)}-a)|<1,$$

so that $PSL(3,2)$ is the group of Trinks' equation.

Exercises

1. (a) Find approximate values for $\eta_1,\ldots,\eta_7$ and determine natural numbers b, c, d serving as upper estimates for the maximum, the next highest and the second highest among the $|\eta_i|$ of Trinks' equation.
   (b) Compute the six ζ's to the required amount of accuracy and determine which of them is a rational integer.
   (c) Choose a prime number suitable to test the group of Trinks' equation and determine the possible outcomes of $\sum_{i=1}^{7}\eta_i\eta_{i+1}\eta_{i+3}$ modulo p depending on the 6 possible arrangements of $\eta_1\eta_2\eta_4$, assuming already that $(1\,234\,567)\in G=G_{168}$.
   Verify in this way that G is the group of Trinks' equation.
   (d) Which method (a) + (b) or (c) is faster?

2.11. The cyclotomic equation

Any finite group G defines the group ring RG over a given unital commutative ring R with the elements of G as R-basis and the multiplication table of G as basic multiplication rule. The trace of the regular representation (relative to the basis $G=\{g_1,\ldots,g_{|G|}\}$, see (3.25c) and (12.2))

$$M: RG \to R^{|G|\times|G|}: g \mapsto M(g) \tag{11.1}$$

is given by

$$\mathrm{Tr}(g) = \delta_{1,g}|G|, \tag{11.2}$$

so that the discriminant is

$$\begin{aligned} d(RG/R) &= \det(\mathrm{Tr}(g_i g_k)) \\ &= |G|^{|G|}. \end{aligned} \tag{11.3}$$

In particular for finite cyclic groups

$$G = \langle g_0 \rangle = C_{|G|}, \tag{11.4a}$$

we find that

$$RG \underset{R}{\cong} R[t]/(t^{|G|}-1), \tag{11.4b}$$

$$d(RG/R) = d(t^{|G|}-1) = |G|^{|G|}, \tag{11.4c}$$

hence the cyclotomic polynomial

$$t^{|G|}-1 \tag{11.5}$$

is separable over R if and only if $|G|$ is not a zero divisor of R.

The cyclotomic equation

$$x^n - 1 = 0 \quad (n \in \mathbb{Z}^{>0}) \tag{11.6}$$

has been the subject of inquiry since the earliest beginnings of algebra, but has still not revealed all its secrets. Its group over the rational integer ring is abelian. Its importance is demonstrated by L. Kronecker's theorem that any finite abelian extension of the rational number field can be embedded into a suitable cyclotomic extension.

Proposition (11.7)
If the natural number n is not divisible by the characteristic of the field F then the group of the separable equation (11.6) *is abelian.*

Proof
Let $S((t^n-1)/F)$ be the universal splitting ring of the monic separable polynomial t^n-1 and let $x_1,\ldots,x_n$ be the standard root generators of $S((t^n-1)/F)$ so that

$$t^n - 1 = \prod_{i=1}^{n}(t-x_i).$$

Then for any primitive idempotent e we obtain the n distinct projections ex_i $(1 \leqslant i \leqslant n)$ such that

$$t^n - e = \prod_{i=1}^{n} (t - ex_i)$$

and the elements of

$$\bar{G} = \operatorname{Stab}(e/\mathfrak{S}_n)$$

permute the root projections ex_i $(1 \leqslant i \leqslant n)$ generating the minimal splitting field $E = eS((t^n - 1)/F)$ of $t^n - e$ over eF.

We observe that $ex_1, \ldots, ex_n$ are the only roots of $t^n - e$ in E. Since the equations

$$(ex_i)^n = (ex_j)^n = e$$

imply upon multiplication that

$$(ex_i ex_j)^n = ee = e,$$

it follows that the roots of $t^n - e$ in E form a group X of order n.

Since any finite subgroup of the multiplicative group of a field is cyclic it follows that we have after suitable numbering of $x_1, \ldots, x_n$

$$ex_i = (ex_1)^i \quad (1 \leqslant i \leqslant n),$$

so that there holds the multiplicative to additive isomorphism

$$\tau\colon X \to \mathbb{Z}/n\colon ex_i \mapsto i + n\mathbb{Z} \quad (1 \leqslant i \leqslant n).$$

Each permutation automorphism $\bar{\pi} \in \bar{G}$ restricts on X to an automorphism of X. Since X generates E over eF it follows that $\bar{G}$ restricts on X faithfully to a subgroup of

$$\operatorname{Aut}(X) \simeq \operatorname{Aut}(\mathbb{Z}/n) \simeq U(\mathbb{Z}/n).$$

The units of $\mathbb{Z}/n$ form an Abelian group of order $\varphi(n)$ where

$$\varphi\colon \mathbb{Z}^{>0} \to \mathbb{Z}^{>0}\colon n \mapsto \#\{i \in \mathbb{Z}^{>0} \mid 1 \leqslant i \leqslant n, \gcd(i, n) = 1\}$$

is the Euler φ-function, G is isomorphic to the subgroup represented by the integers i_π defined by

$$\bar{\pi}(ex_1) = ex_{i_\pi} \quad (\bar{\pi} \in \bar{G},\ 1 \leqslant i_\pi \leqslant n). \qquad \square$$

Theorem (11.8)

The group of the cyclotomic equation (11.6) *over the rational number field is isomorphic to* $U(\mathbb{Z}/n)$.

Proof

Using the same terms as in the proof of proposition (11.7), but for $F = \mathbb{Q}$, we note that the universal splitting ring $S((t^n - 1)/R_1)$ with $R_1 = \langle \mathbb{Z}, n^{-n} \rangle$ contains the same idempotents as $S((t^n - 1)/\mathbb{Q})$ and that we have an epimorphism

$$\varepsilon: S((t^n - 1)/R_1) \to S((t^n - 1)/(\mathbb{Z}/p)),$$

mapping the standard root generators $x_1, \ldots, x_n$ of $S((t^n - 1)/R_1)$ on the standard root generators $y_1, \ldots, y_n$ of $S((t^n - 1)/(\mathbb{Z}/p))$ for any prime number p not dividing n. It maps the primitive idempotent e of $S((t^n - 1)/R_1)$ on some idempotent $\varepsilon(e)$ of $S((t^n - 1)/(\mathbb{Z}/p))$ such that

$$ex_i = (ex_1)^i,$$
$$\varepsilon(ex_i) = \varepsilon(e)y_i = (\varepsilon(e)y_1)^i \quad (1 \leqslant i \leqslant n),$$
$$\operatorname{Stab}(\varepsilon(e)/\bar{\mathfrak{S}}_n) = \operatorname{Stab}(e/\bar{\mathfrak{S}}_n) = \bar{G},$$

where G is the group of the equation (1.6) over $\mathbb{Q}$. There is a primitive idempotent e' of $S((t^n - 1)/(\mathbb{Z}/p))$ for which $e'\varepsilon(e) = e'$, and the projections

$$e'y_i = e'ey_i = e'ey_1^i = e'y_1^i \quad (1 \leqslant i \leqslant n)$$

are distinct, generating a minimal splitting field $\mathbb{F}_{p^\mu}$ of the equation $x^n - e' = 0$ over $e'\mathbb{F}_p$. The group of the equation $x^n - e = 0$ is generated by a permutation automorphism π which leaves e' fixed and raises every element $e'y_i$ to its pth power:

$$\bar{\pi}(e'y_i) = (e'y_i)^p = e'y_{pi} \quad (1 \leqslant i \leqslant n, y_j = y_k \ \text{ for } j \equiv k \bmod n).$$

Hence G contains the permutation automorphisms $\bar{\pi}_p$ mapping x_i on x_{pi} (p any prime number not dividing n) subject to

$$x_j = x_k \quad \text{for } j \equiv k \bmod n.$$

Using prime factorization of any natural number m that is prime to n it follows that G contains the permutation mapping x_i on x_{mi} $(1 \leqslant i \leqslant n)$. Hence $G \simeq U(\mathbb{Z}/n)$. □

It follows from theorem (11.8) that the minimal splitting field E_n of $t^n - 1$ over $\mathbb{Q}$ is an Abelian extension of degree $\varphi(n)$ over $\mathbb{Q}$ with automorphism group isomorphic to $U(\mathbb{Z}/n)$. It is generated by the n roots $\zeta_n, \zeta_n^2, \ldots, \zeta_n^n$ of $t^n - 1$ where ζ_n is a primitive nth root of unity. Any one of the $\varphi(n)$ powers

$$\zeta_n^i \quad (1 \leqslant i \leqslant n, \gcd(i, n) = 1)$$

is also a primitive nth root of unity. They form a conjugacy class under the automorphism group $\operatorname{Aut}(E_n/\mathbb{Q})$.

Since the complex numbers $\zeta_n, \zeta_n^2, \ldots, \zeta_n^n = 1$ form the vertices of a regular n-gon with the origin as center in the complex plane we speak of E_n as the nth *cyclotomic field* and of the polynomial

$$\phi_n(t) = \prod_{\substack{1 \leqslant i \leqslant n \\ \gcd(i,n) = 1}} (t - \zeta_n^i)$$

as the nth *cyclotomic polynomial*. Its coefficients are fixed by the automorphisms of E_n over $\mathbb{Q}$, hence they are rational. On the other hand, the monic polynomial $\phi_n(t)$ divides $t^n - 1$, hence the coefficients of $\phi_n(t)$ are rational integers. According to theorem (11.8) the polynomial $\phi_n(t)$ is irreducible of degree $\varphi(n)$.

We have $E_n = E_{2n}$ if n is odd, since in that case the roots of $\phi_{2n}(t)$ differ at most by sign from the roots of $\phi_n(t)$, But there are no other equalities between cyclotomic fields. Indeed, if $E_n = E_m$ where n, m are two natural numbers satisfying $n < m$ then E_n contains the products of the nth roots of unity and the mth roots of unity which are simply the lcm(n, m)-th roots of unity so that $E_n = E_{\text{lcm}(n,m)}$,

$$\varphi(n) = \varphi(m) = \varphi(\text{lcm}(n, m)), \quad m = 2n, n \equiv 1 \bmod 2.$$

There hold the factorizations

$$t^n - 1 = \prod_{i=1}^{n} (t - \zeta_n^i) = \prod_{d|n} \phi_d(t). \tag{11.9a}$$

Using the Moebius inversion formula (5.14b) we find that

$$\phi_n(t) = \prod_{d|n} (t^d - 1)^{\mu(n/d)}. \tag{11.9b}$$

In order to find the discriminant of the nth cyclotomic polynomial we make use of the following lemma.

Lemma (11.10)

Let $n > 1$ be a natural number. Let

$$\Lambda = \mathbb{Z}[t]/\phi_n(t) = \mathbb{Z}[\zeta_n] \ (\zeta_n = t/(t^n - 1))$$

be the equation ring of the nth *cyclotomic polynomial. Then the following elements are units of Λ:*

$$1 - \zeta_n^i \quad (1 \leqslant i < n, n/\gcd(i, n) \textit{ is not a prime power}), \tag{11.11a}$$

$$\frac{1 - \zeta_n^{in/p^\nu}}{1 - \zeta_n^{jn/p^\nu}} \quad (\nu \in \mathbb{Z}^{>0}, p \textit{ a prime number}, p^\nu | n, 1 < i < p^\nu,$$

$$p \nmid i, 1 < j < p^\nu, p \nmid j, i \neq j). \tag{11.11b}$$

Those units are said to be the **cyclotomic units.**

Proof

For any prime number p and any natural number ν we have

$$\frac{t^{p^\nu} - 1}{t - 1} = t^{p^\nu - 1} + t^{p^\nu - 2} + \cdots + t + 1$$

$$= \prod_{i=1}^{\nu} \phi_{p^i}(t).$$

Substituting 1 for t we obtain

$$p^{\nu} = \prod_{i=1}^{\nu} \phi_{p^i}(1),$$

hence (by induction on ν)

$$p = \phi_{p^i}(1) = \prod_{|\langle\zeta\rangle| = p^i} (1 - \zeta) \quad (i = 1, 2, \ldots). \tag{11.11c}$$

Suppose

$$n = \prod_{i=1}^{s} p_i^{\nu_i} \quad (s \in \mathbb{Z}^{>0},\ \nu_i \in \mathbb{Z}^{>0},\ 1 \leqslant i \leqslant s)$$

is a factorization of n into the power product of the distinct prime numbers $p_1, \ldots, p_s$. Then we have

$$\begin{aligned} \frac{t^n - 1}{t - 1} &= t^{n-1} + t^{n-2} + \cdots + t + 1 \\ &= \prod_{1 < d | n} \phi_d(t), \end{aligned}$$

and on the other hand

$$n = \prod_{i=1}^{s} \prod_{j=1}^{\nu_i} \phi_{p_i^j}(1)$$

by (11.11c).

Hence $|\phi_d(1)| = 1$ if d is a natural number which is not a prime power, but divides n. On the other hand,

$$\begin{aligned} \phi_d(t) &= \prod_{\substack{1 \leqslant j < d \\ \gcd(j,d) = 1}} (t - \zeta_n^{jn/d}), \\ \phi_d(1) &= \prod_{\substack{1 \leqslant j < d \\ \gcd(j,d) = 1}} (1 - \zeta_n^{jn/d}), \end{aligned}$$

hence (11.11a) is a unit (set $d = n/\gcd(i, n)$, $j = i/\gcd(i, n)$).

In order to show the unit nature of (11.11b) we show that numerator and denominator divide one another (as elements of Λ). But in our case $\zeta_n^{inp^{-\nu}}$ and $\zeta_n^{jnp^{-\nu}}$ are powers of each other, so that the quotients in question can be evaluated as geometrical series. □

Proposition (11.12)

The discriminants of the cyclotomic polynomials $\phi_n(t)$ are given by (11.12a–c):

$$d(\phi_1) = d(\phi_2) = 1, \tag{11.12a}$$

$$d(\phi_{p^k}) = (-1)^{\varphi(p^k)/2} p^{\varphi(p^k)(k - 1/(p-1))} \quad (p \in \mathbb{P},\ k \in \mathbb{N},\ p^k > 2), \tag{11.12b}$$

$$d(\phi_n) = (-1)^{\varphi(n)/2} \left(\prod_{p | n} p^{-\varphi(n)/(p-1)} \right) n^{\varphi(n)} \quad (n \in \mathbb{N},\ n > 2). \tag{11.12c}$$

Proof

(11.12a) is trivial. To show (11.12b) we note that $d(\phi_{p^k})$ is a rational integer.

Hence it suffices to compute its absolute value and its sign. Let ζ be a primitive p^kth root of unity.

$$|d(\phi_{p^k})| = \left| \prod_{i=0}^{p^{k}-1} \prod_{j=1}^{p-1} \prod_{\substack{\mu=0 \\ (i,j)\neq(\mu,\nu)}}^{p^{k}-1} \prod_{\nu=1}^{p-1} (\zeta^{ip+j} - \zeta^{\mu p+\nu}) \right|$$

$$= \left| \prod_{i=0}^{p^{k-1}-2} \prod_{j=1}^{p-1} \prod_{\mu=i+1}^{p^{k-1}-1} (1 - \zeta^{(\mu-i)p}) \right|^2$$

$$\times \left| \prod_{i=0}^{p^{k-1}-1} \prod_{j=1}^{p-1} \prod_{\mu=i}^{p^{k-1}-1} \prod_{\substack{\nu=1 \\ \nu\neq j \\ \nu>j \text{ for } i=\mu}}^{p-1} (1 - \zeta^{(\mu-i)p+(\nu-j)}) \right|^2$$

$$= \left| \prod_{\lambda=1}^{p^{k-1}-1} \prod_{\tau=1}^{p^{k-1}-\lambda} (1 - \zeta^{p\lambda}) \right|^{2(p-1)} \left| \prod_{\lambda=0}^{p^{k-1}-1} \prod_{\rho=1}^{p-1} (1 - \zeta^{\lambda p+\rho}) \right|^{(p-2)p^{k-1}}$$

$$= \left| \prod_{\lambda=1}^{p^{k-1}-1} (1 - \zeta^{p\lambda})^{p^{k-1}} \right|^{p-1} |\phi_{p^k}(1)|^{(p-2)p^{k-1}}$$

$$= \left(\left[\frac{x^{p^{k-1}} - 1}{x - 1} \right]_{x=1} \right)^{(p-1)p^{k-1}} p^{\varphi(p^k)(p-2)/(p-1)}$$

$$= p^{\varphi(p^k)(k-1/(p-1))}$$

(see (11.11c)). For the computation of the sign of

$$d(\phi_n) = (-1)^{(\varphi(n)2)} \prod_{|\langle\zeta\rangle| = |\langle\zeta'\rangle| = n, \zeta\neq\zeta'} (\zeta - \zeta')$$

in case $n = p^k$ we proceed as follows.

The factor $\zeta - \zeta'$ is real if and only if $\zeta\zeta' = -1$. In that case n is even,

$$\zeta - \zeta' = \zeta + \zeta^{-1} = -(\zeta' - \zeta)$$

so that the partial product $(\zeta - \zeta')(\zeta' - \zeta)$ is negative. The number of such partial products is $\varphi(n)/2$, the sign of the product of the partial products is $(-1)^{\varphi(n)/2}$. In all other cases $\zeta - \zeta'$ is complex but not real, so that it is distinct from $\zeta^{-1} - (\zeta')^{-1}$ and $(\zeta - \zeta')(\zeta^{-1} - (\zeta')^{-1})$ is positive real. It follows that

$$\operatorname{sign}(d(\phi_n)) = (-1)^{\varphi(n)/2} \quad \text{for } n = p^k > 2.$$

Hence, (11.12b) is proven.

Finally, (11.12c) is a consequence of (11.12b) and (9.29). □

Exercises

1. Derive (11.9b) from (11.9a).
2. Let $d, n \in \mathbb{N}$ subject to $d|n$. Prove:

$$\phi_d(t) = \prod_{\substack{1\leq i<d \\ \gcd(i,d)=1}} (t - \zeta_n^{in/d}).$$

3. Carry out the necessary computations for proving (11.12) in detail.

2.12. Normal bases

1. Introduction

It was stated in chapter 1, section 5 already that the Galois correspondence between the intermediary fields of a normal extension E of the field F and the subgroups of the group

$$G = \mathrm{Aut}(E/F) \tag{12.1}$$

is constructively set in evidence by means of a *normal basis*, i.e. an F-basis B of E consisting of the G-conjugates of a single element x of E. The existence of such a basis and its construction is the topic of this section.

2. Analysis of the problem

The action of G on the members of a normal basis B is the regular permutation action:

$$g(i(x)) = gi(x) \quad (g, i \in G). \tag{12.2a}$$

We have seen already before that the action of an automorphism of E over F is a linear transformation of E over F. With respect to the F-basis

$$i(x) \quad (i \in G)$$

we associate the matrix

$$P(g) = (\delta_{i,gk}) \quad (i, k \in G) \tag{12.2b}$$

with the action of G on a normal basis.

We observe that we obtain the same action for left-multiplication of g on the basis formed by the elements of G of the group algebra FG of G over F, thus P is the *left-regular matrix representation of* G with respect to the basis G of FG, or the *Cayley matrix representation of* G *over* F.

We have taken here the elements of G as row and column indices. If one prefers natural numbers for indices then we must number the elements of G as

$$g_1, g_2, \ldots, g_n \quad (n = |G|) \tag{12.2c}$$

and set

$$gg_i = g_{\pi(g)(i)} \tag{12.2d}$$

where $\pi(g)$ is a permutation of $1, 2, \ldots, n$ derived from the multiplication table. In this case the Cayley matrix representation is derived from the Cayley permutation representation upon transition from a permutation π of $(1, 2, \ldots, n)$ to the corresponding permutation matrix $(\delta_{i,\pi(k)})$.

On the other hand, if we start with an arbitrary F-basis

$$B' = \{b'_1, b'_2, \ldots, b'_n\} \tag{12.3a}$$

of E then the action of G on B' is defined by means of linear equations of the form

$$g(b_i') = \sum_{k=1}^{n} \gamma_{ik}'(g) b_k' \quad (1 \leqslant i \leqslant n) \quad (\gamma_{ik}'(g) \in F; i, k = 1, 2, \ldots, n; g \in G), \tag{12.3b}$$

where the matrices

$$\Gamma'(g) = (\gamma_{ik}'(g)) \in F^{n \times n} \quad (g \in G) \tag{12.3c}$$

of degree n over F provide a *regular matrix representation*

$$\Gamma': G \to F^{n \times n}, \tag{12.3d}$$

i.e. a certain homomorphism Γ' of G into the $n \times n$ matrices with entries in F.

The task is to choose a suitable new basis

$$B = \{b_1, b_2, \ldots, b_n\}, \tag{12.4a}$$

subject to

$$b_i = \sum_{k=1}^{n} \xi_{ik} b_k', \tag{12.4b}$$

$$b_i' = \sum_{k=1}^{n} \eta_{ik} b_k \quad (1 \leqslant i \leqslant n), \tag{12.4c}$$

and with

$$X = (\xi_{ik}), \quad Y = (\eta_{ik}) \in F^{n \times n}, \tag{12.4d}$$

$$XY = YX = I_n, \tag{12.4e}$$

$$g(b_i) = \sum_{k=1}^{n} \gamma_{ik}(g) b_k, \tag{12.4f}$$

such that the new representation

$$\Gamma: G \to F^{n \times n}: g \mapsto X \Gamma'(g) X^{-1} \tag{12.4g}$$

arises and the action of G on $b_1, b_2, \ldots, b_n$ becomes a mere permutation of the basis elements in accordance with the regular permutation representation.

In order to carry out our task let us discuss matrix representations of groups and rings in greater generality and apply the concepts and statements obtained to our task.

3. Matrix representations

Definition (12.5)

Given a unital ring R and an R-ring Λ, then any homomorphism

$$\Delta: \Lambda \to R^{f \times f} \quad (f \in \mathbb{Z}^{>0}) \tag{12.6a}$$

of Λ into the ring of all matrices $R^{f \times f}$ of degree f over R with the customary rules of matrix calculus is said to be a **matrix representation** *of Λ of degree f over R. It* *is said to be* **faithful** *if the kernel of the representation*

$$\ker \Delta = \{x \in \Lambda \mid \Delta(x) = 0\} \text{ is zero.} \tag{12.6b}$$

This is because

$$\ker \Delta = 0 \tag{12.6c}$$

is the necessary and sufficient condition for Δ of (12.6a) to be a monomorphism. Two matrix representations of Λ over R, say (12.6a) and

$$\Delta' \colon \Lambda \to R^{f' \times f'} \quad (f' \in \mathbb{Z}^{>0}) \tag{12.6d}$$

are said to be *equivalent* over R if $f = f'$ and if there is an element

$$X \in U(R^{f \times f}),$$

for which

$$\Delta'(x) = X^{-1} \Delta(x) X \quad (x \in \Lambda) \tag{12.6e}$$

or, briefly,

$$\Delta' = X^{-1} \Delta X. \tag{12.6f}$$

The unit group $U(R^{f \times f})$ of $R^{f \times f}$ is said to be the *general linear group of degree* f over R:

$$U(R^{f \times f}) = GL(f, R).$$

From the multiplicative property of determinants it follows that the matrix equation

$$XY = I_f \quad (X, Y \in R^{f \times f}) \tag{12.6}$$

implies the determinantal relation

$$\det X \cdot \det Y = 1,$$

so that the determinant of X is a unit of R, hence

$$YX = I_f, \quad Y = X^{-1}, \quad X \in U(R^{f \times f}). \tag{12.6h}$$

Conversely, if the determinant of a matrix X of degree f over R is a unit of R then (12.6g) is solvable in $R^{f \times f}$ by a matrix Y of degree f over R, X is in $U(R^{f \times f})$, (12.6h) holds with Y uniquely determined. For this reason the elements of $GL(f, R)$ are said to be the *unimodular matrices* of degree f over R. They are said to be *proper* if their determinant is 1. The proper unimodular matrices form a normal subgroup $SL(f, R)$ of $GL(f, R)$ with representative subgroup $R(f, R)$ formed by the special diagonal matrices

$$\operatorname{diag}(\varepsilon, 1, \ldots, 1) = \begin{pmatrix} \varepsilon & & & 0 \\ & 1 & & \\ & & \ddots & \\ 0 & & & 1 \end{pmatrix} \quad (\varepsilon \in U(R)).$$

In other words $GL(f, R)$ is a semidirect product of the normal subgroup

$SL(f, R)$ and the representative subgroup $R(f, R)$:

$$GL(f, R) = SL(f, R) \rtimes R(f, R). \tag{12.6i}$$

The normal subgroup

$$SL(f, R) = \{X \in R^{f \times f} | \det(X) = 1\}$$

of $GL(f, R)$ is said to be the *special linear group of degree f over R.*

The equivalence between two matrix representations Δ, Δ' is denoted by

$$\Delta \sim \Delta'.$$

This relation is reflexive, symmetric and transitive so that the matrix representations of Λ over R are partitioned into equivalence classes.

Definition (12.7)
The Λ-module M is said to be a **representation module** *for the matrix representations* (12.5) *if it is an R-module with an R-basis $u_1, u_2, \ldots, u_f$ such that*

$$xu_k = \sum_{i=1}^{f} \xi_{ik}(x) u_i \quad (x \in \Lambda; 1 \leqslant k \leqslant f), \tag{12.8a}$$

where

$$\xi_{ik}(x) \in R \quad (i, k = 1, 2, \ldots, f) \tag{12.8b}$$

and

$$\Delta(x) = (\xi_{ik}(x)). \tag{12.8c}$$

For example, the f-column module $R^{f \times 1}$ turns into a representation module of (12.6) *if we define $R^{f \times 1}$ as a Λ-module via*

$$xu = \Delta(x)u \quad (x \in \Lambda, u \in R^{f \times 1}). \tag{12.8d}$$

The corresponding R-basis is formed by the f unit columns

$$\mathbf{e}_k \quad (1 \leqslant k \leqslant f) \tag{12.8e}$$

which will again be used in chapter 3, *section* 2.

The equivalence of two matrix representations Δ, Δ' with representation modules M, M', respectively, amounts to the same as the existence of an operator-isomorphism

$$\varphi: M \to M' \tag{12.8f}$$

of M on M', i.e. a module-isomorphism satisfying the additional conditions

$$\varphi(\lambda u) = \lambda\varphi(u), \quad \varphi(xu) = x\varphi(u) \quad (\lambda \in R, x \in \Lambda, u \in M). \tag{12.8g}$$

We observe that a representation module M of Λ over R is an R-Λ-module, i.e. a unital R-module as well as a Λ-module with the additional rule

$$(\lambda x)u = \lambda(xu) \quad (\lambda \in R, x \in \Lambda, u \in M). \tag{12.8h}$$

The converse need not happen.

However, if R is a field, Λ is a unital R-module, M is a finite dimensional

R-module, and M is an R–M-module then there is a finite R-basis of M and M is a matrix representation module of Λ over R. In that case we speak of M as *representation space* of Λ over R.

From the constructive viewpoint the existence of an R-basis of M is a very powerful aid. In modern research, under the leadership of O. Taussky, this aspect of representation theory has prevailed in most applications.

Definition (12.9)
A homomorphism

$$\Gamma: G \to R^{f \times f} \quad (f \in \mathbb{Z}^{>0}) \tag{12.10a}$$

of a semigroup G into the multiplicative semigroup of the ring of $f \times f$ matrices over a unital commutative ring R is said to be a **matrix representation of degree** *f over R.*

Thus the mapping (12.10a) is characterized as matrix representation by the condition

$$\Gamma(xy) = \Gamma(x)\Gamma(y) \quad (x, y \in G). \tag{12.10b}$$

The new representation concept turns out to be an extension of the previous one upon introduction of semigroup rings extending the group ring concept of section 11.

Definition (12.11)
The **semigroup ring** *of a semigroup G over a unital commutative ring R is defined as the R-ring RG with G as R-basis and the multiplication table of G as basic multiplication rule.*

Any matrix representation (12.10a) *of G gives rise to the matrix representation*

$$\Delta:\ RG \to R^{f \times f}:\ \sum_{g \in G} \lambda(g) g \mapsto \sum_{g \in G} \lambda(g)\Gamma(g) \tag{12.12a}$$

of RG over R by linear expansion of Γ. (Here $\lambda: G \to R$ denotes a mapping with $\lambda(g) = 0$ for all but a finite number of $g \in G$.)

Conversely, every matrix representation

$$\Delta:\ RG \to R^{f \times f} \quad (f \in \mathbb{Z}^{>0}) \tag{12.12b}$$

yields a matrix representation (12.10a) *of G via restriction*

$$\Gamma:\ G \to R^{f \times f}:\ g \mapsto \Delta(g), \tag{12.12c}$$

such that (12.12a) *holds.*

From this reason matrix representations of semigroups and semigroup rings will be used interchangeably. However, it may happen that Γ is faithful and Δ is not. For example, if G is the cyclic group $C_4 = \langle g \rangle$ of order 4 and

R is a ring with characteristic not 2, then the group representation

$$\Gamma\colon C_4 \to R^{2\times 2}\colon g^k \mapsto \begin{pmatrix} 0 & -1 \\ 1 & 0 \end{pmatrix}^k \quad (0 \leqslant k \leqslant 3)$$

is faithful, but the corresponding matrix representation of RG is not faithful because of

$$\Delta(g+g^3)=0 \quad \text{and} \quad g+g^3 \neq 0.$$

Equivalence of matrix representations amounts to the same concept:

$$\begin{aligned} \Gamma \sim \Gamma' &\Leftrightarrow \exists X \in GL(f,R)\colon \Gamma' = X^{-1}\Gamma X \\ &\Leftrightarrow \exists X \in GL(f,R)\colon \Delta' = X^{-1}\Delta X \Leftrightarrow \Delta \sim \Delta'. \end{aligned}$$

Definition (12.13)

The **sum of two matrix representations**

$$\Delta_i\colon \Lambda \to R^{f_i \times f_i} \quad (i=1,2) \tag{12.14a}$$

of the R-ring Λ of finite degree over R is defined by block diagonal composition

$$\Delta_1 \oplus \Delta_2\colon \Lambda \to R^{(f_1+f_2)\times(f_1+f_2)}\colon x \mapsto \begin{pmatrix} \Delta_1(x) & 0 \\ 0 & \Delta_2(x) \end{pmatrix}. \tag{12.14b}$$

The representation conditions for $\Delta_1 \oplus \Delta_2$ are consequences of the following rules for block diagonal matrix composition

$$A_1 \oplus A_2 = \begin{pmatrix} A_1 & 0_{f_1 \times g_2} \\ 0_{f_2 \times g_1} & A_2 \end{pmatrix} \quad (A_i \in R^{f_i \times g_i}, f_i, g_i \in \mathbb{Z}^{>0};\ i=1,2), \tag{12.14c}$$

$$\left.\begin{aligned} &A_1 \oplus A_2 + B_1 \oplus B_2 = (A_1+B_1) \oplus (A_2+B_2), \\ &\lambda(A_1 \oplus A_2) = \lambda A_1 \oplus \lambda A_2 \ (A_i, B_i \in R^{f_i \times g_i}, f_i, g_i \in \mathbb{Z}^{>0}, \lambda \in R;\ i=1,2), \end{aligned}\right\} \tag{12.14d}$$

$$(A_1 \oplus A_2)(B_1 \oplus B_2) = A_1B_1 \oplus A_2B_2 \ (A_i, B_i \in R^{f_i \times f_i}, f_i \in \mathbb{Z}^{>0};\ i=1,2). \tag{12.14e}$$

If M_i is a matrix representation module for the matrix representation (12.14a) of Λ over R, then $M_1 \oplus M_2$ is a matrix representation module for $\Delta_1 \oplus \Delta_2$. This remark sets in evidence that the addition of matrix representations satisfies the substitutional law, the commutative law, and the associative law up to equivalence:

$$\Delta_i \sim \Delta_i' \quad (i=1,2) \Rightarrow \Delta_1 \oplus \Delta_2 \sim \Delta_1' \oplus \Delta_2', \tag{12.15a}$$

$$\Delta_1 \oplus \Delta_2 \sim \Delta_2 \oplus \Delta_1, \tag{12.15b}$$

$$\Delta_1 \oplus (\Delta_2 \oplus \Delta_3) = (\Delta_1 \oplus \Delta_2) \oplus \Delta_3. \tag{12.15c}$$

For any unital R-ring Λ we distinguish *proper*, *improper*, and *null* matrix representations Δ depending on whether

$$\Delta(1_\Lambda) = I_f, \quad \Delta(1_\Lambda) \neq I_f, \quad \Delta(1_\Lambda) = O_f.$$

In the latter case we will have $\Delta(x) = \Delta(x1_\Lambda) = \Delta(x)\Delta(1_\Lambda) = O_f$. If R is a field then any improper matrix representation Δ which is not null is equivalent to the sum of a proper and a null representation. This is because of the Peirce decomposition of any representation space M of Δ:

$$M = 1M \dot{+} \{x - 1x | x \in M\}.$$

4. Permutation representations and matrix representations

The permutation representations

$$\Delta: G \to \mathfrak{S}_n \tag{12.16a}$$

of a group G by permutations of n letters give rise to matrix representations Δ of degree n over any given unital commutative ring R upon representing $\mathfrak{S}_n$ faithfully by *permutation matrices* of degree n

$$\Gamma: \mathfrak{S}_n \to R^{n \times n}: \pi \mapsto (\delta_{i\pi(j)}), \tag{12.16b}$$

defining the action of the permutation π on the standard basis of $R^{n \times 1}$ by means of permuting the indices j of the basis vectors $\mathbf{e}_j$ $(1 \leqslant j \leqslant n)$. We note that $\Gamma\Delta$ is proper. The sum of two permutation representations corresponds to the sum of the permutation matrix representations. If the two permutation representations Δ_1, Δ_2 of G of degree n are equivalent, then the corresponding permutation matrix representations $\Gamma\Delta_1, \Gamma\Delta_2$ are also equivalent. The converse need not happen (see exercise 1).

For any matrix representation (12.6a) of the R-ring Λ of finite degree f over the unital commutative ring R and for any unital overring R' of R we obtain the matrix representation $1_{R'} \otimes_R \Delta$ of $R' \otimes_R \Lambda$ of degree f over R':

$$1_{R'} \otimes_R \Delta\left(\sum_{i=1}^{s} x_i \otimes y_i \right) = \sum_{i=1}^{s} x_i \Delta(y_i) \quad (s \in \mathbb{Z}^{>0}; x_i \in R', y_i \in \Lambda; 1 \leqslant i \leqslant s). \tag{12.17a}$$

If M is a matrix representation module of Λ, then $R' \otimes_R M$ is a matrix representation module of $1_{R'} \otimes_R \Lambda$. On the other hand, if

$$\Delta': R' \otimes_R \Lambda \to R'^{f \times f} \tag{12.17b}$$

is a matrix representation of $R' \otimes_R \Lambda$ of degree f over R', then, of course, its restriction to $1_{R'} \otimes_R \Lambda$ defines the R-homomorphism

$$\Delta'|_\Lambda: \Lambda \to R'^{f \times f}: x \mapsto \Delta'(1_{R'} \otimes_R x) \tag{12.17c}$$

of Λ into $R'^{f \times f}$, and any such R-homomorphism Δ determines uniquely a matrix representation

$$\Delta' = 1_{R'} \otimes_R \Delta \tag{12.17d}$$

of $R' \otimes_R \Lambda$ of degree f over R' in accordance with (12.17a). By abuse of language we frequently speak of Δ as a *matrix representation of* Λ *of degree*

f *over* R', even though it may be impossible to make Δ equivalent to a matrix representation of degree f over R (see exercise 2).

The equivalence of two matrix representations

$$\Delta_i : \Lambda \to R^{f \times f} \quad (i = 1, 2) \tag{12.18a}$$

of the R-ring Λ of degree f over Λ is tantamount to the existence of a unimodular matrix X of $R^{f \times f}$ satisfying the equation

$$\Delta_2(x) = X^{-1}\Delta_1(x)X \quad (x \in \Lambda),$$

which is equivalent to

$$X\Delta_2(x) = \Delta_1(x)X \quad (x \in \Lambda). \tag{12.18b}$$

Dropping the unimodularity condition we realize that (12.18b) amounts to a system of linear homogeneous equations for the entries of the matrix X.

For any two matrix representations

$$\Delta_i : \Lambda \to R^{f_i \times f_i} \quad (i = 1, 2) \tag{12.18c}$$

of Λ of degree f_i over R the rectangular matrices $X \in R^{f_1 \times f_2}$ satisfying (12.18b) form an R-module $C(\Delta_1, \Delta_2)$ of $R^{f_1 \times f_2}$ which is said to be the *connecting R-module* of Δ_1, Δ_2. The two matrix representations (12.18c) are equivalent precisely if $f_1 = f_2$ and the connecting R-module $C(\Delta_1, \Delta_2)$ contains a unimodular matrix. The connecting R-module $C(\Delta, \Delta)$ of the matrix representation (12.5) is an R-ring which is said to be the *centralizer* of Δ. It contains I_f as unit element.

For any two matrix representations (12.18c) we have

$$C(\Delta_1, \Delta_1)C(\Delta_1, \Delta_2) = C(\Delta_1, \Delta_2) = C(\Delta_1, \Delta_2)C(\Delta_2, \Delta_2), \tag{12.18d}$$

and we note that

$$C(X_1^{-1}\Delta_1 X_1, X_2^{-1}\Delta_2 X_2) = X_1^{-1}C(\Delta_1, \Delta_2)X_2 \quad (X_i \in GL(f_i, R); i = 1, 2). \tag{12.18e}$$

If R is a field then we have

$$C(1_{R'} \otimes_R \Delta_1, 1_{R'} \otimes_R \Delta_2) = R'C(\Delta_1, \Delta_2) \simeq R' \otimes_R C(\Delta_1, \Delta_2) \tag{12.18f}$$

for any extension R' of R. In that case $C(\Delta_1, \Delta_2)$ has an R-basis $X_1, \ldots, X_k$ ($k = k(C(\Delta_1, \Delta_2))$) of at most $f_1 f_2$ elements.

Let us now suppose $f_1 = f_2$. Then we form the polynomial

$$P(t_1, t_2, \ldots, t_k) = \det\left(\sum_{i=1}^{k} t_i X_i\right) \in R[t_1, \ldots, t_k]. \tag{12.18g}$$

If it is zero, then the two representations

$$1_{R'} \otimes_R \Delta_i \quad (i = 1, 2) \tag{12.18h}$$

are not equivalent for any extension R' of R because of (12.18f). On the other hand, for $P \neq 0$ the representations (12.18h) can be equivalent in some extension of R, for example in $R(t_1, t_2, \ldots, t_k)$. As a matter of fact we have

Theorem (12.19)
(Deuring-Noether). *If* $1_{R'} \otimes_R \Delta_1 \sim 1_{R'} \otimes_R \Delta_2$ *for some extension* R' *of* R, *then* $\Delta_1 \sim \Delta_2$.

A proof will be given later on for arbitrary fields R and extensions R'. In the case of R being an infinite field (12.19) is a consequence of

Lemma (12.20)
For any non-zero polynomial $P(t_1, \ldots, t_k)$ *in* k *variables* $t_1, \ldots, t_k$ *over an infinite field* R *there is a non-zero specialization.*

Proof
By induction on k. For $k=1$ we choose $1+\deg(P)$ distinct elements $\xi_0, \xi_1, \ldots, \xi_{\deg(P)}$ of R. Then clearly $P(\xi_i) \neq 0$ for some index i ($0 \leqslant i \leqslant \deg(P)$) since P has at most $\deg(P)$ zeros. For $k>1$ we thus obtain a specialization $t_k \mapsto \lambda_k \in R$ such that $P(t_1, \ldots, t_{k-1}, \lambda_k)$ is not zero. Then we apply the induction hypothesis. □

If R is finite and we take theorem (12.19) for granted, then we know that (12.18g) is not zero and that there is at least one non-zero specialization in R, hence we can simply operate by trial and error to actually find a non-singular matrix transforming Δ_1 into Δ_2 over R.

5. Construction of normal bases

Given a normal extension E with automorphism group G over the field of reference F we have seen already in section 6 that the tensor product ring $A = \bar{E} \otimes_F E$ of an extension $\bar{E}$ of F isomorphic to E over F acts as representation space for the regular representation of G inasmuch as the primitive idempotents of A form an $\bar{E}$-basis of A with regular permutation action Γ of the group $\bar{G} = 1_E \otimes_F G \neq G$.

On the other hand, of course E itself is a representation space of G of degree $n = |G|$ over F for some matrix representation $\Delta: G \to F^{n \times n}$. Also Δ can be interpreted as a matrix representation of G over $\bar{E}$ with A as representation space. The two are known to be equivalent over $\bar{E}$. Hence they are equivalent over F. It follows that there is a normal basis $B = \{g(x) | g \in G\}$ of E over F which can be found upon suitable specialization of the generic solution of a system of linear homogeneous equations.

For every subgroup S of G we obtain the left coset decomposition

$$G = \bigcup_{i=1}^{(G:S)} Sg_i, \tag{12.21a}$$

corresponding to the fix elements

$$\theta_{iS} := \sum_{s\in S} sg_i(x) \quad (1 \leqslant i \leqslant (G:S)), \tag{12.21b}$$

of S which are said to be the *Gauss periods* of B with respect to S. All fix elements of S form an intermediate field E^S with (12.21b) as F-basis:

$$E^S = \{y\in E \mid \forall g\in S: g(y) = y\} = \sum_{i=1}^{(G:S)} F\theta_{iS}. \tag{12.21c}$$

Indeed, the θ_i's are linearly independent over F since they are non-overlapping F-linear combinations of B. Their number equals

$$(G:S) = dim_F E^S. \tag{12.21d}$$

Each of the Gauss periods generates E^S over F since there holds a right coset decomposition

$$G = \bigcup_{i=1}^{(G:S)} g_i'S. \tag{12.21e}$$

The G-conjugates

$$g_j'\theta_{iS} \quad (1 \leqslant j \leqslant (G:S)) \tag{12.21f}$$

are F-linearly independent because they are non-overlapping F-linear combinations of B, hence they are distinct. Thus the Gauss periods turn out to be primitive elements of E^S over F. They are G-conjugates if and only if S is a normal subgroup of G.

This occurs for example for the normal basis of the pth cyclotomic extension

$$E_p := \mathbb{Q}(\zeta_p), \quad \zeta_p^{p-1} + \cdots + \zeta_p + 1 = 0 \tag{12.22a}$$

of $\mathbb{Q}$, which is generated by $x = \zeta_p$ and consists of

$$\zeta_p, \zeta_p^2, \ldots, \zeta_p^{p-1}. \tag{12.22b}$$

This was the normal basis which Gauss studied in the Disquisitiones Arithmeticae. We observe that as a rule the S-conjugates of x form a normal basis of E over E^S. In case of the prime factorization $p-1 = p_1 p_2 \cdot \ldots \cdot p_s$ ($p_1, p_2, \ldots, p_s$ prime numbers) we construct E_p via the sequence of cyclic extensions

$$\mathbb{Q} = \Psi_0 \subset \Psi_1 \subset \cdots \subset \Psi_s = E_p,$$

with $\Psi_i = \Psi_{i-1}(\theta_i)$ of prime degree p_i over Ψ_{i-1}.

Here $\theta_i = \sum_{g\in G_i} g(x)$ with $G = G_0 = \langle g_0 \rangle$ and $g_0(\zeta_p) = \zeta_p^\sigma$, σ a primitive root modulo p, $G_j := \langle g_0^{(p-1)/(p_1\cdot\ldots\cdot p_j)} \rangle$ $(1 \leqslant j \leqslant s)$, $G = G_0 \supset G_1 \supset \cdots \supset G_s = 1$.

A sequence of quadratic extensions suffices to construct E_p, if and only if p is a prime number of the form $p = 1 + 2^s$. Such prime numbers are called *Fermat prime numbers.* If the exponent s is divisible by an odd prime number

q, then $(1+2^{s/q})|(1+2^s)$, hence a Fermat prime number must be of the form $p=1+2^{2^t}$. For $t=0, 1, 2, 3, 4$ we indeed obtain the Fermat prime numbers 3, 5, 17, 257, 65 537 which are the only ones known so far. We note, for example, that $641|(1+2^{2^5})$.

C. F. Gauss answered the question which cyclotomic construction tasks can be solved by ruler and compass by the statement that the nth primitive unit root can be constructed as a point of the Gaussian plane with 0 as origin and $1 \neq 0$ as second reference point if and only if n is a product of a power of 2 and a number of distinct Fermat primes. He anticipated Galois theory in a special setting by about 30 years.

6. The theorem of the primitive element

We have seen already that every finite separable extension E of a field of reference F can be generated by a primitive element. However, in order to find a primitive element generating E over F we must embed E into a minimal splitting field Ψ of E over F, we must construct a normal basis of Ψ over F, and we must form a period with respect to $\mathrm{Aut}(\Psi/E)$. A more direct way is given by the following theorem.

Theorem (12.23)

(J. Sonn and H. Zassenhaus). Let E be an extension of the field F with finite basis B. Then E contains a primitive element over F if and only if at least one of the $2^{|B|}-1$ *non-empty sums over distinct basis elements generates E over F.*

Proof

If the characteristic of F is a prime number p and if there are two elements x, y of E for which

$$\left.\begin{array}{l} x^p, y^p \in \Psi := \mathrm{Sep}(F, E) := \{z \in E \mid z \text{ separable over } F\}, \\ [\Psi(x,y):\Psi] = p^2, \end{array}\right\} \qquad (12.24)$$

then it is impossible to find any single generator ξ of E over ψ. Otherwise there is a natural number μ for which

$$\xi^{\mu} \in \Psi, \quad \xi^{\mu-1} \notin \Psi,$$

hence $\xi^{p^{\mu-1}}$ cannot be contained both in $\Psi(x)$, $\Psi(y)$. Assume $\xi^{p^{\mu-1}} \notin \Psi(x)$, hence $\Psi(x,\xi) \simeq \Psi(x) \otimes_{\Psi} \Psi(\xi)$, $x \notin \Psi(\xi)$, $\Psi(\xi) \subset E$, a contradiction.

On the other hand, if there are no two elements x, y of E satisfying (12.24) and if $\Psi \subset E$, then there is a maximal natural number μ such that $E^{p^{\mu-1}} \nsubseteq \Psi$, $E^{p^{\mu}} \subseteq \Psi$, hence there is an element ξ of E for which

$$\xi^p \in \Psi, \quad \xi^{p-1} \notin \Psi, \quad E_0 = \Psi \subset E_1 = \Psi(\xi^{p^{\mu-1}}) \subset \cdots \subset E_{\mu-1} \subset E_{\mu} = E_{\mu-1}(\xi),$$
$$[E_i : E_{i-1}] = p \quad (1 \leqslant i \leqslant \mu).$$

Let η be an element of E not contained in E_{μ}. There is a natural number

$\nu \leqslant \mu$ for which $\eta^{p^{\nu}} \in \Psi$, $\eta^{p^{\nu-1}} \notin \Psi$. Hence there is a natural number $\lambda \leqslant \nu$ for which $\eta^{p^{\lambda-1}} =: \zeta \notin E_{\mu}$, $\zeta^p \in E_{\mu}$. By assumption $\zeta^p \notin E_0 = \Psi$. Hence, $\zeta^p \in E_i$, $\zeta^p \notin E_{i-1}$ for some number $i \leqslant \mu$, and we have $\zeta^{p^{i+1}} \in \Psi$, $\zeta^{p^i} \notin E_1$, a contradiction. It follows that $E_{\mu} = E = \Psi(\xi)$. □

Because of $E = (F^{p^{-\infty}} \cap E) \otimes_F \Psi$ it follows that there is an element η of $F^{p^{-\infty}} \cap E$ for which $F(\eta) = F^{p^{-\infty}} \cap E$, $\eta^{p^{\mu}} \in F$, $\eta^{p^{\mu-1}} \notin F$. We embed E into a minimal splitting field M of E over F. The fixed subfield of $\mathrm{Aut}(M/F)$ is $F^{p^{-\infty}} \cap E$. Any element ζ of E which is not a primitive element of E over F either is contained in the subfield $\mathrm{Sep}(F(\eta^{p^{\mu-1}}), E) = F(\eta^{p^{\mu-1}}) \otimes_F \Psi$ of degree $p^{\mu-1}[\Psi:F]$ over F or in one of the $[E:\Psi]-1$ subfields

$$E_i = \{x \in E \mid x = g_i(x)\} \quad (1 < i \leqslant [E:\Psi]), \tag{12.25a}$$

where

$$\mathrm{Aut}(M/F) = \bigcup_{i=1}^{[E:\Psi]} g_i \, \mathrm{Aut}(M/E) \quad (g_1 = 1).$$

We note that $\Psi \subseteq E_i \subset E$ so that

$$[E_i:F] \leqslant \tfrac{1}{2}[E:F] \quad (1 < i \leqslant [E:\Psi]) \tag{12.25b}$$

according to the degree theorem.

If E is separable over F, then we again embed E into a minimal splitting field M of E over F. The fixed subfield of $\mathrm{Aut}(M/F)$ is F. Any element ζ of E which is not a primitive element of E over F is contained in one of the $[E:F]-1$ subfields

$$E_i := \{x \in E \mid x = g_i(x)\} \quad (1 < i \leqslant [E:F])$$

where

$$\mathrm{Aut}(M/F) = \bigcup_{i=1}^{[E:F]} g_i \, \mathrm{Aut}(M/E) \quad (g_1 = 1).$$

We note that

$$[E_i:F] \leqslant \tfrac{1}{2}[E:F] \quad (1 < i \leqslant [E:F]) \tag{12.25c}$$

according to the degree theorem. To conclude the proof we need the following combinatorial lemma.

Lemma (12.26)

Let E be a linear space over the field F with a basis B. Any set $\{B_1, B_2, \ldots, B_s\}$ of finitely many disjoint finite subsets $B_i \subseteq B$ $(1 \leqslant i \leqslant s)$ generates the **s-hypercube** *formed by the 2^s sums $\sum_{b \in B} \varepsilon(b) b$ subject to the conditions*

$$\begin{aligned} \varepsilon(b) &= 0 && \text{for } b \notin B_1 \cup B_2 \cup \cdots \cup B_s, \\ \varepsilon(b) &\in \{0, 1\} && \text{for } b \in B_1 \cup B_2 \cup \cdots \cup B_s, \\ \varepsilon(b) &= \varepsilon(b') && \text{for } b, b' \in B_i \quad (1 \leqslant i \leqslant s). \end{aligned}$$

It spans an F-linear subspace X with the s basis elements $\sum_{b \in B_i} b$ $(1 \leqslant i \leqslant s)$

such that the s-hypercube is characterized as the set of all sums over distinct elements of B contained in X. This is the s-hypercube situation. If the characteristic of F is 2 then we also allow all invertible linear transformations of B over $\mathbb{F}_2$. *Then any F-linear subspace X of E of finite F-dimension s contains at most* 2^s *sums over distinct elements of B. Equality is attained only in the s-hypercube situation.*

Proof of (12.26)

Let X be an F-linear subspace of E of finite dimension s over F and with the property that there is a set S of 2^s sums over distinct elements of B contained in X, but the s-hypercube situation does not hold. Let B' be the subset of B formed by the elements of B occurring in one of the 2^s sums. Assume that the number of elements of B' is as small as possible. We want to produce a contradiction.

Without loss of generality we assume that B, B' are finite. We obtain an equivalence relation on B by defining b to be equivalent to b' if and only if either both b, b' or neither occur in any one of the 2^s sums of S. Upon transition from B to the set B' formed by the sums over each equivalence class we find another counterexample. Due to the minimal property of B it follows that each equivalence class consists of a single element of B. If an element b of B belongs to X then upon transition from B to $B\backslash\{b\}$, E to $Y := \sum_{b' \in B\backslash\{b\}} Fb'$, X to $X \cap Y$, and S to the subset of the sums in which b does not occur we find another counterexample with a lesser number of basis elements contrary to the minimal property of B. Hence no element of B belongs to X.

Let $b \in B$, $Y := \dot{\sum}_{b' \in B\backslash\{b\}} Fb'$, $\dim_F(Y \cap X) = s - 1$. If $S \cap (X \cap Y)$ consists of 2^{s-1} or more elements, then it follows from the minimal property of B that we are in the $(s-1)$-hypercube situation, hence $B\backslash\{b\} \subseteq X \cap Y$, $X = E$, a contradiction. It follows that $S \cap (X \cap Y)$ consists of less than 2^{s-1} elements, each of the form $\sum_{b' \in B\backslash\{b\}} \alpha(b')b'$ $(\alpha(b') \in \{0, 1\})$ and $S\backslash(S \cap (X \cap Y))$ consists of more than 2^{s-1} elements of the form $b + \sum_{b' \in B\backslash\{b\}} \beta(b')b'$ $(\beta(b') \in \{0, 1\})$ where never $\alpha(b') = \beta(b')$ for all $b' \in B\backslash\{b\}$. Hence, $\dim_F(Fb + X) = s + 1$, $Fb + X$ contains S as well as the 2^s new sums $b + \sum_{b' \in B\backslash\{b\}} \alpha(b')b', \sum_{b' \in B\backslash\{b\}} \beta(b')b'$. Because of the maximal property of s we are now in the $(s+1)$-hypercube situation. This means that $B \subseteq Fb + X$, $E = Fb + X$, $\dim_F E = s + 1$. Hence S contains all sums $b + b'$ $(b' \in B\backslash\{b\})$ yielding $X = \dot{\sum}_{b' \in B\backslash\{b\}} F(b + b')$.

If F is of characteristic 2, then X itself is in the s-hypercube situation after suitable invertible linear transformations of B over $\mathbb{F}_2$, a contradiction. Hence the characteristic of F is not 2. For $s = 1$ we are in the s-hypercube situation. Let finally $s > 1$. Replacing b by another element b'' of B we find that also $X = \dot{\sum}_{b' \in B\backslash\{b''\}} F(b'' + b')$, which is impossible. □

We apply lemma (12.26) to conclude the proof of (12.23). If E is separable of dimension n over F, then we have to avoid at most $1 + (n-1)2^{n/2} < 2^n$

sums of distinct members of B as a consequence of (12.26). If E is not separable of dimension np^{μ} ($\mu \in \mathbb{Z}^{>0}$) over F, then we have to avoid at most $1+(n-1)2^{np^{\mu}/2}+2^{np^{\mu-1}}<2^{np^{\mu}}$ sums of distinct members of B. In either case this can be done. In fact, the chances are the better the larger n, p, μ are.

□

Exercises

1. Let R be a unital commutative ring with $2 \in U(R)$. Show that the sum of the three non-equivalent permutation representations of degree 2 of Klein's Four Group is equivalent via matrix representation to the sum of the regular permutation representation and two permutation representations of degree 1.
2. Show that the cyclic group of order 4 has no faithful representation of degree 1 over $\mathbb{Q}$ though there is one over $\mathbb{Q}(-1)^{\frac{1}{2}}$.
3. For any two unital commutative overrings R_1, R_2 of the unital ring R and for any semigroup G we have $(R_1 \otimes_R R_2)G = R_1G \otimes_R R_2G$.
4. Let A be known to be an algebraic sum of finitely many separable finite extensions of the field F with F-basis $b_1, \ldots, b_n$. Show that at least one of the 2^n sums $x = \sum_{i=1}^{n} \varepsilon_i b_i$ ($\varepsilon_i = 1$ or 0) is a primitive element of A over F so that $1, x, \ldots, x^{n-1}$ form another F-basis of A. Estimate the likelihood of hitting a primitive element by 'random' choice.
5. Does (12.23) hold generally for unital commutative F-algebras A of finite dimension over F?

3

Methods from the geometry of numbers

3.1. Introduction

The geometry of numbers was introduced by H. Minkowski who 'got the methods which provided arithmetical theorems by spacious perception'. This was about 1900 and since then the geometry of numbers has become an independent discipline of mathematics. We, of course, are mainly interested in those results which can be applied to problems of computational number theory. They will be outlined in the following three sections. In section 2 we consider modules over principal entire rings with regard to a subsequent specialization to lattices in Euclidean n-space. Especially, in connection with the study of the bases of a module and the bases of its submodules we derive the Hermite and Smith normal form of matrices. In section 3 our considerations are confined to lattices. The search for special lattice bases consisting of vectors of small length leads to the concept of reduction. We present several reduction methods such as Minkowski-reduction, a total-ordering-reduction and a new reduction algorithm of Lenstra, Lenstra and Lovasz [10] and also some applications. Closely connected with reduction theory are the successive minima of a lattice, also studied in section 3. Finally, section 4 contains Minkowski's famous Convex Body Theorem and some of its consequences for algebraic number theory.

3.2. Free modules over principal entire rings

In the sequel we consider free unital modules M over principal entire rings R which are finitely generated, i.e. $M = \bigoplus_{i=1}^{n} b_i R$ for suitable elements $n \in \mathbb{N}$, $b_1, \ldots, b_n \in M$. Those elements $b_1, \ldots, b_n$ are then called *(R-)basis* of M, n the *(R-)rank* of M.

Lemma (2.1)

If $b_1, \ldots, b_n$ and $c_1, \ldots, c_n$ are bases of M, then there is a matrix $U \in GL(n, R)$ satisfying $(b_1, \ldots, b_n) = (c_1, \ldots, c_n)U$.

Proof

According to assumption there are $U, V \in R^{n \times n}$ subject to $(b_1, \ldots, b_n) = (c_1, \ldots, c_n)U$, $(c_1, \ldots, c_n) = (b_1, \ldots, b_n)V$ and the uniqueness of the presentation of elements by a basis yields $UV = I_n$, I_n the $n \times n$ unit matrix. □

In order to discuss the connections between the bases of a module and its submodules we need a few tools from matrix theory which are implicitly presented in most algebra courses but which may not be familiar to the reader. However, from a constructive point of view these tools are of considerable value.

We recall that matrix multiplication from the right (left) yields an operation on the columns (rows) of the matrix. Especially we are interested in multiplications by so-called *elementary matrices* of $R^{n \times n}$. For an easy description we introduce the following matrix types $(1 \leqslant i, j \leqslant n)$.

E_{ij} contains exactly one entry 1 in column j and row i, otherwise zeros; hence

$$I_n = \sum_{i=1}^{n} E_{ii}; \tag{2.2a}$$

$$S_{ij} := \sum_{\substack{k=1 \\ k \neq i,j}}^{n} E_{kk} + E_{ij} + E_{ji} \quad (i \neq j); \tag{2.2b}$$

$$T_{ij}(a) := I_n + aE_{ij} \quad (a \in R, i \neq j); \tag{2.2c}$$

$$\mathrm{diag}(a_1, \ldots, a_n) := \sum_{i=1}^{n} a_i E_{ii} \quad (a_i \in R, 1 \leqslant i \leqslant n); \tag{2.2d}$$

$$D_j := I_n - 2E_{jj}. \tag{2.2e}$$

Multiplication of a matrix $A \in R^{n \times n}$ from the right (left) by S_{ij} interchanges the columns (rows) i and j. Multiplication of a matrix A from the right (left) by $T_{ij}(a)$ adds a-times column i (row j) to column j (row i). Moreover, the inverse matrices of S_{ij}, $T_{ij}(a)$, D_j are easily seen to be

$$S_{ij}^{-1} = S_{ij}, \quad T_{ij}(a)^{-1} = T_{ij}(-a), \quad D_j^{-1} = D_j. \tag{2.3}$$

We use these matrices to transform any matrix $A \in R^{m \times n}$ into a suitable normal form. To do this we need two preparatory lemmata.

Lemma (2.4)

Let $a_1, \ldots, a_n$ be elements of the principal entire ring R. Then $R^{n \times n}$ contains a matrix $A = (a_{ij})$ subject to $a_{1j} = a_j$ $(1 \leqslant j \leqslant n)$ and $\det(A) = \gcd(a_1, \ldots, a_n)$.

Proof

By induction on n. The case $n = 1$ is trivial. We therefore assume the existence of $\tilde{A} = (\tilde{a}_{ij}) \in R^{(n-1)\times(n-1)}$ satisfying $\tilde{a}_{1j} = a_j$ $(1 \leqslant j \leqslant n-1)$ and $d_{n-1} := \det(\tilde{A}) = \gcd(a_1, \ldots, a_{n-1})$. For $d_n := \gcd(d_{n-1}, a_n)$ there exist $u, v \in R$ subject to $d_n = ud_{n-1} + va_n$. We set

$$a_{ij} = \tilde{a}_{ij} \quad (1 \leqslant i, j \leqslant n-1), \quad a_{1n} = a_n, \quad a_{in} = 0 \quad (2 \leqslant i \leqslant n-1),$$
$$a_{nn} = u_n, \quad a_{nj} = -(a_j/d_{n-1})v \quad (1 \leqslant j \leqslant n-1).$$

Then $\det(A) = d_n$ is obtained upon expansion according to the last column. □

Example
Let $R = \mathbb{Z}$ and $a_1 = 30$, $a_2 = 42$, $a_3 = 70$, $a_4 = 105$. By calculating

$$d_1 = 30,$$
$$d_2 = 6 = 3 \cdot 30 + (-2)\ 42,$$
$$d_3 = 2 = 12 \cdot 6 + (-1)\ 70,$$
$$d_4 = 1 = 53 \cdot 2 + (-1)\ 105$$

we get

$$A = \begin{pmatrix} 30 & 42 & 70 & 105 \\ 2 & 3 & 0 & 0 \\ 5 & 7 & 12 & 0 \\ 15 & 21 & 35 & 53 \end{pmatrix}.$$

We note that the actual computation of A requires the calculation of a presentation of $d = \gcd(a, b)$ by a and b and is therefore practicable in Euclidean rings only.

Lemma (2.5)
Let $a_1, \ldots, a_n$ be elements of the principal entire ring R and $d_n := \gcd(a_1, \ldots, a_n)$. Then there exists $U \in GL(n, R)$ satisfying $(a_1, \ldots, a_n)\ U = (d_n, 0, \ldots, 0)$.

Proof
Let $A = (a_{ij}) \in R^{n \times n}$ as in (2.4). The matrix $\tilde{A} = (\tilde{a}_{ij})$ given by $\tilde{a}_{ij} := a_{ij}$ $(2 \leqslant i \leqslant n)$ and $\tilde{a}_{1j} := a_{1j}/d_n$ $(1 \leqslant j \leqslant n)$ is clearly in $GL(n, R)$ and satisfies $(d_n, 0, \ldots, 0)\ \tilde{A} = (1, 0, \ldots, 0)\ A = (a_1, \ldots, a_n)$, hence $U := \tilde{A}^{-1}$ does the job. □

The last lemma yields a first normal form for matrices over a principal entire ring R.

Theorem (2.6)
Let R be a principal entire ring and $\mathfrak{R}$ a full system of representatives of $R/U(R)$. For every matrix $A = (a_{ij}) \in R^{m \times n}$ there exists a matrix $U \in GL(n, R)$ such that $H(A) = (h_{ij}) := AU$ is a lower triangular matrix the entries of which satisfy $h_{ii} \in \mathfrak{R}$ $(1 \leqslant i \leqslant \min(m, n))$ and in case $h_{ii} \neq 0$ the entries h_{ij} with $j < i$ are uniquely determined modulo h_{ii}. $H(A)$ is called **Hermite normal form** *of A, respectively,* **(Hermite-)column-reduced**. *If $H(A^t)$ is column-reduced, then A itself is said to be* **(Hermite-)row-reduced**.

Proof

By induction on n. For $n = 1$ let $d \in \mathfrak{R}$ be associated to a_{11}, i.e. there exists $e \in U(R)$ subject to $d = a_{11}e$. In that case $U = (e) \in GL(1, R)$ satisfies $AU = H(A)$.

Now let $n > 1$ and the theorem be true for matrices over R with less than n columns. For $\tilde{d} := \gcd(a_{11}, \ldots, a_{1n})$ let $d \in \mathfrak{R}$ be associate to $\tilde{d}$. According to (2.5) there is a matrix $U \in GL(n, R)$ such that the matrix $\tilde{A} = (\tilde{a}_{ij}) := AUe$ for $e = \tilde{d}/d$ has entries $\tilde{a}_{11} = d$, $\tilde{a}_{1j} = 0$ $(j > 1)$. In case $m = 1$ we are done. For $m > 1$ we apply the induction hypothesis to the matrix $B = (b_{ij}) \in R^{(m-1)\times(n-1)}$ with entries $b_{ij} = \tilde{a}_{i+1,j+1}$. There is $\tilde{V} = (\tilde{v}_{ij}) \in GL(n-1, R)$ such that $H(B) = B\tilde{V}$. We choose $V = (v_{ij}) \in GL(n, R)$ via $v_{11} = 1$, $v_{1j} = 0 = v_{i1}$, $v_{ij} = \tilde{v}_{ij}$ $(i > 1, j > 1)$, and obtain $H(A) = AUeV$. □

Especially, for $R = \mathbb{Z}$ we choose $\mathfrak{R} = \mathbb{Z}^{\geq 0}$ and $h_{ij} \in \{0, 1, \ldots, h_{ii} - 1\}$ for $h_{ii} \neq 0$. As we already mentioned the Euclidean algorithm allows the practical computation of $H(A)$ in that case.

Algorithm for the computation of Hermite normal form (2.7)

Input. $A \in \mathbb{Z}^{m\times n}$

Output. $H \in \mathbb{Z}^{m\times n}$, $U \in GL(n, \mathbb{Z})$ such that $AU = H$ and H is in Hermite normal form.

Step 1. (Initialization). Set $H \leftarrow A$, $U \leftarrow I_n$, $r \leftarrow \min(m, n)$, $i \leftarrow 1$.

Step 2. (Determination of smallest element in row i). Let $S_i := \{h_{ij} \mid j \geq i, h_{ij} \neq 0\}$. For $S_i = \emptyset$ go to 6; else compute $k \in \{i, \ldots, n\}$ minimal with $|h_{ik}| = \min\{|h_{ij}| \mid h_{ij} \in S_i\}$. In case $k = i$ go to 4, else to 3.

Step 3. (Change of columns i and k). Set $U \leftarrow US_{ik}$, $H \leftarrow HS_{ik}$.

Step 4. (Reduction of elements h_{ij} modulo h_{ii} for $j > i$). Set $H \leftarrow HT_{ij}(-\{h_{ij}/h_{ii}\})$, $U \leftarrow UT_{ij}(-\{h_{ij}/h_{ii}\})$ for $j = i+1, \ldots, n$. If $h_{ij} = 0$ $(j = i+1, \ldots, n)$ go to 5, else to 2.

Step 5. (Reduction of elements h_{ij} modulo h_{ii} for $j < i$). For $h_{ii} < 0$ set $H \leftarrow HD_i$, $U \leftarrow UD_i$. Then set $H \leftarrow HT_{ij}(-\lfloor h_{ij}/h_{ii} \rfloor)$, $U \leftarrow UT_{ij}(-\lfloor h_{ij}/h_{ii} \rfloor)$ for $j = 1, \ldots, i-1$.

Step 6. (Increase i). For $i = r$ terminate, else set $i \leftarrow i+1$. For $i = r$ go to 5, else to 2.

If we also need U^{-1}, we can compute it analogously to U. We start from $U^{-1} = I_n$ in step 1 and each time when we multiply U by an elementary matrix from the right we multiply U^{-1} by the inverse of that matrix from the left.

Example

We compute the Hermite normal form of $M = \begin{pmatrix} 3 & 7 & 10 \\ 15 & 20 & 25 \\ 31 & 36 & 40 \end{pmatrix}$

step	index i	elementary matrices for H, U	H	U	elementary matrices for U^{-1}	U^{-1}
1	1	—	$\begin{pmatrix} 3 & 7 & 10 \\ 15 & 20 & 25 \\ 31 & 36 & 40 \end{pmatrix}$	I_3	—	I_3
4	1	$T_{1,2}(-2)$, $T_{1,3}(-3)$	$\begin{pmatrix} 3 & 1 & 1 \\ 15 & -10 & -20 \\ 31 & -26 & -53 \end{pmatrix}$	$\begin{pmatrix} 1 & -2 & -3 \\ 0 & 1 & 0 \\ 0 & 0 & 1 \end{pmatrix}$	$T_{1,2}(2)$, $T_{1,3}(3)$	$\begin{pmatrix} 1 & 2 & 3 \\ 0 & 1 & 0 \\ 0 & 0 & 1 \end{pmatrix}$
3	1	$S_{1,2}$	$\begin{pmatrix} 1 & 3 & 1 \\ -10 & 15 & -20 \\ -26 & 31 & -53 \end{pmatrix}$	$\begin{pmatrix} -2 & 1 & -3 \\ 1 & 0 & 0 \\ 0 & 0 & 1 \end{pmatrix}$	$S_{1,2}$	$\begin{pmatrix} 0 & 1 & 0 \\ 1 & 2 & 3 \\ 0 & 0 & 1 \end{pmatrix}$
4	1	$T_{1,2}(-3)$ $T_{1,3}(-1)$	$\begin{pmatrix} 1 & 0 & 0 \\ -10 & 45 & -10 \\ -26 & 109 & -27 \end{pmatrix}$	$\begin{pmatrix} -2 & 7 & -1 \\ 1 & -3 & -1 \\ 0 & 0 & 1 \end{pmatrix}$	$T_{1,2}(3)$ $T_{1,3}(1)$	$\begin{pmatrix} 3 & 7 & 10 \\ 1 & 2 & 3 \\ 0 & 0 & 1 \end{pmatrix}$
3	2	$S_{2,3}$	$\begin{pmatrix} 1 & 0 & 0 \\ -10 & -10 & 45 \\ -26 & -27 & 109 \end{pmatrix}$	$\begin{pmatrix} -2 & -1 & 7 \\ 1 & -1 & -3 \\ 0 & 1 & 0 \end{pmatrix}$	$S_{2,3}$	$\begin{pmatrix} 3 & 7 & 10 \\ 0 & 0 & 1 \\ 1 & 2 & 3 \end{pmatrix}$

4	2	$T_{2,3}(4)$	$\begin{pmatrix} 1 & 0 & 0 \\ -10 & -10 & 5 \\ -26 & -27 & 1 \end{pmatrix}$	$\begin{pmatrix} -2 & -1 & 3 \\ 1 & -1 & -7 \\ 0 & 1 & 4 \end{pmatrix}$	$T_{2,3}(-4)$	$\begin{pmatrix} 3 & 7 & 10 \\ -4 & -8 & -11 \\ 1 & 2 & 3 \end{pmatrix}$
3	2	$S_{2,3}$	$\begin{pmatrix} 1 & 0 & 0 \\ -10 & 5 & -10 \\ -26 & 1 & -27 \end{pmatrix}$	$\begin{pmatrix} -2 & 3 & -1 \\ 1 & -7 & -1 \\ 0 & 4 & 1 \end{pmatrix}$	$S_{2,3}$	$\begin{pmatrix} 3 & 7 & 10 \\ 1 & 2 & 3 \\ -4 & -8 & -11 \end{pmatrix}$
4	2	$T_{2,3}(2)$	$\begin{pmatrix} 1 & 0 & 0 \\ -10 & 5 & 0 \\ -26 & 1 & -25 \end{pmatrix}$	$\begin{pmatrix} -2 & 3 & 5 \\ 1 & -7 & -15 \\ 0 & 4 & 9 \end{pmatrix}$	$T_{2,3}(-2)$	$\begin{pmatrix} 3 & 7 & 10 \\ 9 & 18 & 25 \\ -4 & -8 & -11 \end{pmatrix}$
5	2	$T_{2,1}(2)$	$\begin{pmatrix} 1 & 0 & 0 \\ 0 & 5 & 0 \\ -24 & 1 & -25 \end{pmatrix}$	$\begin{pmatrix} 4 & 3 & 5 \\ -13 & -7 & -15 \\ 8 & 4 & 9 \end{pmatrix}$	$T_{2,1}(-2)$	$\begin{pmatrix} 3 & 7 & 10 \\ 3 & 4 & 5 \\ -4 & -8 & -11 \end{pmatrix}$
5	3	D_3, $T_{3,1}(1)$	$\begin{pmatrix} 1 & 0 & 0 \\ 0 & 5 & 0 \\ 1 & 1 & 25 \end{pmatrix}$	$\begin{pmatrix} -1 & 3 & -5 \\ 2 & -7 & 15 \\ -1 & 4 & -9 \end{pmatrix}$	D_3, $T_{3,1}(-1)$	$\begin{pmatrix} 3 & 7 & 10 \\ 3 & 4 & 5 \\ 1 & 1 & 1 \end{pmatrix}$

Hence, we obtain the Hermite normal form of M as

$$H = \begin{pmatrix} 1 & 0 & 0 \\ 0 & 5 & 0 \\ 1 & 1 & 25 \end{pmatrix} = \begin{pmatrix} 3 & 7 & 10 \\ 15 & 20 & 25 \\ 31 & 36 & 40 \end{pmatrix} \begin{pmatrix} -1 & 3 & -5 \\ 2 & -7 & 15 \\ -1 & 4 & -9 \end{pmatrix} = MU.$$

Remark

From the Hermite normal form H of a square matrix M the absolute value of its determinant is easily computed as the product of the diagonal elements of H. The sign of the determinant is $(-1)^h$, where h denotes the number of multiplications by matrices S_{ij}, D_j in the course of the computation of H. This follows at once from the properties of the determinant, since obviously $\det T_{ij} = 1$, $\det S_{ij} = -1$, $\det D_j = -1$. In the example we have $h = 4$, hence $\det M = 125$.

Even though Algorithm (2.9) seems to be quite easy a few comments should be made. While the method works well if m, n and the absolute values of the entries of M are small, the growth of the numbers in the course of the computations can be tremendous. Several authors developed strategies to avoid this effect (see, for example, [4]). Upper bounds for the size of the intermediate entries which are polynomial in the input data were obtained for a special elimination scheme by [5]. However, these bounds are huge and by no means realistic so that their computational value is rather limited.

If working with higher dimensions or with larger initial entries of M it will therefore always be necessary to have a multiple precision arithmetic package available. If there is none supplied by the local computer center one can easily write one in FORTRAN following Knuth's book [7] but using only half of each computer word for storing information. In this way arithmetic with the supplied hardware instructions is still possible but – of course – slow.

Because of the slowness and the extensive use of storage one should not compute the unimodular transformation matrix U in the course of calculating $H(M)$ when multiple precision arithmetic is used. In later applications (see chapter 5 (4.10)) we at most need U^{-1} and that can be easily calculated once we know $H(M)$. Namely, from $M = H(M)U^{-1}$ we obtain the following system of linear equations

$$m_{ik} = \sum_{j=1}^{i} h_{ij}\tilde{u}_{jk} \quad (1 \leqslant i, k \leqslant n)$$

for the entries $\tilde{u}_{jk}$ of U^{-1} which can be solved recursively. For $i = 1$ we obtain $\tilde{u}_{1k} = m_{1k}h_{11}^{-1}$ $(1 \leqslant k \leqslant n)$. Once we already know $\tilde{u}_{lk}$ $(1 \leqslant l \leqslant i, 1 \leqslant k \leqslant n)$ we easily calculate

$$\tilde{u}_{i+1,k} = \left(m_{i+1,k} - \sum_{j=1}^{i} h_{i+1,j}\tilde{u}_{j,k} \right) h_{i+1,i+1}^{-1} \quad (1 \leqslant k \leqslant n).$$

We note that for regular matrices $M \in R^{n \times n}$ only intermediate entries can grow large since the (non-negative) entries of $H(M)$ are obviously bounded by $|\det(M)|$. Hence, in that case $H(M)$ can be calculated 'modulo N' ($N = |\det(M)|$ or N the largest elementary divisor of M) by computing $H(M \vdots \operatorname{diag}(N, \ldots, N))$ and deleting the last n columns containing zeros. This is of importance in chapter 6.

So far we only applied column operations to the given matrix $M \in \mathbb{Z}^{m \times n}$. If we also admit row operations it is quite obvious that we can transform M into a diagonal matrix:

Theorem (2.8)

Let R be a principal entire ring and $A \in R^{m \times n}$ of rank r. There exist $U \in GL(n, R)$ and $V \in GL(m, R)$ such that $S(A) = (s_{ij}) := VAU$ is in **Smith normal form (elementary divisor normal form)**, *i.e. $s_{ij} = 0$ for $i \neq j$ and $i = j > r$ as well as $s_{11} | s_{22} | \cdots | s_{rr}$. The diagonal elements of $S(A)$ are uniquely determined modulo $U(R)$* **(elementary divisors)**.

Proof

By induction on n. For $n = 1$ we obtain $S(A) = (s_{11}, 0, \ldots, 0)^t$ with $s_{11} = \gcd(a_{11}, a_{21}, \ldots, a_{m1})$ upon application of (2.5) to A^t. The case $n > 1$, $m = 1$ was already treated in (2.5). Now let n and m both be greater than one. Similarly to the proof of (2.6) we apply (2.5) to the first row of A to obtain $\tilde{A}^{(1)} = (\tilde{a}_{ij}^{(1)})$ with $\tilde{a}_{1j}^{(1)} = 0$ for $j > 1$. Application of (2.5) to $\tilde{A}^{(1)t}$ yields $A^{(1)} = (a_{ij}^{(1)})$ with $a_{i1}^{(1)} = 0$ for $i > 1$. Clearly, $\tilde{a}_{11} = \gcd(a_{11}, a_{12}, \ldots, a_{1n})$, $a_{11}^{(1)} = \gcd(\tilde{a}_{11}, a_{21}, \ldots, a_{m1})$. If $a_{1j}^{(1)} = 0$ for $j > 1$ we can apply our induction hypothesis. Otherwise we repeat that procedure to obtain $A^{(2)}, A^{(3)}, \ldots$ which must finally provide $a_{1j}^{(\nu)} = 0$ for $j > 1$ and some $\nu \in \mathbb{N}$ since the number of prime elements dividing $a_{11}^{(\nu)}$ strictly decreases if ν increases in case there is an element $a_{1j}^{(\nu)}$ $(j > 1)$ which is not divisible by $a_{11}^{(\nu)}$. Therefore we obtain $\tilde{U} \in GL(n, R)$, $\tilde{V} \in GL(m, R)$ such that $(\tilde{a}_{ij}) = \tilde{A} := \tilde{V} A \tilde{U}$ is a diagonal matrix.

Next we obtain the divisibility condition. Let us assume that there are indices $1 \leq i < j \leq \min(m, n)$ such that $\tilde{a}_{ii} \nmid \tilde{a}_{jj}$. (This includes the case $\tilde{a}_{ii} = 0 \neq \tilde{a}_{jj}$, however we stipulate $0|0$.) Let $c = \gcd(\tilde{a}_{ii}, \tilde{a}_{jj})$ and $u, v \in R$ subject to $c = u\tilde{a}_{ii} + v\tilde{a}_{jj}$. Then we compute $\tilde{\tilde{A}} = V_{i,j} \tilde{A} U_{i,j}$ for

$$V_{i,j} = T_{ji}\left(\frac{-\tilde{a}_{ii}}{c}\right) S_{ij} T_{ji}(u) \in GL(m, \mathbb{Z}), \quad U_{i,j} = T_{ji}(v) T_{ij}\left(-\frac{\tilde{a}_{jj}}{c}\right) D_j.$$

The new entries $\tilde{\tilde{a}}_{ij}$ then satisfy $\tilde{\tilde{a}}_{ii} = c$, $\tilde{\tilde{a}}_{jj} = (\tilde{a}_{ii}/c)\tilde{a}_{jj}$, all other entries remain unchanged. A repeated application yields $s_{11} | s_{22} | \cdots | s_{rr}$, $s_{jj} = 0$ for $j > r$ as required. It remains to prove that the Smith normal form S is uniquely determined by the matrix A. For this purpose let $d_i = d_i(A)$ denote the greatest common divisor of all minors of A of order i. Obviously, $d_{i-1} | d_i$ $(2 \leq i \leq r)$ because of Laplace's theorem. For $U \in GL(n, \mathbb{Z})$ we obtain $d_i(A) | d_i(AU)$, since the columns of AU are linear combinations of the columns of A and therefore each minor of AU is a sum of minors of A. This yields $d_i(A) \leqslant d_i(AU) \leqslant d_i(AUU^{-1}) = d_i(A)$. Similarly we see $d_i(A) = d_i(VA)$ for $V \in GL(m, \mathbb{Z})$. Hence, $d_i(S) = d_i(A) = \prod_{j=1}^{i} s_{jj}$ $(1 \leqslant i \leqslant r)$, if $S = S(A)$ is the Smith normal form of A. Therefore also the elementary divisors $s_{ii} = d_i/d_{i-1}$ $(1 \leqslant i \leqslant r,\ d_0 := 1)$ are uniquely determined by A. □

Example

We continue the example from above. Without the transformations for U, U^{-1} (see exercise 2) we obtain

$$\begin{pmatrix} 1 & 0 & 0 \\ 0 & 5 & 0 \\ 1 & 1 & 25 \end{pmatrix} \xrightarrow{t} \begin{pmatrix} 1 & 0 & 1 \\ 0 & 5 & 1 \\ 0 & 0 & 25 \end{pmatrix} \xrightarrow{(4,1)} \begin{pmatrix} 1 & 0 & 0 \\ 0 & 5 & 1 \\ 0 & 0 & 25 \end{pmatrix}$$

$$\xrightarrow{(3,2)} \begin{pmatrix} 1 & 0 & 0 \\ 0 & 1 & 5 \\ 0 & 25 & 0 \end{pmatrix} \xrightarrow[(5,3)]{(4,2)} \begin{pmatrix} 1 & 0 & 0 \\ 0 & 1 & 0 \\ 0 & 25 & 125 \end{pmatrix}$$

$$\xrightarrow{t} \begin{pmatrix} 1 & 0 & 0 \\ 0 & 1 & 25 \\ 0 & 0 & 125 \end{pmatrix} \xrightarrow{(4,2)} \begin{pmatrix} 1 & 0 & 0 \\ 0 & 1 & 0 \\ 0 & 0 & 125 \end{pmatrix}.$$

The numbers in brackets refer to the number of the step and the index i of Algorithm (2.7), respectively; t means transition to the transposed matrix. The elementary divisors of M are $1, 1, 125$.

We apply the preceding results to obtain several useful relations between the bases of a module and its submodules.

Lemma (2.9)

Let $N \subseteq M$ be free modules over a principal entire ring R of rank n, m, respectively.

(i) *For each basis $a_1, \ldots, a_m$ of M there exists a basis $b_1, \ldots, b_n$ of N such that $(b_1, \ldots, b_n) = (a_1, \ldots, a_m)A$, where $A = (a_{ij}) \in R^{m \times n}$ is an upper triangular matrix and the entries a_{ij} $(j > i)$ are uniquely determined modulo a_{ii}. In case $n = m$ the matrix A is regular, i.e. $\det(A) \neq 0$.*

(ii) *For each basis $b_1, \ldots, b_n$ of N there exists a basis $a_1, \ldots, a_m$ of M such that $(b_1, \ldots, b_n) = (a_1, \ldots, a_m)A$, where $A = (a_{ij}) \in R^{m \times n}$ is an upper triangular matrix and the entries a_{ij} $(i < j)$ are uniquely determined modulo a_{ii}.*

(iii) *There are bases $a_1, \ldots, a_m$ of M and $b_1, \ldots, b_n$ of N such that $b_i = \varepsilon_i a_i$ $(1 \leqslant i \leqslant m)$ and $\varepsilon_i | \varepsilon_{i+1}$ $(0 \leqslant i \leqslant n, \varepsilon_0 = 1)$. The elements $\varepsilon_i \in R$ are uniquely determined by M, N.*

Proof

(i) Let $c_1, \ldots, c_n$ be an arbitrary basis of N. There exists $B \in R^{m \times n}$ satisfying $(c_1, \ldots, c_n) = (a_1, \ldots, a_m)B$. Let $U \in GL(n, R)$ subject to $BU = H(B)$. We set

$(b_n,\ldots,b_1):=(c_1,\ldots,c_n)U$ and permute according to

$$(b_1,\ldots,b_n)=(a_1,\ldots,a_n)BU\prod_{1\leqslant i\leqslant n/2} S_{i,n+1-i}.$$

Since $1\otimes b_i$ $(1\leqslant i\leqslant n)$ as well as $1\otimes a_i$ $(1\leqslant i\leqslant n)$ are linearly independent in $\mathfrak{Q}(R)\otimes_R M$ we obtain $\det(A)\neq 0$ for $m=n$.

(ii) Let $c_1,\ldots,c_m$ be an arbitrary basis of M and $A\in R^{m\times n}$ subject to $(b_1,\ldots,b_n)=(c_1,\ldots,c_m)A$. Let $V\in GL(m,R)$ such that $H_r(A):=VA$ is Hermite-row-reduced. Then $(c_1,\ldots,c_m)V^{-1}$ does the job.

(iii) This is an immediate consequence of (2.8). □

Lemma (2.10)

Let M be a free module with basis $b_1,\ldots,b_n$ over the principal entire ring R. Let i be a fixed index $(1\leqslant i\leqslant n)$ and $c_i:=\sum_{j=1}^{n}\gamma_j b_j\in M$. Then $b_1,\ldots,b_{i-1},c_i$ can be extended to an R-basis of M if and only if $\gcd(\gamma_i,\ldots,\gamma_n)=1$.

Proof

$$\gcd(\gamma_i,\ldots,\gamma_n)=1\underset{(2.5)}{\Leftrightarrow}\exists U\in GL(n+1-i,R)\colon\ (\gamma_i,\ldots,\gamma_n)U=(1,0,\ldots,0)$$

$$\Leftrightarrow(b_1,\ldots,b_n)\left(\begin{array}{c|cc} I_{i-1} & \begin{matrix}\gamma_1\\ \vdots\\ \gamma_{i-1}\end{matrix} & 0\\ \hline 0 & \multicolumn{2}{c}{(U^{-1})^t}\end{array}\right)\ \text{basis of } M.$$

□

Exercises

1. Write an algorithm to compute the Smith normal form $S(A)$ of a matrix $A\in\mathbb{Z}^{m\times n}$.
2. Compute $U,V\in GL(3,\mathbb{Z})$ such that

$$V\begin{pmatrix}3 & 7 & 10\\ 15 & 20 & 25\\ 31 & 36 & 40\end{pmatrix}U=\begin{pmatrix}1 & 0 & 0\\ 0 & 1 & 0\\ 0 & 0 & 125\end{pmatrix}.$$

3. Develop a strategy for step 5 of Algorithm (2.7) which keeps the occurring entries as small as possible.

3.3. Lattices and basis reduction

To introduce geometric tools we consider $\mathbb{Z}$-modules in the n-dimensional Euclidean space $\mathbb{R}^n$. Their elements are written as column vectors $\mathbf{x}=(x_1,\ldots,x_n)^t$, and $\mathbf{e}_1,\ldots,\mathbf{e}_n$ denote the canonical basis, i.e. the coordinates e_{ij}

of $\mathbf{e}_i$ satisfy $e_{ij} = \delta_{ij}$ $(1 \leqslant i, j \leqslant n)$. The familiar scalar product

$$\mathbb{R}^n \times \mathbb{R}^n \to \mathbb{R}^{\geqslant 0}: (\mathbf{x}, \mathbf{y}) \mapsto \mathbf{x}^t\mathbf{y} \tag{3.1}$$

provides the *norm* (*length*) of a vector $\mathbf{x} \in \mathbb{R}^n$ via

$$\|\mathbf{x}\| := (\mathbf{x}^t\mathbf{x})^{\frac{1}{2}}. \tag{3.2}$$

The connection to arithmetics is then obtained by considering vectors with integral coordinates (relative to some basis).

Definition (3.3)
A free $\mathbb{Z}$-module $\Lambda \subset \mathbb{R}^n$ of rank $k := \dim_{\mathbb{R}} \mathbb{R} \otimes_{\mathbb{Z}} \Lambda$ is called a **lattice** *of* **dimension (rank)** *k. For a $\mathbb{Z}$-basis $\mathbf{a}_1, \ldots, \mathbf{a}_k$ of Λ the* **discriminant** *$d(\Lambda)$ of Λ is defined by $d(\Lambda) = |\det(\mathbf{a}_i{}^t\mathbf{a}_j)_{1 \leqslant i,j \leqslant k}|^{\frac{1}{2}}$.*

We remark that $\Lambda_0 = \{\mathbf{0}\}$ (the only 0-dimensional lattice) and $\Lambda_k := \bigoplus_{i=1}^{k} \mathbb{Z}\mathbf{e}_i$ $(1 \leqslant k \leqslant n)$ (with $d(\Lambda_k) = 1$) are trivial examples of lattices of $\mathbb{R}^n$. The discriminant of a lattice is also the (k-dimensional) volume of the *fundamental parallelotope*

$$\prod(\Lambda) := \left\{\mathbf{x} \in \mathbb{R}^n \,\middle|\, \mathbf{x} = \sum_{i=1}^{k} x_i\mathbf{a}_i, 0 \leqslant x_i < 1, 1 \leqslant i \leqslant k\right\}. \tag{3.4}$$

It is independent of the choice of basis because of (2.1).

Definition (3.5)
A subset Λ' of a lattice Λ is called a **sublattice** *if Λ' itself is a lattice of $\mathbb{R}^n$. The (module-) index $(\Lambda : \Lambda')$ is said to be the* **index** *of the lattice Λ' in Λ.*

Lemma (3.6)
If Λ' is a sublattice of the lattice Λ of the same rank k, then $(\Lambda : \Lambda') = d(\Lambda')/d(\Lambda)$.

Proof
By definition $d(\Lambda')/d(\Lambda)$ is the absolute value of the determinant of the transition matrix A from a basis of Λ to a basis of Λ'. If we choose bases as in (2.9) (i) we obtain $d(\Lambda')/d(\Lambda) = \prod_{i=1}^{k} a_{ii}$. On the other hand $\{\sum_{i=1}^{k} m_i\mathbf{a}_i \mid m_i \in \mathbb{Z}, 0 \leqslant m_i < a_{ii}, 1 \leqslant i \leqslant k\}$ is a full set of representatives of the residue classes of Λ/Λ'. □

Before we start to discuss how to choose special bases for a lattice which are of computational interest we shortly consider lattices from a different point of view. Namely, for each lattice $\Lambda = \bigoplus_{i=1}^{k} \mathbb{Z}\mathbf{a}_i$ we have the corresponding scalar product matrix

$$A = (a_{ij}) := (\mathbf{a}_i{}^t\mathbf{a}_j)_{1 \leqslant i,j \leqslant k} \in \mathbb{R}^{k \times k}. \tag{3.7}$$

yielding the positive definite quadratic form

$$Q: \mathbb{Z}^k \times \mathbb{Z}^k \to \mathbb{R}^{\geq 0}: \mathbf{x} \mapsto \mathbf{x}^t A \mathbf{x}. \tag{3.8}$$

On the other hand, each positive definite matrix $A \in \mathbb{R}^{k \times k}$ can be viewed as the scalar product matrix of the basis vectors $\mathbf{a}_1, \ldots, \mathbf{a}_k$ of some lattice Λ in $\mathbb{R}^n$ ($n \geq k$). Then we can carry out computations in that lattice without knowing the basis vectors explicitly as will be outlined below. For example, we could view A as the scalar product matrix of $\mathbf{e}_1, \ldots, \mathbf{e}_k$ in $\mathbb{R}^k$ thus dealing with a different scalar product than the standard one. But instead of considering different scalar products we prefer the viewpoint of considering different lattices. By Cholesky's method (compare (3.12)) we can decompose A into a product

$$A = R^t R \quad (R \in \mathbb{R}^{k \times k} \text{ an upper triangular matrix}) \tag{3.9}$$

and then attach that k-dimensional lattice of $\mathbb{R}^k$ to A whose vectors are the columns of R.

Probably the most important property of a lattice is its discreteness. This means that any bounded set of $\mathbb{R}^n$ contains at most finitely many lattice points, or – differently – that there is a positive constant $\delta > 0$ such that each pair of different lattice points $\mathbf{x}, \mathbf{y}$ satisfies $\|\mathbf{x} - \mathbf{y}\| \geq \delta$. We present a constructive proof in form of an algorithm which allows us to compute all $\mathbf{a}$ of bounded norm of a given lattice Λ. Namely, let $\Lambda = \bigoplus_{i=1}^k \mathbb{Z}\mathbf{a}_i \subset \mathbb{R}^n$ and A, Q be given as in (3.7), (3.8), also let C be a positive constant. We then need to determine all $\mathbf{a} \in \Lambda$ subject to $\|\mathbf{a}\|^2 \leq C$ and via the presentation $\mathbf{a} = \sum_{i=1}^k x_i \mathbf{a}_i$ this is tantamount to

$$Q(\mathbf{x}) = \mathbf{x}^t A \mathbf{x} \leq C \quad (\mathbf{x} \in \mathbb{Z}^k). \tag{3.10}$$

The set $E := \{\mathbf{x} \in \mathbb{R}^k \mid Q(\mathbf{x}) \leq C\}$ is an ellipsoid, hence the solutions of (3.10) are just the lattice points of an ellipsoid. We supplement Q to a sum of full squares:

Algorithm for quadratic supplement (3.11)

Input. $A = (a_{ij}) \in \mathbb{R}^{k \times k}$ positive definite.
Output. $(q_{ij}) \in \mathbb{R}^{k \times k}$ an upper triangular matrix satisfying
$\mathbf{x}^t A \mathbf{x} = \sum_{i=1}^k q_{ii}(x_i + \sum_{j=i+1}^k q_{ij} x_j)^2$.
Step 1. (Initialization). Set $q_{ij} \leftarrow a_{ij}$ $(1 \leq i \leq j \leq k)$.
Step 2. (Supplementing). For $i = 1, 2, \ldots, k-1$ set $q_{ji} \leftarrow q_{ij}$, $q_{ij} \leftarrow (q_{ij}/q_{ii})$ $(i+1 \leq j \leq k)$ and then for each i and $\mu = i+1, \ldots, k$: $q_{\mu\nu} \leftarrow q_{\mu\nu} - q_{\mu i} q_{i\nu}$ $(\mu \leq \nu \leq k)$.
Step 3. (Update). Set $q_{ij} \leftarrow 0$ $(1 \leq j < i \leq k)$ and terminate.

We note that this algorithm is similar to a Cholesky decomposition R^tR of A. The entries r_{ij} of R can be easily obtained from the q_{ij} via

$$r_{ii}=(q_{ii})^{\frac{1}{2}}, \quad r_{ij}=r_{ii}q_{ij} \quad (1\leqslant i<j\leqslant k), \tag{3.12}$$

and vice versa, of course.

The mode of operation of (3.11) and its notion 'quadratic supplement' is best explained by an example.

Example (3.13)

Let $A=\begin{pmatrix}2&1&1\\1&2&1\\1&1&2\end{pmatrix}$.

$(q_{ij})=$	$Q(\mathbf{x})=$
$\begin{pmatrix}2&1&1\\ *&2&1\\ *&*&2\end{pmatrix}$	$2{x_1}^2+2x_1x_2+2x_1x_3 + 2{x_2}^2 + 2x_2x_3 + 2{x_3}^2$
$\downarrow$	$\downarrow$
$\begin{pmatrix}2&\frac{1}{2}&\frac{1}{2}\\ 1&\frac{3}{2}&\frac{1}{2}\\ 1&*&\frac{3}{2}\end{pmatrix}$	$2(x_1+\frac{1}{2}x_2+\frac{1}{2}x_3)^2 + \frac{3}{2}{x_2}^2+x_2x_3 + \frac{3}{2}{x_3}^2$
$\downarrow$	$\downarrow$
$\begin{pmatrix}2&\frac{1}{2}&\frac{1}{2}\\ 1&\frac{3}{2}&\frac{1}{3}\\ 1&\frac{1}{2}&\frac{4}{3}\end{pmatrix}$	$2(x_1+\frac{1}{2}x_2+\frac{1}{2}x_3)^2 + \frac{3}{2}(x_2+\frac{1}{3}x_3)^2 + \frac{4}{3}{x_3}^2$

From the output of (3.11) it is now clear how we will solve (3.10). Since $\mathbf{x}^tA\mathbf{x}$ has been transformed into a sum of full squares, we obtain for each item

$$q_{ii}\left(x_i+\sum_{j=i+1}^{k}q_{ij}x_j\right)^2\leqslant C-\sum_{\mu=i+1}^{k}q_{\mu\mu}\left(x_\mu+\sum_{j=\mu+1}^{k}q_{\mu j}x_j\right)^2=:T_i$$

$$(i=k,k-1,\ldots,1). \tag{3.14a}$$

Hence, for each $x_k\in\mathbb{Z}$ subject to

$$|x_k|\leqslant(T_k/q_{kk})^{\frac{1}{2}}=(C/q_{kk})^{\frac{1}{2}}, \tag{3.14b}$$

we determine all possibilities for x_{k-1}. And for fixed $x_{i+1},\ldots,x_k\in\mathbb{Z}$ satisfying $\sum_{\mu=i+1}^{k}q_{\mu\mu}(x_\mu+\sum_{j=\mu+1}^{k}q_{\mu j}x_j)^2\leqslant T_{i+1}$ we obtain all possibilities for x_i from

$$-(T_i/q_{ii})^{\frac{1}{2}}-U_i\leqslant x_i\leqslant(T_i/q_{ii})^{\frac{1}{2}}-U_i \quad \left(U_i:=\sum_{j=i+1}^{k}q_{ij}x_j;\ k-1\geqslant i\geqslant 1\right). \tag{3.14c}$$

From this the following algorithm is immediate.

Algorithm for solving $Q(x) \leqslant C$ (3.15)

Input. The matrix (q_{ij}) from the output of (3.11) and $C > 0$.

Output. All $\mathbf{x} \in \mathbb{Z}^k$, $\mathbf{x} \neq \mathbf{0}$, satisfying (3.10) and additionally each corresponding value $Q(\mathbf{x})$.

Step 1. (Initialization). Set $i \leftarrow k$, $T_i \leftarrow C$, $U_i \leftarrow 0$.

Step 2. (Bounds for x_i). Set $Z \leftarrow (T_i/q_{ii})^{\frac{1}{2}}$, $UB(x_i) \leftarrow \lfloor Z - U_i \rfloor$, $x_i \leftarrow \lceil -Z - U_i \rceil - 1$.

Step 3. (Increase x_i). Set $x_i \leftarrow x_i + 1$; for $x_i \leqslant UB(x_i)$ go to 5.

Step 4. (Increase i). Set $i \leftarrow i + 1$ and go to 3.

Step 5. (Decrease i). For $i = 1$ go to 6; else set $i \leftarrow i - 1$, $U_i \leftarrow \sum_{j=i+1}^{k} q_{ij} x_j$, $T_i \leftarrow T_{i+1} - q_{i+1,i+1}(x_{i+1} + U_{i+1})^2$ and go to 2.

Step 6. (Solution). For $\mathbf{x} = \mathbf{0}$ terminate; else print $\mathbf{x}$, $-\mathbf{x}$ and $Q(\mathbf{x}) = C - T_1 + q_{11}(x_1 + U_1)^2$, then go back to 3.

Remarks (3.16)

(i) Clearly, the algorithm produces at most finitely many solution vectors $\mathbf{x}$.

(ii) The non-vanishing coordinate of highest index of $\mathbf{x}$ is always negative. This can be stipulated because of $Q(\mathbf{x}) = Q(-\mathbf{x})$.

(iii) The necessary computations usually employ floating point numbers (Step 2). Hence, in larger dimensions one must take care of round-off errors. An error analysis of algorithms (3.11), (3.15) is contained in [13].

(iv) Algorithm (3.15) can still be improved. If R^tR is the Cholesky decomposition of A and $\tilde{\mathbf{r}}_i$ denotes the ith column of $(R^{-1})^t$ we obtain for the coordinates x_i of any solution $\mathbf{x}$ of (3.10) by the Cauchy–Schwarz inequality:

$$|x_i|^2 = (\tilde{\mathbf{r}}_i^t(R\mathbf{x}))^2 \leqslant \|\tilde{\mathbf{r}}_i\|^2 \|R\mathbf{x}\|^2 \leqslant \|\tilde{\mathbf{r}}_i\|^2 C \quad (1 \leqslant i \leqslant k).$$

Therefore we shall reduce the norms of $\tilde{\mathbf{r}}_1, \ldots, \tilde{\mathbf{r}}_k$ by the LLL-algorithm discussed at the end of this section. This means to determine U, $U^{-1} \in GL(k, \mathbb{Z})$ such that $(\tilde{\mathbf{r}}_1, \ldots, \tilde{\mathbf{r}}_k)\, U^t$ is reduced. Then A has to be replaced by $(RU^{-1})^t(RU)$ and the obtained solutions must be multiplied by U^{-1} (see exercise 2). It is also recommendable to order the coordinates such that the bounds for $|x_i|$ in terms of $\|\tilde{\mathbf{r}}_i\|^2$ do increase. For details and a complexity analysis of the algorithms see [4].

We note, that (3.15) is appropriate for solving a few other tasks as well.

Computation of a vector of minimal length in a lattice. (3.17a)
For that purpose we set $C \leftarrow \min_{1 \leqslant i \leqslant k} a_{ii}$ (length of shortest basis vector) and each time a shorter vector $\mathbf{x}$ is found C and all T_i are decreased appropriately.

Computation of all $\mathbf{x} \in \mathbb{Z}^k$ *satisfying* $0 < \tilde{C} \leqslant \mathbf{x}^t A \mathbf{x} \leqslant C$. (3.17b)
In that case the bounds for x_1 have to be changed suitably in (3.15). Unfortunately this has barely any effect on the running time.

Computation of all $\mathbf{a} \in \Lambda$ *subject to* $\|\mathbf{y} - \mathbf{a}\| \leqslant C$ *for some fixed* $\mathbf{y} \in \mathbb{R} \otimes_{\mathbb{Z}} \Lambda$. (3.17c)
This means to determine short lattice vectors in the residue class $\mathbf{y} + \Lambda$. From the presentation $\mathbf{y} = \sum_{i=1}^{k} y_i \mathbf{a}_i$ ($y_i \in \mathbb{R}$, $1 \leqslant i \leqslant k$) we obtain the inequalities (3.14a–c) for $\eta_i := y_i - x_i$ ($1 \leqslant i \leqslant k$) instead of x_i. This requires minor changes of (3.15), the algorithm will then especially terminate if i becomes larger than k in Step 4.

Numerous computed examples ([4], [13]) suggest that (3.15) is indeed very efficient. For larger dimensions k (say $k > 10$) it should be applied only in connection with LLL-reduction, however (compare (3.16)(iv)).

In the sequel we consider the construction of special bases of lattices Λ. For the applications and for geometrical reasons (form of the fundamental parallelotope) we are interested in bases consisting of vectors of small norm. Ordering the vectors of $\mathbb{R}^n$ according to the size of their norm defines a partial ordering on $\mathbb{R}^n$:

$$\mathbf{a} < \mathbf{b} :\Leftrightarrow \|\mathbf{a}\| < \|\mathbf{b}\| \quad (\mathbf{a}, \mathbf{b} \in \mathbb{R}^n). \tag{3.18a}$$

This induces a partial ordering on the set $\mathfrak{S}_\Lambda$ of all bases of a k-dimensional lattice Λ via

$$(\mathbf{b}_1 \cdots \mathbf{b}_k) < (\mathbf{a}_1 \cdots \mathbf{a}_k) :\Leftrightarrow \exists j \in \{1, \ldots, k\} \ \forall i \in \{1, \ldots, j-1\}: \|\mathbf{b}_i\| = \|\mathbf{a}_i\| \wedge \mathbf{b}_j < \mathbf{a}_j. \tag{3.18b}$$

Definition (3.18c)
A minimal element of $\mathfrak{S}_\Lambda$ *with respect to* $<$ *is called a* **Minkowski reduced basis** *of* Λ.

Such a reduced basis is by no means unique. For example, for every $\pi \in \mathfrak{S}_n$ (the symmetric group of n elements) $\mathbf{e}_{\pi(1)}, \ldots, \mathbf{e}_{\pi(n)}$ is a reduced basis of $\mathbb{Z}^n$ in the sense of (3.18c). Moreover, the automorphism group of Λ need not even operate transitively on the set of Minkowski reduced bases. This is best illustrated by the hexagonal lattice in 3-dimensional space $\mathbb{R}^3$. Let

$$\Lambda = \mathbb{Z}\mathbf{x} + \mathbb{Z}\mathbf{y} + \mathbb{Z}\mathbf{z} \quad \text{for} \quad \mathbf{x}^t = (1, 0, 0), \quad \mathbf{y}^t = (\tfrac{1}{2}, \tfrac{3^{1/2}}{2}, 0), \quad \mathbf{z}^t = (\tfrac{1}{2}, \tfrac{3^{1/2}}{6}, \tfrac{6^{1/2}}{3}).$$

Then $\|\mathbf{x}\| = \|\mathbf{y}\| = \|\mathbf{z}\| = 1$ and $\mathbf{x}^t\mathbf{y} = \mathbf{x}^t\mathbf{z} = \mathbf{y}^t\mathbf{z} = \frac{1}{2}$. The basis $\mathbf{x}, \mathbf{y}, \mathbf{z}$ of Λ is Minkowski reduced. Namely, we can easily compute all $\mathbf{u} \in \Lambda$ with $\|\mathbf{u}\| \leqslant 1$. For $\mathbf{u} = \xi\mathbf{x} + \eta\mathbf{y} + \zeta\mathbf{z}$ we obtain (also compare (3.13))

$$\begin{aligned} \|\mathbf{u}\|^2 &= \xi^2 + \eta^2 + \zeta^2 + \xi\eta + \xi\zeta + \eta\zeta \\ &= (\xi + \tfrac{1}{2}\eta + \tfrac{1}{2}\zeta)^2 + \tfrac{3}{4}(\eta + \tfrac{1}{3}\zeta)^2 + \tfrac{2}{3}\zeta^2, \end{aligned}$$

from which all non-zero vectors $\mathbf{u}$ of length $\leqslant 1$ are easily calculated by (3.15). There are 12 of them, all of length 1: $\pm\mathbf{x}$, $\pm\mathbf{y}$, $\pm\mathbf{z}$, $\pm(\mathbf{x}-\mathbf{y})$, $\pm(\mathbf{x}-\mathbf{z})$, $\pm(\mathbf{y}-\mathbf{z})$. Hence, all bases of Λ consisting only of vectors of length 1 are Minkowski reduced. But whereas the scalar products of the basis vectors of $(\mathbf{x},\mathbf{y},\mathbf{z})$ are all $\frac{1}{2}$, we obtain for the reduced basis $(\mathbf{x},\mathbf{y}-\mathbf{z},\mathbf{z})$ that $\mathbf{x}^t(\mathbf{y}-\mathbf{z})=0$, $\mathbf{x}^t\mathbf{z}=-(\mathbf{y}-\mathbf{z})^t\mathbf{z}=\frac{1}{2}$. Therefore both bases cannot be isomorphic. Uniqueness of reduction can be obtained by ordering vectors of equal norm lexicographically. For most applications this is not necessary. However, the theory becomes much easier if we stipulate a total ordering. Then (3.18a)–(3.18c) have to be replaced by

$$\mathbf{a}<_T\mathbf{b}:\Leftrightarrow\|\mathbf{a}\|<\|\mathbf{b}\|\vee(\|\mathbf{a}\|=\|\mathbf{b}\|\wedge\exists j\in\{1,\ldots,n\}\ \forall i=1,\ldots,j-1:a_i=b_i\wedge a_j>b_j)\quad(\mathbf{a},\mathbf{b}\in\mathbb{R}^n). \tag{3.19a}$$

(The somewhat strange condition $a_j>b_j$ implies that $-\mathbf{e}_i$ is greater than $\mathbf{e}_i$ $(1\leqslant i\leqslant n)$, i.e. the canonical basis of $\mathbb{Z}^n$ is totally reduced which we would naturally expect.)

$$(\mathbf{b}_1,\ldots,\mathbf{b}_k)<_T(\mathbf{a}_1,\ldots,\mathbf{a}_k):\Leftrightarrow\exists j\in\{1,\ldots,k\}\ \forall i=1,\ldots,j-1:\ \mathbf{b}_i=\mathbf{a}_i\wedge\mathbf{b}_j<_T\mathbf{a}_j. \tag{3.19b}$$

A minimal element of $\mathfrak{S}_\Lambda$ *with respect to* $<_T$ *is called a* **totally reduced** *basis of* Λ. (3.19c)

Hence, totally reduced bases are uniquely determined and are of course Minkowski reduced. (We note that in the hexagonal lattice above the basis $\mathbf{x},\mathbf{y},\mathbf{z}$ is totally reduced.) However, their computation can be rather time consuming. Therefore in many cases we shall use LLL-reduction as a substitute. LLL-reduced bases can be easily calculated but don't have such nice properties as totally reduced bases. (Their basis vectors need not even be ordered with respect to the norm.) The concept of LLL-reduction is discussed at the end of this section.

In the sequel we develop an algorithm for the computation of a totally reduced basis of a lattice Λ. It will be calculated inductively. Let $\mathbf{b}_1,\ldots,\mathbf{b}_k$ be the initial basis of Λ. We define r-dimensional sublattices via $\Lambda_r:=\mathbb{Z}\mathbf{b}_1+\cdots+\mathbb{Z}\mathbf{b}_r$ $(1\leqslant r\leqslant k)$. A totally reduced basis of Λ_1 is either $\mathbf{b}_1$ or $-\mathbf{b}_1$. Hence, we can assume that we know a basis $\mathbf{b}_1,\ldots,\mathbf{b}_k$ of Λ such that $\mathbf{b}_1,\ldots,\mathbf{b}_r$ is a totally reduced basis of Λ_r for some r $(1\leqslant r<k)$. Then we need a criterion, whether $\mathbf{b}_1,\ldots,\mathbf{b}_{r+1}$ also form a totally reduced basis for Λ_{r+1}.

If the above is not the case, there must be a vector $\mathbf{c}\in\Lambda_{r+1}$ with the properties

$$\mathbf{c}<_T\mathbf{b}_{r+1}, \tag{3.20a}$$

$$\mathbf{c} \text{ is linearly independent of } \mathbf{b}_1, \ldots, \mathbf{b}_r, \tag{3.20b}$$

$$\mathbf{c} \text{ can be supplemented to a basis of } \Lambda_{r+1} \text{ which is smaller than the basis } \mathbf{b}_1, \ldots, \mathbf{b}_{r+1}. \tag{3.20c}$$

Hence, lemma (2.10) suggests we consider the set

$$S_{r+1} := \left\{ \mathbf{x} \in \Lambda_{r+1} \,\middle|\, \mathbf{x} = \sum_{i=1}^{r+1} \xi_i \mathbf{b}_i, \xi_{r+1} \neq 0, \gcd(\xi_1, \ldots, \xi_{r+1}) = 1 \right\}. \tag{3.21}$$

There are two possibilities:

$$\forall \mathbf{x} \in S_{r+1} \ \forall j \in \{1, \ldots, r+1\}: \gcd(\xi_j, \ldots, \xi_{r+1}) = 1 \Rightarrow \mathbf{x} = \mathbf{b}_j \vee \mathbf{b}_j <_T \mathbf{x}. \tag{3.22a}$$

In this case $\mathbf{b}_1, \ldots, \mathbf{b}_{r+1}$ is a totally reduced basis for Λ_{r+1} according to lemma (2.10). The other possibility is

$$\exists \mathbf{x} \in S_{r+1} \exists j \in \{1, \ldots, r+1\}: \gcd(\xi_j, \ldots, \xi_{r+1}) = 1 \wedge \mathbf{x} <_T \mathbf{b}_j. \tag{3.22b}$$

In this case $(\xi_j, \ldots, \xi_{r+1})$ can be made the first column of a matrix $U \in GL(r+2-j, \mathbb{Z})$, and

$$(\mathbf{b}_1 \cdots \mathbf{b}_{r+1}) \left(\begin{array}{c:c} I_{j-1} & \begin{matrix} \xi_1 & \\ \vdots & 0 \\ \xi_{j-1} & \end{matrix} \\ \hdashline 0 & U \end{array} \right)$$

is a basis of Λ_{r+1} which is smaller with respect to $<_T$. Because of (3.16)(*i*) possibility (3.22b) can occur only a finite number of times. The problem will be solved if we show that we must consider only a finite subset of vectors of S_{r+1}. Let $V_r := \mathbb{R}\mathbf{b}_1 + \cdots + \mathbb{R}\mathbf{b}_r \simeq \mathbb{R} \otimes_{\mathbb{Z}} \Lambda_r$ and $\mathbf{b}_p$ the orthogonal projection of $\mathbf{b}_{r+1}$ into V_r, $\mathbf{b}^*_{r+1}$ the corresponding orthogonal complement:

$$\mathbf{b}_{r+1} = \mathbf{b}_p + \mathbf{b}^*_{r+1}. \tag{3.23}$$

To obtain a presentation of $\mathbf{b}_p, \mathbf{b}^*_{r+1}$ we compute the orthogonal basis $\mathbf{b}^*_1, \ldots, \mathbf{b}^*_k$ of V_r belonging to $\mathbf{b}_1, \ldots, \mathbf{b}_k$ by the usual Gram-Schmidt orthogonalization procedure:

$$\mathbf{b}^*_i := \mathbf{b}_i - \sum_{j=1}^{i-1} \mu_{ij} \mathbf{b}^*_j \quad \text{for } \mu_{ij} := \frac{\mathbf{b}_i^t \mathbf{b}^*_j}{\mathbf{b}^{*t}_j \mathbf{b}^*_j} \quad (1 \leqslant j < i \leqslant k), \tag{3.24}$$

which provides

$$\mathbf{b}_p = \sum_{j=1}^{r} \mu_{r+1,j} \mathbf{b}^*_j. \tag{3.25}$$

We note that this also yields a presentation of $\mathbf{b}_p$ in terms of $\mathbf{b}_1, \ldots, \mathbf{b}_r$ (see exercise 1).

Since we are only interested in those vectors $\mathbf{x}$ of S_{r+1} with $\mathbf{x} <_T \mathbf{b}_{r+1}$

the equation

$$\|\mathbf{x}\|^2 = \left\|\sum_{i=1}^{r+1} \xi_i \mathbf{b}_i\right\|^2 = \left\|\sum_{i=1}^{r} \xi_i \mathbf{b}_i + \xi_{r+1}\mathbf{b}_p\right\|^2 + \xi_{r+1}^2 \|\mathbf{b}_{r+1}^*\|^2 \tag{3.26}$$

yields the following restriction for ξ_{r+1}:

$$0 < |\xi_{r+1}| \leqslant \frac{\|\mathbf{b}_{r+1}\|}{\|\mathbf{b}_{r+1}^*\|} =: B_{r+1}. \tag{3.27}$$

(Since $\mathbf{b}_{r+1}$ is linearly independent from $\mathbf{b}_1, \ldots, \mathbf{b}_r$ we must have $\mathbf{b}_{r+1}^* \neq \mathbf{0}$.)

Thus we need consider only finitely many possibilities for ξ_{r+1}. For a fixed value of ξ_{r+1} each $\mathbf{x} \in \Lambda_{r+1}$ subject to $\|\mathbf{x}\| \leqslant \|\mathbf{b}_{r+1}\|$ satisfies $\|\mathbf{x}_p\|^2 \leqslant \|\mathbf{b}_{r+1}\|^2 - \xi_{r+1}^2 \|\mathbf{b}_{r+1}^*\|^2 =: B(\xi_{r+1})$. But this means we have to determine vectors of small length in the residue class of $\xi_{r+1}\mathbf{b}_p + \Lambda_r$, a problem which is also solved by (3.15) as was discussed in (3.17c).

Before we start to describe a general reduction algorithm we note the following. The crucial point will be possibility (3.22b). If it occurs we may end up with a reduced basis $\mathbf{b}_1, \ldots, \mathbf{b}_j$ with j much smaller than r before. Thus the process of building up a reduced basis can become rather lengthy. Of course, we would prefer to be quite sure that (3.22b) will not occur. Hence, we apply a weaker type of reduction (see (3.40)) to the basis vectors first. The algorithm below concerns total reduction of a lattice basis, the corresponding one for Minkowski reduction is derived from it by replacing $<_T$ by $<$.

General reduction algorithm (3.28)

Input. A basis $\mathbf{b}_1, \ldots, \mathbf{b}_k$ of a lattice Λ of a rank $k \geqslant 2$ which has been LLL-reduced (see (3.40)) and the basis vectors have been totally ordered.

Output. A totally reduced basis $\mathbf{b}_1, \ldots, \mathbf{b}_k$ of Λ.

Step 1. (Initialization). Set $r \leftarrow 1, \Lambda_r \leftarrow \mathbb{Z}\mathbf{b}_1, \mathbf{b}_1^* \leftarrow \mathbf{b}_1$.

Step 2. (Orthogonal decomposition). Compute $\mu_{r+1,j} \leftarrow (\mathbf{b}_{r+1}^t \mathbf{b}_j^* / \|\mathbf{b}_j^*\|^2)$ $(1 \leqslant j \leqslant r)$ (compare (3.24)) and $\beta_{r+1,j}$ of $\mathbf{b}_{r+1}^* = \mathbf{b}_{r+1} - \sum_{j=1}^{r} \beta_{r+1,j}\mathbf{b}_j = \mathbf{b}_{r+1} - \mathbf{b}_p$ (see exercise 1), as well as $B_{r+1} \leftarrow \lfloor \|\mathbf{b}_{r+1}\| / \|\mathbf{b}_{r+1}^*\| \rfloor$, $\xi_{r+1} \leftarrow 1$.

Step 3. (Computation of short vectors in $\xi_{r+1}\mathbf{b}_p + \Lambda_r$). Set $B(\xi_{r+1}) \leftarrow \|\mathbf{b}_{r+1}\|^2 - \xi_{r+1}^2 \|\mathbf{b}_{r+1}^*\|^2$ and compute by (3.15) (subject to the changes outlined in (3.17c) the set

$$\mathfrak{M} := \left\{ \mathbf{x} \in \Lambda_r + \mathbb{Z}\mathbf{b}_{r+1} \mid \mathbf{x} = \sum_{i=1}^{r+1} \xi_i \mathbf{b}_i, \mathbf{x} - \xi_{r+1}\mathbf{b}_{r+1}^* \in \xi_{r+1}\mathbf{b}_p + \Lambda_r, \right.$$

$$\left. \|\mathbf{x} - \xi_{r+1}\mathbf{b}_{r+1}^*\|^2 \leqslant B(\xi_{r+1}) \right\}.$$

For $\mathfrak{M} = \emptyset$ go to 5.

Step 4. (Reduction criterion). For $j = 1, \ldots, r+1$ let $\mathbf{x}_j \in \mathfrak{M}$ be minimal with respect to $<_T$ satisfying $\gcd(\xi_j, \ldots, \xi_{r+1}) = 1$ and $\mathbf{x}_j <_T \mathbf{b}_j$; if $\mathfrak{M}$ contains no $\mathbf{x}_j$ satisfying these conditions set $\mathbf{x}_j \leftarrow \mathbf{0}$. For each j $(1 \leqslant j \leqslant r+1)$ with $\mathbf{x}_j \neq \mathbf{0}$ transform (if possible) $\mathbf{b}_1, \ldots, \mathbf{b}_{r+1}$ into a new basis $\mathbf{b}_1, \ldots, \mathbf{b}_{j-1}, \mathbf{x}_j, \tilde{\mathbf{b}}_{j+1}, \ldots, \tilde{\mathbf{b}}_{r+1}$ by an application of (2.10). Then set $\mathbf{b}_j \leftarrow \mathbf{x}_j$, $\mathbf{b}_i \leftarrow \tilde{\mathbf{b}}_i$ $(j < i \leqslant r+1)$ and update the $\mathbf{b}_i^*$ $(j \leqslant i \leqslant r+1)$ and the coordinates of $\mathbf{x}_\nu$ $(\nu > j)$. If no transformation is carried out go to 5. Otherwise let j_0 be the smallest index for which a transformation occurred. For $j_0 \leqslant r$ set $r \leftarrow j_0$, $\Lambda_r \leftarrow \mathbb{Z}\mathbf{b}_1 + \cdots + \mathbb{Z}\mathbf{b}_r$ and go to 2.

Step 5. (Increase ξ_{r+1}). Set $\xi_{r+1} \leftarrow \xi_{r+1} + 1$. For $\xi_{r+1} \leqslant B_{r+1}$ go to 3.

Step 6. (Increase r). Set $r \leftarrow r+1$, $\Lambda_r \leftarrow \Lambda_{r-1} + \mathbb{Z}\mathbf{b}_r$. For $r < k$ go to 2, otherwise terminate.

It is obvious that (3.28) is not suited for computations by hand. Also, for small values of k it is usually easier to calculate all lattice vectors of Λ of norm $\leqslant \|\mathbf{b}_k\|$ and to choose a reduced basis from those.

Besides the Minkowski reduced bases there are other important sets of maximally many independent vectors of small norm in a lattice. Namely, if we only require (3.20a), (3.20b), but not (3.20c), we obtain Minkowski's famous successive minima.

Definition (3.29)

Let Λ be a k-dimensional lattice in $\mathbb{R}^n$. For $i = 1, 2, \ldots, k$ let $M_i \in \mathbb{R}^{>0}$ be minimal with the property that there exist linearly independent vectors $\mathbf{y}_1, \ldots, \mathbf{y}_i$ in Λ satisfying $\|\mathbf{y}_j\|^2 \leqslant M_i$ $(1 \leqslant j \leqslant i)$. $M_1, \ldots, M_k$ are called **successive minima of** *Λ (with respect to the function $\|\ \ \|$).*

The following chain of inequalities is a trivial consequence of the definition:

$$M_1 \leqslant M_2 \leqslant \cdots \leqslant M_k. \tag{3.30}$$

Since we use a different notation, the reader will already surmise that $\mathbf{y}_1, \ldots, \mathbf{y}_k$ will not be a reduced basis for Λ, generally speaking. Fortunately, we can be more precise and describe the relationship between reduced basis vectors $\mathbf{b}_i$ and successive minima vectors $\mathbf{y}_i$ in greater detail $(1 \leqslant i \leqslant k)$.

The following example was already known to Minkowski:

Example (3.31)

Let $\mathbf{b}_i = \mathbf{e}_i$ $(i = 1, \ldots, 4)$, $\mathbf{b}_5 = \frac{1}{2}\sum_{i=1}^{5} \mathbf{e}_i$ in $\mathbb{R}^5$, $\Lambda = \sum_{i=1}^{5} \mathbb{Z}\mathbf{b}_i$. It is easily seen that $\mathbf{b}_1, \ldots, \mathbf{b}_5$ is a totally reduced basis for Λ. (We recommend this as an exercise). On the other hand $M_1 = \cdots = M_5 = 1$ with $\mathbf{y}_i = \mathbf{b}_i$ $(i = 1, \ldots, 4)$, $\mathbf{y}_5 = 2\mathbf{b}_5 - \mathbf{b}_4 - \mathbf{b}_3 - \mathbf{b}_2 - \mathbf{b}_1$.

Hence, successive minima vectors will not necessarily yield a reduced basis in dimensions greater than or equal to five. What about smaller dimensions? Here we have

Theorem (3.32)

Let Λ be a k-dimensional lattice of $\mathbb{R}^n$ and $M_i = \|\mathbf{y}_i\|^2$ $(i = 1, \ldots, k)$ the successive minima of Λ. Then $\mathbf{y}_1, \mathbf{y}_2, \mathbf{y}_3$ can be supplemented to a (reduced) basis of Λ and there is always a reduced basis $\mathbf{b}_1, \ldots, \mathbf{b}_k$ of Λ such that $M_i = \|\mathbf{b}_i\|^2$ for $i = 1, 2, \ldots, \min(k, 4)$.

We see from (3.31) that this result is best possible. On the other hand there are lattices with $M_i = \|\mathbf{b}_i\|^2$ $(i = 1, \ldots, k;\ k > 4)$, and the proof of (3.32) gives some idea, under which additional premises we can extend the last statement of the theorem to indices $i > 4$.

Proof

We already noticed that the first vector $\mathbf{b}_1$ of a reduced basis satisfies $M_1 = \|\mathbf{b}_1\|^2$. Hence, for a reduced basis $\mathbf{b}_1, \ldots, \mathbf{b}_k$ of Λ we can assume $M_i = \|\mathbf{b}_i\|^2$ for $i = 1, \ldots, r$, $r \geq 1$. Each $\mathbf{x} \in \Lambda$ has a presentation $\mathbf{x} = \sum_{i=1}^{k} x_i \mathbf{b}_i$ $(x_i \in \mathbb{Z}, 1 \leq i \leq k)$ and $\|\mathbf{x}\|^2$ is a positive definite quadratic form

$$\|\mathbf{x}\|^2 = Q(x_1, \ldots, x_k) = \sum_{1 \leq i,j \leq k} b_{ij} x_i x_j \quad (b_{ij} := \mathbf{b}_i{}^t \mathbf{b}_j)$$

in the variables $x_1, \ldots, x_k$.

As in (3.11) we compute $q_{ij} \in R$ $(i = 1, \ldots, k; j \geq i)$ such that

$$Q(x_1, \ldots, x_k) = \sum_{i=1}^{k} Q_i(x_i, \ldots, x_k),$$

with

$$Q_i(x_i, \ldots, x_k) := q_{ii} \left(x_i + \sum_{j=i+1}^{k} q_{ij} x_j \right)^2, \tag{3.33a}$$

and we define

$$\tilde{Q}_i(x_i, \ldots, x_k) := \sum_{j=i}^{k} Q_j(x_j, \ldots, x_k). \tag{3.33b}$$

All these quadratic forms are positive definite again. From our assumptions we know that

$$b_{ii} = M_i \geq q_{ii} > 0 \quad (1 \leq i \leq r), \tag{3.33c}$$

and we shall test if $M_{r+1} = \|\mathbf{b}_{r+1}\|^2$. For $M_{r+1} = \|\mathbf{x}\|^2$ we must have $|x_{r+1}| + \cdots + |x_k| > 0$, hence we will consider $m := \gcd(x_{r+1}, \ldots, x_k)$. In case $m = 1$ the vector $\mathbf{x}$ can be taken as new basis vector $\mathbf{b}_{r+1}$ because of (2.10).

So, let us assume $m > 1$.

From (3.33a), (3.33b) we obtain

$$m^2 \tilde{Q}_{r+1}\left(\frac{x_{r+1}}{m}, \ldots, \frac{x_k}{m} \right) \leq M_{r+1}. \tag{3.33d}$$

Therefore we can determine $\mathbf{v} = \sum_{i=1}^{k} v_i \mathbf{b}_i \in \Lambda$ in the following way:

$$v_i := \frac{x_i}{m} \quad (r+1 \leqslant i \leqslant k), \tag{3.33e}$$

$$\left| v_i + \sum_{j=i+1}^{k} q_{ij} v_j \right| \leqslant \tfrac{1}{2} \quad (i = r, r-1, \ldots, 1). \tag{3.33f}$$

Since $\mathbf{b}_1, \ldots, \mathbf{b}_r$, $\mathbf{v}$ are linearly independent we must have

$$M_{r+1} \leqslant \sum_{i=1}^{r} \frac{q_{ii}}{4} + \tilde{Q}_{r+1}(v_{r+1}, \ldots, v_k). \tag{3.33g}$$

Now (3.33d) and (3.33g) yield

$$(m^2 - 1)\tilde{Q}_{r+1}(v_{r+1}, \ldots, v_k) \leqslant \sum_{i=1}^{r} \frac{q_{ii}}{4}. \tag{3.33h}$$

Applying (3.33g) and (3.33h) we obtain

$$\begin{aligned} M_{r+1} &\leqslant \sum_{i=1}^{r} \frac{q_{ii}}{4}\left(1 + \frac{1}{m^2 - 1}\right) \\ &\leqslant \frac{m^2 r}{(m^2 - 1)4} \max\{q_{ii} \mid 1 \leqslant i \leqslant r\} \\ &\leqslant \frac{m^2 r}{(m^2 - 1)4} M_r \\ &\leqslant \frac{r}{3} M_r. \end{aligned} \tag{3.33i}$$

For $r = 1, 2$ this yields a contradiction to (3.30), and we get the first statement of the theorem.

In case $r = 3$ we still get a contradiction except for $m = 2$ and $M_1 = \cdots = M_4 = q_{11} = q_{22} = q_{33}$. But then $q_{12} = q_{13} = q_{23} = 0$ and we have $M_4 = \|\mathbf{v}\|^2$ for $\mathbf{v} = (0, 0, 0, x_4/m, \ldots, x_k/m)$, hence $\mathbf{v}$ is a candidate for $\mathbf{b}_4$. □

In case the first successive minima are known and if they are not equal, (3.33) can usually be improved, i.e. (3.33i) yields a contradiction also for $r \geqslant 4$.

The importance of the successive minima hinges on the estimates stated in the following theorem.

Theorem (3.34)

Let Λ be an n-dimensional lattice with successive minima $M_1, \ldots, M_n$. There exists a constant $\mu_n \in \mathbb{R}^{>0}$ depending only on n such that

$$M_1{}^n \leqslant \mu_n d(\Lambda)^2, \tag{3.34a}$$

and even

$$M_1 \cdot \ldots \cdot M_n \leqslant \mu_n d(\Lambda)^2. \tag{3.34b}$$

Proof

(a) We consider balls $B(\mathbf{x}, r) := \{\mathbf{y} \in \mathbb{R}^n \mid \|\mathbf{x} - \mathbf{y}\| \leqslant r\}$ for $\mathbf{x} \in \mathbb{R}^n$. It is obvious that the intersection of balls centered at different lattice points contains only boundary points for $r = M_1^{\frac{1}{2}}/2$.

We define parallelotopes ($1 \leqslant i_1 < \ldots < i_k \leqslant n$; $0 \leqslant k \leqslant n$)

$$\Pi(i_1, \ldots, i_k) := \{\mathbf{x} \in \mathbb{R}^n \mid 0 \leqslant x_j < 1 \text{ for } j \notin \{i_1, \ldots, i_k\},\\ 0 > x_j \geqslant -1 \text{ for } j \in \{i_1, \ldots, i_k\}\}.$$

There are 2^n disjoint parallelotopes, each containing $\mathbf{0}$ on its boundary.

The idea of the proof is to partition $B(\mathbf{0}, M_1^{\frac{1}{2}}/2)$ into 2^n subsets which can be transformed into the fundamental parallelotope $\Pi = \Pi(\emptyset)$ by translations such that the images are disjoint except for boundary points. Hence, the volume of $B(\mathbf{0}, M_1^{\frac{1}{2}}/2)$ is less than the mesh of Λ yielding (3.34a).

Namely, for $B(i_1, \ldots, i_k) := \Pi(i_1, \ldots, i_k) \cap B(\mathbf{0}, M_1^{\frac{1}{2}}/2)$ the translated set

$$\tilde{B}(i_1, \ldots, i_k) := B(i_1, \ldots, i_k) + \mathbf{e}_{i_1} + \cdots + \mathbf{e}_{i_k}$$

is of equal volume and is contained in $\Pi \cap B(\mathbf{e}_{i_1} + \cdots + \mathbf{e}_{i_k}, M_1^{\frac{1}{2}}/2)$. Therefore we obtain

$$V(B(\mathbf{0}, M_1^{\frac{1}{2}}/2)) \leqslant V(\Pi)$$

which yields

$$\pi^{n/2} \Gamma\left(\frac{n}{2} + 1\right)^{-1} (M_1^{\frac{1}{2}}/2)^n \leqslant d(\Lambda).$$

Setting $\mu_n := (4/\pi)^n \Gamma(n/2 + 1)^2$ we get (3.34a).

(b) To obtain (3.34b) let $\mathbf{y}_1, \ldots, \mathbf{y}_n$ be independent lattice points with $M_i = \|\mathbf{y}_i\|^2$ $(1 \leqslant i \leqslant n)$. For $\mathbf{x} \in \mathbb{R}^n$ we have $\mathbf{x} = \xi_1 \mathbf{y}_1 + \cdots + \xi_n \mathbf{y}_n$ ($\xi_i \in \mathbb{R}$, $1 \leqslant i \leqslant n$) and $\|\mathbf{x}\|^2$ is a positive definite quadratic form in $\xi_1, \ldots, \xi_n$. Applying quadratic supplement we obtain $\|\mathbf{x}\|^2 = z_1^2 + \cdots + z_n^2$ with $z_i = z_i(\xi_i, \ldots, \xi_n)$. Then we introduce a new norm using weights by $\|\mathbf{x}\|_n^2 := (1/M_1) z_1^2 + \cdots + (1/M_n) z_n^2$.

For each $\mathbf{x} \in \Lambda$, $\mathbf{x} \neq \mathbf{0}$, there is a maximal index k $(1 \leqslant k \leqslant n)$ such that $\xi_k \neq 0$, $\xi_{k+1} = \cdots = \xi_n = 0$, and we get $\|\mathbf{x}\|_n^2 \geqslant (1/M_k)(z_1^2 + \cdots + z_k^2) = (1/M_k)\|\mathbf{x}\|^2 \geqslant 1$ because of the definition of M_k.

Hence, for Λ (with $\| \ \|_n) M_{1_n} \geqslant 1$ and $d(\Lambda)_n^2 = (M_1 \cdot \ldots \cdot M_n)^{-1} d(\Lambda)^2$ (because of the weighted scalar product) and an application of (3.34a) gives (3.34b). □

The constants μ_n obtained in the proof are not very good. The best possible values ('Hermite's constants') are known only for $n \leqslant 8$, they are denoted by γ_n^n.

n	1	2	3	4	5	6	7	8
γ_n^n	1	4/3	2	4	8	64/3	64	256

(3.35a)

The best upper bounds known for $n > 8$ were given by Blichfeldt [2]:

$$\gamma_n^{\,n} \leqslant \left(\frac{2}{\pi}\right)^n \Gamma\left(1 + \frac{n+2}{2}\right)^2. \tag{3.35b}$$

Usually (3.34) is applied when we know a few successive minima and look for a lower bound of some invariant occurring in the determinant of the lattice. Therefore it is important that our general reduction algorithm (3.28) also can be used to compute $M_1, \ldots, M_k$.

Algorithm for the computation of successive minima (3.36)
The following changes in (3.28) are necessary. Add to step 1: Set $\mathbf{y}_i \leftarrow \mathbf{b}_i$ $(i = 1, \ldots, k)$. Replace step 3 by: 'Set $B(\xi_{r+1}) \leftarrow \|\mathbf{b}_{r+1}\|^2 - \xi_{r+1}^2 \|\mathbf{b}_{r+1}^*\|^2$ and compute $\mathbf{y} \in \xi_{r+1}\mathbf{b}_p + \Lambda_r$ of shortest length and set $\mathbf{x} = \sum_{i=1}^{r+1} m_i \mathbf{b}_i \leftarrow \mathbf{y} + \xi_{r+1}\mathbf{b}_{r+1}^*$. For $\|\mathbf{x}\| \geqslant \|\mathbf{y}_{r+1}\|$ go to step 5.' Step 4 is to be replaced by: 'Let $j \in \{1, \ldots, r+1\}$ be minimal with the property $\|\mathbf{x}\| < \|\mathbf{y}_j\|$ and set $a \leftarrow \gcd(\xi_j, \ldots, \xi_{r+1})$. Transform $\mathbf{b}_1, \ldots, \mathbf{b}_{r+1}$ into a basis $\mathbf{b}_1, \ldots, \mathbf{b}_{j-1}, \tilde{\mathbf{b}}_j, \ldots, \tilde{\mathbf{b}}_{r+1}$ of $\Lambda_r + \mathbb{Z}\mathbf{b}_{r+1}$ such that $\tilde{\mathbf{b}}_j = \frac{1}{a}(\mathbf{x} - \sum_{i=1}^{j-1} m_i \mathbf{b}_i)$. Set $r \leftarrow j$, $\mathbf{y}_j \leftarrow \mathbf{x}$, $\mathbf{b}_i \leftarrow \tilde{\mathbf{b}}_i$ $(j \leqslant i \leqslant r+1)$, present $\mathbf{y}_j, \ldots, \mathbf{y}_n$ in terms of the new basis and go to 2.'

The output consists of the successive minima $M_i := \|\mathbf{y}_i\|^2$ $(1 \leqslant i \leqslant k)$.

In (3.34a), (3.34b) we gave upper bounds for the successive minima of a lattice (with respect to the norm). It is clear that we cannot derive lower bounds for M_1 alone, depending only on $d(\Lambda)$. However, for $M_1 \cdot \ldots \cdot M_n$ it is not difficult to determine such a lower bound.

Theorem (3.37)
Let Λ be an n-dimensional lattice of $\mathbb{R}^n$ with successive minima $M_1, \ldots, M_n$. Then $M_1 \cdot \ldots \cdot M_n \geqslant d(\Lambda)^2$.

Proof
There exist linearly independent lattice vectors $\mathbf{y}_1, \ldots, \mathbf{y}_n$ for which $M_i = \|\mathbf{y}_i\|^2$. Those span a sublattice $\Lambda_1 := \mathbb{Z}\mathbf{y}_1 + \cdots + \mathbb{Z}\mathbf{y}_n$ of Λ of index $m \in \mathbb{N}$. Hence,

$$d(\Lambda)^2 \leqslant |\det(\mathbf{y}_1, \ldots, \mathbf{y}_n)|^2 \leqslant \prod_{i=1}^{n} \|\mathbf{y}_i\|^2$$

by Hadamard's inequality. □

As already mentioned the computation of a totally or Minkowski reduced basis of a lattice can be very time consuming. Hence, in many cases one is satisfied with constructing bases of lattices which are reduced in a much weaker sense. Those bases either have sufficiently good properties themselves (see example (3.42)) or they are used as initial bases of algorithms (3.15), (3.28). Besides the well-known notion of pair-reduction (see exercise 3) the most important reduction procedure now in use is LLL-reduction which was introduced in 1982 by Lenstra, Lenstra & Lovasz in a paper on the factorization of polynomials with rational integral coefficients [10].

Definition (3.38)
A basis $\mathbf{b}_1,\ldots,\mathbf{b}_k$ *of a lattice* Λ *is called* **LLL-reduced** *if the vectors* $\mathbf{b}_1^*,\ldots,\mathbf{b}_k^*$ *of the corresponding orthogonal basis (see (3.24)) satisfy*

$$|\mu_{ij}| \leqslant \tfrac{1}{2} \quad (1 \leqslant j < i \leqslant k), \tag{3.38a}$$

$$\| \mathbf{b}_i^* + \mu_{i,i-1}\mathbf{b}_{i-1}^* \|^2 \geqslant \tfrac{3}{4} \| \mathbf{b}_{i-1}^* \|^2. \tag{3.38b}$$

We note, that the constant $\frac{3}{4}$ in (3.38b) can be replaced by any constant α belonging to the open interval $(\frac{1}{4}, 1)$. The estimates of (3.39), (3.41) must then be changed appropriately. The larger the constant α is the better are the properties of a LLL-reduced basis but the larger is also the amount of computational steps for its calculation. Inequality (3.38b) means that the ellipsoid $\mathbf{x}^t\mathbf{A}\mathbf{x} \leqslant C$ derived from the basis vectors as in (3.10) is roughly shaped like a ball.

We state without proof several useful properties of LLL-reduced bases. The proof is analogous to that given in [10] for n-dimensional lattices (see also exercise 4).

Lemma (3.39)
Let $\mathbf{b}_1,\ldots,\mathbf{b}_k$ *be a LLL-reduced basis of a lattice* $\Lambda \subset \mathbb{R}^n$ *with corresponding orthogonal basis* $\mathbf{b}_1^*,\ldots,\mathbf{b}_k^*$ *and constants* μ_{ij} *of* (3.24). *Then the following estimates hold:*

(i) $\|\mathbf{b}_j\|^2 \leqslant 2^{i-1}\|\mathbf{b}_i^*\|^2$ $(1 \leqslant j \leqslant i \leqslant k)$;
(ii) $d(\Lambda) \leqslant \prod_{i=1}^k \|\mathbf{b}_i\| \leqslant 2^{k(k-1)/4} d(\Lambda)$;
(iii) $\|\mathbf{b}_1\| \leqslant 2^{(k-1)/4} d(\Lambda)^{1/k}$;
(iv) $\|\mathbf{b}_1\|^2 \leqslant 2^{k-1}\|\mathbf{x}\|^2$ *for all* $\mathbf{x} \in \Lambda$, $\mathbf{x} \neq \mathbf{0}$;
(v) $\|\mathbf{b}_j\|^2 \leqslant 2^{k-1}\max\{\|\mathbf{x}_1\|^2,\ldots,\|\mathbf{x}_t\|^2\}$ $(1 \leqslant j \leqslant t)$, where $t \in \mathbb{N}$, $1 \leqslant t \leqslant k$, *and* $\mathbf{x}_1,\ldots,\mathbf{x}_t$ *are linearly independent vectors of* Λ.

Before we present an algorithm which transforms a given basis of a lattice into a LLL-reduced one we shortly explain the underlying ideas. At the start the constants μ_{ij} and the orthogonal basis vectors $\mathbf{b}_i^*$ are calculated by (3.24). Then a LLL-reduced basis is constructed inductively. The induction is on the number of reduced basis vectors. The initial value of the induction parameter is $m = 2$, in case of $m > k$ the procedure terminates. There are three major steps:

(A) Reduce $\mu_{m,m-1}$ to $|\mu_{m,m-1}| \leqslant \frac{1}{2}$ by subtracting a suitable multiple of $\mathbf{b}_{m-1}$ from $\mathbf{b}_m$. (All $\mathbf{b}_i^*$ remain unchanged!)

(B) If (3.38b) holds for $i = m$ proceed to (C), else interchange $\mathbf{b}_{m-1}$ and $\mathbf{b}_m$. In case $m > 2$ also replace m by $m - 1$. Then go on with (A).

(C) Reduce (similar to (A)) μ_{mj} to $|\mu_{mj}| \leqslant \frac{1}{2}$ for $j = m-2, m-3, \ldots, 1$. Then increase m by 1. For $m > k$ terminate, else go on with (A).

Remark

In the algorithm the vectors $\mathbf{b}_i^*$ are not used explicitly but only the squares of their norms $B_i := \mathbf{b}_i^{*\prime}\mathbf{b}_i^*$.

LLL-reduction algorithm (3.40)

Input. Basis vectors $\mathbf{b}_1, \ldots, \mathbf{b}_k$ of a k-dimensional lattice Λ.

Output. A basis $\mathbf{b}_1, \ldots, \mathbf{b}_k$ of Λ which is LLL-reduced.

Step 1. (Initialization). For $i = 1, \ldots, k$ set:

$$\mu_{ij} \leftarrow \mathbf{b}_i'\mathbf{b}^*{}_j/B_j \quad (1 \leqslant j \leqslant i-1),$$

$$\mathbf{b}_i^* \leftarrow \mathbf{b}_i - \sum_{j=1}^{i-1} \mu_{ij}\mathbf{b}_j^*, \quad B_i \leftarrow \mathbf{b}_i^{*\prime}\mathbf{b}^*{}_i.$$

Then set $m \leftarrow 2$.

Step 2. (Set l). Set $l \leftarrow m - 1$.

Step 3. (Change μ_{ml} in case $|\mu_{ml}| > \frac{1}{2}$). If $|\mu_{ml}| > \frac{1}{2}$, set

$$r \leftarrow \{\mu_{ml}\}, \quad \mathbf{b}_m \leftarrow \mathbf{b}_m - r\mathbf{b}_l,$$

$$\mu_{mj} \leftarrow \mu_{mj} - r\mu_{lj} \quad (1 \leqslant j \leqslant l-1), \quad \mu_{ml} \leftarrow \mu_{ml} - r.$$

For $l = m - 1$ go to 4, else to 5.

Step 4. ((3.38b) violated on level m?) For $B_m < (\frac{3}{4} - \mu_{m,m-1}^2)B_{m-1}$ go to 6.

Step 5. (Decrease l). Set $l \leftarrow l - 1$. For $l > 0$ go to 3. If $m = k$, terminate; else set $m \leftarrow m + 1$ and go to 2.

Step 6. (Interchange $\mathbf{b}_{m-1}, \mathbf{b}_m$). Set $\mu \leftarrow \mu_{m,m-1}, B \leftarrow B_m + \mu^2 B_{m-1}$, $\mu_{m,m-1} \leftarrow \mu B_{m-1}/B, B_m \leftarrow B_{m-1}B_m/B, B_{m-1} \leftarrow B$; then set

$$\begin{pmatrix}\mathbf{b}_{m-1}\\ \mathbf{b}_m\end{pmatrix} \leftarrow \begin{pmatrix}\mathbf{b}_m\\ \mathbf{b}_{m-1}\end{pmatrix}, \quad \begin{pmatrix}\mu_{m-1,j}\\ \mu_{mj}\end{pmatrix} \leftarrow \begin{pmatrix}\mu_{mj}\\ \mu_{m-1,j}\end{pmatrix} \quad (1 \leqslant j \leqslant m-2),$$

$$\begin{pmatrix}\mu_{i,m-1}\\ \mu_{im}\end{pmatrix} \leftarrow \begin{pmatrix}1 & \mu_{m,m-1}\\ 0 & 1\end{pmatrix}\begin{pmatrix}0 & 1\\ 1 & -\mu\end{pmatrix}\begin{pmatrix}\mu_{i,m-1}\\ \mu_{im}\end{pmatrix} \quad (m+1 \leqslant i \leqslant k).$$

For $m > 2$ decrease m by 1. Then go to 2.

The transformation formulae of steps 3, 6 are easily derived from the ideas in (A), (B), (C) (see also [10]). It remains to show that the algorithm terminates. For this purpose let

$$D_i := \det(\mathbf{b}_\mu'\mathbf{b}_\nu)_{1 \leqslant \mu,\nu \leqslant i} \quad (1 \leqslant i \leqslant k)$$

be the principal minors of $d(\Lambda)^2$ ($= D_k$) and

$$D := \prod_{j=1}^{k-1} D_j.$$

Because of (3.3), (3.24) we also have

$$D_i = \prod_{j=1}^{i} \|\mathbf{b}_j^*\|^2 \quad (1 \leqslant i \leqslant k).$$

Each time algorithm (3.40) passes step 6 the value D_{m-1} is diminished by a factor $\frac{3}{4}$ whereas all other D_i remain unchanged. Hence, also D decreases by a factor $\frac{3}{4}$. But – as exercise 5 shows – there are positive lower bounds S_i for the D_i, i.e.

$$D_i \geqslant S_i > 0 \quad (1 \leqslant i \leqslant k)$$

independently of the chosen basis $\mathbf{b}_1, \ldots, \mathbf{b}_k$ of Λ. Therefore algorithm (3.40) can pass step 6 at most $\lfloor -\log(\frac{1}{D}\prod_{j=1}^{k-1} S_j)/\log\frac{3}{4} \rfloor$ times, the algorithm terminates after a finite number of steps.

In case the lattice Λ under consideration is of rank n and *integral* (i.e. contained in $\mathbb{Z}^n$) one can show [10]:

Lemma (3.41)

Let $\Lambda = \bigoplus_{i=1}^{n} \mathbb{Z}\mathbf{b}_i, \mathbf{b}_i \in \mathbb{Z}^n$, $\|\mathbf{b}_i\|^2 \leqslant B$ $(1 \leqslant i \leqslant n; B \geqslant 2)$. *Then the number of arithmetical operations of algorithm* (3.40) *is bounded by* $O(n^4 B)$, *all occurring numbers (integers!) have at most* $O(n \log B)$ *binary digits.*

Before we present some important applications of algorithm (3.40) we illustrate its disadvantages by two examples. Especially, the first example shows that the unsatisfying estimate (3.39) (iv) cannot be improved by much.

Examples (3.42)

(i) Let $B = (\beta_{ij}) \in \mathbb{R}^{n \times n}$ be given by

$$\beta_{ij} := \begin{cases} (3^{1/2}/2)^{i-1} & \text{for } j = i \\ 0 & \text{for } j < i \\ \frac{1}{2}(3^{\frac{1}{2}}/2)^{i-1} & \text{for } j > i \end{cases}; \quad 1 \leqslant i \leqslant n.$$

Let $\Lambda_k := \bigoplus_{i=1}^{k} \mathbb{Z}\mathbf{b}_i$ for fixed k $(1 \leqslant k \leqslant n)$, where $\mathbf{b}_1, \ldots, \mathbf{b}_n$ denote the columns of B. Then $\mathbf{b}_1, \ldots, \mathbf{b}_k$ is a LLL-reduced basis of Λ_k and $\min\{\|\mathbf{x}\| \mid \mathbf{x} \in \Lambda_k, \mathbf{x} \neq \mathbf{0}\} = (3^{1/2}/2)^{k-2}$ in case of $k \geqslant 2$ ([1], see also exercise 6).

(ii) In this example the basis vectors of the lattice under consideration are again given by the columns of the matrices shown below. A basis (corresponding to the first matrix) is LLL-reduced, then Minkowski-reduced and finally LLL-reduction is applied again. The listed squares of the norms of the basis vectors show that LLL-reduction can even spoil a short basis. This example was found in [14].

Initial basis:

9	6	0	1	3	6	4	3	4	1	7	7	7	7	6	7	8	7	1	8
9	0	1	4	7	7	9	3	8	9	1	7	0	8	2	8	0	8	1	4
7	5	2	1	1	6	6	8	4	5	1	4	2	8	8	6	8	4	7	2
2	1	5	2	3	1	8	9	2	3	5	3	0	7	8	1	6	5	7	2
0	9	7	8	0	1	8	0	8	1	1	3	1	9	0	6	9	1	7	5
0	8	5	6	4	7	1	1	0	3	3	4	3	0	2	0	7	3	1	3
10	4	3	1	6	7	1	8	9	9	7	3	2	4	6	4	7	4	1	6
8	9	1	7	7	0	8	6	3	0	1	5	7	1	5	8	9	5	2	10
1	0	1	5	7	3	5	4	7	3	6	7	3	6	6	6	9	5	6	0
8	4	2	2	4	3	5	3	2	7	4	6	2	6	1	8	6	0	8	7
9	6	6	6	0	2	2	8	7	3	2	5	6	3	4	1	2	2	9	3
9	8	2	3	0	5	5	10	3	3	9	7	8	3	8	0	4	7	8	5
7	8	5	0	4	9	7	9	2	3	4	3	7	8	9	8	6	8	7	9
0	1	6	10	3	2	2	4	5	3	4	4	9	0	5	9	1	9	6	9
5	3	7	8	6	4	1	2	0	7	6	3	4	9	0	4	0	1	5	8
0	1	7	0	0	9	9	7	3	8	8	9	0	0	8	4	8	5	0	1
5	1	8	5	5	1	2	4	1	5	7	7	2	5	5	3	6	0	5	0
8	2	3	2	8	0	1	1	9	6	4	2	5	3	8	8	9	2	5	6
2	2	1	0	5	9	0	9	7	1	9	6	3	2	8	3	0	6	2	7
9	7	9	4	1	4	5	3	3	6	3	10	9	0	2	4	5	2	7	9

Squares of norms of columns:

854 553 473 455 410 548 571 710 543 512 560 645 494 597 677 642 804 498 613 734

LLL-reduced basis:

3	3	3	3	2	1	6	5	1	1	1	1	1	3	2	3	2	2	2	1
−4	−1	0	−4	0	−1	−2	−4	6	−6	0	2	−5	0	−4	−1	1	−5	2	3
4	−1	−2	4	0	−1	2	−5	−4	0	−2	−3	−3	−2	−1	2	−1	3	−3	−1
−2	0	−4	1	1	−6	2	−1	−3	−4	5	−1	3	1	2	3	3	−3	0	−3
−2	3	−2	−3	3	−2	4	−1	1	−1	−2	−2	−2	2	−3	4	−3	3	2	1
1	4	−3	2	1	5	−4	−2	1	−5	−2	−4	0	0	−2	−5	−6	4	3	1
3	1	−4	−4	0	0	−2	1	−2	−2	0	−1	−1	−1	2	0	0	−1	−5	−2
2	4	2	2	0	−1	−3	3	0	1	−3	7	1	5	−4	−2	0	7	−2	6
−3	1	0	−4	−2	3	0	−1	−1	−3	0	0	0	−5	3	−2	−1	2	3	4
−2	0	−1	1	4	−3	−1	−2	−1	−2	−4	−2	−3	−2	0	1	−1	2	−2	0
1	3	−3	−1	−1	−2	1	−1	−2	3	2	1	−3	1	5	5	6	−2	−2	0
0	4	2	−2	5	4	4	3	−1	3	−1	−2	−4	0	−1	6	0	2	2	1
4	−4	−4	1	4	5	1	−2	−1	0	2	−2	6	2	6	1	1	−3	2	−5
−3	1	−2	3	−6	0	−1	3	4	−1	−3	−1	−4	−2	1	−2	2	−3	2	4
−1	−2	4	−1	−4	0	2	2	1	0	1	3	−2	6	3	4	2	−4	−2	1
−1	−3	−3	−2	2	−1	4	−5	−3	−6	2	−1	−1	−7	1	3	−2	1	2	−7
−1	3	0	0	−3	1	3	0	−5	3	2	3	3	−2	4	2	7	−2	−3	−3
−1	0	−1	−1	1	−3	2	−3	−6	−2	−2	6	3	4	−1	−5	3	−2	−5	−1
1	2	2	−2	1	−1	−1	−1	−2	−1	−7	−2	5	−1	6	0	0	−2	7	−1
3	1	−2	−4	0	0	−2	2	0	−1	5	−4	0	1	0	−1	5	4	−5	−5

Squares of norms of columns:

116	123	130	133	144	144	151	153	167	167	168	174	184	189	193	194	194	201	204	206

Minkowski reduced basis:

3	3	3	1	3	3	2	4	2	1	1	3	0	3	6	3	4	3	1	4
−1	−4	−1	2	0	−4	−4	2	0	2	−1	0	2	2	−2	−1	−1	−2	4	0
0	4	−1	5	−2	4	−3	−5	0	−2	−1	−3	0	−4	2	0	2	−1	0	−1
−1	−2	0	1	−4	1	3	3	1	3	−6	0	−3	0	2	0	4	3	2	−2
0	−2	3	5	−2	−3	1	0	3	−1	−2	−5	−2	3	4	−1	2	4	4	0
1	1	4	1	−3	2	1	3	1	5	5	−4	4	4	−4	−3	−1	0	0	−5
−2	3	1	−2	−4	−4	5	3	0	−3	0	−2	3	1	−2	−7	1	−4	4	−1
−1	2	4	−4	2	2	1	2	0	1	−1	−2	−3	−2	−3	1	1	−5	3	1
2	−3	1	1	0	−4	−1	2	−2	1	3	1	3	−4	0	3	2	1	−3	5
−2	−2	0	1	−1	1	−1	0	4	−5	−3	2	6	−1	−1	−1	3	−3	5	−1
−4	1	3	−2	−3	−1	2	−4	−1	1	−2	−1	2	1	1	−1	−3	4	−3	−3
−2	0	4	−3	2	−2	1	0	5	0	4	5	2	2	4	0	2	−5	−3	−2
0	4	−4	3	−4	1	2	−2	4	0	5	3	3	0	1	4	−2	−1	2	−3
1	−3	1	−1	−2	3	5	4	−6	−1	0	2	1	1	−1	0	−4	−2	0	3
−2	−1	−2	−1	4	−1	−2	2	−4	2	0	−3	−3	−2	2	−2	6	−3	4	−2
1	−1	−3	−3	−3	−2	−2	1	2	−2	−1	1	1	1	4	4	−3	0	1	−2
1	−1	3	−2	0	0	0	−3	−3	−3	1	−2	−4	1	3	−1	−1	0	−2	3
7	−1	0	0	−1	−1	−2	−1	1	5	−3	4	2	−5	2	−4	−3	−1	1	−1
−1	1	2	−3	2	−2	−3	0	1	−4	−1	2	2	1	−1	3	2	−1	4	−5
−1	3	1	0	−2	−4	4	3	0	2	0	2	1	6	−2	3	2	−3	0	−3

Squares of norms of columns:

103	116	123	125	130	133	139	140	144	144	144	144	149	149	150	151	152	153	156	156	157

LLL-reduced basis obtained from Minkowski reduced basis:

3	3	3	1	3	3	2	1	5	5	1	1	5	3	2	2	1	1	4	5
−1	−4	−1	2	0	−4	0	2	−4	2	1	2	−3	−4	−3	0	2	−3	−2	−3
0	4	−1	5	−2	4	0	−2	−5	2	1	2	1	−1	−1	8	−2	−1	−1	−2
−1	−2	0	1	−4	1	1	3	−1	−2	4	1	−2	6	−1	−3	−2	0	1	−4
0	−2	3	5	−2	−3	3	−1	−1	−3	1	−1	3	−1	−3	−1	−5	4	4	1
1	1	4	1	−3	2	1	5	−2	3	−2	−5	−2	2	−4	−2	2	−1	2	3
−2	3	1	−2	−4	−4	0	−3	1	1	−5	0	3	5	−2	1	−6	2	4	−1
−1	2	4	−4	2	2	0	1	3	3	2	−1	−5	2	−4	0	−2	1	2	3
2	−3	1	1	0	−4	−2	1	−1	0	0	−1	0	3	−4	3	6	−5	0	4
−2	−2	0	1	−1	1	4	−5	−2	−1	−2	−4	1	1	−1	−1	−4	0	3	1
−4	1	3	−2	−3	−1	−1	1	−1	−5	1	−1	0	−2	−3	−1	−3	−4	0	1
−2	0	4	−3	2	−2	5	0	3	−1	4	2	0	1	1	−5	2	−1	−6	0
0	4	−4	3	−4	1	4	0	−2	0	−2	1	3	−1	1	3	−6	3	−7	2
1	−3	1	−1	−2	3	−6	−1	3	−4	−1	−1	0	0	−6	−3	−1	−6	3	2
−2	−1	−2	−1	4	−1	−4	2	2	3	−4	−3	6	1	3	−5	0	3	0	0
1	−1	−3	−3	−3	−2	2	−2	−5	2	2	−1	2	4	4	2	2	−1	−3	−1
1	−1	3	−2	0	0	−3	−3	0	−4	5	1	−2	2	3	2	−1	−4	−2	−3
7	−1	0	0	−1	−1	1	5	−3	0	5	−5	−3	5	0	1	0	−6	5	−9
−1	1	2	−3	2	−2	1	−4	−1	4	3	2	−4	2	3	−3	−1	−4	−1	2
−1	3	1	0	−2	−4	0	2	2	2	0	−8	−2	3	3	−2	−1	−1	0	4

Squares of norms of columns:

103 116 123 125 130 133 144 144 153 157 158 165 169 171 176 184 191 199 204 211

In many applications LLL-reduction is used to show the polynomial time behavior of suitable algorithms. Actually, that reduction method was presented for the first time to show that polynomials of $\mathbb{Q}[t]$ can be factored into irreducible ones in a number of arithmetical operations which is polynomial in the input data. Since we shall need such factorization procedures later on (see chapter 5, section 4) we shall now discuss them in some detail.

It is convenient to restrict factorization in $\mathbb{Q}[t]$ to polynomials of the form

$$g(t) = \sum_{i=0}^{m} g_i t^i \in \mathbb{Z}[t], \quad \gcd(g_0, g_1, \ldots, g_m) = 1. \tag{3.43}$$

(The transition from an arbitrary polynomial of $\mathbb{Q}[t]$ to a corresponding one of the form (3.43) is possible by Gauss's Lemma, see [8], for example.) Then common factorization methods proceed in four steps:

(I) Compute a (small) prime number p which does not divide the discriminant $d(g)$ of $g(t)$.
(II) Factorize $g(t)$ modulo $p\mathbb{Z}[t]$.
(III) Lift that factorization to a factorization of $g(t)$ modulo $p^k\mathbb{Z}[t]$ for a sufficiently large exponent $k \in \mathbb{N}$.
(IV) Search for products of lifted factors which are potential factors of $g(t)$ in $\mathbb{Z}[t]$.

Step I is elementary, the computation of $d(g)$ was discussed in chapter 2, section 3. Step II is carried out by Berlekamp's method presented in chapter 2, section 5. Step III is done by Hensel's method which will be explained in detail in chapter 4, section 5, subsection 10. The exponent k is to be chosen in a way such that p^k is larger than twice the absolute value of the coefficients of any factor of $g(t)$ in $\mathbb{Z}[t]$. Upper bounds for those coefficients are easily derived from the root estimates in chapter 2, section 10, subsection 7. (For other bounds which are frequently used see [11].)

It remains to consider step IV. That step can cause a problem since there are infinitely many positive integers m such that there are polynomials $g(t)$ of $\mathbb{Z}[t]$ of degree m which have at least $m/2$ factors modulo $p^k\mathbb{Z}[t]$ for each prime number p and all $k \in \mathbb{N}$ [12]. (For example, this phenomenon occurs for $g(t) = t^4 + 1$, see exercise 5 of chapter 2, section 5.) Hence, a simple search through all possible factors of $g(t)$ modulo $p^k\mathbb{Z}[t]$ to exhibit a factor of $g(t)$ in $\mathbb{Z}[t]$ can yield a number of candidates which is exponential in $\deg(g)$.

This is where LLL-reduction comes in. After carrying out steps I, II, III we assume that we know a prime number p, a sufficiently large exponent k and (at most $\deg(g)$) polynomials $h(t) \in \mathbb{Z}[t]$ with the properties:

$h(t)$ is monic with coefficients bounded by p^k in absolute

value; in $\mathbb{Z}/p\mathbb{Z}[t]$ the image $\bar{h}(t)$ of $h(t)$ is irreducible,
$\bar{h}(t)|\bar{g}(t)$, $\bar{h}(t)^2 \nmid \bar{g}(t)$;
$h(t)$ divides $g(t)$ in $\mathbb{Z}/p^k\mathbb{Z}[t]$.

We develop a method to find out, whether $g(t)$ has an irreducible divisor $h_0(t)$ in $\mathbb{Z}[t]$ such that $\bar{h}(t)|\bar{h}_0(t)$. We set $l := \deg(h)$, $l_0 := \deg(h_0)$, $m := \deg(g)$. For each $\tilde{l} \in \mathbb{N}$, $l \leqslant \tilde{l} \leqslant m$ we choose $k \in \mathbb{N}$ (of step III) sufficiently large such that

$$p^{kl} > 2^{l_0 m/2} \binom{2l_0}{m}^{m/2} \|g\|^{l_0+m}. \tag{3.44}$$

Here the norm $\|g\|$ of a polynomial $g(t) \in \mathbb{Z}[t]$ is defined as the square root of the sum of the squares of its coefficients. (In this way bounds for the coefficients of $h_0(t)$ in terms of $\|g\|$ are obtained.) Especially, we know from our assumptions: $\|h\|^2 \leqslant 1 + lp^k$.

Then we consider the set of polynomials

$$L := \{u(t) \in \mathbb{Z}[t] \,|\, \deg(u) \leqslant l_0, \bar{h}(t)|\bar{u}(t)\}$$

together with the injective mapping

$$\varphi: L \to \mathbb{Z}^{l_0+1}: \sum_{i=0}^{l_0} u_i t^i \mapsto (u_0, u_1, \ldots, u_{l_0})^t.$$

Obviously, $\varphi(L)$ is an integral lattice with basis

$$\varphi(\{p^k t^i | 0 \leqslant i \leqslant l\} \cup \{h(t)t^j | 0 \leqslant j \leqslant l_0 - 1\}),$$

and discriminant $d(\varphi(L)) = p^{kl}$. Upon calculating a LLL-reduced basis $\mathbf{b}_1, \ldots, \mathbf{b}_{l_0+1}$ of $\varphi(L)$ one obtains the following two lemmata which we only cite from [10]:

Lemma (3.45)

(i) *In $\mathbb{Z}[t]$ $g(t)$ has a divisor $h_0(t)$ with $\deg(h_0) \leqslant l_0$ and $\bar{h}(t)|h_0(t)$ if and only if*

$$\|\mathbf{b}_1\| < (p^{kl}/\|\mathbf{g}\|^{l_0})^{1/m}. \tag{3.45a}$$

(ii) *If (3.45a) is satisfied and $t \in \{1, \ldots, l_0 + 1\}$ is maximal subject to $\|\mathbf{b}_t\| < (p^{kl}/\|g\|^{l_0})^{1/m}$, then*

$$h_0(t) = \gcd(\varphi^{-1}(\mathbf{b}_1), \ldots, \varphi^{-1}(\mathbf{b}_t)) \quad \text{with } \deg(h_0) = l_0 + 1 - t. \tag{3.45b}$$

Lemma (3.46)

The method described above determines an irreducible factor of a given polynomial $g(t) \in \mathbb{Z}[t]$ in the form (3.43) in $O(\deg(g)(\deg(g)^5 + \deg(g)^4 \log\|g\| + \deg(g)^3 \log p))$ arithmetical operations. The occurring integers have at most $O(\deg(g)^3 + \deg(g)^2 \log\|g\| + m \log p)$ binary digits.

Remark (3.47)
In case $\bar{g}(t)$ has only a few irreducible factors in $\mathbb{Z}/p\mathbb{Z}[t]$, then a direct search in step IV is preferable, of course.

In chapter 5, section 4 we shall extend this method to the factoring of polynomials over number fields. It is used there for the calculation of the torsion subgroup of the unit group of an order. Another application is a method to decide whether two algebraic number fields $F = \mathbb{Q}(\alpha)$, $E = \mathbb{Q}(\beta)$ of the same degree are isomorphic. To solve this problem if suffices to test whether the minimal polynomial $m_\beta(t) \in \mathbb{Q}[t]$ of β has a zero in F (see exercise 5 of chapter 5, section 4).

Another application of LLL-reduction is in connection with enlarging sublattices. The problem which we want to solve is the following: Let $\Lambda_1 \subseteq \Lambda$ be two integral k-dimensional lattices, $\mathbf{b}_1, \ldots, \mathbf{b}_{k+1} \in \Lambda$ subject to $\Lambda_1 = \dot{\sum}_{i=1}^{k} \mathbb{Z}\mathbf{b}_i$. We want to exhibit a basis of the lattice $\tilde{\Lambda} = \sum_{i=1}^{k+1} \mathbb{Z}\mathbf{b}_i \subseteq \Lambda$. Clearly, $\mathbf{b}_1, \ldots, \mathbf{b}_{k+1}$ are linearly dependent, hence algorithm (3.40) needs to be modified.

MLLL-reduction algorithm (3.48)

Input. Vectors $\mathbf{b}_1, \ldots, \mathbf{b}_{k+1}$ of a k-dimensional lattice $\Lambda \subset \mathbb{R}^n$, $\mathbf{b}_1, \ldots, \mathbf{b}_k$ being linearly independent.

Output. Integers $m_1, \ldots, m_{k+1}$ subject to $\sum_{i=1}^{k+1} m_i \mathbf{b}_i = \mathbf{0}$, $\sum_{i=1}^{k+1} |m_i| > 0$, and vectors $\hat{\mathbf{b}}_1, \ldots, \hat{\mathbf{b}}_k$ satisfying $\sum_{i=1}^{k+1} \mathbb{Z}\mathbf{b}_i = \sum_{i=1}^{k} \mathbb{Z}\hat{\mathbf{b}}_i$.

Step 1. (Initialization). For $i = 1, 2, \ldots, k+1$ set: $\hat{\mathbf{b}}_i \leftarrow \mathbf{b}_i$, $\mu_{ij} \leftarrow \mathbf{b}_i^t \mathbf{b}_j^* / B_j$ $(1 \leqslant j \leqslant i-1)$, $\mathbf{b}_i^* \leftarrow \mathbf{b}_i - \sum_{j=1}^{i-1} \mu_{ij} \mathbf{b}_j^*$, $B_i \leftarrow \mathbf{b}_i^{*t} \mathbf{b}_i^*$. Then set $m \leftarrow 2$, $H \leftarrow I_{k+1}$ and denote the columns of H by $\mathbf{h}_1, \ldots, \mathbf{h}_{k+1}$.

Step 2. (Set l). Set $l \leftarrow m-1$.

Step 3. (Change μ_{ml} in case $|\mu_{ml}| > \frac{1}{2}$). For $|\mu_{ml}| > \frac{1}{2}$ set $r \leftarrow \{\mu_{ml}\}$, $\hat{\mathbf{b}}_m \leftarrow \hat{\mathbf{b}}_m - r\hat{\mathbf{b}}_l$, $\mathbf{h}_m \leftarrow \mathbf{h}_m - r\mathbf{h}_l$, $\mu_{ml} \leftarrow \mu_{ml} - r$, $\mu_{mj} \leftarrow \mu_{mj} - r\mu_{lj}$ $(1 \leqslant j \leqslant l-1)$. In case $\hat{\mathbf{b}}_m = \mathbf{0}$ set $\hat{\mathbf{b}}_i \leftarrow \hat{\mathbf{b}}_{i+1}$ $(m \leqslant i \leqslant k)$, $(m_1, \ldots, m_{k+1})^t \leftarrow \mathbf{h}_m$ and terminate. For $\hat{\mathbf{b}}_m \neq \mathbf{0}$ check whether $l < m-1$. In that case go to 5.

Step 4. ((3.38b) violated on level m?). For $B_m < (\frac{3}{4} - \mu_{m,m-1}^2)\, B_{m-1}$ go to 6.

Step 5. (Decrease l). Set $l \leftarrow l-1$. For $l > 0$ go to 3. Else set $m \leftarrow m+1$ and go to 2.

Step 6. ($B_m = \mu = 0$?). Set $\mu \leftarrow \mu_{m,m-1}$, $B \leftarrow B_m + \mu^2 B_{m-1}$. For $B = 0$ go to 7. Else set $\mu_{m,m-1} \leftarrow \mu B_{m-1}/B$, $B_m \leftarrow B_m B_{m-1}/B$,

$$\begin{pmatrix} \mu_{i,m-1} \\ \mu_{i,m} \end{pmatrix} \leftarrow \begin{pmatrix} 1 & \mu_{m,m-1} \\ 0 & 1 \end{pmatrix} \begin{pmatrix} 0 & 1 \\ 1 & -\mu \end{pmatrix} \begin{pmatrix} \mu_{i,m-1} \\ \mu_{i,m} \end{pmatrix} \quad (m+1 \leqslant i \leqslant k+1).$$

Step 7. (Interchange $\hat{\mathbf{b}}_{m-1}, \hat{\mathbf{b}}_m$). Set $B_{m-1} \leftarrow B$,

$$\begin{pmatrix}\mathbf{h}_{m-1}\\ \mathbf{h}_m\end{pmatrix}\leftarrow\begin{pmatrix}\mathbf{h}_m\\ \mathbf{h}_{m-1}\end{pmatrix},\quad \begin{pmatrix}\hat{\mathbf{b}}_{m-1}\\ \hat{\mathbf{b}}_m\end{pmatrix}\leftarrow\begin{pmatrix}\hat{\mathbf{b}}_m\\ \hat{\mathbf{b}}_{m-1}\end{pmatrix},$$
$$\begin{pmatrix}\mu_{m-1,j}\\ \mu_{m,j}\end{pmatrix}\leftarrow\begin{pmatrix}\mu_{m,j}\\ \mu_{m-1,j}\end{pmatrix}\quad (1\leqslant j\leqslant m-2).$$

For $m>2$ decrease m by 1. Then go to 2.

We need to add some explanations about the changes made in (3.40). In step 1 we observe that $B_{k+1}=0$. Thus steps 6, 7 become most important. The possibility of $B\neq 0$ can occur only a finite number of times (for $m=k+1$, in the beginning) since in that case B_{m-1} is multiplied by a factor $\frac{1}{4}$. Hence, $B_{m-1}=0$ after some computational steps which means that then the vectors $\hat{\mathbf{b}}_1,\ldots,\hat{\mathbf{b}}_{m-1}$ are already linearly dependent. In this way we finally obtain a linear dependency of only two vectors yielding $\hat{\mathbf{b}}_m=\mathbf{0}$ in step 3. At that stage the coefficients $m_1,\ldots,m_{k+1}$ are the entries of the mth column of H, the vectors $\hat{\mathbf{b}}_1,\ldots,\hat{\mathbf{b}}_{m-1}$, $\hat{\mathbf{b}}_{m+1},\ldots,\hat{\mathbf{b}}_{k+1}$ form a basis of $\sum_{i=1}^{k+1}\mathbb{Z}\mathbf{b}_i$, and those vectors are obtained from $\mathbf{b}_1,\ldots,\mathbf{b}_{k+1}$ by $(\mathbf{b}_1\cdots\mathbf{b}_{k+1})\,H'=(\hat{\mathbf{b}}_1\cdots\hat{\mathbf{b}}_{m-1}\hat{\mathbf{b}}_{m+1}\cdots\hat{\mathbf{b}}_{k+1})$ where the accent at H means that the mth column is removed from H.

We illustrate the performance of algorithm (3.48) by a simple example. It was chosen so that all computations can be done by hand.

Example (3.49)

Let

$$(\mathbf{b}_1,\mathbf{b}_2,\mathbf{b}_3)=\begin{pmatrix}1 & 2 & 1\\ 1 & 1 & -1\end{pmatrix},\quad k=2.$$

Applying (3.48) m is immediately increased to $m=3$. Then we get $B\neq 0$ in step 6, $\hat{\mathbf{b}}_2$ and $\hat{\mathbf{b}}_3$ are interchanged. Next $\hat{\mathbf{b}}_1$ is subtracted from $\hat{\mathbf{b}}_2$, $\hat{\mathbf{b}}_1$ and $\hat{\mathbf{b}}_2$ are interchanged. Again $\hat{\mathbf{b}}_1$ is subtracted from $\hat{\mathbf{b}}_2$, $\hat{\mathbf{b}}_2$ is added to $\hat{\mathbf{b}}_3$. Now $B=0$, $\hat{\mathbf{b}}_2$ and $\hat{\mathbf{b}}_3$ are interchanged and we have a linear dependency of $\hat{\mathbf{b}}_1$ and $\hat{\mathbf{b}}_2$. In step 3 the vector $\hat{\mathbf{b}}_2$ becomes $\mathbf{0}$. We list the result:

$$(\hat{\mathbf{b}}_1,\hat{\mathbf{b}}_2,\hat{\mathbf{b}}_3)=\begin{pmatrix}1 & 0 & 0\\ 0 & 0 & 1\end{pmatrix},$$

$$H=(\mathbf{h}_1,\mathbf{h}_2,\mathbf{h}_3)=\begin{pmatrix}-1 & 2 & 2\\ 0 & 1 & 0\\ 1 & -1 & -1\end{pmatrix}.$$

We note that (3.48) can be used to advantage for solving systems of linear equations over the integers (see also exercise 10).

Concluding this section we remark that it is very easy to compute generators $\mathbf{b}_1,\ldots,\mathbf{b}_k$ of $\sum_{i=1}^{k+1}\mathbb{Z}\mathbf{b}_i$ once we know a relation $\sum_{i+1}^{k+1}m_i\mathbf{b}_i=\mathbf{0}$ ($m_i\in\mathbb{Z}$, $\gcd(m_1,\ldots,m_{k+1})=1$). Then we only need to apply division with remainder to the m_i as the following algorithm demonstrates.

Algorithm for enlarging sublattices (3.50)

Input. Vectors $\mathbf{b}_1, \ldots, \mathbf{b}_{k+1}$ of a k-dimensional lattice Λ such that $\mathbf{b}_1, \ldots, \mathbf{b}_k$ are $\mathbb{R}$-linearly independent. Integers $m_1, \ldots, m_{k+1} \in \mathbb{Z}$ subject to $\mathbf{0} = \sum_{i=1}^{k+1} m_i \mathbf{b}_i$ and $\gcd(m_1, \ldots, m_{k+1}) = 1$.

Output. Vectors $\mathbf{c}_1, \ldots, \mathbf{c}_k$ of Λ satisfying $\sum_{i=1}^{k} \mathbb{Z}\mathbf{c}_i = \sum_{i=1}^{k+1} \mathbb{Z}\mathbf{b}_i$.

Step 1. (Initialization). Set $\mathbf{c}_i \leftarrow \mathbf{b}_i$ $(1 \leqslant i \leqslant k+1)$.

Step 2. (Determine non-zero m_i of minimal absolute value). Set $p \leftarrow \min\{|m_i| \mid 1 \leqslant i \leqslant k+1, m_i \neq 0\}$ and choose $j \in \{1, \ldots, k+1\}$ maximal with $|m_j| = p$.

Step 3. (Division with remainder). For $p \mid m_i$ $(1 \leqslant i \leqslant k+1)$ go to 4. Else set $m_{ij} \leftarrow \{m_i/m_j\}$, $m_i \leftarrow m_i - m_{ij} m_j$, $\mathbf{c}_j \leftarrow \mathbf{c}_j + m_{ij}\mathbf{c}_i$ for all i $(1 \leqslant i \leqslant k+1)$ with $p \nmid m_i$. Then return to 2.

Step 4. (Done). For $i = j+1, \ldots, k+1$ set $\mathbf{c}_{i-1} \leftarrow \mathbf{c}_i$ and terminate.

We illustrate (3.50) by a simple example.

Example

Let $\mathbf{b}_1, \mathbf{b}_2, \mathbf{b}_3, \mathbf{b}_4 \in \mathbb{R}^3$ subject to $3\mathbf{b}_1 - 4\mathbf{b}_2 + 2\mathbf{b}_3 + 2\mathbf{b}_4 = \mathbf{0}$. This equation is immediately reduced to

$$\mathbf{b}_1 - 4\mathbf{b}_2 + 2\mathbf{b}_3 + 2(\mathbf{b}_4 + \mathbf{b}_1) = \mathbf{0}$$

yielding

$$\mathbf{c}_1 = \mathbf{b}_2, \quad \mathbf{c}_2 = \mathbf{b}_3, \quad \mathbf{c}_3 = \mathbf{b}_4 + \mathbf{b}_1.$$

Exercises

1. Let $\mathbf{b}_1, \ldots, \mathbf{b}_k \in \mathbb{R}^n$ be linearly independent and $\mathbf{b}_i^*, \mu_{ij}$ $(1 \leqslant j < i \leqslant k)$ as in (3.24). Write an algorithm which determines the (rational) coefficients β_i of the presentation $\mathbf{b}_p = \sum_{i=1}^{k} \beta_i \mathbf{b}_i$, where $\mathbf{b}_p$ is defined by (3.25).
2. Incorporate the refined techniques suggested in (3.16) (iv) into algorithm (3.15) (compare [4]).
3. Two linearly independent vectors $\mathbf{b}_1, \mathbf{b}_2 \in \mathbb{R}^n$ are called *pairwise reduced* in case of $\|\mathbf{b}_1\| \leqslant \|\mathbf{b}_2\|$ and $|\mathbf{b}_1^t \mathbf{b}_2| \leqslant \frac{1}{2}\|\mathbf{b}_1\|^2$.

   (i) Show that two linearly independent vectors $\mathbf{a}, \mathbf{b} \in \mathbb{R}^n$ are transformed into pairwise reduced vectors upon repeated application of

$$\begin{pmatrix} \mathbf{a} \\ \mathbf{b} \end{pmatrix} \leftarrow \begin{pmatrix} \mathbf{a} \\ \mathbf{b} - \left\{ \dfrac{\mathbf{a}^t \mathbf{b}}{\|\mathbf{a}\|^2} \right\} \mathbf{a} \end{pmatrix}$$

   and in case $\|\mathbf{b}\| > \|\mathbf{a}\|$ interchanging $\mathbf{a}$ and $\mathbf{b}$.

   (ii) Interpret the reduction step in (i) geometrically.

   (iii) Show that two pairwise reduced vectors are also LLL-reduced and reduced in the sense of Minkowski.

4. Prove Lemma (3.39).
5. Let $\Lambda = \bigoplus_{i=1}^{k} \mathbb{Z}\mathbf{b}_i$ be a k-dimensional lattice of $\mathbb{R}^n$ and $D_i := \det(\mathbf{b}_\mu' \mathbf{b}_\nu)_{1 \leq \mu, \nu \leq i}$ $(1 \leq i \leq k)$. Prove:

$$M_{i1} := \min\left\{ \|\mathbf{x}\|^2 \,|\, \mathbf{x} \in \Lambda_i := \bigoplus_{j=1}^{i} \mathbb{Z}\mathbf{b}_j, \mathbf{x} \neq \mathbf{0} \right\} \leq (\tfrac{4}{3})^{(i-1)/2} D_i^{1/i}.$$

(Hint: For a reduced basis $\mathbf{c}_1, \ldots, \mathbf{c}_i$ of Λ_i $(M_{i1} = \|\mathbf{c}_1\|^2)$ decompose the corresponding quadratic form $Q_i(\mathbf{x}) = \sum_{\mu,\nu=1}^{i} \gamma_{\mu\nu} x_\mu x_\nu$ $(\gamma_{\mu\nu} := \mathbf{c}_\mu' \mathbf{c}_\nu)$ into $\gamma_{11}(x_1 + (\gamma_{12}/\gamma_{11})x_2 + \cdots + (\gamma_{1i}/\gamma_{11})x_i)^2 + g(x_2, \ldots, x_i)$ and apply induction.) Show that the result also yields $\gamma_i^i \leq (\tfrac{4}{3})^{i(i-1)/2}$.
6. Prove the statements in example (3.42) (i). (Hint: Consider $\mathbf{b}_k - \mathbf{b}_{k-1}$.)
7. Prove: $\gamma_2^2 = \tfrac{4}{3}$.
8. Modify (3.15) such that the output consists of all vectors $\mathbf{y} \in \Lambda$ of given square length $\|\mathbf{y}\|^2 = T$.
9. Use (3.15) to show that lattices are *discrete*, i.e. for each lattice Λ there is a positive constant δ such that all $\mathbf{x}, \mathbf{y} \in \Lambda$, $\mathbf{x} \neq \mathbf{y}$, satisfy $\|\mathbf{x} - \mathbf{y}\| \geq \delta$.
10. Generalize algorithm (3.48) in the following sense. The input consists of vectors $\mathbf{b}_1, \ldots, \mathbf{b}_u$ of a k-dimensional lattice $\Lambda \subset \mathbb{R}^n$. The output is to consist of vectors $\hat{\mathbf{b}}_1, \ldots, \hat{\mathbf{b}}_g$ which are linearly independent and satisfy $\sum_{i=1}^{u} \mathbb{Z}\mathbf{b}_i = \dot{\sum}_{i=1}^{g} \mathbb{Z}\hat{\mathbf{b}}_i$. Use the result to develop an algorithm for solving systems of linear Diophantine equations.

3.4. Minkowski's convex body theorem

Making use of the connection between quadratic forms and lattices we can interpret theorem (3.34) in another way. If we fix a basis $\mathbf{b}_1, \ldots, \mathbf{b}_n$ of the lattice Λ, then $\|\mathbf{x}\|^2$ $(\mathbf{x} \in \Lambda)$ becomes a positive definite quadratic form $Q(\mathbf{x})$ with coefficients $\mathbf{b}_i'\mathbf{b}_j$ $(1 \leq i, j \leq n)$ and determinant $d(\Lambda)^2$. Hence, the set $E := \{\mathbf{x} \in \mathbb{R}^n | Q(\mathbf{x}) \leq r^2\}$ $(r > 0)$ is an ellipsoid which contains non-zero lattice points if and only if $r^2 \geq M_1$. Because of (3.34) $r^2 \geq \gamma_n d(\Lambda)^{2/n}$ suffices.

For more general point sets it is much more difficult to determine whether they contain non-zero lattice points. A very useful result in this context is Minkowski's famous Convex Body Theorem.

Definition (4.1)
A set $C \subseteq \mathbb{R}^n$ is called **convex**, *if for every $\mathbf{x}, \mathbf{y} \in C$, $\lambda \in [0, 1]$ the vector $\lambda\mathbf{x} + (1 - \lambda)\mathbf{y}$ (i.e. the line segment between $\mathbf{x}$ and $\mathbf{y}$) belongs to C.*

Examples
Let $A = (a_{ij}) \in \mathbb{R}^{n \times n}$, and $C_1, \ldots, C_n$ positive constants. Then (4.1a)

$$S := \left\{ \mathbf{x} \in \mathbb{R}^n \,\middle|\, \left| \sum_{j=1}^{n} a_{ij} x_j \right| \underset{(\leq)}{<} C_i \quad (1 \leq i \leq n) \right\}$$

is convex.

Namely, *for* $\mathbf{x}, \mathbf{y} \in S$, $\lambda \in [0,1]$ we obtain

$$\left|\sum_{j=1}^{n} a_{ij}(\lambda x_j + (1-\lambda)y_j)\right| \leqslant \lambda \left|\sum_{j=1}^{n} a_{ij}x_j\right| + (1-\lambda)\left|\sum_{j=1}^{n} a_{ij}y_j\right|$$
$$\underset{(\leqslant)}{<} \lambda C_i + (1-\lambda)C_i = C_i \quad (1 \leqslant i \leqslant n).$$

Let Q be a positive definite quadratic form and $C > 0$. Then (4.1b)

$$S := \{\mathbf{x} \in \mathbb{R}^n \mid Q(\mathbf{x})^{1/2} \underset{(\leqslant)}{<} C\}$$

is convex.

For the proof we assume that $Q(\mathbf{x}) = \mathbf{x}^t A \mathbf{x}$ for $A \in R^{n \times n}$. Then

$$Q(\lambda\mathbf{x} + (1-\lambda)\mathbf{y})^{1/2} = (\lambda^2 Q(\mathbf{x}) + (1-\lambda)^2 Q(\mathbf{y}) + 2\lambda(1-\lambda)\mathbf{x}^t A\mathbf{y})^{1/2}$$
$$\leqslant (\lambda^2 Q(\mathbf{x}) + (1-\lambda)^2 Q(\mathbf{y}) + 2\lambda(1-\lambda)(Q(\mathbf{x})Q(\mathbf{y}))^{1/2})^{1/2}$$

(by the Cauchy–Schwarz inequality)

$$= \lambda Q(\mathbf{x})^{1/2} + (1-\lambda)Q(\mathbf{y})^{1/2}$$
$$\underset{(\leqslant)}{<} \lambda C + (1-\lambda)C = C.$$

The intersection of convex sets is convex. (4.1c)

For a convex set C its closure $\bar{C}$ and the set of its interior points $\mathring{C}$ are convex. (4.1d)

We note that convex subsets C of $\mathbb{R}^n$ are Jordan-measurable (for a proof see [9]). The corresponding volume is denoted by $V(C)$.

***Theorem** (Minkowski's Convex Body Theorem)* (4.2)
Let $C \subseteq \mathbb{R}^n$ be a convex, $\mathbf{0}$-symmetric (i.e. $\mathbf{x} \in C$ *implies* $-\mathbf{x} \in C$) *set and Λ an n-dimensional lattice. In case $V(C) > 2^n d(\Lambda)$ or $V(C) = 2^n d(\Lambda)$ and C compact the set C contains a lattice point $\mathbf{x} \in \Lambda$, $\mathbf{x} \neq \mathbf{0}$.*

Proof
For $A \subseteq \mathbb{R}^n$, $\mathbf{y} \in \mathbb{R}^n$, $\alpha \in \mathbb{R}$ we write $\alpha A := \{\alpha\mathbf{x} \mid \mathbf{x} \in A\}$, $\mathbf{y} + A := \{\mathbf{y} + \mathbf{x} \mid \mathbf{x} \in A\}$ for abbreviation.

Let us assume $V(C) > 2^n d(\Lambda)$ at first. Using the fundamental parallelotope Π of the lattice Λ (which was defined in (3.4)) we obtain

$$\tfrac{1}{2}C = \bigcup_{\mathbf{x} \in \Lambda} (\tfrac{1}{2}C \cap (\mathbf{x} + \Pi)),$$

hence,

$$d(\Lambda) = V(\Pi) < \tfrac{1}{2^n} V(C) = V(\tfrac{1}{2}C) = \sum_{\mathbf{x}\in\Lambda} V(\tfrac{1}{2}C \cap (\mathbf{x} + \Pi))$$
$$= \sum_{\mathbf{x}\in\Lambda} V((-\mathbf{x} + \tfrac{1}{2}C) \cap \Pi).$$

This implies that there exist $\mathbf{x}, \mathbf{y} \in \Lambda$, $\mathbf{x} \neq \mathbf{y}$, satisfying $-\mathbf{x} + \frac{1}{2}C \cap -\mathbf{y} + \frac{1}{2}C \neq \varnothing$. As a consequence there are $\mathbf{c}_1, \mathbf{c}_2 \in C$ subject to $-\mathbf{x} + \frac{1}{2}\mathbf{c}_1 = -\mathbf{y} + \frac{1}{2}\mathbf{c}_2$, hence, $\mathbf{0} \neq \mathbf{x} - \mathbf{y} = \frac{1}{2}\mathbf{c}_1 + \frac{1}{2}(-\mathbf{c}_2) \in C$.

If $V(C) = 2^n d(\Lambda)$ and C is compact then we use a sequence $\varepsilon_n \searrow 0$ $(n \in \mathbb{N}, \varepsilon_n > 0)$ in order to obtain non-zero lattice points $\mathbf{x}_n$ in $(1 + \varepsilon_n)C$. Since the sequence $(\mathbf{x}_n)_{n\in\mathbb{N}}$ is bounded, it contains a convergent subsequence which will be constant for n large enough because of the discreteness of Λ (see (3.15) and exercise 9 of chapter 3). Hence, $\mathbf{0} \neq \mathbf{x}_n = \dot{\mathbf{x}} \in \Lambda \cap (1 + \varepsilon_n)C$ for large n, i.e. $\mathbf{x} \in C$ because C is compact. □

Corollary (4.3)

Under the same premises for C, Λ as in (4.2) and additionally $V(C) > m2^n d(\Lambda)$ or $V(C) = m2^n d(\Lambda)$ and C compact ($m \in \mathbb{N}$), there exist m distinct pairs of non-zero lattice points in C.

The proof is recommended as an easy exercise.

Remark

Considering $C := \{\mathbf{x} \in \mathbb{R}^n \mid |x_i| < 1\}$ for $\Lambda = \mathbb{Z}^n$ shows at once that (4.2) cannot be improved in general.

We give a few applications of Minkowski's Convex Body Theorem which will turn out to be useful for later sections.

Theorem (4.4)

Let Λ be an n-dimensional lattice in $\mathbb{R}^n$, $A = (a_{ij}) \in \mathbb{R}^{n\times n}$ a matrix, and C_j $(1 \leqslant j \leqslant n)$ positive constants such that $C_1 \cdots C_n \geqslant |\det A| d(\Lambda)$. Then Λ contains a non-zero lattice point $\mathbf{u} = \sum_{i=1}^n u_i \mathbf{e}_i$ with

$$\left|\sum_{j=1}^n a_{1j}u_j\right| \leqslant C_1, \quad \left|\sum_{j=1}^n a_{ij}u_j\right| < C_i \quad (2 \leqslant i \leqslant n).$$

Proof

We consider the set $S \subset \mathbb{R}^n$ given by

$$S := \left\{\mathbf{x} \in \mathbb{R}^n \,\middle|\, \left|\sum_{j=1}^n a_{1j}x_j\right| \leqslant C_1, \quad \left|\sum_{j=1}^n a_{ij}x_j\right| < C_i \quad (2 \leqslant i \leqslant n)\right\}.$$

S is convex (by (4.1a)) and $\mathbf{0}$-symmetric. The matrix A describes a linear transformation of $\mathbb{R}^n$:

$$A\colon \mathbb{R}^n \to \mathbb{R}^n\colon \mathbf{x} \mapsto A\mathbf{x} =: \mathbf{X}.$$

Hence,

$$V(S) = \int_S d\mathbf{x} = \int_{A(S)} \left|\frac{\partial \mathbf{x}}{\partial \mathbf{X}}\right| d\mathbf{X} = (\det A)^{-1} \int_{A(S)} d\mathbf{X}.$$

Therefore $V(S) = \infty$, if A is singular, and in that case S contains non-zero lattice points of Λ according to (4.2). For regular A the set $A(S)$ has volume $2^n C_1 \cdot \ldots \cdot C_n$. If $C_1 \cdot \ldots \cdot C_n > |\det A| d(\Lambda) = d(A(\Lambda))$, the theorem follows again from (4.2). Finally, for $C_1 \cdot \ldots \cdot C_n = d(A(\Lambda))$ we obtain non-zero lattice points $\mathbf{X}_\varepsilon \in A(\Lambda)$ for each $\varepsilon > 0$ which satisfy $|X_{\varepsilon_1}| \leqslant C_1 + \varepsilon$, $|X_{\varepsilon_i}| < C_i$ $(2 \leqslant i \leqslant n)$. Because of (3.15) there are only finitely many $\mathbf{X}_\varepsilon$ at all, if we bound ε by 1 from above, and for $\varepsilon \to 0$ one of these $\mathbf{X}_\varepsilon$ must already satisfy $|X_{\varepsilon_1}| \leqslant C_1$.

□

In most applications, however, the pointset S under consideration is by no means convex. In that case it can be helpful to apply Minkowski's Convex Body Theorem to a convex subset C of S.

For example, in algebraic number fields the non-zero integers are of absolute norm greater than or equal to one. For the Minkowski-mapping of those integers into $\mathbb{R}^n$ (see chapter 6 (3.6)) the set S,

$$S := \left\{ \mathbf{x} \in \mathbb{R}^n \,\middle|\, \prod_{i=1}^{s} |x_i| \prod_{\substack{i=s+1 \\ i-s \text{ odd}}}^{n-1} (x_i^2 + x_{i+1}^2) < 1 \right\} \tag{4.5}$$

for suitable $s \in \mathbb{Z}^{\geqslant 0}$, $n - s = 2t$, cannot contain images of integers different from $\mathbf{0}$. Since S is not convex in general we consider the subset C instead:

$$C := \left\{ \mathbf{x} \in \mathbb{R}^n \,\middle|\, \sum_{i=1}^{s} |x_i| + 2 \sum_{\substack{i=s+1 \\ i-s \text{ odd}}}^{n-1} (x_i^2 + x_{i+1}^2)^{1/2} < n \right\}. \tag{4.6}$$

($C \subseteq S$ is an immediate consequence of the inequality between geometric and arithmetic means.)

Then C cannot contain images of non-zero integers either, and as C is convex (by (4.1a), (4.1b)), Theorem (4.2) yields $V(C) \leqslant 2^n d(\Lambda)$, where Λ denotes the lattice of the image of the algebraic integers. This yields an important estimate for $d(\Lambda)$ in terms of $V(C)$.

Computation of $V(C)$.

Let

$$C_{s,t}(\lambda) := \left\{ \mathbf{x} \in \mathbb{R}^n \,\middle|\, \sum_{i=1}^{s} |x_i| + 2 \sum_{\substack{i=s+1 \\ i-s \text{ odd}}}^{n-1} (x_i^2 + x_{i+1}^2)^{1/2} \leqslant \lambda \right\}$$

$$(s \in \mathbb{Z}^{\geqslant 0},\ n - s = 2t,\ \lambda \in \mathbb{R}^{\geqslant 0}).$$

Obviously,

$$V(C_{s,t}(\lambda)) = \lambda^n V(C_{s,t}(1)). \tag{4.7a}$$

For $s>0$ we have

$$V(C_{s,t}(1)) = \int_{-1}^{1} V(C_{s-1,t}(1-|x|))\,dx$$
$$= 2V(C_{s-1,t}(1)) \int_0^1 (1-x)^{n-1}\,dx$$
$$= \frac{2}{n} V(C_{s-1,t}(1)) \tag{4.7b}$$

and

$$V(C_{1,0}(1)) = 2. \tag{4.7c}$$

For $t>0$ we obtain

$$V(C_{0,t}(1)) = \iint_{x^2+y^2\leq 1/4} V(C_{0,t-1}(1-2(x^2+y^2)^{1/2}))\,dx\,dy.$$

Introducing polar coordinates $x = r\cos\varphi$, $y = r\sin\varphi$ we compute

$$V(C_{0,t}(1)) = 2\pi \int_0^{\frac{1}{2}} V(C_{0,t-1}(1-2r))r\,dr$$
$$= 2\pi V(C_{0,t-1}(1)) \int_0^{\frac{1}{2}} (1-2r)^{2(t-1)} r\,dr.$$

Substituting $\tilde{r} = \frac{1}{2} - r$ yields

$$V(C_{0,t}(1)) = 2\pi V(C_{0,t-1}(1)) \left(2^{2t-3}\frac{1}{2t-1}2^{1-2t} - 2^{2(t-1)}\frac{1}{2t}2^{-2t} \right)$$
$$= \frac{\pi}{2}\frac{1}{2t(2t-1)} V(C_{0,t-1}(1)) \tag{4.7d}$$

and

$$V(C_{0,1}(1)) = \pi/4. \tag{4.7e}$$

Combining (4.7a–e) we finally obtain

$$V(C_{s,t}(\lambda)) = \lambda^n V(C_{s,t}(1)) = \lambda^n \frac{2}{n}\cdot\frac{2}{n-1}\cdot\ldots\cdot\frac{2}{n-s+1} V(C_{0,t}(1))$$
$$= \lambda^n 2^s \frac{(n-s)!}{n!} \left(\frac{\pi}{2}\right)^{t-1} \frac{1}{(2t)!} 2\frac{\pi}{4}$$
$$= \lambda^n \frac{2^s}{n!}\left(\frac{\pi}{2}\right)^t. \tag{4.8}$$

We will make use of the results obtained so far in all later chapters. There they will be applied to the integers of algebraic number fields F, more precisely to homomorphic images of the ring of integers which are lattices. Theorems (3.34), (4.2), (3.37) are used to derive bounds for discriminants, regulators and the coefficients of a suitable generating polynomial of F.

Exercises

1. Let C be a convex subset of $\mathbb{R}^n$. Prove:

   $\forall m \in \mathbb{N} \quad \forall \mathbf{x}_1, \ldots, \mathbf{x}_m \in C \quad \forall t_1, \ldots, t_m \in \mathbb{R}^{\geq 0}$:

$$\sum_{i=1}^{m} t_i = 1 \Rightarrow \sum_{i=1}^{m} t_i \mathbf{x}_i \in C.$$

2. Let C be a convex subset of $\mathbb{R}^n$. Then show either C is contained in a hyperplane or C contains interior points.
3. Prove Corollary (4.3).
4. Let Λ be an n-dimensional lattice and $\mathbf{x}_1, \ldots, \mathbf{x}_{n+1} \in \Lambda$. Use (4.4) to prove that there exist $u_1, \ldots, u_{n+1} \in \mathbb{Z}$ satisfying $\sum_{j=1}^{n+1} |u_j| > 0$ and $\sum_{j=1}^{n+1} u_j \mathbf{x}_j = \mathbf{0}$. (Hint: For $\mathbf{x}_i = (x_{i_1}, \ldots, x_{i_n})^t$ set $x_{i_{n+1}} = 0$ and consider the system of linear inequalities $|\sum_{i=1}^{n+1} x_{ij} u_j| \leq M_1^{\frac{1}{2}}/(2n)$, where M_1 denotes the first successive minimum of Λ.)

4

Embedding of commutative orders into the maximal order

4.1. Introduction

Solving algebraic equations

$$x^n + a_1 x^{n-1} + \cdots + a_n = 0 \quad (a_i \in \mathbb{Q}, 1 \leqslant i \leqslant n) \tag{1.1a}$$

over the rational number field $\mathbb{Q}$ establishes algebraic numbers. Following the ideas of E. Galois, number theory began to occupy itself with the algebraic relations between several algebraic numbers. It was shown in chapters 1 and 2 how to build up systems of algebraic numbers by associating the $\mathbb{Q}$-algebra

$$\begin{aligned} A_f &= \mathbb{Q}[t]/f(t)\mathbb{Q}[t] \\ &= \mathbb{Q} + \mathbb{Q}\alpha_f + \cdots + \mathbb{Q}\alpha_f^{n-1} \end{aligned} \tag{1.1b}$$

with (1.1a). Here

$$f(t) = t^n + a_1 t^{n-1} + \cdots + a_n \in \mathbb{Q}[t] \tag{1.1c}$$

is the monic polynomial associated with (1.1a) and

$$\alpha_f = t + f(t)\mathbb{Q}[t]$$

is an element of A_f solving (1.1a) and generating the algebra A_f over $\mathbb{Q}$.

Applying the construction repeatedly as needed one is able to carry out the program of E. Galois constructively in terms of commutative and associative algebras with a finite basis over $\mathbb{Q}$. Of course the structures which are built in this way are not always fields but only rings, but we have learned also how to solve the basic problems (like finding the group of an equation, or factorising a polynomial) with rings rather than with finite field extensions.

What does that mean for arithmetics?

The basic question of multiplicative algebraic number theory is the following: Given a system Σ of algebraic numbers, describe the multiplicative semigroup of the number ring $\mathbb{Z}[\Sigma]$ generated by Σ and $\mathbb{Z}$ in general terms. Are there prime elements, how do the units behave, is it possible to factorize any non-zero divisor into a product of prime factors, is there uniqueness up to the order and equivalence?

We have seen already that $\mathbb{Z}[i]$ is a unique factorization domain. Its multiplicative structure is quite similar to that of $\mathbb{Z}$.

The ring $\mathbb{Z}[1/2]$ has the same prime elements 3, 5, 7, 11,... as $\mathbb{Z}$, apart from 2 which turns into a unit so that the multiplicative number theory of $\mathbb{Z}[1/2]$ becomes a copy of the multiplicative number theory of $\mathbb{Z}$ with the exception of one prime element turning into a unit. Note that $\mathbb{Z}[1/2]$ has no $\mathbb{Z}$-basis. Similarly, joining an element of the form λ/v $(\lambda, v \in R\backslash\{0\})$ to a factorial ring R the prime elements of $R[\lambda/v]$ also are prime elements of R, but there are a finite number of additional prime elements of R that are non-equivalent among each other in R, but turn into units in $R[\lambda/v]$. We call the transition from R to $R[\lambda/v]$ (and its iterations) a *localization*. It has the effect of possibly diminishing the prime structure of R, and focussing on the multiplicative behavior of the remainder. In general there will be no basis. In terms of the equation ring $\mathbb{Z}[\alpha_f]$ derived from (1.1a–c) its multiplicative structure will be enriched upon transition to $\mathbb{Z}[v\alpha_f]$ for any non-zero rational integer $v > 1$. The monic polynomial corresponding to $v\alpha_f$ is $g(t) = t^n + va_1t^{n-1} + \cdots + v^na_n$. Those transitions are particularly important for which the coefficients of $g(t)$ are in $\mathbb{Z}$. Suppose $a_i = b_i/N_i$, $b_i \in \mathbb{Z}$, $N_i \in \mathbb{Z}^{>0}$, $\gcd(b_i, N_i) = 1$ then the necessary and sufficient condition for $va_i \in \mathbb{Z}$ is $N \mid v$ where N is the least common multiple of the numbers $(N_i)_i$ denoting the largest divisor of N_i such that the ith power divides N_i $(1 \leq i \leq n)$.

The $\mathbb{Z}$-equation ring $\mathbb{Z}[2i]$ is not a factorial ring because $4 = 2\cdot 2 = -2i\cdot 2i$ and 2, $2i$ are two non-equivalent primes. But the reason for this behavior is obvious: $\mathbb{Z}[2i]$ is properly contained in the equation ring $\mathbb{Z}[i]$ where 2 and $2i$ are equivalent. The defect is remedied by transition to a larger number ring where both of them have a $\mathbb{Z}$-basis with the same number of basis elements. It is left to the reader as an exercise to show that the equation ring $\mathbb{Z}[(-5)^{\frac{1}{2}}]$ cannot be enlarged into a larger subring of its algebraic background $\mathbb{Q}((-5)^{\frac{1}{2}})$ which still has a $\mathbb{Z}$-basis. But yet there hold the two factorizations $6 = 2\cdot 3 = (1 + (-5)^{\frac{1}{2}})(1 - (-5)^{\frac{1}{2}})$ with non-equivalent prime elements $2, 3, 1 + (-5)^{\frac{1}{2}}, 1 - (-5)^{\frac{1}{2}}$. This is because there are no elements of norm 2, 3, in the ring, so that the numbers of norm 4, 9, 6 cannot split any further.

This type of non-uniqueness of prime factorization was first dealt with by E. Kummer after French mathematicians had pointed out that in dealing with elements of the equation order $R = \mathbb{Z}[\zeta_p] = \mathbb{Z}[t]/(t^{p-1} + t^{p-2} + \cdots + 1)$ (p-cyclotomic numbers, p a prime number) he could not always assume that a gcd exists in such a way that two products with factors mutually prime to one another are themselves prime to each other. Correspondingly Kummer remedied the defect of his earlier treatment by introducing other algebraic integers for which he stipulated that they are equivalent to gcd's even though those are not contained in R. These gcd's are the *ideal numbers* of R. They form a semigroup with unique factorization. Its prime elements are the *ideal prime elements* of R.

Clearly Kummer's treatment introduces outside elements in order to clarify an internal structural question. Also the construction of the ideal numbers is not unique though the emerging ideal arithmetic of R itself will be unique. For example in the case of $\mathbb{Z}[(-5)^{\frac{1}{2}}]$ any one of the algebraic numbers $2^{\frac{1}{2}}, (-2)^{\frac{1}{2}}, (-3)^{\frac{1}{2}}, 6^{\frac{1}{2}}, (-6)^{\frac{1}{2}}$, etc. can be used as an ideal number. In any case one will have $2 \sim \Pi_1\Pi_2, 3 \sim \Pi_3\Pi_4, 1+(-5)^{\frac{1}{2}} \sim \Pi_1\Pi_3, 1-(-5)^{\frac{1}{2}} \sim \Pi_2\Pi_4$, where $\Pi_1, \Pi_2, \Pi_3, \Pi_4$ are ideal prime elements.

R. Dedekind invented the ideal concept for the same purpose. For example all numbers of $\mathbb{Z}[(-5)^{\frac{1}{2}}]$ having the ideal prime divisor Π_1 form a subset of $\mathbb{Z}[(-5)^{\frac{1}{2}}]$ which is closed under addition and under multiplication by arbitrary elements of $\mathbb{Z}[(-5)^{\frac{1}{2}}]$. In this case it is the set of linear combinations of $2, 1+(-5)^{\frac{1}{2}}$ over $\mathbb{Z}[(-5)^{\frac{1}{2}}]$, or as we say with Dedekind, it is the *ideal* generated by 2 and $1+(-5)^{\frac{1}{2}}$. Dedekind showed that the ideals of $\mathbb{Z}[(-5)^{\frac{1}{2}}]$ (more generally of the maximal order of an algebraic number field) behave under multiplication the same way as the rational integers up to equivalence: they form an abelian semigroup with unique prime factorization (up to the order). A number ring with those properties is said to be a Dedekind domain over $\mathbb{Z}$.

Dedekind showed that every algebraic number field $\mathbb{Q}(\alpha)$ contains precisely one Dedekind domain over $\mathbb{Z}$ generating $\mathbb{Q}(\alpha)$ in the field-theoretic sense, the maximal order.

It so happens that its elements are the algebraic integers. But a true characterization of algebraic integers is arrived at only by localization using the language and the concepts of valuation theory created by Hensel, Kuershak and Ostrowski. Upon localization the true meaning of Eisenstein's ingenious construction of irreducible polynomials became clear: actually he constructed local maximal orders. In another context, viz. the theory of algebraic curves, the algebraic background is a finite extension E of the rational function field $F(x)$ in one variable x over a finite field F. By a suitable choice of the independent variable it is possible to make E a separable extension of $F(x)$ such that $E = F(x, y)$ where y satisfies the irreducible equation $y^n + a_1 y^{n-1} + \cdots + a_n = 0$, with coefficients $a_1, \ldots, a_n$ in the polynomial ring $R = F[x]$ in x over F. The equation ring $\Lambda_{f/R} = R[t]/f(t)R[t] = F[x, y]$ of the monic polynomial $f(t) = t^n + a_1 t^{n-1} + \cdots + a_n$ will not be a factorial ring, not even a Dedekind domain. But again there is precisely one subring $\bar{\Lambda}$ of E such that R is contained in $\bar{\Lambda}$, any two elements of R are equivalent in $\bar{\Lambda}$ if and only if they are equivalent in R and every ideal $\neq 0$ of $\bar{\Lambda}$ is the product of prime ideals of $\bar{\Lambda}$. It consists of all elements that are integral with respect to R i.e. they are zeros of monic polynomials of $R[t]$ in E. In order to find $\bar{\Lambda}$, quite the same methods are used as in algebraic number theory. Actually the algorithm used in section 6 will do the work for embedding $\Lambda_{f/R}$ into $\bar{\Lambda}$.

This example shows that the methods used in algebraic number theory can be usefully generalized. In the example given the residue class rings modulo prime ideals always turn out to be finite. The algorithm developed in section 6 will make essential use of this property which characterizes the so-called global fields. In this chapter we do not follow the historic order of development but we begin in section 2 with the study of the algebraic background which includes quotient ring formation and localization. This is followed in section 3 and section 4 by an intensive study of valuation theory, actually extending beyond the immediate needs of algebraic number theory, responding also to the more complex needs of algebraic geometry (Krull–Zariski valuations). Only in section 5 are Dedekind domains dealt with, but by local methods so that we are prepared to introduce an efficient algorithm for computing the maximal order of any separable equation ring over $\mathbb{Z}$ in its quotient ring in section 6.

4.2. The algebraic background

1. Quotient rings

In this section we study the quotient ring.

As we remarked already in chapter 1 the quotient ring $\mathfrak{Q}(R)$ of a commutative ring R containing zero divisors is defined as the unital commutative overring formed by the formal quotients x/v of any element x of R by any non-zero divisor v of R subject to the rules

$$\left.\begin{aligned} &\frac{x}{v}=\frac{x'}{v'}\Leftrightarrow v'x=vx' \\ &\frac{x}{v}+\frac{x'}{v'}=\frac{v'x+vx'}{vv'} \\ &\frac{x}{v}\cdot\frac{x'}{v'}=\frac{xx'}{vv'}, \end{aligned}\right\} \tag{2.1a}$$

$$1_{\mathfrak{Q}(R)}=\frac{v}{v}, \tag{2.1b}$$

$$\frac{x}{v}=x'\Leftrightarrow vx'=x \tag{2.1c}$$

($x, x'\in R$; v, v' two non-zero divisors of R), such that

$$\mathfrak{Q}(\mathfrak{Q}(R))=\mathfrak{Q}(R), \tag{2.2}$$

$$\mathfrak{Q}(R_1\oplus R_2)=\mathfrak{Q}(R_1)\oplus\mathfrak{Q}(R_2), \tag{2.3}$$

$$NR(\mathfrak{Q}(R)) = \mathfrak{Q}(R)\,NR(R), \tag{2.4}$$

$$\mathfrak{Q}(R)/NR(\mathfrak{Q}(R)) \simeq \mathfrak{Q}(R/NR(R)). \tag{2.5}$$

If R is subring of the unital commutative ring Λ containing 1_Λ then the canonical monomorphism $\iota: R \to \mathfrak{Q}(\Lambda): a \mapsto \frac{a}{1}$ extends uniquely to the monomorphism $\iota': \mathfrak{Q}(R,\Lambda) \to \mathfrak{Q}(\Lambda): \frac{a}{v} \mapsto \frac{a}{v}$ of the subring $\mathfrak{Q}(R,\Lambda)$ of $\mathfrak{Q}(R)$ formed by those formal quotients $\frac{a}{v}$ where $a \in R$ and $v \in R$ is a non-zero divisor of Λ.

In constructive algebraic number theory we must solve one algebraic equation or a system of finitely many equations in a constructively given commutative ring. The smallest ring available is the ring generated by the coefficients of the equations to be solved. In any case our constructions employ only finitely many algebraic quantities. Moreover, it is usually not easy and often not even necessary to test the irreducibility of the given equations. That is the reason why we pay much attention to finitely generated commutative rings in this presentation.

Lemma (2.6)

Let Λ be a unital commutative ring with zero nilradical which is finitely generated over the subring R, an entire ring containing 1_Λ such that every non-zero element of R is a non-zero divisor of Λ.

Then we have

$$\mathfrak{Q}(R,\Lambda) = \mathfrak{Q}(R),$$

and $\mathfrak{Q}(\Lambda)$ is the algebraic sum of finitely many fields, each of which is isomorphic to a finitely generated extension of $\mathfrak{Q}(R)$.

Each component has non-zero intersection with Λ.

Proof

By assumption

$$\Lambda = \langle R, x_1, \ldots, x_s \rangle,$$

where $s \in \mathbb{Z}^{>0}$, $x_1 = 1_\Lambda = 1_R$. If $s = 1$ then $\Lambda = R$, $\mathfrak{Q}(\Lambda) = \mathfrak{Q}(R)$.

Apply induction over s.

If $s > 1$ then by induction hypothesis applied to the subring

$$\Lambda' = \langle R; x_1, \ldots, x_{s-1} \rangle$$

of Λ we find that

$$\mathfrak{Q}(\Lambda') = \bigoplus_{i=1}^{\tau} e_i \Lambda' \supseteq \mathfrak{Q}(R) = \bigoplus_{i=1}^{\tau} e_i \mathfrak{Q}(R),$$

where $e_1, \ldots, e_\tau$ are the finitely many primitive idempotents of $\mathfrak{Q}(\Lambda')$, such that $e_i \Lambda'$ is a field generated by the projections $e_i x_j$ $(1 \leqslant j < s)$ over the subfield $e_i \mathfrak{Q}(R)$ $(1 \leqslant i < \tau)$.

There is the canonical monomorphism

$$\theta' : \mathfrak{Q}(\Lambda', \Lambda) \to \mathfrak{Q}(\Lambda'),$$
$$\theta'(x_i) = x_i \quad (1 \leqslant i < s),$$
$$\theta'(\lambda) = \lambda \quad (\lambda \in R),$$

and by assumption

$$\mathfrak{Q}(R) \subseteq \theta' \mathfrak{Q}(\Lambda', \Lambda).$$

Moreover, $\mathfrak{Q}(\Lambda)$ contains no nilpotent element and there is an element y_i of Λ for which

$$0 \neq y_i \in e_i \mathfrak{Q}(\Lambda').$$

For any polynomial $f(t)$ of $\Lambda'[t]$ satisfying

$$f(x_s) = 0,$$

we have

$$\theta' f(t) = \sum_{i=1}^{\tau} e_i f(t).$$

All polynomials $e_i f$ generate an ideal of $e_i \mathfrak{Q}(\Lambda')[t]$ which is either 0 or it is generated by a polynomial $m_i(t)$ of $(\mathfrak{Q}(\Lambda', \Lambda) \cap e_i \mathfrak{Q}(\Lambda'))[t]$. If there is a monic polynomial d of $(\mathfrak{Q}(\Lambda', \Lambda) \cap e_i \mathfrak{Q}(\Lambda'))[t]$ properly dividing m_i such that m_i divides d^2 then we find

$$m'(t) = m_i(y_i^{-1} t), \quad d'(t) = d(y_i^{-1} t) \in e_i \mathfrak{Q}(\Lambda')[t]$$

so that

$$\Lambda \ni d'(y_i x_s) \neq 0, \quad m'(y_i x_s) = 0,$$

hence

$$d'(y_i x_s)^2 = 0,$$

a contradiction.

Hence m_i is a square free polynomial of $e_i \mathfrak{Q}(\Lambda')[t]$. Hence

$$e_i \mathfrak{Q}(\Lambda')[t]/m_i = \bigoplus_{j=1}^{n_i} E_{ij} = e_i \mathfrak{Q}(\Lambda')[x_{si}],$$

where

$$x_{si} = t/m_i,$$
$$e_i = \sum_{j=1}^{n_i} e_{ij},$$
$$0 \neq e_{ij} = e_{ij}^2 \in E_{ij} = e_{ij} \mathfrak{Q}(\Lambda')[e_{ij} x_{si}],$$

and E_{ij} is a finite extension of $e_{ij} \mathfrak{Q}(\Lambda')$ $(1 \leqslant j \leqslant n_i)$. If all polynomials $e_i f(t)$ are 0 for some index i then we set $n_i = 1$ and form the rational function field E_{i1} in one variable x_{si} over $e_i \mathfrak{Q}(\Lambda')$.

Thus we obtain the algebraic sum

$$E = \bigoplus_{i=1}^{\tau} \bigoplus_{j=1}^{n_i} E_{ij}$$

of finitely many fields and a monomorphism θ of Λ into E which maps

$$x \text{ on } \bigoplus_{i=1}^{\tau} \bigoplus_{j=1}^{n_i} e_{ij}\theta(x) \quad \text{for } x \text{ of } \Lambda',$$

$$x_s \text{ on } \bigoplus_{i=1}^{\tau} \bigoplus_{j=1}^{n_i} e_{ij}x_{si}$$

such that

$$E = \mathfrak{Q}(\theta\Lambda).$$

Moreover, in each case we find a non-zero element y_{ij} of Λ belonging to E_{ij}.

□

Lemma (2.7)

The quotient ring of a finitely generated commutative ring Λ contains only finitely many idempotents.

Proof

Without loss of generality we assume that Λ is unital and that Λ contains no nilpotent elements $\neq 0$. By assumption there is an epimorphism η of the polynomial ring $\mathbb{Z}[t_1, \ldots, t_s]$ on Λ for some natural number s.

Let m be the least common multiple of the *characteristics* of $\eta(t_i)$, i.e. the orders of the additive groups generated by the $\eta(t_i)$s $(1 \leqslant i \leqslant s)$.

If it is 0 then let $m' = 2$, if $m > 0$ then let m' be the product of the distinct prime numbers dividing m. In any case, $\mathfrak{Q}(\Lambda/m'\Lambda)$ contains the same number of idempotents as $\mathfrak{Q}(\Lambda)$.

Without loss of generality we assume that Λ is unital with non-zero nilradical and of characteristic

$$m' = \prod_{i=1}^{\sigma} p_i,$$

where $\sigma \in \mathbb{Z}^{>0}$ and $p_1, \ldots, p_\sigma$ are distinct prime numbers. Hence,

$$\Lambda = \bigoplus_{i=1}^{\sigma} (m'/p_i)\Lambda,$$

$$\mathfrak{Q}(\Lambda) = \bigoplus_{i=1}^{\sigma} \mathfrak{Q}((m'/p_i)\Lambda),$$

and $(m'/p_i)\Lambda_i$ is a finitely generated unital commutative ring of prime characteristic p_i whose quotient ring contains only finitely many idempotents according to lemma (2.6).

□

Concerning non-constructive extensions of the quotient ring concept see exercises 1–4 at the end of this section.

2. Localization

The construction (2.1a) of a ring using a given commutative ring R and certain elements of R as denominators can be applied to any ring R containing a subsemigroup S of the (multiplicative) semigroup of the center of R. It yields a unital ring R/S, said to be the *S-localization* of R. Its unit element is

$$1_{R/S} = \frac{\nu}{\nu} \quad (\nu \in S), \tag{2.8a}$$

and the mapping

$$\varphi\colon R \to \frac{R}{S}\colon x \mapsto \frac{\nu x}{\nu} \quad (x \in R, \nu \in S), \tag{2.8b}$$

suggested by (2.1b) is unique. The image ring $\varphi(R)$ is a subring of R/S such that the elements of $\varphi(S)$ are non-zero divisors of R/S forming a subsemigroup of the center of R/S.

This mapping is injective (monomorphic) if and only if the elements of S are non-zero divisors of R. In that case we consider φ as the canonical embedding of R in R/S, and interpret R/S as an overring of R.

For example

$$\frac{R}{S} = \frac{\varphi(R)}{\varphi(S)} = \frac{\dfrac{\varphi(R)}{\varphi(S)}}{\varphi(S)} \tag{2.8c}$$

is an overring of $\varphi(S)$.

It is because of (2.8c) that we confine our consideration to localization by means of semigroups of central non-zero divisors.

Any ideal $\mathfrak{a}$ of R generates the ideal

$$\frac{\mathfrak{a}}{S} = \mathfrak{a}\frac{R}{S} = \frac{R}{S}\mathfrak{a} \tag{2.9a}$$

of R/S such that

$$\left.\begin{aligned} \frac{\mathfrak{a}+\mathfrak{b}}{S} &= \frac{\mathfrak{a}}{S} + \frac{\mathfrak{b}}{S}, \\ \frac{\mathfrak{a}\mathfrak{b}}{S} &= \frac{\mathfrak{a}}{S}\frac{\mathfrak{b}}{S}, \\ \frac{\mathfrak{a}\cap\mathfrak{b}}{S} &= \frac{\mathfrak{a}}{S} \cap \frac{\mathfrak{b}}{S} \end{aligned}\right\} \tag{2.9b}$$

($\mathfrak{a}, \mathfrak{b}$ any two ideals of R). Conversely, for any ideal $\mathfrak{A}$ of R/S we find that the intersection with R is an ideal of R generating $\mathfrak{A}$ over S:

$$\frac{\mathfrak{A} \cap R}{S} = \mathfrak{A}, \tag{2.9c}$$

the intersection of $\mathfrak{A}$ with R is the largest ideal of R generating $\mathfrak{A}$ over S. In other words, localization sifts out certain ideals $\mathfrak{A} \cap R$ of R, the S-ideals in one-to-one correspondence with the ideals of R/S such that the ideal structure in regard to addition, multiplication and intersection is preserved. For example for any prime ideal $\mathfrak{p}$ of any commutative ring R the elements of R not belonging to $\mathfrak{p}$ form a multiplicative semigroup $R \backslash \mathfrak{p}$. It defines a unital overring $R/(R \backslash \mathfrak{p})$ by localization precisely if R contains no non-zero divisors and in that case $R/(R \backslash \mathfrak{p})$ is an entire ring. This particular localization is said to be the $\mathfrak{p}$-*localization* of R though, of course, the denominators used are precisely those elements of R which do not belong to $\mathfrak{p}$.

In fact $\mathfrak{p}$ generates the maximal ideal $\mathfrak{p}/(R \backslash \mathfrak{p})$ of $R/(R \backslash \mathfrak{p})$ and this is the only maximal ideal of $R/(R \backslash \mathfrak{p})$. This is because the units of $R/(R \backslash \mathfrak{p})$ are precisely the elements of $R/(R \backslash \mathfrak{p})$ which do not belong to $\mathfrak{p}/(R \backslash \mathfrak{p})$:

$$U\left(\frac{R}{R \backslash \mathfrak{p}}\right) = \frac{R}{R \backslash \mathfrak{p}} \Big/ \frac{\mathfrak{p}}{R \backslash \mathfrak{p}}. \tag{2.9d}$$

Definition (2.10)

A ring with only one maximal ideal is said to be a **local ring.**

A *local domain* is characterized as an entire ring with precisely one maximal ideal. It is a ring providing its own localization with respect to its maximal ideal.

For example the 2-localization of $\mathbb{Z}$, i.e. the localization of $\mathbb{Z}$ with respect to the prime ideal $2\mathbb{Z}$ is obtained as the ring $\dfrac{\mathbb{Z}}{\mathbb{Z} \backslash 2\mathbb{Z}}$ of all rational numbers with odd denominators. The 2-ideals of $\mathbb{Z}$ are the ideals generated by powers of 2 and 0.

Similarly, for any prime number p of $\mathbb{Z}$ the p-localization of $\mathbb{Z}$ emphasizes the powers of p and 0 among the ideals of $\mathbb{Z}$.

Thus p-localization appears as a tool for detailed aspects of the arithmetical structure of $\mathbb{Z}$. Upon gathering the aspects obtained by many p-localizations we obtain a richer global view of it. For any ideal $\mathfrak{a}$ of a commutative ring R the elements x of R for which the residue class $x/\mathfrak{a}$ modulo $\mathfrak{a}$ represented by x is a non-zero divisor of $R/\mathfrak{a}$ form a subsemigroup $S_{\mathfrak{a}}$ of R. The $\mathfrak{a}$-localization of R is defined as the ring $R/S_{\mathfrak{a}}$. If R is an entire ring then $R/S_{\mathfrak{a}}$ is a subring of $\mathfrak{Q}(R)$ containing R as a subring.

For any non-zero ideal of $\mathbb{Z}$, say the ideal $m\mathbb{Z}$ generated by the natural number m, the m-localization $\mathbb{Z}/S_m$ consists of the rational numbers with denominator prime to m.

We have $\mathbb{Z}/S_1 = \mathbb{Z}$. For $m > 1$ there are only finitely many maximal ideals of $\mathbb{Z}/S_m$, viz. the principal ideals p/S_m where p runs through the prime numbers dividing m.

Definition (2.11)
A ring with only a finite number of maximal ideals is said to be **semilocal.**

The semilocal localizations of $\mathbb{Z}$ and of generalizations of $\mathbb{Z}$ (Dedekind orders) become important in the detailed study of finitely many distinct prime ideals and their interaction.

The localization concept extends to modules over rings. Suppose R is a unital commutative ring and M is an R-module, in other words M is a module with a binary multiplication of the elements λ of R (scalars) by elements x of M defining the homomorphism

$$\theta: R \to End(M),$$
$$\theta(\lambda)(x) = \lambda x \quad (\lambda \in R,\ x \in M)$$

of R into the endomorphism ring End(M) with

$$\theta(1_R) = 1_M.$$

For any subsemigroup S of R the S-localization M/S is the R/S-module formed by the formal quotients $\frac{x}{\lambda}$ $(x \in M, \lambda \in S)$ with the operational rules

$$\left.\begin{aligned} &\frac{x}{\lambda} = \frac{x'}{\lambda'} \Leftrightarrow \lambda' x = \lambda x', \\ &\frac{x}{\lambda} + \frac{x'}{\lambda'} = \frac{\lambda' x + \lambda x'}{\lambda\lambda'}, \\ &\frac{\mu}{\nu} \cdot \frac{x}{\lambda} = \frac{\mu x}{\nu\lambda} \quad (x, x' \in M;\ \mu \in R;\ \lambda, \lambda', \nu \in S) \end{aligned}\right\} \tag{2.12}$$

as is easily verified by the reader, who also realizes that the mapping

$$\phi: M \to \frac{M}{S}: c \mapsto \frac{\nu x}{\nu} \quad (x \in M, \nu \in S) \tag{2.13a}$$

provides an additive homomorphism satisfying

$$\phi(\lambda x) = \varphi(\lambda)\phi(x) \quad (\lambda \in R,\ x \in M), \tag{2.13b}$$

where φ is as in (2.8b). However, ϕ is injective if and only if S acts *torsion free* on M:

$$\forall \lambda \in S \quad \forall x \in M: \lambda x = 0 \Rightarrow x = 0. \tag{2.13c}$$

The kernel of ϕ is the *S-torsion submodule*

$$\operatorname{Tor}(M/S) = \{x \in M \mid \exists \lambda \in S : \lambda x = 0\},$$

an R-submodule of M such that ϕ lifts to an injective mapping of $M/\operatorname{Tor}(M/S)$ into $(M/\operatorname{Tor}(M/S))/S$ so that $M/\operatorname{Tor}(M/S)$ is S-torsion free.

If M is S-torsion free then the S–R-submodules of M are defined as those R-submodules which are obtained by intersecting an R/S-submodule of M/S with M. They are the largest R-submodules of M among those generating the same R/S-submodule of M/S. They are in 1–1-correspondence with the R/S-submodules of M/S such that addition and intersection of modules is preserved. Note also that for any ideal $\mathfrak{a}$ of R and any R-submodule $\mathfrak{m}$ of M we have

$$\frac{\mathfrak{a}\mathfrak{m}}{S} = \frac{\mathfrak{a}}{S}\frac{\mathfrak{m}}{S}.$$

Exercises

1. Let R be a unital commutative ring. An ideal $\mathfrak{a}$ of R is said to be *large* ('non-zero divisor ideal') if 0 is the only annihilator of $\mathfrak{a}$. For example, R itself is large. Show:
   (a) Any ideal of R containing a large ideal is large.
   (b) The sum of two large ideals is large.
   (c) The product of two large ideals is large.
   (d) The intersection of two large ideals is large.
   (e) Any ideal of R containing a non-zero divisor of R is large.

2. (Lambek) *A partial endomorphism* of a unital commutative ring R is defined as a mapping $\varphi:\mathfrak{a}\to R$ of a large ideal $\mathfrak{a}$ of R into R satisfying $\varphi(a+b)=\varphi(a)+\varphi(b)$, $\varphi(\lambda a)=\lambda\varphi(a)$ $(a,b\in\mathfrak{a},\lambda\in R)$. For example, the fraction $\frac{2}{3}$ is associated with the partial endomorphism $\varphi: 3\mathbb{Z}\to\mathbb{Z}: 3x\mapsto 2x$. Two partial endomorphisms $\varphi_i:\mathfrak{a}_i\to R$ $(i=1,2)$ of R are said to be *equivalent*, if their restrictions to the intersection of the large ideals $\mathfrak{a}_1$, $\mathfrak{a}_2$ coincide: $\varphi_1|_{\mathfrak{a}_1\cap\mathfrak{a}_2}=\varphi_2|_{\mathfrak{a}_1\cap\mathfrak{a}_2}$. For example, the partial endomorphisms of $\mathbb{Z}$ associated with the fractions $\frac{4}{6}$, $\frac{6}{9}$ coincide on $6\mathbb{Z}\cap 9\mathbb{Z}=18\mathbb{Z}$. Both are equivalent to $\frac{2}{3}$. Show:
   (a) The equivalence of partial endomorphisms is reflexive, symmetric, and transitive.
   (b) If $\varphi:\mathfrak{a}\to R$, $\psi:\mathfrak{b}\to R$ are two partial endomorphisms of R, then also the sum $\varphi+\psi:\mathfrak{a}\cap\mathfrak{b}\to R: x\mapsto\varphi(x)+\psi(x)$ and the product $\varphi\psi:\mathfrak{a}\mathfrak{b}\to R: x\mapsto\varphi(\psi(x))$ are partial endomorphisms.
   (c) Both addition and multiplication of partial endomorphisms satisfy the substitutional law so that the equivalence of partial endomorphisms $\varphi_i (i=1,2)$, $\psi_j (j=1,2)$ implies the equivalence of $\varphi_1+\psi_1$ and $\varphi_2+\psi_2$ as well as of $\varphi_1\psi_1$ and $\varphi_2\psi_2$.
   (d) The equivalence classes of the partial endomorphisms of the unital commutative

ring R define a unital commutative ring $\bar{\mathfrak{Q}}(R)$, the *Lambek–Ushida quotient ring.*

(e) There is the canonical monomorphism $\tau\colon \mathfrak{Q}(R)\to\bar{\mathfrak{Q}}(R)$ subject to $\tau(x/v)\colon vR\to R\colon vy\mapsto xy$ ($x\in R$, v a non-zero divisor of R) of the standard quotient ring $\mathfrak{Q}(R)$ into the Lambek–Ushida quotient ring.

(f) For entire rings R we have $\mathfrak{Q}(R)=\bar{\mathfrak{Q}}(R)$.

(g) There are unital commutative rings R satisfying $\mathfrak{Q}(R)\subset\bar{\mathfrak{Q}}(R)$.

(h) $\bar{\mathfrak{Q}}(\bar{\mathfrak{Q}}(R))=\bar{\mathfrak{Q}}(R)$.

3. Let R be a unital commutative ring and M an R-module.

(a) There is the R-module homomorphism $\eta\colon M\to\mathfrak{Q}(R)\otimes_R M\colon u\mapsto 1\otimes u$ such that $\mathfrak{Q}(R)\otimes_R M$ is a $\mathfrak{Q}(R)$-module according to the rule $x(y\otimes u)=xy\otimes u$ ($x, y\in\mathfrak{Q}(R)$, $u\in M$) and $\mathfrak{Q}(R)\otimes_R M=\mathfrak{Q}(R)\eta(M)$.

(b) For entire rings R there holds $\ker\eta=\operatorname{Tor}(M/R)$, where $\operatorname{Tor}(M/R):=\operatorname{Tor}(M/S_R)$ and S_R denotes the non-zero divisors of R.

(c) If we define $\operatorname{Tor}(M/R)$ as $\ker\eta$ then $\operatorname{Tor}((M/\operatorname{Tor}(M/R)/R)=0$.

4. (a) Under the assumption of 3. show that there is the R-module homomorphism $\bar{\eta}\colon M\to\bar{\mathfrak{Q}}(R)\otimes_R M\colon u\mapsto 1\otimes u$ such that $\bar{\mathfrak{Q}}(R)\otimes_R M$ is a $\mathfrak{Q}(R)$-module according to the rule $x(y\otimes u)=xy\otimes u$ ($x, y\in\bar{\mathfrak{Q}}(R)$, $u\in M$).

(b) $\ker\bar{\eta}=\ker\eta$.

4.3. Valuation theory

1. Pseudo-valuations

The mapping of a complex number on its absolute value is a mapping φ of $\mathbb{C}$, the complex number field, into the real number field $\mathbb{R}$ for which

$$\varphi(a)\geqslant 0, \tag{3.1a}$$
$$\varphi(a)=0\Leftrightarrow a=0, \tag{3.1b}$$
$$\varphi(a\pm b)\leqslant\varphi(a)+\varphi(b), \tag{3.1c}$$
$$\varphi(ab)=\varphi(a)\varphi(b), \tag{3.1d}$$
$$\varphi(\pm 1)=1 \tag{3.1e}$$

for all complex numbers a, b. The real number field $\mathbb{R}$ provides an example of an *algebraically ordered ring* which is defined as a ring Φ with a total ordering relation such that

$$\alpha>0\wedge\beta>0\Rightarrow\alpha+\beta>0, \tag{3.2a}$$
$$\alpha>0\wedge\beta>0\Rightarrow\alpha\beta>0, \tag{3.2b}$$
$$\text{if } \alpha\neq 0, \text{ and if not } \alpha>0 \text{ then } -\alpha>0, \tag{3.2c}$$
$$\alpha>\beta\Leftrightarrow\alpha-\beta>0 \text{ for all elements } \alpha, \beta \text{ of } \Phi. \tag{3.2d}$$

Definition (3.3)

The mapping φ of any unital ring R into the algebraically ordered unital ring Φ satisfying (3.1a)–(3.1c), (3.1e) *for a, b of R and the generalization*

$$\varphi(ab) \leqslant \varphi(a)\varphi(b) \quad (a, b \in R) \tag{3.3a}$$

is said to be a **pseudo-valuation** *of R in Φ.*

The pseudo-valuation is said to be *multiplicative* or just a *valuation* if (3.1d) is satisfied for a, b of R.

For example, the rational number field $\mathbb{Q}$ is algebraically ordered according to the rule

$$\frac{a}{b} > \frac{a'}{b'} \Leftrightarrow abb'^2 > a'b'b^2 \quad (a, a', b, b' \in \mathbb{Z},\ 0 \neq bb'). \tag{3.4}$$

For any prime number p we have the *p-adic valuation*

$$\varphi_p : \mathbb{Q} \to \mathbb{Q}^{\geqslant 0} : \frac{p^{\nu} a}{b} \mapsto p^{-\nu} \quad (\nu \in \mathbb{Z};\ a, b \in \mathbb{Z},\ p \nmid ab).$$
$$0 \mapsto 0$$

Similarly we have for any composite natural number n the non-multiplicative n-adic valuation:

$$\varphi_n : \mathbb{Q} \to \mathbb{Q}^{\geqslant 0},$$
$$\varphi_n\left(\frac{n^{\nu} a}{b}\right) = n^{-\nu} \quad (\nu \in \mathbb{Z};\ a, b \in \mathbb{Z},\ n \nmid a, \gcd(b, n) = 1),$$
$$\varphi_n(0) = 0.$$

It is a pseudo-valuation but no valuation. The pseudo-valuations $= \varphi_n$ satisfy the rule

$$\varphi(a + b) \leqslant \max(\varphi(a), \varphi(b)), \tag{3.5}$$

which implies the triangle inequality (3.1c). Pseudo-valuations φ satisfying (3.5) are said to be *non-archimedean.* On the other hand the ordinary absolute value valuation

$$\varphi_{\infty} : \mathbb{Q} \to \mathbb{Q}^{\geqslant 0} : \frac{a}{b} \mapsto \left|\frac{a}{b}\right| \quad (a \in \mathbb{Z}, 0 \neq b \in \mathbb{Z}) \tag{3.6}$$

does not satisfy the stronger condition (3.5). It is said to be *archimedean.*

2. Algebraically ordered rings and semirings

Let us now study the pseudo-valuations for the purpose of embedding orders into maximal orders.

Firstly we observe that only the non-negative elements $\Phi^{\geqslant 0}$ of Φ occur as values of the valuation function φ.

They form an *algebraically ordered semiring* as is implied by the following definitions.

Definition (3.7)
A **semiring** *is defined as a set with two associative binary operations, addition and multiplication, linked by the two distributive laws*

$$a(b+c) = ab + ac, \quad (b+c)a = ba + ca.$$

For example, the natural numbers form a semiring.

Definition (3.8)
A total order relation on a semiring S is said to be an **algebraic ordering** *if it satisfies the rules*

$$\alpha > \beta \Rightarrow \gamma + \alpha > \gamma + \beta \wedge \alpha + \gamma > \beta + \gamma \tag{3.8a}$$

and

$$\alpha > \beta \wedge \gamma > \delta \Rightarrow \alpha\gamma + \beta\delta > \beta\gamma + \alpha\delta \quad \textit{for all } \alpha, \beta, \gamma, \delta \textit{ of } S. \tag{3.8b}$$

The reader will find it an easy exercise to deduce (3.8a), (3.8b) from (3.2a–d).

On the other hand we note that addition in a semiring need not be commutative. Let us show that algebraically ordered semirings have commutative addition. Certainly both cancellation laws of addition are satisfied:

$$\gamma + \alpha = \gamma + \beta \Rightarrow \alpha = \beta, \tag{3.9a}$$

$$\alpha + \gamma = \beta + \gamma \Rightarrow \alpha = \beta. \tag{3.9b}$$

This is because of (3.8a) and the trichotomy of total ordering which means that for any two elements α, β of a totally ordered set one and only one of the 3 relations: $\alpha > \beta$, $\beta > \alpha$, $\alpha = \beta$ takes place.

But in a semiring S satisfying both cancellation laws of addition the elements of S^2 commute under addition since

$$\begin{aligned}(\alpha+\beta)(\gamma+\delta) &= (\alpha+\beta)\gamma + (\alpha+\beta)\delta = \alpha\gamma + \beta\gamma + \alpha\delta + \beta\delta \\ &= \alpha(\gamma+\delta) + \beta(\gamma+\delta) = \alpha\gamma + \alpha\delta + \beta\gamma + \beta\delta\end{aligned}$$

implies upon cancellation of $\alpha\gamma$ on the left, $\beta\delta$ on the right that

$$\beta\gamma + \alpha\delta = \alpha\delta + \beta\gamma.$$

If an algebraically ordered semiring S contains a neutral element n of addition (zero element) at all then n is characterized by the equation $n + n = n$. Indeed, if $n + n = n$ then we have $n + n + \alpha = n + \alpha$, $\alpha + n + n = \alpha + n$ by uniqueness and $n + \alpha = \alpha$, $\alpha + n = \alpha$ by the cancellation laws of addition. Conversely any one of the equations $n + \alpha = \alpha$, $\alpha + n = \alpha$ implies the equation $n + n = n$.

For an algebraically ordered semiring S we also have the cancellation laws of multiplication

$$\alpha\beta = \alpha\gamma \Rightarrow \beta = \gamma \tag{3.9c}$$

and

$$\beta\alpha = \gamma\alpha \Rightarrow \beta = \gamma \tag{3.9d}$$

for three elements α, β, γ of S for which $\alpha + \alpha \neq \alpha$. Indeed, if $\alpha + \alpha > \alpha$, $\beta > \gamma$ then $(\alpha + \alpha)\beta + \alpha\gamma > \alpha\beta + (\alpha + \alpha)\gamma$ by (3.8b), hence $\alpha\beta + \alpha\beta + \alpha\gamma > \alpha\beta + \alpha\gamma + \alpha\gamma$ and upon cancellation of $\alpha\beta$ on the left, $\alpha\gamma$ on the right $\alpha\beta > \alpha\gamma$. Similarly we conclude that $\alpha\beta < \alpha\gamma$ from $\alpha > \alpha + \alpha$, $\beta > \gamma$.

There is just one semiring with one element only, the null ring.

In an algebraically ordered semiring S which is not a null ring the addition is commutative because there is an element α of S for which $\alpha + \alpha \neq \alpha$. For any two elements β, γ of S we derive from the commutativity $\alpha\beta + \alpha\gamma = \alpha\gamma + \alpha\beta$ for the addition of the products $\alpha\beta$, $\alpha\gamma$ upon cancellation by α on the left the commutative rule $\beta + \gamma = \gamma + \beta$.

A semiring S with commutative law of addition and the cancellation laws of addition also is said to be a *half-ring*. This is because it is embedded into the ring $\Phi = S - S$ formed by the formal difference elements $\alpha - \beta(\alpha, \beta \in S)$ subject to the operational rules:

$$\alpha - \beta = \gamma - \delta \Rightarrow \alpha + \delta = \beta + \gamma, \tag{3.10a}$$

$$(\alpha - \beta) + (\gamma - \delta) = (\alpha + \gamma) - (\beta + \delta), \tag{3.10b}$$

$$(\alpha - \beta)(\gamma - \delta) = (\alpha\gamma + \beta\delta) - (\alpha\delta + \beta\gamma), \tag{3.10c}$$

$$\alpha - \beta = \gamma \Leftrightarrow \alpha = \beta + \gamma \ (\alpha, \beta, \gamma, \delta \in S). \tag{3.10d}$$

The ring properties of Φ and the embedding of S in Φ may be verified by the reader without difficulty. Note that $(\alpha - \beta) + \beta = \alpha$ as it should be in a ring.

The algebraic ordering of an algebraically ordered semiring S extends to a total ordering of S–S upon defining

$$\alpha - \beta > \gamma - \delta \Leftrightarrow \alpha + \delta > \beta + \gamma \quad (\alpha, \beta, \gamma, \delta \in S) \tag{3.11a}$$

and verifying that

$$\begin{aligned} &\alpha - \beta > \gamma - \delta \quad \text{and} \quad \alpha - \beta = \alpha' - \beta', \quad \gamma - \delta = \gamma' - \delta' \\ &\text{implies that } \alpha' - \beta' > \gamma' - \delta', \end{aligned} \tag{3.11b}$$

and that

$$\alpha - \beta > \gamma - \delta, \alpha - \beta = \varepsilon, \gamma - \delta = \eta \Rightarrow \varepsilon > \eta \tag{3.11c}$$

$$(\alpha, \alpha', \beta, \beta', \gamma, \gamma', \delta, \delta', \varepsilon, \eta \in S).$$

It is left to the reader to verify that the total ordering of S–S provided by (3.11a) satisfies the rules (3.2a–d) for an algebraic ordering of S–S extending the algebraic ordering of S. Indeed this is the only way to do so because for an algebraic ordering of S–S any time we have $\alpha - \beta > \gamma - \delta$ $(\alpha, \beta, \gamma, \delta \in S)$ we gather upon addition of $\beta + \delta$ on both sides that $\alpha + \delta > \gamma + \beta = \beta + \gamma$.

Note that in any algebraically ordered ring Φ, the positive elements, i.e. the elements >0, form an algebraically ordered semiring $\Phi^{>0} = \{x \in \Phi \mid x > 0\}$, which generates $\Phi = \Phi^{>0} - \Phi^{>0}$ as described above.

Also note that the squares of non-zero elements and their sums are positive.

If the square sums of the non-zero elements of a ring Φ form a half-ring S with the property that Φ is the disjoint union of S, 0, $-S$ then S is the positivity half-ring of a totally ordering relation of Φ that is uniquely determined by Φ itself.

For example $\mathbb{Z}^{>0}$ consists of the sums of $1 = 1^2$ so that there is only one algebraic ordering of the rational integer ring. Similarly $\mathbb{R}^{>0}$ consists of the squares of non-zero real numbers so that there is only one algebraic ordering of the real number field $\mathbb{R}$.

The null ring is embedded into the algebraically ordered ring $\mathbb{Z}$. Any other algebraically ordered ring R is embedded into the unital algebraically ordered ring $R \oplus \mathbb{Z}$ formed by the formal sums $x \oplus \lambda$ $(x \in R, \lambda \in \mathbb{Z})$ with operational rules

$$x \oplus \lambda = y \oplus \mu \Leftrightarrow \forall \alpha \in R: (x-y)\alpha = \alpha(x-y) = (\mu - \lambda)\alpha, \tag{3.12a}$$

$$(x \oplus \lambda) + (y \oplus \mu) = (x+y) \oplus (\lambda + \mu), \tag{3.12b}$$

$$(x \oplus \lambda)(y \oplus \mu) = (xy + \lambda y + \mu x) \oplus \lambda\mu \tag{3.12c}$$

$$x \oplus \lambda > 0 \Leftrightarrow \exists \alpha \in R: \alpha > 0 \wedge x\alpha > -\lambda\alpha \quad (x, y \in R; \lambda, \mu \in \mathbb{Z}) \tag{3.12d}$$

as may be verified by the reader.

Thus it comes that in valuation theory the values are taken from a unital algebraically ordered ring Φ though not all elements of Φ occur as values.

We remark that Φ values itself multiplicatively by the *absolute valuation*

$$|\alpha| = \begin{cases} \alpha & \text{if} \quad \alpha > 0 \\ 0 & \text{if} \quad \alpha = 0 \quad (\alpha \in \Phi), \\ -\alpha & \text{if} \quad 0 > \alpha \end{cases} \tag{3.13}$$

in as much as

$$\begin{aligned} |a| &\geqslant 0, \\ |a| = 0 &\Leftrightarrow a = 0, \\ |a+b| &\leqslant |a| + |b|, \\ |ab| &= |a| \cdot |b|, \\ |\pm 1| &= 1 \quad (a, b \in \Phi). \end{aligned}$$

The sign of any element of Φ is defined by

$$\operatorname{sign}(\alpha) = \begin{cases} 1 & \text{if} \quad \alpha > 0 \\ 0 & \text{if} \quad \alpha = 0 \\ -1 & \text{if} \quad 0 > \alpha. \end{cases} \tag{3.14a}$$

It is multiplicative:

$$\operatorname{sign}(\alpha\beta) = \operatorname{sign}(\alpha)\operatorname{sign}(\beta) \quad (\alpha, \beta \in \Phi). \tag{3.14b}$$

3. General remarks on pseudo-valuations

The restriction (3.1b) on pseudo-valuations is merely for the sake of convenience in as much as for a *generalized pseudo-valuation*, i.e. a mapping φ of R into Φ satisfying merely (3.1a), (3.1c), (3.1d), (3.1e), all elements of R with φ-value zero form an ideal $I(\varphi, 0)$ of R such that the mapping

$$\bar{\varphi}: R/I(\varphi, 0) \to \Phi: a/I(\varphi, 0) \mapsto \varphi(a)$$

defines a pseudo-valuation $\bar{\varphi}$ on $R/I(\varphi, 0)$.

The strength of valuation theory rests in the fact that it provides an evaluation of complex algebraic structures like those provided by the algebraic structure of a ring R by means of a totally ordered structure like the algebraically ordered unital ring Φ.

Therefore we say that two pseudo-valuations of R, say φ in Φ, φ' in Φ', are *equivalent* if

$$\varphi(a) > \varphi(b) \Leftrightarrow \varphi'(a) > \varphi'(b) \quad (a, b \in R).$$

This equivalence relation between pseudo-valuations is reflexive, symmetric, and transitive.

For example the n-adic pseudo-valuation φ_n of $\mathbb{Q}$ ($n \in \mathbb{Z}^{>1}$) is equivalent to any one of the pseudo-valuations

$$\varphi: \mathbb{Q} \to \mathbb{R}: x \mapsto \varphi_n(x)^{\beta}$$

with positive exponent β.

The strength of our particular choice of φ_p (p a prime number) and of φ_∞ rests in the *product formula*

$$\prod_p \varphi_p(x) = 1 \quad (0 \neq x \in \mathbb{Q}), \tag{3.15}$$

where the product on the left extends over all prime numbers p as well as over ∞. Moreover, for each x only a finite number of factors is not 1. The product is taken only over those factors. The product formula is an elegant equivalent to the fundamental theorem of number theory. It suggests to look at φ_∞, i.e. the absolute value valuation of $\mathbb{Q}$, as of the infinite prime p_∞ of $\mathbb{Q}$. As this example shows, valuation theory provides for a proper analysis of the complexities of an algebraic structure by considering many valuations.

The non-archimedean valuations of a *division ring* (= skew field) which are also called *Krull valuations* have their values only in a multiplicative group and in 0. The restriction of an algebraic ordering of a unital ring Φ to a subgroup G of the unit group of Φ establishes an *algebraic ordering of* G. It is a total ordering of G subject to the rules

$$\alpha > 1, \quad \beta > 1 \Rightarrow \alpha\beta > 1, \tag{3.16a}$$

$$\alpha > 1 \Rightarrow \xi\alpha\xi^{-1} > 1, \tag{3.16b}$$

$$\alpha \in G, \quad \alpha \not> 1, \quad \alpha \neq 1 \Rightarrow \alpha^{-1} > 1, \tag{3.16c}$$

$$\alpha > \beta \Leftrightarrow \alpha\beta^{-1} > 1 \quad (\alpha, \beta, \xi \in G). \tag{3.16d}$$

Conversely, if a group G contains a subset $G^{>1}$ satisfying (3.16a–c) then by means of (3.16d) a total ordering of G is established that is extended to an algebraic ordering of the group ring $\mathbb{Z}G$ by defining

$$\sum_{g \in G} \lambda(g)g > 0 \Leftrightarrow (\exists g_0 \in G : \lambda(g_0) > 0 \wedge (\forall g \in G : g > g_0 \Rightarrow \lambda(g) = 0)). \tag{3.16e}$$

It follows that the total ordering of G determined by means of (3.16d) is an algebraic ordering of G.

This brief excursion shows that even in case the non-zero values belong to an algebraically ordered group it is still acceptable to speak of a valuation into an algebraically ordered unital ring.

For any non-archimedean pseudo-valuation

$$\varphi : R \rightarrow \Phi \tag{3.17a}$$

of the unital ring R we have the modules

$$I(\varphi, \rho) = \{x \in R \mid \varphi(x) \leqslant \rho\}. \tag{3.17b}$$

of R defined for each ρ of $\Phi^{\geqslant 0}$. They are said to be the *valuation modules* corresponding to φ and they have the properties

$$I(\varphi, \rho) \subseteq I(\varphi, \rho') \quad \text{if } 0 \leqslant \rho \leqslant \rho', \tag{3.17c}$$

$$I(\varphi, \rho)I(\varphi, \rho') \subseteq I(\varphi, \rho\rho') \ (\rho, \rho' \in \Phi^{\geqslant 0}), \tag{3.17d}$$

$$I(\varphi, 0) = 0, \tag{3.17e}$$

$$\bigcup_{\rho \in \Phi^{\geqslant 0}} I(\varphi, \rho) = R, \tag{3.17f}$$

$$\forall x \in R : x \in I(\varphi, \varphi(x)) \wedge \forall \rho \in \Phi^{<\varphi(x)} : x \notin I(\varphi, \rho). \tag{3.17g}$$

The valuation module $I(\varphi, 1)$ is a ring said to be the *valuation ring* of φ.

Conversely, if $I(\rho)$ are submodules of R defined for each element ρ of $\Phi^{\geqslant 0}$ and subject to the conditions (3.17c–f) with $I(\rho)$ in place of $I(\varphi, \rho)$ then there is the non-archimedean pseudo-valuation (3.17a) defined by (3.17g) so that $I(\rho) = I(\varphi, \rho)$ $(\rho \in \Phi^{\geqslant 0})$.

From now on we simply speak of 'valuations' in place of 'non-archimedean pseudo-valuations'.

4. Extending valuations I

A central problem of valuation theory is the task of extending a given valuation (3.17a) of R to a valuation of a given unital overring Λ.

If 1_R can be extended to an R-basis B of Λ such that

$$\alpha b_i = b_i \alpha \quad (\alpha \in R, \ b_i \in B), \tag{3.18a}$$

$$b_i b_j = \sum_{b_k \in B} \gamma_{ijk} b_k \quad (\gamma_{ijk} \in C(R)), \tag{3.18b}$$

then we have the inequalities

$$\bar{\varphi}(b_i b_j) \leqslant \bar{\varphi}(b_i)\bar{\varphi}(b_j) \quad (b_i, b_j \in B), \tag{3.18c}$$

$$\bar{\varphi}(b_i b_j) \leqslant \max\{\varphi(\gamma_{ijk})\bar{\varphi}(b_k) \mid b_k \in B\} \quad (b_i, b_j \in B), \tag{3.18d}$$

for every valuation

$$\bar{\varphi}: \Lambda \to \Phi \tag{3.18e}$$

of Λ extending φ.

Suppose now, we can solve the inequalities

$$\max\{\varphi(\gamma_{ijk})K_{b_k} \mid b_k \in B\} \leqslant K_{b_i}K_{b_j} \ (b_i, b_j \in B) \tag{3.18f}$$

by positive constants $K_{b_i} \in \Phi$ $(b_i \in B)$ subject to

$$K_1 = 1, \tag{3.18g}$$

then we define an extension $\bar{\varphi}$ of φ to Λ by setting

$$\bar{\varphi}\left(\sum_{b_k \in B} \xi_k b_k\right) := \max\{\varphi(\xi_k)K_{b_k} \mid b_k \in B\} \tag{3.18h}$$

($\xi_k \in R$, $\xi_k = 0$ for all but a finite number of b_k). If B is finite, then (3.18f, g) are always solvable, viz.:

$$K_1 = 1, \quad K_{b_k} = \max\{\varphi(\gamma_{ijk}) \mid b_i, b_j \in B\} \quad (1 \neq b_k \in B). \tag{3.18i}$$

The extensions $\bar{\varphi}$ obtained in this way satisfy the rule

$$\bar{\varphi}\left(\sum_{b_k \in B} \xi_k b_k\right) = \max\{\varphi(\xi_k)\bar{\varphi}(b_k) \mid b_k \in B\} \tag{3.18j}$$

which characterizes the *normalized bases* of Λ over R. Of course, in general, it can happen that Λ has no normalized R-basis.

The advantage offered by a normalized R-basis B of Λ is the possibility of defining the *product valuation*

$$\bar{\varphi}' \otimes_R \bar{\varphi}: \Lambda' \otimes_R \Lambda \to \Phi \tag{3.19a}$$

of the tensor product R-ring $\Lambda' \otimes_R \Lambda$ for two unital R-rings Λ, Λ' with valuations $\bar{\varphi}$, $\bar{\varphi}'$ extending φ, respectively. It has the properties

$$\bar{\varphi}' \otimes_R \bar{\varphi}(x' \otimes x) \leqslant \bar{\varphi}'(x')\bar{\varphi}(x) \quad (x' \in \Lambda', x \in \Lambda), \tag{3.19b}$$

$$\bar{\varphi}' \otimes_R \bar{\varphi}(y' \otimes y) = \varphi(y')\varphi(y) \quad (y, y' \in R), \tag{3.19c}$$

$\bar{\varphi}' \otimes_R \bar{\varphi}$ *is maximal among the generalized valuations of* $\Lambda' \otimes_R \Lambda$ *satisfying* (3.19b, c). (3.19d)

Indeed, every element z of $\Lambda' \otimes_R \Lambda$ is uniquely presented in the form

$$z = \sum_{b_k \in B} x'_k \otimes b_k \quad (x'_k \in \Lambda', x'_k = 0 \text{ for all but a finite number of } b_k). \tag{3.19e}$$

Then (3.19b–d) are satisfied by setting

$$\bar{\varphi}' \otimes_R \bar{\varphi}(z) = \max\{\bar{\varphi}'(x'_k)\bar{\varphi}(b_k) \mid b_k \in B\}. \tag{3.19f}$$

We observe that for any system S of generalized valuations $\varphi: X \to \Phi$ of a unital ring X in Φ also the maximized mapping

$$\varphi_S: X \to \Phi: x \mapsto \max\{\varphi(x) \mid \varphi \in S\} \tag{3.19g}$$

is a generalized valuation of X in Φ provided that the maximum exists for all $x \in X$. For example, the possibility of defining $\bar{\varphi}' \otimes_R \bar{\varphi}$ in general depends on the conditions

(i) $1_{\Lambda'} \otimes \rho = 0 \Rightarrow \rho = 0$ for all $\rho \in R$,
(ii) the existence of some generalized valuation satisfying (3.19c),
(iii) the existence of the maximization demanded by (3.19d).

Before we continue to discuss the extension problem a few simple remarks on valuations φ of R are in order.

$$\varphi(-a) = \varphi(a) \quad (a \in R). \tag{3.20a}$$

For any two elements a, b of R we know that

$$\varphi(a \pm b) \leqslant \max(\varphi(a), \varphi(b)).$$

However, if $\varphi(a) \neq \varphi(b)$ then we know more precisely

$$\varphi(a \pm b) = \max(\varphi(a), \varphi(b)). \tag{3.20b}$$

Indeed, for $\varphi(a) < \varphi(b)$ we obtain

$$\varphi(a \pm b) \leqslant \varphi(b) = \varphi((b \pm a) \mp a) \leqslant \max(\varphi(b \pm a), \varphi(\mp a)) = \varphi(a \pm b).$$

Furthermore, if $ab = 1_R$ then we have

$$1 = \varphi(1_R) = \varphi(ab) \leqslant \varphi(a)\varphi(b).$$

If $\varphi(a)\varphi(b) = 1$ then we have for every element x of R

$$\begin{aligned} \varphi(x) &= \varphi(x\,ab) \\ &\leqslant \varphi(xa)\varphi(b) \leqslant \varphi(x)\varphi(a)\varphi(b) = \varphi(x), \end{aligned}$$

hence

$$\varphi(xa) = \varphi(x)\varphi(a). \tag{3.20c}$$

Similarly,

$$\varphi(bx) = \varphi(b)\varphi(x). \tag{3.20d}$$

The unit a of R is said to be a *φ-unit* if

$$\varphi(a)\varphi(a^{-1}) = \varphi(a^{-1})\varphi(a) = 1.$$

For φ-units a we have

$$\varphi(a^{-1}) = \varphi(a)^{-1},$$
$$\varphi(ax) = \varphi(a)\varphi(x), \ \varphi(xa) = \varphi(x)\varphi(a) \quad (x \in R).$$

The φ-units form a subgroup of the unit group of R.

Similarly, we see that the elements a, b of R satisfying (3.20c, d) for all x of R form unital semigroups (of *right φ-multipliers, left-φ-multipliers*).

The pseudo-valuation φ is multiplicative if and only if every element of R is a right-φ-multiplier (left-φ-multiplier).

This implies the condition that

$$I(\varphi, \rho)I(\varphi, \sigma) = I(\varphi, \rho\sigma) \tag{3.20e}$$

for any two ρ, σ of $\Phi \cap \varphi(R)$, though the converse need not hold.

If R is a division ring then the valuation φ of R is multiplicative (a Krull valuation) if and only if

$$\varphi(R \backslash \{0\}) \subseteq U(\Phi),$$

and for any element of R not belonging to $I(\varphi, 1)$ its inverse does belong to $I(\varphi, 1)$.

Lemma (3.21)

A subring R_0 of a division ring R is a valuation ring of a Krull valuation **(Krull valuation ring)** *of R precisely if $1_R \in R_0$, if R_0 is invariant under transformation by non-zero elements of R ($\xi R_0 \xi^{-1} \subseteq R_0\ \forall \xi \in R, \xi \neq 0$), and if R_0 contains the inverse of any element of R that does not belong to R_0.*

Proof

"$\Rightarrow$" Any valuation ring $I(\varphi, 1)$ clearly has those 3 properties. "$\Leftarrow$" Two elements a, b of R are said to be R_0-equivalent if either both are zero or both are non-zero with ab^{-1}, ba^{-1} in R_0. This equivalence is reflexive because for $a \neq 0$ we have $aa^{-1} = 1_R \in R_0$. It is obviously symmetric and it is transitive because for any three non-zero elements a, b, c of R with ab^{-1}, ba^{-1}, bc^{-1}, cb^{-1} in R_0 we have $ac^{-1} = ab^{-1} \cdot bc^{-1} \in R_0$, $ca^{-1} = cb^{-1} \cdot ba^{-1} \in R_0$.

R_0-equivalence also is multiplicative because for any four non-zero elements a, b, c, d of R with ab^{-1}, ba^{-1}, cd^{-1}, dc^{-1} in R_0 we have

$$\begin{aligned}(ac)(bd)^{-1} &= acd^{-1}b^{-1} \\ &= a(cd^{-1})a^{-1} \cdot ab^{-1} \in R_0, \\ (bd)(ac)^{-1} &= bdc^{-1}a^{-1} \\ &= b(dc^{-1})b^{-1} \cdot ba^{-1} \in R_0.\end{aligned}$$

It follows that the non-zero R_0-equivalence classes form a group G. The group G is totally ordered by defining the R_0-equivalence class of a to be greater than the R_0-equivalence class of b if ab^{-1} is in R_0.

Indeed, if $a \neq 0$, $b \neq 0$, $ab^{-1} \in R_0$, $a' \neq 0$, $b' \neq 0$, aa'^{-1}, $a'a^{-1}$, bb'^{-1}, $b'b^{-1} \in R_0$ then we have $a'b'^{-1} = a'a^{-1} \cdot ab^{-1} \cdot bb'^{-1} \in R_0$, so that the order definition depends only on R_0-equivalence classes, not merely on the individual elements.

For any two non-zero elements a, b of R we have either $ab^{-1} \in R_0$, or $ab^{-1} \notin R_0$, $ba^{-1} = (ab^{-1})^{-1} \in R_0$ by assumption. Thus, for any two elements α, β of G one and only one of the 3 relations $\beta > \alpha$, $\alpha = \beta$, $\alpha > \beta$ takes place.

And if $\beta > \alpha$, $\gamma > \beta$ then we have for the representatives a, b, c of α, β, γ in R that $ab^{-1} \in R_0$, $ba^{-1} \notin R_0$, $bc^{-1} \in R_0$, $cb^{-1} \notin R_0$, hence

$$ac^{-1} = ab^{-1} \cdot bc^{-1} \in R_0.$$

If ca^{-1} is in R_0 then also

$$cb^{-1} = ca^{-1} \cdot ab^{-1}$$

is in R_0, a contradiction. Hence $ca^{-1} \notin R_0$. It follows that $\gamma > \alpha$. G is totally ordered. Let us show that G also is algebraically ordered.

If $\alpha > \beta$ then we have for the representatives a, b of α, β in R the relations $ba^{-1} \in R_0, ab^{-1} \notin R_0$, so that $\alpha\beta^{-1} > 1$. Conversely, if $\alpha\beta^{-1} > 1$, then $ba^{-1} \in R_0$, $\alpha > \beta$.

If $\alpha > 1$ and if a, x are representatives of the R_0-equivalence classes α, ξ in R then we have $a^{-1} \in R_0$, $xa^{-1}x^{-1} = (xax^{-1})^{-1} \in R_0$, $\xi\alpha\xi^{-1} > 1$.

If $\alpha > 1$, $\beta > 1$ then we have for the representatives a, b of α, β in R that

$$a \notin R_0, \quad a^{-1} \in R_0, \quad b \notin R_0, \quad b^{-1} \in R_0$$

hence $b^{-1}a^{-1} \in R_0$. If $ab \in R_0$ then $a = ab \cdot b^{-1} \in R_0$, a contradiction. Hence $ab \notin R_0, \alpha\beta > 1$. We extend the algebraic ordering of G to an algebraic ordering of $\Phi = \mathbb{Z}G$, the group ring of G.

If follows that there is the multiplicative valuation

$$\varphi: R \to \Phi,$$

defined by

$$\varphi(0) = 0, \quad \varphi(a) = \alpha \quad (0 \neq a \in R, \alpha \text{ the } R_0\text{-equivalence class of } a),$$

and that $R_0 = I(\varphi, 1)$. We remember that any algebraically ordered group can be embedded into an algebraically ordered ring so that φ also can be interpreted as a pseudo-valuation in the former sense. □

Let us remark that any valuation φ of a ring R in Φ for which

$$\varphi(C(R)) \subseteq C(U(\Phi)) \cup \{0\} \tag{3.22a}$$

is extended uniquely to a valuation $\bar{\varphi}$ of the central quotient ring $\mathfrak{Q}(R)$ of R by setting

$$\bar{\varphi}\left(\frac{x}{v}\right) = \varphi(x)\varphi(v)^{-1} \tag{3.22b}$$

for the formal quotient of the element x of R and the non-zero divisor v of R belonging to $C(R)$.

The reader verifies quickly that (3.22b) indeed is unique either way and that the rules for a valuation $\bar{\varphi}$ of $\mathfrak{Q}(R)$ extending φ are satisfied.

Similarly it is shown that any algebraic ordering of a ring Φ is extended uniquely to an algebraic ordering of the central quotient ring $\mathfrak{Q}(\Phi)$ upon defining

$$\frac{x}{v} > \frac{x'}{v'} \Leftrightarrow vv'^2x > v^2v'x'$$

for any two formal quotients x/v, x'/v' of elements x, x' of Φ by non-zero elements v, v' of $C(\Phi)$. We observe that Φ is an entire ring in case Φ is unital commutative and that $\mathfrak{Q}(\Phi)$ is a field.

If Φ is an algebraically ordered field then the non-zero elements of Φ form a commutative algebraically ordered group.

Furthermore, we remark that the valuation ring $I(\varphi, 1)$ of a field R with a multiplicative valuation

$$\varphi: R \to \Phi$$

is a local entire ring. Its single maximal ideal $\mathfrak{m}$ is the union of the valuation modules $I(\varphi, \rho)$ with $\rho \in \Phi^{\geq 0} \cap \Phi^{<1}$.

Indeed, any element of $I(\varphi, 1)$ not belonging to that ideal $\mathfrak{m}$ of $I(\varphi, 1)$ has φ-value 1 so that it is a unit of $I(\varphi, 1)$. Hence

$$\mathfrak{m} = \bigcup_{\rho \in \Phi^{\geq 0} \cap \Phi^{<1}} I(\varphi, \rho). \tag{3.23}$$

Lemma *(Chevalley)* (3.24)

For any field Λ, any subring R of Λ containing 1_Λ and any prime ideal $\mathfrak{p}$ of R there is a Krull valuation ring of Λ with maximal ideal intersecting R in $\mathfrak{p}$.

Proof

Without loss of generality we may assume already that R is a local entire ring.

Indeed, any Krull valuation ring of Λ with maximal ideal intersecting $R/(R\backslash\mathfrak{p})$ in $\mathfrak{p}/(R\backslash\mathfrak{p})$ has the property that its maximal ideal intersects R in $\mathfrak{p}$.

According to Zorn's Lemma there is a largest overring I of R among the overrings I' of R in Λ with the property that $\mathfrak{p}I'$ does not contain 1. Hence, the ideal $\mathfrak{p}I$ of the unital ring I is contained in a maximal ideal $\mathfrak{P}$ of I. Because of

$$1 \notin \mathfrak{P} = I \cap \frac{\mathfrak{P}}{I\backslash\mathfrak{P}} \supseteq I \cap \mathfrak{p}\frac{I}{I\backslash\mathfrak{P}}$$

it follows from the maximal property of I that $I = \dfrac{I}{I\backslash\mathfrak{P}}$. Hence, I is a local ring with $\mathfrak{P}$ as its single maximal ideal.

We want to show that I is a Krull valuation ring of Λ. If that is not true then there is an element x of Λ such that neither x nor x^{-1} is contained in I. Hence, the overring $I[x]$ of I is larger than I yielding $1 \in \mathfrak{P}I[x]$,

$$1 = -\sum_{i=0}^{m} a_i x^i \quad (m \in \mathbb{Z}^{>0}; a_i \in \mathfrak{P}, \quad 0 \leq i \leq m). \tag{3.25}$$

Among the elements $x' \in \Lambda$ with $x', x'^{-1} \notin I$ there is one, say x, for which the value of m in (3.25) is minimal. It follows that $a_m \neq 0, 1 + a_0 \equiv 1 \bmod \mathfrak{P}$ and therefore

$$(1+a_0)^{-1} \in I,\ 1 + \sum_{i=1}^{m} b_i x^i = 0 \quad (b_i := a_i(1+a_0)^{-1} \in \mathfrak{P}, 1 \leqslant i \leqslant m),$$

$$(x^{-1})^m + \sum_{i=1}^{m} b_i (x^{-1})^{m-i} = 0, \quad I[x^{-1}] = \sum_{i=0}^{m-1} I x^{-i}.$$

By assumption we have $x^{-1} \notin I$, hence

$$\mathfrak{P} I[x^{-i}] \ni 1, \quad 1 \in \sum_{i=0}^{m-1} \mathfrak{P} I x^{-i}, \quad 1 = \sum_{i=0}^{m-1} c_i x^{-i} \quad (c_i \in \mathfrak{P}, 0 \leqslant i < m).$$

Now the minimal property of m yields the contradiction $(x^{-1})^{-1} = x \in I$. Hence, I is a Krull valuation ring of Λ with maximal ideal $\mathfrak{P}$ intersecting the given subring R in $\mathfrak{p}$. □

Corollary (3.26)

For any unital commutative ring Λ and for any subring $R \ni 1_\Lambda$ with prime ideal $\mathfrak{p}$ there is a generalized valuation ring of Λ with maximal ideal intersecting R in $\mathfrak{p}$.

Proof

By Zorn's Lemma there is a largest ideal $\mathfrak{a}$ among all ideals of Λ intersecting R in the ideal contained in $\mathfrak{p}$. If $\mathfrak{a}$ is not a prime ideal then there exist two ideals $\mathfrak{a}_1, \mathfrak{a}_2$ of Λ for which $\mathfrak{a} \subset \mathfrak{a}_1$, $\mathfrak{a} \subset \mathfrak{a}_2$, $\mathfrak{a}_1\mathfrak{a}_2 \subseteq \mathfrak{a}$, hence $(\mathfrak{a}_1 \cap R)(\mathfrak{a}_2 \cap R) \subseteq \mathfrak{a} \cap R \subseteq \mathfrak{p}$. Since $\mathfrak{p}$ is a prime ideal either $\mathfrak{a}_1 \cap R \subseteq \mathfrak{p}$ or $\mathfrak{a}_2 \cap R \subseteq \mathfrak{p}$, a contradiction to the maximality of $\mathfrak{a}$. Therefore $R/\mathfrak{a}$ is a unital subring of the entire ring $\Lambda/\mathfrak{a}$ and $\mathfrak{p}/\mathfrak{a}$ is a prime ideal of $R/\mathfrak{a}$. An application of Lemma (3.24) to $\mathfrak{Q}(\Lambda/\mathfrak{a})$, $R/\mathfrak{a}$, $\mathfrak{p}/\mathfrak{a}$ yields a Krull valuation of $\mathfrak{Q}(\Lambda/\mathfrak{a})$ with values <1 on $\mathfrak{p}/\mathfrak{a}$ but with value 1 on $R/\mathfrak{a} \backslash \mathfrak{p}/\mathfrak{a}$. Upon restriction of that Krull valuation to $\Lambda/\mathfrak{a}$ we find a valuation with $\mathfrak{p}/\mathfrak{a}$ equalling the union of all proper valuation ideals of $R/\mathfrak{a}$. Hence, there is a generalized valuation of Λ with R in its valuation ring and with $\mathfrak{p}$ as the intersection of R and the union of the proper valuation ideals of Λ. □

5. Divisible groups

Suppose we want to extend a valuation φ of the field R to a valuation $\bar{\varphi}$ of the extension Λ of R which is generated by a solution of the pure equation $x^n = a \in R$. This is possible only if there is a solution of the equation $\eta^n = \varphi(a)$ in Φ.

Definition (3.27)
A group G is said to be **divisible** *if the equation $\eta^n = \alpha$ is solvable by $\eta \in G$ for any given $\alpha \in G$, $n \in \mathbb{N}$.*

For example, the multiplicative group of $\mathbb{R}$ (respectively $\mathbb{C}$) is divisible. The positive real numbers $\mathbb{R}^{>0}$ form a uniquely divisible commutative group. The same is true for the additive group of $\mathbb{Q}$ as well as of any field of zero characteristic.

Lemma (3.28)
Any commutative group G without elements $\neq 1$ of finite order (torsion free abelian group) has a unique torsion-free divisible commutative overgroup $G^{1/\mathbb{N}}$ formed by all symbols a^r ($a \in G$, $r \in \mathbb{Q}$) subject to the operational rules

$$a^{p/q} = b^{p'/q'} \Leftrightarrow a^{pq'} = b^{p'q}, \quad a^{p/q}b^{p'/q'} = (a^{pq'}b^{p'q})^{1/(qq')} \quad (a, b \in G;\, p, p' \in \mathbb{Z};\, q, q' \in \mathbb{N}).$$

The proof is left as an easy exercise to the reader.

Any algebraically ordered commutative group G is torsion-free. The algebraic ordering of G is uniquely extendable to an algebraic ordering of $G^{1/\mathbb{N}}$ by defining

$$a^{p/q} > b^{p'/q'} :\Leftrightarrow a^{pq'} > b^{p'q} \quad (a, b \in G;\, p, p' \in \mathbb{Z};\, q, q' \in \mathbb{N}). \tag{3.29}$$

It will be shown in the next subsection that for any algebraically ordered field Φ there is an algebraically ordered extension with divisible positivity group. At any rate, any valuation φ of a unital ring R on Φ is equivalent to the valuation $\varphi': R \to \mathfrak{Q}(\mathbb{Z}(\Phi^{>0})^{1/\mathbb{N}})$ of R in terms of an algebraically ordered field such that the non-zero values of φ' belong to a divisible algebraically ordered commutative group.

6. Extending valuations II

Any valuation $\varphi: R \to \Phi$ of a unital commutative ring R is equivalent to a valuation in an algebraically ordered field. Indeed, we have $\varphi(x)\varphi(y) = \varphi(xy) = \varphi(yx) = \varphi(y)\varphi(x)$ for all x, y of R so that $\varphi(R)$ generates a commutative unital subring $\langle \varphi(R) \rangle$ of Φ. Since any subsemiring of an algebraically ordered semiring is also algebraically ordered we find that $\langle \varphi(R) \rangle$ is an algebraically ordered commutative unital ring. Its quotient field Φ' is algebraically ordered so that the ordering relation of $\langle \varphi(R) \rangle$ is extended to Φ'. Now we have the multiplicative valuation $\varphi': R \to \Phi'$ with the same values on R as φ. It is equivalent to φ.

***Theorem** (Chevalley)* (3.30)
Any Krull valuation $\varphi: R \to \Phi$ of the subfield R of a field Λ in an algebraically

ordered field Φ with divisible multiplicative group can be extended to a Krull valuation of Λ in Φ.

Proof

Among the extensions Λ'' of R in Λ with a multiplicative valuation φ'' in Φ extending φ there is a maximal one, say Λ' with multiplicative valuation $\varphi':\Lambda'\to\Phi$ such that $\varphi'|_R=\varphi$, by Zorn's Lemma. If there is an element x of Λ such that $\Lambda'[x]$ is a polynomial ring in x over Λ' then φ' can be extended to the multiplicative valuation

$$\varphi^*:\Lambda'[x]\to\Phi:\sum_{i=0}^{m}a_ix^i\mapsto\max_{0\leqslant i\leqslant m}\varphi'(a_i),$$

and φ^* extends uniquely to a multiplicative valuation of the subfield $\Lambda'(x)=\mathfrak{Q}(\Lambda'[x])$ of Λ, contradicting the maximal property of Λ'.

So, if there is $x\in\Lambda\backslash\Lambda'$, then x satisfies an equation $f(x)=0$, where $f(t)\in\Lambda'[t]$ is irreducible of degree $n>1$. By Lemma (3.24) there is a Krull valuation ring I of $\Lambda'(x)$ with maximal ideal $\mathfrak{P}$ intersecting Λ' in the maximal ideal

$$\mathfrak{p}=\bigcup_{\rho\in\Phi^{\geqslant 0}\cap\Phi^{<1}}I(\varphi',\rho)\quad\text{of } I(\varphi',1).$$

Hence, the I-equivalence classes of $\Lambda'(x)$ form a totally ordered group G with the property that the subgroup G' of those equivalence classes represented by the elements of $\Lambda'\backslash\{0\}$ is isomorphic to the algebraically ordered group $\varphi'(\Lambda'\backslash\{0\})$. Here the isomorphism θ preserves the ordering. Replacing each element of G' by its θ-image in $\varphi'(\Lambda'\backslash\{0\})$ we obtain a new totally ordered group $\bar{G}$ isomorphic to the totally ordered group of the I-equivalence classes of $\Lambda'(x)$ so that each non-zero element y is valued by an element $\bar{\varphi}(y)$ of $\bar{G}$ and $\bar{\varphi}|_{\Lambda'\backslash\{0\}}=\varphi'|_{\Lambda'\backslash\{0\}}$. We set $\bar{\varphi}(0)=0$.

We want to show that the index of $\varphi'(\Lambda'\backslash\{0\})$ in $\bar{G}$ is finite. Indeed, for any finite number of cosets of $\bar{G}$ modulo $\varphi'(\Lambda'\backslash\{0\})$ we have representatives, say $\alpha_1,\alpha_2,\dots,\alpha_s$ and elements $y_1,y_2,\dots,y_s\in\Lambda'(x)$, satisfying $\bar{\varphi}(y_i)=\alpha_i$ $(1\leqslant i\leqslant s)$. If a linear relation, $\sum_{i=1}^{s}a_iy_i=0$ $(a_i\in\Lambda';\ 1\leqslant i\leqslant s)$, holds, then the non-zero terms a_iy_i lead to distinct cosets $\bar{\varphi}(a_iy_i)$ modulo $\varphi'(\Lambda'\backslash\{0\})$. Hence, the non-zero values $\bar{\varphi}(a_iy_i)$ are all distinct. If there is one at all, then we have $\bar{\varphi}(\sum_{i=1}^{s}a_iy_i)=\max_{1\leqslant i\leqslant s}\bar{\varphi}(a_iy_i)\neq 0$. But this is a contradiction to $0=\bar{\varphi}(0)=\bar{\varphi}(\sum_{i=1}^{s}a_iy_i)$. It follows that $0=a_1=a_2=\cdots=a_s$ so that $y_1,\dots,y_s$ are linearly independent over Λ', hence $s\leqslant[\Lambda'(x):\Lambda']=n$.

Therefore the index $(\bar{G}:\varphi'(\Lambda'\backslash\{0\}))$ is finite. Because of the divisibility property of the abelian group $\Phi^{>0}$ it follows that the identity mapping of $\varphi'(\Lambda'\backslash\{0\})$ can be uniquely extended to an order preserving isomorphism ψ of $\bar{G}$ on an overgroup $\bar{\bar{G}}$ of $\varphi'(\Lambda'\backslash\{0\})$ in $\Phi^{>0}$. Hence, we have the Krull valuation $\tilde{\varphi}:\Lambda'(x)\to\Phi$ with $\tilde{\varphi}(y)=\psi(\bar{\varphi}(y))$ for $y\in\Lambda'(x)$, $y\neq 0$, and $\tilde{\varphi}(0)=0$, contradicting the maximal property of Λ'.

It follows that $\Lambda = \Lambda'$ so that φ' is an extension of φ to a Krull valuation of Λ. □

From now on we only discuss non-archimedean valuations.

Corollary (3.31)
Any generalized valuation

$$\varphi: R \to \Phi \tag{3.31a}$$

of the subring $R \ni 1_\Lambda$ of the unital commutative ring Λ with values in the algebraically ordered field Φ with divisible group $\Phi^{>0}$ can be extended to a generalized valuation of Λ in Φ provided that the necessary condition

$$I(\varphi, 0)\Lambda \cap R = I(\varphi, 0) \tag{3.31b}$$

is satisfied.

Proof
As in the proof of (3.26) we form a prime ideal $\mathfrak{a}$ of Λ which intersects R in the prime ideal $I(\varphi, 0)$. This is possible under the premise of (3.31b). We observe that $I(\varphi, 0)$ is contained in the prime ideal $\mathfrak{p}_\varphi := \bigcup_{\rho \in \Phi^{\geq 0} \cap \Phi^{<1}} I(\varphi, \rho)$ of R. Hence, $R/\mathfrak{a}$ is a subring of the entire ring $\Lambda/\mathfrak{a}$ which contains both $1_\Lambda/\mathfrak{a}$ and the prime ideal $\mathfrak{p}_\varphi/\mathfrak{a}$. Moreover, the valuation $\varphi': R/\mathfrak{a} \to \Phi: x/\mathfrak{a} \mapsto \varphi(x)$ can be uniquely extended to a Krull valuation φ'' of the subfield $\mathfrak{Q}(R/\mathfrak{a})$ of $\mathfrak{Q}(\Lambda/\mathfrak{a})$. According to theorem (3.30) φ'' can be extended to a Krull valuation $\bar{\varphi}$ of $\mathfrak{Q}(\Lambda/\mathfrak{a})$ in Φ. Hence, $\bar{\varphi}: \Lambda \to \Phi: x \mapsto \bar{\varphi}(x/\mathfrak{a})$ is a generalized valuation of Λ extending φ. □

7. Integral closure

Definition (3.32)
Let Λ be a unital commutative overring of the ring R. The intersection of all generalized multiplicative valuation rings of Λ containing R is said to be the **integral closure** *of R in Λ; it is an overring of R in Λ which we denote by* $\mathrm{Cl}(R, \Lambda)$.

The definition implies that $\mathrm{Cl}(R, \Lambda)$ is its own integral closure in Λ:

$$\mathrm{Cl}(\mathrm{Cl}(R, \Lambda), \Lambda) = \mathrm{Cl}(R, \Lambda). \tag{3.33a}$$

Moreover, we note that

$$\mathrm{Cl}(R_1, \Lambda) \subseteq \mathrm{Cl}(R_2, \Lambda) \quad \text{for } R_1 \subseteq R_2 \subseteq \Lambda, \tag{3.33b}$$

$$\mathrm{Cl}(R, \Lambda_1) \subseteq \mathrm{Cl}(R, \Lambda_2) \quad \text{for } R \subseteq \Lambda_1 \subseteq \Lambda_2, \tag{3.33c}$$

$$\mathrm{Cl}(R/\mathfrak{A}, \Lambda/\mathfrak{A}) \supseteq \mathrm{Cl}(R, \Lambda)/\mathfrak{A} \quad \text{for any proper ideal } \mathfrak{A} \text{ of } \Lambda, \tag{3.33d}$$

$$\mathrm{Cl}(R_1 \oplus R_2, \Lambda_1 \oplus \Lambda_2) = \mathrm{Cl}(R_1, \Lambda_1) \oplus \mathrm{Cl}(R_2, \Lambda_2) \tag{3.33e}$$

for unital overrings Λ_i of R_i $(i = 1, 2)$.

Definition (3.34)
The unital commutative ring R is said to be **integrally closed,** *if it coincides with its integral closure relative to its quotient ring:* $R = \mathrm{Cl}(R, \mathfrak{Q}(R))$.

We observe that $\mathrm{Cl}(R, \mathfrak{Q}(R))$ is always integrally closed.

Definition (3.35)
Let Λ be a unital overring of the ring R. An element x of Λ is said to be **integral** *over R, if x satisfies a monic algebraic equation over R:*

$$x^n + \sum_{i=1}^{n} a_i x^{n-i} = 0 \quad (n \in \mathbb{Z}^{>0};\ a_i \in R,\ 1 \leqslant i \leqslant n). \tag{3.35a}$$

For example, the Gaussian integers are integral over $\mathbb{Z}$. In fact, they are the only elements of $\mathbb{Q}(i)$ which are integral over $\mathbb{Z}$.

The elements of an extension E of $\mathbb{Q}$ which are integral over $\mathbb{Z}$ are also called the *algebraic integers* of E. We denote them by o_E.

Lemma (3.36)
Every integral element of Λ over R belongs to $\mathrm{Cl}(R, \Lambda)$.

Proof
Let $x \in \Lambda$ and (3.35a) be satisfied. Then for any generalized multiplicative valuation $\varphi: \Lambda \to \Phi$ of Λ containing R in its valuation ring we have

$$\begin{aligned}\varphi(x^n) = \varphi(x)^n = \varphi\left(-\sum_{i=1}^{n} a_i x^{n-i}\right) &\leqslant \max_{1 \leqslant i \leqslant n} \varphi(a_i x^{n-i}) \\ = \max_{1 \leqslant i \leqslant n} \varphi(a_i)\varphi(x)^{n-i} &\leqslant \max_{1 \leqslant i \leqslant n} \varphi(x)^{n-i}\end{aligned}$$

hence $\varphi(x) \leqslant 1$, i.e. $x \in I(\varphi, 1)$. □

Lemma (3.37)
If Λ is an entire ring, then every element of $\mathrm{Cl}(R, \Lambda)$ *is integral over R.*

Proof
Let x be an element of $\mathrm{Cl}(R, \Lambda)$. By assumption x belongs to the valuation ring of every generalized multiplicative valuation of Λ for which R is contained in its valuation ring. Any Krull valuation of $\mathfrak{Q}(\Lambda)$ containing R in its valuation ring restricts on Λ to a valuation containing R in its valuation ring.

If $x = 0$, then x is integral over R. If there exists an equality $1 = -\sum_{i=1}^{m} a_i x^{-i}$ ($m \in \mathbb{Z}^{>0}$; $a_i \in R$, $1 \leqslant i \leqslant m$; $a_m \neq 0$), then we obtain $x^m + \sum_{i=1}^{m} a_i x^{m-i} = 0$ and x is integral over R, too. Finally, we assume that $\sum_{i=1}^{\infty} R x^{-i}$ is a proper ideal of the overring $R[x^{-1}]$ of R in $\mathfrak{Q}(\Lambda)$. It is contained in a maximal ideal $\mathfrak{m} \ni x^{-1}$. By theorem (3.24) there is a Krull valuation φ of $\mathfrak{Q}(\Lambda)$ with maximal ideal of its valuation ring intersecting $R[x^{-1}]$ in $\mathfrak{m}$. Since the valuation ring of φ contains $R[x^{-1}]$ (and therefore R) it follows that $\varphi(x^{-1}) < 1$. On the other hand our construction yields $\varphi(x^{-1}) = \varphi(x)^{-1} \geqslant 1$, a contradiction. □

Exercise 9 shows that not every element of the integral closure of R in the unital overring Λ needs to be integral over R. However, if the ring R is finitely generated that property can be shown using (2.6) and the preceding theory. Thus we can characterize the elements of Λ which are integral over R as those elements of Λ which belong to the integral closure in Λ of some finitely generated subring of R containing 1_Λ. Thereby, of course, we establish the ring property of the elements of Λ which are integral over R.

The following constructive confirmation is sufficient for our purposes.

Kronecker's criterion (3.38)

An element x of Λ is integral over R, if and only if there are finitely many elements $\omega_1, \ldots, \omega_n$ of Λ such that

(i) *there hold equations*

$$x\omega_k = \sum_{i=1}^{n} \xi_{ik}\omega_i \quad (1 \leqslant k \leqslant n;\ \xi_{ik} \in R), \tag{3.39a}$$

(ii) *for given $y \in \Lambda$ the equations*

$$y\omega_k = 0 \quad (1 \leqslant k \leqslant n) \tag{3.39b}$$

imply that $y = 0$.

Proof

If $x \in \Lambda$ is an integral element over R, there holds an equation (3.35a). Then the elements $\omega_i := x^{i-1}$ $(1 \leqslant i \leqslant n)$ of Λ satisfy (3.39a, b).

Conversely, if x satisfies Kronecker's conditions (3.39a, b), then there holds the matrix equation $(xI_n - (\xi_{ik}))\boldsymbol{\omega} = \mathbf{0}$ for $\boldsymbol{\omega} = (\omega_1, \ldots, \omega_n)^t \in \Lambda^{n \times 1}$. This yields $\det(xI_n - (\xi_{ik}))\omega_k = 0$ $(1 \leqslant k \leqslant n)$, hence $\det(xI_n - (\xi_{ik})) = 0$ because of (3.39b). Therefore x is a zero of the characteristic polynomial of the matrix (ξ_{ik}) and because of (3.35) an integral element of Λ over R. □

If α_i $(i = 1, 2)$ are elements of Λ satisfying monic equations of degree n_i over R, then the application of (3.39) to the $n_1 n_2$ elements $\omega_{\nu_1,\nu_2} := \alpha_1^{\nu_1}\alpha_2^{\nu_2}$ $(0 \leqslant \nu_i < n_i;\ i = 1, 2)$ yields equations of degree $n_1 n_2$ for $\alpha_1 \pm \alpha_2$, $\alpha_1\alpha_2$. Moreover, the integral elements of Λ over R are closed in the sense that every element x of Λ satisfying a monic equation $x^n + b_1 x^{n-1} + \cdots + b_n = 0$

with coefficients $b_i \in \Lambda (1 \leqslant i \leqslant n)$ satisfying monic equations

$$b_i^{m_i} + a_{i_1} b_i^{m_i - 1} + \cdots + a_{i_{m_i}} = 0 \quad (a_{i_k} \in R, 1 \leqslant k \leqslant m_i; m_i \in \mathbb{Z}^{>0}, 1 \leqslant i \leqslant n)$$

over R satisfies itself a monic equation of degree $nm_1 \cdot \ldots \cdot m_n$ over R according to an application of Kronecker's criterion to the elements $x^\nu a_1{}^{\nu_1} \cdot \ldots \cdot a_n{}^{\nu_n}$ $(0 \leqslant \nu < n;\ 0 \leqslant \nu_i < m_i,\ 1 \leqslant i \leqslant n)$.

8. Exponential valuations and rank

Definition. *A mapping* (3.40)

$$\eta: R \to M \cup \{\infty\} \tag{3.40a}$$

of a unital ring R into an algebraically ordered additive group M is called an **exponential valuation** *if it satisfies*

$$\eta(x) = \infty \Leftrightarrow x = 0, \tag{3.40b}$$

$$\eta(x+y) \geqslant \min(\eta(x), \eta(y)), \tag{3.40c}$$

$$\eta(xy) \geqslant \eta(x) + \eta(y), \tag{3.40d}$$

$$\eta(\pm 1) = 0 \quad (x, y \in R). \tag{3.40e}$$

We observe that the valuation ring of η is $I(\eta, 0) = \{x \in R \mid \eta(x) \geqslant 0\}$ with corresponding maximal ideal $I(\eta, >0) = \mathfrak{p}_\eta = \{x \in R \mid \eta(x) > 0\}$.

For example, let R_0 be a unital commutative ring, then the mapping $\eta: R_0[t] \to \mathbb{Z} \cup \{-\infty\}: f(t) \mapsto -\deg(f)$ is an exponential valuation, the so-called *degree valuation*. Since an algebraically ordered additive group is just an algebraically ordered group with addition as binary operation, exponential valuations are special non-archimedean valuations. All that we require for interpreting a non-archimedean valuation in an algebraically ordered unital ring Φ as exponential valuation is the condition that the non-zero values generate a subgroup of the unit group of Φ. For example, Krull valuations of fields may be interpreted as exponential valuations. They are additive in the sense that we have

$$\eta(xy) = \eta(x) + \eta(y) \quad (x, y \in R). \tag{3.40f}$$

Definition (3.41)
An element $a \in M^{\geqslant 0}$ is said to be **infinitely larger** *than the element $b \in M^{\geqslant 0}$: $a \gg b$ (respectively $b \ll a$), if $a > nb$ for any $n \in \mathbb{N}$.*

For example, $a > 0$ implies $a \gg 0$.

Elements $a, b \in M^{\geqslant 0}$ are said to be *comparable*, if neither $a \gg b$ nor $b \gg a$ holds, i.e. there exist $m, n \in \mathbb{N}$ such that $ma \geqslant b$, $nb \geqslant a$. Especially, a is comparable to zero, if and only if $a = 0$. The relation $\gg$ is transitive but neither reflexive nor symmetric. However, comparability yields an equivalence relation on $M^{\geq 0}$.

If a and a', b and b' are comparable and if $a \gg b$, then also $a' \gg b'$. If $a, b \in M^{>0}$ satisfy $a > b$, then either a, b are comparable or we have $a \gg b$. Hence, the relation $\gg$ induces a total ordering relation of the comparability classes. The number of comparability classes not containing 0 is said to be the *order rank* $\rho(M)$ of the algebraic ordering of M.

Definition (3.42)

A rank one algebraically ordered additive group is called **archimedean ordered.**

In the foundations of analysis it is shown that archimedean ordered additive groups are modules and that there is a largest one among them, say $\bar{M}$, with the property that for every archimedean ordered module M there is an order preserving monomorphism of M into $\bar{M}$ which is unique up to order preserving automorphisms of $\bar{M}$. The order preserving automorphisms of $\bar{M}$ form an algebraically ordered abelian group $\mathbb{R}^{>0}$ in accordance with

$$\alpha > 1 \Leftrightarrow \forall x \in \bar{M}^{>0} : \alpha(x) > x.$$

The inversion automorphism $-\mathbf{1}$ generates together with $\mathbb{R}^{>0}$ an algebraically ordered subfield $\mathbb{R} = \mathbb{R}^{>0} \cup \{0\} \cup \mathbb{R}^{<0}, \mathbb{R}^{<0} := -\mathbf{1}\mathbb{R}^{>0}$, of the endomorphism ring of $\bar{M}$. The field $\mathbb{R}$ is called the real number field. For any element $\xi \in \bar{M}^{>0}$ there is the order preserving isomorphism of $\mathbb{R}$ on $\bar{M}$ which maps ρ on $\rho(\xi)$ $(\rho \in \mathbb{R})$. For any algebraically ordered additive group M and any element α of $M^{>0}$ the elements of $M^{>0}$ which are infinitely larger than α form an additive half-module $H(\alpha) = \{\beta \in M^{>0} \mid \beta \gg \alpha\}$ such that $G(\alpha) := H(\alpha) \cup \{0\} \cup -H(\alpha)$ is a normal additive subgroup of M. Also the comparability class of α and $G(\alpha)$ generate an additive subgroup $G'(\alpha)$ of M which is normal in M such that the additive factor group $G'(\alpha)/G(\alpha)$ is archimedean ordered in accordance with $x/G(\alpha) > y/G(\alpha) \Leftrightarrow x > y$ $(x, y \in G'(\alpha))$. Similarly, $M/G(\alpha)$ is algebraically ordered.

For any exponential valuation (3.40a) the elements x of R with $\eta(x) \in H(\alpha) \cup \{\infty\}$ form an ideal $I^{**}(R, \alpha)$ of the ring $I^*(R, \alpha) := \{x \in R \mid \eta(x) > -H(\alpha)\}$ such that η induces the exponential valuation $\eta(R, \alpha) : I^*(R, \alpha)/I^{**}(R, \alpha) \to M/G(\alpha)$.

Definition (3.43)

An exponential valuation $\eta : D \to M \cup \{\infty\}$ corresponding to a Krull valuation of the division ring D such that $\eta(D \backslash \{0\}) = M$ is called a **Krull exponential valuation** *of D.*

The comparability classes of $M^{\geq 0}$ form a totally ordered set $\tilde{M}^{\geq 0}$ with 0 as smallest element. The initial segments of $\tilde{M}^{\geqslant 0}$ corresponding to those subsets X of $M^{\geqslant 0}$ which have the properties

$$0 \in X, \tag{3.44a}$$

$$\text{if } a \in X,\quad b \in M^{\geqslant 0},\quad b > a,\quad \text{then } b \in X, \tag{3.44b}$$

$$\text{if } a \in X,\ b \in M^{\geqslant 0},\ b \text{ comparable to } a, \text{ then } b \in X, \tag{3.44c}$$

are in one-to-one correspondence with the prime ideals $\mathfrak{p}$ of the valuation ring $I(\eta, 0) := \{x \in D \mid \eta(x) \geqslant 0\}$, inasmuch as $\mathfrak{p}$ is of the form $\mathfrak{p}_X := \{x \in D \mid \eta(x) \in X\}$ for some X. Conversely, $\eta(\mathfrak{p})$ is an initial segment of $\tilde{M}^{\geqslant 0}$.
In the converse direction we have

Lemma (3.45)

Let R be an entire ring with quotient field F and let X be a non-empty set of prime ideals of R which is well ordered by set theoretic inclusion. Then there is a Krull exponential valuation $\eta: F \to M \cup \{\infty\}$ such that $R \subseteq I(\eta, 0)$, and for each $\mathfrak{p} \in X$ the intersection of all prime ideals of $I(\eta, 0)$ containing $\mathfrak{p}$ is a prime ideal $\bar{\mathfrak{p}}$ of $I(\eta, 0)$ satisfying $\bar{\mathfrak{p}} \cap R = \mathfrak{p}$.

Proof

The union of all members of X is a prime ideal $\mathfrak{m}$ of R. Without loss of generality we may assume $\mathfrak{m} \in X$. There is the one-to-one correspondence between the elements $\mathfrak{p} \in X$ and the prime ideals $\mathfrak{p}^* := \dfrac{\mathfrak{p}}{R \backslash \mathfrak{m}}$ of the $\mathfrak{m}$-localization $R^* := \dfrac{R}{R \backslash \mathfrak{m}}$ such that the $\mathfrak{p}^*$s form a set X^* of prime ideals of R^* which is well ordered by set theoretic inclusion and contains $\mathfrak{m}^*$. Since all $\mathfrak{p} \in X$ satisfy $\mathfrak{p}^* \cap R = \mathfrak{p}$ we can assume without loss of generality that $\mathfrak{m}$ is a maximal ideal of R.

Among the local subrings $\tilde{R}$ of F with the property that for each $\mathfrak{p} \in X$ the intersection of all prime ideals of $\tilde{R}$ containing $\mathfrak{p}$ is a prime ideal $\tilde{\mathfrak{p}}$ of $\tilde{R}$ intersecting R in $\mathfrak{p}$ there is a maximal one R' by Zorn's lemma. Without loss of generality we can assume $R = R'$. For each $\mathfrak{p} \in X$ there is an additive exponential valuation $\eta_{\mathfrak{p}}: F \to M \cup \{\infty\}$ such that $R \subseteq I(\eta_{\mathfrak{p}}, 0)$, $I(\eta_{\mathfrak{p}}, > 0) \cap R = \mathfrak{p}$ according to Chevalley's lemma. If there is a first element $\mathfrak{p} \in X$ with $\mathfrak{p}_{\eta_{\mathfrak{p}}} \supset \mathfrak{p}$, then $\tilde{R} := R + \mathfrak{p}_{\eta_{\mathfrak{p}}}$ has the property described above yielding $\tilde{R} \supset R$ contrary to our assumption. Hence, we must have $\mathfrak{p}_{\eta_{\mathfrak{p}}} = \mathfrak{p}$ for all $\mathfrak{p} \in X$, and therefore R is a Krull valuation ring. □

Corollary (3.46)

(a) *Under the assumption of lemma (3.45) let $X = \{\mathfrak{p}, \tilde{\mathfrak{p}}\}$ such that $\tilde{\mathfrak{p}} \subset \mathfrak{p}$, then for each Krull exponential valuation $\eta: F \to M \cup \{\infty\}$ for which $I(\eta, 0) \supseteq R$, $I(\eta, > 0) \cap R = \mathfrak{p}$ we derive the Krull exponential valuation*

$\tilde{\eta}: F \to M \cup \{\infty\}$ *such that* $I(\eta,0) \supset I(\tilde{\eta},0) \supseteq R$, $I(\tilde{\eta},>0) \cap R = \tilde{\mathfrak{p}}$. *If the rank of* η *is finite then the rank of* $\tilde{\eta}$ *is smaller.*

(b) *If every non-zero prime ideal of the entire ring R is maximal, then every non-trivial Krull valuation of* $\mathfrak{Q}(R)$ *containing R in its valuation ring is of rank one, and vice versa.*

Among the rank one valuations we distinguish the *discrete valuations.* They are characterized as Krull valuations with infinite cyclic value group:

$$\varphi: F \to \Phi: a \mapsto \xi_0^{\eta(a)} \quad (0 < \xi_0 < 1), \tag{3.47a}$$

or, in terms of an exponential valuation,

$$\eta: F \to \mathbb{Z} \cup \{\infty\}, \quad \eta \text{ surjective}. \tag{3.47b}$$

For example, the p-adic valuations of the rational number field $\mathbb{Q}$ are discrete valuations. Conversely, we have

Lemma (3.48)

Every non-trivial Krull valuation of $\mathbb{Q}$ *is equivalent to a p-adic valuation for some prime number p.*

Proof

Let $\varphi: \mathbb{Q} \to \Phi$ be a non-trivial Krull valuation. Then we have $\mathbb{Z} \subseteq I(\varphi, 1)$ and since φ is non-trivial on $\mathbb{Q}$, also $\varphi|_{\mathbb{Z}}$ is non-trivial. Hence, $\mathfrak{p}_\varphi \cap \mathbb{Z}$ is a prime ideal of $\mathbb{Z}$, i.e. there is a prime number p subject to $\mathfrak{p}_\varphi \cap \mathbb{Z} = p\mathbb{Z}$ with the consequences $0 < \varphi(p) = \xi_0 < 1$; $\varphi(m) = 1$ for $p \nmid m$ $(m \in \mathbb{Z})$, $\varphi(p^\nu m n^{-1}) = \xi_0^\nu$ $(y, m, n \in \mathbb{Z}, p \nmid m, p \nmid n)$. Clearly, φ is equivalent to φ_p. □

The fundamental theorem of number theory implies the (*strong*) *independence of the p-adic valuations* (p running through all prime numbers) in the following sense:

Definition (3.49)

A system S of non-archimedean valuations $\varphi: R \to \Phi$ *of the unital ring R in the algebraically ordered ring* Φ *is said to be* **independent**, *if for every finite subset* $\{\varphi_1, \ldots, \varphi_m\}$ *of S and any m elements* $\varepsilon_i \in \varphi_i(R \backslash \{0\})$ $(1 \leqslant i \leqslant m)$ *there is an element x of R satisfying* $\varphi_i(x) = \varepsilon_i (1 \leqslant i \leqslant m)$. *If for any finite subset* $\{\varphi_1, \ldots, \varphi_m\}$ *of S and* $\varepsilon_i \in \varphi_i(R \backslash \{0\}) \cap \Phi^{\leqslant 1} (1 \leqslant i \leqslant m)$ *there is even an element* $x \in R_{S \leqslant 1} := \{y \in R \mid \varphi(y) \leqslant 1 \, \forall \varphi \in S\}$ *satisfying* $\varphi_i(x) = \varepsilon_i$ $(1 \leqslant i \leqslant m)$, *then S is called* **strongly independent**.

Unique factorization rings are characterized as entire subrings $R_{S \leqslant 1}$ formed according to (3.49) for a strongly independent set S of discrete Krull valuations of a field F such that the conditions of (3.49) can be met by an element

$x \in R_{S \leqslant 1}$ which additionally satisfies

$$\varphi(x) = 0 \quad \text{for all } \varphi \in S \setminus \{\varphi_1, \ldots, \varphi_m\}. \tag{3.50}$$

We note that F is the quotient field of $R_{S \leqslant 1}$ and that the members of S are in one-to-one correspondence with the equivalence classes of prime elements π of $R_{S \leqslant 1}$ such that π corresponds to the discrete Krull valuation

$$\varphi_\pi : F \to \Phi : \pi^r m n^{-1} \mapsto \xi_\pi^r \quad (r \in \mathbb{Z}; m, n \in R_{S \leqslant 1}, \pi \nmid mn; \xi_\pi \in \Phi, 0 < \xi_\pi < 1). \tag{3.51}$$

We also note that there are only finitely many members of S with value < 1 for any given non-zero element of F.

For example, the polynomial ring $R_0 = F_0[X]$ in the polynomial variables belonging to a given non-empty set X of symbols over the field F_0 is a unique factorization ring. Let Π be a representative set of the equivalence classes of irreducible polynomials in X over F_0. For each member π of Π we have the discrete Krull valuation

$$\varphi_\pi : F \to \mathbb{R} : \pi^r m n^{-1} \mapsto q^{-r \deg(\pi)} \quad (r \in \mathbb{Z}; m, n \in R_0, \pi \nmid mn)$$
$$0 \mapsto 0$$

of the quotient field $F = \mathfrak{Q}(R_0) = F_0(X)$, where q is some fixed real number greater than one. The system of all such φ_π constitutes a strongly independent set S of discrete Krull valuations of F such that $R = R_{S \leqslant 1}$.

We observe that there is also the degree valuation

$$\varphi_\infty : F \to \mathbb{R} : f g^{-1} \mapsto q^{\deg(f) - \deg(g)} \quad (f, g \in R_0, fg \neq 0),$$
$$0 \mapsto 0 \tag{3.52}$$

again a discrete Krull valuation of F. The uniqueness of prime factorization in R finds its equivalent expression in the product formula

$$\prod_{\pi \in \Pi \cup \{\infty\}} \varphi_\pi(x) = 1 \quad (x \in F^\times). \tag{3.53}$$

It follows that the system $S \cup \{\varphi_\infty\}$ is not strongly independent. Indeed, $\bigcap_{\varphi \in S \cup \{\varphi_\infty\}} I(\varphi, 1) = F_0$.

If X contains more than one element, then there are also non-discrete Krull valuations of F containing R in their valuation ring. In fact, there are such Krull valuations of rank greater than one.

But if X consists only of one element, say t, then every non-trivial Krull valuation φ of F with R contained in its valuation ring is equivalent to precisely one member of S. Indeed, the elements of R with value less than one form a non-zero prime ideal $\mathfrak{p}$. Since R is a principal ideal ring it follows that $\mathfrak{p} = R\pi$ for some $\pi \in \Pi$, hence $\varphi(\pi^r m n^{-1}) = \varphi(\pi)^r$ $(r \in \mathbb{Z}; m, n \in R, \pi \nmid mn)$ so that φ is equivalent to φ_π.

Moreover, every Krull valuation φ of F satisfying

$$I(\varphi, 1) \supseteq F_0, \quad I(\varphi, 1) \not\supseteq R \tag{3.54}$$

is equivalent to φ_∞. This is because (3.54) implies that $\varphi(t) > 1$, hence $\varphi(t^{-1}) < 1$ and $F_0[t^{-1}] \subseteq I(\varphi, 1)$.

The rational integers $\mathbb{Z}$ and the polynomial ring in one variable over a field form the prime examples of entire rings R with the following valuation theoretic properties:

R is integrally closed; (3.55a)

every non-trivial Krull valuation of $F := \mathfrak{Q}(R)$ (3.55b)
containing R in its valuation ring is discrete;

any system of infinitely many Krull valuation prime (3.55c)
ideals over R intersects in zero.

Kummer and Dedekind discovered in the nineteenth century that the integral closure of $\mathbb{Z}$ in any finite extension of $\mathbb{Q}$ again has the properties (3.55a–c). Today entire rings with the properties (3.55a–c) are called *Dedekind rings.* We are going to study their properties, in particular with regard to subrings defined by a monic equation in section 5.

9. Order rank, rational rank, and degree of transcendency

For any Krull exponent valuation

$$\eta: F \to M \cup \{\infty\} \tag{3.56a}$$

of a field F on the join of an algebraically ordered module M and the symbol ∞ we have three invariants, the order rank $\rho(M)$ of M as defined in subsection 8, the rational rank $r(M/\mathbb{Q})$ which will be discussed below, and the degree of transcendency $d(F)$ of F over the prime field F_0, where the transcendency concept is assumed to be familiar to the reader. The three invariants satisfy

$$\rho(M) \leqslant r(M/\mathbb{Q}) \leqslant d(F) \tag{3.56b}$$

in case the restriction of η to F_0 is trivial. Otherwise F_0 is the rational number field, and $\eta|_{F_0}$ is equivalent to a p-adic valuation (p a prime number), and we have the relation

$$\rho(M) \leqslant r(M/\mathbb{Q}) < d(F). \tag{3.56c}$$

Let us observe that an algebraically ordered module M is torsion-free: $\forall a \in M \forall n \in \mathbb{Z}$: $na = 0 \Rightarrow a = 0 \vee n = 0$. Hence, M is embedded into its $\mathbb{Z}$-quotient module $M/[\mathbb{Z}\setminus\{0\}]$ formed by the formal quotients u/n ($u \in M, n \in \mathbb{Z}^{>0}$) with operational rules:

$$\frac{u}{n} = \frac{u'}{n'} \Leftrightarrow n'u = nu', \quad \frac{u}{n} + \frac{u'}{n'} = \frac{n'u + nu'}{nn'},$$

and

$$\frac{u}{n} = u' \Leftrightarrow nu' = u \quad (u, u' \in M; n, n' \in \mathbb{Z}^{>0}).$$

$M/[\mathbb{Z}\backslash\{0\}]$ is an algebraically ordered $\mathbb{Q}$-module according to the operational rules:

$$\frac{p}{q}\cdot\frac{u}{n} = \frac{pu}{qn}, \quad \frac{u}{n} > \frac{u'}{n'} \Leftrightarrow n'u > nu' \quad (p \in \mathbb{Z}; q, n, n' \in \mathbb{Z}^{>0}; \ u, u' \in M).$$

The order rank of $M/[\mathbb{Z}\backslash\{0\}]$ is the same as that of M: $\rho(M) = \rho(M/[\mathbb{Z}\backslash\{0\}])$. The dimension of the $\mathbb{Q}$-linear space $M/[\mathbb{Z}\backslash\{0\}]$ is defined as the *rational rank* of M: $r(M/\mathbb{Q}) = \dim_{\mathbb{Q}}(M/[\mathbb{Z}\backslash\{0\}]) =: r(M/[\mathbb{Z}\backslash\{0\}])$. The analysis carried out in subsection 8 yields the inequality $\rho(M) \leqslant r(M/\mathbb{Q})$. By construction M contains a $\mathbb{Q}$-basis B of $M/[\mathbb{Z}\backslash\{0\}]$. For each $b \in B$ there is an element $\xi(b) \in F$ satisfying $\eta(\xi(b)) = b$.

If $\eta|_{F_0}$ is trivial, then the values

$$\eta(\xi(b_1)^{\nu_1}\xi(b_2)^{\nu_2}\cdot\ldots\cdot\xi(b_s)^{\nu_s}) = \sum_{i=1}^{s} \nu_i b_i$$

$$(s \in \mathbb{Z}^{>0}; b_i \in B, \nu_i \in \mathbb{Z}, \nu_i \neq 0, 1 \leqslant i \leqslant s; b_1 < b_2 < \cdots < b_s)$$

are all distinct so that the non-trivial linear combinations of the monomials in $\xi(B)$ over the prime field F_0 of F have η-values in M which means that they are not zero. Therefore we get $d(F) \geqslant r(M\backslash\mathbb{Q})$. If $\eta|_{F_0}$ is not trivial, then the same argument applies to $B\backslash\{b_0\}$, where $b_0 = \eta(p)$ and (3.56c) is obtained.

Let us observe that for any purely transcendental extension $F = F_0(x_1, \ldots, x_d)$ we find that M is free abelian. Any module M with finite $\mathbb{Z}$-basis $u_1, \ldots, u_n (n \in \mathbb{Z}^{>0})$ has the rank n algebraical ordering based on lexicographic ordering:

$$\sum_{j=1}^{n} \lambda_j u_j > 0 \Leftrightarrow \lambda_i > 0 \quad \text{for } i = \min\{j \mid 1 \leqslant j \leqslant n, \lambda_j \neq 0\}.$$

Also all rank n algebraical orderings of M are lexicographic with respect to a suitable $\mathbb{Z}$-basis.

Algebraic number theory utilizes rational rank 1 valuations. Algebraic function theory in n variables utilizes rational valuations of rank $\leqslant n$.

Exercises

1. Deduce (3.8a, b) from (3.2a–d).
2. Prove lemma (3.28).
3. For every natural number $n > 1$ and for every non-negative rational integer x there holds a unique presentation

$$x = \sum_{i=0}^{\infty} a_i(x,n)n^i \quad (a_i(x,n)\in\mathbb{Z}, 0 \leqslant a_i(x,n) < n,$$
$$0 = a_i(x,n) \text{ for } n^i > x).$$

Develop an algorithm for that 'n-adic presentation' of x.

4. (Ostrowski)

   (a) If $\varphi:\mathbb{Z}\to\mathbb{R}^{>0}$ is a valuation satisfying $\varphi(n) \leqslant 1$ for some integer $n > 1$, then show we have $\varphi(x) \leqslant (n-1)(1+\log_n x)$, $\varphi(x^j) \leqslant (n-1)(1+j\log_n x)\,(j\in\mathbb{Z}^{>0})$ for every natural number x.

   (b) If φ is multiplicative, then show we have $\varphi(x) \leqslant (n-1)^{1/j}(1+j\log_n x)^{1/j}$ $(j\in\mathbb{Z}^{>0})$, $\varphi(x)\leqslant 1$ for every natural number x, hence φ is non-archimedean.

5. (Ostrowski) Show that every non-trivial Krull valuation of the rational number field $\mathbb{Q}$ is equivalent to a p-adic valuation for some prime number p.

6. (Hölder) Show that the mapping $\varphi:\mathbb{C}\to\mathbb{R}^{\geqslant 0}: z\mapsto |z|^\alpha = \exp(\alpha \log|z|)$ is a multiplicative archimedean valuation of the complex number field $\mathbb{C}$ for every fixed positive real exponent $\alpha \leqslant 1$. All of those valuations are equivalent.

7. (Ostrowski) Let $\varphi:\mathbb{Q}\to\mathbb{R}^{\geqslant 0}$ be a multiplicative valuation satisfying $\varphi(2) > 1$. Show that

   (a) $\varphi(n) > 1$ for $n\in\mathbb{Z}^{>1}$.

   (b) $\varphi(x) \leqslant (n-1)\varphi(n)^{1+\log_n x}$, $\varphi(x^j) \leqslant (n-1)\varphi(n)^{1+j\log_n x}$, $\varphi(x) \leqslant (n-1)^{1/j}\varphi(n)^{1/j \log_n x}$ $(j\in\mathbb{Z}^{>0})$, $\varphi(x) \leqslant \varphi(n)^{\log_n x}$, $\varphi(x)^{1/\log x} = \varphi(n)^{1/\log n}$ for any two integers $x, n\in\mathbb{Z}^{>1}$.

   (c) $\varphi(x) = |x|^\alpha$ for all $x\in\mathbb{Q}$, where $\varphi(2) = 2^\alpha$, $0 < \alpha \leqslant 1$.

8. (Banach) Let $\varphi:F\to\mathbb{R}^{\geqslant 0}$ be a multiplicative valuation of the field F and let $\Phi:L\to\mathbb{R}^{\geqslant 0}$ be a mapping of the F-linear space L into the non-negative real numbers subject to the condition on a φ-norm:

$$\Phi(u+v) \leqslant \Phi(u) + \Phi(v), \quad \Phi(\lambda u) = \varphi(\lambda)\Phi(u) \quad (u, v\in L, \lambda\in F).$$

   Then show that the F-linear transformations $T:L\to L$ of L subject to the Φ-boundedness condition $\Phi(Tu) \leqslant M\Phi(u)$ $(u\in L)$ for some positive real number M form a unital F-algebra $B(L/\Phi)$. Also show that the Banach algebra $B(L/\Phi)$ of L over Φ has the valuation

$$\Psi: B(L/\varphi)\to\mathbb{R}^{\geqslant 0}: T\mapsto \text{glb}\,\{\Phi(Tu)/\Phi(u) \mid 0 \neq u\in L\},$$

   such that $\Psi(\lambda T) = \varphi(\lambda)\Psi(T)$ $(T\in B(L/\Phi), \lambda\in F)$.

9. The sequences of rational integers form a unital commutative integrally closed ring for the operational rules:

$$(a_n) = (b_n) \Leftrightarrow a_n = b_n \quad (n\in\mathbb{N}),$$
$$(a_n) + (b_n) = (a_n + b_n),$$
$$(a_n)(b_n) = (a_n b_n).$$

   Its prime ring is formed by the constant sequences and is isomorphic to $\mathbb{Z}$. But show that the integral sequences that are algebraic integers over the prime ring form the proper subring of those sequences which have only finitely many distinct entries.

4.4. Eisenstein polynomials

Given a unital commutative ring R with a non archimedean pseudo-valuation

$$\varphi: R \to \Phi \tag{4.1}$$

in an algebraically ordered field Φ with divisible positivity group $\Phi^{>0}$ and a monic polynomial

$$f(t) = t^n + a_1 t^{n-1} + \cdots + a_n \in R[t] \tag{4.2a}$$

with factorization

$$f(t) = \prod_{i=1}^{n} (t - \xi_i) \tag{4.2b}$$

into linear factors over R, let us study the connection between the values of the coefficients and the values of the roots of f.

For simplicity's sake we assume that the roots are ordered by non-increasing magnitude:

$$\varphi(\xi_1) \geqslant \varphi(\xi_2) \geqslant \cdots \geqslant \varphi(\xi_n). \tag{4.3}$$

The Vieta formulae

$$(-1)^i a_i = \sum_{1 \leqslant j_1 < \cdots < j_i \leqslant n} \xi_{j_1} \xi_{j_2} \cdots \xi_{j_i} \tag{4.4}$$

imply that

$$\varphi(a_i) \leqslant \varphi(\xi_1)\varphi(\xi_2) \cdots \varphi(\xi_i) \leqslant \varphi(\xi_1)^i. \tag{4.5}$$

Setting

$$\psi(f) := \max_{1 \leqslant i \leqslant n} \varphi(a_i)^{1/i} \tag{4.6}$$

it follows that

$$\varphi(\xi_1) = \max_{1 \leqslant i \leqslant n} \varphi(\xi_i) \geqslant \psi(f). \tag{4.7}$$

If φ is multiplicative then an estimate of $\varphi(\xi_1)$ from above is derived from the equation

$$\xi_1^n = -a_1 \xi_1^{n-1} - \cdots - a_n,$$

in accordance with the estimate

$$\varphi(\xi_1)^n = \varphi(\xi_1^n) \leqslant \max_{1 \leqslant i \leqslant n} \varphi(a_i \xi_1^{n-i}) \leqslant \varphi(a_j)\varphi(\xi_1)^{n-j}$$

for some $j \in \{1, 2, \ldots, n\}$. Hence,

$$\varphi(\xi_1) \leqslant \varphi(a_j)^{1/j} \leqslant \psi(f),$$

and

$$\max_{1 \leqslant i \leqslant n} \varphi(\xi_i) = \psi(f). \tag{4.8}$$

A generalization of (4.8) is obtained by counting the number s of distinct values $\varphi(\xi_i)$ and setting

$$\varphi(\xi_1) = \cdots = \varphi(\xi_{\nu_1}) > \varphi(\xi_{\nu_1+1}) = \cdots = \varphi(\xi_{\nu_2}) > \cdots > \varphi(\xi_{\nu_{s-1}+1}) = \cdots = \varphi(\xi_n), \tag{4.9a}$$

where

$$1 \leqslant \nu_1 < \nu_2 < \cdots < \nu_{s-1} < \nu_s = n, \tag{4.9b}$$

in accordance with (4.3), (4.4), viz.

$$\varphi(a_{\nu_i}) = \varphi(\xi_1) \cdots \varphi(\xi_{\nu_i}) \quad (1 \leqslant i \leqslant s), \tag{4.9c}$$

hence

$$\varphi(a_{\nu_i}) = \varphi(a_{\nu_{i-1}}) \varphi(\xi_{\nu_i})^{\nu_i - \nu_{i-1}} \quad (1 \leqslant i \leqslant s), \tag{4.9d}$$

with

$$\nu_0 := 0, \quad a_0 := 1. \tag{4.9e}$$

In other words,

$$\varphi(a_{\nu_{i-1}}) > 0, \quad \varphi(\xi_{\nu_i}) = (\varphi(a_{\nu_i}) \varphi(a_{\nu_{i-1}})^{-1})^{(\nu_i - \nu_{i-1})^{-1}} \quad (1 \leqslant i \leqslant s). \tag{4.9f}$$

On the other hand, (4.4) implies for $\nu_{i-1} < \nu < \nu_i$ that

$$\varphi(a_\nu) \leqslant \varphi(a_{\nu_{i-1}}) \varphi(\xi_{\nu_i})^{\nu - \nu_{i-1}}, \tag{4.9g}$$

or

$$\varphi(a_\nu) \leqslant \varphi(a_{\nu_{i-1}})^{[1 - (\nu - \nu_{i-1})/(\nu_i - \nu_{i-1})]} \varphi(a_{\nu_i})^{(\nu - \nu_{i-1})/(\nu_i - \nu_{i-1})}. \tag{4.9h}$$

Note that (4.9a) yields

$$(\varphi(a_{\nu_i}) \varphi(a_{\nu_{i-1}})^{-1})^{(\nu_i - \nu_{i-1})^{-1}} > (\varphi(a_{\nu_{i+1}}) \varphi(a_{\nu_i})^{-1})^{(\nu_{i+1} - \nu_i)^{-1}} \quad (1 \leqslant i < s). \tag{4.9i}$$

In other words, for multiplicative valuations φ the φ-values of the roots of a monic polynomial are uniquely determined by the φ-values of the coefficients in the non-archimedean case.

For an important application of (4.9h, i) we introduce an exponent valuation

$$\eta : R \to H \cup \{\infty\}$$

(H an algebraically ordered divisible module) which corresponds to φ translating the formulae to

$$\eta(a_\nu) \geqslant \left(1 - \frac{\nu - \nu_{i-1}}{\nu_i - \nu_{i-1}}\right) \eta(a_{\nu_{i-1}}) + \left(\frac{\nu - \nu_{i-1}}{\nu_i - \nu_{i-1}}\right) \eta(a_{\nu_i}) \quad (1 \leqslant i \leqslant s), \tag{4.9j}$$

$$\frac{\eta(a_{\nu_i}) - \eta(a_{\nu_{i-1}})}{\nu_i - \nu_{i-1}} < \frac{\eta(a_{\nu_{i+1}}) - \eta(a_{\nu_i})}{\nu_{i+1} - \nu_i}. \tag{4.9k}$$

Following Newton, we associate the $n + 1$ points

$$P_{f,i} := (i, \eta(a_i)) \quad (0 \leqslant i \leqslant n) \tag{4.9l}$$

of the affine plane over $H \cup \{\infty\}$ with f. Then we can interpret (4.9j, k) as the statement that the points

$$P_{f,\nu_0}, P_{f,\nu_1}, \ldots, P_{f,\nu_s}$$

are the vertices of the lower convex hull of the pointset

$$\{(i, x) \mid \eta(a_i) \leqslant x, 0 \leqslant i \leqslant n\}.$$

The chain of connecting straight segments $\overline{P_{f,\nu_0} P_{f,\nu_1}}, \ldots, \overline{P_{f,\nu_{s-1}} P_{f,\nu_s}}$ forms the *Newton polygon*. For example, let $R = \mathbb{Q}$, $\varphi = \varphi_2$, $f(t) = t^3 - 2t^2 - t + 2$. Then we have

$$P_{f,0} = (0,0), \quad P_{f,1} = (1,1), \quad P_{f,2} = (2,0), \quad P_{f,3} = (3,1)$$

with Newton polygon

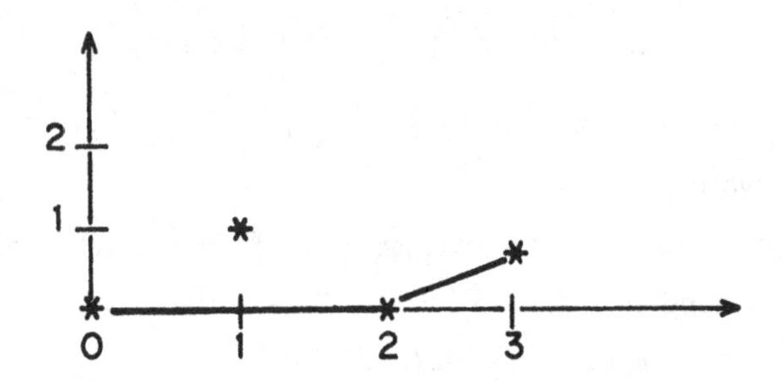

corresponding to

$$s = 2, \quad \nu_0 = 0, \quad \nu_1 = 2, \quad \nu_2 = 3,$$
$$\varphi(\xi_1) = \varphi(\xi_2) = 1, \quad \varphi(\xi_3) = \tfrac{1}{2},$$
$$\eta(\xi_1) = \eta(\xi_2) = 0, \quad \eta(\xi_3) = 1.$$

Definition (4.10)

The monic polynomial (4.2a) *is said to be an* **Eisenstein polynomial** *over the field F with Krull valuation*

$$\varphi: F \to \Phi \tag{4.10a}$$

if

(1) $0 \leqslant \varphi(a_n) < 1$,

(2) $0 \leqslant \varphi(a_i) \leqslant \varphi(a_n) (1 \leqslant i < n)$,

(3) *for any prime divisors q of n the equation* $\varphi(a_n) = \xi^q$ *cannot be solved by a* φ*-value.*

Historically speaking Eisenstein introduced the polynomials named after him as monic polynomials (4.2a) over $\mathbb{Z}$ subject to the conditions

$$p \mid a_i \ (1 \leqslant i \leqslant n), \quad p^2 \nmid a_n$$

for some prime number p. They are the Eisenstein polynomials over $\mathbb{Z}$ with valuation $\varphi = \varphi_p$ and $\varphi(a_n) = 1/p$.

For example, the pth cyclotomic polynomial

$$\Phi_p(t) = \frac{t^p - 1}{t - 1} = t^{p-1} + t^{p-2} + \cdots + 1$$

is turned into an Eisenstein polynomial

$$\frac{(t'+1)^p-1}{t'} = t'^{p-1} + \binom{p}{1}t'^{p-2} + \binom{p}{2}t'^{p-3} + \cdots + \binom{p}{p-1}$$

by the Tschirnhausen transformation

$$t \mapsto t'+1.$$

This is because all binomial coefficients $\binom{p}{i}(1 \leqslant i \leqslant p)$ are divisible by p. It is left as an exercise for the reader to prove that the same transformation also turns the p^νth cyclotomic polynomial

$$\Phi_{p^\nu}(t) = \frac{t^{p^\nu}-1}{t^{p^{\nu-1}}-1} = \Phi_p(t^{p^{\nu-1}})$$

into an Eisenstein polynomial for $\nu > 1$.

Note that according to definition (4.10) also polynomials like

$$t^3 - 250t + 25$$

are φ_5-Eisenstein, not merely the polynomials like

$$t^3 - 250t + 5$$

with last coefficient not divisible by 5^2.

The importance of the Eisenstein polynomials is primarily derived from their assured irreducibility which is shown below, though we will learn to know their significance from another point of view in section 6.

Lemma (4.11)

Every Eisenstein polynomial is irreducible.

Proof

Let $f(t) \in F[t]$ of the form (4.2a) be an Eisenstein polynomial relative to the Krull valuation (4.10a) of F, and let E be a finite extension of F generated by a root ξ of f over F. According to the Chevalley theorem (3.30) there is an extension

$$\bar{\varphi}: E \to \Phi$$

of φ to a Krull valuation $\bar{\varphi}$ of E. According to (4.9g, i) we have $s = 1$, $\nu_1 = n$, $\bar{\varphi}(\xi) = \varphi(a_n)^{1/n}$. We will show that $1, \xi, \ldots, \xi^{n-1}$ are linearly independent over F implying $[E:F] = n$ and therefore the irreducibility of f. In any non-trivial linear combination

$$\zeta = \sum_{i=0}^{n-1} \lambda_i \xi^i$$

of $1, \xi, \ldots, \xi^{n-1}$ over F there are at least two non-zero summands on the right-hand side. For any two non-zero terms, say

$$\lambda_i \xi^i, \quad \lambda_k \xi^k \quad (0 \leqslant i < k < n;\ \lambda_i,\ \lambda_k \in F,\ \lambda_i \lambda_k \neq 0),$$

it is impossible that

$$\bar{\varphi}(\lambda_i \xi^i) = \bar{\varphi}(\lambda_k \xi^k)$$

because such an equality yields $0 < \bar{\varphi}(\xi^{k-i}) = \varphi(\lambda_i \lambda_k^{-1}) \in \varphi(F)$. But then both $\bar{\varphi}(\xi)^{k-i}$ and $\bar{\varphi}(\xi)^n = \varphi(a_n)$ are in $\varphi(F)$ implying $\bar{\varphi}(\xi)^d \in \varphi(F)$ for $d = \gcd(n, k-i)$, hence, $\varphi(a_n) = (\bar{\varphi}(\xi)^d)^{n/d}$, contradicting property (3) of the Eisenstein polynomial f. It follows that the $\bar{\varphi}$-value of ζ equals the maximum of the values $\bar{\varphi}(\lambda_i \xi^i)$, a positive element of Φ. Hence $\zeta = 0$ if and only if $\lambda_1 = \cdots = \lambda_n = 0$, and the elements $1, \xi, \ldots, \xi^{n-1}$ are linearly independent over F. □

An application of the Eisenstein lemma (4.11) yields another proof of the irreducibility of the p^νth cyclotomic polynomials over $\mathbb{Z}(\nu \geqslant 0)$. We observe that the index of the value group of φ in the value group of $\bar{\varphi}$ is n, the degree of E over F. According to the proof of (3.30) the index is optimal.

Definition (4.12)

The Krull valuation

$$\bar{\varphi}: E \to \Phi$$

of the finite extension E of the field F is said to be **totally ramified** *over its restriction to F:*

$$\varphi: F \to \Phi,$$

if the index of the value group of φ in the value group of $\bar{\varphi}$ equals the degree of E over F. We say that E is an **Eisenstein extension** *of F if E is totally ramified over F and if the factor group of the two value groups is cyclic.*

Our second observation regarding the Eisenstein polynomials states that they define an Eisenstein extension of the field of reference. Conversely we have

Lemma (4.13)

Let E be an Eisenstein extension of F. Then the minimal polynomial f of every non-zero element ξ of E with the property that the order of $\bar{\varphi}(\xi)$ modulo $\varphi(F \backslash \{0\})$ equals the degree n of E over F and that $\bar{\varphi}(\xi) < 1$ is an Eisenstein polynomial of degree $[E:F]$.

Proof

Lemma (4.13) follows from the fact that the $\bar{\varphi}$-value of one of the roots of f has order n modulo $\varphi(F \backslash \{0\})$ according to (4.9i). □

Finally we observe that Eisenstein polynomials f determine the integral closure of the entire ring

$$R = I(\varphi, 1)$$

in the Eisenstein extension

$$F[t]/f(t)F[t] = E = F(\xi) \quad (\xi = t/f(t)F[t]), \tag{4.14}$$

as follows:

$$\mathrm{Cl}(R, E) = \sum_{i=0}^{n-1} I(\varphi, \bar{\varphi}(\xi)^{-i})\xi^i = I(\bar{\varphi}, 1). \tag{4.15}$$

Indeed, every element of E is of the form

$$x = \sum_{i=0}^{n-1} \lambda_i \xi^i, \tag{4.16a}$$

with coefficients λ_i in F. As we saw already in the proof of lemma (4.11), we have

$$\bar{\varphi}(x) = \max_{0 \leqslant i < n} \varphi(\lambda_i)\bar{\varphi}(\xi)^i, \tag{4.16b}$$

hence the second equation in (4.15). Since f is an Eisenstein polynomial it follows from (4.9i) that for every Krull valuation

$$\varphi^*: E \to \Phi$$

of E extending φ we have $\varphi^*(\xi) = \bar{\varphi}(\xi)$ and, as before, $\varphi^*(x) = \max_{0 \leqslant i < n} \varphi(\lambda_i)\varphi^*(\xi)^i = \bar{\varphi}(x)$ $(x \in E)$, hence $\varphi^* = \varphi$. Hence,

$$\mathrm{Cl}(R, E) = I(\bar{\varphi}, 1) \quad \text{(see lemma (3.34))}. \qquad \square$$

For any non-archimedean valuation (4.1) of the unital commutative ring R and for every polynomial (4.2a) with factorization (4.2b) over R it is possible, of course, to determine the natural numbers s, ν_1, $\nu_2, \ldots, \nu_s$ in accordance with (4.9g, h) and to determine the non-negative elements

$$\rho_{f,1} = (\varphi(a_{\nu_i})\varphi(a_{\nu_{i-1}})^{-1})^{(\nu_i - \nu_{i-1})^{-1}} \quad (1 \leqslant i \leqslant s) \tag{4.17}$$

of Φ in analogy to (4.9a–d). But now we have merely the inequalities,

$$\prod_{j=1}^{i} \varphi(\xi_j) \geqslant \left(\prod_{k=1}^{h} \rho_{f,k}^{\nu_k - \nu_{k-1}} \right) \rho_{f,h+1}^{i-\nu_h} \quad (\nu_h < i \leqslant \nu_{h+1}, 1 \leqslant h < s), \tag{4.18}$$

implied by (4.5).

However, let

$$a_\mu \neq 0 \tag{4.19a}$$

be the last non-zero coefficient of f so that

$$0 \leqslant \mu \leqslant n, \quad a_{\mu+i} = 0 \quad (\mu < i \leqslant n), \tag{4.19b}$$

so that

$$f(t) = f_1(t)t^{n-\mu}, \quad f_1(t) = t^\mu + a_1 t^{\mu-1} + \cdots + a_\mu. \tag{4.19c}$$

Assuming that

$$f_1(t) = \prod_{i=1}^{\mu} (t - \xi_i), \quad \xi_{\mu+i} = 0 \quad (0 < i \leqslant n - \mu) \tag{4.19d}$$

and also that a_μ is a φ-multiplier such that

$$\mu > 0, \tag{4.19e}$$

then we have the formula

$$a_\mu = \prod_{i=1}^{\mu} \xi_i \tag{4.19f}$$

showing that the roots $\xi_1, \ldots, \xi_\mu$ are φ-multipliers, the first μ of which are non-zero. Thus we have the equations

$$\varphi(\xi_{j_1}\xi_{j_2}\cdots\xi_{j_i}) = \prod_{k=1}^{i} \varphi(\xi_{j_k}) \quad (1 \leqslant i \leqslant n) \tag{4.19g}$$

which are required for the proof of (4.9f, h, i).

In case $R = \mathbb{Z}$, $\Phi = \mathbb{R}$, $\varphi = \varphi_m$ (the m-adic pseudo-valuation for some natural number $m > 1$) we have the following alternative:

(I) a_μ is not a φ_m-unit. This is tantamount to

$$1 < \gcd(a_\mu, m) < m, \tag{4.20}$$

hence the proper factorization $m = \gcd(a_\mu, m)(m/\gcd(a_\mu, m))$ is implied.

(II) a_μ is a φ_m-unit. In this case (4.9f, h, i) hold.

It is not required to test whether the other coefficients $a_1, \ldots, a_{\mu-1}$ are φ-units. However, it is suggested to do so anyway inasmuch as those tests increase the likelihood of finding a proper factorization of m in case m is not prime.

The following observation is appropriate at this point. Let

$$m = \prod_{i=1}^{s} p_i^{v_i} \quad (s \in \mathbb{Z}^{>0};\ v_i \in \mathbb{Z}^{>0},\ 1 \leqslant i \leqslant s)$$

be a factorization of m into the power product of distinct prime numbers $p_1, \ldots, p_s$, then we have the equation

$$\varphi_m(x) = \max_{1 \leqslant i \leqslant s} \varphi_{p_i}^{v_i}(x)$$

for all φ_m-units x of $\mathbb{Q}$. (Note that

$$\varphi_{p_i}^{v_i} : \mathbb{Q} \to \mathbb{R}$$

is a Krull valuation of $\mathbb{Q}$.) For any finite extension E of $\mathbb{Q}$ there are only finitely many distinct extensions

$$\varphi : E \to \mathbb{R}$$

of the $\varphi_{p_i}^{v_i}$ to Krull valuations of E (see section 5). We denote by $\bar{\varphi}_m(y)$ the maximum of the $\bar{\varphi}(y)$ for y of E so that

$$\bar{\varphi}_m: E \to \mathbb{R}$$

is a non-archimedean pseudo-valuation of E which coincides with φ_m precisely on those rational numbers x which are φ_m-units.

The monic polynomial (4.2) over $\mathbb{Z}$ may be said to be pseudo-Eisenstein if

(1) all coefficients $a_1, a_2, \ldots, a_n$ are φ_n-multipliers,
(2) $0 < \varphi_m(a_n) < 1$,
(3) $0 \leqslant \varphi_m(a_i) \leqslant \varphi_m(a_n)\,(1 \leqslant i < n)$,
(4) there is no prime number q dividing n such that

$$\varphi_m(a_n)^{-1/q} \in \mathbb{Z}^{>0}.$$

In this case it follows again by the arguments of the proof of lemma (4.11) applied to $\bar{\varphi}_m$ that f is irreducible. Let us observe that the test of (4) is relatively easy since one must try out only those prime divisors q of n which are less than or equal to $\log m/\log 2$.

In general we have the following:

Definition (4.21)
Let F be a field with a finite number of Krull valuations, say $\psi_1, \psi_2, \ldots, \psi_s$, in Φ and let

$$\varphi := \max_{1 \leqslant i \leqslant s} \psi_i$$

be the intermediate pseudo-valuation of F derived by maximum formation from $\psi_1, \ldots, \psi_s$. Then the monic polynomial (4.2a) *over F is said to be* **pseudo-Eisenstein**, *if*

(1) *all coefficients $a_1, \ldots, a_n$ are φ-multipliers,*
(2) $0 < \varphi(a_n) < 1$,
(3) $0 \leqslant \varphi(a_i) \leqslant \varphi(a_n)\,(1 \leqslant i < n)$,
(4) *there is no prime divisor q of n such that*

$$\varphi(a_n)^{1/q} \in \bigcap_{i=1}^{s} \psi_i(F \backslash \{0\}).$$

Again it is shown by the arguments given in the proof of the Eisenstein lemma (4.11) *that f is irreducible over F.*

Exercises

1. Let R be a Krull valuation ring of its quotient field F with maximal ideal $\mathfrak{p}$ and perfect residue class field $R/\mathfrak{p}$. Then show that the finite extension E of F is

Eisenstein relative to R if and only if the minimal polynomial of every element of $\mathrm{Cl}(R, E)$ is congruent to a power of a linear factor modulo $\mathfrak{p}[t]$.

4.5. Dedekind rings and orders

1. Fractional ideals

It is useful to introduce fractional ideals over a ring.

Definition (5.1)
For any commutative unital ring R the R-modules $v^{-1}\mathfrak{a}$ of $\mathfrak{Q}(R)$ derived from the ideals $\mathfrak{a}$ of R upon multiplication by the inverse of a non-zero divisor v of R are said to be the **R-fractional ideals** *of $\mathfrak{Q}(R)$ with denominator v.*

Clearly, the ideals of R are fractional ideals with denominator 1.

The fractional R-ideals form an abelian semigroup with R as unit element and 0 as null element. The intersection and the sum of two R-fractional ideals are again R-fractional ideals. If $\mathfrak{a}$ is an R-fractional ideal containing a non-zero divisor of R and if $\mathfrak{b}$ is any R-fractional ideal then also the quotient module

$$[\mathfrak{b}/\mathfrak{a}] = \{x \in \mathfrak{Q}(R) \mid x\mathfrak{a} \subseteq \mathfrak{b}\}$$

is an R-fractional ideal.

A fractional R-ideal $\mathfrak{a}$ is said to be *invertible* with respect to R if there is an R-fractional ideal $\mathfrak{b}$ satisfying

$$\mathfrak{a}\mathfrak{b} = \mathfrak{b}\mathfrak{a} = R.$$

From the elements of group theory we know that $\mathfrak{b}$ is uniquely determined if it exists at all. It is customary to denote $\mathfrak{b}$ as $\mathfrak{a}^{-1}$ and to speak of $\mathfrak{a}^{-1}$ as the *inverse* of $\mathfrak{a}$ or as the *inverse R-fractional ideal* of $\mathfrak{a}$ so that we have the defining equation

$$\mathfrak{a}\mathfrak{a}^{-1} = \mathfrak{a}^{-1}\mathfrak{a} = R.$$

They imply that

$$[R/\mathfrak{a}] = \mathfrak{a}^{-1},$$
$$[\mathfrak{a}/\mathfrak{a}] = R.$$

They also imply that $\mathfrak{a}^{-1}$ is invertible with respect to R and that

$$\mathfrak{a} = (\mathfrak{a}^{-1})^{-1}.$$

Moreover, if $\mathfrak{a}, \mathfrak{b}$ are both invertible with respect to R, then also $\mathfrak{a}\mathfrak{b}$ is invertible with respect to R, and we have

$$(\mathfrak{a}\mathfrak{b})^{-1} = \mathfrak{b}^{-1}\mathfrak{a}^{-1}.$$

In particular, the principal R-fractional ideal ξR generated by an element ξ

of $\mathfrak{Q}(R)$ is invertible with respect to R, if an only if ξ is a unit of $\mathfrak{Q}(R)$ and in that case we have

$$(\xi R)^{-1} = \xi^{-1}R.$$

Hence, the fractional R-ideals that are invertible with respect to R form an abelian group.

2. Dedekind rings

Naturally, the question arises when every non-zero R-fractional ideal is invertible with respect to R.

Suppose that is the case. Then every non-zero element of $\mathfrak{Q}(R)$ is invertible so that $\mathfrak{Q}(R)$ is a field, R is entire.

Moreover, every non-zero prime ideal $\mathfrak{p}$ of R is maximal. Namely, otherwise $\mathfrak{p}$ is properly contained in a maximal ideal $\mathfrak{m}$ of R so that

$$\mathfrak{p} \subset \mathfrak{m} \subset R,$$
$$\mathfrak{p}\mathfrak{m}^{-1} \subset \mathfrak{m}\mathfrak{m}^{-1} = R,$$

and

$$\mathfrak{p} = (\mathfrak{p}\mathfrak{m}^{-1})\mathfrak{m},$$

hence $\mathfrak{p}$ is not a prime ideal.

More generally, if $\mathfrak{p}$, $\mathfrak{m}$ are any two fractional ideals satisfying

$$\mathfrak{p} \subseteq \mathfrak{m} \neq 0,$$

then we have

$$\mathfrak{p}\mathfrak{m}^{-1} \subseteq \mathfrak{m}\mathfrak{m}^{-1} \subseteq R,$$
$$\mathfrak{p} = (\mathfrak{p}\mathfrak{m}^{-1})\mathfrak{m},$$

so that $\mathfrak{p}$ is a multiple of $\mathfrak{m}$ by an ideal of R, i.e.

$$\mathfrak{m} \mid \mathfrak{p}.$$

Conversely, if $\mathfrak{m} \mid \mathfrak{p}$ then $\mathfrak{p} \subseteq \mathfrak{m}$ because of $\mathfrak{a}^{-1} \supseteq R$ for non-zero ideals $\mathfrak{a}$ of R. We have used a special case of

Lemma (5.2)

If the R-invertible ideal $\mathfrak{m}$ of R contains an ideal $\mathfrak{p}$ of R then $\mathfrak{p}$ is a multiple of $\mathfrak{m}$ in accordance with

$$\mathfrak{p} = (\mathfrak{p}\mathfrak{m}^{-1})\mathfrak{m}.$$

Conversely, if the ideal $\mathfrak{p}$ is a multiple of any ideal $\mathfrak{m}$ of R then $\mathfrak{p}$ is contained in $\mathfrak{m}$.

Thus set theoretic containment is equivalent to divisibility in the opposite direction for our rings.

Also the ring R is Noetherian. Indeed, every non-zero ideal $\mathfrak{a}$ of R is invertible so that

$$\mathfrak{a}\mathfrak{a}^{-1} = R,$$

implying an equation

$$\sum_{i=1}^{n} a_i b_i = 1,$$

with elements a_i of $\mathfrak{a}$, b_i of $\mathfrak{a}^{-1}$, hence we have

$$1 \in \left(\sum_{i=1}^{n} a_i R\right)\mathfrak{a}^{-1} \subseteq \mathfrak{a}\mathfrak{a}^{-1} = R,$$

$$\left(\sum_{i=1}^{n} a_i R\right)\mathfrak{a}^{-1} = R,$$

$$\sum_{i=1}^{n} a_i R = \mathfrak{a},$$

so that $\mathfrak{a}$ is finitely generated.

Finally we show that R is integrally closed. Indeed, if the element ξ of $\mathfrak{Q}(R)$ satisfies the monic equation

$$\xi^n + a_1 \xi^{n-1} + \cdots + a_n = 0,$$

with coefficients $a_1, \ldots, a_n$ in R, then the fractional non-zero ideal

$$\mathfrak{a} = \sum_{i=0}^{n-1} \xi^i R$$

satisfies the Kronecker condition

$$\xi \mathfrak{a} \subseteq \mathfrak{a},$$

implying that

$$\xi R = \xi \mathfrak{a}\mathfrak{a}^{-1} \subseteq \mathfrak{a}\mathfrak{a}^{-1} = R,$$

hence $\xi \in R$.

A special example of the rings we are studying here are the *principal ideal domains* (PID), i.e. the entire rings with the property that every ideal is principal, e.g. the rational integer ring $\mathbb{Z}$.

For such rings an ideal is a prime ideal, if and only if it is generated by a prime element. The statement that every non-zero element of a PID is factorizable into a product of finitely many prime elements, where the factors are uniquely determined up to order and equivalence, is equivalent to the statement that every non-zero ideal of the ring is the product of prime ideals which is unique up to the order of the factors. In this form we generalize the statement as follows:

Lemma (5.3)

If the non-zero R-fractional ideals of the entire ring R form a group G, then every element $\mathfrak{g}$ *of* G *is a power product of prime ideals*

$$\mathfrak{g} = \prod_{i=1}^{s} \mathfrak{p}_i^{\nu_i}, \tag{5.3a}$$

with distinct non-zero prime ideals $\mathfrak{p}_1, \mathfrak{p}_2, \ldots, \mathfrak{p}_s$ *of* R *and non-zero integral multiplicities* $\nu_1, \ldots, \nu_s$. *For any non-zero prime ideal* $\mathfrak{p} \notin \{\mathfrak{p}_1, \ldots, \mathfrak{p}_s\}$ *the multiplicity of* $\mathfrak{p}$ *is said to be zero. The multiplicity of each non-zero prime ideal with respect to the given ideal* $\mathfrak{g}$ *is unique, it is independent of the particular prime factorization* (5.3a).

Proof

If there is a non-zero $\mathfrak{g}$ of R for which no factorization (5.3a) exists with non-negative multiplicities then we choose $\mathfrak{g}$ to be maximal among the counter examples. Since R itself has the empty factorization ($s = 0$) with all multiplicities equal to zero it follows that $0 \subset \mathfrak{g} \subset R$, and therefore $\mathfrak{g}$ is contained in a maximal ideal $\mathfrak{p}_1$ of R, hence $\mathfrak{g} = \mathfrak{p}_1 \tilde{\mathfrak{g}}$, where $\mathfrak{g} \subset \tilde{\mathfrak{g}} \subseteq R$ so that

$$\tilde{\mathfrak{g}} = \prod_{i=1}^{s} \mathfrak{p}_i^{\tilde{\nu}_i},$$

where $\mathfrak{p}_1, \mathfrak{p}_2, \ldots, \mathfrak{p}_s$ are distinct non-zero prime ideals of R and the multiplicities $\tilde{\nu}_i$ are non-negative integers such that

$$\tilde{\nu}_i > 0 \quad \text{for } 2 \leqslant i \leqslant s.$$

Hence,

$$\mathfrak{p} = \mathfrak{p}_1 \tilde{\mathfrak{g}} = \prod_{i=1}^{s} \mathfrak{p}_i^{\nu_i}$$

for

$$\nu_i = \begin{cases} 1 + \tilde{\nu}_1 & \text{if } i = 1 \\ \tilde{\nu}_i & \text{if } 1 < i \leqslant s, \end{cases}$$

$$\nu_i > 0 \quad (1 \leqslant i \leqslant s).$$

Thus it is shown that every non-zero ideal of R is a prime ideal product with non-negative multiplicities.

Any non-zero R-fractional ideal is of the form

$$g^{-1}\mathfrak{g} = (gR)^{-1}\mathfrak{g},$$

with $\mathfrak{g}$ a non-zero ideal of R and g a non-zero element of R. Hence also gR is a prime ideal product with non-negative multiplicities, say

$$gR = \prod_{i=1}^{t} \mathfrak{p}_i^{\mu_i},$$

$$\mu_i \geqslant 0 \quad (1 \leqslant i \leqslant t),$$

$$\mu_i > 0 \quad (s < i \leqslant t),$$

where also $\mathfrak{p}_{s+1}, \ldots, \mathfrak{p}_t$ are distinct non-zero prime ideals of R, and

$$(gR)^{-1} = \prod_{i=1}^{t} \mathfrak{p}_i^{-\mu_i},$$

$$g^{-1}\mathfrak{g} = \prod_{i=1}^{t} \mathfrak{p}_i^{\nu_i - \mu_i},$$

where we set

$$\nu_i = 0 \quad \text{for } s < i \leqslant t.$$

In order to prove uniqueness we discuss the equation

$$\prod_{i=1}^{s} \mathfrak{p}_i^{\mu_i} = \prod_{i=1}^{s} \mathfrak{p}_i^{\nu_i}, \tag{5.4}$$

where $\mathfrak{p}_1, \mathfrak{p}_2, \ldots, \mathfrak{p}_s$ are distinct non-zero prime ideals and μ_i, ν_i are rational integers.

Upon moving the factors with negative exponents to the other side, respectively, we must show that there is no equation (5.4) with distinct non-zero prime ideals $\mathfrak{p}_1, \ldots, \mathfrak{p}_s$ $(s > 0)$ and with rational integers μ_i, ν_i subject to the condition that either

$$\mu_i > 0, \quad \nu_i = 0 \quad \text{or} \quad \mu_i = 0, \quad \nu_i > 0 \quad (1 \leqslant i \leqslant s).$$

Indeed, if $\mu_1 > 0$ then the left-hand side is contained in $\mathfrak{p}_1$, hence the right-hand side also is contained in $\mathfrak{p}_1$.

It follows that at least one of the prime ideal factors with positive multiplicity on the right-hand side is contained in and distinct from $\mathfrak{p}_1$, say

$$\mathfrak{p}_i \subset \mathfrak{p}_1 \quad \text{for some index } i \quad (1 < i \leqslant s).$$

But that is impossible, since every prime ideal is maximal. Thus (5.2) is demonstrated. □

The multiplicity $\nu_{\mathfrak{p}}(\xi)$ of the non-zero prime ideal $\mathfrak{p}$ of R in the non-zero element ξ of $\mathfrak{Q}(R)$ defines the $\mathfrak{p}$-*adic Krull exponential valuation*

$$\nu_{\mathfrak{p}}: \mathfrak{Q}(R) \to \mathbb{Z} \cup \{\infty\}. \tag{5.5}$$

It is defined for ξ of R as follows:

$$\nu_{\mathfrak{p}}(0) = \infty,$$

and for $0 \neq \xi \in R$ we set $\nu_{\mathfrak{p}}(\xi)$ to be the unique rational integer satisfying

$$\xi \in \mathfrak{p}^{\nu_{\mathfrak{p}}(\xi)}, \quad \xi \notin \mathfrak{p}^{\nu_{\mathfrak{p}}(\xi)+1}.$$

For

$$\xi = \eta\tau^{-1} \in \mathfrak{Q}(R) \quad (0 \neq \tau \in R, 0 \neq \eta \in R)$$

we set

$$\nu_{\mathfrak{p}}(\xi) := \nu_{\mathfrak{p}}(\eta) - \nu_{\mathfrak{p}}(\tau).$$

The $\mathfrak{p}$-adic Krull exponential valuation is discrete with $\dfrac{R}{R \backslash \mathfrak{p}}$ as its valuation ring. Hence there is precisely one Krull valuation ring of $\mathfrak{Q}(R)$ containing R such that its maximal ideal intersects R in $\mathfrak{p}$ according to (3.46).

Since R is integrally closed, it follows that R is the intersection of the Krull valuation rings of $\mathfrak{Q}(R)$ in which it is contained. From lemma (5.2) it follows that any non-zero element of R is contained only in finitely many maximal ideals of distinct Krull valuation rings of $\mathfrak{Q}(R)$ containing R. In other words R is a Dedekind ring.

The previous observations are extended by the next theorem.

Theorem (5.6)

Let R be an entire ring. Then the following conditions are equivalent.

(I) *R is a Dedeking ring.*
(II) *The fractional R-ideals $\neq 0$ form a group under multiplication.*
(III) *R is Noetherian, integrally closed and each non-zero prime ideal of R is maximal.*
(IV) *Every ideal of R is a product of prime ideals.*

Proof

We have shown already that

$$(\mathrm{II}) \Rightarrow (\mathrm{III}), \quad (\mathrm{II}) \Rightarrow (\mathrm{IV}), \quad (\mathrm{II}) \Rightarrow (\mathrm{I}).$$

We are going to show that

(a) (III) $\rightarrow$ (II),
(b) (I) $\rightarrow$ (III),
(c) (IV) $\Rightarrow$ (II).

(a) Let R be an entire ring satisfying (III). In order to show the group property of the non-zero R fractional ideals it suffices to show that every non-zero ideal of R is invertible with respect to R.

Let us assume that there are non-zero ideals of R which are not invertible with respect to R. Then among those there is a maximal one, say $\mathfrak{p}$. Because of $R^{-1} = R$ it follows that $\mathfrak{p}$ is a proper ideal.

If $\mathfrak{p}$ is contained in a maximal ideal $\mathfrak{m}$ and $\mathfrak{m}$ is invertible with respect to R then we have seen in (5.2) that $\mathfrak{p}$ is a multiple of $\mathfrak{m}$ and hence it is a prime ideal only if $\mathfrak{p} = \mathfrak{m}$.

It remains to prove that every maximal non-zero ideal $\mathfrak{p}$ of R is invertible with respect to R. There is a non-zero element π of $\mathfrak{p}$. We want to show the existence of a prime ideal product $\neq 0$ contained in the ideal πR, say

$$0 \subset \mathfrak{p}_1 \mathfrak{p}_2 \cdots \mathfrak{p}_s \subseteq \pi R. \tag{5.7}$$

If that is wrong then among the ideals of R not containing a non-zero prime ideal product there is a largest one, say $\mathfrak{a}$. The ideal $\mathfrak{a}$ is proper and not a prime ideal. Hence, there are two elements a_1, a_2 of R for which

$$a_1 \notin \mathfrak{a}, \quad a_2 \notin \mathfrak{a}, \quad a_1 a_2 \in \mathfrak{a},$$

so that

$$\mathfrak{a} \subset \mathfrak{a}_1 = \mathfrak{a} + Ra_1, \quad \mathfrak{a} \subset \mathfrak{a}_2 = \mathfrak{a} + Ra_2, \quad \mathfrak{a}_1 \mathfrak{a}_2 \subseteq \mathfrak{a}.$$

According to our assumption there are non-zero prime ideal products contained in $\mathfrak{a}_i$, say

$$0 \subset \mathfrak{p}_{i1} \mathfrak{p}_{i2} \cdots \mathfrak{p}_{in_i} \subseteq \mathfrak{a}_i \quad (i = 1, 2),$$

implying the relation

$$0 \subset \prod_{i=1}^{2} \prod_{j=1}^{n_i} \mathfrak{p}_{ij} \subseteq \mathfrak{a}_1 \mathfrak{a}_2 \subseteq \mathfrak{a}.$$

Thus (5.7) holds for suitable non-zero prime ideals $\mathfrak{p}_1, \ldots, \mathfrak{p}_s$.

We stipulate that s is as small as possible, hence

$$\prod_{\substack{i=l \\ i \neq j}}^{s} \mathfrak{p}_i \nsubseteq \pi R \quad (j = 1, 2, \ldots, s).$$

Since we have $\pi \in \mathfrak{p}$ it follows that

$$\prod_{i=1}^{s} \mathfrak{p}_i \subseteq \pi R \subseteq \mathfrak{p},$$

so that at least one of the prime ideals $\mathfrak{p}_i$ is contained in $\mathfrak{p}$, say

$$\mathfrak{p}_1 \subseteq \mathfrak{p}.$$

Because of the maximal property of every non-zero prime ideal of R it follows that

$$\mathfrak{p}_1 = \mathfrak{p}.$$

Moreover,

$$\prod_{i=2}^{s} \mathfrak{p}_i \nsubseteq \pi R,$$

$$\pi^{-1} \prod_{i=2}^{s} \mathfrak{p}_i \nsubseteq R,$$

but

$$\pi^{-1} \prod_{i=1}^{s} \mathfrak{p}_i \subseteq R,$$

so that

$$\pi^{-1} \prod_{i=2}^{s} \mathfrak{p}_i \subseteq [R/\mathfrak{p}],$$

hence

$$R \subset [R/\mathfrak{p}].$$

Because of the maximal property of $\mathfrak{p}$ and $\mathfrak{p} = \mathfrak{p}R \subseteq \mathfrak{p}[R/\mathfrak{p}] \subseteq R$ we have either

$$\mathfrak{p} = \mathfrak{p}[R/\mathfrak{p}],$$

or

$$R = \mathfrak{p}[R/\mathfrak{p}].$$

In the first case it follows from the Kronecker criterion that $[R/\mathfrak{p}]$ belongs to the integral closure of R in $\mathfrak{Q}(R)$ contradicting the integral closedness of R.

Hence, we find that

$$[R/\mathfrak{p}] = \mathfrak{p}^{-1},$$

demonstrating (II).

(b) Now let R be a Dedekind ring. We want to establish (III). By definition R is integrally closed.

If $\mathfrak{p}$ is a non-zero prime ideal of R, then there is a Krull valuation ring $\bar{R}$ of $\mathfrak{Q}(R)$ containing R such that the maximal ideal of $\bar{R}$ intersects R in $\mathfrak{p}$. Since the corresponding Krull valuation is discrete, it follows that $\mathfrak{p}$ cannot properly contain another prime ideal $\neq 0$ of R according to (3.46). Thus it is seen that every non-zero prime ideal of R is maximal.

Finally we show that any non-zero ideal $\mathfrak{a}$ of R is a finitely generated R-module. There is an element $\alpha \neq 0$ of $\mathfrak{a}$, and there are only finitely many Krull valuation rings of $\mathfrak{Q}(R)$, say $R_1, R_2, \ldots, R_s$ such that $R \subseteq R_i$, $\alpha R_i \subset R_i (1 \leqslant i \leqslant s)$. The integral closure property of R yields

$$R = R_1 \cap R_2 \cap \cdots \cap R_s.$$

By assumption there are discrete Krull exponential valuations

$$\eta_i : \mathfrak{Q}(R) \to \mathbb{Z} \cup \{\infty\},$$

with R_i as valuation rings such that the intersections

$$I(\eta_i > 0) \cap R = \mathfrak{m}_i$$

are s maximal ideals,

$$\eta_i(\alpha) > 0 \quad (1 \leqslant i \leqslant s)$$

$$\prod_{i=1}^{s} \mathfrak{m}_i^{\eta_i(\alpha)} \subseteq \alpha R \subseteq \mathfrak{a}.$$

It suffices to show that the ideal $\bar{\mathfrak{a}} = \mathfrak{a}\bar{R}$ of the localization $\bar{R} = R/(\prod_{i=1}^{s} R \backslash \mathfrak{m}_i)$ is a finitely generated $\bar{R}$-module.

In other words, we can assume without loss of generality that R contains only a finite number of distinct maximal ideals, R is a semilocal ring, hence $R = R_1 \cap \cdots \cap R_s$.

Moreover, $\mathfrak{Q}(R)$, $R_1, \ldots, R_s$ are the only Krull valuation rings of $\mathfrak{Q}(R)$ containing R.

We observe that the set $Y(\mathfrak{a})$ of all s-tuples $(\eta_1(y), \eta_2(y), \ldots, \eta_s(y))$ for non-zero y of $\mathfrak{a}$ is bounded from below by $(\eta_1(\alpha), \eta_2(\alpha), \ldots, \eta_s(\alpha))$ in the partial ordering of $\mathbb{Z}^{1\times s}$ provided by

$$(x_1, x_2, \ldots, x_s) \leqslant (y_1, y_2, \ldots, y_s) \Leftrightarrow x_i \leqslant y_i \quad (1 \leqslant i \leqslant s).$$

Applying the subsequent lemma (5.8) to the set $Y(\mathfrak{a})$ of the s-tuples $(\eta_1(x), \ldots, \eta_s(x))$ $(0 \neq x \in \mathfrak{a})$ it follows that there are finitely many elements of $\mathfrak{a}$, say $x_1, \ldots, x_\sigma$, with the property that for any non-zero element x of $\mathfrak{a}$ there is an index $i \in \{1, \ldots, \sigma\}$ for which $\eta_j(x_i) \leqslant \eta_j(x)$ $(1 \leqslant j \leqslant s)$, hence $\eta_j(xx_i^{-1}) \geqslant 0$ $(1 \leqslant j \leqslant s)$, hence $xx_i^{-1} \in R_1 \cap R_2 \cap \cdots \cap R_s = R$. This of course implies $\mathfrak{a} = Rx_1 + \cdots + Rx_\sigma$.

(c) To conclude (II) from (IV) it suffices to show that every non-zero ideal of R is invertible. Since every ideal is a product of prime ideals we only need to prove the invertibility of non-zero prime ideals $\mathfrak{p}$ of R. We do this in two steps.

Firstly we show $(\mathfrak{p} + Ra)^2 = \mathfrak{p} + Ra^2$ for all $a \in R \backslash \mathfrak{p}$.

There are (not necessarily unique) presentations of $\mathfrak{p} + Ra$, $\mathfrak{p} + Ra^2$ as products of prime ideals:

$$\mathfrak{p} + Ra = \prod_{i=1}^{m} \mathfrak{p}_i^{\mu_i}, \quad \mathfrak{p} + Ra^2 = \prod_{j=1}^{n} \mathfrak{y}_j^{\nu_j} \quad (\mu_i, \nu_j \in \mathbb{N}).$$

Clearly, $\mathfrak{p} \subset \mathfrak{p} + Ra \subseteq \mathfrak{p}_i$, $\mathfrak{p} \subset \mathfrak{p} + Ra^2 \subseteq \mathfrak{y}_j$ for $1 \leqslant i \leqslant m$, $1 \leqslant j \leqslant n$. Transition to the (entire!) residue class ring $\bar{R} := R/\mathfrak{p}$ yields $\bar{R}\bar{a} = \prod_{i=1}^{m} \bar{\mathfrak{p}}_i^{\mu_i}$, $\bar{R}\bar{a}^2 = \prod_{j=1}^{n} \bar{\mathfrak{y}}_j^{\nu_j}$. Because of $a \notin \mathfrak{p}$ the elements $\bar{a}$, $\bar{a}^2$ are different from $\bar{0}$ and therefore the ideals $\bar{\mathfrak{p}}_i$, $\bar{\mathfrak{y}}_j$ invertible with respect to $\bar{R}$ via

$$\bar{\mathfrak{p}}_i^{-1} = \bar{a}^{-1} \bar{\mathfrak{p}}_i^{\mu_i - 1} \prod_{\substack{\rho=1 \\ \rho \neq i}}^{m} \bar{\mathfrak{p}}_\rho^{\mu_\rho}, \quad \mathfrak{y}_j^{-1} = \bar{a}^{-2} \bar{\mathfrak{y}}_j^{\nu_j - 1} \prod_{\substack{\tau=1 \\ \tau \neq j}}^{n} \bar{\mathfrak{y}}_\tau^{\nu_\tau} \quad (1 \leqslant i \leqslant m,\ 1 \leqslant j \leqslant n).$$

But then the equation

$$\prod_{i=1}^{m} \bar{\mathfrak{p}}_i^{2\mu_i} = \prod_{j=1}^{n} \bar{\mathfrak{y}}_j^{\nu_j}$$

yields $m = n$ and – after reordering the $\bar{\mathfrak{y}}_j$, if necessary – $\bar{\mathfrak{p}}_i = \bar{\mathfrak{y}}_i$, $2\mu_i = \nu_i$ $(1 \leqslant i \leqslant n)$ as we saw in the proof of lemma (5.3). Because of $\mathfrak{p} \subset \mathfrak{p}_i$, $\mathfrak{p} \subset \mathfrak{y}_j (1 \leqslant i \leqslant m,\ 1 \leqslant j \leqslant n)$ we then obtain $\mathfrak{p}_i = \mathfrak{y}_i (1 \leqslant i \leqslant m)$.

In the second step we show that every non-zero prime ideal $\mathfrak{p}$ of R is maximal and invertible with respect to R.

There is $b \in \mathfrak{p}$, $b \neq 0$, and there exist prime ideals $\mathfrak{p}_1, \ldots, \mathfrak{p}_r$ of R such that $\mathfrak{p}_1 \cdot \ldots \cdot \mathfrak{p}_r = Rb \subseteq \mathfrak{p}$. Without loss of generality we can assume $\mathfrak{p}_1 \subseteq \mathfrak{p}$. As above, $\mathfrak{p}_1$ is invertible with respect to R and therefore maximal as we will see below. Hence, $\mathfrak{p}_1 = \mathfrak{p}$, and also $\mathfrak{p}$ is invertible with respect to R.

It remains to show that $\mathfrak{p}_1 + Ra = R$ for all $a \in R \backslash \mathfrak{p}_1$. For any elements $p \in \mathfrak{p}_1$ and $a \in R \backslash \mathfrak{p}_1$, we have according to the first step: $p \in \mathfrak{p}_1 + Ra^2 = (\mathfrak{p}_1 + Ra)^2 = \mathfrak{p}_1^2 + a\mathfrak{p}_1 + Ra^2 \subset \mathfrak{p}_1^2 + Ra$. Hence, there are elements $q \in \mathfrak{p}_1^2$, and $r \in R$ such that $p = q + ra$. But this implies $p - q = ra$ and therefore $r \in \mathfrak{p}_1$, $\mathfrak{p}_1 \subseteq \mathfrak{p}_1^2 + \mathfrak{p}_1 a$, and the invertibility of $\mathfrak{p}_1$ yields

$$R = \mathfrak{p}_1 \mathfrak{p}_1^{-1} \subseteq (\mathfrak{p}_1^2 + \mathfrak{p}_1 a)\mathfrak{p}_1^{-1} \subseteq \mathfrak{p}_1 + Ra \subseteq R. \qquad \square$$

For the proof of (b) we made use of the following lemma.

Lemma (5.8)

Let $\mathbb{Z}^n$ be partially ordered according to

$$\mathbf{x} \leqslant \mathbf{y} :\Leftrightarrow x_i \leqslant y_i \quad (1 \leqslant i \leqslant n).$$

Then any non-empty subset Y of $\mathbb{Z}^n$ which is bounded from below has only finitely many minima.

Proof

By induction on n. The case $n = 1$ is trivial. For $n > 1$ we use the projections

$$\Pi_1 : \mathbb{Z}^n \to \mathbb{Z} : \mathbf{x} = (x_1, \ldots, x_n)^t \mapsto x_1,$$
$$\Pi_2 : \mathbb{Z}^n \to \mathbb{Z}^{n-1} : \mathbf{x} = (x_1, \ldots, x_n)^t \mapsto (x_2, \ldots, x_n)^t = : \tilde{\mathbf{x}}.$$

Let Y be a non-empty subset of $\mathbb{Z}^n$ which is bounded from below. According to the induction hypothesis $\Pi_2(Y)$ has only finitely many minima, say $\tilde{\mathbf{x}}_1, \ldots, \tilde{\mathbf{x}}_s$. In $\Pi_2^{-1}(\tilde{\mathbf{x}}_i) \cap Y$ we choose $\mathbf{x}_i$ with minimal first coordinate $(1 \leqslant i \leqslant s)$. Let m be the maximum of the first coordinates of $\mathbf{x}_1, \ldots, \mathbf{x}_s$. Clearly, $\Pi_1(Y)$ contains only finitely many integers x satisfying $x \leqslant m$, say $\xi_1, \ldots, \xi_k$. Thus we are led to consider the subsets $Y_i := \{\mathbf{y} \in Y \mid \Pi_1(y) = \xi_i\}$ of Y $(1 \leqslant i \leqslant k)$. Again, $\Pi_2(Y_i)$ has at most finitely many minima, say $\tilde{\mathbf{y}}_{i_1}, \ldots, \tilde{\mathbf{y}}_{i_{s_i}}$ with (unique) preimages $\mathbf{y}_{i_1}, \ldots, \mathbf{y}_{i_{s_i}}$ in Y_i. Now all minima of Y are contained in the finite set $\{\mathbf{y}_{i_\nu} \mid 1 \leqslant \nu \leqslant s_i, 1 \leqslant i \leqslant k\}$. $\square$

Historically speaking, the theory of Dedekind rings originated from the remark made by certain astute French mathematicians that in the ring of cyclotomic integers $\mathbb{Z}[\zeta_n]$ (ζ_n a primitive nth root of unity) there does not always exist a greatest common divisor of two non-zero elements α, β, as Dirichlet pointed out to E.E. Kummer who had made the assumption implicitly. Thus E.E. Kummer was inspired to introduce 'ideal' numbers (i.e. algebraic integers outside the field $\mathbb{Q}(\zeta_n)$) playing the role of the greatest common divisor of α, β. R. Dedekind introduced the ideal generated by α, β as a substitute for the greatest common divisor which does not always exist in algebraic number fields, i.e. finite extensions of $\mathbb{Q}$. Dedekind showed that the ideals of the algebraic integer ring $\mathrm{Cl}(\mathbb{Z}, E)$ form a semigroup with unique factorization into prime ideal products for any algebraic number field E.

He also observed that the non-zero fractional ideals of $\mathrm{Cl}(\mathbb{Z}, E)$ form a group and that this property implies both (III) and (IV). He already suggested to take (III) as defining property for a more axiomatic treatment. This program was carried out in a famous 1926 paper by E. Noether [2].

The valuation theoretic treatment leading to (I) goes back to the work of W. Krull. It is seen from our treatment that (I), (III) run parallel inasmuch as

(a) integral closure,
(b) the Noetherian property,
(c) the maximality of prime ideals,

have equivalent valuation theoretic and ordinary ring theoretic definitions.

The characterizations (I), (III) both can be used to show that the integral closure of a Dedekind ring in a finite extension of the quotient field is again a Dedekind ring. Actually (I) will be used here since it does not require to distinguish the case of a separable and a non-separable extension.

The characterization (IV) given by Matusita appears to be the most natural one but it is not as useful as the other three.

Theorem (5.9)
The integral closure Λ of a Dedekind ring R in a finite extension E of the quotient field $F = \mathfrak{Q}(R)$ is a Dedekind ring.

Proof
The classical argument (of E. Noether) demonstrates (III) for Λ. This must be done separately for separable and for inseparable extensions. Since the latter do not occur in algebraic number theory we shall give the classical demonstration only in the case that E is a separable finite extension of F. This demonstration is then followed by a demonstration of (I) for Λ without separability condition on E. However, we shall combine the latter demonstration with the proof of a weak independence which is of intrinsic value.

(i) Classical argument
By definition Λ is integrally closed. In order to show the maximality of the non-zero prime ideals of Λ we must show that every Krull exponential valuation

$$\eta: E \to M \cup \{\infty\}$$

(M an algebraically ordered module) of E satisfying $\eta(\Lambda) \geqslant 0$ is discrete. But the restriction of η to F is a Krull exponential valuation of F satisfying

$$\eta(R) \geqslant 0.$$

Since R is a Dedekind ring it follows that $\eta|_F$ is discrete.

We have seen in the proof of (3.30) that any subset X of E with the property that $\eta(X)$ is a representative set of $\eta(E\backslash\{0\})$ modulo $\eta(F\backslash\{0\})$ is linearly independent over F. Hence, the index of

$$(\eta(E\backslash\{0\}):\eta(F\backslash\{0\})),$$

which is also called the *ramification index* of $\eta|_F$ in E, is finite. Since the module $\eta(F\backslash\{0\})$ is cyclic of order 1 or ∞ and since the module $\eta(E\backslash\{0\})$ is torsion-free and of finite index over $\eta(F\backslash\{0\})$ it follows that $\eta(E\backslash\{0\})$ is cyclic of order 1 or ∞. Hence, η is discrete.

Finally, it must be shown that Λ is Noetherian. For this purpose we need the assumption of separability which implies, firstly, the existence of a primitive element ξ of E so that $1, \xi, \ldots, \xi^{n-1}$ is an F-basis of E and, secondly, the non-vanishing of the discriminant $d(f)$ of the irreducible polynomial

$$f(t) = t^n + a_1 t^{n-1} + \cdots + a_n,$$

with coefficients in R of which ξ is a root. This in turn implies that Λ is contained in the R-module with basis

$$d(f)^{-1}, \quad d(f)^{-1}\xi, \ldots, d(f)^{-1}\xi^{n-1},$$

so that

$$R[t]/f \simeq \sum_{i=0}^{n-1} R\xi^i \subseteq \mathrm{Cl}(R, E) = \Lambda \subseteq \sum_{i=0}^{n-1} d(f)^{-1} R\xi^i.$$

(See also (5.17).) Hence, the Noetherian property of R implies the Noetherian property of the R-module Λ. *A fortiori*, Λ is a Noetherian ring.

(*ii*) *Valuation theoretic demonstration*

Making no separability assumption on E we will demonstrate (5.6) (I) for Λ. It has already been shown above that Λ is integrally closed and that every Krull valuation of E with Λ in its valuation ring is discrete. It remains to prove that for every non-zero element ξ of Λ there are only finitely many Krull valuation rings of E containing Λ such that ξ is contained in the corresponding maximal ideals.

We know that ξ satisfies an irreducible monic equation

$$\xi^m + a_1 \xi^{m-1} + \cdots + a_m = 0,$$

with coefficients $a_1, \ldots, a_m$ in the integrally closed ring R such that $a_m \neq 0$. If we show that

$$a_m = \xi(-\xi^{m-1} - a_1 \xi^{m-2} - \cdots - a_{m-1})$$

belongs to only finitely many maximal ideals of Krull valuation rings of E containing Λ, then the same is true for the element ξ dividing a_m in Λ. Without loss of generality we can therefore assume that ξ belongs to R.

We know already that there are only finitely many Krull valuation rings of F, say $R_1, R_2, \ldots, R_s$ with the property that the maximal ideal $\mathfrak{m}_i$ of R_i contains $\xi (1 \leqslant i \leqslant s)$. Thus it suffices to show that every Krull valuation ring

$\bar{R}$ of F is contained only in finitely many Krull valuation rings of E intersecting F in $\bar{R}$. More sharply, there are at most as many distinct extension rings

$$\Lambda_1, \quad \Lambda_2, \ldots \text{ of } \bar{R}$$

to Krull valuation rings of E intersecting F in $\bar{R}$ as the degree n of E over F.

Suppose we have the Krull exponential valuation

$$\psi: F \to M \cup \{\infty\}$$

of F with $\bar{R}$ as valuation ring and with divisible algebraically ordered value module M and the distinct extensions

$$\Psi_i: E \to M \cup \{\infty\},$$

with Krull valuation rings Λ_i $(i = 1, 2, \ldots, s)$ such that $\Psi_i|_F = \psi$.

We establish the *weak independence* of $\Psi_1, \ldots, \Psi_s$ by the subsequent lemma (5.10), where we show the existence of elements $\varepsilon_i \in \Lambda_i$ $(1 \leqslant i \leqslant s)$ satisfying

$$0 < \Psi_i(\varepsilon_i - 1), \quad 0 < \Psi_i(\varepsilon_k) \in \psi(F) \quad (1 \leqslant i,\, k \leqslant s,\, i \neq k).$$

We now go on to prove that those $\varepsilon_1, \ldots, \varepsilon_s$ are linearly independent over F implying $s \leqslant n$ thus concluding the proof of the theorem.

Namely, let $\lambda_1, \ldots, \lambda_\sigma \in F \backslash \{0\} (1 \leqslant \sigma \leqslant s)$ be given subject to $\psi(\lambda_1) \geqslant \psi(\lambda_2) \geqslant \cdots \geqslant \psi(\lambda_\sigma)$. We show that

$$\xi := \sum_{i=1}^{\sigma} \lambda_i \varepsilon_{\pi(i)} \quad (\pi \in \mathfrak{S}_s) \quad \text{satisfies } \Psi_{\pi(\sigma)}(\xi) = \psi(\lambda_\sigma) \neq \infty,$$

hence there is no non-trivial representation of 0 as a linear combination of $\varepsilon_1, \ldots, \varepsilon_s$ over F. The last equation is obtained by the following easy computations:

$$\Psi_{\pi(\sigma)}(\xi) = \Psi_{\pi(\sigma)}\left(\lambda_\sigma \varepsilon_{\pi(\sigma)} + \sum_{i=1}^{\sigma-1} \lambda_i \varepsilon_{\pi(i)}\right),$$

$$\Psi_{\pi(\sigma)}(\lambda_\sigma \varepsilon_{\pi(\sigma)}) = \psi(\lambda_\sigma) + \Psi_{\pi(\sigma)}(\varepsilon_{\pi(\sigma)}) = \psi(\lambda_\sigma) + \Psi_{\pi(\sigma)}(1 + (\varepsilon_{\pi(\sigma)} - 1)) = \psi(\lambda_\sigma),$$

$$\Psi_{\pi(\sigma)}\left(\sum_{i=1}^{\sigma-1} \lambda_i \varepsilon_{\pi(i)}\right) \geqslant \min_{1 \leqslant i \leqslant \sigma-1} (\psi(\lambda_i) + \Psi_{\pi(\sigma)}(\varepsilon_{\pi(i)})) > \min_{1 \leqslant i \leqslant \sigma-1} \psi(\lambda_i) \geqslant \psi(\lambda_\sigma). \qquad \square$$

Lemma (5.10)

Under the premises of (5.9) *let* ψ *be any Krull exponential valuation of* F *and* $\Psi_1, \ldots, \Psi_s$ *finitely many distinct Krull exponential valuations of* E *such that their valuation rings* Λ_i $(1 \leqslant i \leqslant s)$ *intersect* F *in the valuation ring* $\bar{R}$ *of* ψ. *Then there exist elements* $\varepsilon_i \in \Lambda_i$ $(1 \leqslant i \leqslant s)$ *satisfying*

$$0 < \Psi_i(\varepsilon_i - 1), \quad 0 < \Psi_i(\varepsilon_k) \in \Psi(F \backslash \{0\}) \quad (1 \leqslant i,\, k \leqslant s;\, i \neq k).$$

Proof

The case $s = 1$ is trivial, choose $\varepsilon_1 = 1$.

For $s=2$ we choose an element $0 \neq \alpha \in E$ such that $\Psi_1(\alpha) \neq \Psi_2(\alpha)$. For $\kappa_1 := (\Psi_1(E\backslash\{0\}):\Psi(F\backslash\{0\}))$ the element $\beta = \alpha^{\kappa_1}$ satisfies $\Psi_1(\beta) \in \Psi(F\backslash\{0\})$ and $\Psi_1(\beta) = \kappa_1 \Psi_1(\alpha) \neq \kappa_1 \Psi_2(\alpha) = \Psi_2(\beta)$. Hence, there exists $0 \neq a \in F$ with $\Psi_1(\beta) = \psi(a)$, and we obtain $\Psi_1(\gamma) = 0 \neq \Psi_2(\gamma)$ for $\gamma = a^{-1}\beta$. In case of $\Psi_2(\gamma) < 0$ we replace γ by γ^{-1}. In this way we always find an element $0 \neq \gamma \in E$ with $\Psi_1(\gamma) = 0 < \Psi_2(\gamma)$.

Then the element $\delta := \gamma^{\kappa_2}$, $\kappa_2 := (\Psi_2(E\backslash\{0\}):\Psi(F\backslash\{0\}))$, satisfies $\Psi_1(\delta) = 0 < \Psi_2(\delta) \in \psi(F\backslash\{0\})$, hence $\Psi_2(\delta) = \psi(b)$ for a suitable $b \in F\backslash\{0\}$. Finally, we set $\varepsilon_1 = (1 + b\delta^{-2})^{-1}$. We claim that ε_1 has the properties stated in the lemma. This follows by some easy calculations:

$$\Psi_2(b\delta^{-2}) = -\Psi_2(\delta) < 0, \quad \text{hence } \psi(F\backslash\{0\}) \ni \Psi_2(\delta) = \Psi_2(\varepsilon_1) > 0;$$

$$\Psi_1(\varepsilon_1 - 1) = \Psi_1\left(\frac{-b}{\delta^2 + b}\right) = \psi(b) - \Psi_1(\delta^2 + b) = \psi(b) > 0.$$

Analogously we find an element ε_2 for which $0 < \Psi_2(\varepsilon_2 - 1)$, $0 < \Psi_1(\varepsilon_2) \in \Psi(F\backslash\{0\})$ hold. We note that $\Psi_i(\varepsilon_i - 1) > 0$ implies $\Psi_i(\varepsilon_i) = 0$, hence $\varepsilon_i \in \Lambda_i$ $(i = 1, 2)$.

For $s > 2$ we apply induction on s. Hence, we assume that there is an element $\tilde{\varepsilon}_1$ of Λ_1 such that $0 < \Psi_1(\tilde{\varepsilon}_1 - 1)$, $0 < \Psi_i(\tilde{\varepsilon}_1) \in \psi(F\backslash\{0\})$ $(1 < i < s)$. Clearly, $\varepsilon := \tilde{\varepsilon}_1^{\kappa_s}$, $\kappa_s := (\Psi_s(E\backslash\{0\}):\psi(F\backslash\{0\}))$, also satisfies these inequalities. If $\Psi_s(\varepsilon) > 0$ we set $\varepsilon_1 = \varepsilon$ and obtain $0 < \Psi_i(\varepsilon_i) \in \psi(F\backslash\{0\})$ for $i = s$. If $\Psi_s(\varepsilon) \leqslant 0$, then we choose $b \in F\backslash\{0\}$ such that
$0 < \psi(b) = \min\{|\Psi_i(\varepsilon)| \mid 1 \leqslant i \leqslant s, \Psi_i(\varepsilon) \neq 0\}$.

In case of $\Psi_s(\varepsilon) < 0$ we set $\varepsilon_1 = (\varepsilon^{-2} + b\varepsilon^2)^{-1}$. Clearly, $\Psi_i(b\varepsilon^2) = \psi(b) + 2\Psi_i(\varepsilon) > 0$ for $1 < i < s$, hence, $\Psi_i(\varepsilon_1) = -\Psi_i(\varepsilon^{-2}) > 0$ according to assumption. For $i = s$ we obtain $\Psi_s(\varepsilon^{-2}) > 0$, $\Psi_i(b\varepsilon^2) = \psi(b) + 2\Psi_i(\varepsilon) \leqslant \Psi_i(\varepsilon) < 0$, hence $\Psi_s(\varepsilon_1) > 0$. Finally,

$$\Psi_1\left(\frac{1}{\varepsilon^{-2} + \pi\varepsilon^2} - 1\right) = \Psi_1\left(\frac{\varepsilon^2 - 1 - \pi\varepsilon^4}{1 + \pi\varepsilon^4}\right) = \Psi_1(\varepsilon^2 - 1 - \pi\varepsilon^4) > 0$$

because of $\Psi_1(\varepsilon) = 0$, $\psi(\pi) > 0$, $\Psi_1(\varepsilon^2 - 1) > 0$.

In case of $\Psi_s(\varepsilon) = 0$ we additionally employ the existence of $\eta \in \Lambda_1$ such that $\Psi_1(\eta - 1) > 0$, $0 < \Psi_s(\eta)$, $\Psi_i(\eta) \in \psi(F\backslash\{0\})$ $(1 < i \leqslant s)$. Such an element exists according to our considerations in the case $s = 2$. Now we choose $b \in F$ satisfying $0 < \psi(b) = \min\{|\Psi_i(\eta)| \mid 1 \leqslant i \leqslant s, \Psi_i(\eta) \neq 0\}$ and set
$\varepsilon_1 = \varepsilon(\eta^2 + b\eta^{-2})^{-1}$.

As above we easily see that $\Psi_i(\eta^2 + b\eta^{-2}) \leqslant 0$ for $(1 < i < s)$, and $\Psi_s(\eta^2 + b\eta^{-2}) < 0$, hence $\Psi_i(\varepsilon_1) > 0$ for $1 < i \leqslant s$. Finally,

$$\Psi_1(\varepsilon_1 - 1) = \Psi_1\left(\frac{\varepsilon\eta^2 - \eta^4 - b}{\eta^4 + b}\right) = \Psi_1(\eta^2(\varepsilon - \eta^2) - b) > 0$$

because of $\Psi_1(b) = \psi(b) > 0$, $\Psi_1(\eta) = 0$ and

$$\Psi_1(\varepsilon - \eta^2) = \Psi_1((\varepsilon - 1) + (1 - \eta^2)) > 0.$$

Similarly, we obtain elements $\varepsilon_2, \ldots, \varepsilon_s$ satisfying the conditions of the lemma. □

3. Orders

The solution of a monic irreducible equation

$$x^n + a_1 x^{n-1} + \cdots + a_n = 0 \tag{5.11}$$

over the rational integer ring is formally achieved by forming the equation ring

$$\Lambda_f = \mathbb{Z}[t]/f \tag{5.12a}$$

for the polynomial

$$f(t) = t^n + a_1 t^{n-1} + \cdots + a_n \tag{5.12b}$$

of $\mathbb{Z}[t]$. But important arithmetic invariants like the field discriminant, the fundamental units and the ideal class group need to be known in terms of the integral closure $\mathrm{Cl}(\mathbb{Z}, A)$ of $\mathbb{Z}$ in the algebraic number field

$$A = \mathbb{Q}[t]/f = \mathfrak{Q}(\Lambda_f).$$

The question arises how to find $\mathrm{Cl}(\mathbb{Z}, A)$ when f is given.

Definition (5.13)

Given a Dedekind ring R, the overring Λ of R is said to be an **R-order** *if*

(a) *Λ is unital;*

(b) *Λ is a finitely generated R-module, in other words there are finitely many elements $a_1, a_2, \ldots, a_s$ of Λ such that $\Lambda = Ra_1 + Ra_2 + \cdots + Ra_s$;*

(c) *the R-module Λ is torsion-free: If $\lambda a = 0$, $0 \neq \lambda \in R$, $a \in \Lambda$, then $a = 0$ which means that the central quotient ring $A = \mathfrak{Q}(\Lambda)$ of Λ is an (associative) algebra over $F = \mathfrak{Q}(R)$;*

(d) *Λ contains no nilpotent ideal $\neq 0$.*

In this book we consider only commutative orders. An example of a commutative order is the order of a monic separable equation (5.11) with coefficients in the Dedekind ring R. It is the equation order (5.12).

Definition (5.14)

The R-order Λ is said to be a **maximal R-order** *if it is maximal among the R-orders of $\mathfrak{Q}(\Lambda)$.*

Lemma (5.15)

Any commutative maximal R-order is the integral closure of R in its central quotient ring.

Proof

Let Λ be any commutative R-order. By Kronecker's criterion applied to Λ any element of Λ belongs to the integral closure of R in $\mathfrak{Q}(\Lambda)$. □

On the other hand, it is possible that a commutative R-order is not contained in a maximal order of its central quotient ring. The reader is asked to realize this occurrence by going through exercises 1–3.

If the quotient field of a Dedekind ring R is perfect, then every finite extension E of $\mathfrak{Q}(R)$ can be generated by the adjunction of an element x of E satisfying a monic separable equation (5.11) over R, so that the discriminant of the polynomial f is not zero. We have seen before that $1, x, \ldots, x^{n-1}$ is an R-basis of the equation order Λ_f and that the trace bilinear form

$$\operatorname{Tr}(ab) = \sum_{i=1}^{n} \sum_{k=1}^{n} \alpha_i \beta_k \operatorname{Tr}(x^{i-1+k-1}),$$

$$a = \sum_{i=1}^{n} \alpha_i x^{i-1}, \quad b = \sum_{k=1}^{n} \beta_k x^{k-1} \quad (\alpha_i, \beta_k \in R; i, k = 1, 2, \ldots, n) \tag{5.16}$$

has the determinant

$$\det(\operatorname{Tr}(x^{i+k-2})) = d(f) \neq 0,$$

so that we find for any element

$$\xi = \sum_{i=1}^{n} \xi_i x^{i-1} \quad (\xi_i \in R, 1 \leqslant i \leqslant n)$$

of $\mathrm{Cl}(R, E)$ that

$$\operatorname{Tr}(\xi x^{k-1}) = \sum_{i=1}^{n} \xi_i \operatorname{Tr}(x^{i+k-2}) \in R.$$

By Cramer's rule we have

$$\xi_i d(f) \in R,$$

hence

$$\mathrm{Cl}(R, E) \subseteq d(f)^{-1} \Lambda_f. \tag{5.17}$$

Since R is a Noetherian ring and since by (5.17) the integral closure $\mathrm{Cl}(R, E)$ is contained in the finitely generated R-module

$$\sum_{i=1}^{n} R d(f)^{-1} x^{i-1},$$

it follows that $\mathrm{Cl}(R, E)$ itself is finitely generated so that in this case $\mathrm{Cl}(R, E)$ is itself an R-order.

The absence of nilpotent ideals $\neq 0$ from $\mathrm{Cl}(R, E)$ follows from the absence of such ideals from $\mathfrak{Q}(\Lambda_f)$ which in turn follows from their absence from Λ_f. The same argument holds also in case f is separable, but reducible. One must use the result that $\mathfrak{Q}(\Lambda_f)$ is the algebraic sum of finite extensions of F (see exercises 4–11, for example).

In order to prove the next theorem we need a preparatory lemma.

Lemma (5.18)

Let R be a unital commutative ring which is integrally closed in $\mathfrak{Q}(R)$. Let A be a unital commutative $\mathfrak{Q}(R)$-ring with matrix representation Δ of degree $n > 0$ over $\mathfrak{Q}(R)$. Then for every element x of $\mathrm{Cl}(R, A)$ the characteristic polynomial of $\Delta(x)$ is monic of degree n over R.

Proof

For any generalized multiplicative valuation $\varphi: A \to \Phi$ the elements of A of φ-value 0 form a maximal ideal $\mathfrak{M}$ such that the replacement of elements of $\mathfrak{Q}(R)$ by the residue classes of $\mathfrak{Q}(R)/\mathfrak{M}$ carries Δ to a matrix representation $\bar{\Delta}$ of $\bar{A} = A/\mathfrak{M}$ of degree n over $\overline{\mathfrak{Q}(R)} = \mathfrak{Q}(R)/\mathfrak{M}$ satisfying $\bar{\Delta}(\bar{x}) \in \mathrm{Cl}(\bar{R}, \bar{\Delta}(\bar{A}))$ $(\bar{R} = R/\mathfrak{M}, \bar{x} = x/\mathfrak{M})$.

The matrix representation $\bar{\Delta}$ restricts on the subalgebra generated by $\bar{x}$ over $\mathfrak{Q}(\bar{R})$ to a matrix representation $\tilde{\Delta}$ of degree n over $\overline{\mathfrak{Q}(R)}$ with the property that for each irreducible component Γ the characteristic polynomial of $\Gamma(\bar{x})$ is a power of the minimal polynomial of $\Gamma(\bar{x})$. The latter is a monic irreducible polynomial over $\bar{R}$. This is because $\bar{\Delta}(\bar{x})$ belongs to $\mathrm{Cl}(\bar{R}, \bar{\Delta}(\bar{A}))$ and because $\bar{R}$ is integrally closed in the quotient field $\mathfrak{Q}(\bar{R}) = \overline{\mathfrak{Q}(R)}$, according to our assumption on R. Hence, the characteristic polynomial of $\bar{\Delta}(\bar{x})$ is a monic polynomial of degree n over $\bar{R}$.

It follows by assumption that the characteristic polynomial of $\Delta(x)$ is monic of degree n over R. □

Theorem (5.19)

The integral closure of a Dedekind ring R in a separable commutative $\mathfrak{Q}(R)$-algebra A of finite dimension n over $\mathfrak{Q}(R)$ is an R-order.

Proof

We know that there is a primitive element ξ of A such that the powers $1, \xi, \ldots, \xi^{n-1}$ form a $\mathfrak{Q}(R)$-basis of A. The $n+1$ coefficients of the (monic) minimal polynomial of ξ have a common denominator $D \neq 0$ in R. Then $D\xi$ is also a primitive element of A with monic minimal polynomial of degree n over R. Without loss of generality we can assume that $A = \mathfrak{Q}(R)[t]/f(t)\mathfrak{Q}(R)[t]$ is the algebraic background of the equation order $\Lambda_f = R[t]/f(t)R[t]$ of the monic separable polynomial f of degree $n > 0$ over R.

The *regular trace* of an element x of A is defined as the negative second highest coefficient of the characteristic polynomial of the regular representation of x, a matrix representation of degree n over $\mathfrak{Q}(R)$. Hence, $\mathrm{Tr}(x)$ belongs to R for every x of $\mathrm{Cl}(R, A)$ according to lemma (5.18). It follows that the regular trace bilinear form $B: A \times A \to \mathfrak{Q}(R): (a_1, a_2) \mapsto \mathrm{Tr}(a_1 a_2)$ restricts on $\mathrm{Cl}(R, A)$ to a symmetric bilinear form with values in R. Thus it follows that

$$\Lambda_f \subseteq \mathrm{Cl}(R, A) \subseteq \Lambda_f^\perp,$$

where the B-dual $\Lambda_f^\perp$ is defined by

$$\Lambda_f^\perp = \{y \in A \mid \mathrm{Tr}(y\Lambda_f) \subseteq R\}.$$

Because of $d(f) = \det((\mathrm{Tr}(\xi^{i+k-2}))_{1 \leqslant i,k \leqslant n}) \neq 0$ it follows that the system of linear equations

$$\mathrm{Tr}\left(\xi^{i-1}\left(\sum_{k=1}^{n} y_{jk}\xi^{k-1}\right)\right) = \delta_{ij} \quad (1 \leqslant i \leqslant n) \tag{5.20a}$$

for the unknowns $y_{j1}, \ldots, y_{jn}$ $(1 \leqslant j \leqslant n)$ has a unique solution so that there are elements

$$b_j = \sum_{k=1}^{n} y_{jk}\xi^{k-1} \quad (y_{jk} \in \mathfrak{Q}(R), 1 \leqslant j \leqslant n) \tag{5.20b}$$

of A, satisfying

$$\mathrm{Tr}(\xi^{i-1}b_j) = \delta_{ij} \quad (1 \leqslant i, j \leqslant n). \tag{5.20c}$$

Those elements form an R-basis of $\Lambda_f^\perp$, a so-called *dual basis*. Indeed, for any element x of $\Lambda_f^\perp$ we have

$$\mathrm{Tr}(\xi^{i-1}x) = \mathrm{Tr}\left(\xi^{i-1}\left(\sum_{k=1}^{n} \mathrm{Tr}(\xi^{k-1}x)b_k\right)\right) \quad (1 \leqslant i \leqslant n) \tag{5.20d}$$

so that the difference

$$x' = x - \sum_{k=1}^{n} \mathrm{Tr}(\xi^{k-1}x)b_k$$

has the property

$$\mathrm{Tr}(\xi^{i-1}x') = 0 \quad (1 \leqslant i \leqslant n),$$

which implies

$$x' = 0$$

as we had already observed in chapter 2. It follows that every element x of $\Lambda_f^\perp$ is uniquely presented as the R-linear combination

$$x = \sum_{k=1}^{n} \mathrm{Tr}(\xi^{k-1}x)b_k \tag{5.20e}$$

of $b_1, \ldots, b_n$ and that

$$\Lambda_f^\perp = \sum_{k=1}^{n} Rb_k. \tag{5.20f}$$

For the more general theory of duality and orthogonal complementation see exercises 12–14.

Since R is a Noetherian ring it follows that every R-submodule of the finitely generated R-module $\Lambda_f^{\perp}$ is finitely generated over R (see exercises 15, 16). In particular $\mathrm{Cl}(R, A)$ is finitely generated over R.

Since A is separable over $\mathfrak{Q}(R)$ it follows that A is isomorphic to the algebraic sum of finitely many fields so that A contains no nilpotent element besides zero. Hence, $\mathrm{Cl}(R, A)$ is an R-order. □

4. Finitely generated modules over Dedekind rings

In the remainder of this section we prepare an algorithm for embedding the equation order Λ_f of the monic separable polynomial $f(t)$ of degree $n > 0$ over R into its maximal order $\mathrm{Cl}(R, A) = \mathrm{Cl}(R, \mathfrak{Q}(R)[t]/f(t)\mathfrak{Q}(R)[t])$. The algorithm itself is developed in section 6.

We begin with the task of characterizing a finitely presented module over a Dedekind ring by means of a full system of invariants.

Let R be a unital commutative ring. Then for any module M with finitely many generators $v_1, \ldots, v_n$ over R there is the standard R-epimorphism η: $R^{1\times n} \to M$ of the n-row module over R on M which sends the unit row $\mathbf{e}_j'$ on the generator v_j $(1 \leqslant j \leqslant n)$. The kernel of that epimorphism is formed by the n-rows $(\lambda_1, \ldots, \lambda_n)$ for which there holds the linear relation $\lambda_1 v_1 + \cdots + \lambda_n v_n = 0$ between the R-generators $v_1, \ldots, v_n$ of M. Thus $\ker(\eta)$ is the *relation* or (first) *syzygy-module* of M relative to the generating set $v_1, \ldots, v_n$ over R, and the factor module is R-isomorphic to M: $R^{1\times n}/\ker(\eta) \simeq M$.

Definition (5.21)

The **elementary ideals** $\mathfrak{E}_i = \mathfrak{E}_i(M/R)$ *of* M *over* R *are defined as the ideals of* R *generated by the* $(n-i) \times (n-i)$ *minors of the* $(n-i) \times n$ *matrices formed from any* $n-i$ *rows of* $\ker(\eta)$. *For* $i \in \mathbb{Z}^{\geqslant n}$ *we define* $\mathfrak{E}_i(M/R) := R$.

The reader will easily verify that the elementary ideals are independent of the choice of the finite generator set of M over R (see exercise 17). In case the relation module is finitely generated over R the finitely many generators of it form the rows of a rectangular matrix, the so-called *relation matrix* of M relative to the finitely many generators of M over R.

Conversely, every $s \times n$-matrix $\Lambda = (\lambda_{ik})$ over R is associated with the R-submodule $\mathfrak{R}(\Lambda)$ of $R^{1\times n}$ generated by the s rows of Λ. That R-submodule is said to be the *row module* of Λ over R, the factor module $R^{1\times n}/\mathfrak{R}(\Lambda)$ is an R-module with the generators $\mathbf{e}_j'/\mathfrak{R}(\Lambda)$ $(1 \leqslant j \leqslant n)$ over R and Λ as relation matrix. The elementary ideals are easily computable (see exercise 18).

The definition can be extended to arbitrary R-modules (see exercise 19) and behaves constructively for algebraic sums (see exercise 20). As the definition shows the elementary ideals form an ascending sequence

$$\mathfrak{E}_0 \subseteq \mathfrak{E}_1 \subseteq \mathfrak{E}_2 \subseteq \cdots. \tag{5.22}$$

For any ideal $\mathfrak{a}$ of R we find that the factor module $M/\mathfrak{a}M$ is an $R/\mathfrak{a}$-module according to the definition

$$(\lambda/\mathfrak{a})(u/\mathfrak{a}M) := \lambda u/\mathfrak{a}M \quad (\lambda \in R, u \in M), \tag{5.23a}$$

and that

$$\mathfrak{E}_i((M/\mathfrak{a}M)/(R/\mathfrak{a})) = \mathfrak{E}_i(M/R)/\mathfrak{a} \quad (i \in \mathbb{Z}^{\geq 0}). \tag{5.23b}$$

For any commutative overring Λ of R with

$$1_\Lambda = 1_R, \tag{5.24a}$$

it follows that there is the R-monomorphism

$$\eta_\Lambda : M \to \Lambda \otimes_R M : u \mapsto 1_\Lambda \otimes u \tag{5.24b}$$

of M into $\Lambda \otimes_R M$ such that

$$\mathfrak{E}_i(\Lambda \otimes_R M) = \Lambda \otimes_R \mathfrak{E}_i((M/\ker(\eta_\Lambda))/R) \quad (i \in \mathbb{Z}^{\geq 0}). \tag{5.24c}$$

In particular, we have

$$\mathfrak{Q}(R) \otimes_R M \simeq_{\mathfrak{Q}(R)} \mathfrak{Q}(R) \otimes_R (M/\mathrm{Tor}(M/R)). \tag{5.24d}$$

If R is an entire ring then we have

$$\mathfrak{E}_i(\mathfrak{Q}(R) \otimes_R M) = \mathfrak{E}_i((\mathfrak{Q}(R) \otimes_R M/\mathrm{Tor}(M/R))/\mathfrak{Q}(R)) = \begin{cases} 0 & \text{for } i < r(M/R) \\ \mathfrak{Q}(R) & \text{otherwise} \end{cases}, \tag{5.24e}$$

where $r(M/R) := \dim_{\mathfrak{Q}(R)}(\mathfrak{Q}(R) \otimes_R M)$ is the rank of M over R. Two rectangular matrices $\Lambda_i \in R^{s_i \times n}$ $(s_i \in \mathbb{Z}^{>0}, i = 1, 2)$ with the same number of columns are said to be *row equivalent*, if their row modules over R coincide. The constructive decision on equivalence of Λ_1 and Λ_2 hinges on the assumption that there is an algorithm known to decide whether a given element λ of R belongs to the ideal of R generated by finitely many given elements $\lambda_1, \lambda_2, \ldots, \lambda_\rho$ and, in case the decision is affirmative, to exhibit a presentation of the form $\lambda = \xi_1\lambda_1 + \cdots + \xi_\rho\lambda_\rho$ with $\xi_1, \ldots, \xi_\rho$ in R (see exercise 22).

A theoretical concept for the solution of that task is to develop suitable normal forms for matrices over R. In case of R being a Euclidean ring this was done already in chapter 3, section 2. In the sequel we generalize the definition of an elementary divisor normal form to arbitrary unital commutative rings R.

Definition (5.25)

Let R be a unital commutative ring. The R-module $M = \bigoplus_{i=1}^{s} R/\mathfrak{n}_i$, where $\mathfrak{n}_1, \ldots, \mathfrak{n}_s$ $(s \in \mathbb{Z}^{>0})$ are ideals of R subject to $0 \subseteq \mathfrak{n}_1 \subseteq \mathfrak{n}_2 \subseteq \cdots \subseteq \mathfrak{n}_s \subset R$, is said *to be in* **elementary divisor form presentation,** *and the ideals* $\mathfrak{n}_1,\ldots,\mathfrak{n}_s$ *are said to be the* **elementary divisor ideals** of M.

We observe that under the premise of (5.25)

$$\mathfrak{E}_0(M/R)=\prod_{i=1}^{s}\mathfrak{n}_i,\quad \mathfrak{E}_j(M/R)=\prod_{i=j+1}^{s}\mathfrak{n}_i\ \ (0<j<s),\quad \mathfrak{E}_{s+j}(M/R)=R\ \ (j\in\mathbb{Z}^{\geqslant 0}).$$

It can be shown in general that $\mathfrak{n}_1,\mathfrak{n}_2,\ldots,\mathfrak{n}_s$ are uniquely determined by M. Because of the uniqueness of the elementary ideals certainly $\mathfrak{n}_1,\ldots,\mathfrak{n}_s$ are uniquely determined if the non-zero elementary ideals are invertible with respect to R.

A particular elementary divisor form presentation is associated with the relation matrix

$$\mathfrak{R}=\operatorname{diag}(e_1,e_2,\ldots,e_s)\quad (e_i\in R,\ e_i\,|\,e_{i+1}\quad (1\leqslant i<s), e_s\in R\backslash U(R)).\quad (5.26)$$

In that case the elementary divisor ideals are the principal ideals $\mathfrak{n}_i=e_{n+1-i}R$ $(1\leqslant i\leqslant s)$ generated by the elementary divisors e_i, and vice versa.

For constructive purposes we restrict the rings R under consideration.

Definition (5.27)

The unital commutative ring R is said to be a **Prüfer ring,** *if every ideal of R which is generated by two elements is a principal ideal.*

For example, every principal ideal ring and every Krull valuation ring is a Prüfer ring (see also exercise 23). We note that in a Prüfer ring every finitely generated ideal is principal.

For Prüfer rings the row normal form of a rectangular matrix can be simplified (see exercises 24, 25) provided we can assume that there is an algorithm to exhibit a generator $\lambda=\xi_1\lambda_1+\xi_2\lambda_2$ $(\xi_1,\xi_2\in R)$ of the ideal $R\lambda_1+R\lambda_2$ for arbitrary λ_1,λ_2 of R. In general this is possible only for Euclidean rings, however.

Also there must be an algorithm for deciding whether the element λ of R divides the element η of R and in case the decision is positive producing an element ξ of R satisfying $\lambda\xi=\eta$.

Any finitely presented R-module over a Prüfer ring can be carried into the form (5.26) (see exercises 25, 26). Using exercises 28, 22 we obtain

Theorem (5.28)

Every finitely generated torsion module over a Dedekind ring R has an elementary divisor form presentation.

For the rational integer ring $\mathbb{Z}$ as base ring every finitely generated torsion module M has a set of generators $v_1,\ldots,v_s$ with defining relators

$$e_i v_i = 0, \tag{5.29a}$$

where the numbers

$$e_i \in \mathbb{Z}^{>1} \quad (1 \leqslant i \leqslant s) \tag{5.29b}$$

are the elementary divisors of M. As we saw in chapter 3 they are uniquely determined by (5.29a, b) and by

$$e_{i+1} | e_i \quad (1 \leqslant i < s) \tag{5.29c}$$

so that

$$M \simeq \bigoplus_{i=1}^{s} \mathbb{Z}/e_i\mathbb{Z}, \quad |M| = \prod_{i=1}^{s} e_i, \quad |M|\mathbb{Z} = \mathfrak{E}_0(M/\mathbb{Z}). \tag{5.29d}$$

(This is in principle the basis theorem for finite abelian groups.)
We see that the order of any finite module M generates the ideal $\mathfrak{E}_0(M/\mathbb{Z})$. Correspondingly we define

Definition (5.30)
The **order ideal** $(M:_R 0)$ *of the module M over the unital commutative ring R is defined as $\mathfrak{E}_0(M/R)$.*

The **exponent** *of a finite abelian group (5.29d) is defined as the smallest natural number annihilating M. It turns out to be e_s.*

Correspondingly we define

Definition (5.31)
The **exponent ideal** $\mathfrak{n}(M/R)$ *of a module M over the unital commutative ring R is defined as the ideal of R consisting of all elements λ of R which annihilate M: $\lambda M = 0$.*

In analogy to Fermat's theorem we have

Lemma (5.32)
$\mathfrak{E}_0(M/R) \subseteq \mathfrak{n}(M/R)$ *for any finitely generated R-module M.*

Proof
Let η be an R-epimorphism of $R^{1\times n}$ on M. Then $\mathfrak{E}_0(M/R)$ is the ideal of R generated by the determinants of all $n\times n$-matrices $A = (\alpha_{ik})$ over R with the property that $\sum_{k=1}^{n} \alpha_{ik}\eta(\mathbf{e}_k^t) = 0$ $(1 \leqslant i \leqslant n)$. Denoting by A_{ik} the algebraic complement of α_{ik} in A we have the formulae

$$\sum_{i=1}^{n} A_{ih}\alpha_{ik} = \delta_{hk}\det(A) \quad (h,k = 1,2,\ldots,n)$$

of determinantal calculus which imply that

$$\det(A)\eta(e'_n) = \sum_{i=1}^{n} A_{ih}\alpha_{ih}\eta(e'_h) = \sum_{i=1}^{n} A_{ih} \sum_{k=1}^{n} \alpha_{ik}\eta(e'_k) = 0$$

for $h = 1, 2, \ldots, n$ so that $\det(A)M = 0$, $\mathfrak{E}_0(M/R) \subseteq \mathfrak{n}(M/R)$. □

For the generalization of lemma (5.32) to arbitrary R-modules see exercise 28.

We observe that

$$\mathfrak{n}(M/R) = \mathfrak{n}_1(M/R) \tag{5.33}$$

in case that the R-module M can be presented in elementary divisor form. Moreover,

$$\mathfrak{n}(M_1 \oplus M_2) = \mathfrak{n}(M_1) \cap \mathfrak{n}(M_2). \tag{5.34}$$

For finite groups G there holds the index formula $(G:T) = (G:S)(S:T)$ for any two subgroups S, T of G satisfying $S \supseteq T$. Correspondingly we have

Lemma (5.35)

For any unital commutative ring R and for any R-module M with R-submodules M_1, M_2 subject to $M_1 \supseteq M_2$ we have

$$(M:_R M_2) = (M:_R M_1)(M_1:_R M_2), \tag{5.35a}$$

where the **index ideals** *are defined via the order ideal by*

$$(M:_R M_i) = (M/M_i:_R 0)\ (i = 1, 2), \quad (M_1:_R M_2) = (M_1/M_2:_R 0), \tag{5.35b}$$

provided that the occurring factormodules can be finitely presented so that there are not more defining syzygies than there are generators.

Proof

Without loss of generality, we can assume that $M_2 = 0$ and that M_1 can be finitely generated over R. Hence M can be finitely generated. It follows from our assumption (which remains unchanged under elementary transformations) that $M = \sum_{i=1}^{n} Rv_i$ for some $n \in \mathbb{N}$ so that $M_1 = \sum_{i=1}^{n} Rw_i$ with

$$w_k = \sum_{i=1}^{n} \alpha_{ik} v_i \quad (A = (\alpha_{ik}) \in R^{n \times n}) \tag{5.36a}$$

such that the syzygies $0 = \sum_{i=1}^{n} \alpha_{ik} v_i / M_1$ $(1 \leqslant k \leqslant n)$ are the defining syzygies of M/M_1 over R. We have to show that

$$\mathfrak{E}_0(M/R) = \mathfrak{E}_0((M/M_1)/R)\mathfrak{E}_0(M_1/R). \tag{5.36b}$$

Since A is a relation matrix of M/M_1 over R it follows that $\mathfrak{E}_0((M/M_1)/R) = \det(A)\,R$.

The order ideal of M_1 over R is the ideal of R generated by the determinants of those matrices $B=(\beta_{ik})\in R^{n\times n}$ for which there hold the syzygies

$$\sum_{i=1}^{n}\beta_{ik}w_i=0 \tag{5.36c}$$

of M_1 relative to the generators $w_1,\ldots,w_n$. Upon substitution of (5.36a) in (5.36c) we obtain the syzygies

$$\sum_{i=1}^{n}\sum_{j=1}^{n}\alpha_{ij}\beta_{jk}v_i=0 \quad (1\leqslant k\leqslant n) \tag{5.36d}$$

of M corresponding to the matrix AB of determinant $\det(A)\det(B)$. By assumption the syzygies $\sum_{i=1}^{n}\alpha_{ij}\beta_{jk}v_i=0$ of M derived from the syzygies $\sum_{i=1}^{n}\beta_{ik}w_i=0$ of M_1 over R form a system of defining syzygies. Thus (5.36b) is shown. □

For any semilocal Dedekind ring R the assumption of lemma (5.35) is satisfied whenever M/M_1 is finitely generated over R. It follows that (5.36b) holds in all cases in which M/M_1 can be finitely generated over R. By the localization argument (see exercise 21) this remains true for all Dedekind rings. We remark that for any unital ring R and any ideal $\mathfrak{a}$ of R we have

$$\mathfrak{E}_0((R/\mathfrak{a})/R)=\mathfrak{a}. \tag{5.37}$$

This is because $R/\mathfrak{a}$ is an R-module with one generator $1/\mathfrak{a}$ and defining relation $\alpha 1/\mathfrak{a}=0$ $(\alpha\in\mathfrak{a})$. It follows that any equation of the form

$$\mathfrak{E}_0(R/\mathfrak{a})\mathfrak{E}_0(\mathfrak{a}/\mathfrak{a}')=\mathfrak{E}_0(R/\mathfrak{a}') \tag{5.38}$$

for two ideals $\mathfrak{a},\mathfrak{a}'$ of R satisfying $\mathfrak{a}\supseteq\mathfrak{a}'$ implies that $\mathfrak{a}$ divides $\mathfrak{a}'$. If R is entire then (5.38) is universally true if and only if R is a Dedekind ring as we have seen before.

Theorem (5.39)

Let R be a Dedekind ring.

(a) *The finitely generated R-module M is R-isomorphic to a non-zero R-fractional ideal precisely if $\mathfrak{E}_0(M/R)=0$, $\mathfrak{E}_1(M/R)=R$.*
(b) *Two R-fractional ideals $\mathfrak{a}_1,\mathfrak{a}_2$ are R-isomorphic R-modules precisely if they are equivalent in the sense that there is a unit λ of $\mathfrak{Q}(R)$ satisfying $\lambda\mathfrak{a}_1=\mathfrak{a}_2$.*
(c) *The equivalence classes of the non-zero R-fractional ideals form an abelian group Cl_R, the* **class group** *of R.*
(d) *(E. Landau) Every non-zero R-fractional ideal can be generated by two elements.*
(e) *(E. Steinitz) Every R-module M splits over every R-submodule M' with torsion-free finitely generated factormodule.*
(*f*) *(E. Steinitz) For every finitely generated torsion-free R-module M the*

R-rank $r = r(M/R) = \dim_{\mathfrak{Q}(R)} \mathfrak{Q}(R) \otimes_R M$ is connected with the elementary ideals by the relations

$$\mathfrak{E}_i(M/R) = \begin{cases} 0 & \text{if } 0 \leqslant i < r \\ R & \text{if } r \leqslant i \end{cases}. \tag{5.39a}$$

(g) *For every set of r independent elements $v_1, v_2, \ldots, v_r$ of the finitely generated torsion-free R-module M of R-rank r the equivalence class $\mathfrak{N}(M/R)$ of the index ideal $(M:_R \sum_{i=1}^r Rv_i) = \mathfrak{E}_0((M/\sum_{i=1}^r Rv_i)/R)$ is independent of the choice of $v_1, \ldots, v_r$. It is the* **Steinitz class** *of M over R.*

(h) *Two finitely generated R-modules M_1, M_2 are R-isomorphic precisely if*

$$\mathfrak{E}_i(M_1/R) = \mathfrak{E}_i(M_2/R) \quad (i \in \mathbb{Z}^{\geqslant 0}), \tag{5.39b}$$

$$\mathfrak{N}(M_1/R) = \mathfrak{N}(M_2/R), \tag{5.39c}$$

where the Steinitz class of a torsion module is defined as the equivalence class of the principal R-fractional ideals and the Steinitz class of a non-torsion finitely generated R-module M is defined as

$$\mathfrak{N}(M/R) = \mathfrak{N}((M/\mathrm{Tor}_R(M))/R). \tag{5.39d}$$

Proof

(a) If M is a non-zero R-fractional ideal of the entire ring R then we have $\mathfrak{n}(M/R) = 0$, hence $\mathfrak{E}_0(M/R) = 0$ in case M is finitely generated over R. Moreover, $\mathfrak{E}_1(M/R)$ is not zero because of $0 \subset M \subseteq \mathfrak{Q}(R) \otimes_R M$.

Now let R be a Dedekind ring. We want to show that $\mathfrak{E}_1(M/R) = R$. Since there is a non-zero element n of R satisfying $nM \subseteq R$ and $M \simeq_R nM$ it suffices to deal with the case that $M \subseteq R$. Upon $\mathfrak{p}$-localization for any non-zero prime ideal $\mathfrak{p}$ of R the ideal M is always carried into the non-zero principal ideal

$\dfrac{M}{R\backslash\mathfrak{p}}$ of $\dfrac{R}{R\backslash\mathfrak{p}}$ so that $\mathfrak{E}_1\left(\dfrac{M}{R\backslash\mathfrak{p}} \Big/ \dfrac{R}{R\backslash\mathfrak{p}}\right) = \dfrac{R}{R\backslash\mathfrak{p}} \mathfrak{E}_1(M/R)$, hence $\mathfrak{E}_1(M/R) = R$. Conversely, if $\mathfrak{E}_0(M/R) = 0$, $\mathfrak{E}_1(M/R) = R$ for the finitely generated R-module M then it follows from (5.32) that M is torsion-free, hence $M \subseteq \mathfrak{Q}(R) \otimes_R M$, M is $\mathfrak{Q}(R)$-isomorphic to a non-zero fractional ideal of R.

(b) Clearly, for any R-fractional ideal $\mathfrak{a}$ of the unital commutative ring R multiplication of $\mathfrak{a}$ by a unit ε of $\mathfrak{a}(R)$ constitutes an R-isomorphism of the R-module $\mathfrak{a}$ with the R-module $\varepsilon\mathfrak{a}$ defined by the R-fractional ideal $\varepsilon\mathfrak{a}$. Conversely, let $\eta: \mathfrak{a}_1 \to \mathfrak{a}_2$ be an R-isomorphism of the R-fractional ideals $\mathfrak{a}_1, \mathfrak{a}_2$ and let R be entire. Then either $1\mathfrak{a}_1 = \mathfrak{a}_1 = \mathfrak{a}_2 = 0$ or $\mathfrak{a}_1$ contains an element $x \neq 0$ and we have $\mathfrak{a}_1 \neq 0$, $\mathfrak{a}_2 \neq 0$ and $\eta(x) = \varepsilon x$, where $\varepsilon = \eta(x)x^{-1}$ is an element of $\mathfrak{Q}(R)$. Moreover, for any element y of $\mathfrak{a}_1$ there are elements x', y' of R satisfying $x' \neq 0$, $v'/x' = v/x$, $y'x = x'y$, $\eta(y'x) = y'\eta(x) = \eta(x'y) = x'\eta(y)$, $\eta(y)/\eta(x) = y'/x'$, $\eta(y) = \varepsilon x(y'/x') = \varepsilon y$. Because of $\mathfrak{a}_2 \neq 0$ it follows that $\varepsilon \neq 0$. There are non-zero elements $n, \varepsilon_1, \varepsilon_2$ of R satisfying $n\mathfrak{a}_1 \subseteq R$, $\varepsilon = \varepsilon_1\varepsilon_2^{-1}$, hence $n\varepsilon_2\mathfrak{a}_2 = n\varepsilon_2\varepsilon\mathfrak{a}_1 = \varepsilon_1(n\mathfrak{a}_1) \subseteq R$ so that $\mathfrak{a}_2$ is an R-fractional ideal.

(c) It is clear that the equivalence of R-fractional ideals is reflexive, symmetric, and transitive for any unital commutative ring R. Moreover, the equivalence is multiplicative in the sense that $\mathfrak{a}_i \sim_R \mathfrak{a}'_i \leftrightarrow \exists \lambda_i \in U(\mathfrak{Q}(R)) : \mathfrak{a}'_i = \lambda_i \mathfrak{a}_i$ $(i = 1, 2)$ implies $\mathfrak{a}'_1 \mathfrak{a}'_2 = \lambda_1 \lambda_2 \mathfrak{a}_1 \mathfrak{a}_2$ with $\lambda_1 \lambda_2$ in $U(\mathfrak{Q}(R))$ so that $\mathfrak{a}'_1 \mathfrak{a}'_2 \sim_R \mathfrak{a}_1 \mathfrak{a}_2$. It follows that the equivalence classes of R-fractional ideals with zero exponent ideal form a unital commutative semigroup $\mathfrak{C}(\mathfrak{R})$, the so-called *class semigroup*. Its unit element is the equivalence class formed by the principal ideals $R\varepsilon$ generated by a unit ε of $\mathfrak{Q}(R)$. The class semigroup is a group if and only if the R-fractional ideals with zero exponent ideal form a group. In case R is entire that means that R is a Dedekind ring.

(d) It suffices to deal with the case that $0 \subset \mathfrak{a} \subseteq R$. There is a non-zero element a of $\mathfrak{a}$, so that $0 \subset Ra \subseteq \mathfrak{a}$. By Ra-localization we obtain the semilocal ring $\bar{R} = R/(R \backslash \mathfrak{p}_1 \cup \mathfrak{p}_2 \cup \cdots \cup \mathfrak{p}_s)$, where $\mathfrak{p}_1, \mathfrak{p}_2, \ldots, \mathfrak{p}_s$ are the finitely many distinct prime ideals containing a. It is a principal ideal ring and contains the ideal $\mathfrak{a}\bar{R}$ such that $\mathfrak{a}\bar{R} = bc^{-1}\bar{R}$ with $0 \neq b \in R$, $0 \neq c \in R$, $cR + aR = R$, $\mathfrak{a}\bar{R} \cap R = \mathfrak{a} = aR + bR$.

(e) If the R-rank of M/M' is 1 then we know already from (a) that $M/M' \simeq_R \mathfrak{a}$, where $\mathfrak{a}$ is a non-zero ideal of R. Also we always have $M = R \otimes_R M$, $\mathfrak{a} \otimes_R \mathfrak{a}^{-1} \simeq_R R$, hence $M = \mathfrak{a} \otimes_R (\mathfrak{a}^{-1} \otimes_R M)$, $M' = \mathfrak{a} \otimes_R (\mathfrak{a}^{-1} \otimes_R M')$, $\mathfrak{a}^{-1} \otimes_R M/\mathfrak{a}^{-1} \otimes_R M' = \mathfrak{a}^{-1} \otimes_R M/M' \simeq_R \mathfrak{a}^{-1} \otimes_R \mathfrak{a} \simeq_R R$. Thus there is an element u of $\mathfrak{a}^{-1} \otimes_R M$ satisfying
$Ru/\mathfrak{a}^{-1} \otimes_R M' \simeq_R \mathfrak{a}^{-1} \otimes_R M/\mathfrak{a}^{-1} \otimes_R M'$, so that $\mathfrak{a}^{-1} \otimes_R M = Ru \dotplus \mathfrak{a}^{-1} \otimes_R M'$, $M = \mathfrak{a} \otimes_R Ru \dotplus M'$. If the R-rank of M/M' is greater than one then there is a maximal one among the R-submodules $\tilde{M}$ of M containing M' for which $\tilde{M}$ splits over M' and $M/\tilde{M}$ is non-zero torsion-free. It follows that $M/\tilde{M}$ is of rank one and thus M splits over $\tilde{M}$, $\tilde{M}$ splits over M'.

Note in particular that

$$M = \mathrm{Tor}_R(M) \dotplus M_0 \tag{5.40}$$

for some R-torsion free submodule M_0 of M in case M is an R-module with finitely generated factormodule over $\mathrm{Tor}_R(M)$.

(f) It follows from (e) that

$$M \simeq \mathfrak{a}_1 \oplus \mathfrak{a}_2 \oplus \cdots \oplus \mathfrak{a}_r, \tag{5.41}$$

where r is the R-rank of M and $\mathfrak{a}_1, \ldots, \mathfrak{a}_r$ are non-zero ideals of R. Hence, $\mathfrak{E}_0(\mathfrak{a}_i) = 0$, $\mathfrak{E}_1(\mathfrak{a}_i) = R$ $(1 \leqslant i \leqslant r)$ and (f) follows from exercise 20.

We add the remark that in case $r > 1$ it is always possible to find a decomposition

$$M = Rv_1 \dotplus Rv_2 \dotplus \cdots \dotplus Rv_{r-1} \dotplus M_r \tag{5.42}$$

of M into the direct sum of $r - 1$ cyclic R-submodules $Rv_i \neq 0$ $(1 \leqslant i < r)$ and one R-submodule M_r which is R-isomorphic to a non-zero ideal of R. For this purpose it suffices to study the case that $M = \mathfrak{a}_1 \oplus \mathfrak{a}_2$ is the algebraic sum of two non-zero ideals $\mathfrak{a}_1, \mathfrak{a}_2$ of R and to find an element v of M such that $Rv \neq 0$ and M/Rv is torsion free.

But there is a non-zero element v_1 of $\mathfrak{a}_1$ and a non-zero element v_2' of $v_1\mathfrak{a}_1^{-1}\mathfrak{a}_2$, hence by (d) there is an element v_2 of $\mathfrak{a}_2$ for which $\mathfrak{a}_2 = Rv_2' + Rv_2$. Setting $v = v_1 \oplus v_2$ we attain our purpose.

(g), (h) follow from (5.28). □

5. Extension of the index ideal concept

In case M_1, M_2 are two R-submodules of the finitely generated R-module M over the Dedekind ring R and that the module $M_2/M_1 \cap M_2$ is R-torsion we extend the definition of the *index ideal* as follows:

$$(M_1 :_R M_2) := (M_1 :_R M_1 \cap M_2)(M_2 :_R M_1 \cap M_2)^{-1}. \tag{5.43a}$$

Thus the index ideal is defined as an R-fractional ideal also in many cases for which M_2 is not contained in M_1. In case of $M_2 \subseteq M_1$ the new definition coincides with the earlier one. We observe that for any R-submodule M' of $M_1 \cap M_2$ for which M_2/M' is R-torsion we have

$$(M_1 :_R M_2) = (M_1 :_R M')(M_2 :_R M')^{-1}. \tag{5.43b}$$

This is because of $(M_1 :_R M') = (M_1 :_R M_1 \cap M_2)(M_1 \cap M_2 :_R M')$, $(M_2 :_R M') = (M_2 :_R M_1 \cap M_2)(M_1 \cap M_2 :_R M')$, $0 \subset (M_1 \cap M_2 :_R M') \subseteq R$. It follows that

$$(M_1 :_R M_2)(M_2 :_R M_3) = (M_1 :_R M_3) \tag{5.43c}$$

in case M_1, M_2, M_3 are three R-submodules of M for which the R-factormodules $M_1/M_1 \cap M_3$, $M_2/M_1 \cap M_2$ and $M_3/M_2 \cap M_3$ are R-torsion. Indeed, in that case we have the R-isomorphism $M_2 \cap M_3/M_1 \cap M_2 \cap M_3 \simeq (M_1 \cap M_2) + (M_2 \cap M_3)/M_1 \cap M_2$. The right-hand side is contained in $M_2/M_1 \cap M_2$ so that it is R-torsion. Hence both $M_3/M_2 \cap M_3$ and $M_2 \cap M_3/M_1 \cap M_2 \cap M_3$ are R-torsion. Hence $M_3/M_1 \cap M_2 \cap M_3$ is R-torsion. *A fortiori*, $M_3/M_1 \cap M_3$ is R-torsion. Therefore all three index ideals of (5.43c) are defined. Moreover, $M_2/M_1 \cap M_2 \cap M_3$ is also R-torsion so that upon using the rule (5.43b) with $M' = M_1 \cap M_2 \cap M_3$ we obtain

$$(M_1 :_R M_2)(M_2 :_R M_3) = (M_1 :_R M')(M_3 :_R M')^{-1} = (M_1 :_R M_3).$$

From (5.43c) we gather the rule

$$(M_1 :_R M_2)(M_2 :_R M_1) = R \tag{5.43d}$$

in case both $M_1/M_1 \cap M_2$ and $M_2/M_1 \cap M_2$ are R-torsion.

6. Norms of ideals

Definition (5.44)

Let R be a Dedekind ring and let Λ be a commutative R-order of R-rank n.

Then the **norm** *of any Λ-fractional ideal $\mathfrak{a}$ is defined as the index ideal of Λ over $\mathfrak{a}$:* $N_{\Lambda/R}(\mathfrak{a}) := (\Lambda :_R \mathfrak{a})$.

We observe that the norm of a Λ-fractional ideal $\mathfrak{a}$ over R is an R-fractional ideal which is not zero precisely if $\mathfrak{a}$ is of maximal R-rank n. If $\mathfrak{a}$ is principal, say $\mathfrak{a} = \alpha\Lambda$, then we have $N_{\Lambda/R}(\alpha\Lambda) = N_{\mathfrak{Q}(\Lambda)/\mathfrak{Q}(R)}(\alpha)R$. Hence we have the multiplicativity

$$N_{\Lambda/R}(\mathfrak{a}\mathfrak{b}) = N_{\Lambda/R}(\mathfrak{a})N_{\Lambda/R}(\mathfrak{b}) \tag{5.45}$$

of the norm ideal for Λ-fractional ideals $\mathfrak{a}, \mathfrak{b}$ in case $\mathfrak{a}$ is principal. More generally speaking (5.45) holds in case the Λ-fractional ideal $\mathfrak{a}$ is invertible relative to Λ (see exercises 29–33). The same is also trivially true if $\mathfrak{a}$ is of R-rank less than n since in that case both $N_{\Lambda/R}(\mathfrak{a}\mathfrak{b})$ and $N_{\Lambda/R}(\mathfrak{a})$ vanish. On the other hand (5.45) does not hold for all Λ-fractional ideals $\mathfrak{b}$ in case $\mathfrak{a}$ is a Λ-fractional ideal of R-rank n which is not invertible relative to R and Λ can be properly embedded into an R-maximal order of $\mathfrak{Q}(\Lambda)$. This is because by a theorem of Dade, Taussky and Zassenhaus (see exercise 34) the Λ-fractional ideal $\mathfrak{a}^{n-1}$ is invertible relative to its order $\Lambda_1 = [\mathfrak{a}^{n-1}/\mathfrak{a}^{n-1}]$, an R-overorder of Λ in $\mathfrak{Q}(\Lambda)$ that contains Λ properly. Namely, in that case we have

$$\begin{aligned} N_{\Lambda/R}(\mathfrak{a}^{2(n-1)}) &= (\Lambda :_R \mathfrak{a}^{2(n-1)}) = (\Lambda :_R \Lambda_1)(\Lambda_1 :_R \mathfrak{a}^{2(n-1)}) \\ &= (\Lambda :_R \Lambda_1)N_{\Lambda_1/R}((\mathfrak{a}^{n-1})^2) = (\Lambda :_R \Lambda_1)N_{\Lambda_1/R}(\mathfrak{a}^{n-1})^2. \end{aligned}$$

On the other hand, (5.45) would imply that

$$\begin{aligned} N_{\Lambda/R}(\mathfrak{a}^{2(n-1)}) &= N_{\Lambda/R}(\mathfrak{a}^{n-1}\mathfrak{a}^{n-1}) = (N_{\Lambda/R}(\mathfrak{a}^{n-1}))^2 = (\Lambda :_R \mathfrak{a}^{n-1})^2 \\ &= ((\Lambda :_R \Lambda_1)(\Lambda_1 :_R \mathfrak{a}^{n-1}))^2 = (\Lambda :_R \Lambda_1)^2(N_{\Lambda_1/R}(\mathfrak{a}^{n-1}))^2, \end{aligned}$$

hence,

$$(\Lambda :_R \Lambda_1) = (\Lambda :_R \Lambda_1)^2, \quad (\Lambda :_R \Lambda_1) = R, \quad \Lambda = \Lambda_1,$$

a contradiction.

7. The arithmetic radical of an order

The Jacobson radical of a Dedekind ring R with infinitely many prime ideals is zero. It follows that the Jacobson radical of any order Λ over such a Dedekind ring is zero, too. On the other hand the Jacobson radical of a semilocal Dedekind ring is not zero.

It follows that the Jacobson radical of a Dedekind ring is not localizable.

For orders over Dedekind rings we also have to consider the effect of inseparability which will prevent localizability of the Jacobson radical even over some orders over local Dedekind rings.

We assume from now on that for the orders Λ considered the regular trace bilinear form of $\mathfrak{Q}(\Lambda)$ over $\mathfrak{Q}(R)$ is non-degenerate.

Definition (5.46)

The **discriminant ideal** $\mathfrak{D}(\Lambda/R)$ *of* Λ *is defined as the index ideal of* $\Lambda^\perp$ *over* R:

$$\mathfrak{D}(\Lambda/R) := (\Lambda^\perp :_R \Lambda). \tag{5.46a}$$

The **arithmetic radical** $\mathrm{AR}(\Lambda)$ *of* Λ *is defined as the intersection of the maximal ideals of* Λ *containing the discriminant ideal. The* **reduced discriminant ideal** $\mathfrak{D}_0(\Lambda/R)$ *of* Λ *is defined as the exponent ideal of the R-factormodule* $\Lambda^\perp/\Lambda$:

$$\mathfrak{D}_0(\Lambda/R) := \mathfrak{n}((\Lambda^\perp/\Lambda)/R). \tag{5.46b}$$

We shall see in (5.49) that the new definition of discriminant ideals generalizes the one given in chapter 2. All three concepts: discriminant ideal, arithmetic radical and reduced discriminant ideal are invariant under localization. For any subsemigroup S of $R\backslash\{0\}$ we have

$$\mathfrak{D}\left(\frac{\Lambda}{S}\Big/\frac{R}{S}\right) = \frac{\mathfrak{D}(\Lambda/R)}{S}, \quad \mathrm{AR}\left(\frac{\Lambda}{S}\Big/\frac{R}{S}\right) = \frac{\mathrm{AR}(\Lambda/R)}{S}, \quad \mathfrak{D}_0\left(\frac{\Lambda}{S}\Big/\frac{R}{S}\right) = \frac{\mathfrak{D}_0(\Lambda/R)}{S}, \tag{5.47}$$

as the reader will easily verify.

We generalize (5.46a). For any R-submodule M of $\mathfrak{Q}(\Lambda)$ the discriminant ideal of M over R is defined as

$$\mathfrak{D}(M/R) := (M^\perp :_R M). \tag{5.48a}$$

It follows invariance under localization

$$\mathfrak{D}\left(\frac{M}{S}\Big/\frac{R}{S}\right) = \frac{\mathfrak{D}(M/R)}{S}, \tag{5.48b}$$

as well as the *index discriminant rule*

$$\mathfrak{D}(M_1/R) = (M_2 :_R M_1)^2 \mathfrak{D}(M_2/R) \tag{5.48c}$$

for any two R-submodules M_1, M_2 of $\mathfrak{Q}(\Lambda)$ of maximal R-rank. It is implied by the *dual-index rule*

$$(M_1 :_R M_2) = (M_2^\perp :_R M_1^\perp), \tag{5.48d}$$

in accordance with

$$\mathfrak{D}(M_2/R) = (M_2^\perp :_R M_2) = (M_2^\perp :_R M_1^\perp)(M_1^\perp :_R M_1)(M_1 :_R M_2).$$

The dual-index rule itself is a consequence of the *involutory rule*

$$(M^\perp)^\perp = M, \tag{5.48e}$$

and the *index multiplicativity*

$$(M_1 :_R M_3) = (M_1 :_R M_2)(M_2 :_R M_3), \tag{5.48f}$$

both of which were demonstrated before. As a consequence of (5.48f) we must show (5.48d) merely for the case that M_2 is a maximal R-submodule of M_1. But in that case it follows from (5.48e) that $M_1^\perp$ is a maximal R-submodule of $M_2^\perp$. Moreover the exponent ideal of M_1/M_2 is a maximal ideal $\mathfrak{p}$ of R. Because of $\mathfrak{p}M_1 \subseteq M_2$ it follows that $\mathfrak{p}M_2^\perp \subseteq M_1^\perp$, hence $\mathfrak{n}((M_2^\perp/M_1^\perp)/R) = \mathfrak{p}$. Now the maximality of M_2 in M_1, $M_1^\perp$ in $M_2^\perp$ implies $M_1/M_2 \simeq_R R/\mathfrak{p} \simeq_R M_2^\perp/M_1^\perp$, hence $(M_1 :_R M_2) = (M_2^\perp :_R M_1^\perp)$.

Lemma (5.49)
The discriminant ideal $\mathfrak{D}(M/R)$ *of any* R*-submodule* M *of* $\mathfrak{Q}(\Lambda)$ *over* R *equals the* R*-fractional ideal generated by the determinants*

$$\det((\mathrm{Tr}(u_i v_k))_{1 \leqslant i,k \leqslant n}), \tag{5.50}$$

where Tr *denotes the regular trace of* $\mathfrak{Q}(\Lambda)$ *over* $\mathfrak{Q}(R)$, $n = \dim_{\mathfrak{Q}(R)} \mathfrak{Q}(\Lambda)$, *and* $u_1, \ldots, u_n, v_1, \ldots, v_n$ *are any* $2n$ *elements of* M.

Proof
Using localization it suffices to prove the lemma only for local Dedekind rings. But for such Dedekind rings there is an R-basis $b_1, \ldots, b_n$ of M so that the ideal generated by the determinants given in (5.50) equals the principal R-fractional ideal $\det((\mathrm{Tr}(b_i b_k))_{1 \leqslant i,k \leqslant n})R$. Moreover, there is the R-basis $c_1, \ldots, c_n$ of $M^\perp$ that is dual to $b_1, \ldots, b_n$ in accordance with the equations $\mathrm{Tr}(b_i c_k) = \delta_{ik}$ $(i, k = 1, \ldots, n)$. Hence, we have equations

$$b_k = \sum_{i=1}^{n} \sigma_{ik} c_i \quad \text{with} \quad \sigma_{ik} \in \mathfrak{Q}(R) \quad (1 \leqslant i, k \leqslant n)$$

so that

$$\mathrm{Tr}(b_i b_k) = \mathrm{Tr}\left(b_i \sum_{j=1}^{n} \sigma_{jk} c_j\right) = \sum_{j=1}^{n} \sigma_{jk} \mathrm{Tr}(b_i c_j) = \sigma_{ik}, \quad \det((\mathrm{Tr}(b_i b_k))) = \det((\sigma_{ik})),$$

$$\mathfrak{D}(M/R) = \det((\mathrm{Tr}(b_i b_k)))R = \det((\sigma_{ik}))R = (M^\perp :_R M). \qquad \square$$

The arithmetic radical $\mathrm{AR}(\Lambda)$ of the order Λ over R contains per definition the ideal $\mathfrak{D}(\Lambda/R)\Lambda$ such that the factorring is the Jacobson radical of the factorring of Λ over $\mathfrak{D}(\Lambda/R)\Lambda$:

$$\mathrm{AR}(\Lambda)/\mathfrak{D}(\Lambda/R)\Lambda = J(\Lambda/\mathfrak{D}(\Lambda/R)\Lambda). \tag{5.51}$$

In case Λ is commutative, or even if it is only known that the factorring of Λ over $\mathrm{AR}(\Lambda)$ is commutative it follows that the factorring $\Lambda/\mathrm{AR}(\Lambda)$ is the algebraic sum of finitely many fields each of which is a finite extension of a field $R/\mathfrak{p}$ where $\mathfrak{p}$ is one of the prime ideal divisors of $\mathfrak{D}(\Lambda/R)$. If R is local and if the maximal ideal $\mathfrak{p}$ of R is known, then $\mathrm{AR}(\Lambda)$ can be computed from the formula

$$\mathrm{AR}(\Lambda) = \{x \in \Lambda \mid \forall y \in \Lambda : \mathrm{Tr}(xy) \in \mathfrak{p}\} \tag{5.52}$$

provided that the characteristic of $R/\mathfrak{p}$ is either zero or greater than n. This follows from the remark that the regular representation of Λ over R leads to a representation of $\Lambda/\mathrm{AR}(\Lambda)$ of degree n over $R/\mathfrak{p}$ for which none of its multiplicities is divisible by the characteristic of $R/\mathfrak{p}$.

The arithmetic radical of a commutative order Λ is an indicator of the maximality or non-maximality of Λ according to

Lemma (5.53)

The commutative order Λ over the Dedekind ring R with discriminant ideal $\mathfrak{D}(\Lambda/R) \neq 0$ is maximal if and only if $[\mathrm{AR}(\Lambda)/\mathrm{AR}(\Lambda)] = \Lambda$.

Proof

We note that for any order Λ over R the discriminant ideal $\mathfrak{D}(\Lambda/R)$ is defined as the ideal of R generated by all expressions (5.50). But $\mathfrak{D}(\Lambda/R)$ is zero unless the regular trace bilinear form of $\mathfrak{Q}(\Lambda)$ over $\mathfrak{Q}(R)$ is non degenerate and in the latter case we have $\mathfrak{D}(\Lambda/R) \neq 0$.

Now let Λ be commutative. Then Λ is embedded into the maximal R-order

$$\Lambda_1 = \mathrm{Cl}(R, \mathfrak{Q}(\Lambda)) \tag{5.54a}$$

of $\mathfrak{Q}(\Lambda)$ so that $\Lambda \subseteq \Lambda' := [\mathrm{AR}(\Lambda)/\mathrm{AR}(\Lambda)] \subseteq \Lambda_1$. We assume that

$$\Lambda \subset \Lambda_1, \tag{5.54b}$$

and want to show that

$$\Lambda \subset \Lambda'. \tag{5.54c}$$

Since all concepts used are localizable we can assume without loss of generality that R has precisely one non-zero maximal ideal, say $\mathfrak{p}$. It follows that $\mathrm{AR}(\Lambda) = J(\Lambda)$ and that the exponent ideal of the R-module $\Lambda_1/J(\Lambda)$ is a $\mathfrak{p}$-power: $\mathfrak{n}((\Lambda_1/J(\Lambda))/R) = \mathfrak{p}^\lambda$, $\lambda \in \mathbb{Z}^{>0}$, $0 \subset \mathfrak{p}^\lambda\Lambda_1 \subseteq J(\Lambda)$, hence, $J(\Lambda/\mathfrak{p}^\lambda\Lambda_1) = J(\Lambda)/\mathfrak{p}^\lambda\Lambda_1$, the factoring of $J(\Lambda)$ over $\mathfrak{p}^\lambda\Lambda_1$ is nilpotent, hence there is a natural number μ satisfying $J(\Lambda)^\mu \subseteq \mathfrak{p}^\lambda\Lambda_1$, $J(\Lambda)^\mu\Lambda_1 \subseteq \mathfrak{p}^\lambda\Lambda_1^2 = \mathfrak{p}^\lambda\Lambda_1 \subseteq J(\Lambda)$. For $J(\Lambda)\Lambda_1 \subseteq J(\Lambda)$ we have $\Lambda_1 \subseteq \Lambda'$, $\Lambda' = \Lambda_1$, $\Lambda \subset \Lambda'$. For $J(\Lambda)\Lambda_1 \nsubseteq J(\Lambda)$ there is an index $\mu' \in \mathbb{Z}^{>1}$ satisfying $J(\Lambda)^{\mu'}\Lambda_1 \subseteq J(\Lambda)$, $J(\Lambda)^{\mu'-1}\Lambda_1 \nsubseteq J(\Lambda)$, so that there is an element x of $J(\Lambda)^{\mu'-1}\Lambda_1$ for which $x \notin J(\Lambda)$, $x/\mathfrak{p}^\lambda\Lambda_1$ is nilpotent, $x \notin \Lambda$, $xJ(\Lambda) \subseteq J(\Lambda)^{\mu'}\Lambda_1 \subseteq J(\Lambda)$, $x \in \Lambda'$, $\Lambda' \supset \Lambda$. □

Of course the lemma also yields a method of embedding the commutative order Λ into its maximal order (5.54a): Either $[\mathrm{AR}(\Lambda)/\mathrm{AR}(\Lambda)] = \Lambda$ and Λ itself is the maximal order Λ_1 or $\Lambda' = [\mathrm{AR}(\Lambda)/\mathrm{AR}(\Lambda)] \supset \Lambda$, in which case we continue with Λ' in place of Λ. However, in general the computation of $\mathrm{AR}(\Lambda)$ is too time and storage consuming. Even if $\mathrm{AR}(\Lambda)$ is already at hand, the computation of $[\mathrm{AR}(\Lambda)/\mathrm{AR}(\Lambda)]$ consumes more time and storage than the method presented in the next section as experience has shown. On the other hand, in the special case $\Lambda = \Lambda_f$ lemma (5.53) yields a very useful criterion which was already known to R. Dedekind.

Criterion (Dedekind) (5.55)

Let R be a local Dedekind ring with $\mathfrak{p} = \pi R$ *as its only non-zero maximal ideal and let f be a monic separable polynomial of* $R[t]$ *with* $n = \deg(f) > 0$. *Let*

$$f \equiv \prod_{i=1}^{n} g_i^i \bmod (\mathfrak{p}R[t]) \tag{5.55a}$$

be the congruence factorization into a power product of monic polynomials $g_1, \ldots, g_n$ *over R that are mutually prime and separable* mod$(\mathfrak{p}R[t])$. *Then we have*

$$J(\Lambda_f) = \mathfrak{p}\Lambda_f + \prod_{i=1}^{n} g_i(\xi)\Lambda_f, \tag{5.55b}$$

where

$$\xi = t/f(t)R[t], \quad \Lambda_f = \sum_{i=1}^{n} R\xi^{i-1}, \tag{5.55c}$$

and

$$f(t) - \prod_{i=1}^{n} g_i(t)^i = \pi h(t) \quad (h(t) \in R[t]). \tag{5.55d}$$

The equation order Λ_f *is maximal precisely if* $h(t)$ *is prime to* $\prod_{i=2}^{n} g_i(t)$ *modulo* $\mathfrak{p}R[t]$.

Proof

Because of (5.55a) it follows that the element $\eta = \prod_{i=1}^{n} g_i(\xi)$ satisfies the congruence $\eta^n \equiv 0 \bmod (\mathfrak{p}\Lambda_f)$ so that $\eta \in J(\Lambda)$. Hence the right-hand side of (5.55b) is contained in the left-hand side so that we have $g(\xi) \equiv 0 \bmod J(\Lambda_f)$ for $g = \prod_{i=1}^{n} g_i$. Since $\Lambda_{f/\mathfrak{p}R[t]} = \Lambda_f/\mathfrak{p}\Lambda_f$ it follows that the minimal polynomial of ξ modulo $\mathfrak{p}\Lambda_f$ equals $f \bmod (\mathfrak{p}R[t])$. On the other hand, it divides mod$(\mathfrak{p}R[t])$ any polynomial $j(t) \in R[t]$ satisfying $j(\xi) \equiv 0 \bmod \mathfrak{p}\Lambda_f$. Because of the nilpotency of $J(\Lambda_f)$ modulo $\mathfrak{p}\Lambda_f$ it follows that some power of the minimal polynomial of $\xi \bmod J(\Lambda_f)[t]$ will be contained in $\mathfrak{p}R[t]$. Hence, that minimal polynomial equals $g \bmod \mathfrak{p}R[t]$. Thus (5.55b) is established.

Any element x of $[J(\Lambda_f)/J(\Lambda_f)]$ satisfies the condition $x\mathfrak{p}\Lambda \subseteq J(\Lambda_f)$, therefore it is congruent to an element of the form $y = \pi^{-1}g(\xi)j_1(\xi)$ with $j_1(t) \in R[t]$, $\deg(j_1) < \deg(f) - \deg(g)$, modulo the R-module Λ_f. But it also satisfies the condition $yg(\xi) = \pi^{-1}g^2(\xi)j_1(\xi) \in J(\Lambda_f)$. Hence, $j_1 \equiv j_2 \prod_{i=3}^{n} g_i^{i-2} \bmod \mathfrak{p}[t]$, where j_2 is a polynomial of $R[t]$ of degree less than $\sum_{i=2}^{n} \deg(g_i)$ so that y is congruent to an element z of the form $z = \pi^{-1}j_2(\xi)g_1(\xi)g_2(\xi)\prod_{i=3}^{n} g_i(\xi)^{i-1}$ modulo the R-submodule Λ_f of $[J(\Lambda_f)/J(\Lambda_f)]$. Now we obtain the condition

$$zg(\xi) = \pi^{-1}j_2(\xi)g_1(\xi)\prod_{i=1}^{n} g_i(\xi)^i = j_2(\xi)g_1(\xi)h(\xi) \in J(\Lambda_f)$$

which can always be satisfied in a non-trivial way unless h is prime to g_i

modulo $\mathfrak{p}[t]$ for all $i = 2, 3, \ldots, n$. □

The Dedekind criterion turns out to be extremely useful, since experience has shown (see [3], [4], for example) that most local equation orders are maximal and the application of the criterion weeds out all those happenings with little computational effort (see also example (6.8)).

For the remaining cases it was first discovered by R. Land experimentally that the reduced discriminant can be used in many situations in place of the discriminant with much less computations involved. For example, the arithmetic radical of the R-order Λ can also be defined as the intersection of all maximal ideals of Λ containing $\mathfrak{D}_0(\Lambda/R)\Lambda$. This is because the exponent ideal of finitely generated modules over a commutative ring R is contained in the same maximal ideals of R as the order ideal, due to the existence of the elementary divisor normal form of finitely generated modules over Krull valuation rings.

The computation of the reduced discriminant usually requires a greater effort than the computation of the discriminant ideal itself. However, in the important case of equation orders we have

Lemma (5.56)

Let R be a Dedekind ring and $f(t) \in R[t]$ be monic, separable of degree $n > 1$. Then the Euclidean algorithm in $\mathfrak{Q}(R)[t]$ yields an equation

$$Xf + YD_t(f) = 1, \tag{5.57a}$$

with polynomials

$$X = X(f, D_t(f)) \in \mathfrak{Q}(R)[t], \quad Y = Y(f, D_t(f)) \in \mathfrak{Q}(R)[t], \tag{5.57b}$$

uniquely determined by the degree conditions

$$\deg(X) < \deg(D_t(f)), \quad \deg(Y) < n = \deg(f). \tag{5.57c}$$

Then the R-fractional ideal generated by the coefficients of X, Y is the inverse of the reduced discriminant ideal.

Proof

Because of the localizability of the concepts used it suffices to deal with the case that R has just one non-zero maximal ideal, say $\mathfrak{p} = \pi R$. Since both polynomials f, $D_t(f)$ have coefficients in R it follows that the coefficients of X, Y generate an ideal of the form $\mathfrak{p}^{-\lambda}$ with $\lambda \in \mathbb{Z}^{\geq 0}$, thus we have an equation

$$X_1 f + Y_1 D_t(f) = \pi^\lambda, \tag{5.57d}$$

where X_1, Y_1 are polynomials of $R[t]$ with coefficients that are not all divisible by π. The equation (5.57d) implies that

$$Y_1(\xi)D_t(f)(\xi) = \pi^\lambda \quad (\xi = t/f(t)R[t] \in \Lambda_f). \tag{5.57e}$$

We have already seen earlier that $D_t(f)(\xi)\Lambda_f^{\perp} \subseteq \Lambda_f$, hence $\pi^\lambda\Lambda_f^{\perp} \subseteq \Lambda_f$, $\pi^\lambda \in \mathfrak{D}_0(\Lambda_f/R)$.

Conversely, if $\pi^{\lambda'} \in \mathfrak{C}_0(\Lambda_f/R)$ ($\lambda' \in \mathbb{Z}^{\geq 0}$, $\lambda' \leq \lambda$), then we have $\pi^{\lambda'}\ \Lambda_f^{\perp} \subseteq \Lambda_f$. But we also saw that $D_t(f)(\xi)^{-1} \in \Lambda_f^{\perp}$, hence $\pi^{\lambda'} D_t(f)(\xi)^{-1} \in \Lambda_f$, so that there holds an equation $Y_2(\xi)D_t(f)(\xi) = \pi^{\lambda'}$, where $Y_2(t) \in R[t]$ is of degree less than n. Hence, there also holds an equation $X_2(t)f(t) + Y_2(t)D_t(f)(t) = \pi^{\lambda'}$ with $X_2(t) \in R[t]$ of degree less than $\deg(D_t(f))$. Since (5.57a, b) are unique under the degree condition (5.57c) it follows that $X_2 = X_1$, $Y_2 = Y_1$, $\lambda = \lambda'$. □

As a consequence of lemma (5.56) we obtain the reduced discriminant of a separable monic polynomial f by the usual Euclidean division algorithm applied to $f, D_t(f)$ over $\mathfrak{Q}(R)$ and a simple computation with the coefficients of $X(f, D_t(F))$, $Y(f, D_t(f))$ whereas the discriminant computation of f needs pseudo-division, hence a much more careful inspection of each division step.

8. Structural stability

Lemma (5.58)

The embedding of an order Λ over a Dedekind ring R into an R-overorder Λ_1 of the same R-rank is stable in the following sense: Let $\mathfrak{a}$ be an ideal of R contained in the square of the exponent ideal $\mathfrak{n} = \mathfrak{n}((\Lambda_1/\Lambda)/R) \neq 0$ of Λ_1/Λ over R. Let $\tilde{\Lambda}$ be an R-order and let $\sigma: \Lambda \to \tilde{\Lambda}$ be an R-isomorphism of the R-module Λ on the R-module $\tilde{\Lambda}$ satisfying the congruence condition $\sigma(xy) \equiv \sigma(x)\sigma(y) \bmod (\mathfrak{a}\tilde{\Lambda})$ $(x, y \in \Lambda)$. Then there is a unique extension $\tau: A \to \tilde{A}$ of σ to a $\mathfrak{Q}(R)$-isomorphism of the $\mathfrak{Q}(R)$-module $A = \mathfrak{Q}(R) \otimes_R \Lambda$ on the $\mathfrak{Q}(R)$-module $\tilde{A} = \mathfrak{Q}(R) \otimes_R \tilde{\Lambda}$ such that the restriction of τ to Λ_1 yields an R-overorder $\tilde{\Lambda}_1 = \tau(\Lambda_1)$ of $\tilde{\Lambda}$ in its central quotient ring $\tilde{A}$.

Proof

Since Λ is a torsion-free R-module the unique extendability of σ to τ is obvious. Let $\tilde{\Lambda}_1 = \tau(\Lambda_1)$. Because of $\mathfrak{n}\Lambda_1 \subseteq \Lambda$ it follows that

$$\mathfrak{n}\tilde{\Lambda}_1 \subseteq \tilde{\Lambda}.$$

Let $\tilde{x}_1, \tilde{y}_1$ be any two elements of $\tilde{\Lambda}_1$. By definition there are elements x_1, y_1 of Λ_1 satisfying $\tau(x_1) = \tilde{x}_1$, $\tau(y_1) = \tilde{y}_1$. Now $\mathfrak{n}\tilde{\Lambda}_1 \subseteq \tilde{\Lambda}$ implies for any two elements λ, μ of $\mathfrak{n}$:

$$x := \lambda x_1 \in \Lambda, \quad y := \mu y_1 \in \Lambda, \quad \sigma(x) = \tau(x) = \lambda\tilde{x}_1 \in \tilde{\Lambda}, \quad \sigma(y) = \tau(y) = \mu\tilde{y}_1 \in \tilde{\Lambda}$$

and by assumption

$$\begin{aligned}\mathfrak{a}\tilde{\Lambda} \ni \sigma(x)\sigma(y) - \sigma(xy) &= \tau(x)\tau(y) - \tau(xy) = \tau(\lambda x_1)\tau(\mu y_1) - \tau((\lambda x_1)(\mu y_1)) \\ &= \lambda\mu(\tau(x_1)\tau(y_1) - \tau(x_1 y_1)) = \lambda\mu(\tilde{x}_1\tilde{y}_1 - \tilde{z}_1),\end{aligned}$$

where $x_1 y_1 = z_1 \in \Lambda_1$, $\tau(z_1) = \tilde{z}_1 \in \tau(\Lambda_1) = \tilde{\Lambda}_1$. Hence,

$$\mathfrak{a}\tilde{\Lambda} \supseteq \mathfrak{n}^2(\tilde{x}_1\tilde{y}_1 - \tilde{z}_1), \quad \mathfrak{n}^{-2}\mathfrak{a}\tilde{\Lambda} \ni \tilde{x}_1\tilde{y}_1 - \tilde{z}_1.$$

By assumption we have $\mathfrak{a} \subseteq \mathfrak{n}^2 \neq 0$, hence

$$\mathfrak{n}^{-2}\mathfrak{a} \subseteq R, \quad \tilde{\Lambda} \supseteq \mathfrak{n}^{-2}\mathfrak{a}\tilde{\Lambda}, \quad \tilde{x}_1\tilde{y}_1 - \tilde{z}_1 \in \tilde{\Lambda}, \quad \tilde{x}_1\tilde{y}_1 \in \tilde{\Lambda}_1 + \tilde{\Lambda} = \tilde{\Lambda}_1. \qquad \square$$

The application of lemma (5.58) is as follows. Let Λ_f be the equation order of the monic separable polynomial $f(t)$ of positive degree n over the Dedekind ring R and let $\tilde{f}(t) \in R[t]$ be another monic separable polynomial satisfying

$$d_0(f)/d_0(\tilde{f}) \in U(R), \tag{5.59a}$$

$$\tilde{f} \equiv f \bmod (d_0(f)^2 R). \tag{5.59b}$$

We note that the reduced discriminant ideal $d_0(f)R$ is contained in the exponent ideal of the factormodule of the integral closure $\mathrm{Cl}(R, A_f)$ in the central quotient ring $\mathfrak{Q}(\Lambda_f) = \mathfrak{Q}(R)[t]/f(t)\mathfrak{Q}(R)[t] = A_f$ over Λ_f. We remark that $\mathrm{Cl}(R, A_f)$ consists of certain linear combinations $\sum_{i=1}^{n} \lambda_i \xi^{i-1} \in \mathfrak{Q}(R)[\xi]$ for $\xi = t/f(t)R[t]$. According to the lemma applied in both directions (from $\Lambda = \Lambda_f$ to $\tilde{\Lambda} = \Lambda_{\tilde{f}}$ and from $\tilde{\Lambda}$ to Λ) it follows that the maximal order $\mathrm{Cl}(R, A_{\tilde{f}})$ of $\Lambda_{\tilde{f}}$ consists of the very same linear combinations $\sum_{i=1}^{n} \lambda_i \tilde{\xi}^{i-1} \in \mathfrak{Q}(R)[\tilde{\xi}]$ for $\tilde{\xi} = t/\tilde{f}(t)R[t]$ as are used to compute $\mathrm{Cl}(R, A_f)$. In other words: the task of embedding Λ_f into its maximal order is equivalent to the task of embedding $\Lambda_{\tilde{f}}$ into its maximal order. Regarding the condition (5.59a) on the reduced discriminants of $f, \tilde{f}$ it is a consequence of (5.59b) in case R is semilocal so that $J(R) \ni d_0(f)$. In fact, we have the stronger statement:

Proposition (5.60)
Let $f(t)$ be a monic separable polynomial of positive degree n over the semilocal Dedekind ring R such that $d_0(f)$ is contained in the Jacobson radical $J(R)$ of R. Then any monic polynomial $\tilde{f}(t) \in R[t]$ satisfying the congruence condition

$$\tilde{f} \equiv f \bmod (d_0(f)J(R)) \tag{5.60a}$$

is separable over R such that (5.59a) *holds.*

Proof
We observe that

$$d_0(f)R = (R[t]f(t) + R[t]D_t(f)(t)) \cap R. \tag{5.61}$$

This is because by definition $d_0(f)$ is uniquely presentable as a linear combination

$$d_0(f) = Xf + YD_t(f), \tag{5.62a}$$

with polynomials $X, Y \in R[t]$, such that

$$\deg(X) < \deg(D_t(f)), \quad \deg(Y) < n, \tag{5.62b}$$

and the greatest common divisor of the coefficients of X, Y is one. Hence the element $d_0(f)$ of R is contained in the ideal of $R[t]$ generated by $f, D_t(f)$.

On the other hand, any presentation $a = X_1 f + Y_1 D_t(f)$ $(X_1, Y_1 \in R[t])$ of an element $a \in R$ gives rise to a division with remainder $Y_1 = Q(Y_1, f)f + R(Y_1, f)$ $(Q(Y_1, f), R(Y_1, f) \in R[t], \deg(R(Y_1, f)) < n)$ and to the equation

$$a = X_2 f + Y_2 D_t(f), \quad X_2 = Q(Y_1, f)D_t(f) + X_1, \quad Y_2 = R(Y_1, f) \qquad (5.63a)$$

with $\deg(Y_2) < n$, hence

$$\deg(X_2) < \deg(D_t(f)). \qquad (5.63b)$$

Because of the uniqueness of the presentation

$$1 = d_0(f)^{-1} X f + d_0(f)^{-1} Y D_t(f) \qquad (5.64a)$$

derived from (5.62a) in $\mathfrak{Q}(R)[t]$ and in view of the degree conditions

$$\deg(d_0(f)^{-1} X) < \deg(D_t(f)), \quad \deg(d_0(f)^{-1} Y) < n \qquad (5.64b)$$

derived from (5.62b) it follows from (5.63a, b) that $X_2 = \lambda X$, $Y_2 = \lambda Y$ with λ in $\mathfrak{Q}(R)$. In fact λ is the greatest common divisor of the coefficients of X_2, Y_2 over R. Since by construction both X_2, Y_2 are in $R[t]$ it follows that λ is in R, $d_0(f)$ divides a, (5.61) is established.

Because of the invariance of the concepts used in proposition (5.60) under localization it suffices to prove it only for the case that R is a local Dedekind ring with just one non-zero maximal ideal $\mathfrak{p}$. According to (5.60a) there holds an equation

$$f = \bar{f} + d_0(f) g, \qquad (5.65)$$

where the polynomial $g(t)$ is in $\mathfrak{p}[t]$ with $\deg(g) < n$. Hence

$$D_t(f) = D_t(\bar{f}) + d_0(f) D_t(g), \qquad (5.66)$$

and upon substitution of (5.65), (5.66) in (5.62a) we obtain the equation

$$X\bar{f} + Y D_t(\bar{f}) = d_0(f)(1 - Xg - Y D_t(g)). \qquad (5.67)$$

If $\bar{f}$ is inseparable, then the greatest common divisor of $\bar{f}, D_t(\bar{f})$ is a non-constant monic polynomial of $R[t]$ dividing $d_0(f)$ modulo $(d_0(f)\mathfrak{p}[t])$, obviously a contradiction. Hence, $\bar{f}$ must be separable, say $d_0(\bar{f})R = \mathfrak{p}^\lambda R$ for some $\lambda \in \mathbb{Z}^{\geq 0}$,

$$d_0(\bar{f}) = \tilde{X}\bar{f} + \tilde{Y} D_t(\bar{f}) \qquad (5.68)$$

for some polynomials $\tilde{X}$, $\tilde{Y} \in R[t]$ with $\gcd(\tilde{X}, \tilde{Y}) = 1$. For $\lambda = 0$ we obtain $d_0(\bar{f}) | d_0(f)$. For $\lambda > 0$ we derive from (5.67) upon multiplication by $1 + \sum_{i=1}^{\lambda-1} (Xg + YD_t(g))^i$ and substitution of (5.68) an equation of the form $X_1 \bar{f} + Y_1 D_t(\bar{f}) = d_0(f)$ $(X_1, Y_1 \in R[t])$ which shows in view of (5.61) (for $\bar{f}$ in place of f) that $d_0(\bar{f}) | d_0(f)$. Similarly we show that $d_0(f) | d_0(\bar{f})$. (Note that (5.59b) also holds with f and $\bar{f}$ reversed.) Hence (5.59a) is proved. □

9. Reducible polynomials

If the monic separable polynomial $f(t)$ over the Dedekind ring R permits a factorization $f = f_1 f_2$ into the product of two monic non-constant polynomials $f_1, f_2 \in \mathfrak{Q}(R)[t]$, then both factors already belong to $R[t]$. This is because for the universal splitting ring $A = S(f_1/S(f_2/\mathfrak{Q}(R)))$ the generating root symbols x_{hi} $(1 \leqslant i \leqslant \deg(f_h) =: n_h, h = 1,2)$ entering the defining factorizations $f_h(t) = \prod_{i=1}^{n_h}(t - x_{hi})$ $(h = 1,2)$ satisfy the monic equation $f(x_{hi}) = 0$ over R so that they belong to the integral closure $\mathrm{Cl}(R, A)$ and hence the coefficients of f_h belong to the intersection of $\mathrm{Cl}(R, A)$ with $\mathfrak{Q}(R)$ which is R because of the integral closure property of Dedekind rings.

Due to the separability of f it follows that both f_1, f_2 are separable and mutually prime in $\mathfrak{Q}(R)[t]$ so that we have the algebraic decomposition $A_f = \mathfrak{Q}(R)[t]/\!/f(t)\,\mathfrak{Q}(R)[t] = A_{f_1} \oplus A_{f_2}$ with $A_{f_h} = \mathfrak{Q}(R)[t]/\!/f_h(t)\,\mathfrak{Q}(R)[t]$ $(h = 1,2)$. As was shown before we obtain the idempotents $e_h = 1_{A_{f_h}}$ $(h = 1,2)$ serving to define $A_{f_h} = e_h A_f$ by means of the Euclidean division algorithm applied to f_1, f_2 over $\mathfrak{Q}(R)$ leading to an equation $a_1 f_1 + a_2 f_2 = 1$ with $a_h \in \mathfrak{Q}(R)[t]$ and to $e_h = a_{3-h}(\xi) f_{3-h}(\xi)$ for $\xi = t/\!/f(t)\mathfrak{Q}(R)[t]$ $(h = 1,2)$. It follows that $\mathrm{Cl}(R, A_f) = \mathrm{Cl}(Re_1, A_{f_1}) \oplus \mathrm{Cl}(Re_2, A_{f_2})$.

Let us suppose that we have solved the embedding problem of the R-order Λ_{f_h} into the maximal order $\bar{\Lambda}_{f_h} := \mathrm{Cl}(R, A_{f_h})$ already, say by means of establishing an R-basis of the form $\omega_{h1}, \ldots, \omega_{hn_h}$ such that

$$\omega_{hi} = \sum_{j=1}^{k} \beta_{hij} \xi_h^{j-1} (\beta_{hij} \in \mathfrak{Q}(R), 1 \leqslant j \leqslant k, 1 \leqslant i \leqslant n_h, h = 1,2), \tag{5.69}$$

with $\xi_h = t/\!/f_h(t)\,\mathfrak{Q}(R)[t]$. Then we set $\xi_h = e_h \xi$ $(h = 1,2)$ and use the $n_1 + n_2 = n = \deg(f)$ elements ω_{hi} $(1 \leqslant i \leqslant n_h, h = 1,2)$ as R-basis of $\bar{\Lambda}_f = \mathrm{Cl}(R, A_f)$. Of course, the new R-basis does not have the canonical form of an R-basis $\omega_1, \ldots, \omega_n$ of $\bar{\Lambda}_f$ for which we demand that

$$\omega_i = \sum_{j=1}^{i} \beta_{ij} \xi^{j-1} \quad (\beta_{ij} \in \mathfrak{Q}(R), 1 \leqslant j \leqslant k, 1 \leqslant i \leqslant n). \tag{5.70}$$

But, provided that R is a principal ideal ring, it is always possible by means of presenting $\xi_h = a_{3-h}(\xi) f_{3-h}(\xi) \xi$ in the normal form

$$\xi_h = \sum_{j=1}^{n} \alpha_{hj} \xi^{j-1} \quad (\alpha_{hj} \in \mathfrak{Q}(R), 1 \leqslant j \leqslant n, h = 1,2) \tag{5.71}$$

and substitution of (5.71) into (5.69) to present the basis $\omega_i' = \omega_{1i}$ $(1 \leqslant i \leqslant n_1)$, $\omega_{n_1+i}' = \omega_{2i}$ $(1 \leqslant i \leqslant n_2)$ in the normal form $\omega_i' = \sum_{j=1}^{n} \beta_{ij}' \xi^{j-1}$. By means of Hermite row reduction of the quadratic matrix (β_{ij}') we obtain a reduced matrix (β_{ij}) with all entries above the diagonal being zero. This leads to a canonical R-basis (5.70) of $\bar{\Lambda}_f$.

This construction will be tautly adopted in step 4 of section 6.

10. The Hensel lemma

The structural stability lemma (5.58) of course applies to any situation in which the monic separable polynomial $f(t)$ of positive degree n over the semilocal ring R is modified modulo a suitable ideal $\mathfrak{a}$ of R contained in $J(R)$ to a monic polynomial $\tilde{f}$, say $\tilde{f} \equiv f \bmod \mathfrak{a}[t]$. Whenever $\mathfrak{a}$ is contained in $d_0(f)^2R$ we are entitled to use a canonical R-basis $\tilde{\omega}_i$ $(1 \leqslant i \leqslant n)$ of $\mathrm{Cl}(R, A_{\tilde{f}})$ of the form

$$\tilde{\omega}_i = \sum_{j=1}^{i} \beta_{ij}\tilde{\xi}^{j-1} \quad (A_{\tilde{f}} = \mathfrak{Q}(R)[t]/\tilde{f}(t)\mathfrak{Q}(R)[t], \tilde{\xi} = t/\tilde{f}(t)\mathfrak{Q}(R)[t],$$
$$\beta_{ij} \in \mathfrak{Q}(R), 1 \leqslant j \leqslant i, 1 \leqslant i \leqslant n) \tag{5.72}$$

in order to produce the canonical R-basis ω_i $(1 \leqslant i \leqslant n)$ of $\mathrm{Cl}(R, A_f)$ of the form

$$\omega_i = \sum_{j=1}^{i} \beta_{ij}\xi^{j-1} \quad (1 \leqslant i \leqslant n, A_f = \mathfrak{Q}(R)[t]/f(t)\mathfrak{Q}(R)[t], \xi = t/f(t)\mathfrak{Q}(R)[t]). \tag{5.73}$$

The main application is made in case a congruence factorization

$$f \equiv f_{10}f_{20} \bmod \mathfrak{b}[t] \tag{5.74}$$

of f is known modulo an ideal $\mathfrak{b}$ of R contained in $J(R)$ such that f_{10}, f_{20} are two non-constant monic polynomials for which an equation

$$a_{10}f_{10} + a_{20}f_{20} = 1 + a_{00} \quad (a_{i0} \in R[t], 0 \leqslant i \leqslant 2, a_{00} \in \mathfrak{b}[t]) \tag{5.75}$$

is given which expresses in a constructive manner the idea that f_{10}, f_{20} are mutually prime modulo $\mathfrak{b}[t]$.

It is evident from the description that an immediate application of the structural stability lemma and of the results of subsection 7 on reducible polynomials is out of the question since the ideal $\mathfrak{b}$ may not be contained in $d_0(f)^2R$. It becomes therefore necessary first to raise the congruence modulo $\mathfrak{b}$ to a suitable power of $\mathfrak{b}$ contained in $d_0(f)^2R$. That this can always be done is the assertion of

Hensel's lemma (5.76)

Let R be a unital commutative ring, $\mathfrak{b}$ an ideal of R, and $f, f_{10}, f_{20} \in R[t]$ monic non-constant subject to (5.74), (5.75). Then for every $k \in \mathbb{N}$ there holds a congruence factorization

$$f \equiv f_{1k}f_{2k} \bmod \mathfrak{b}^{2^k}[t] \tag{5.76a}$$

($f_{1k}, f_{2k} \in R[t]$ monic non-constant) satisfying the coherence condition

$$f_{ik} \equiv f_{i0} \bmod \mathfrak{b}[t] \quad (i = 1, 2) \tag{5.76b}$$

and an equation

$$a_{1k}f_{1k} + a_{2k}f_{2k} = 1 + a_{ok}$$
$$(a_{ik} \in R[t], \deg(a_{ik}) < \deg(f_{3-i,k}), i = 1, 2, a_{ok} \in \mathfrak{b}^{2^k}[t]). \tag{5.76c}$$

Proof

We show how to obtain the result for $k=1$. The rest is done by induction on k.

We try to obtain f_{i1} in the form

$$f_{i1} = f_{i0} + d_{i0} \quad (d_{i0} \in \mathfrak{b}[t], \deg(d_{i0}) < \deg(f_{i0}), \quad i = 1,2) \tag{5.77}$$

which already meets the coherence condition (5.76b) for $k=1$. The congruence condition (5.76a) for $k=1$ then becomes $d_0 := f - f_{10}f_{20} \equiv f_{10}d_{20} + f_{20}d_{10} \bmod \mathfrak{b}^2[t]$, which is essentially met by setting $d_{i0}^* := a_{3-i,0}d_0$ $(i=1,2)$ because of (5.74), (5.75). However, the degree condition requires that we replace d_{i0}^* by its remainder upon division by f_{i0}: $d_{i0} := R(d_{i0}^*, f_{i0})$. We note that both quotient $Q(d_{i0}^*, f_{i0})$ and remainder d_{i0} are in $\mathfrak{b}[t]$ $(i=1,2)$. Thus we get

$$\begin{aligned} d_1 &:= f - f_{11}f_{21} = f - (f_{10}+d_{10})(f_{20}+d_{20}) = d_0 - f_{10}d_{20} - f_{20}d_{10} - d_{10}d_{20} \\ &= d_0 - f_{10}d_{20}^* - f_{20}d_{10}^* - d_{10}d_{20} + f_{10}f_{20}(Q(d_{10}^*, f_{10}) + Q(d_{20}^*, f_{20})), \end{aligned}$$

where we already know that the first term on the right-hand side is in $\mathfrak{b}^2[t]$. On the other hand, the left-hand side is of degree less than $\deg(f) = \deg(f_{11}) + \deg(f_{12})$. *A fortiori*, the left-hand side is of degree less than $\deg(f_{11}) + \deg(f_{12})$ modulo $\mathfrak{b}^2[t]$. Hence the same is true for the right-hand side. But $f_{10}f_{20}$ is monic of degree $\deg(f)$ modulo $\mathfrak{b}^2[t]$ yielding $Q(d_{10}^*, f_{10}) + Q(d_{20}^*, f_{20}) \in \mathfrak{b}^2[t]$. Hence (5.76a) is satisfied for $k=1$.

Analogously we try to solve (5.76c) for $k=1$ by setting

$$a_{i1} = a_{i0} + b_{i0} \quad (b_{i0} \in R[t], i = 1,2). \tag{5.78}$$

This leads to the condition

$$\begin{aligned} a_{01} &:= (a_{10} + b_{10})(f_{10} + d_{10}) + (a_{20} + b_{20})(f_{20} + d_{20}) - 1 \\ &= (a_{00} + b_{10}f_{10} + b_{20}f_{20} + a_{10}d_{10} + a_{20}d_{20}) \\ &\quad + (b_{10}d_{10} + b_{20}d_{20}) \in \mathfrak{b}^2[t] \end{aligned}$$

which is solved by setting

$$b_{i0} = -a_{i0}(a_{00} + a_{10}d_{10} + a_{20}d_{20}) \in \mathfrak{b}[t] \quad (i = 1,2). \tag{5.79}$$

Of course, the solution a_{i1} $(0 \leqslant i \leqslant 2)$ of (5.76c) obtained in this way will in general not yet satisfy the degree condition. Hence we replace a_{i1} by $R(a_{i1}, f_{3-i,1}) =: a'_{i1}$ $(i=1,2)$. Upon substitution into (5.76c) for $k=1$ we obtain

$$(Q(a_{11}, f_{21}) + Q(a_{21}, f_{11}))f_{11}f_{21} + (a'_{11}f_{11} + a'_{21}f_{21} - 1) \in \mathfrak{b}^2[t],$$

hence $Q(a_{11}, f_{21}) + Q(a_{21}, f_{11}) \in \mathfrak{b}^2[t]$, also implying $a'_{11}f_{11} + a'_{21}f_{21} - 1 \in \mathfrak{b}^2[t]$. Therefore we meet the degree condition by substituting $a_{i1} \leftarrow a'_{i1}$, $a_{01} \leftarrow a'_{11}f_{11} + a'_{21}f_{21} - 1$ $(i=1,2)$. □

11. Localization

Throughout subsections 1–10 of this section we have used the localization argument which is based on the observation that for transition from a Dedekind ring R as base ring to the Dedekind ring R/S for any subsemigroup S of $R\backslash\{0\}$ the concept of orders, arithmetic radicals, discriminant ideals, reduced discriminant ideals etc. remains invariant. A slightly different form of localization is introduced by means of

Definition (5.80)
Let R be a Dedekind ring, Λ an R-order, and $\mathfrak{a}$ a non-zero ideal of R. The R-overorder Λ_1 of Λ is said to be an $\mathfrak{a}$-**overorder** *if $\mathfrak{a}^\lambda\Lambda_1 \subseteq \Lambda$ for some $\lambda \in \mathbb{N}$.*

It follows that an $\mathfrak{a}$-overorder of Λ has the same R-rank as Λ. Both the order and the exponent ideal of the R-module Λ_1/Λ contain some power of $\mathfrak{a}$. The intersection of the members of any system of $\mathfrak{a}$-overorders is an $\mathfrak{a}$-overorder. If the subring generated by two $\mathfrak{a}$-overorders of Λ is itself an R-order (as is the case if Λ is commutative) then it is an $\mathfrak{a}$-overorder of Λ. If the non-zero ideal $\mathfrak{b}$ of R is contained in every prime ideal of R containing $\mathfrak{a}$ then every $\mathfrak{a}$-overorder of Λ also is a $\mathfrak{b}$-overorder. For $\mathfrak{D}(\Lambda/R) \neq 0$ every R-overorder of Λ of the same R-rank is a $\mathfrak{D}(\Lambda/R)$-overorder and also a $\mathfrak{D}_0(\Lambda/R)$-overorder.

The connection with localization is established by

Lemma (5.81)
Let R be a Dedekind ring, $\mathfrak{a}$ a non-zero ideal of R, $S_\mathfrak{a}$ the subsemigroup of all elements $x \in R$ satisfying $xR + \mathfrak{a} = R$ (i.e. $x/\mathfrak{a}$ is a unit of $R/\mathfrak{a}$). Let Λ be an R-order. Then for every $\mathfrak{a}/S_\mathfrak{a}$-overorder $\bar\Lambda_1$ of $\Lambda/S_\mathfrak{a}$ of the same rank there is the $\mathfrak{a}$-overorder $\bar\Lambda_1 \wedge \Lambda = \Lambda_1 := \{x \in \bar\Lambda_1 \mid \exists \lambda \in \mathbb{Z}^{>0}: \mathfrak{a}^\lambda x \subseteq \Lambda\}$ such that

$$\bar\Lambda_1 = \frac{\Lambda_1}{S_\mathfrak{a}}. \tag{5.82}$$

Conversely, for every $\mathfrak{a}$-overorder Λ_1 of Λ we find the $\mathfrak{a}/S_\mathfrak{a}$-overorder $\Lambda_1/S_\mathfrak{a}$ of $\Lambda/S_\mathfrak{a}$ such that $\Lambda_1/S_\mathfrak{a} \wedge \Lambda = \Lambda_1$.

Proof
The order property of $\bar\Lambda_1 \wedge \Lambda$ follows from the remark that the exponent ideal of the $R/S_\mathfrak{a}$-module $\bar\Lambda_1 \wedge \Lambda/S_\mathfrak{a}$ contains a power of $\mathfrak{a}/S_\mathfrak{a}$, say $(\mathfrak{a}/S_\mathfrak{a})^\mu\bar\Lambda_1 \subseteq \Lambda/S_\mathfrak{a}$ for some $\mu \in \mathbb{Z}^{>0}$. Hence for any x of $\bar\Lambda_1 \wedge \Lambda$ we have $\mathfrak{a}^\lambda x \subseteq \Lambda$ for some $\lambda \in \mathbb{Z}^{>0}$ and $(\mathfrak{a}^\mu/S_\mathfrak{a})x \subseteq \Lambda/S_\mathfrak{a}$ implying $\mathfrak{a}^\mu x \subseteq \Lambda$. It follows that $\mathfrak{a}^\mu\Lambda_1 \subseteq \Lambda$. For x, y of Λ_1 we have $\mathfrak{a}^\mu x \subseteq \Lambda$, $\mathfrak{a}^\mu y \subseteq \Lambda$, hence $\mathfrak{a}^\mu(x+y) \subseteq \Lambda$, $\mathfrak{a}^{2\mu}xy = (\mathfrak{a}^\mu x)(\mathfrak{a}^\mu y) \subseteq \Lambda$, $\mathfrak{a}^\mu xy \subseteq \Lambda$. Therefore Λ_1 is an $\mathfrak{a}$-overorder of Λ. By Landau's theorem (5.39d) there is an element $\alpha \neq 0$ of $\mathfrak{a}^\mu$ satisfying $\alpha\mathfrak{a}^{-\mu} + \mathfrak{a}^\mu = R$. For any element x of $\bar\Lambda_1$ we have $\mathfrak{a}^\mu x \subseteq \Lambda/S_\mathfrak{a}$, hence $\alpha x \in \Lambda/S_\mathfrak{a}$. Hence there is $\beta \in S_\mathfrak{a}$ satisfying $\alpha\beta x \in \Lambda$. By Landau's theorem there is an element γ of $\alpha\mathfrak{a}^{-\mu}$ for which

$\gamma\alpha\mathfrak{a}^{-\mu} + \alpha R = \alpha\mathfrak{a}^{-\mu}$, hence, $\gamma \in S_{\mathfrak{a}}$, $\mathfrak{a}^{\mu}\gamma\beta x \subseteq \Lambda$, $\gamma\beta \in S_{\mathfrak{a}}$, $\gamma\beta x \in \Lambda_1$, $x \in \Lambda_1/S_{\mathfrak{a}}$ so that the first part of (5.81) is established.

Conversely, let Λ_1 be an $\mathfrak{a}$-overorder of Λ. Then $\bar{\Lambda}_1 = \Lambda_1/S_{\mathfrak{a}}$ is an $\mathfrak{a}/S_{\mathfrak{a}}$-overorder of $\Lambda/S_{\mathfrak{a}}$. There is $\mu \in \mathbb{N}$ satisfying $\mathfrak{a}^{\mu}\Lambda_1 \subseteq \Lambda$, $(\mathfrak{a}^{\mu}/S_{\mathfrak{a}})\bar{\Lambda}_1 \subseteq \Lambda/S_{\mathfrak{a}}$. If for some element x of $\bar{\Lambda}_1$ and for some $\lambda \in \mathbb{Z}^{>0}$ we have $\mathfrak{a}^{\lambda}x \subseteq \Lambda$ then we obtain $\mathfrak{a}^{\mu}x \subseteq \Lambda$ and also $x = y/\alpha$ with $y \in \Lambda_1$, $\alpha \in S_{\mathfrak{a}}$. Hence, $\alpha x \in \Lambda_1$, $\mathfrak{a}^{\mu}x \subseteq \Lambda \subseteq \Lambda_1$, $Rx = (\alpha R + \mathfrak{a}^{\mu})x = \alpha Rx + \mathfrak{a}^{\mu}x \subseteq \Lambda_1, x \in \Lambda$. □

Definition (5.83)

Let R be a Dedekind ring, $\mathfrak{a}$ a non-zero ideal of R. Then the R-overorder Λ_1 of the R-order Λ is said to be **$\mathfrak{a}$-maximal,** *if Λ_1 is an $\mathfrak{a}$-overorder of Λ and if any $\mathfrak{a}$-overorder of Λ containing Λ_1 coincides with Λ_1.*

It follows from lemma (5.81) that the $\mathfrak{a}$-overorder Λ_1 of the R-order Λ is $\mathfrak{a}$-maximal precisely if $\Lambda_1/S_{\mathfrak{a}}$ is an $\mathfrak{a}/S_{\mathfrak{a}}$-maximal order.

If the discriminant ideal $\mathfrak{D}(\Lambda/R)$ of the R-order Λ is not zero then the R-overorder Λ_1 is maximal precisely if it is a $\mathfrak{D}(\Lambda/R)$-maximal R-overorder of Λ. If in that case $\mathfrak{a}$ is any non-zero ideal of R then any $\mathfrak{a}$-overorder Λ_1 of Λ also is an $(\mathfrak{a} + \mathfrak{D}(\Lambda/R))$-overorder of Λ. Among the $\mathfrak{a}$-overorders of Λ contained in the R-overorder Λ_1 of Λ precisely one is maximal, viz. $\Lambda_1/S_{\mathfrak{a}} \wedge \Lambda$.

Let μ be a natural number. For any collection of μ non-zero ideals $\mathfrak{a}_1, \ldots, \mathfrak{a}_{\mu}$ of R the overorders $\Lambda_1/S_{\mathfrak{a}_i} \wedge \Lambda (1 \leqslant i \leqslant \mu)$ generate the R-overorder $\Lambda_1/S_{\mathfrak{a}_1 \cdots \mathfrak{a}_{\mu}}$. If Λ_1 is a maximal R-overorder of Λ then Λ_1 contains a maximal R-overorder Λ_2 of Λ of the same R-rank as the R-rank of Λ. Furthermore, if $\mathfrak{a}_1, \ldots, \mathfrak{a}_{\mu}$ are comaximal ideals of R with the property that some power of $\mathfrak{a}_1 \cdots \mathfrak{a}_{\mu}$ is contained in $(\Lambda_2 :_R \Lambda)$ then we have

$$\Lambda_2 = \sum_{i=1}^{\mu} \frac{\Lambda_1}{S_{\mathfrak{a}_i}} \wedge \Lambda, \tag{5.84}$$

where there holds the direct R-module decomposition

$$\Lambda_2/\Lambda = \bigoplus_{i=1}^{\mu} \left(\frac{\Lambda_1}{S_{\mathfrak{a}_i}} \wedge \Lambda \right) \Big/ \Lambda. \tag{5.85}$$

This relation is the strongest expression of the localization argument in terms of the concept of $\mathfrak{a}$-overorders. It is used within the embedding algorithm of section 6 as follows.

Let a monic separable non-constant polynomial $f(t)$ be given over the Dedekind ring R. For the purpose of embedding the equation order Λ_f into its maximal order $\bar{\Lambda}_f = \mathrm{Cl}(R, \mathfrak{Q}(R)[t]/f(t)\mathfrak{Q}(R)[t])$ one determines certain non-zero elements $\delta_1, \ldots, \delta_{\mu}$ of R such that the principal ideals $\mathfrak{a}_j = \delta_j R$ $(1 \leqslant j \leqslant \mu)$ of R are comaximal and that $(\bar{\Lambda}_f :_R \Lambda_f)$ contains a power of $\mathfrak{a}_1 \cdots \mathfrak{a}_{\mu}$. The algorithm provides a canonical R-basis

$$\omega_{ij} = \sum_{k=1}^{i} \beta_{ijk}\xi^{k-1} \; (\beta_{ijk} \in \mathfrak{Q}(R), 1 \leqslant k \leqslant i, 1 \leqslant i \leqslant n, \xi = t/f(t)\mathfrak{Q}(R)[t]) \quad (5.86)$$

of the $\mathfrak{a}_j$-maximal R-overorder $\bar{\Lambda}_f/S_{\mathfrak{a}_j} \wedge \Lambda_f$ of Λ_f $(1 \leqslant j \leqslant \mu)$. The final task is then to provide a canonical R-basis $\omega_i = \sum_{k=1}^{i} \beta_{ik}\xi^{k-1}$ $(\beta_{ik} \in \mathfrak{Q}(R), 1 \leqslant k \leqslant i, 1 \leqslant i \leqslant n)$ of $\bar{\Lambda}_f$. But according to (5.85) the μn elements ω_{ij} of (5.86) provide a system of generators of $\bar{\Lambda}_f$ over R. Since ξ^{i-1} is contained in $\bar{\Lambda}_f$ as well as in $\bar{\Lambda}_f/S_{\mathfrak{a}_j} \wedge \Lambda_f$ for $j = 1, \ldots, \mu$, it follows that $\beta_{ii} \neq 0, \beta_{iji} \neq 0, \beta_{ii}^{-1} \in R, \beta_{iji}^{-1} \in R$, moreover $\beta_{iji}R$ contains some power of $\mathfrak{a}_i$ and β_{ii}^{-1} is equivalent to $\prod_{j=1}^{\mu} \beta_{iji}^{-1}$. Hence, it suffices to find a linear combination

$$\sum_{j=1}^{\mu} \lambda_{ij}\beta_{iji}^{-1} = \prod_{j=1}^{\mu} \beta_{iji}^{-1} \quad (5.87)$$

with coefficients λ_{ij} in R and to set $\omega_i = \sum_{j=1}^{\mu} \lambda_{ij}\omega_{ij}$ $(1 \leqslant i \leqslant n)$ in order to obtain a canonical R-basis of $\bar{\Lambda}_f$. For the purpose of solving (5.87) we form the elements $\gamma_{ij} = \prod_{\substack{h=1 \\ h \neq j}}^{\mu} \beta_{ihi}^{-1}$ of R and observe that

$$\sum_{j=1}^{\mu} \gamma_{ij}R = R. \quad (5.88)$$

Assuming the existence of a Euclidean division algorithm in R we use it to obtain elements λ_{ij} of R satisfying the equation

$$\sum_{j=1}^{\mu} \lambda_{ij}\gamma_{ij} = 1 \quad (5.89)$$

corresponding to (5.88). The equation (5.89) is tantamount with (5.87).

In the next section we shall refer to the construction just described as *amalgamation of canonical R-bases of the* $\mathfrak{a}_j$*-maximal overorders* $(1 \leqslant j \leqslant \mu)$ *to a canonical R-basis of* Λ_f.

Exercises

1. Let F_0 be a field. The *formal power series ring* in one variable t over F_0 is defined as the ring $F_0[[t]]$ of all formal sums $\sum_{i=0}^{\infty} a_i t^i$, where $(a_i | i \in \mathbb{Z}^{\geqslant 0})$ is any sequence of elements of F_0, with operational rules

$$\sum_{i=0}^{\infty} a_i t^i = \sum_{i=0}^{\infty} b_i t^i \Leftrightarrow a_i = b_i \forall i \in \mathbb{Z}^{\geqslant 0},$$

$$\sum_{i=0}^{\infty} a_i t^i + \sum_{i=0}^{\infty} b_i t^i = \sum_{i=0}^{\infty} (a_i + b_i) t^i,$$

$$\left(\sum_{i=0}^{\infty} a_i t^i\right)\left(\sum_{i=0}^{\infty} b_i t^i\right) = \sum_{i=0}^{\infty} \left(\sum_{j=0}^{i} a_j b_{i-j}\right) t^i.$$

(a) Show that $F_0[[t]]$ is an entire ring.
(b) Which sequence of elements of F_0 corresponds to the unit element, which to the zero element, which to the negative of an arbitrary element of $F_0[[t]]$?
(c) Show that the mapping $\iota: F_0[t] \to F_0[[t]]: \sum_{i=0}^{n} a_i t^i \mapsto \sum_{i=0}^{\infty} a_i t^i$ subject to $a_i = 0$ for $i > n$ is a monomorphism. It is said to be the canonical monomorphism of $F_0[t]$ into $F_0[[t]]$.
(d) Show that the unit group of $F_0[[t]]$ is formed by all power series $\sum_{i=0}^{\infty} a_i t^i$ with non-zero constant term a_0.
(e) The quotient field $F_0((t))$ of $F_0[[t]]$ consists of all *formal Laurent series* $L = \sum_{i=-\infty}^{\infty} a_i t^i$ ($a_i \in F_0, i \in \mathbb{Z}$); there is an index $\eta(L)$ for each Laurent series different from $0 = \sum_{i=-\infty}^{\infty} 0 t^i$ such that $\eta(L) \in \mathbb{Z}$ and $a_i = 0$ for $i < \eta(L)$. The operational rules for the Laurent series are obtained from those for the power series by substituting $-\infty$ for 0 below the summation symbol. Show that the mapping $\tilde{\iota}: F_0[[t]] \to F_0((t)): \sum_{i=0}^{\infty} a_i t^i \mapsto \sum_{i=-\infty}^{\infty} a_i t^i$ subject to $a_i = 0$ for $i < 0$ is a canonical monomorphism of $F_0[[t]]$ into $F_0((t))$.
(f) Show that $F_0[[t]]$ is a discrete Krull valuation ring of $F_0((t))$ for the exponent valuation $\eta: F_0((t)) \to \mathbb{Z} \cup \{\infty\}$ of part (e).

2. Let R be a Krull valuation ring of the field F. Let E be a finite extension of $\mathfrak{Q}(R)$.

(a) Show that the maximal ideal $\mathfrak{p}$ of R generates a proper ideal of $\mathrm{Cl}(R, E)$, and there is a finite basis $b_1, b_2, \ldots, b_{n'}$ of $\mathrm{Cl}(R, E)$ modulo $\mathfrak{p}\mathrm{Cl}(R, E)$.
(b) Show that for any such basis the elements $b_1, b_2, \ldots, b_{n'}$ are linearly independent over F.
(c) Show that if R is discrete and if $\mathrm{Cl}(R, E)$ is finitely generated over R, then the number n' equals the degree of E over F.

3. Let $F_0 = \mathbb{F}_p$ be the prime field of prime characteristic $p > 0$. Let F be the subfield of $\mathbb{F}_p((t))$ generated by t and by $x = \sum_{i=0}^{\infty} t^{pi^2}$. Let $R = \mathbb{F}_p((t)) \cap F$.

(a) Show that x is algebraically independent of t and that R is a Dedekind ring with F as quotient field.
(b) Let E be the subfield of $\mathbb{F}_p((t))$ generated by t and by $y = \sum_{i=0}^{\infty} t^{i^2}$. Show that E is a purely inseparable extension of degree p over F such that $E = F(y)$, $y^p = x$.
(c) Show that $\mathfrak{p} := tR$ is the maximal ideal of the local ring R.
(d) Show that $\mathrm{Cl}(R, E) = \mathfrak{p}\mathrm{Cl}(R, E) + R\mathfrak{p}$.
(e) Show that $\mathrm{Cl}(R, E)$ is not finitely generated over R.

4. Two ideals $\mathfrak{a}_1, \mathfrak{a}_2$ of the unital ring Λ are said to be *comaximal* if they satisfy $\mathfrak{a}_1 + \mathfrak{a}_2 = \Lambda$. Show that

(a) Any two distinct maximal ideals of Λ are comaximal.
(b) Two ideals $\mathfrak{a}_1, \mathfrak{a}_2$ of Λ are comaximal, if and only if there are elements $a_i \in \mathfrak{a}_i$ $(i = 1, 2)$ satisfying $a_1 + a_2 = 1$.
(c) Any two comaximal ideals $\mathfrak{a}_1, \mathfrak{a}_2$ of Λ satisfy

$$\mathfrak{a}_1\mathfrak{a}_2 = \mathfrak{a}_1 \cap \mathfrak{a}_2 = \mathfrak{a}_2\mathfrak{a}_1, \quad \Lambda/(\mathfrak{a}_1 \cap \mathfrak{a}_2) \simeq \Lambda/\mathfrak{a}_1 \oplus \Lambda/\mathfrak{a}_2.$$

(d) If the finitely many ideals $\mathfrak{a}_1, \ldots, \mathfrak{a}_s$ of Λ are pairwise comaximal then we have

$$\mathfrak{a}_1\mathfrak{a}_2\cdots\mathfrak{a}_s = \mathfrak{a}_1 \cap \mathfrak{a}_2 \cap \cdots \cap \mathfrak{a}_s, \quad \Lambda/(\mathfrak{a}_1 \cap \mathfrak{a}_2 \cap \cdots \cap \mathfrak{a}_s) \simeq \bigoplus_{i=1}^{s} \Lambda/\mathfrak{a}_i.$$

5. The intersection $J(\Lambda)$ of a ring Λ with its maximal ideals is called the *Jacobson radical* of Λ. Show that every ring epimorphism of Λ onto another ring Λ' maps $J(\Lambda)$ on $J(\Lambda')$.
6. If the Jacobson radical of the unital ring Λ is already the intersection of finitely many maximal ideals $\mathfrak{a}_1, \ldots, \mathfrak{a}_s$ of Λ then show that $\mathfrak{a}_1, \mathfrak{a}_2, \ldots, \mathfrak{a}_s$ are the only maximal ideals of Λ. In that case we have $\Lambda/J(\Lambda) \simeq \bigoplus_{i=1}^{s} \Lambda/\mathfrak{a}_i$, and the 2^s distinct ideals $\mathfrak{a}_{i_1} \cap \mathfrak{a}_{i_2} \cap \cdots \cap \mathfrak{a}_{i_r}$ $(0 \leqslant r \leqslant s,\ 1 \leqslant i_1 < i_2 < \cdots < i_r \leqslant s)$ of Λ are the only ideals containing $J(\Lambda)$.
7. For any two rings Λ_1, Λ_2 we have $J(\Lambda_1 \oplus \Lambda_2) = J(\Lambda_1) \oplus J(\Lambda_2)$.
8. A commutative ring is said to be *simple* if it is not nilpotent and if it contains no ideal other than itself and zero. Show that a commutative ring is simple if and only if it is a field.
9. A commutative ring is said to be *semisimple* if it contains no nilpotent ideal different from zero and if the intersection of finitely many maximal ideals is zero. Show that

   (a) A commutative ring $R \neq 0$ is semisimple if and only if it is isomorphic to the algebraic sum of finitely many fields.
   (b) A commutative F-algebra over the field F is semisimple if and only if it is the algebraic sum of finitely many extensions of F.
   (c) A unital commutative F-algebra of finite F-dimension is semisimple if and only if its Jacobson radical is zero.

10. (E. Noether) Let $\mathfrak{r} \neq 0$ be a non-nilpotent minimal right-ideal of the ring Λ. Show that

    (a) $\mathfrak{r}^2 = \mathfrak{r}$.
    (b) The left-annihilators of $\mathfrak{r}$ in $\mathfrak{r}$, i.e. the elements ρ of $\mathfrak{r}$ satisfying $\rho\mathfrak{r} = 0$, form a right-ideal of Λ which is contained in $\mathfrak{r}$.
    (c) The only left-annihilator of $\mathfrak{r}$ in $\mathfrak{r}$ is 0.
    (d) For any non-zero element ρ of $\mathfrak{r}$ we have $\rho\mathfrak{r} = \mathfrak{r}$.
    (e) For any non-zero element ρ of $\mathfrak{r}$ there is an element e of $\mathfrak{r}$ satisfying $\rho e = \rho$.
    (f) For the elements ρ, e of (e) we find that ρ is a left-annihilator of $e^2 - e$.
    (g) For the element e of (e) we find that $(e^2 - e)\Lambda \subset \mathfrak{r}$.
    (h) $\mathfrak{r} = e\Lambda = e\mathfrak{r} \ni e^2 = e \neq 0$.
    (i) What is the corresponding ('dual') statement for left-ideals?

11. Let F be a field. Show that

    (a) Every F-algebra H of finite F-dimension which is not nilpotent contains an idempotent. (Hint: use exercise 10.)
    (b) If the F-algebra H of finite F-dimension over F is a nilring then it is nilpotent. (Hint: use exercise 10.)

(c) The maximal nilradical $\mathrm{NR}(H)$ of an F-algebra H of finite F-dimension is nilpotent; it contains the Jacobson radical. Moreover, there are only finitely many maximal ideals of H containing $\mathrm{NR}(H)$, say $\mathfrak{a}_1,\ldots,\mathfrak{a}_s$, and we have $J(H/\mathrm{NR}(H)) = \mathrm{NR}(H/\mathrm{NR}(H)) = 0$,

$$H/\mathrm{NR}(H) \simeq H/\mathfrak{a}_1 \oplus \cdots \oplus H/\mathfrak{a}_s,$$
$$J(H/\mathfrak{a}_i) = \mathrm{NR}(H/\mathfrak{a}_i) = 0 \quad (1 \leqslant i \leqslant s).$$

(d) If H is a unital F-algebra of finite F-dimension then $J(H) = \mathrm{NR}(H)$ is the maximal nilpotent ideal of H.

(e) If H is a commutative F-algebra of finite F-dimension then $H/\mathrm{NR}(H)$ is isomorphic to an algebraic sum of finitely many fields.

12. Let R be a unital commutative ring and A, M two R-modules. The most general *R-bilinear form* is defined as an R-linear mapping $B: A \otimes_R A \to M$. The *orthogonal right B-complement* of any subset X of A is defined as the set $X^{\perp}$ of all elements y of A satisfying $B(x \otimes y) = 0$ for all x of X (notation: $B(X \otimes y) = 0$). The *orthogonal left B-complement* of X is defined as the set ${}^{\perp}X$ of all elements z of A satisfying $B(z \otimes x) = 0$ for all z *of* A (notation: $B(z \otimes X) = 0$). Show that

(a) $X^{\perp}, {}^{\perp}X$ are R-submodules of A.

(b) $X^{\perp} = (RX)^{\perp}$, ${}^{\perp}X = {}^{\perp}(RX)$.

(c) ${}^{\perp}(X^{\perp}) \supseteq X, ({}^{\perp}X)^{\perp} \supseteq X$.

(d) $({}^{\perp}(X^{\perp}))^{\perp} = X^{\perp}$, ${}^{\perp}(({}^{\perp}X)^{\perp}) = {}^{\perp}X$.

(e) $X \subseteq Y \Rightarrow X^{\perp} \supseteq Y^{\perp}$, ${}^{\perp}X \supseteq {}^{\perp}Y$.

(f) $(X \cup Y)^{\perp} = (RX + RY)^{\perp}$, ${}^{\perp}(X \cup Y) = {}^{\perp}(RX + RY)$.

(g) $(X \cap Y)^{\perp} \supseteq X^{\perp} + Y^{\perp}$, ${}^{\perp}(X \cap Y) \supseteq X^{\perp} + Y^{\perp}$.

(h) ${}^{\perp}(X^{\perp} \cap Y^{\perp}) = {}^{\perp}(X^{\perp}) + {}^{\perp}(Y^{\perp})$, $({}^{\perp}X \cap {}^{\perp}Y)^{\perp} = ({}^{\perp}X)^{\perp} + ({}^{\perp}Y)^{\perp}$.

(i) B is said to be *non-degenerate* if $A^{\perp} = {}^{\perp}A = 0$. B is said to be *symmetric* if $B(a \otimes b) = B(b \otimes a)$ for all a, b of A. B is said to be *antisymmetric* (*skew symmetric*) if $B(a \otimes b) = -B(b \otimes a)$ for all a, b of A. Prove: if B is symmetric or antisymmetric then we have $X^{\perp} = {}^{\perp}X$ for any subset X of A, and B induces a non-degenerate bilinear form $\bar{B}$ on $A/A^{\perp}$ upon setting $\bar{B}(a/A^{\perp} \otimes b/A^{\perp}) = B(a \otimes b)$ $(a, b \in A)$.

(j) If $b_1, b_2, \ldots, b_n$ is an R-generator set of A then B is symmetric if and only if the matrix $(B(b_i \otimes b_k))_{1 \leqslant i,k \leqslant n}$ is symmetric. B is antisymmetric if and only if that matrix is antisymmetric.

(k) If $b_1, \ldots, b_n$ is a finite R-basis of A and $M = R$ then B is non-degenerate if and only if $\det(B(b_i \otimes b_k))$ is a non-zero divisor of R.

(l) If M_1 is an R-submodule of M then B induces the R-bilinear from $\tilde{B}: A \otimes_R A \to M/M_1 : a \otimes b \mapsto B(a \otimes b)/M_1$.

(m) If A is a $\mathfrak{Q}(R)$-module then B is a $\mathfrak{Q}(R)$-bilinear form.

(n) Let A be a $\mathfrak{Q}(R)$-module with finite $\mathfrak{Q}(R)$-basis $b_1, \ldots, b_n$. Then the $\mathfrak{Q}(R)$-bilinear form B is non-degenerate if and only if $\det(B(b_i \otimes b_k))$ is a unit of $\mathfrak{Q}(R)$.

(o) Let A be a $\mathfrak{Q}(R)$-module with finite $\mathfrak{Q}(R)$-basis $b_1, \ldots, b_n$ such that $\det(B(b_i \otimes b_k))$ is a unit of $\mathfrak{Q}(R)$. Then there is the uniquely determined dual

$\mathfrak{Q}(R)$-basis $b_1^\perp, \ldots, b_n^\perp$ of A satisfying $B(b_i \otimes b_k^\perp) = \delta_{ik}$ $(1 \leqslant i, k \leqslant n)$, and we have $\Lambda^\perp = \sum_{k=1}^n Rb_k^\perp$ for $\Lambda = \sum_{k=1}^n Rb_k$. Note that ${}^\perp(\Lambda^\perp) = \Lambda$.

13. The same notations as in 12 are used. Let A be an R-ring.
    (a) Let $\Delta: A \to \mathfrak{Q}(R)^{d \times d}$ be an R-homomorphism of A into the ring of matrices of degree d over $\mathfrak{Q}(R)$. Show that the Δ-trace $\mathrm{Tr}_\Delta: A \to \mathfrak{Q}(R): x \mapsto \mathrm{Tr}(\Delta(x))$ is an R-linear form giving rise to the symmetric R-bilinear form $B_\Delta: A \otimes_R A \to \mathfrak{Q}(R): x \otimes y \mapsto \mathrm{Tr}_\Delta(xy)$ satisfying the admissibility condition $B_\Delta(x \otimes yz) = B_\Delta(xy \otimes z)$ $(x, y, z \in A)$.
    (b) Let $B: A \otimes_R A \to \mathfrak{Q}(R)$ be an admissible symmetric R-bilinear form on A. Then show that for any R-submodule Λ of A we find that $[\Lambda \backslash \Lambda]\Lambda^\perp \subseteq \Lambda^\perp$, ${}^\perp\Lambda[\Lambda/\Lambda] \subseteq {}^\perp\Lambda$.
14. Let R be a unital commutative ring. The R-module M is said to be *cyclic* if it can be generated by one element over R. Show that M is cyclic over R if and only if it is an R-epimorphic image of R.
15. Let R be a unital commutative ring. The R-module M is said to be *Noetherian* if every R-submodule of M can be finitely generated over R. Show that
    (a) Every R-submodule and every R-factormodule of a Noetherian R-module is Noetherian.
    (b) If both the R-submodule m of the R-module M and the R-factormodule M/m are Noetherian then M is Noetherian, too.
    (c) (Lasker–McCaulay) If R is Noetherian then every finitely generated R-module is Noetherian.
    (d) Show the converse of (c).
16. (Hilbert) Let R be a unital commutative ring. An ascending sequence $M_0 \subseteq M_1 \subseteq M_2 \subseteq \cdots$ of R-submodules of the R-module M is said to be a *filtration* of M if $M = \bigcup_{i=0}^\infty M_i$. The corresponding *grading* of M is defined as the algebraic sum $\bigoplus_{i=0}^\infty M_i'$ with $M_0' = M_0$, $M_i' = M_i/M_{i-1}$ $(i \in \mathbb{N})$.
    (a) Show that for any R-submodule m of M there is the filtration $m \cap M_0 \subseteq m \cap M_1 \subseteq m \cap M_2 \subseteq \cdots$ and the grading $\bigoplus_{i=0}^\infty m_i'$ with $m_0' = m \cap M_0$, $m_i' = m \cap M_i / m \cap M_{i-1} \subseteq M_i' (i \in \mathbb{N})$ induced by the given filtration and grading of M.
    (b) Show that there are the grading epimorphisms $\eta_0: M_0 \to M_0': u \mapsto u$, $\eta_i: M_i \to M_i': u \mapsto u/M_{i-1}$ $(i \in \mathbb{N})$.
    (c) Show that two submodules X, Y of M satisfying $X \subseteq Y$ coincide if and only if the given grading of M induces the same grading of X and of Y.
    (d) For every R-submodule m of M show there is the filtration $M_0/m \subseteq M_1/m \subseteq \cdots$ of M/m induced by the given filtration of M. Describe the corresponding grading.
    (e) The polynomial ring $R[t]$ in one variable t has the filtration $M_0 \subseteq M_1 \subseteq \cdots$, where M_i denotes the R-module formed by all polynomials of degree $\leqslant i$. Show that the filtration splits inasmuch as $M = R[t] = \bigoplus_{i=0}^\infty M_i'$ for $M_i' = Rt^i$ so that $\eta_i(\sum_{k=0}^i b_k t^k) = b_i t^i$.
    (f) For any ideal $\mathfrak{a}$ of $R[t]$ show we have $\eta_i(\mathfrak{a} \cap M_i) = \mathfrak{a}_i t^i$, where $\mathfrak{a}_i$ is an ideal of R

such that there is the filtration $\mathfrak{a}_0 \subseteq \mathfrak{a}_1 \subseteq \mathfrak{a}_2 \subseteq \cdots$ of the ideal $\bigcup_{i=0}^{\infty} \mathfrak{a}_i$ of R.

(g) If R is Noetherian then $R[t]$ is Noetherian.

17. (a) Show that the transition from one finite generator set $v_1, \ldots, v_n$ of the R-module M to another one can be done by a finite number of *elementary changes*:
    (i) Increase the generator set to $v_1, \ldots, v_n, v_{n+1}$, where $v_{n+1} = \lambda_1 v_1 + \cdots + \lambda_n v_n$ is presented as a linear combination of $v_1, \ldots, v_n$ with coefficients $\lambda_1, \ldots, \lambda_n$ of R.
    (ii) Decrease the generator set to $v_1, \ldots, v_{n-1}$, if $v_n = \lambda_1 v_1 + \cdots + \lambda_{n-1} v_{n-1}$ is presented as a linear combination of $v_1, \ldots, v_{n-1}$ with coefficients $\lambda_1, \ldots, \lambda_{n-1}$ of R.
    (iii) Permute the generator set in any one of the $n!$ ways.
    (b) Suppose there is a relation matrix $\Lambda \in R^{s \times n}$. Produce a relation matrix for any one of the generator sets obtained by an elementary change.
    (c) Show that the elementary ideals of M remain unchanged for any elementary change.
    (d) For any epimorphism ε of the finitely generated R-module M on the R-module M' show that it follows that M' is finitely generated over R and that $\mathfrak{E}_i(M'/R) \supseteq \mathfrak{E}_i(M/R)$ $(i \in \mathbb{Z}^{\geq 0})$.

18. (a) If the ring R is Noetherian and η is an R-epimorphism of $R^{1 \times n}$ on the R-module M then show that $\ker(\eta)$ is finitely generated, say $\ker(\eta) = \sum_{i=1}^{s} R\mathbf{r}_i'$, $\mathbf{r}_i' = (\lambda_{i1}, \ldots, \lambda_{in})$, so that the matrix $\Lambda = (\lambda_{ik}) \in R^{s \times n}$ is a relation matrix of M relative to the R-generators $v_i = \eta(\mathbf{r}_i')$ $(1 \leq i \leq n)$.
    (b) If there is a relation matrix $\Lambda = (\lambda_{ik}) \in R^{s \times n}$ of M relative to the R-generators $v_1, \ldots, v_n$ then show that the elementary ideal $\mathfrak{E}_g(M/R)$ is the ideal of R which is generated by the finitely many determinants $\det(\lambda_{h_i j_k})_{1 \leq i,k \leq n-g}$ subject to $1 \leq h_1 < h_2 < \cdots < h_{n-g} \leq s$, $1 \leq j_1 < j_2 < \cdots < j_{n-g} \leq n$ in case $n > g \geq \min(s, n)$. Moreover, we have

$$\mathfrak{E}_g(M/R) = \begin{cases} 0 & \text{for } 0 \leq g < \min(s, n) \\ R & \text{otherwise} \end{cases}$$

19. (a) If M is a finitely generated R-module then $\mathfrak{E}_i(M/R)$ is the intersection of all $\mathfrak{E}_i((M/m)/R)$ for the R-submodules m of M.
    (b) For any R-module M define $\mathfrak{E}_i(M/R)$ as the intersection of all $\mathfrak{E}_i((M/m)/R)$ with m any R-submodule of M with finitely generated factormodule M/m over R.
    (c) Show that $\mathfrak{E}_i(M'/R) \supseteq \mathfrak{E}_i(M/R)$ $(i \in \mathbb{Z}^{\geq 0})$ for any R-epimorphic image M' of M.

20. Show that $\mathfrak{E}_i((M_1 \oplus M_2)/R) = \sum_{j=0}^{i} \mathfrak{E}_j(M_1/R)\mathfrak{E}_{i-j}(M_2/R)$ for any two R-modules M_1, M_2 and $i \in \mathbb{Z}^{\geq 0}$.

21. Show that $\mathfrak{E}_0(R/\mathfrak{a}) = \mathfrak{a}, \mathfrak{E}_i(R/\mathfrak{a}) = R$ $(i \in \mathbb{N})$ for any ideal $\mathfrak{a}$ of R.

22. (a) For the R-module $R^{s \times n}$ there is the filtration $0 = X_0(R^{s \times n}) \subset X_1(R^{s \times n}) \subset \cdots \subset X_n(R^{s \times n}) = R^{s \times n}$ of $R^{s \times n}$, where $X_j(R^{s \times n})$ consists of all matrices $(\lambda_{ik}) \in R^{s \times n}$ satisfying $\lambda_{ik} = 0$ for $1 \leq i \leq s$, $1 \leq k \leq n - j$ $(0 < j < n)$. This filtration splits

inasmuch as $X_j(R^{s\times n}) = X_{j-1}(R^{s\times n}) \dot{+} Y_j(R^{s\times n})$, where $Y_j(R^{s\times n})$ is the R-module formed by the matrices $(\lambda_{ik})\in R^{s\times n}$ satisfying $\lambda_{ik}=0$ for $1\leqslant i\leqslant s$, $1\leqslant k\leqslant n-j$ or $n-j+1<k\leqslant n$ $(0<j<n)$.

(b) Let $\varepsilon_j: X_j(R^{s\times n})\to Y_j(R^{s\times n}): (\lambda_{ik})\mapsto(\delta_{n-j+1,k}\lambda_{ik})$ so that $\ker(\varepsilon_j)=X_{j-1}(R^{s\times n})$ $(0<j\leqslant n)$. For any matrix $\Lambda=(\lambda_{ik})\in R^{s\times n}$ there is the filtration $0=X_0(\Lambda)\subseteq X_1(\Lambda)\subseteq\cdots\subseteq X_n(\Lambda)=\mathfrak{R}(\Lambda)$. Show that the ideal $\mathfrak{a}_j(\Lambda)$ of R which is generated by the coefficients of the $(n-j+1)$th column of $X_j(\Lambda)$ is the same for any matrix Λ' that is row equivalent to Λ.

(c) Let $k_1,k_2,\ldots,k_r$ be the natural numbers satisfying $0<k_1<k_2<\cdots<k_r\leqslant n$, $\mathfrak{a}_{n-k_i}(\Lambda)\neq 0$ $(1\leqslant i\leqslant r)$, $\mathfrak{a}_j(\Lambda)=0$ if $n-j$ is distinct from $k_1,\ldots,k_r$. Develop an algorithm which carries the matrix $\Lambda\in R^{s\times n}$ into its row equivalent row normal form

$$\Lambda'=\begin{pmatrix}\Lambda'_1\\ \Lambda'_2\\ \vdots\\ \Lambda'_r\\ \Lambda'_{r+1}\end{pmatrix}\quad\text{subject to } \Lambda'_i\in R^{s_i\times n},\ s_i\in\mathbb{Z}^{>i},$$

$$X_{n-k_i}(\Lambda'_i)=\mathfrak{R}(\Lambda'_i),\ \mathfrak{a}_{n-k_i}(\Lambda'_i)=\mathfrak{a}_{n-k_i}(\Lambda)=\mathfrak{a}_{n-k_i}(\Lambda')\ (1\leqslant i\leqslant r),$$

$$\Lambda'_{r+1}=0\in R^{s_{r+1}\times n},\quad \sum_{i=1}^{r+1}s_i=s.$$

(d) Let both rectangular matrices $\Lambda\in R^{s\times n}, \Lambda'\in R^{s'\times n}$ be in normal form. Develop an algorithm to decide an equivalence of Λ,Λ' and to exhibit the equivalence in case the decision is affirmative.

23. Show that every semilocal Dedekind ring is a Prüfer ring.
24. If R is a Prüfer ring then show that any rectangular matrix $\Lambda=(\lambda_{ik})\in R^{s\times n}$ is row equivalent to one in *Hermite normal form* which is defined as a row normal form in terms of 22. Satisfying $s'=s$ and $s_i=1$ $(1\leqslant i\leqslant r)$.

25. (a) The units of $R^{s\times s}$ form a multiplicative group $GL(s,R)$ said to be the *general linear group* of degree s over R (or the *unimodular group* of degree s over R). Show that it contains the *special linear group* $SL(s,R)$ of degree s over R formed by all matrices of $GL(s,R)$ of determinant 1 as normal subgroup such that the diagonal matrices $\mathrm{diag}(\varepsilon,1,\ldots,1)$ $(\varepsilon\in U(R))$ form a representative group isomorphic to the unit group of $R: GL(s,R)=SL(s,R)\rtimes U(R)$.

(b) The centre of $GL(s,R)$ is formed by all scalar matrices εI_s with $\varepsilon\in U(R)$ so that $C(GL(s,R))\simeq U(R)$. Show that it intersects $SL(s,R)$ in its centre:

$$C(GL(s,R))\cap SL(s,R)=C(SL(s,R))\simeq\{\varepsilon\in U(R)\mid\varepsilon^s=1\}.$$

(c) For any rectangular matrix Λ of $R^{s\times n}$ and for any element P of $GL(s,R)$ show that the matrix $P\Lambda$ is row equivalent to Λ.

(d) If R is a Prüfer domain (i.e. an entire Prüfer ring) then show that for any two elements λ_1, λ_2 of R and for any generator λ of $\lambda_1 R + \lambda_2 R$ there is a matrix $\begin{pmatrix} \xi_1 & \eta_1 \\ \xi_2 & \eta_2 \end{pmatrix}$ of $SL(2, R)$ satisfying $(\lambda_1, \lambda_2)\begin{pmatrix} \xi_1 & \eta_1 \\ \xi_2 & \eta_2 \end{pmatrix} = (\lambda, 0)$.

(e) If R is a Prüfer domain then show that two rectangular matrices $\Lambda, \Lambda' \in R^{s \times n}$ are row equivalent if and only if there is a unimodular matrix P of degree s over R satisfying $P\Lambda = \Lambda'$.

26. (a) Show that any semilocal Dedekind ring R is a principal ideal ring. (Hint: Let $\mathfrak{a}$ be a non-zero ideal of R and let $\mathfrak{p}_1, \ldots, \mathfrak{p}_s$ be the finitely many non-zero maximal ideals of R. Then there are elements a_i of $\mathfrak{a}\prod_{j=1, j \neq i}^{s} \mathfrak{p}_j$ not belonging to $\mathfrak{a}\mathfrak{p}_i$. Show that $\mathfrak{a} = (\sum_{i=1}^{s} a_i)R$.)
    (b) For any semilocal Dedekind ring R there is a Euclidean division algorithm.

27. (a) Let R be a unital commutative ring, $\eta: R^{1 \times n} \to M$ an epimorphism on the R-module M with relation matrix $\mathfrak{R}$, $\kappa: R^{1 \times n} \to R^{1 \times n}: \mathbf{u}' \mapsto \mathbf{u}'K$ ($K \in GL(n, R)$ a non-singular R-linear transformation), then there is also the R-epimorphism $\eta\kappa: R^{1 \times n} \to M$ with relation matrix $\mathfrak{R}K^{-1}$.
    (b) Suppose the relation matrix $\mathfrak{R} \in R^{s \times n}$ of the finitely presented R-module M contains the nth unit row $\mathbf{e}_n^t$, then we have $M = R\eta(\mathbf{e}_1^t) + R\eta(\mathbf{e}_2^t) + \cdots + R\eta(\mathbf{e}_{n-1}^t)$ with relation matrix $\mathfrak{R}'$ derived from $\mathfrak{R}$ by removing the nth column.

28. Let M be a module over the unital commutative ring R. Then show that the intersection of all R-submodules M' of M with finitely generated R-factormodule is an R-submodule M_0 of M satisfying $\mathfrak{n}((M/M_0)/R) \supseteq \mathfrak{E}_i((M/M_0)/R) = \mathfrak{E}_i(M/R)$ $(i \in \mathbb{Z}^{\geq 0})$.

29. (a) For any non-zero polynomial P in n variables $t_1, \ldots, t_n$ over the infinite field F show there is a specialization $t_i \mapsto \tau_i$ $(1 \leq i \leq n)$ in F such that $P(\tau_1, \ldots, \tau_n) \neq 0$.
    (b) Show that the polynomials in n variables $t_1, \ldots, t_n$ over the finite field $\mathbb{F}_q$ that vanish for all specializations $t_i \mapsto \tau_i \in \mathbb{F}_q$ $(1 \leq i \leq n)$ form an ideal of $\mathbb{F}_q[t_1, \ldots, t_n]$. It is generated by the monic polynomials $t_i^q - t_i$ $(1 \leq i \leq n)$.
    (c) Construct for any non-zero polynomial P of $\mathbb{F}_q[t_1, \ldots, t_n]$ a finite extension E and n elements $\tau_1, \ldots, \tau_n$ of E such that $P(\tau_1, \ldots, \tau_n) \neq 0$.

30. (a) Let Λ be an order over the Dedekind ring R and let E be a finite separable extension of $\mathfrak{Q}(R)$. Then show that $\bar{\Lambda} = \mathrm{Cl}(R, E) \otimes_R \Lambda$ is an order over the Dedekind ring $\bar{R} = \mathrm{Cl}(R, E)$.
    (b) If $\mathfrak{a}$ is a Λ-fractional ideal then show that $\bar{\mathfrak{a}} = \bar{R} \otimes_R \mathfrak{a}$ is a $\bar{\Lambda}$-fractional ideal.
    (c) Show that $(\bar{\Lambda} :_{\bar{R}} \bar{\mathfrak{a}}) = \bar{R} \otimes_R (\Lambda :_R \mathfrak{a})$.
    (d) If under the assumption of (b) Λ is commutative and if the Λ-fractional ideal $\mathfrak{a}$ is invertible relative to Λ then show that $\bar{\Lambda}$ is commutative and $\bar{\mathfrak{a}}$ is invertible relative to $\bar{\Lambda}$ such that $\bar{\mathfrak{a}}^{-1} = \bar{R} \otimes_R \mathfrak{a}^{-1}$.

31. If R is a Dedekind ring and Λ is an R-order then show that for any semigroup S of non-zero divisors of R the S-localization Λ/S is an R-order. Moreover, show that we have $(\mathfrak{a}/S :_{R/S} \mathfrak{b}/S) = (\mathfrak{a} :_R \mathfrak{b})/S$ for the Λ-fractional ideals $\mathfrak{a}, \mathfrak{b}$.

32. If R is a Dedekind ring and Λ is a commutative R-order then show that any Λ-fractional ideal $\mathfrak{a}$ of maximal R-rank is principal, if and only if $\mathfrak{a}$ contains an element α satisfying $(\Lambda:_R\mathfrak{a}) = (\Lambda:_R\alpha\Lambda)$. In that case we have $\mathfrak{a} = \alpha\Lambda$.
33. Let Λ be a commutative order over the local Dedekind ring R and let $\mathfrak{a}$ be a Λ-fractional ideal which is invertible with respect to Λ.
    (a) Show that there is a finite extension E of $\mathfrak{Q}(R)$ such that $\mathrm{Cl}(R, E) \otimes_R \mathfrak{a}$ is a principal $\mathrm{Cl}(R, E) \otimes_R \Lambda$-fractional ideal.
    (b) Show that (5.45) holds for all Λ-fractional ideals $\mathfrak{b}$ in case the Λ-fractional ideal $\mathfrak{a}$ is invertible.

34. (E.C. Dade, O. Taussky, H. Zassenhaus)
    (a) Let M be a submodule of a unital ring A containing 1_A. Show that $M^i = M^{i+1}$ ($i \in \mathbb{N}$) implies that M^i is the subring of A generated by M.
    (b) Let R be a Dedekind ring and Λ an R-module with n basis elements. Let $\mathfrak{a}$ be a non-zero ideal of R. Then show that for any R-submodule M of Λ containing $\mathfrak{a}\Lambda$ there is an R-basis $b_1, \ldots, b_n$ of Λ and there are elements $e_1, \ldots, e_n$ of R such that $0 \subset \mathfrak{a} + e_1 R \subseteq \mathfrak{a} + e_2 R \subseteq \cdots \subseteq \mathfrak{a}_n + e_n R \subseteq R$ and $M = \sum_{i=1}^n (\mathfrak{a} + e_i R) b_i$.
    (c) Let R be a Dedekind ring and A a unital R-ring with n basis elements over R. Let M be an R-submodule of A containing 1_A. Show that M^{n-1} is the subring of A generated by M. (Hint: Use induction over the natural number λ in case $M \supseteq \mathfrak{p}^\lambda A$ for some prime ideal $\mathfrak{p} \neq 0$ of R, then apply a localization argument.)
    (d) Let R be a local Dedekind ring. Then show that every commutative maximal R-order is a principal ideal ring.
    (e) Let R be a local Dedekind ring and Λ_1 a commutative maximal R-order with n basis elements over R. Let Λ be an R-suborder of Λ_1 of the maximal R-rank n so that $A = \mathfrak{Q}(\Lambda) = \mathfrak{Q}(\Lambda_1)$ is a semisimple commutative $\mathfrak{Q}(R)$-algebra of dimension n. Let $\mathfrak{a}$ be a Λ-fractional ideal of maximal R-rank n. Show that $\mathfrak{A}$ contains a unit α of A such that $\mathfrak{A}\Lambda_1 = \alpha\Lambda_1$ and that $\Lambda' = (\alpha^{-1}\mathfrak{A})^{n-1}$ is an R-overorder of Λ. Hence, $\mathfrak{a}^{n-1}$ is invertible with respect to its order.
    (f) Let R be a Dedekind ring and Λ be a commutative R-order of R-rank n. Then show that for any Λ-fractional ideal $\mathfrak{A}$ of maximal R-rank n the power ideal $\mathfrak{A}^{n-1}$ is invertible relative to $[\mathfrak{A}^{n-1}/\mathfrak{A}^{n-1}]$.

4.6. Embedding algorithm

In this section we describe an algorithm for embedding the equation order

$$\Lambda_f = R[t]/f(t)R[t] = \sum_{i=1}^{n} Rx^{i-1} \quad (x = t/f(t)R[t]) \tag{6.1a}$$

of the monic separable polynomial

$$f(t) = t^n + a_1 t^{n-1} + \cdots + a_n \quad (n \in \mathbb{N}) \tag{6.1b}$$

with coefficients $a_1, \dots, a_n$ belonging to the Dedekind ring R into the maximal order $\mathrm{Cl}(R, A)$ of the algebraic background

$$A = A_f = \mathfrak{Q}(R)[t]/f(t)\mathfrak{Q}(R)[t] = \sum_{i=1}^{n} \mathfrak{a}(R)x^{i-1}. \tag{6.1c}$$

We are especially interested in the two cases

$$R = \mathbb{Z}, \tag{6.2a}$$

$$R = \mathbb{F}_q[\xi] \quad (q = p^\nu, p \in \mathbb{P}, \nu \in \mathbb{N}, \xi \text{ an independent variable over } \mathbb{F}_q) \tag{6.2b}$$

in which R is a PID. The output will be obtained as an *integral* (=minimal) *basis*

$$\omega_i = \left(\sum_{k=1}^{i} a_{ik}x^{k-1} \right) \Big/ N_i \tag{6.3a}$$

of $\mathrm{Cl}(R, A)$ over R, where

$$a_{ik} \in R, \quad 0 \neq N_i \in R, \quad \gcd(a_{i1}, a_{i2}, \dots, a_{ii}) = 1 \tag{6.3b}$$

and – as we already know –

$$a_{ii} \neq 0, \quad N_1 = 1 = a_{11}, N_{i-1} | N_i \quad (1 \leqslant i \leqslant n). \tag{6.3c}$$

If it is desired we also attain uniqueness by making the additional demands

$$a_{ii} > 0 (1 \leqslant i \leqslant n), \quad 0 \leqslant a_{ij} < a_{ii} (1 \leqslant j < i \leqslant n), \quad N_i > 0 (1 \leqslant i \leqslant n) \tag{6.4a}$$

in case (6.2a), whereas we demand in case (6.2b) that

$$a_{ii}, N_i \text{ are monic } (1 \leqslant i \leqslant n), \quad \deg(a_{ij}) < \deg(a_{ii}) \quad (1 \leqslant j < i \leqslant n). \tag{6.4b}$$

A rational integer x is said to be *reduced* modulo the natural number $\delta > 1$ in case $-\delta < 2x \leqslant \delta$. For any rational integer x there is precisely one reduced rational integer $R(x, \delta)$ modulo δ which is congruent to x modulo δ. It is found by means of division with remainder of x by δ (compare chapter 1 (1.6)).

Analogously the polynomial x of $\mathbb{F}_q[t]$ is said to be *reduced* modulo the monic non-constant polynomial δ of $\mathbb{F}_q[t]$ in case the degree of x is less than the degree of δ. For any polynomial x of $\mathbb{F}_q[t]$ there is precisely one polynomial which is reduced modulo δ and congruent to x modulo δ, viz. $R(x, \delta)$. It is found by means of division with remainder of x by δ.

In the sequel we shall use certain elements $\delta_1, \dots, \delta_\mu$ of R which are mutually prime divisors of the discriminant $d(f)$ of the polynomial f such that δ_i is not in $U(R)$ and $\delta_i^{\kappa_i}$ divides $d(f)$ for some natural number κ_i, but the quotient $d(f)/\delta_i^{\kappa_i}$ is prime to δ_i:

$$d(f) \Big/ \left(\prod_{i=1}^{\mu} \delta_i^{\kappa_i} \right) \in U(R) \quad (\kappa_i \in \mathbb{N},\ 1 \leqslant i \leqslant \mu). \tag{6.5}$$

In general we only know about some of δ_i's whether they are prime elements or at least square free elements of R. In any event certain computations modulo $\delta_i R[t]$ will have to be made in order to perform the embedding algorithm.

If δ_i is not known to be a prime element of R then the task of dividing an element α of R modulo δ_i by $\beta \in R$, $\beta \not\equiv 0 \bmod \delta_i$, can be carried out uniquely if and only if the Euclidean division algorithm of R for β, δ_i yields $1 = \gcd(\beta, \delta_i) = X(\beta, \delta_i)\beta + Y(\beta, \delta_i)\delta_i$ in which case $\alpha/\beta \equiv X(\beta, \delta_i)\alpha \bmod \delta_i$. Otherwise we find $\gcd(\beta, \delta_i)$ to be a proper divisor of δ_i which is not a unit of R. In that case we obtain by divisor cascading of β, δ_i a factorization

$$\delta_i = \prod_{j=1}^{\sigma} \delta_j'^{\kappa_j'} \quad (\sigma \in \mathbb{Z}^{>1}, \delta_1', \ldots, \delta_\sigma' \text{ mutually prime elements of } R \backslash U(R), \\ \kappa_j' \in \mathbb{N}, 1 \leqslant j \leqslant \sigma)$$

and we replace

$$\delta_i \leftarrow \delta_1', \ \delta_{\mu+j} \leftarrow \delta_{j+1}' \quad (1 \leqslant j < \sigma),$$
$$\kappa_i \leftarrow \kappa_i \kappa_1', \ \kappa_{\mu+j} \leftarrow \kappa_i \kappa_{j+1}' \quad (1 \leqslant j < \sigma), \quad \mu \leftarrow \mu + \sigma - 1$$

so that (6.5) still holds but μ has increased.

Examples (6.6)

(i) $\mu = 1, \alpha = 2, \beta = 6, \delta_1 = 15, R = \mathbb{Z}$. We compute $\gcd(\beta, \delta_1) = 3, \delta_1 = 3 \cdot 5$, $\sigma = 2$, and set $\delta_1 \leftarrow 3$, $\delta_2 \leftarrow 5$, $\mu \leftarrow 2$.

(ii) $\mu = 1$, $\alpha = 2$, $\beta = 7$, $\delta_1 = 15$, $R = \mathbb{Z}$. It follows that $\gcd(7, 15) = 1 = -2 \cdot 7 + 1 \cdot 15$, $\alpha/\beta \equiv -2 \cdot 2 \equiv -4 \bmod 15$.

(iii) $\mu = 1$, $\alpha = 2$, $\beta = t^2 + t$, $\delta_1 = t^3 + 2t$, $R = \mathbb{F}_5[t]$. We obtain $\gcd(t^2 + t, t^3 + 2t) = t$, and set $\delta_1 \leftarrow t$, $\delta_2 \leftarrow t^2 + 2$, $\mu \leftarrow 2$.

(iv) $\mu = 1, \alpha = 2, \beta = t^2 + t + 1, \delta_1 = t^3 + 2t, R = \mathbb{F}_5[t]$. We get $\gcd(t_2 + t + 1, t^3 + 2t) = 1 = (-t^2 - 2t + 1)(t^2 + t + 1) + (t - 2)(t^3 + 2t)$, $\alpha/\beta \equiv (-t^2 - 2t + 1)2 \equiv -2t^2 + t + 2 \bmod (t^3 + 2t)$.

For the application in the subsequent algorithm we always assume that the division algorithms yield positive results as otherwise we obtain proper non-unit divisors of some δ_i which lead to an increase of μ as explained above. In that case we have to start anew with that increased value of μ.

In the sequel we describe the embedding algorithm step by step. Most of the underlying theory was already developed in the preceding section.

Input
The input consists of the degree n and the coefficients $a_1, \ldots, a_n$ of the polynomial $f(t)$ of (6.1b). (Of course, we make the implicit assumption that we can do arithmetic in R, i.e. that R is given constructively.)

Output
An integral basis $\omega_1, \ldots, \omega_n$ of $\mathrm{Cl}(R, A_f)$ subject to (6.3a–c).

Initialization
Set $a_{ik} \leftarrow \delta_{ik}$, $N_i \leftarrow 1$ $(1 \leqslant k \leqslant i,\ 1 \leqslant i \leqslant n)$.

Step 1 (Square test)
Compute $d(f)$. If $d(f)$ is square-free, terminate. Else go to step 2.

Remarks
In case $R = \mathbb{Z}$ it may happen that $|d(f)|$ is so large that the lion's share of the computation time is consumed by testing the existence of a square divisor $\delta^2 > 1$ of $d(f)$. This is because no method to test for a square factor greater than one is known which is polynomial time in terms of $\log |d(f)|$. For that reason the following compromise is suggested:

For a suitable natural number $M > 1$ the first M prime numbers $p_1 = 2$, $p_2 = 3, \ldots, p_M$ are added to the input data such that we have at least $n \leqslant p_M$. Then we determine the prime factorization $d(f) = p_1^{\nu_1} p_2^{\nu_2} \cdots p_M^{\nu_M} p_0$ with $\nu_i \in \mathbb{Z}^{\geqslant 0}$ $(1 \leqslant i \leqslant M)$ and $p_0 \in \mathbb{N}$ not divisible by any p_i $(1 \leqslant i \leqslant M)$. We set

$$\delta_0' := \prod_{i=1}^{M} p_i^{\lfloor \nu_i/2 \rfloor}, \quad \delta_1' := \begin{cases} 1 & \text{if} \quad p_0 < (p_M + 2)^2 \\ p_0 & \text{if} \quad p_0 \geqslant (p_M + 2)^2 \end{cases}.$$

In case of $\delta_0' = \delta_1' = 1$ the test is affirmative, i.e. $d(f)$ is square free, and we terminate. If $\delta_0' = 1$, $\delta_1' > 1$ then no decision is made and we proceed to Step 2 with entries $\mu = 1$, $\delta_1 = \delta_1' > 1$, $\kappa_1 = 1$. For $\delta_0' > 1$ the test is negative. We assume that there are $\mu' \in \mathbb{N}$ and indices $1 \leqslant j_1 < j_2 < \cdots < j_{\mu'} \leqslant M$ such that $\nu_{j_i} > 1$ for $1 \leqslant i \leqslant \mu'$ and $0 \leqslant \nu_k \leqslant 1$ in case of $k \notin \{j_i \mid 1 \leqslant i \leqslant \mu'\}$. We set $\delta_i \leftarrow p_{j_i}$, $\kappa_i \leftarrow \nu_{j_i}$ $(1 \leqslant i \leqslant \mu')$ and $\mu \leftarrow \mu'$ for $p_0 = \pm 1$ but $\mu \leftarrow \mu' + 1$, $\delta_\mu \leftarrow |p_0|$ for $|p_0| > 1$. Then we proceed to step 2.

In case $R = \mathbb{F}_q[t]$ we apply the four operations D_t, gcd, division without rest (if applicable), pth root extraction in case of $D_t(x) = 0$ (which amounts to extended divisor cascading, see 1 (6.9), (6.12), (6.13)) repeatedly to $d(f)$ in order to obtain a factorization $d(f) = \prod_{i=1}^{n} g_i^i$ $(g_1, \ldots, g_n \in \mathbb{F}_q[t]$ monic and mutually prime). In case $g_2 = g_3 = \cdots = g_n = 1$ we know that $d(f)$ is square free, the R-order Λ_f is maximal, and we terminate. Otherwise let μ be the number of non-constant factors, say $g_{j_1}, g_{j_2}, \ldots, g_{j_\mu}$ among the polynomials $g_2, \ldots, g_n$ so that $1 < j_1 < j_2 < \cdots < j_\mu \leqslant n$, $\nu_{j_k} = j_k$ $(1 \leqslant k \leqslant \mu)$, $g_i = 1$ for $1 \leqslant i \leqslant n$ and $i \notin \{j_k \mid 1 \leqslant k \leqslant \mu\}$. In this case we set $\delta_i \leftarrow g_{j_i}$, $\kappa_i \leftarrow \nu_{j_i}$ $(1 \leqslant i \leqslant \mu)$ and proceed to step 2.

Step 2. (Dedekind test).
Set $i \leftarrow 1$. If it is known that δ_i is square-free then we form the congruence factorization $f(t) = \prod_{j=1}^{n} g_{ij}^{j} \bmod (\delta_i R[t])$ with monic polynomials g_{ij} of $R[t]$ that are reduced modulo δ_i, and both separable and mutually prime modulo $(\delta_i R[t])$. It is obtained via extended divisor cascading of $f(t)$, $D_t(F(t))$ modulo $(\delta_i R[t])$ (compare chapter 1 (6.9), (6.12) and for $\operatorname{char}(R/\delta_i R) \in \mathbb{P}$ also (6.13)).

Then we compute a modulo δ_i reduced polynomial $h_i(t)$ of $R[t]$ satisfying $\deg(h_i) < n$ and the congruence $(1/\delta_i)(f - \prod_{j=1}^{n} g_{ij}^{j}) \equiv h_i \bmod (\delta_i R[t])$. From (5.55) and the results of subsection 10 of section 5 we obtain the criterion:

Λ_f *is* δ_i*-maximal if and only if* $\gcd(h_i, \prod_{j=2}^{n} \sigma_{ij}^{j})$ *modulo* $(\delta_i R[t])$ *is one.* (6.7)

Hence, we proceed as follows:

If Λ_f is not δ_i-maximal we set $i \leftarrow i+1$ in case $i < \mu$ and repeat the Dedekind test, for $i = \mu$ the test is over. The same provisions obtain in case it should not yet be known whether δ_i is square-free.

If Λ_f is δ_i-maximal we set $\delta_j \leftarrow \delta_{j+1}$ $(i \leqslant j < \mu)$, $\mu \leftarrow \mu - 1$. For $i > \mu$ the Dedekind test is over, for $i \leqslant \mu$ we continue testing.

At the end of step 2 we either obtain $\mu = 0$ indicating that Λ_f is maximal in which case we terminate, or we have $\mu > 0$ in which case Λ_f is not δ_i-maximal or δ_i not known to be square free $(1 \leqslant i \leqslant \mu)$.

Examples (6.8)

(a) $R = \mathbb{Z}, f(t) = t^6 + 3t^4 + 3t^2 + t + 3$, $d(f) = -7^2 41^2 43$. We have $\mu = 2$, $\delta_1 = 7$, $\delta_2 = 41$ and compute $g_{11}(t) = t^2 - 2t - 2$, $g_{12}(t) = t^2 + t - 3$, $g_{1j}(t) = 1$ $(3 \leqslant j \leqslant 6)$, $h_1(t) = t^4 + t^2 - t$, Λ_f is δ_1-maximal, $g_{21}(t) = t^3 + 21t^2 + 10t + 8$, $g_{23}(t) = t - 7$, $g_{22}(t) = g_{2j}(t) = 1$ for $4 \leqslant j \leqslant 6$, $h_2(t) = 7t^4 + 20t^3 - 20t^2 + 14t - 14$, Λ_f is δ_2-maximal, hence $\Lambda_f = \mathrm{Cl}(R, A_f)$.

(b) $R = \mathbb{Z}, f(t) = t^3 - t^2 - 2t - 8$, $d(f) = -2^2 503$, We obtain $\mu = 1$, $\delta_1 = 2$, $g_{11}(t) = t + 1$, $g_{12}(t) = t$, $g_{13}(t) = 1$, $h_1(t) = -t^2 - t$, Λ_f is not maximal.

(c) *Integral bases of quadratic number fields.* A quadratic extension of the rational number field $\mathbb{Q}$ is usually given in the form $F = \mathbb{Q}(m^{\frac{1}{2}})$ ($m \in \mathbb{Z} \setminus \{0, 1\}$ square-free), the minimal polynomial of the generator $m^{\frac{1}{2}}$ being $f(t) = t^2 - m$. For the construction of a $\mathbb{Z}$-basis of $o_F := \mathrm{Cl}(\mathbb{Z}, F)$ we apply the Dedekind test. We compute $d(f) = 4m$ and since m was assumed to be square-free we obtain $\mu = 1$, $\delta_1 = 2$. We distinguish several cases:

(i) $2|m$ (implying $m/2 \equiv 1 \bmod 2$). We find $g_{11}(t) = 1$, $g_{12}(t) = t$, $h_1(t) = 1$, $o_F = \mathbb{Z}1 + \mathbb{Z}m^{\frac{1}{2}}$.

(ii) $m \equiv 3 \bmod 4$. We get $g_{11}(t) = 1$, $g_{12}(t) = t + 1$, $h_1(t) = t$, hence again $o_F = \mathbb{Z}1 + \mathbb{Z}m^{\frac{1}{2}}$.

(iii) $m \equiv 1 \bmod 4$. In that case $g_{11}(t) = 1$, $g_{12}(t) = t + 1$, $h_1(t) = t + 1$, the Dedekind test is negative, hence $\mathbb{Z}1 + \mathbb{Z}m^{\frac{1}{2}} \subset o_F$. As a consequence we must have $N_2 = 2$ in (5.3) and therefore $\omega_1 = 1$, $\omega_2 = (1 + m^{\frac{1}{2}})/2$ according to (6.4) (since $m^{\frac{1}{2}}/2$ is not in $\mathrm{Cl}(\mathbb{Z}, F)$).

We note that in case (iii) we can replace the generating polynomial $f(t)$ for F by $\tilde{f}(t) = t^2 - t - (m-1)/4 \in \mathbb{Z}[t]$. Its discriminant is $d(\tilde{f}) = m$, hence $o_F = \mathbb{Z} + \mathbb{Z}\rho$ for a zero $\rho = (1 \pm m^{\frac{1}{2}})^2$ of $\tilde{f}(t)$ by the Dedekind test. Thus we have shown:

Proposition (6.9)

Let $F = \mathbb{Q}(m^{\frac{1}{2}})$ be a quadratic number field. Then an integral basis for $o_F = \mathrm{Cl}(\mathbb{Z}, F)$ is given by $\omega_1 = 1$, $\omega_2 = \begin{Bmatrix} m^{\frac{1}{2}} & \text{for } m \not\equiv 1 \bmod 4 \\ (1 + m^{\frac{1}{2}})/2 & \text{for } m \equiv 1 \bmod 4 \end{Bmatrix}$.

(If we denote the discriminant of o_F by d_F, then we can set $\omega_2 = (d_F + d_F^{\frac{1}{2}})/2$ in each case.)

(d) *Integral bases of cyclotomic fields.* As in chapter 2, section 11, let $\phi_n(t)$ be the nth cyclotomic polynomial whose roots are precisely the primitive nth roots of unity ζ_n^ν $(1 \leqslant \nu \leqslant n,\ \gcd(\nu, n) = 1,\ \zeta_n = e^{2\pi i/n})$. We want to find a $\mathbb{Z}$-basis of $o_{F_n} = \mathrm{Cl}(\mathbb{Z}, F_n)$ for $F_n = \mathbb{Q}(\zeta_n)$ in case $n > 2$, i.e. $F_n \supset \mathbb{Q}$. If n is a power product of the distinct prime numbers $p_1, \ldots, p_\mu$, say $n = \prod_{i=1}^{\mu} p_i^{m_i}$ $(m_i \in \mathbb{N})$, then also the discriminant $d(\phi_n)$ of the polynomial ϕ_n is a power product of $p_1, \ldots, p_\mu$ according to 2 (11.12). Hence, we apply the Dedekind test for all $\delta_i = p_i$ $(1 \leqslant i \leqslant \mu)$. Unfortunately this requires a deeper knowledge of the properties of cyclotomic polynomials so that we state the following proposition without proof. (But see exercise 5 for the case $\mu = 1$.)

Proposition (6.10)

The maximal order o_{F_n} of $F_n = \mathbb{Q}(e^{2\pi i/n})$ has the $\mathbb{Z}$-basis ζ_n^{j-1} $(1 \leqslant j \leqslant \varphi(n)$, $\zeta_n = e^{2\pi i/n}, n \in \mathbb{N})$.

Remark

Even though the order Λ_f was not always maximal in example (6.8) (c) in each case we could find a polynomial $\tilde{f}$ satisfying $A_f = A_{\tilde{f}}$ such that $\Lambda_{\tilde{f}}$ was maximal. However, this is not possible in general. For example, exercise 4 yields that for each monic cubic polynomial $g(t) \in \mathbb{Z}[t]$ satisfying $A_g = A_f$ for $f(t) = t^3 - t^2 - 2t - 8$ (compare (6.8) (b)) we find that Λ_g is not 2-maximal. (2 is a so-called *common inessential discriminant divisor.*)

Step 3 (Reduced discriminant)

In subsection 7 of section 5 a subroutine for the computation of the reduced discriminant $d_0(g)$ of any monic separable polynomial $g(t)$ over the Dedekind ring R was developed. It forms step 3 of the embedding algorithm and is used for the computation of the reduced discriminant of the polynomial $f(t)$.

Before we present the three final steps of the algorithm we need an

Introduction to the core steps of the algorithm

At this point we are given a non-unit δ of the Dedekind ring R for which it is known that

$$d_0(f) = \delta^\kappa \hat{\delta} \ (\kappa \in \mathbb{Z}^{>0},\ \delta R + \hat{\delta} R = R). \tag{6.11}$$

It is our task (compare (5.85)) to construct a canonical R-basis $\omega_{1\delta}, \ldots, \omega_{n\delta}$

of the maximal δ-overorder $\bar{\Lambda}_f$ of Λ_f such that

$$\omega_{i\delta} = \sum_{k=1}^{i} \beta_{ik\delta}\xi^{k-1} \quad (\beta_{ik\delta}\in\mathfrak{Q}(R),\ 1 \leqslant k \leqslant i,\ 1 \leqslant i \leqslant n). \tag{6.12a}$$

It suffices to construct any R-basis $\omega'_{1\delta},\ldots,\omega'_{n\delta}$ satisfying

$$\omega'_{i\delta} = \sum_{k=1}^{n} \beta'_{ik\delta}\xi^{k-1} \quad (\beta'_{ik\delta}\in\mathfrak{Q}(R),\ 1 \leqslant k \leqslant n,\ 1 \leqslant i \leqslant n). \tag{6.12b}$$

The coefficients $\beta_{ik\delta}$ of a canonical R-basis (6.12a) are then obtained by Hermite row reduction of the $n \times n$-matrix $(\beta'_{ik\delta})$.

Any element ζ of $\mathfrak{Q}(R)\Lambda_f$ is uniquely presented in the form

$$\zeta = \sum_{i=1}^{n} \lambda_i(\zeta)\xi^{i-1} \quad (\lambda_i(\zeta)\in\mathfrak{Q}(R),\ 1 \leqslant i \leqslant n). \tag{6.13a}$$

For a better understanding of the situation we interpret ζ as an n-vector with components $\lambda_1(\zeta),\ldots,\lambda_n(\zeta)$ over $\mathfrak{Q}(R)$. The minimal polynomial m_ζ of ζ is obtained e.g. by straightforward computations using linear algebra, namely by looking for the smallest natural number μ_ζ for which the n-vector corresponding to ζ^{μ_ζ} is a $\mathfrak{Q}(R)$-linear combination of the n-vectors corresponding to $1,\zeta,\ldots,\zeta^{\mu_\zeta-1}$, say

$$\zeta^{\mu_\zeta} = -\sum_{i=1}^{\mu_\zeta} \gamma_i(\zeta)\zeta^{\mu_\zeta-i} \quad (\gamma_i(\zeta)\in\mathfrak{Q}(R),\ 1 \leqslant i \leqslant \mu_\zeta), \tag{6.13b}$$

so that

$$m_\zeta(t) = t^{\mu_\zeta} + \sum_{i=1}^{\mu_\zeta} \gamma_i(\zeta)t^{\mu_\zeta-i}. \tag{6.13c}$$

(For an alternative see Collins' method given in exercise 1.) We use the procedure indicated as a further subroutine of the embedding algorithm. The element ζ belongs to the maximal order $\mathrm{Cl}(R, \mathfrak{Q}(R)\Lambda_f)$ precisely if m_ζ belongs to $R[t]$, i.e.

$$\gamma_i(\zeta)\in R \quad (1 \leqslant i \leqslant \mu_\zeta). \tag{6.13d}$$

The element ζ of $\mathrm{Cl}(R, \mathfrak{Q}(R)\Lambda_f)$ belongs to the δ-maximal overorder $\bar{\Lambda}_f$ of Λ_f precisely if

$$\delta^\kappa\zeta\in\Lambda_f. \tag{6.13e}$$

For any monic non-constant polynomial $m(t)$ over R we compute, just as in step 2, a congruence factorization

$$m \equiv \prod_{i=1}^{s(m,\delta)} g_{im\delta}^{\alpha_{im\delta}} \operatorname{mod}(\delta R[t]) \tag{6.14a}$$

of m into the power product of monic non-constant polynomials $g_{im\delta}(t)$ over R of which it is known that

$$g_{im\delta}(t)R[t] + g_{jm\delta}(t)R[t] = R[t] \quad (1 \leqslant i < j \leqslant s(m,\delta)). \tag{6.14b}$$

We also stipulate that the coefficients of the $g_{im\delta}$ are reduced modulo δ. Then calculation of $s(m,\delta)$, $\alpha_{im\delta}, g_{im\delta}$ $(1 \leqslant i \leqslant s(m,\delta))$ is assumed to be done by another subroutine of the embedding algorithm.

Definition (6.15)
*The element ζ of $\bar{\Lambda}_f$ is said to be δ-**split** if we have $s(m_\zeta,\delta) > 1$ for its minimal polynomial m_ζ.*

If ζ is δ-split then we apply a Hensel lift (see (5.76)) to the congruence factorization (6.14a) of $m = m_\zeta$ in order to produce the congruence factorization

$$m_\zeta \equiv \prod_{i=1}^{s(\zeta,\Delta)} g_{i\zeta\Delta}^{\alpha_{i\zeta\Delta}} \bmod (\Delta R[t]) \tag{6.16a}$$

$$(\Delta = \delta^{2\kappa}, s(\zeta,\Delta) = s(m,\delta), \alpha_{i\zeta\Delta} = \alpha_{im\delta}, g_{i\zeta\Delta} \equiv g_{im\delta} \bmod (\delta R[t]))$$

and a presentation

$$\sum_{i=1}^{s(\zeta,\Delta)} a_{i\zeta\Delta}(\zeta)\hat{g}_{i\zeta\Delta}(\zeta) \equiv 1 \bmod (\Delta R[t]) \quad (a_{i\zeta\Delta}(t) \in R[t], \deg(a_{i\zeta\Delta}) < \deg(g_{im\delta}),$$

$$\hat{g}_{i\zeta\Delta} = \prod_{\substack{j=1\\ j\neq i}}^{s(\zeta,\Delta)} g_{j\zeta\Delta}, 1 \leqslant i \leqslant s(\zeta,\Delta)), \tag{6.16b}$$

which evolves from (6.14b). In this way we compute the set of $s(\zeta,\Delta)$ orthogonal Δ-idempotents

$$e_{i\zeta\Delta} := a_{i\zeta\Delta}(\zeta)\hat{g}_{i\zeta\Delta}(\zeta) \quad (1 \leqslant i \leqslant s(\zeta,\Delta)), \tag{6.16c}$$

characterized by the congruences:

$$e_{i\zeta\Delta}e_{j\zeta\Delta} \equiv 0 \bmod (\Delta R[\zeta]), \quad 0 \not\equiv e_{i\zeta\Delta}^2 \equiv e_{i\zeta\Delta} \bmod (\Delta R[\zeta])$$
$$(1 \leqslant i \leqslant s(\zeta,\Delta)),$$
$$\sum_{i=1}^{s(\zeta,\Delta)} e_{i\zeta\Delta} \equiv 1 \bmod (\Delta R[\zeta]). \tag{6.16d}$$

The R-order $\Lambda_f^* = \Lambda_f R[\zeta]$ has the property that $\Delta\Lambda_f^* \subseteq \Lambda_f$, hence Λ_f^* contains the suborder $\Lambda_f^{**} = \Delta\Lambda_f^* + \sum_{i=1}^{s(\zeta,\Delta)} e_{i\zeta\Delta}\Lambda_f$ with the property that $f/\Delta\Lambda_f^{**}$ is the minimal polynomial of $\xi/\Delta\Lambda_f^{**}$ over R/Δ and that $\Lambda_f^{**}/\Delta\Lambda_f^{**} = \bigoplus_{i=1}^{s(\zeta,\Delta)} e_{i\zeta\Delta}\Lambda_f/\Delta\Lambda_f^{**}$ so that $f \equiv \prod_{i=1}^{s(\zeta,\Delta)} f_i \bmod (\Delta R[t])$, where $f_i(t)$ is a monic non-constant polynomial of $R[t]$ for which $f_i/\Delta R[t]$ is the minimal polynomial of $e_{i\zeta\Delta}\xi/\Delta\Lambda_f^{**}$ over R/Δ and $\xi = t/f(t)R[t]$ as before. For the computation of the f_i it suffices to form the R-orders $\Lambda_{f_i}^{**}$ generated by ξ and $e_{i\zeta\Delta}$ and to compute the minimal polynomial of $e_{i\zeta\Delta}\xi$ modulo Δ. Using the remarks made at the end of subsection 8 of section 5 we realize that the task of embedding Λ_f into its δ-maximal overorder is reduced to the task of embedding $\Lambda_{f_1 f_2 \cdots f_{s(\zeta,\Delta)}}$ into its δ-maximal overorder. That task is reduced to the tasks of embedding the R-equation orders Λ_{f_i} into their δ-maximal overorder $(1 \leqslant i \leqslant s(\zeta,\Delta))$. It follows that any δ-split leads to a reduction of the embedding task for f to similar embedding tasks for finitely many polynomials of degree lower than the degree of f.

Example (6.17)
For $f(t) = t^3 - t^2 - 2t - 8$, $d(f) = -2^2 503$ we compute $\delta = 2$, $d_0(f) = 2 \cdot 503$, $\kappa = 2$, $\Delta = 2^2 = 4$, $f_1(t) = t^2 - 2t$, $f_2(t) = t + 1$.

It suffices therefore to make the assumption throughout the core algorithm that every element ζ of $\bar{\Lambda}_f$ that is brought to the test turns out not to be δ-split.

Definition (6.18)
*The element ζ of $\bar{\Lambda}_f$ is said to be δ-**uniform** if $s(\zeta, \Delta) = 1$, $m_\zeta(0)R + \delta R = R$.*

The δ-uniform elements of $\bar{\Lambda}_f$ are special units of $\bar{\Lambda}_f$ as follows from the equation $m_\zeta(\zeta) = 0$. An element ζ of $\bar{\Lambda}_f$ which is neither δ-split nor δ-uniform is characterized by the congruence

$$m_\zeta(t) = t^{\mu_\zeta} \bmod \Delta R[t] \tag{6.19}$$

or else we find a factorization of δ.

In order to deal with such elements we introduce the δ-adic exponential valuation $\eta: \mathfrak{Q}(R) \to \mathbb{Z} \cup \{\infty\}$, $\eta(\delta) = 1$, $\eta(0) = 0$, $\eta(\delta^i x y^{-1}) = i$ ($i \in \mathbb{Z}$, $x, y \in R$, $\delta \nmid x$, $yR + \delta R = R$). As was stated before, it suffices to assume that $\gamma_i(\zeta)\delta^{-\eta(\gamma_i(\zeta))}R + \delta R = R$ whenever $\gamma_i(\zeta)$ is not zero and is brought to the test ($1 \leqslant i \leqslant \mu_\zeta$). Hence the non-zero coefficients of m_ζ are η-multipliers. According to assumption we have $\gamma_i(\zeta) \in \delta R$ ($1 \leqslant i \leqslant \mu_\zeta$).

It was pointed out for the special case $R = \mathbb{Z}$ at the end of section 4 that the δ-adic exponential valuation η is the minimum of certain additive exponential valuations $\eta_1, \eta_2, \ldots, \eta_c$ of $\mathfrak{Q}(R)$ in $\mathbb{Z} \cup \{\infty\}$ corresponding to the c prime ideals of R containing δ.

Suppose that E is a minimal splitting field of m_ζ over $\mathfrak{Q}(R)$ then there are finitely many extensions of $\eta_1, \ldots, \eta_c$ to additive exponential valuations of E, which define an exponential valuation $\bar{\eta}: E \to \mathbb{Q} \cup \{\infty\}$ upon taking their minimum. This exponential valuation restricts to η on $\mathfrak{Q}(R)$. Since the constant term of m_ζ is an η-multiplier it follows that the roots $\zeta_1, \ldots, \zeta_{\mu_\zeta}$ of m_ζ in E are $\bar{\eta}$-multipliers. Let $\bar{\eta}(\zeta_1) \geqslant \bar{\eta}(\zeta_2) \geqslant \cdots \geqslant \bar{\eta}(\zeta_{\mu_\zeta}) > 0$ and let λ_2 be the denominator of the positive rational number $\bar{\eta}(\zeta_{\mu_\zeta}) = \lambda_1/\lambda_2$ ($\lambda_1, \lambda_2 \in \mathbb{Z}^{>0}$, $\gcd(\lambda_1, \lambda_2) = 1$) then it follows that $\bar{\eta}(\zeta_i^{\lambda_2}\delta^{-\lambda_1}) \geqslant \bar{\eta}(\zeta_{\mu_\zeta}^{\lambda_2}\delta^{-\lambda_1}) = 0$ so that for the element $\zeta^* := \zeta^{\lambda_2}\delta^{-\lambda_1} \in \bar{\Lambda}_f$ the corresponding minimal polynomial m_{ζ^*} has the roots $\zeta_i^{\lambda_2}\delta^{-\lambda_1}$, though perhaps with a multiplicity which is not as large.

For elements ζ of $\mathfrak{Q}(R)\Lambda_f$ satisfying

$$\bar{\eta}(\zeta_1) = \bar{\eta}(\zeta_2) = \cdots = \bar{\eta}(\zeta_{\mu_\zeta}) \tag{6.20a}$$

it is safe to define $\bar{\eta}(\zeta)$ as the rational number occurring in (6.20a). Using the $\bar{\eta}$-Newton polygon method of section 3 the statement (6.20a) is equivalent to the inequalities

$$\eta(\gamma_{\mu_\zeta}(\zeta)) \leqslant \frac{i}{\mu_\zeta}\eta(\gamma_i(\zeta)) \quad (1 \leqslant i < \mu_\zeta) \tag{6.20b}$$

which can be easily tested. If the test is positive then we have

$$\bar{\eta}(\zeta) = \eta(\gamma_{\mu_\zeta}(\zeta))/\mu_\zeta. \tag{6.20c}$$

Definition (6.21)
*The element ζ of $\bar{\Lambda}_f$ is called a **δ-element** if it satisfies (6.20b) and if $\bar{\eta}(\zeta) \neq 0$, $\bar{\eta}(\zeta)^{-1} \in \mathbb{Z}^{>0}$.*

Starting from an arbitrary non-zero element ζ of $\bar{\Lambda}_f$ we produce a δ-element of $R[\zeta]$ as follows. Test whether ζ is δ-split. If that is not the case form $\zeta^* = g_{1m\delta}(\zeta) \neq 0$. Test whether $\eta(\gamma_{\mu_{\zeta^*}}(\zeta^*)) \leqslant (i/\mu_{\zeta^*})\,\eta(\gamma_i(\zeta^*))$ $(1 \leqslant i \leqslant \mu_{\zeta^*})$. If that is the case form $\eta(\zeta^*) = \lambda_1/\lambda_2$ $(\lambda_1, \lambda_2 \in \mathbb{Z}^{>0}, \gcd(\lambda_1, \lambda_2) = 1)$, $k_1\lambda_1 + k_2\lambda_2 = 1$ $(k_1, k_2 \in \mathbb{Z})$, $\pi(\zeta) := (\zeta^*)^{k_1}\delta^{k_2}$, hence $0 < \bar{\eta}(\pi(\zeta)) = 1/\lambda_2$. In particular, the use of δ-elements permits to give a new criterion for the δ-maximality of Λ_f.

Criterion (6.22)
The equation order Λ_f is δ-maximal precisely if $\pi(\xi)$ can be formed and satisfies

$$\pi(\xi) = g_{1\xi\delta}(\xi) \quad (g_{1\xi\delta}(t) := g_{1m_\xi\delta}(t)), \tag{6.22a}$$

$$\mu_\xi \eta(\pi(\xi)) = 1. \tag{6.22b}$$

Proof
We use localization. It suffices to assume that R is local with δR as maximal ideal. Then Λ_f is maximal precisely if the elements $\xi^i\xi^{*k}$ $(0 \leqslant i < \deg(g_{1\xi\delta})$, $0 \leqslant k < \eta(\pi(\xi))^{-1})$ form an R-basis of Λ_f as follows upon projecting Λ_f on the simple components of $\mathfrak{Q}(R)\Lambda_f$. Hence, (6.22a, b). □

The criterion (6.22) yields a useful test for δ-maximality:

Criterion (6.23)
*Let (6.22a) be satisfied (in which case we call ξ **normalized**) and assume that*

$$\pi(\xi)^{\eta(\pi(\xi))^{-1}} \equiv \sum_{i=0}^{\eta(\pi(\xi))^{-1}-1} c_i(\xi)\pi(\xi)^i \bmod (\Delta\Lambda_f)$$

$$(c_i(t) \in R[t], \deg(c_i) < \deg(g_{1\xi\delta}), 0 \leqslant i < \eta(\pi(\xi))^{-1}).$$

Then Λ_f is δ-maximal precisely if $c_0(\xi)/\delta$ is a δ-unit.

After the preceding introductory remarks we proceed to expound the last three computational steps yielding the embedding of Λ_f into a δ-maximal overorder. They constitute the

Core algorithm

Step 4
(Normalization of ξ). If $g_{1\xi\delta}(\xi)=\pi(\xi)$ is already satisfied then proceed to step 5, else set $\xi' \leftarrow \xi + \pi(\xi)$.

It follows that $g_{1\xi'\delta}=g_{1\xi\delta}$ and that ξ' is not δ-split. This is because by assumption f has n distinct roots $\xi_1, \xi_2, \ldots, \xi_n$ in any minimal splitting extension E of f over $\mathfrak{Q}(R)$ and either $\bar{\eta}(\xi_i-\xi_j)=0$ or $\bar{\eta}(\xi_i-\xi_j)\geqslant \bar{\eta}(g_{1\xi\delta}(\pi(\xi)))>\bar{\eta}(\pi(\xi))$ $(1\leqslant i<j\leqslant n)$. Hence, the algebraic conjugates of $g_{1\xi'\delta}(\xi')$ have the same value in any exponential valuation extending η to E. Clearly, ξ' belongs to $\bar{\Lambda}_f$ and we have $\deg(m_{\xi'})=n$. Because of the separability of $g_{1\xi\delta}$ we find that $(D_t(g_{1\xi\delta}))(\pi(\xi))$ is a unit of $\bar{\Lambda}_f$ and hence $g_{1\xi\delta}(\xi')=g_{1\xi'\delta}(\xi')$, $\eta(g_{1\xi'\delta}(\xi'))=\eta(\pi(\xi))$, $\bar{\Lambda}_{m_{\xi'}}\subseteq\bar{\Lambda}_f$, $\bar{\Lambda}_f=\Lambda_f+\bar{\Lambda}_{m_{\xi'}}$. The transition from $\bar{\Lambda}_{m_{\xi'}}$ to $\bar{\Lambda}_f$ is routine. Substituting ξ' for ξ, $m_{\xi'}$ for f we can therefore assume that $\pi(\xi)=g_{1\xi\delta}(\xi)$ so that (6.22a) holds. Then we proceed to

Step 5 (Initial term of $\delta^{-1}\pi(\xi)^{\eta(\pi(\xi))^{-1}}$).
In case (6.22b) holds we terminate since Λ_f is δ-maximal. Else we form $\bar{c}_0=\delta^{-1}\pi(\xi)^{\eta(\pi(\xi))^{-1}}$ so that $\bar{\eta}(\bar{c}_0)=0$, $\bar{c}_0\in\bar{\Lambda}_f$. For $\bar{c}_0\in\Lambda_f$ we proceed to step 6. Else we form expressions $\xi':=\xi+P_1(\bar{c}_0)\,(P_1(t)\in R[t], \deg(P_1)<\deg(g_1\bar{c}_0\delta))$ beginning with $P_1(t)=t$, $P_1(\bar{c}_0)=\bar{c}_0$, $\xi'=\xi+\bar{c}_0$ such that

$$\deg(g_{1\xi'\delta})>\deg(g_{1\xi\delta}). \tag{6.24}$$

This must be achievable in a finite number of trials and errors since the residue class fields under consideration are finite and there holds the theorem of the primitive element in the strong form that a finite field $\mathbb{F}$ generated by two elements $\tilde{\xi}, \tilde{c}_0$ is already generated by one element of the form $\tilde{\xi}+\tilde{P}(\tilde{c}_0)$, where $\tilde{P}$ is some polynomial with coefficients in the ground field.

Now suppose that (6.24) is satisfied. If $\deg(m_{\xi'})$ is less than n we form expressions

$$\xi''=\xi'+\delta P_2(\xi) \quad (P_2(t)\in R[t] \quad \delta\text{-reduced with } \deg(P_2)<n) \tag{6.25}$$

beginning with $P_2(t)=t$, $\xi''=\xi'+\delta\xi$. After a finite number of trials and errors we will have

$$\deg(m_{\xi''})=n. \tag{6.26}$$

We find that $\bar{\Lambda}_{m_{\xi''}}\subseteq\bar{\Lambda}_f$, $\bar{\Lambda}_f=\Lambda_f+\bar{\Lambda}_{m_{\xi''}}$. We replace ξ by ξ''. The transition from $\bar{\Lambda}_{m_{\xi''}}$ to $\bar{\Lambda}_f$ is routine. We go back to step 4.

We note that the degree of $g_{1\xi\delta}$ increases each time we carry out step 5. That can happen only a finite number of times.

Step-6 (development of $\pi(\xi)^{\eta(\pi(\xi))^{-1}}$).
At this point we have

$$\pi(\xi)=g_{1\xi\delta}(\xi), \quad \bar{c}_0=\delta^{-1}\pi(\xi)^{\nu}\in\Lambda_f, \quad \bar{\eta}(\bar{c}_0)=0, \quad \nu=\eta(\pi(\xi))^{-1}<\mu_{\xi}. \tag{6.27}$$

It follows that there holds a congruence development $\bar{c}_0 \equiv \sum_{i=0}^{v-1} c_i(\xi)\pi(\xi)^i \bmod (\pi(\xi)^v \Lambda_f)$ with δ-reduced polynomials $c_i(t)$ of $R[t]$ satisfying

$$\deg(c_i) < \deg(g_{1\xi\delta}) \quad (0 \leqslant i < v), \quad c_0 \neq 0, \tag{6.28}$$

which is easily obtained by calculations in $R/\delta R[t]$ modulo $\pi(\xi)^v R/\delta R[t]$. Since $\pi(\xi)^v$ is contained in $\delta\bar{\Lambda}_f$ it follows that there holds the congruence development $\bar{c}_0 \equiv \sum_{i=0}^{v-1} c_i(\xi)\pi(\xi)^i \bmod (\delta\bar{\Lambda}_f)$. We extend it as follows. Assuming that there already holds a congruence development

$$\bar{c}_0 \equiv \sum_{i=0}^{v-1} c_i(\xi)\pi(\xi)^i \bmod (\delta^j \bar{\Lambda}_f) \quad (j \in \mathbb{N}) \tag{6.29}$$

with polynomials $c_i(t)$ of $R[t]$ satisfying (6.28), the expression $\rho = \delta^{-j}(\bar{c}_0 - \sum_{i=0}^{v-1} c_i(\xi)\pi(\xi)^i)$ is not zero since the minimal polynomial of ξ over R is f, and we have $\bar{\eta}(\rho) \geqslant 0$, hence

$$\rho = \rho'\pi(\xi)^\lambda \rho''^{-1} \quad (\rho' \in \bar{\Lambda}_f, \lambda \in \mathbb{Z}, \rho'' \in R, \delta R + \rho'' R = R). \tag{6.30}$$

For $\bar{\eta}(\rho') > 0$ we form $\xi' = \xi + \rho'$ so that $\deg(m_{\xi'}) = n$, $g_{1\xi\delta} = g_{1\xi'\delta}$, $\mu_\xi = \mu_{\xi'}$, $\pi(\xi') = g_{1\xi'\delta}(\xi')$, $\eta(\pi(\xi')) = \eta(\rho') < \bar{\eta}(\pi(\xi))$. Now we replace ξ by ξ' and go back to step 5. (We observe that $\bar{\Lambda}_{m_{\xi'}} \subseteq \bar{\Lambda}_f$, $\bar{\Lambda}_f = \Lambda_f + \bar{\Lambda}_{m_{\xi'}}$ so that the transition from $\bar{\Lambda}_{m_{\xi'}}$ to $\bar{\Lambda}_f$ is routine.)

For $\bar{\eta}(\rho') = 0$, $\rho' \in \Lambda_f$ we proceed as in step 5 with ρ' in place of $\bar{c}_0$. There holds a congruence presentation $\rho' \equiv \sum_{i=0}^{v-1} c_i'(\xi)\pi(\xi)^i \bmod (\pi(\xi)^v \Lambda_f)$ with δ-reduced polynomials $c_0'(t), \ldots, c_{v-1}'(t)$ of $R[t]$ of degree less than μ_ξ. It implies the congruence presentation

$$\rho' \equiv \sum_{i=0}^{v-1} c_i'(\xi)\pi(\xi)^i \bmod (\delta\bar{\Lambda}_f). \tag{6.31}$$

Upon substitution of (6.31) in (6.30), (6.29), (6.28) we obtain a new congruence presentation for $\bar{c}_0$ of the same form as (6.29), but with j increased by one.

If necessary this procedure is applied repeatedly. We note that $j = \kappa$ is impossible because of $\delta^\kappa \bar{\Lambda}_f \subseteq \Lambda_f$ so that in case $j = \kappa$ we would obtain a congruence $\pi(\xi)^v \equiv \sum_{i=0}^{v-1} \delta c_i(\xi)\pi(\xi)^i \bmod (\delta\Lambda_f)$ which is not possible since f also is the minimal polynomial of $\xi \bmod (\delta\Lambda_f)$.

Hence, the algorithm collapses at some stage $j < \kappa$ yielding the δ-maximal overorder of Λ_f.

We remark that the algorithm works even if the prime ideals dividing δ do not all divide the same prime number. Supposing we allow factorization of rational integers into prime numbers then we reduce the core algorithm to the case that all prime ideals containing δ also contain the same prime number p. In that case one can bring to bear the special structure of finite fields. This is always possible in the analogous situation for algebraic function fields of characteristic p [5].

In the case that the polynomial f is irreducible over R we deduce from the core algorithm

Theorem *(Hensel, Krasner)* (6.32)
Let R be a local Dedekind ring. Then for any separable finite extension E of $F = \mathfrak{Q}(R)$ there is a separable finite extension E_1 of degree n_1 over F with the following properties:

(I) *$J(R)$ splits completely in E_1 so that $\mathrm{Cl}(R, E_1) = \bigcap_{i=1}^{n_1} R_i$, where $R_1, \ldots, R_{n_1}$ are n_1 local Dedekind rings intersecting F in R.*

(II) *$E_1 \otimes_F E = \bigoplus_{i=1}^{s} E_{1i}$, where the E_{1i} are finite extensions of E_1 $(1 \leqslant i \leqslant s, s \in \mathbb{N})$.*

(III) *The field E_{1i} of (II) contains an intermediary field Ψ_i over E_1 such that $J(R_j)$ is unramified in Ψ_i, moreover the degree of the prime ideal $J(\mathrm{Cl}(R_j, \Psi_i))$ of $\mathrm{Cl}(R_j, \Psi_i)$ over $J(R_j)$ equals the degree of Ψ_i over E_1.*

(IV) *The finite extension E_{1i} of (II) is an Eisenstein extension of Ψ_i relative to the Dedekind ring $\mathrm{Cl}(R_j, \Psi_i)$ $(1 \leqslant j \leqslant n_1, 1 \leqslant i \leqslant s)$.*

Exercises

1. (Collins) Let R be a Dedekind ring, $f(t), g(t) \in R[t]$ and $f(t)$ monic and separable with $n = \deg(f) > 0$. Then show that the resultant of the polynomials $f(t)$, $y - g(t)$ relative to $R[y]$ provides a monic polynomial $m(y)$ over R such that $m(g(\xi)) = 0$ for $\xi = t/f(t)R[t] \in \Lambda_f$ and $m(t) = m_{g(\xi)}(t)$.
2. (Collins) Let $p_1, p_2, \ldots, p_k \in \mathbb{Z}^{>1}$ be mutually prime $(k \in \mathbb{N})$. Let $f(x, y), h(x, y)$ be two non-zero polynomials of $\mathbb{Z}[x, y]$. Let m be the degree of f in x over $\mathbb{Z}[y]$, n the degree of h in x over $\mathbb{Z}[y]$. Let R_i be the m, n-resultant of f, h modulo $\mathbb{Z}/p_i\mathbb{Z}[y]$, and let R be the $p_1 p_2 \cdots p_k$-reduced polynomial in y congruent to R_i modulo $p_i\mathbb{Z}[y]$ $(1 \leqslant i \leqslant k)$. Assuming that the coefficients of f also are reduced modulo $p_1 \cdots p_k$ show that $R = \mathrm{Res}_y(f, h)$. (For an improved version see [1].)
3. Apply the Dedekind test to the polynomials $t^3 - 2t^2 - 9t + 2, t^4 + 5t^2 + 1, t^3 - 3t + 6$ of $\mathbb{Z}[t]$.
4. Let $F = \mathbb{Q}(\beta)$ for a zero β of $f(t) = t^3 - t^2 - 2t - 8$ of $\mathbb{Z}[t]$. Prove that $o_F = \mathrm{Cl}(\mathbb{Z}, F)$ has no $\mathbb{Z}$-basis of the form $1, \rho, \rho^2$.
   (Hint: *A* $\mathbb{Z}$-basis of o_F is $1, \beta, (\beta + \beta^2)/2$. Then any presentation $\rho = a_0 + a_1\beta + a_2(\beta + \beta^2)/2$ $(a_0, a_1, a_2 \in \mathbb{Z})$ yields $a_2 \equiv 1 \bmod 2$, hence $\beta^2 \notin \mathbb{Z} + \mathbb{Z}\rho + \mathbb{Z}\rho^2$.)
5. Prove (6.10) in the case where n is the power of a prime number.

5

Units in algebraic number fields

5.1. Introduction

In this chapter we consider unital ring extensions of $\mathbb{Z}$ which have a finite $\mathbb{Z}$-basis. These structures were already analyzed in some detail in chapter 4. As rings of an equation they were introduced in chapter 1, section 4 in the form $\mathbb{Z}[t]/f\mathbb{Z}[t]$ for non-constant monic polynomials of $\mathbb{Z}[t]$. In the sequel we shall always assume f to be irreducible, since the general case can be derived from this case without difficulties by the methods of the preceding chapter (but see also [13]). Hence, if $\rho \in \mathbb{C}$ denotes a zero of $f(t)$, then $F = \mathbb{Q}(\rho)$ is an *algebraic number field* of *degree* $n = \deg(f)$ over $\mathbb{Q}$. The integral closure of $\mathbb{Z}$ in F is denoted by o_F. We shall consider the invertible elements of o_F and of those unital subrings R of finite index in o_F (see also section 2), shortly called *orders of F*.

The importance of the units of such rings R is best demonstrated by a simple example. If we try to find all solutions of

$$a^2 - 5b^2 = 11 \tag{1.1}$$

in $\mathbb{Z}$, then we see at once one solution: $a = 4$, $b = 1$. (With every solution a, b also $\pm a$, $\pm b$ is one, hence we can assume $a, b \in \mathbb{N}$). It is clear that the solvability of (1.1) is closely connected to the ring $R = \mathbb{Z}[5^{\frac{1}{2}}]$ because of

$$a^2 - 5b^2 = \mathrm{N}(a + b5^{\frac{1}{2}}). \tag{1.2}$$

Thus two solutions a, b and $\tilde{a}$, $\tilde{b}$ at once yield

$$1 = \mathrm{N}\left(\frac{a + b5^{\frac{1}{2}}}{\tilde{a} + \tilde{b}5^{\frac{1}{2}}}\right), \tag{1.3}$$

and also $\varepsilon = (a + b5^{\frac{1}{2}})/(\tilde{a} + \tilde{b}5^{\frac{1}{2}})$, ε^{-1} are in $\mathbb{Z}[5^{\frac{1}{2}}]$ as will be shown in chapter 6. Hence for a unit $\varepsilon = e_1 + e_2 5^{\frac{1}{2}}$ and (a, b) a solution of (1.1) also

$$(e_1 a + 5e_2 b, e_1 b + e_2 a) \tag{1.4}$$

is a solution. The structure of the unit group of $\mathbb{Z}[5^{\frac{1}{2}}]$ then shows that all solutions of (1.1) are of the form (1.4) for $(a, b) = (4, 1)$, ε a power of $9 + 4(5)^{\frac{1}{2}}$. The option that one can compute all units of R is therefore a premise for solving nonlinear Diophantine equations.

We give another instructive example for units in quadratic number fields. Clearly, each unit has norm ± 1 and lemma (2.2) shows that also each algebraic integer of norm ± 1 is a unit. For $F = \mathbb{Q}(d^{\frac{1}{2}})$, $d \in \mathbb{Z}$ square-free, this leads to the norm equation

$$x^2 - dy^2 = \pm 1 \quad (\text{for } d \equiv 2, 3 \bmod 4), \quad x^2 - dy^2 = \pm 4 \quad (\text{for } d \equiv 1 \bmod 4), \tag{1.5}$$

the so-called *Pell's equation*. An obvious approach for solving (1.5) would be to check for $y = 1, 2, 3, \ldots$, whether $\pm 1 + dy^2$ (respectively $\pm 4 + dy^2$) is a square. The solution for which y is smallest is called the *fundamental unit* of F. For small d (say $d < 50$) this seems to work quite well. But let us compare the fundamental units of $\mathbb{Q}(93^{\frac{1}{2}})$, $\mathbb{Q}(94^{\frac{1}{2}})$, $\mathbb{Q}(95^{\frac{1}{2}})$: They are

$$\varepsilon_1 = \frac{29 + 3(93)^{\frac{1}{2}}}{2},$$

$$\varepsilon_2 = 2\,143\,295 + 221\,064(94)^{\frac{1}{2}},$$

$$\varepsilon_3 = 39 + 4(95)^{\frac{1}{2}},$$

respectively. Hence, our first idea for solving (1.5) was certainly not good enough even for quite small values of d. For example, the first coefficient of the fundamental unit of $\mathbb{Q}(9199^{\frac{1}{2}})$ has already 88 digits.

If we try to solve $x^2 - 39y^2 = -1$ and write a simple computer program for the approach made above most computers will indeed provide a solution, even though that equation can have no integral solution since it is already insolvable in $\mathbb{Z}/4\mathbb{Z}$. The reason is that most computers give no warning when integer overflow occurs. Assuming a wordsize of 32 bits the result for $39y^2 - 1$ will already become false in case of $y = 7421$. This little exercise is quite instructive in a course on constructive number theory.

In the following sections we develop the theory of units of number fields (section 2), respectively orders R in such fields, and also the tools for a constructive approach. A first step in that direction is done in section 3. There we present a new method for solving norm equations which will be frequently applied later on.

In section 4 we determine methods for the computation of the units of finite order, i.e. the roots of unity of R. Then we discuss how to produce arbitrarily many integers of R of bounded (small) absolute norm, from which we derive units by forming quotients whenever possible. From there it is easy to construct subgroups of finite index in the whole unit group $U(R)$ of R. But for the final step, the enlarging of such a subgroup of finite index to $U(R)$

we need an upper bound for that index. Such bounds are derived in section 6. In section 7 we carry out the enlarging in detail. Finally, section 8 contains some remarks on the computerization of those methods, especially on necessary subroutines and error estimates.

We note that the methods for computing units which we present in this chapter were developed for arbitrary algebraic number fields. So far, no other methods of similar efficiency for the calculation of fundamental units have been published. However, for special number fields there are several efficient methods. We only mention the work of Shanks and H.W. Lenstra for quadratic fields, of Buchmann, Godwin, Gras, Williams for cubic fields, S. Mäki for special sextic fields and of Bernstein, Stender for parametrized fields. Some of their papers are listed in the additional references for this chapter.

5.2. The Dirichlet Theorem

In the sequel we assume $F = \mathbb{Q}(\rho)$ for a zero $\rho \in \mathbb{C}$ of the monic irreducible polynomial $f(t) \in \mathbb{Z}[t]$ of degree $n \geqslant 2$. In $\mathbb{C}[t]$ the polynomial f splits into $f(t) = \prod_{j=1}^{n}(t - \rho^{(j)})$ $(\rho^{(1)} = \rho)$, where the zeros $\rho^{(1)}, \ldots, \rho^{(s)}$ are assumed to be real $(0 \leqslant s \leqslant n)$, the remaining $2t = n - s$ zeros $\rho^{(s+1)}, \ldots, \rho^{(n)}$ to be complex and non-real. They are ordered such that $\rho^{(k)} = \overline{\rho^{(k+t)}}$ $(s+1 \leqslant k \leqslant s+t)$. The mapping $\rho \mapsto \rho^{(j)}$ provides a $\mathbb{Q}$-isomorphism of F into the complex numbers. The image of F (an element β of F) is called the jth *conjugate* of F (β) and denoted by $F^{(j)}$ $(\beta^{(j)})$. In particular, F is called *totally real* for $s = n$, *totally complex* for $s = 0$. In case $n = 2$ we speak of *real quadratic* $(s = 2)$ and *complex quadratic* $(s = 0)$ *fields*.

We want to compute the units of $\mathbb{Z}$-orders R contained in F (and therefore in $o_F = \mathrm{Cl}(\mathbb{Z}, F)$). The discriminant of such an order is denoted by $d(R)$, the discriminant of o_F by d_F (*discriminant* of F). $\omega_1, \ldots, \omega_m$ denotes a $\mathbb{Z}$-basis of R, m the $\mathbb{Z}$-rank of R. We show that $\mathfrak{Q}(R) = \mathbb{Q}\omega_1 + \cdots + \mathbb{Q}\omega_m$.

Let $0 \neq \beta \in R$ and $f_\beta(t) = t^n + \sum_{i=1}^{n} b_i t^{n-i} \in \mathbb{Z}[t]$ be the characteristic polynomial of β. Because of $f_\beta(\beta) = 0$ we obtain

$$(-1)^{n+1} N(\beta) = -b_n = \beta(\beta^{n-1} + b_1 \beta^{n-2} + \cdots + b_{n-1}) \in \beta R, \qquad (2.1)$$

hence $\alpha/\beta = (\alpha N(\beta)/\beta)/N(\beta) \in (1/N(\beta))R$ for every $\alpha \in R$.

The degree $[\mathfrak{Q}(R):\mathbb{Q}]$ is therefore the number of $\mathbb{Q}$-independent elements among $\omega_1, \ldots, \omega_m$, for $m < n$ we have $\mathfrak{Q}(R) \subset F$. Hence, in the sequel (and also in chapter 6) we assume $m = n, \mathfrak{Q}(R) = F$. The group of units of R (*unit group*) is denoted by $U(R)$.

Lemma (2.2)

An element $\varepsilon \in R$ is a unit, iff $\mathrm{N}(\varepsilon) = \pm 1$.

Proof
In case $N(\varepsilon) = \pm 1$ (2.1) yields $1 \in \varepsilon R$, hence $\varepsilon \in U(R)$. For $\varepsilon \in U(R)$ we apply the norm function to $\varepsilon(1/\varepsilon) = 1$. This yields $N(\varepsilon) | 1$ in $\mathbb{Z}$. □

We recall from chapter 1, section 6 that two elements α, β of R are called *associate*, $\alpha \sim \beta$, if there is a unit $\varepsilon \in U(R)$ with $\alpha = \beta\varepsilon$.

Lemma (2.3)
R contains only a finite number of non-associate elements of bounded norm.

Proof
It suffices to prove that R contains only finitely many non associate elements γ with $|N(\gamma)| = c$ ($c \in \mathbb{Z}^{\geq 2}$). For arbitrary $\alpha \in R$ we have $\alpha \equiv a_1\omega_1 + \cdots + a_n\omega_n \bmod cR$ ($0 \leq a_i < c$; $1 \leq i \leq n$), and R contains c^n congruence classes modulo c. We prove that there are at most c^n non-associate elements γ of R satisfying $|N(\gamma)| = c$. For this purpose let $\alpha, \beta \in R$, $\alpha \equiv \beta \bmod cR$ and $|N(\alpha)| = |N(\beta)| = c$. Then there exists $\gamma \in R$ with $\alpha - \beta = c\gamma$ which yields $\alpha/\beta = 1 \pm (N(\beta)/\beta)\gamma \in R$. Analogously, $\beta/\alpha \in R$, and therefore α and β must be associate. □

Lemma (2.4)
There are only finitely many elements α in R for which all of their conjugates are bounded: $|\alpha^{(j)}| \leq C$ $(1 \leq j \leq n)$, for some constant $C \in \mathbb{R}^{>0}$.

Proof
For the coefficients of the characteristic polynomial Φ_α of α, $\Phi_\alpha(t) = \sum_{i=0}^{n} a_i t^{n-i} \in \mathbb{Z}[t]$ $(a_0 = 1)$, we have

$$|a_i| = \left| \sum_{1 \leq j_1 < \cdots < j_i \leq n} \alpha^{(j_1)} \cdot \ldots \cdot \alpha^{(j_i)} \right| \leq \binom{n}{i} C^i \quad (1 \leq i \leq n).$$

Hence, there are only finitely many possibilities for Φ_α. □

Lemma (2.5)
Let $\zeta \in R$. Then ζ is a root of unity, iff $|\zeta^{(j)}| = 1$ $(1 \leq j \leq n)$.

Proof
Each root of unity naturally satisfies $|\zeta^{(j)}| = 1$ $(1 \leq j \leq n)$. Now let $\zeta \in R$ with $|\zeta^{(j)}| = 1$ $(1 \leq j \leq n)$. According to (2.4) there can be only finitely many different elements among ζ^k $(k \in \mathbb{N})$. If, for example, $\zeta^k = \zeta^l$ $(k < l)$, we obtain $\zeta^{l-k} = 1$, which shows that ζ is a root of unity. □

Lemma (2.6)
The roots of unity of R from a finite cyclic group which we denote by $TU(R)$ (torsion subgroup of $U(R)$).

Proof
Let $TU(R)$ be the set of all roots of unity of R. Clearly, $TU(R)$ is a subgroup of $U(R)$. Each $\zeta \in TU(R)$ is of finite order $\mathrm{ord}(\xi) = \min\{m \in \mathbb{N} \mid \zeta^m = 1\}$. (This

is a consequence of the proof of (2.5).) The group is finite because of (2.4).

The proof of $TU(R)$ – and similarly of any finite multiplicative subgroup of a (commutative) field – being cyclic was given in chapter 2, section 5. □

Example (2.7)

(a) For complex quadratic fields we have $U(F) = TU(F)$. Let $d < 0$ be the discriminant of a complex quadratic field F. Then $1, (d + d^{\frac{1}{2}})/2$ is an integral basis of F. Each unit $\varepsilon \in o_F$ must satisfy $\mathrm{N}(\varepsilon) = \pm 1$ because of (2.2). For $\varepsilon = e_1 + e_2(d + d^{\frac{1}{2}})/2$ we obtain $\pm 1 = (2e_1 + e_2 d)^2/4 - d(e_2^2/4)$. (Because of $d < 0$ the norm of each element of F is non-negative.) For $e_2 = 0$ the only solutions are $\varepsilon = \pm 1$. For $e_2 \neq 0$ we obtain (because of $d \leqslant -3$)

$$e_2 = \pm 1, \quad e_1 = 2e_2 \qquad \text{in case } d = -4$$

and

$$e_2 = \pm 1, \quad e_1 = e_2, 2e_2 \quad \text{in case } d = -3.$$

Hence, the unit group of a complex quadratic field consists of ± 1 except for $\mathbb{Q}(-1)^{\frac{1}{2}}$, $\mathbb{Q}(-3)^{\frac{1}{2}}$ where it consists of the 4th, 6th roots of unity respectively.

(b) For real quadratic fields there are units which are not roots of unity. For example, in $R = \mathbb{Z}[2^{\frac{1}{2}}]$ the element $\varepsilon = 1 + 2^{\frac{1}{2}}$ has $\mathrm{N}(\varepsilon) = -1$, hence it belongs to $U(R)$ but obviously it is not in $TU(R)$ because of (2.5). Later on we shall see that $TU(R) = U(R)$, if and only if $F = \mathbb{Q}$ or F is a complex quadratic field (compare (2.16)).

To determine the structure of $U(R)$ completely, we need

Definition (2.8)

Units $\varepsilon_1, \ldots, \varepsilon_k \in U(R)$ $(k \in \mathbb{N})$ *are called* **independent**, *if an equation* $\varepsilon_1^{m_1} \cdot \ldots \cdot \varepsilon_k^{m_k} = 1$ $(m_i \in \mathbb{Z},\ 1 \leqslant i \leqslant k)$ *yields* $m_1 = m_2 = \cdots = m_k = 0$. *Otherwise they are called* **dependent.**

If $\varepsilon_1, \ldots, \varepsilon_k \in U(R)$ are independent units, obviously none of them can be a root of unity. On the other hand, $\varepsilon = 1 + 2^{\frac{1}{2}}$ is an independent unit, and so is every power of ε in $\mathbb{Z}[2^{\frac{1}{2}}]$. To exhibit a maximal set of independent units we make use of Minkowski's Convex Body Theorem. For fixed $k \in \{1, 2, \ldots, s + t\}$ we choose

$$c_{i,k} = c_{i,k}(1) \in \mathbb{R}^{>0} \quad (1 \leqslant i \leqslant n) \tag{2.9a}$$

subject to

$$c_{i,k} < 1 \quad (1 \leqslant i \leqslant s + t,\ i \neq k), \quad c_{s+t+j,k} = c_{s+j,k} \quad (1 \leqslant j \leqslant t), \quad \prod_{i=1}^{n} c_{i,k} = |d(R)|^{\frac{1}{2}}.$$

(Here $d(R)$ denotes the discriminant of R.) Since

$$\Lambda := \bigoplus_{i=1}^{n} \mathbb{Z}(\omega_i^{(1)}, \ldots, \omega_i^{(s)},\ \operatorname{Re}\omega_i^{(s+1)},\ \operatorname{Im}\omega_i^{(s+1)}, \ldots, \operatorname{Re}\omega_i^{(s+t)},\ \operatorname{Im}\omega_i^{(s+t)})^t$$

is an n-dimensional lattice in $\mathbb{R}^n$ of discriminant $2^{-t}|d(R)|^{1/2}$, R contains an element $\alpha_k = \alpha_k(1) \neq 0$ satisfying

$$\left.\begin{array}{ll} |\alpha_k^{(j)}| \leqslant c_{j,k} & (1 \leqslant j \leqslant s), \\ |\operatorname{Re} \alpha_k^{(j)}| \leqslant c_{j,k}/2^{\frac{1}{2}}, & \\ |\operatorname{Im} \alpha_k^{(j)}| \leqslant c_{j,k}/2^{\frac{1}{2}} & (s+1 \leqslant j \leqslant s+t), \end{array}\right\} \tag{2.9b}$$

by chapter 3 (4.4). This, of course, implies

$$|\alpha_k^{(j)}| \leqslant c_{j,k} \quad (1 \leqslant j \leqslant n).$$

Thus we obtain a sequence $(\alpha_k(l))_{l\in\mathbb{N}}$ of non-zero elements of R by choosing $c_{i,k}(l+1) \in \mathbb{R}^{>0}$ subject to

$$\left.\begin{array}{r} c_{i,k}(l+1) < |\alpha_k(l)^{(i)}| \quad (1 \leqslant i \leqslant s+t, i \neq k), \\ c_{s+t+j,k}(l+1) = c_{s+j,k}(l+1) \quad (1 \leqslant j \leqslant t), \\ \prod_{i=1}^{n} c_{i,k}(l+1) = |d(R)|^{\frac{1}{2}} \end{array}\right\} \tag{2.9c}$$

for $l = 1, 2, \ldots$. The elements $\alpha_k(l)$ then satisfy

$$1 > |\alpha_k(l)^{(j)}| > |\alpha_k(l+1)^{(j)}| \quad (l \in \mathbb{N}; 1 \leqslant j \leqslant s+t, j \neq k). \tag{2.10}$$

All elements of this sequence have a norm whose absolute value is bounded by $|d(R)|^{\frac{1}{2}}$. According to (2.3) there must be indices $l < \tilde{l}$ for which $\alpha_k(l)$ and $\alpha_k(\tilde{l})$ are associate: $\varepsilon_k \alpha_k(l) = \alpha_k(\tilde{l})$ for some $\varepsilon_k \in U(R)$. The unit ε_k satisfies the inequalities

$$|\varepsilon_k^{(i)}| < 1, \quad |\varepsilon_k^{(k)}| > 1 \quad (1 \leqslant i \leqslant s+t, i \neq k) \tag{2.11}$$

because of (2.10) and $|\mathrm{N}(\varepsilon_k)| = 1$.

Lemma (2.12)

Let $r = s+t-1$ and $\varepsilon_k \in R$ $(1 \leqslant k \leqslant r)$ be units satisfying (2.11). *Then $\varepsilon_1, \ldots, \varepsilon_r$ are independent.*

Proof

Let us assume that $\varepsilon_1, \ldots, \varepsilon_r$ are dependent. Then there exist $m_i \in \mathbb{Z}$ $(1 \leqslant i \leqslant r)$ satisfying $|m_1| + \cdots + |m_r| > 0$ and

$$1 = \prod_{i=1}^{r} \varepsilon_i^{m_i}. \tag{2.12a}$$

Let $I := \{i \in \{1, \ldots, r\} \mid m_i \geqslant 0\}$ and $J := \{1, \ldots, r\} \backslash I$. We order the factors of the right-hand side of (2.12a) correspondingly:

$$1 = \prod_{i\in I} \varepsilon_i^{m_i} \left(\prod_{i \in J} \varepsilon_i^{-m_i} \right)^{-1}, \tag{2.12b}$$

which remains valid by transition to conjugates. In case $J = \varnothing$ we obtain the contradiction

$$1 = \prod_{i \in I} |\varepsilon_i^{(s+t)}|^{m_i} < 1 \tag{2.12c}$$

because of (2.11) and $\sum_{i \in I} m_i = \sum_{i=1}^{r} |m_i| > 0$. A similar argument excludes the possibility $I = \varnothing$. In case $I \neq \varnothing$, $J \neq \varnothing$ (2.11) yields

$$1 < \prod_{j \in \tilde{J}} |\varepsilon_i^{(j)}| \text{ for each } i \in J \text{ and } \tilde{J} := J \cup \{j \mid s+t < j < n, j-t \in J\}. \tag{2.12d}$$

Since $\sum_{i \in J} |m_i|$ and $\sum_{i \in I} m_i$ are both positive, we obtain

$$1 < \prod_{i \in J} \prod_{j \in \tilde{J}} |\varepsilon_i^{(j)}|^{|m_i|} = \prod_{j \in \tilde{J}} \prod_{i \in I} |\varepsilon_i^{(j)}|^{m_i} < 1 \tag{1.12e}$$

because of (2.12d), (2.12b), (2.11) respectively. Hence, we get a contradiction in each case, the units $\varepsilon_1, \ldots, \varepsilon_r$ must be independent. □

Lemma (2.13)

Let $\varepsilon_1, \ldots, \varepsilon_r$ *as in* (2.12). *Then for every* $\varepsilon \in U(R)$ *the units* $\varepsilon, \varepsilon_1, \ldots, \varepsilon_r$ *are dependent.*

Proof

Let $\alpha \in R$. For abbreviation we set

$$w(\alpha) := \prod_{j=1}^{n} \max\{1, |\alpha^{(j)}|\}. \tag{2.13a}$$

It is obvious that

$$|\alpha^{(j)}| \leqslant w(\alpha) \quad (1 \leqslant j \leqslant n), \tag{2.13b}$$

$$w(\alpha)^{-1} = \prod_{j=1}^{n} \min\{1, |\alpha^{(j)}|\} \text{ for } \alpha \in U(R) \text{ because of } |N(\alpha)| = 1. \tag{2.13c}$$

According to (2.4) there are only finitely many elements $\beta \in R$ satisfying $w(\beta) \leqslant w(\alpha)$. We claim that we can multiply each $\varepsilon \in U(R)$ with suitable powers $\varepsilon_i^{m_i}$ $(1 \leqslant i \leqslant r)$ to obtain

$$\eta = \varepsilon \varepsilon_1^{m_1} \cdot \ldots \cdot \varepsilon_r^{m_r}, \tag{2.13d}$$

subject to

$$|\varepsilon_i^{(i)}|^{-1} \leqslant |\eta^{(i)}| \leqslant |\varepsilon_i^{(i)}| \quad (1 \leqslant i \leqslant r). \tag{2.13e}$$

Namely, let us assume that the unit η of (2.13d) violates (2.13e), i.e. there is $\delta \in \{1, -1\}$ such that

$$|\eta^{(i)}|^{\delta} > |\varepsilon_i^{(i)}|$$

for some $i \in \{1, \ldots, r\}$. Then

$$\tilde{\eta} := \eta \varepsilon_i^{-\delta}$$

satisfies the inequalities

$$|\tilde{\eta}^{(i)}|^{\delta} > 1, \quad |\tilde{\eta}^{(j)}|^{\delta} > |\eta^{(j)}|^{\delta}$$

for $j=1,\ldots,r+1, j\neq i$; therefore we obtain for $\delta=1$:

$$w(\tilde{\eta})^{-1}=\prod_{j=1}^{n}\min\{1,|\tilde{\eta}^{(j)}|\}>\prod_{j=1}^{n}\min\{1,|\eta^{(j)}|\}=w(\eta)^{-1},$$

and, for $\delta=-1$:

$$w(\tilde{\eta})=\prod_{j=1}^{n}\max\{1,|\tilde{\eta}^{(j)}|\}<\prod_{j=1}^{n}\max\{1,|\eta^{(j)}|\}=w(\eta).$$

Hence, in each case we obtain a unit $\tilde{\eta}$ such that $w(\tilde{\eta})<w(\eta)$. Since there are only finitely many possibilities for $\tilde{\eta}$, as noted in the beginning of the proof, a unit η of the form (2.13d) with

$$w(\eta)=\min\{w(\varepsilon\varepsilon_1^{m_1}\cdot\ldots\cdot\varepsilon_r^{m_r})|\, m_i\in\mathbb{Z},\ 1\leqslant i\leqslant r\}$$

must satisfy (2.13e)

According to (2.13e) and $|\mathrm{N}(\eta)|=1$ all conjugates of η are bounded, and because of (2.4) η belongs to a finite set of units

$$\mathfrak{U}:=\{\eta_1,\ldots,\eta_v\}. \tag{2.13f}$$

More precisely, for each $\varepsilon\in U(R)$ there are rational integers $m_1,\ldots,m_r$ such that

$$\varepsilon\varepsilon_1{}^{m_1}\cdot\ldots\cdot\varepsilon_r{}^{m_r}\in\mathfrak{U}.$$

(Note: $TU(R)\subset\mathfrak{U}$.) Since $\mathfrak{U}$ is finite, for each $\varepsilon\in U(R)$ there are exponents $l>m>0$ such that ε^l and ε^m yield the same element of $\mathfrak{U}$, say η_u, i.e. there are m, $l\in\mathbb{N}$, $m_1,\ldots,m_r$, $l_1,\ldots,l_r\in\mathbb{Z}$ such that

$$\varepsilon^m\varepsilon_1{}^{m_1}\cdot\ldots\cdot\varepsilon_r{}^{m_r}=\varepsilon^l\varepsilon_1^{l_1}\cdot\ldots\cdot\varepsilon_r{}^{l_r}=\eta_u\in\mathfrak{U}.$$

Hence,

$$\varepsilon^{l-m}\varepsilon_1{}^{l_1-m_1}\cdot\ldots\cdot\varepsilon_r{}^{l_r-m_r}=1, \tag{2.13g}$$

and the units ε, $\varepsilon_1,\ldots,\varepsilon_r$ are dependent. □

From the preceding lemmata we easily derive:

Theorem *(Dirichlet)* (2.14)

Let R be a subring of the integral closure of $\mathbb{Z}$ in an algebraic number field F of degree n. Let the $\mathbb{Z}$-rank of R be n, and let F have $s+2t$ conjugates ordered in the usual way. Let $r=s+t-1$. Then the unit group $U(R)$ of R is the direct product of its torsion subgroup, generated by a root of unity ζ, and r infinite cyclic groups, generated by so-called **fundamental units** *$E_1,\ldots,E_r$: $U(R)=\langle\zeta\rangle\times\langle E_1\rangle\times\cdots\times\langle E_r\rangle$.*

Proof

The torsion subgroup $TU(R)$ of R was already determined in (2.6). From the proof of (2.13) we know that each $\varepsilon \in U(R)$ is of a form

$$\left.\begin{array}{l}\varepsilon = \eta_l \varepsilon_1^{m_1} \cdot \ldots \cdot \varepsilon_r^{m_r} \quad (\eta_l \in \mathfrak{U} \text{ (compare (2.13f)),} \\ m_i \in \mathbb{Z},\ \varepsilon_i \in U(R) \text{ subject to (2.11), } 1 \leqslant i \leqslant r),\end{array}\right\} \tag{2.15a}$$

and that each unit is dependent from $\varepsilon_1, \ldots, \varepsilon_r$. In particular, there are minimal exponents $n_l \in \mathbb{N}$ such that $\eta_l^{n_l}$ is a power product of $\varepsilon_1, \ldots \varepsilon_r$ $(1 \leqslant l \leqslant v)$. Let

$$M := \operatorname{lcm}(n_1, \ldots, n_v). \tag{2.15b}$$

Then for each $\varepsilon \in U(R)$ the Mth power ε^M belongs to the subgroup $\langle \varepsilon_1, \ldots, \varepsilon_r \rangle$ of $U(R)$, i.e. this subgroup is of index at most M in $U(R)$ by chapter 3 (2.9),

$$\varepsilon^M = \varepsilon_1^{m_1} \cdot \ldots \cdot \varepsilon_r^{m_r} \quad (\varepsilon \in U(R);\ m_i \in \mathbb{Z},\ 1 \leqslant i \leqslant r). \tag{2.15c}$$

The rest of the proof is an easy consequence of the principal theorem on finitely generated abelian groups. However, for the reader who is not familiar with this theorem we also prove the remaining part.

As a system of generators of $U(R)$ modulo $TU(R)$ we shall obtain Mth roots of suitable power products $\varepsilon_1^{m_1} \cdot \ldots \cdot \varepsilon_r^{m_r}$. For this purpose we consider sets of units $\mathfrak{M}_i$ defined by

$$\mathfrak{M}_i := \{\varepsilon \in U(R) \mid \varepsilon^M \in \langle \varepsilon_i, \ldots, \varepsilon_r \rangle\} \quad (1 \leqslant i \leqslant r). \tag{2.15d}$$

In the presentation (2.15c) of ε^M for $\varepsilon \in \mathfrak{M}_i$ we have $m_1 = \cdots = m_{i-1} = 0$ and the occurring m_i form a $\mathbb{Z}$-ideal $f_i\mathbb{Z}$. We choose $E_i \in \mathfrak{M}_i$ for which the exponent of ε_i in (2.15c) is f_i $(1 \leqslant i \leqslant r)$. Hence, given $\varepsilon \in U(R)$ we successively get rational integers $a_1, \ldots, a_r$ by

$$\begin{aligned}\varepsilon^M = \varepsilon_1^{m_1} \cdot \ldots \cdot \varepsilon_r^{m_r} &= E_1^{Ma_1} \varepsilon_2^{\tilde{m}_2} \cdot \ldots \cdot \varepsilon_r^{\tilde{m}_r} \\ &= \cdots \\ &= E_1^{Ma_1} \cdot \ldots \cdot E_r^{Ma_r}.\end{aligned} \tag{2.15e}$$

Therefore $(\varepsilon E_1^{-a_1} \cdot \ldots \cdot E_r^{-a_r})^M = 1$, i.e. $\varepsilon E_1^{-a_1} \cdot \ldots \cdot E_r^{-a_r}$ belongs to $TU(R)$, and we have shown that a suitable $\zeta \in TU(R)$ and $E_1, \ldots, E_r$ generate $U(R)$. It remains to prove that $E_1, \ldots, E_r$ are independent. Let us assume

$$E_1^{m_1} \cdot \ldots \cdot E_r^{m_r} = 1 \quad (m_i \in \mathbb{Z},\ 1 \leqslant i \leqslant r). \tag{2.15f}$$

This implies

$$E_1^{m_1 M} \cdot \ldots \cdot E_r^{m_r M} = 1,$$

and we can substitute $\varepsilon_1, \ldots, \varepsilon_r$ and obtain $\varepsilon_1^{h_1} \cdot \ldots \cdot \varepsilon_r^{h_r} = 1$ for suitable $h_i \in \mathbb{Z}$ $(1 \leqslant i \leqslant r)$, hence $h_1 = \cdots = h_r = 0$. But, $h_1 = m_1 f_1$ which yields $m_1 = 0$. Successively we obtain $m_2 = \cdots = m_r = 0$. This and exercise 1 complete the proof.

□

Remark

The preceding lemmata and the Dirichlet theorem are valid also for $r = 0$. Especially (2.14) yields $U(R) = TU(R)$ iff $s + t = 1$, i.e. $s = 1, t = 0$ $(F = \mathbb{Q})$ or $s = 0, t = 1$ $(F = \mathbb{Q}((-d)^{\frac{1}{2}}),\ d \in \mathbb{N},\ d$ squarefree).

Unfortunately, the deduction of the Dirichlet theorem given in this section is not constructive. Of course, looking through the proofs carefully we could derive a method for determining $\zeta, E_1, \ldots, E_r$ in a finite number of steps. But this number is too large for practical computations. Therefore we shall develop better ways for the computation of $TU(R)$ and a system of fundamental units in the next sections.

We note that a system of fundamental units $E_1, \ldots, E_r$ is not at all unique for $U(R)$. For example, we can multiply each E_i by arbitrary powers of ζ. Also, for $E_1, \ldots, E_r$ being fundamental units, the units $\eta_1, \ldots, \eta_r$ satisfying

$$\eta_i := \prod_{j=1}^{r} E_j^{m_{ij}}, \tag{2.17}$$

subject to

$$M = (m_{ij}) \in GL(r, \mathbb{Z})$$

are a system of fundamental units.

Exercises

1. Let $E_1, \ldots, E_r$ be independent units such that $U(R) = TU(R) \times \langle E_1 \rangle \times \cdots \times \langle E_r \rangle$ and $TU(R) = \langle \zeta \rangle$. Prove that for every $\varepsilon \in U(R)$ the presentation $\varepsilon = \zeta^{m_0} E_1^{m_1} \cdot \cdots \cdot E_r^{m_r}$ $(m_i \in \mathbb{Z}, 0 \leqslant i \leqslant r)$ is unique.
2. Determine a fundamental unit $\varepsilon > 1$ in $\mathbb{Z}[7^{\frac{1}{2}}]$, $\mathbb{Z}[37^{\frac{1}{2}}]$ using (2.12)–(2.14).
3. Let p be an odd prime number and $R = \mathbb{Z}[\zeta]$ for $\zeta = e^{2\pi i/p}$. Prove:

    (i) For $\varepsilon \in U(R)$ $\varepsilon/\bar{\varepsilon}$ is also in $U(R)$.
    (ii) For $\varepsilon \in U(R)$ the quotient $\varepsilon/\bar{\varepsilon}$ satisfies $\varepsilon/\bar{\varepsilon} = \zeta^k$ for some $k \in \mathbb{N}$, $1 \leqslant k \leqslant p$.
    (iii) Each $\varepsilon \in U(R)$ is the product of an element of $U(R) \cap \mathbb{R}$ and an element of $TU(R)$.
    (iv) $U(R) = U(\mathbb{Z}[\zeta + \zeta^{-1}]) \times \langle \zeta \rangle$.

5.3. On solving norm equations I

In this section we develop a new method for solving norm equations which will be frequently employed in unit group and class group computations. As usual, F denotes an algebraic number field of degree $n = s + 2t$ over $\mathbb{Q}$ and o_F its ring of algebraic integers.

For $\mathbb{Q}$-linearly independent elements $\alpha_1, \ldots, \alpha_n \in o_F$ let

$$A = \mathbb{Z}\alpha_1 + \cdots + \mathbb{Z}\alpha_n \tag{3.1a}$$

be a free $\mathbb{Z}$-module of o_F and R_A the corresponding ring of coefficients

$$R_A = \{\alpha \in o_F \mid \alpha A \subseteq A\}. \tag{3.1b}$$

This concept includes two special cases, namely

$$A = R_A \text{ an order of } o_F, \tag{3.1c}$$

$$A \text{ an ideal of } o_F \text{ with } R_A = o_F. \tag{3.1d}$$

For given $k \in \mathbb{N}$ we consider the task of solving the equation

$$|\mathrm{N}(x)| = k, \tag{3.2}$$

by elements $x \in A$. If R_A contains infinitely many units and (3.2) is solvable, then there are infinitely many solutions since for $x \in A$ subject to $|\mathrm{N}(x)| = k$ also $|\mathrm{N}(x\varepsilon)| = k$ holds for all $\varepsilon \in U(R_A)$ (the unit group of R_A) because of (2.2).

Therefore we require additional restrictions for possible solutions x of (3.2) which make the set of solutions finite and also allow an efficient computation of all solutions. From the applications made in later sections it will become clear that those restrictions are connected with the problem in a natural way. From (2.4) we already know that the conditions

$$R_j \leqslant |x^{(j)}| \leqslant S_j \quad (R_j \in \mathbb{R}^{\geqslant 0},\ S_j \in \mathbb{R}^{>0};\ 1 \leqslant j \leqslant n) \tag{3.3}$$

for the conjugates of $x \in A$ guarantee that there are at most finitely many $x \in A$ satisfying both (3.2) and (3.3).

The standard method of determining all solutions of (3.2), (3.3) is the following. We calculate a *dual basis*

$$\alpha_1^*, \ldots, \alpha_n^* \in F, \tag{3.4a}$$

for A subject to

$$\mathrm{Tr}(\alpha_i \alpha_j^*) = \delta_{ij} \quad (1 \leqslant i, j \leqslant n). \tag{3.4b}$$

This is usually done by floating point arithmetic up to a prescribed accuracy (compare section 8). Then the coefficients x_i $(1 \leqslant i \leqslant n)$ of $x = x_1\alpha_1 + \cdots + x_n\alpha_n \in A$ satisfy

$$x_i = \mathrm{Tr}(x\alpha_i^*), \tag{3.5a}$$

respectively,

$$|x_i| \leqslant \sum_{j=1}^{n} |x^{(j)}|\,|\alpha_i^{*(j)}|. \tag{3.5b}$$

If the conjugates of x are bounded by (3.3), then (3.5b) yields numerical bounds for x_i:

$$|x_i| \leqslant \sum_{j=1}^{n} S_j |\alpha_i^{*(j)}| =: T_i \quad (1 \leqslant i \leqslant n). \tag{3.5c}$$

The bounds T_i strongly depend on the size of the absolute values of the conjugates of the elements of the dual basis. Hence, we recommend that we operate with a basis $\alpha_1, \ldots, \alpha_n$ of A which is reduced with respect to the length $\sum_{j=1}^{n} |\alpha^{(j)}|^2$. Given the box

$$Q := \{\mathbf{x} \in \mathbb{R}^n \mid |x_i| \leqslant T_i\}, \tag{3.6a}$$

we just enumerate $Q \cap \mathbb{Z}^n$ and test for each of its

$$M_Q = \prod_{i=1}^{n} (2\lfloor T_i \rfloor + 1) \tag{3.6b}$$

elements $\mathbf{x}$ whether $x := \sum_{i=1}^{n} x_i \alpha_i$ satisfies (3.2). Actually it suffices to choose $\mathbf{x}$ lexicographically positive and to omit $\mathbf{x} = \mathbf{0}$. Hence, there remain

$$\tfrac{1}{2}(M_Q - 1) \tag{3.6c}$$

possibilities for $\mathbf{x}$ for which the norms of the corresponding elements $x \in A$ must be computed.

Though norm computations can be done by $O(n^2)$ arithmetical operations (see section 8) this method cannot be recommended if M_Q is large since very few lattice points of $\mathbb{Z}^n \cap Q$ lie on the norm surface

$$\left| \prod_{j=1}^{n} \left(\sum_{i=1}^{n} x_i \alpha_i^{(j)} \right) \right| = k \quad (\mathbf{x} \in \mathbb{R}^n). \tag{3.6d}$$

The following example will illustrate this.

Example (3.7)
Let $F = \mathbb{Q}(14^{\frac{1}{2}})$, $A = R_A = o_F$, $\alpha_1 = 1$, $\alpha_2 = 14^{\frac{1}{2}}$, $\alpha_1^* = \frac{1}{2}$, $\alpha_2^* = 14^{\frac{1}{2}}/28$. We want to compute all solutions of $|\mathrm{N}(x)| = k$, $x = x_1 + x_2 14^{\frac{1}{2}} (x_1, x_2 \in \mathbb{Z})$, $k \in \{67, 69\}$, subject to $S_i = 44.81$ for $k = 67$, $S_i = 45.47$ for $k = 69$ $(i = 1, 2)$. (The bounds S_i are derived from the fundamental unit $\varepsilon = 15 + 4(14)^{\frac{1}{2}}$, compare chapter 6 (5.2).) Then $Q \cap \mathbb{Z}^2$ contains 1023, respectively 1137, lexicographically positive vectors $\mathbf{x}$. Among them $(9, \pm 1)^t$ yield the only solutions in case of $k = 67$, for $k = 69$ there is no solution at all.

Hence, in the sequel we do not consider all lattice points of Q but cover that part of the surface described by (3.6d) which is contained in Q by suitable ellipsoids centered at the origin. Solutions of (3.2), (3.3) then correspond to certain lattice points of those ellipsoids which can be easily calculated by the methods used in chapter 3. It has been proved in [1] by a worst case analysis that this new method requires only the square-root of the number of the arithmetical operations which the old standard method uses on the average provided that the box Q is large.

We had the idea of this new approach when we read the paper [2] by K. Mahler in which he develops algebraic number theory from a new concept. He considers functions on the normalized valuations of F with values in $\mathbb{Q}^{>0}$, so-called ceilings. This will be discussed in some detail in chapter 6, section 5. However, the result in which we are interested can be formulated and proved without the use of ceilings. The main result is contained in the following theorem.

Theorem (3.8)
Let (3.2), (3.3) *be solvable and* $\gamma \in \mathbb{R}^{>0}$. *Let* $\lambda, L_j, U_j \in \mathbb{R}$ $(1 \leqslant j \leqslant n)$ *be defined by*

$$h \colon \mathbb{R}^{>1} \to \mathbb{R} \colon t \mapsto \frac{t}{t-1} - \frac{1}{\log t}, \tag{3.8a}$$

$$g: \mathbb{R}^{>1} \to \mathbb{R}: t \mapsto (1-h(t))t^{h(t)} + h(t)t^{h(t)-1}, \tag{3.8b}$$

$$\lambda \in \mathbb{R}^{>1} \textit{ is the unique zero of } g(t) - \left(1+\frac{\gamma}{k}\right)^{2/n} = 0, \tag{3.8c}$$

$$L_j := \left\lfloor \frac{-2}{\log \lambda}\left(\log S_j - \frac{\log k}{n}\right)\right\rfloor,$$

$$U_j := \left\lceil \frac{-2}{\log \lambda}\left(\log R_j - \frac{\log k}{n}\right)\right\rceil \quad (1 \leqslant j \leqslant n;\ R_j,\ S_j \textit{ as in } (3.3)). \tag{3.8d}$$

Then (3.2) *has a solution* $x = \sum_{i=1}^{n} x_i \alpha_i$ *subject to* (3.3) *and*

$$\sum_{j=1}^{n} \lambda^{r_j} |x^{(j)}|^2 \leqslant n(k+\gamma)^{2/n}, \tag{3.8e}$$

where $\mathbf{r} = (r_1, \ldots, r_n)^t \in \mathbb{Z}^n$ *satisfies*

$$\sum_{j=1}^{n} r_j = 0, \tag{3.8f}$$

$$L_j \leqslant r_j \leqslant U_j \quad (1 \leqslant j \leqslant n), \tag{3.8g}$$

$$r_{s+t+j} - r_{s+j} \in \{0,1\}\ (1 \leqslant j \leqslant t) \textit{ and}$$
$$\#\{j \mid 1 \leqslant j \leqslant t,\ r_{s+t+j} - r_{s+j} = 1\} \in \{0,1\}. \tag{3.8h}$$

Proof

We start to show (3.8c). For all $t \in \mathbb{R}^{>1}$ we have $0 < h(t) < 1$ since the left-hand side is equivalent to $0 < t\log t - (t-1)$ and the right-hand side to $0 < t - 1 - \log t$. But the last two inequalities are valid for $t > 1$ by the mean value theorem. Namely, there holds an equality for $t = 1$, and the derivatives of both right-hand sides are positive for $t > 1$. The function $f: (0,1) \times \mathbb{R}^{>1} \to \mathbb{R}$: $(y,z) \mapsto (1-y)z^y + yz^{y-1}$ is strictly increasing in z for fixed y, and for fixed z it has precisely one maximum assumed for $y_z := h(z) = z/(z-1) - 1/\log z$. This is immediate from the corresponding partial derivatives of f. Therefore the function g of (3.8b) is strictly increasing in t and because of $\lim_{t \searrow 1} g(t) = 1$ we obtain a unique solution $\lambda \in \mathbb{R}^{>1}$ of $g(t) = (1+\gamma/k)^{2/n}$.

Let $\alpha = a_1\alpha_1 + \cdots + a_n\alpha_n \in A$ be a solution of (3.2), (3.3) which exists according to our premise. We set

$$y_j = k^{-2/n} |\alpha^{(j)}|^2 \quad (1 \leqslant j \leqslant n), \tag{3.9a}$$

which implies

$$\prod_{j=1}^{n} y_j = 1. \tag{3.9b}$$

Hence, we can write y_j in the form

$$y_j = \lambda^{-r_j + \varepsilon_j} \quad (1 \leqslant j \leqslant n), \tag{3.9c}$$

where $r_j \in \mathbb{Z}$, $\varepsilon_j \in \mathbb{R}$ satisfy

$$\left.\begin{aligned} &0 = \sum_{j=1}^{n} r_j = \sum_{j=1}^{n} \varepsilon_j, \quad 0 \leqslant \varepsilon_k < 1 \quad (1 \leqslant k < n), \\ &r_j = r_{j+t}, \quad \varepsilon_j = \varepsilon_{j+t} \quad (s < j < s+t), \end{aligned}\right\} \tag{3.9d}$$

and therefore

$$r_n = -\sum_{j=1}^{n-1} r_j, \quad \varepsilon_n = -\sum_{j=1}^{n-1} \varepsilon_j \tag{3.9e}$$

because of (3.9b). Let us assume that

$$\varepsilon_n = -m + v \quad (m \in \mathbb{Z}^{\geqslant 0}, 0 \leqslant v < 1). \tag{3.9f}$$

We show how to change the r_i, ε_i successively until they accomplish (3.9c) as well as (3.8f, h).

In the case of $m = 0$ we are done. Therefore let us assume $m > 0$, and let $j \in \{1, 2, \ldots, n-1\}$ be the smallest index for which $\varepsilon_j = \max\{\varepsilon_i \mid 1 \leqslant i < n\}$. There are two possibilities:

(i) $v > \varepsilon_j$. In that case we replace v by $v - 1$, m by $m - 1$.
(ii) $\varepsilon_j \geqslant v$. In that case we replace ε_j by $\varepsilon_j - 1$, r_j by $r_j - 1$, m by $m - 1$ (i.e. ε_n by $\varepsilon_n + 1$), r_n by $r_n + 1$. For $s < j < s + t$ and $m > 0$ we also replace ε_{j+t} by $\varepsilon_{j+t} - 1$, r_{j+t} by $r_{j+t} - 1$, m by $m - 1$, r_n by $r_n + 1$.

We note that those replacements have no effect on (3.9c), (3.8f). We repeatedly apply this procedure as long as m is positive. Thus we obtain the result

$m = 0$ and all *ε_j $(1 \leqslant j \leqslant n)$ are in the closed interval $[a-1, a]$* (3.9g)
for $a := \max\{\varepsilon_j \mid 1 \leqslant j \leqslant n\}$, $0 \leqslant a < 1$.

Obviously (3.8f) is still satisfied as well as $r_j = r_{j+t}$ for $s + 1 \leqslant j \leqslant s + t$ with at most two exceptions. Namely, in case (ii) $r_{j+t} - r_j = 1$ may occur for one index j $(s < j < s + t)$ and we don't know about the difference $r_n - r_{s+t}$. Because of (3.9c) we have $r_{j+t} - r_j = \varepsilon_{j+t} - \varepsilon_j$ and $r_n - r_{s+t} = \varepsilon_n - \varepsilon_{s+t}$ and (3.9g) yields that those differences are $0, 1$ or -1. In the case of $r_{j+t} - r_j = 0$ or of $r_n - r_{s+t} = 0$ (3.8h) is already satisfied. So let us assume $r_{j+t} - r_j = |r_n - r_{s+t}| = 1$. This implies $\varepsilon_{j+t} = a = \varepsilon_j + 1$ and either $\varepsilon_n = a = \varepsilon_{s+t} + 1$ or $\varepsilon_n = a - 1 = \varepsilon_{s+t} - 1$. Hence, we replace ε_{j+t} by $\varepsilon_{j+t} - 1$, r_{j+t} by $r_{j+t} - 1$ and either ε_{s+t} by $\varepsilon_{s+t} + 1$, r_{s+t} by $r_{s+t} + 1$ or ε_n by $\varepsilon_n + 1$, r_n by $r_n + 1$. This yields $r_k = r_{k+t}$ $(s < k \leqslant s + t)$. By these considerations we get (3.8h) in each case.

Since α satisfies (3.3) we conclude by (3.9a, c) that
$2(\log R_j - \log k/n) \leqslant (-r_j + \varepsilon_j)\log\lambda = \log y_j \leqslant 2(\log S_j - \log k/n)$ $(1 \leqslant j \leqslant n)$,
hence (3.8g) because of $-1 < \varepsilon_j < 1$.

It remains to show (3.8e). Using the same notation as before we obtain for any solution α of (3.2), (3.3):

$$\begin{aligned}\sum_{j=1}^{n} \lambda^{r_j}|\alpha^{(j)}|^2 &= k^{2/n} \sum_{j=1}^{n} \lambda^{\varepsilon_j} \\ &\leqslant k^{2/n} \sum_{j=1}^{n} ((1-(a-\varepsilon_j)\lambda^a + (a-\varepsilon_j)\lambda^{a-1}) \\ &= k^{2/n} \sum_{j=1}^{n} ((1-a)\lambda^a + a\lambda^{a-1}) \\ &\leqslant k^{2/n} n g(\lambda) \\ &= n(k+\gamma)^{2/n}.\end{aligned}$$

Here, we made use of (3.9a, c), then of the convexity of the exponential function in the interval $[a-1, a]$, applied (3.9d) and then the inequality $f(a,\lambda) \leqslant f(h(\lambda),\lambda)$ together with (3.8c) shown at the beginning of the proof. □

Remarks

(i) The parameter γ seems to be somewhat artificial. However it helps to reduce the number of arithmetical operations drastically. The appropriate choice of γ is discussed at the end of this section.

(ii) If we actually want to solve norm equations (3.2) subject to (3.3) by using (3.8), we begin by calculating λ, for example with Newton's method. Then we need to generate all $\mathbf{r} \in \mathbb{Z}^n$ subject to (3.8f–h). For each $\mathbf{r}$ we then compute all $x \in A$ ($x = \sum_{i=1}^{n} x_i \alpha_i$, $x_i \in \mathbb{Z}$, $1 \leqslant i \leqslant n$) satisfying (3.8e) by chapter 3 (3.15). Since $r_{j+t} = r_j$ ($s < j \leqslant s+t$) is in general not fulfilled because of (3.8h), we must still prove that the left-hand side of (3.8e) is indeed a positive definite quadratic form in $x_1, \ldots, x_n$. This is the subject of the next lemma.

(iii) If $\alpha \in A$ solves (3.2), (3.3), then (3.9a, c, d) also yield

$$\begin{aligned}\sum_{j=1}^{n} \lambda^{r_j}|\alpha^{(j)}|^2 &= nk^{2/n}\left(\frac{1}{n}\sum_{j=1}^{n} \lambda^{\varepsilon_j}\right) \\ &\geqslant nk^{2/n}\left(\prod_{j=1}^{n} \lambda^{\varepsilon_j}\right)^{1/n} = nk^{2/n}.\end{aligned} \tag{3.10a}$$

This can be used in applications to speed up the inner loop of the algorithm (3.15) in chapter 3.

(iv) Every solution $x \in A$ of (3.8e) is of bounded norm:

$$\begin{aligned}|N(x)| &= \left(\prod_{j=1}^{n} \lambda^{r_j}|x^{(j)}|^2\right)^{1/2} \\ &\leqslant \left(\frac{1}{n}\sum_{j=1}^{n} \lambda^{r_j}|x^{(j)}|^2\right)^{n/2} \leqslant k+\gamma\end{aligned} \tag{3.10b}$$

because of (3.9d).

Lemma (3.11)

Let $\lambda \in \mathbb{R}^{>1}$ and $\mathbf{r} \in \mathbb{Z}^n$ subject to (3.8f–h). Then (3.8e) describes a positive definite quadratic form, namely,

$$\sum_{j=1}^{n} \lambda^{r_j} \left| \sum_{i=1}^{n} x_i \alpha_i^{(j)} \right|^2 = \mathbf{x}^t B^t B \mathbf{x}$$

for $\mathbf{x} = (x_1, \ldots, x_n)^t \in \mathbb{Z}^n$ and $B = (b_{ji}) \in \mathbb{R}^{n \times n}$ subject to

$$b_{ji} = \begin{cases} \lambda^{r_j/2} \alpha_i^{(j)} & 1 \leqslant j \leqslant s \\ (\lambda^{r_j} + \lambda^{r_{j+t}})^{1/2} \operatorname{Re} \alpha_i^{(j)} & s+1 \leqslant j \leqslant s+t \\ (\lambda^{r_j} + \lambda^{r_{j-t}})^{1/2} \operatorname{Im} \alpha_i^{(j-t)} & s+t+1 \leqslant j \leqslant n \end{cases} \quad (1 \leqslant i \leqslant n).$$

Proof

We easily calculate:

$$\begin{aligned} &\sum_{j=1}^{n} \lambda^{r_j} \left| \sum_{i=1}^{n} x_i \alpha_i^{(j)} \right|^2 \\ &= \sum_{j=1}^{s} \lambda^{r_j} \left(\sum_{i=1}^{n} x_i \alpha_i^{(j)} \right)^2 + \sum_{j=s+1}^{s+2t} \lambda^{r_j} \left(\left(\sum_{i=1}^{n} x_i \operatorname{Re} \alpha_i^{(j)} \right)^2 + \left(\sum_{i=1}^{n} x_i \operatorname{Im} \alpha_i^{(j)} \right)^2 \right) \\ &= \sum_{j=1}^{s} \left(\lambda^{r_j/2} \sum_{i=1}^{n} x_i \alpha_i^{(j)} \right)^2 \\ &\quad + \sum_{j=s+1}^{s+t} (\lambda^{r_j} + \lambda^{r_{j+t}}) \left(\left(\sum_{i=1}^{n} x_i \operatorname{Re} \alpha_i^{(j)} \right)^2 + \left(\sum_{i=1}^{n} x_i \operatorname{Im} \alpha_i^{(j)} \right)^2 \right) \\ &= \mathbf{x}^t B^t B \mathbf{x}. \end{aligned}$$

□

Finally, we discuss the appropriate choice of γ for applications of (3.8). Naturally (3.10b) suggests to choose γ small. But this results in small values of λ and therefore in a large number of quadratic forms to be considered because of (3.8c, d, g). Hence, we rather choose γ such that the total number of solutions $\mathbf{x}$ of (3.8e) for all $\mathbf{r}$ subject to (3.8f–h) becomes small. The number of quadratic forms to be considered is roughly proportional to $(\log \lambda)^{1-s-t}$ because of (3.8d, f, g, h). For each quadratic form the number of solutions $\mathbf{x}$ of (3.8e) is proportional to $k + \gamma$. Hence, we compute

$$y^* := \min \{ (\log \lambda)^{1-s-t} g(\lambda)^{n/2} \mid \lambda > 1 \}, \tag{3.12a}$$

where $g(\lambda)$ was defined in (3.8b). With the argument λ^* of a minimal solution of (3.12a) we calculate

$$\gamma^* := (y^* (\log \lambda^*)^{s+t-1} - 1) k \tag{3.12b}$$

according to (3.8c). For small values $n = s + 2t$ we present a list of values γ^*/k:

n, s	2,0	2,1	3,1	3,3	4,0	4,2	4,4	5,1
γ^*/k	0.0	0.7	0.7	1.9	0.7	1.9	4.1	1.8

(3.12c)

n, s	5,3	5,5	6,0	6,2	6,4	6,6	7,1	7,3
γ^*/k	4.0	7.9	1.8	3.9	7.6	14.6	3.8	7.4

n, s	7,5	7,7	8,0	8,2	8,4	8,6	8,8
γ^*/k	14.0	26.3	3.8	7.3	13.6	25.2	46.7

These values strongly agree with those for which the computation time in the examples of [1] was minimal. In some examples the choice of γ^* instead of $\gamma = 1$ reduced the computation time even by a factor of 10^{-2}. However, we note that decreasing γ^* need not effect the number $A(\gamma^*)$ of quadratic forms (3.8e), i.e. of vectors $\mathbf{r} \in \mathbb{Z}^n$ subject to (3.8f–h). Hence, we can save additional computation time by choosing

$$\tilde{\gamma} := \min\{\gamma \in \mathbb{Z}^{\geq 0} \mid A(\gamma) = A(\gamma^*)\}. \tag{3.12d}$$

Those numbers $A(\gamma)$ are easily calculated from (3.8a–d, g–h).

Exercises

1. Develop an algorithm which generates all $\mathbf{r} \in \mathbb{Z}^n$ subject to (3.8f–h) for given input data L_j, U_j $(1 \leq j \leq n)$.
2. Solve (3.7) by using (3.8).

5.4. Computation of roots of unity

From (2.6) we already know that the torsion subgroup $TU(R)$ of the unit group of the order R is a finite cyclic group. We denote its order by g. Obviously, g must be even since $TU(R)$ contains the subgroup $\langle -1 \rangle$ of order 2. Indeed, ± 1 are the only roots of unity of R in most cases.

Lemma (4.1)
If the number s of real conjugates of $F = \mathbb{Q}(R)$ is positive, then ± 1 are the only torsion units of R.

The proof is obvious and left as an exercise to the reader. For the case of $s = 0$ we present two different methods for determining $TU(R)$.

Method I.
By (2.5) $\zeta \in R$ is a root of unity, precisely if all conjugates of ζ are of absolute value one. In that case we clearly have $|\mathrm{N}(\zeta)| = 1$, a result we already knew from (2.2). Hence, it suffices to compute all $\zeta \in R$ subject to

$$|\mathrm{N}(\zeta)| = 1, \quad |\zeta^{(j)}| = 1 \quad (1 \leq j \leq n). \tag{4.2}$$

This can be easily done by the methods of the preceding section. Using the same notation we obtain for $A = R_A = R$ that $k = 1$ in (3.2) and $R_j = S_j = 1$ $(1 \leqslant j \leqslant n)$ in (3.3). Hence, for small n we can avoid the refined approach of (3.8) and just compute T_i $(1 \leqslant i \leqslant n)$ according to (3.5c) and enumerate $Q \cap \mathbb{Z}^n$ for the corresponding box as described in (3.5), (3.6). Of course, for larger n we shall apply (3.8).

Either way we obtain a finite set $\mathfrak{M}$ of elements of R which contains all $\zeta \in TU(R)$. We note that $\zeta \in \mathfrak{M}$ is in $TU(R)$, precisely if it satisfies (4.2) which can be easily checked.

Method II

The second method is based on the fact that $TU(R)$ is cyclic of order g. Therefore it suffices to exhibit a generating element ζ of $TU(R)$, a so-called *primitive gth root of unity*. Then we obtain $TU(R) = \langle \zeta \rangle$, a better result than by the first method which only provides all group elements and therefore the order of $TU(R)$.

If ζ is a primitive gth root of unity, then the same holds for ζ^i $(1 \leqslant i \leqslant g$, $\gcd(i, g) = 1)$ since ζ^i has order g, too. Therefore the number of primitive gth roots of unity is given by the value $\varphi(g)$ of Euler's φ-function. The primitive gth roots of unity are exactly the roots of

$$\Phi_g(t) = \prod_{d \mid g} (t^d - 1)^{\mu(g/d)} \in \mathbb{Z}[t] \tag{4.3}$$

of degree $\varphi(g)$, as we already saw in chapter 2, section 11.

Example (4.4)

Obviously, $\Phi_1(t) = t - 1$. We give a list of $\Phi_m(t)$ for even m, $m \leqslant 24$ and $m = 30$ which can be easily computed from (4.3).

m	$\Phi_m(t)$
2	$\dfrac{t^2 - 1}{t - 1} = t + 1$
4	$\dfrac{t^4 - 1}{t^2 - 1} = t^2 + 1$
6	$\dfrac{(t^6 - 1)(t - 1)}{(t^2 - 1)(t^3 - 1)} = t^2 - t + 1$
8	$\dfrac{t^8 - 1}{t^4 - 1} = t^4 + 1$
10	$\dfrac{(t^{10} - 1)(t - 1)}{(t^5 - 1)(t^2 - 1)} = t^4 - t^3 + t^2 - t + 1$

12	$\dfrac{(t^{12}-1)(t^2-1)}{(t^6-1)(t^4-1)}=t^4-t^2+1$
14	$\dfrac{(t^{14}-1)(t-1)}{(t^7-1)(t^2-1)}=t^6-t^5+t^4-t^3+t^2-t+1$
16	$\dfrac{t^{16}-1}{t^8-1}=t^8+1$
18	$\dfrac{(t^{18}-1)(t^3-1)}{(t^9-1)(t^6-1)}=t^6-t^3+1$
20	$\dfrac{(t^{20}-1)(t^2-1)}{(t^{10}-1)(t^4-1)}=t^8-t^6+t^4-t^2+1$
22	$\dfrac{(t^{22}-1)(t-1)}{(t^{11}-1)(t^2-1)}=t^{10}-t^9+t^8-t^7+t^6-t^5+t^4-t^3+t^2-t+1$
24	$\dfrac{(t^{24}-1)(t^4-1)}{(t^{12}-1)(t^8-1)}=t^8-t^4+1$
30	$\dfrac{(t^{30}-1)(t^5-1)(t^3-1)(t^2-1)}{(t^{15}-1)(t^{10}-1)(t^6-1)(t-1)}=t^8+t^7-t^5-t^4-t^3+t+1$

To determine $TU(R)$ it therefore suffices to compute m with $\varphi(m)$ maximal such that $\Phi_m(t)$ has a root ζ in R. This means ζ generates a subfield of F, hence we need to determine those $m \in \mathbb{N}$, m even, for which $\varphi(m)|n$. Since φ is multiplicative (see exercise 5) this is very easy, of course.

List of even $m \in \mathbb{N}$ satisfying $\varphi(m)|n$ for given $n \in \{2, 4, 6, 8, 10\}$. (4.5)
The corresponding polynomials $\Phi_m(t)$ where listed in (4.4).

n	2	4	6	8	10
m	2,4,6	2,4,6, 8,10,12	2,4,6, 14,18	2,4,6, 8,10,12, 16,20,24,30	2,4,6,22

For n odd we have $m = 2$, since $\varphi(p^k)$ is even in case $p^k > 2$.

Now, the determination of $TU(R)$ is no longer a problem, whence we know how to decide whether $\Phi_m(t)$ splits in $R[t]$ for the finitely many possible values of m for which $\varphi(m)|n$. This can, for example, be done by p-adic methods (see [12]). We present a different approach which is in general very efficient and gives a complete factorization of a polynomial of $\mathbb{Z}[t]$ in $F[t]$, F a finite extension field of $\mathbb{Q}$. We note that $F[t]$ is still Euclidean whereas $R[t]$ is usually not even a unique factorization domain. Of course, we must test afterwards whether the obtained factors in $F[t]$ are already contained in $R[t]$.

The idea of this method goes back to van der Waerden. It was subsequently improved by B. Trager [10]. We present it in a version refined for our purposes. To factor a polynomial $g(t)$ of $F[t]$ we make use of the fact that it is easy to compute greatest common divisors in $F[t]$. We use this to make $g(t)$ square-free (by computing $g(t)/\gcd(g(t), g'(t))$) and then to compute its irreducible factors by computation of gcds of $g(t)$ with suitable polynomials of $\mathbb{Q}[t]$. Also a transition from polynomials of $F[t]$ to ones of $\mathbb{Q}[t]$ via the norm is needed. In $\mathbb{Q}[t]$ it is no problem to factor polynomials applying Berlekamp's method which is implemented in many higher-level program packages like SAC-2 or MACSYMA (compare chapter 3 (3.43) ff.).

Definition (4.6)

Let $g(t) \in F[t]$ and $g^{(j)}(t) \in F^{(j)}[t]$ $(1 \leqslant j \leqslant n)$ be the corresponding polynomials over the conjugate fields obtained by applying conjugation to the coefficients of $g(t)$ only. Then the **norm of g(t)** *is defined by $N(g(t)) = \prod_{j=1}^{n} g^{(j)}(t)$.*

We note that $N(g(t)) \in \mathbb{Q}[t]$ and that

$$N(g_1(t)g_2(t)) = N(g_1(t))N(g_2(t)) \quad \text{for all } g_1(t), g_2(t) \in F[t]. \tag{4.7}$$

Lemma (4.8)

Let $g(t) \in F[t]$ be irreducible. Then $N(g(t))$ is the power of an irreducible polynomial of $\mathbb{Q}[t]$.

Proof

Let $c(t)$ be the irreducibe factor of $N(g(t))$ in $\mathbb{Q}[t]$ for which $g(t) | c(t)$ in $F[t]$. This implies $g^{(j)}(t) | c(t)$ in $F^{(j)}[t]$ $(1 \leqslant j \leqslant n)$ and therefore $N(g(t)) | c(t)^n$ in $\mathbb{Q}[t]$. □

Lemma (4.9)

Let $g(t) \in F[t]$ and $N(g(t))$ both be square-free. Let $q_1(t), \ldots, q_k(t)$ be the irreducible factors of $N(g(t))$ in $\mathbb{Q}[t]$. Then $\prod_{i=1}^{k} \gcd(g(t), q_i(t))$ is a factorization of $g(t)$ into irreducible factors in $F[t]$.

Proof

Let $g_1(t), \ldots, g_l(t)$ be the irreducible factors of $g(t)$ in $F[t]$. Since $N(g(t))$ is square-free we have $N(g_i(t)) = q_j(t)$ $(1 \leqslant i \leqslant l$ and suitable $j = j(i)$, $1 \leqslant j \leqslant k)$ according to (4.8). An assumption $N(g_i(t)) = N(g_j(t))$ for $1 \leqslant i < j \leqslant l$ yields a contradiction because of $N(g_i(t))\,N(g_j(t)) = N(g_i(t)g_j(t)) \mid N(g(t))$ and $N(g(t))$ is square-free. Hence, we obtain $q_i(t) = N(g_i(t))$ $(1 \leqslant i \leqslant l)$ by reordering the factors of $N(g(t))$, if necessary. Finally, (4.7) yields $k = l$ and we are done because of $\gcd(g_j(t)), N(g_i(t)) = 1$ for $i \neq j$. □

We have already noted that it is no problem to make $g(t) \in F[t]$ square-free. But even then, $N(g(t))$ is in general not square-free, take for example $g(t) \in \mathbb{Q}[t]$. Therefore we substitute t by $t - k\alpha$ in $g(t)$, $\alpha \in R$ a generating element for F, k a suitable rational integer.

Lemma (4.10)

Let $g(t) \in F[t]$ be square-free and $F = \mathbb{Q}(\alpha)$. Then there exists $k \in \mathbb{Z}$ such that $N(g(t - k\alpha))$ is square-free.

Proof

Let $\beta_i^{(j)}$ $(1 \leqslant j \leqslant n,\ 1 \leqslant i \leqslant m = \deg(g(t)))$ be the zeros of $g^{(j)}(t)$. Then the zeros of $g^{(j)}(t - k\alpha^{(j)})$ are $\beta_i^{(j)} + k\alpha^{(j)}$. Hence, $N(g(t - k\alpha))$ has multiple roots if for some indices $(i_1, j_1), (i_2, j_2), j_1 \neq j_2$: $\beta_{i_1}^{(j_1)} + k\alpha^{(j_1)} = \beta_{i_2}^{(j_2)} + k\alpha^{(j_2)}$, or

$$k = \frac{\beta_{i_2}^{(j_2)} - \beta_{i_1}^{(j_1)}}{\alpha^{(j_1)} - \alpha^{(j_2)}}. \tag{4.11}$$

Obviously, there are only finitely many possibilities for k such that $N(g(t - k\alpha))$ is not square-free. □

In practice a few trials $k = \pm 1, \pm 2, \ldots$ suffice to obtain a polynomial $g(t - k\alpha)$ with square-free norm. Then we can factor this norm in $\mathbb{Q}[t]$ and obtain the irreducible factors of $g(t - k\alpha)$ by calculating polynomial gcds in $F[t]$. This splits $g(t - k\alpha)$ into irreducible factors in $F[t]$, and, finally, by the reverse substitution $t \mapsto t + k\alpha$ we obtain the desired factorization of $g(t)$.

So we are done except for two things. One is the computation of the norm of a polynomial. It can be obtained by calculating $\mathrm{Res}(M_\alpha(x), g_x(t))$ with respect to x, where $M_\alpha(x)$ denotes the minimal polynomial of α and $g_x(t)$ is obtained from $g(t)$ by substituting x for α (see exercise 2).

There is a modular version for the computation of resultants by Collins (see [1] of chapter 4) which is very fast. However, we suggest a different method. For the computation of $U(R)$ it turns out to be of advantage to compute all conjugates of α up to machine precision (compare also section 5 of this chapter). Then the norm of $g(t) \in R[t]$ is calculated by floating point operations, and since we know $N(g(t)) \in \mathbb{Z}[t]$ the result is obtained by choosing the nearest integers for the coefficients. (For error estimates see section 8.) This is very easy and has the additional advantage that we need only one substitution $t \mapsto t - k\alpha$. Namely, from the proof of (4.10) we know that it suffices to choose k different from all possible quotients (4.11), where the numerator is the difference of roots of conjugates of $g(t)$ and the denominator is the difference of conjugates of α. A lower bound for the absolute value of the denominator is easily obtained once we have computed all conjugates of α. An upper bound for the absolute value of the numerator is also easily derived from the coefficients of the polynomials $g^{(j)}(t)$ $(1 \leqslant j \leqslant n)$ (compare exercise 4). Hence, for $g^{(j)}(t) = \sum_{i=0}^{m} g_i^{(j)} t^{m-i}$ we compute

$$T := \frac{2\max\{|g_i^{(j)}/g_0^{(j)}| + 1 \,|\, 1 \leqslant i \leqslant m,\ 1 \leqslant j \leqslant n\}}{\min\{|\alpha^{(i)} - \alpha^{(j)}| \,|\, 1 \leqslant i < j \leqslant n\}}. \tag{4.12}$$

Then every $k \in \mathbb{N}$ greater than T does the job.

The second remark is that we only obtain a factorization of $\Phi_m(t)$ over F, the factors need not be in $R[t]$. That we must check in each case separately. Before we give an algorithm for the computation of $TU(R)$ a simple example is given to illustrate the method.

Example (4.13)

Let $F = \mathbb{Q}((-3)^{\frac{1}{2}})$. Here we already know $TU(R)$ in case R is the maximal order of F from (2.7). (4.5) yields $\#TU(R) \in \{2, 4, 6\}$. We already computed $\Phi_6(t) = t^2 - t + 1$ in (4.4). Hence, let $g(t) = \Phi_6(t)$, $\alpha = (-3)^{\frac{1}{2}}$. From (4.11) we conclude $k = 1$ and compute

$$\begin{aligned} g(t-\alpha) &= t^2 - 2\alpha t + \alpha^2 - t + \alpha + 1 \\ &= t^2 - (1 + 2\alpha)t + (\alpha - 2), \\ N(g(t-\alpha)) &= (t^2 - (1+2\alpha)t + (\alpha - 2))(t^2 - (1 + 2\bar{\alpha})t + (\bar{\alpha} - 2)) \\ &= t^4 - 2t^3 + 9t^2 - 8t + 7 \\ &= (t^2 - t + 1)(t^2 - t + 7), \end{aligned}$$

$$\gcd(g(t-\alpha), t^2 - t + 1) = t - \frac{1 + (-3)^{\frac{1}{2}}}{2},$$

$$\gcd(g(t-\alpha), t^2 - t + 7) = t - \frac{1 + 3(-3)^{\frac{1}{2}}}{2}.$$

Therefore $\Phi_6(t)$ splits in F, but – for example – not in $\mathbb{Z}[(-3)^{\frac{1}{2}}]$. Hence, $\#TU(o_F) = 6, \#TU(\mathbb{Z}[(-3)^{\frac{1}{2}}]) = 2$.

Remarks (4.14)

In case R is the maximal order of F, a splitting of $\Phi_m(t)$ over F is also one over o_F, since the roots of $\Phi_m(t)$ are algebraic integers. In any case it suffices to produce one linear factor of $\Phi_m(t)$ in $R[t]$ as the primitive roots of unity are powers of each other.

After the preceding considerations the following algorithm for the determination of $TU(R)$ is immediate.

Algorithm for the computation of TU(R) (4.15)

Input. The rank n of $R/\mathbb{Z}$, a generating equation $f(t) = 0$ with root $\alpha \in R$

such that $\mathbb{Q}(R) = \mathbb{Q}(\alpha)$, and a module basis $\omega_1, \ldots, \omega_n$ of R over $\mathbb{Z}$. n splits into $n = s + 2t$, the number of real, respectively complex, conjugates of α.

Output. $m \in \mathbb{N}$ and $\Phi_m(t)$ such that $TU(R)$ is generated by the primitive mth roots of unity ζ with $\Phi_m(\zeta) = 0$.

Step 1. (R not totally complex). For $s > 0$ set $m \leftarrow 2$, $\Phi_2(t) \leftarrow t + 1$ and terminate, else go to 2.

Step 2. (Determine possible m). Compute a complete list $\mathfrak{L}$ of candidates for m (see exercise 6).

Step 3. (Computation of $TU(R)$). Determine $m \in \mathfrak{L}$ maximal such that $\Phi_m(t)$ has a linear factor in $R[t]$ and terminate.

Remarks (4.16)

(i) For $s > 0$ we must necessarily have $m = 2$, since any primitive mth root of unity is not in $\mathbb{R}$ in case $m > 2$.

(ii) The successive powers of a primitive mth root of unity ζ form an integral basis for the ring of integers in $\mathbb{Q}(\zeta)$. That was shown in chapter 4 (5.10). Hence, in many cases the discriminant composition formula, chapter 2 (9.29a), can be used to remove some elements m of $\mathfrak{L}$ from the list of candidates in addition to step 2.

(iii) The tools for step 3 were developed in (4.6)–(4.14).

Exercises

1. Prove lemma (4.1).
2. Let $F = \mathbb{Q}(\alpha)$ be an algebraic number field of degree n, and let $M_\alpha(t) \in \mathbb{Q}[t]$ be the minimal polynomial of α. Then there are three different methods for the computation of the norm of

$$g = \sum_{i=0}^{m} g_i \alpha^{m-i} \quad (m \leqslant n-1,\ g_i \in \mathbb{Q},\ 0 \leqslant i \leqslant m).$$

Set $g(t) := \sum_{i=0}^{m} g_i t^{m-i}$ and let M_g be the right regular representation of g with respect to the basis $1, \alpha, \ldots, \alpha^{n-1}$ of F. Using $N(g) = \prod_{j=1}^{n} g^{(j)}$ as definition prove $N(g) = \det M_g = \operatorname{Res}(M_\alpha(t), g(t))$. Convince yourself that this result remains valid if $g = g(y)$ is a polynomial of $F[y]$.

3. Compute $TU(R)$ with both methods of this section for $R = \mathbb{Z}[\rho]$, ρ a zero of $f(t) = t^4 + 4t^3 + 5t^2 + 2t + 1$. (For example, $\rho = (-2 + 3^{\frac{1}{2}} + (-1)^{\frac{1}{2}})/2$ is a zero of f.)
4. Let $f(t) = t^n + a_1 t^{n-1} + \cdots + a_n \in \mathbb{C}[t]$. The *companion matrix* $M_f = (m_{ij}) \in \mathbb{C}^{n \times n}$ of f is defined via

$$m_{ij} = \left\{ \begin{array}{lll} 1 & \text{for } j = i-1 & (2 \leqslant i \leqslant n) \\ -a_{n+1-i} & \text{for } j = n & (1 \leqslant i \leqslant n) \\ 0 & \text{otherwise} & \end{array} \right\}.$$

Prove:

(a) $\det(tI_n - M_f) = f(t)$.

(b) The eigenvalues of M_f are precisely the zeros of $f(t)$.

(c) Any zero x of $f(t)$ satisfies the estimate $|x| \leqslant \max\{|a_i| + 1 - \delta_{in} | 1 \leqslant i \leqslant n\}$.

5. Prove:

(a) $\varphi(p^k) = p^k - p^{k-1}$ for $k \in \mathbb{N}$, p a prime number.

(b) $\varphi(ap^k) = \varphi(a)\varphi(p^k)$ for $a, k \in \mathbb{N}$, p a prime number with $p \nmid a$. (Hint: The natural numbers x subject to $1 \leqslant x \leqslant p^k a$ which are prime to a and divisible by p are in 1–1-correspondence to the natural numbers y subject to $1 \leqslant y \leqslant p^{k-1} a$ which are prime to a.)

(c) $\varphi(ab) = \varphi(a)\varphi(b)$ for $a, b \in \mathbb{N}$ subject to $\gcd(a, b) = 1$. Hence, φ is said to be *multiplicative*.

6. Use the result of exercise 5 to develop an algorithm which determines for given $x \in \mathbb{N}$ all $m \in \mathbb{N}$ subject to m even and $\varphi(m) | x$.
7. Develop an algorithm to decide whether two algebraic number fields F_1, F_2 are isomorphic. (Hint: Try to factorize a generating polynomial for F_2 over F_1.)

5.5. Computation of independent units

In this section we describe procedures for computing a maximal set of $r = s + t - 1$ independent units in R. We derive them from lattice points in suitable convex regions of $\mathbb{R}^n$. This process is similar to the one which we used to prove the existence of r independent units in section 2. But Minkowski's Convex Body Theorem (and – as a consequence – chapter 3, theorem (4.6)) just guarantees the existence of non-trivial lattice points and does not yield an efficient method of computation. For the latter we must choose the convex point sets in an appropriate way. We suggest consideration of special parallelotopes or ellipsoids centered at the origin. This also guarantees that the lattice points found in this way correspond to elements of R of bounded norm.

Performing division on these elements in R whenever possible we get elements of small absolute norm very rapidly and thus necessarily associate elements among them. These provide units which then need to be tested for independency.

In the sequel we fix a $\mathbb{Z}$-basis $\omega_1, \ldots, \omega_n$ of R. Then there is the bijective mapping

$$\varphi: R \to \mathbb{Z}^n: m_1\omega_1 + \cdots + m_n\omega_n \mapsto (m_1, \ldots, m_n)^t. \tag{5.1}$$

It can be extended to an injective mapping of F into $\mathbb{R}^n$, if necessary. We discuss the cases of parallelotopes and of ellipsoids as suitable point sets separately.

Method I: *parallelotopes.*
The *basic parallelotope*

$$\prod := \{\mathbf{x} = (x_1, \ldots, x_n)^t \in \mathbb{R}^n \mid -1 \leqslant x_i \leqslant 1, 1 \leqslant i \leqslant n\} \tag{5.2}$$

obviously contains non-trivial lattice points of $\mathbb{Z}^n$ and also satisfies the premises of Minkowski's Convex Body Theorem. For each $\mathbf{x} \in \mathbb{Z}^n \cap \Pi$ we have

$$|N(\varphi^{-1}(\mathbf{x}))| \leqslant \max\left\{\prod_{j=1}^{n} |y_1 \omega_1^{(j)} + \cdots + y_n \omega_n^{(j)}| \,|\, \mathbf{y} \in \Pi\right\} =: B. \tag{5.3}$$

B is the upper bound for the absolute norms of all elements of R which will be constructed. Usually $\varphi^{-1}(\Pi \cap \mathbb{Z}^n)$ will not provide a maximal set of independent units. (But compare example (5.11) and exercise 3 of section 6.) Therefore we also consider suitable transforms of Π. The transformations Ψ have to be chosen in a way such that the image $\Psi(\Pi)$ still satisfies the premises of Minkowski's Convex Body Theorem, that $\Psi(\Pi) \cap \mathbb{Z}^n$ can easily be computed, and that the absolute norms of elements of $\varphi^{-1}(\Psi(\Pi) \cap \mathbb{Z}^n)$ stay bounded.

To satisfy the first two conditions we choose Ψ to be linear of determinant ± 1. To fulfill the third we take Ψ as the regular representation matrix M_ω of an element $\omega \in R \backslash \mathbb{Z}$ multiplied by a suitable constant. Namely, for $\omega \in R$ its right regular representation $M_\omega \in \mathbb{Z}^{n \times n}$ is defined by

$$(\omega_1, \ldots, \omega_n)\omega = (\omega_1, \ldots, \omega_n) M_\omega \quad \text{(compare chapter 2 (3.25a)).} \tag{5.4}$$

As a trivial consequence we obtain (see exercise 2 of section 4)

$$N(\omega) = \det M_\omega. \tag{5.5}$$

The linear transformation $\Psi = \Psi_\omega$ is then given by $|N(\omega)|^{-1/n} M_\omega$ with respect to the basis $\omega_1, \ldots, \omega_n$. Obviously, Ψ_ω satisfies the first property required, i.e. Ψ_ω is linear of $\det \Psi_\omega = \pm 1$, hence $\Psi_\omega(\Pi)$ is an $\mathbf{0}$-symmetric parallelotope of volume 2^n. Also the absolute norms of elements $\alpha \in \varphi^{-1}(\Psi_\omega(\Pi) \cap \mathbb{Z}^n)$ are bounded by B because of

$$\prod(\omega) := \Psi_\omega\left(\prod\right) = \{|N(\omega)|^{-1/n} M_\omega \mathbf{x} \in \mathbb{R}^n \mid -1 \leqslant x_i \leqslant 1, 1 \leqslant i \leqslant n\}, \tag{5.6}$$

and

$$\begin{aligned}
&\prod_{j=1}^{n} (|N(\omega)|^{-1/n} |(\omega_1^{(j)}, \ldots, \omega_n^{(j)}) M_\omega \mathbf{x})| \\
&\quad = |N(\omega)|^{-1} \prod_{j=1}^{n} |\omega^{(j)} (\omega_1^{(j)}, \ldots, \omega_n^{(j)}) \mathbf{x}| \\
&\quad = \prod_{j=1}^{n} |\mathbf{x}^t (\omega_1^{(j)}, \ldots, \omega_n^{(j)})^t| \leqslant B \quad \text{for } -1 \leqslant x_i \leqslant 1, \quad 1 \leqslant i \leqslant n.
\end{aligned}$$

It remains to show how we can easily determine $\Psi_\omega(\Pi) \cap \mathbb{Z}^n$. By chapter 3

(2.7) we compute unimodular matrices U_ω, U_ω^{-1} such that $M_\omega^t U_\omega =: N_\omega$ is in Hermite normal form. We recall that N_ω is a lower triangular matrix, hence the product of its diagonal elements n_{ii} $(1 \leqslant i \leqslant n)$ is – up to sign – $N(\omega)$. Elementary integral calculations yield a lower triangular matrix $B_\omega \in \mathbb{Z}^{n \times n}$ such that

$$M_\omega^t U_\omega B_\omega = \operatorname{diag}(|N(\omega)|, \ldots, |N(\omega)|). \tag{5.7}$$

Namely, if we denote the entries of N_ω, B_ω by n_{ij}, b_{ij}, respectively, (5.7) is equivalent to

$$\delta_{ij}|N(\omega)| = \sum_{k=1}^{n} n_{ik} b_{kj} = \sum_{k=j}^{i} n_{ik} b_{kj} \quad (1 \leqslant i, j \leqslant n). \tag{5.8}$$

Let us assume that we have already computed b_{ij} subject to $(\prod_{k=i_0+1}^{n} n_{kk}) | b_{i_0 j}$ for fixed $i_0 \in \mathbb{Z}^{\geqslant 0}$ and $1 \leqslant j \leqslant n$. Then (5.8) yields in case $i = i_0 + 1 \leqslant n$:
for $j = i$: $b_{ii} = |N(\omega)|/n_{ii} \in \mathbb{Z} \backslash \{o\}$,
for $j < i$:

$$b_{ij} = \left(- \sum_{k=j}^{i-1} n_{ik} b_{kj} \right) \Big/ n_{ii} \in \mathbb{Z},$$

for $j > i$: $b_{ij} = 0$; .

hence, we can compute b_{ij} subject to $b_{ij}(\prod_{k=i+1}^{n} n_{kk})^{-1} \in \mathbb{Z}$ for $i = i_0 + 1$, $1 \leqslant j \leqslant n$. Thus $B_\omega \in \mathbb{Z}^{n \times n}$ is obtained successively.

Let $\mathbf{c} = (c_1, \ldots, c_n)^t$ be a lattice point of $\Pi(\omega)$. Then there is $\mathbf{x} = (x_1, \ldots, x_n)^t \in \mathbb{R}^n$ subject to $-1 \leqslant x_i \leqslant 1$ $(1 \leqslant i \leqslant n)$ such that $\mathbf{c} = |N(\omega)|^{-1/n} M_\omega \mathbf{x}$, and also $\mathbf{d} = U_\omega^t \mathbf{c}$ is in $\mathbb{Z}^n$. Multiplication by B_ω^t yields $B_\omega^t \mathbf{d} = |N(\omega)|^{(n-1)/n} \mathbf{x}$, hence each $\mathbf{c} \in \Pi(\omega) \cap \mathbb{Z}^n$ is obtained upon multiplication by $(U_\omega^{-1})^t$ from a solution $\mathbf{d} \in \mathbb{Z}^n$ of

$$-|N(\omega)|^{(n-1)/n} \leqslant \sum_{j=i}^{n} d_j b_{ji} \leqslant |N(\omega)|^{(n-1)/n} \quad (i = 1, \ldots, n). \tag{5.9}$$

Since the ith inequality of (5.9) contains only the coordinates $d_i, \ldots, d_n$, all integral solutions of (5.9) can easily be computed by determining all integers d_i solving the ith inequality for each $(n-i)$ – tuple $(d_{i+1}, \ldots, d_n)$ already obtained $(i = n, n-1, \ldots, 1)$. Each solution $\mathbf{d}$ of (5.9) then is multiplied by $(U_\omega^{-1})^t$ to obtain all lattice points $\mathbf{c}$ of $\Pi(\omega) \cap \mathbb{Z}^n$:

$$\mathbf{c} = (U_\omega^{-1})^t \mathbf{d}. \tag{5.10}$$

Before we discuss the processing of the integers $\varphi^{-1}(\Pi(\omega) \cap \mathbb{Z}^n)$ we should consider the preceding computations more thoroughly. Let us remember that we started fixing an integral basis $\omega_1, \ldots, \omega_n$ of R. The choice of $\omega_1, \ldots, \omega_n$ is of strong influence on the amount of necessary computations, since the size of B of (5.3) is directly affected by it. Let us demonstrate this by a simple but impressive example.

Example (5.11)
Let $R = \mathbb{Z}[6^{\frac{1}{4}}]$. For $\omega_1 = 1, \omega_2 = 6^{\frac{1}{4}}$ we easily compute $B = 6$ (see also exercise 5).

But if we choose $\omega_1 = 1$, $\omega_2 = 2 + 6^{\frac{1}{2}}$, we obtain $B = 5$. The corresponding parallelotope contains the lattice point $(3, 1)^t$. We fix ω_2 and take $\varphi^{-1}((3, 1)^t) = 3 + 6^{\frac{1}{2}}$ as new basis element ω_1. This not only yields $B = 3$, a much better bound, but also the new basic parallelotope contains $\varphi(\varepsilon) = \varphi(\omega_1 + \omega_2) = \varphi(5 + 2(6)^{\frac{1}{2}})$ as a lattice point, ε being the fundamental unit of R.

Of course, we would like to choose $\omega_1, \ldots, \omega_n$ to make B as small as possible. Unfortunately there is no solution for that task as far as we know. Even for the easiest case of a real quadratic number field that problem is about as difficult as solving Pell's equation directly which just means determining the fundamental unit (see exercise 5). From that result we conclude that it will be rather hopeless to look for an optimal $\mathbb{Z}$-basis of R such that B of (5.3) is minimal. On the other hand it suggests to search for basis elements ω_i $(1 \leqslant i \leqslant n)$ of small norm. Since such a basis is also difficult to determine we instead take a reduced $\mathbb{Z}$-basis of R with respect to the length

$$\|\alpha\|^2 := \left(\sum_{j=1}^{n} |\alpha^{(j)}|^2 \right) \quad (\alpha \in R). \tag{5.12}$$

Because of the inequality between arithmetic and geometric means we obtain

$$|N(\alpha)| = \left| \prod_{j=1}^{n} |\alpha^{(j)}|^2 \right|^{1/2} \leqslant \left(\frac{1}{n} \sum_{j=1}^{n} |\alpha^{(j)}|^2 \right)^{n/2} = \left(\frac{\|\alpha\|^2}{n} \right)^{n/2}, \tag{5.13}$$

from which we conclude that elements $\alpha \in R$ of small length also have small norm. We note that a $\mathbb{Z}$-basis of R which is only pairwise or LLL-reduced (see chapter 3) in general suffices for our purposes. Such a basis can be computed very quickly starting from an arbitrary $\mathbb{Z}$-basis of R. The use of such bases was of great advantage in [6]. Not only was the amount of computation time drastically reduced but also the coefficients of the obtained fundamental units became much smaller, sometimes by several powers of ten. Even for some totally real sextic fields we obtained all five fundamental units from the basis parallelotope Π when we used a reduced basis (compare table 6.1 of the appendix).

Another comment must be made about the choice of the transforming element $\omega \in R$. It is clear that $\omega \in \mathbb{Z}$ would yield $M_\omega = \mathrm{diag}(\omega, \ldots, \omega)$ and therefore $\Pi(\omega) = \Pi$. But within $R \backslash \mathbb{Z}$ the choice of ω is completely free. We would like to choose ω such that $\Pi(\omega)$ contains many new lattice points. Unfortunately we don't know how to determine ω for that purpose. From our experience in computing units we suggest the choice of an element ω of small absolute norm and then a few consecutive powers of that element for transforming Π and then to switch to another ω. This method has several advantages. Elements ω of small absolute norm are stored anyway and are therefore always at hand. If ω is of small absolute norm, usually the entries of M_ω are small, too, and the entries of the first few powers of M_ω still fit into one computer word. Also the use of powers $M_\omega^k = M_{\omega^k}$ diminishes the calculations necessary for computing $U_{\omega^k}, B_{\omega^k}$ (see [5]). We note that the choice of transforming elements ω should still be investigated in greater detail. See also exercise 6.

A final remark concerns the computation of $\Pi(\omega)\cap\mathbb{Z}^n$. The method discussed in (5.5)–(5.10) is indeed very simple. All computations (except for $|N(\omega)|^{(n-1)/n}$) are done in the rational integers and are easily programmable. However, solving (5.9) recursively is not always optimal. In many cases the matrix $M_\omega^t U_\omega$ and therefore B_ω are sparse matrices. Namely, quite often we find for N_ω: $n_{ii}=1$ $(1\leqslant i<n)$, $n_{nn}=|N(\omega)|$ and therefore the entries of N_ω off the diagonal are zero except for the last row. It is easy to see that the same holds for the matrix B_ω. Hence, in this case (5.9) becomes

$$\left.\begin{aligned} -|N(\omega)|^{(n-1)/n} &\leqslant d_i|N(\omega)|+d_nb_{ni}\leqslant |N(\omega)|^{(n-1)/n} \quad (1\leqslant i<n),\\ -|N(\omega)|^{(n-1)/n} &\leqslant d_n\leqslant |N(\omega)|^{(n-1)/n}. \end{aligned}\right\} \tag{5.14}$$

In case $|N(\omega)|$ is large the recursive solution of (5.14) requires a lot of computation time since only very few of the d_n in the interval $[0,|N(\omega)|^{(n-1)/n}]$ can be extended to a vector $\mathbf{d}$ satisfying all inequalities. We set $k:=|N(\omega)|^{(n-1)/n}$ and note that we can assume $d_n\geqslant 0$ since we need consider only one solution vector of each pair $\pm\mathbf{d}$.

Straightforward algorithm for solving (5.14) (5.15)

Input. k, b_{ni} $(1\leqslant i\leqslant n)$.

Output. All solutions $\mathbf{d}\in\mathbb{Z}^n$ of (5.14) with $d_n\geqslant 0$.

Step 1. (Initialization). Set $d_n\leftarrow 0$, $L\leftarrow -k/|N(\omega)|$, $U\leftarrow -L$, $E\leftarrow b_{nn-1}/|N(\omega)|$.

Step 2. (Does solution d_{n-1} exist?). In case $\lfloor U\rfloor\geqslant L$ go to 4.

Step 3. (Increase d_n). Set $d_n\leftarrow d_n+1$. For $d_n>k$ terminate, else set $L\leftarrow L-E$, $U\leftarrow U-E$ and go to 2.

Step 4. $((d_n,d_{n-1})$ extendable?). For $i=1,\ldots,n-1$, compute $L_i\leftarrow(-k-d_nb_{ni})/|N(\omega)|$, $U_i\leftarrow(k-d_nb_{ni})/|N(\omega)|$ and for $\lfloor U_i\rfloor<L_i$ go to 3. Otherwise print $d_1,\ldots,d_n$ for all possibilities of d_i, $\lceil L_i\rceil\leqslant d_i\leqslant\lfloor U_i\rfloor$ $(1\leqslant i\leqslant n-1)$. Then go to 3.

Remark

Algorithm (5.15) requires about $3\lfloor k\rceil$ arithmetic operations (in steps 3,4) if (5.14) has few solutions and k is much larger than n.

However, for large k we can do better. To solve (5.14) efficiently in that case we transform the problem suitably. The crucial part is of course calculating the solutions of one pair of inequalities, for example the possibilities for (d_{n-1},d_n) corresponding to steps 1–3 of algorithm (5.15). We therefore fix the index i $(1 \leqslant i \leqslant n-1)$ in the sequel. We note that $b_{ni} \leqslant 0$ because of the matrix N_ω being in Hermite normal form. Our assumption $d_n \geqslant 0$ also yields $d_i \geqslant 0$. Setting $k_0 := \lfloor |N(\omega)|^{(n-1)/n} \rfloor$, $y := k_0 - d_n$ we obtain the equivalent inequalities

$$(|b_{ni}|-1)k_0 \leqslant |N(\omega)|d_i + |b_{ni}|y \leqslant (|b_{ni}|+1)k_0, \quad 0 \leqslant y \leqslant k_0. \tag{5.16}$$

Without loss of generality we can assume $|b_{ni}| > 0$ since the case of B_ω being a diagonal matrix is trivial and occurs rarely. (But see exercise 1.) Because of $d_i \geqslant 0$ we can drop the condition $y \leqslant k_0$ in case $|b_{ni}| > k_0$. Finally, for $|b_{ni}| \leqslant 2k_0$ all solutions of (5.16) are given explicitly in the following lemma.

Lemma (5.17)
Let (5.16) *be given with* $|b_{ni}| \leqslant 2k_0$. *For each* $d_i \in \mathbb{Z}^{\geqslant 0}$ *with* $d_i \leqslant (|b_{ni}|+1)k_0/|N(\omega)|$ *there are solutions* y *subject to* (5.16) *and all solutions of* (5.16) *are obtained in this way.*

The proof is straightforward and left as an exercise to the reader. It remains to solve a pair of inequalities of type

$$j \leqslant ax + by \leqslant k \tag{5.18}$$

for given $a, b, j, k \in \mathbb{Z}^{\geqslant 0}$ subject to $a > b > k - j \geqslant 0$ in $x, y \in \mathbb{Z}^{\geqslant 0}$. Obviously, x, y are bounded from above by $\lfloor k/a \rfloor$, $\lfloor k/b \rfloor$, respectively. We can divide a by b to obtain analogous inequalities with smaller coefficients:

$$j \leqslant b\left(y + \left\lfloor \frac{a}{b} \right\rfloor x\right) + \left(a - \left\lfloor \frac{a}{b} \right\rfloor b\right)x = \tilde{a}\tilde{x} + \tilde{b}\tilde{y} \leqslant k.$$

We do this until one of the coefficients becomes less than or equal to $k - j$. Then the solutions of the last pair of inequalities are computed as in (5.17). The following two lemmata show that this process is correct, i.e. the solutions of the different inequalities are in 1–1-correspondence. The underlying idea for this is to consider the range for possible solutions x.

Lemma (5.19)
Let $j, k, a, b, s, t \in \mathbb{Z}^{\geqslant 0}$, $a > b$, $r := a - \lfloor a/b \rfloor b$, $s \leqslant t \leqslant \lfloor k/a \rfloor$, $r > k - j \geqslant 0$. *Then the solutions* $(x, y)^t \in (\mathbb{Z}^{\geqslant 0})^2$ *of* $j \leqslant ax + by \leqslant k$ *subject to* $s \leqslant x \leqslant t$ *and the solutions* $(u, v)^t \in (\mathbb{Z}^{\geqslant 0})^2$ *of* $j \leqslant bu + rv \leqslant k$ *subject to* $\lceil (j - rt)/b \rceil \leqslant u \leqslant \lfloor (k - rs)/b \rfloor$ *are in* 1–1-*correspondence.*

Proof
It is easily seen that each solution (x, y) with $s \leqslant x \leqslant t$ yields a solution (u, v) with $0 \leqslant \tilde{s} := \lceil (j - rt)/b \rceil \leqslant u \leqslant \lfloor (k - rs)/b \rfloor =: \tilde{t}$ upon setting $u = y + \lfloor a/b \rfloor x$, $v = x$. If, on the other hand, $(u, v)^t \in (\mathbb{Z}^{\geqslant 0})^2$ satisfies $j \leqslant bu + rv \leqslant k$ and

$\tilde{s} \leqslant u \leqslant \tilde{t}$, we set $x = v$, $y = u - \lfloor a/b \rfloor v$ and get

$$x = v \geqslant \left\lceil \frac{j - bu}{r} \right\rceil \geqslant \left\lceil \frac{j - b\left\lfloor \frac{k - rs}{b} \right\rfloor}{r} \right\rceil \geqslant \left\lceil \frac{j - k}{r} \right\rceil + s = s,$$

$$x = v \leqslant \left\lfloor \frac{k - bu}{r} \right\rfloor \leqslant \left\lfloor \frac{k - b\left\lceil \frac{j - rt}{b} \right\rceil}{r} \right\rfloor \leqslant \left\lfloor \frac{k - j}{r} \right\rfloor + t = t,$$

$$y \geqslant \left\lceil \frac{j - rt}{b} \right\rceil - \left\lfloor \frac{a}{b} \right\rfloor t = \left\lceil \frac{j - \left(a - \left\lfloor \frac{a}{b} \right\rfloor b\right) t}{b} \right\rceil - \left\lfloor \frac{a}{b} \right\rfloor t$$

$$\geqslant \left\lceil \frac{j - a\left\lfloor \frac{k}{a} \right\rfloor}{b} \right\rceil \geqslant \left\lceil \frac{j - k}{b} \right\rceil = 0.$$ □

Therefore we can apply Euclid's algorithm to the pair (a, b) of (5.18) as long as the remainder is larger than the difference $k - j$. What happens when it finally becomes smaller?

Lemma (5.20)

Let $a > b > 0$, $k \geqslant j \geqslant 0$, $t \geqslant s \geqslant 0$ be integers subject to $\lfloor (j - at)/b \rfloor \geqslant 0$, $b > k - j$, and $r := a - \lfloor a/b \rfloor b \leqslant k - j$. Then for each $u \in \mathbb{Z}$ satisfying $\lceil (j - rt)/b \rceil \leqslant u \leqslant \lfloor (k - rs)/b \rfloor$ there is a $v \in \mathbb{Z}^{\geqslant 0}$ subject to $j \leqslant bu + rv \leqslant k$, $s \leqslant v \leqslant t$, and each such pair (u, v) yields a solution $x = v$, $y = u - \lfloor a/b \rfloor v \geqslant 0$ of $j \leqslant ax + by \leqslant k$ satisfying $s \leqslant x \leqslant t$.

Proof

Because of $k - b\lfloor (k - rs)/b \rfloor \geqslant rs$, $j - b\lceil (j - rt)/b \rceil \leqslant rt$, we obtain for every u in the interval $[\lceil (j - rt)/b \rceil, \lfloor (k - rs)/b \rfloor]$: $k \geqslant bu + rs$, $bu + rt \geqslant j$, and because of $r \leqslant k - j$ for each such u there exists (at least one) v, $s \leqslant v \leqslant t$, satisfying $j \leqslant bu + rv \leqslant k$. The rest of the proof is by similar arguments as in (5.19). □

Before we now develop an algorithm solving (5.18) we need to be a little more explicit about the necessary computations. At each step i we assume to have an inequality $j \leqslant a_i x_i + b_i y_i \leqslant k$ together with bounds s_i, t_i, $s_i \leqslant x_i \leqslant t_i \leqslant \lfloor k/a_i \rfloor$. According to (5.19) we compute $q_i := \lfloor a_i/b_i \rfloor$, $r_i := a_i - q_i b_i$, $a_{i+1} = b_i$, $b_{i+1} = r_i$,

$$\begin{pmatrix} x_{i+1} \\ y_{i+1} \end{pmatrix} = \begin{pmatrix} q_i & 1 \\ 1 & 0 \end{pmatrix} \begin{pmatrix} x_i \\ y_i \end{pmatrix},$$

$$s_{i+1} := \left\lceil \frac{j - r_i t_i}{b_i} \right\rceil, \quad t_{i+1} := \left\lfloor \frac{k - r_i s_i}{b_i} \right\rfloor.$$

Finally, if $b_n \leqslant k - j$ for the first time we must compute x_1, y_1 for all solutions

x_n, y_n. This is done efficiently in the following way. We set $\tilde{U}_i := \binom{q_i \;\; 1}{1 \;\; 0}$ $(1 \leqslant i \leqslant n-1)$ for abbreviation and define $U_k := \tilde{U}_{k-1}\tilde{U}_{k-2} \cdot \ldots \cdot \tilde{U}_1$ $(2 \leqslant k \leqslant n)$. Obviously, U_n is unimodular and satisfies $\binom{x_n}{y_n} = U_n \binom{x_1}{y_1}$. For $U_n = \binom{u_1 \;\; u_2}{u_3 \;\; u_4}$ we have

$$U_n^{-1} = \frac{1}{\det(U_n)} \begin{pmatrix} u_4 & -u_2 \\ -u_3 & u_1 \end{pmatrix} = (-1)^{n-1} \begin{pmatrix} u_4 & -u_2 \\ -u_3 & u_1 \end{pmatrix}.$$

Algorithm solving $j \leqslant ax + by \leqslant k$ for $x, y \in \mathbb{Z}^{\geqslant 0}$ (5.21)

Input: Integers a, b, j, k satisfying $k \geqslant j \geqslant 0$, $a > b > 0$.

Output: All pairs $(x, y)^t \in (\mathbb{Z}^{\geqslant 0})^2$ satisfying $j \leqslant ax + by \leqslant k$, respectively, 'No solution' if none exists.

Step 1: (Initialization). Set $i \leftarrow 1$, $a_i \leftarrow a$, $b_i \leftarrow b$, $s_i \leftarrow 0$, $t_i \leftarrow \lfloor k/a \rfloor$, $U_i \leftarrow \binom{1 \;\; 0}{0 \;\; 1}$.

Step 2: ($b_i > k - j$?). In case $b_i \leqslant k - j$ go to 4.

Step 3: (Long division of a, b). Set $q_i \leftarrow \lfloor a_i/b_i \rfloor$, $r_i \leftarrow a_i - q_i b_i$. Set $i \leftarrow i + 1$, $a_i \leftarrow b_{i-1}$, $b_i \leftarrow r_{i-1}$ $U_i \leftarrow \binom{q_{i-1} \;\; 1}{1 \;\; 0} U_{i-1}$, $s_i \leftarrow \lceil (j - b_i t_{i-1})/a_i \rceil$, $t_i \leftarrow \lfloor (k - b_i s_{i-1})/a_i \rfloor$. In case $s_i > t_i$ terminate with 'No solution', otherwise go to 2.

Step 4: (Print solutions). For each $u \in \mathbb{Z}$, $s_i \leqslant u \leqslant t_i$ compute all $v \in \mathbb{Z}$ such that $s_{i-1} \leqslant v \leqslant t_{i-1}$ and $j \leqslant a_i u + b_i v \leqslant k$. For each such pair $\binom{u}{v}$ print solution $\binom{x}{y} = U_i^{-1} \binom{u}{v}$.

Remarks (5.22)

(i) In step 4 solutions always exist according to (5.17) and (5.20).

(ii) This algorithm is an improvement of the one given in [5]. It requires only one-third of the arithmetic operations of the latter to proceed from level i to $i + 1$.

(iii) To exclude the superfluous solution $(0, k_0)$ of (5.16) it is advisable to consider the case $s_0 = 0$ in step 1 separately and then to proceed with $s_0 = 1$.

Method II: *ellipsoids*

Again we apply (3.8) in the case $A = R_A = R$, $k = 1$. However, there is the problem that we do not know realistic bounds (3.3) for the conjugates of the elements of bounded norm which we are looking for. Hence, we omit (3.8d) and modify (3.8g) to:

$$|r_j| \leqslant m \quad (1 \leqslant j \leqslant n) \quad \text{and there is an index } j_0 \text{ such that } |r_{j_0}| = m, \tag{5.23}$$

where m denotes a positive integer. The initial value is $m = 1$ of course. If all lattice points of all ellipsoids (3.8e) have been determined for a fixed value of m, then we increase m by 1 and proceed until a maximal set of independent units has been determined.

We remark that the condition $|r_{j_0}| = m$ guarantees that no ellipsoid is considered twice. From (3.10b) we know that the norms of the elements found as lattice points are bounded by $1+\gamma$ in absolute value. The appropriate choice of γ was discussed at the end of section 3.

On the other hand, we can also choose γ in such a way that the obtained ellipsoids always contain non-zero lattice points. By Minkowski's Convex Body Theorem we find by an easy calculation that a choice of

$$\gamma \geqslant -1 + \left(\frac{4}{\pi n}\right)^{n/2} |d_R^{\frac{1}{2}}| \begin{cases} (n/2)! & \text{for } n \text{ even} \\ \dfrac{\pi^{\frac{1}{2}} n!}{2^n \left(\dfrac{n-1}{2}\right)!} & \text{for } n \text{ odd} \end{cases} \tag{5.24}$$

is sufficient for that purpose (see exercise 4). Of course, this can not be recommended if the absolute value of the discriminant of R is large.

Method II has the advantage that it proceeds in a systematic way. Hence, it is guaranteed to provide a maximal set of independent units. Moreover, the procedure of increasing m makes it likely that not only independent but fundamental units are detected.

After we showed how to produce sufficiently many elements of R of bounded norm we need to consider the processing of such elements once they have been computed. Let $x \in R$ be the last element obtained from the parallelotope under consideration. We assume that the elements determined earlier are stored in some array X which contains n_X elements of R of bounded norm at the moment. The corresponding norms – respectively their absolute values – are stored in an array X_N.

Moreover, we need auxiliary arrays $\tilde{X}$, $\tilde{X}_N$ of $\tilde{n}_X$ elements each. The initial values are $n_X = \tilde{n}_X = 0$, of course.

Algorithm for comparing x with stored elements of small absolute norm (5.25)

Input. $x \in R$ of absolute norm $N_x > 1$, arrays X, X_N of length n_X as described above.

Output. X, X_N, n_X and/or units $\varepsilon_1, \ldots, \varepsilon_p$.

Step 1. (Initialization). Set $\tilde{n}_X \leftarrow 0$, $k \leftarrow 0$, $p \leftarrow 0$.

Step 2. (X completely searched?). Set $k \leftarrow k+1$. For $k > n_X$ go to 6.

Step 3. (Next element of X). Set $\alpha \leftarrow X(k)$, $N_\alpha \leftarrow X_N(k)$. For $N_\alpha > N_x$ go to 5.

Step 4. (Compare $X(k)$, x for $X_N(k) \leqslant N_x$). For $m := N_x/N_\alpha \notin \mathbb{Z}$ go to 2. For $\beta := x/\alpha \notin R$ go to 2. For $m = 1$ set $p \leftarrow p+1$, $\varepsilon_p \leftarrow \beta$ and go to 7. For $m > 1$ set $x \leftarrow \beta$, $N_x \leftarrow m$, $k \leftarrow 0$ and go to 2.

Step 5. (Compare $X(k)$, x for $X_N(k) > N_x$). For $m := N_\alpha/N_x \notin \mathbb{Z}$ go to 2. For $\beta := \alpha/x \notin R$ go to 2. For $l = k, \ldots, n_X - 1$ set $X(l) \leftarrow X(l+1)$, $X_N(l) \leftarrow X_N(l+1)$, $n_X \leftarrow n_X - 1$. Then set $\tilde{n}_X \leftarrow \tilde{n}_X + 1$, $\tilde{X}(\tilde{n}_X) \leftarrow \beta$, $\tilde{X}_N(\tilde{n}_X) \leftarrow m$, and go to 4.

Step 6. (Insert x into X). Set $n_X \leftarrow n_X + 1$, $X(n_X) \leftarrow x$, $X_N(n_X) \leftarrow N_X$.

Step 7. (Decrease $\tilde{X}$) For $\tilde{n}_X = 0$ terminate. Else set $x \leftarrow \tilde{X}(\tilde{n}_X)$, $N_X \leftarrow \tilde{X}_N(\tilde{n}_X)$, $\tilde{n}_X \leftarrow \tilde{n}_X - 1$, $k \leftarrow 0$ and go to 2.

The processing of the units obtained from (5.25) will be discussed in section 7 of this chapter. At this stage it is enough to know that we can produce as many units as we may need by this method. Of course, all units of the output belonging to $TU(R)$ will be eliminated. Extensive computations have shown that it is not recommendable to process all x found via the parallelotopes to (5.25) but only those among them whose absolute norm is less than 10000. (The bound B of (5.3) can of course be much larger.)

More advanced techniques also involving the ideal factorization of the involved algebraic integers are generally too complicated for unit computations. They are recommended of course, if also the class number of F is to be determined. These refined methods will be discussed in chapter 6.

Exercises

1. Compute $\Pi(\omega) \cap \mathbb{Z}^n$ in $R = \mathbb{Z}[\alpha]$ for $\omega = \alpha^6$ and α a zero of $t^3 - 7t - 7 = 0$.
2. Let α be a zero of $t^3 - t - 1 = 0$ and $R = \mathbb{Z}[\alpha]$. Compute $\Pi(\omega) \cap \mathbb{Z}^n$ for $\omega = (\alpha + 2)^5$ with algorithms (5.15) and (5.21) and compare the number of arithmetic operations of both methods. (It cannot be recommended to do this exercise without a pocket calculator – preferably a programmable one.)
3. Compute an independent unit in $R = \mathbb{Z}[14^{\frac{1}{3}}]$ with the methods of this section.
4. Prove that a choice of γ according to (5.24) guarantees that each ellipsoid (3.8e) contains a non-zero lattice point.
5. Let $R = \mathbb{Z} + \mathbb{Z}m^{\frac{1}{2}}$, $m \in \mathbb{Z}^{\geq 2}$ square-free. To choose a basis $\omega_1 = a + bm^{\frac{1}{2}}$, $\omega_2 = c + dm^{\frac{1}{2}}$ for R which minimizes B of (5.3) means to solve

$$\min \{\max \{|(x_1(a + bm^{\frac{1}{2}}) + x_2(c + dm^{\frac{1}{2}}))(x_1(a - bm^{\frac{1}{2}}) + x_2(c - dm^{\frac{1}{2}}))| \\ -1 \leq x_i \leq 1,\ i = 1, 2\} \mid a, b, c, d \in \mathbb{Z},\ ad - bc = \pm 1\}.$$

Show that this requires us to solve

$$\min \{\max \{|(-a + c)^2 - m(-b + d)^2|, |(a + c)^2 - m(b + d)^2|, K_1, K_2\} \mid \\ a, b, c, d \in \mathbb{Z},\ ad - bc = \pm 1\}$$

for

$$K_i := \begin{cases} |1 - m/(N(\omega_i))| & \text{in case } |(ac - mbd)/N(\omega_i)| \leq 1 \\ 1 & \text{otherwise} \end{cases}, \quad i = 1, 2.$$

6. Show that for any $\varepsilon \in U(R)$ there is an element $\omega \in R$ such that $\psi(e_1) = \varphi(\varepsilon)$, i.e. $\varphi(\varepsilon)$ is in the transformed parallelotope. Hence, in principle each unit can be obtained as a lattice point in a suitable transform of the basic parallelotope.

5.6. Regulator bounds and index estimates

In order to obtain an upper bound for the index of the computed subgroup $U_\varepsilon(R)$ in the whole unit group $U(R)$ we interpret this index as a $\mathbb{Z}$-module index. Namely, by taking logarithms the multiplicative structure of $U(R)$ becomes an additive one. This obvious procedure is used in most textbooks already in the course of the proof of Dirichlet's Theorem.

We consider two mappings of $U(R)$ into *logarithmic space*:

$$\left.\begin{array}{l} L_1\colon U(R)\to \mathbb{R}^{s+t}\colon \varepsilon\mapsto (c_1\log|\varepsilon^{(1)}|,\ldots,c_{s+t}\log|\varepsilon^{(s+t)}|)^t,\\ L_2\colon U(R)\to \mathbb{R}^{r}\ \ \colon \varepsilon\mapsto (c_1\log|\varepsilon^{(1)}|,\ldots,c_{r}\log|\varepsilon^{(r)}|)^t,\end{array}\right\} \tag{6.1}$$

where $c_j = \begin{cases} 1 & \text{for } 1\leqslant j\leqslant s \\ 2 & \text{else}\end{cases}$.

The following properties of L_i $(i=1,2)$ are obvious:

(a) $L_i(\varepsilon\eta) = L_i(\varepsilon) + L_i(\eta)$ for all $\varepsilon,\eta\in U(R)$,

(b) $\operatorname{Ker}(L_i(U(R))) = TU(R)$, (6.2)

(c) $c_{s+t}\log|\varepsilon^{(s+t)}| = -\sum_{j=1}^{r} c_j\log|\varepsilon^{(j)}|$ for the coordinates of $L_1(\varepsilon)$ $(\varepsilon\in U(R))$.

The next result is not quite so obvious but it will turn out to be of great use later on.

Lemma (6.3)

Let $U_\varepsilon(R) := TU(R)\times\langle\varepsilon_1\rangle\times\cdots\times\langle\varepsilon_r\rangle$ be a subgroup of finite index in $U(R)$. Then $L_i(\varepsilon_j)$ $(1\leqslant j\leqslant r)$ are $\mathbb{R}$-linearly independent $(i=1,2)$. $L_1(\varepsilon_1)^t,\ldots,L_1(\varepsilon_r)^t$ and $(c_1,\ldots,c_{s+t})^t$ form a basis of $\mathbb{R}^{s+t}$.

Proof

Let $E_1,\ldots,E_r$ be a system of fundamental units of $U(R)$. There is $M\in\mathbb{N}$ and $a_{\mu\nu}\in\mathbb{Z}$ $(1\leqslant\mu,\nu\leqslant r)$ such that $|E_\mu^M| = |\prod_{\nu=1}^r \varepsilon_\nu^{a_{\mu\nu}}|$. Obviously, the matrix $A := (1/M)(a_{\mu\nu})_{1\leqslant\mu,\nu\leqslant r}$ is invertible.

On the other hand $U(R)$ contains (independent) units $\eta_1,\ldots,\eta_r$ for which $|\eta_\mu^{(\nu)}|<1$, $|\eta_\mu^{(\mu)}|>1$ $(1\leqslant\nu\leqslant s+t, 1\leqslant\mu\leqslant r, \mu\neq\nu)$. Analogously there is a matrix $B\in GL(r,\mathbb{Z})$ transforming $E_1,\ldots,E_r$ into $\eta_1,\ldots,\eta_r$ up to roots of unity. Therefore we obtain

$$\begin{pmatrix} L_2(\eta_1)^t \\ \vdots \\ L_2(\eta_r)^t\end{pmatrix} = BA\begin{pmatrix} L_2(\varepsilon_1)^t \\ \vdots \\ L_2(\varepsilon_r)^t\end{pmatrix} =: C$$

with a regular matrix BA. Hence, it suffices to show that $L_2(\eta_1),\ldots,L_2(\eta_r)$ are $\mathbb{R}$-linearly independent which is equivalent to $\det(C)\neq 0$. Let us assume that the columns of C are linearly dependent. Then there exist $t_1,\ldots,t_r\in\mathbb{R}$ subject to $t_k := \max\{t_i \mid 1\leqslant i\leqslant r\} > 0$ and $\sum_{j=1}^r t_j(c_j\log|\eta_i^{(j)}|)_{1\leqslant i\leqslant r} = 0$. Considering the kth coordinate we get the contradiction

$$0 = \sum_{j=1}^{r} t_jc_j\log|\eta_k^{(j)}| = \sum_{\substack{j=1\\ j\neq k}}^{r} t_jc_j\log|\eta_k^{(j)}| + t_kc_k\log|\eta_k^{(k)}|$$

$$\geqslant t_k c_k \log|\eta_k^{(k)}| + \sum_{\substack{j=1\\ j\neq k}}^{r} t_k c_j \log|\eta_k^{(j)}| = t_k(-c_{s+t}\log|\eta_k^{(s+t)}|) > 0.$$

Therefore $L_2(\eta_1),\dots,L_2(\eta_r)$ are $\mathbb{R}$-linearly independent implying that $L_2(\varepsilon_1),\dots,L_2(\varepsilon_r)$ and also $L_1(\varepsilon_1),\dots,L_1(\varepsilon_r)$ are $\mathbb{R}$-linearly independent.

To prove the second statement of the lemma it suffices to show that the vector $(c_1,\dots,c_{s+t})^t$ is $\mathbb{R}$-linearly independent from $L_1(\varepsilon_1),\dots,L_1(\varepsilon_r)$. If it were not, there would be a presentation $(c_1,\dots,c_{s+t})^t = \sum_{i=1}^r t_i L_1(\varepsilon_i)$ $(t_i \in \mathbb{R}, 1 \leqslant i \leqslant r, \max\{|t_i| \mid 1 \leqslant i \leqslant r\} > 0)$. But then addition of the coordinates yields the contradiction

$$s + 2t = \sum_{j=1}^{s+t} c_j = \sum_{j=1}^{s+t}\sum_{i=1}^{r} t_i c_j \log|\varepsilon_i^{(j)}|$$

$$= \sum_{i=1}^{r} t_i \sum_{j=1}^{s+t} c_j \log|\varepsilon_i^{(j)}| = \sum_{i=1}^{r} t_i \cdot 0 = 0. \qquad \square$$

Corollary 1 (6.4)

Let $U_\varepsilon(R)$ be a subgroup of $U(R)$ of finite index. Then $L_i(U_\varepsilon(R))$ is a free $\mathbb{Z}$-module of rank r $(i = 1, 2)$.

Corollary 2 (6.5)

Let $U_\varepsilon(R)$ be a subgroup of $U(R)$ of finite index. Then $(U(R){:}\,U_\varepsilon(R)) = d(L_2(U_\varepsilon(R)))/d(L_2(U(R)))$.

Proof

(6.4) is obvious. For the proof of (6.5) we note that $(L_2(U(R)){:}\,L_2(U_\varepsilon(R))) = (U(R){:}\,U_\varepsilon(R))$ follows from the homomorphism theorems of group theory. Then chapter 3 (3.6) is applied. $\square$

Definition (6.6)

Let $U_\varepsilon(R) := TU(R) \times \langle\varepsilon_1\rangle \times \cdots \times \langle\varepsilon_r\rangle$ be a subgroup of $U(R)$ of finite index. Then the mesh $d(L_2(U_\varepsilon(R)))$ is called the **regulator** $\mathrm{Reg}(U_\varepsilon(R))$ *of $U_\varepsilon(R)$. In case $R = \mathrm{Cl}(\mathbb{Z}, \mathfrak{Q}(R))$ the regulator of $U(R)$ is also called* **the regulator of the field** *$F = \mathfrak{Q}(R)$. We denote it by Reg_F, or in short by R_F.*

At the present state of our computations of $U(R)$ we assume that we already calculated independent units $\varepsilon_1,\dots,\varepsilon_r$ generating a subgroup $U_\varepsilon(R)$ of finite index up to roots of unity. Now we can calculate $\mathrm{Reg}(U_\varepsilon(R))$ from $\mathrm{Reg}(U_\varepsilon(R)) = \mathrm{abs}(\det(c_j \log|\varepsilon_i^{(j)}|)_{1\leqslant i,j\leqslant r})$ and obtain an upper bound for the index

$$(U(R){:}\,U_\varepsilon(R)) = \frac{\mathrm{Reg}(U_\varepsilon(R))}{\mathrm{Reg}(U(R))}$$

once we know a lower bound for $\mathrm{Reg}(U(R))$. To derive such a lower bound is the goal of the rest of this section.

We apply the tools of chapter 3, namely, following Remak [7] we consider

$$\sum_{j=1}^{n} (\log|\varepsilon^{(j)}|)^2 \quad (\varepsilon \in U(R)). \tag{6.7}$$

Representing ε by fundamental units this becomes a positive definite quadratic form. The determinant of this quadratic form is essentially $\mathrm{Reg}(U(R))$. Thus we get a lower bound for the regulator of $U(R)$ by chapter 3 (3.34), as soon as we have derived a lower bound for (6.7).

Let us fix a system of fundamental units $E_1, \ldots, E_r$ of $U(R)$. Each $\varepsilon \in U(R)$ then has a (unique) representation by $E_1, \ldots, E_r$ and some element of $TU(R)$. Hence, for $|\varepsilon^{(j)}|$ $(1 \leqslant j \leqslant n)$ we obtain

$$\log|\varepsilon^{(j)}| = \sum_{i=1}^{r} x_i \log|E_i^{(j)}| \quad (x_i \in \mathbb{Z},\ 1 \leqslant i \leqslant r,\ 1 \leqslant j \leqslant n). \tag{6.8}$$

Using the constants c_j $(1 \leqslant j \leqslant s+t)$ of (6.1) we convert (6.7):

$$\begin{aligned}
\sum_{j=1}^{n} (\log|\varepsilon^{(j)}|)^2 &= \sum_{j=1}^{s+t} c_j (\log|\varepsilon^{(j)}|)^2 \\
&= \sum_{j=1}^{r} c_j (\log|\varepsilon^{(j)}|)^2 + c_{s+t}^{-1}\left(-\sum_{j=1}^{r} c_j \log|\varepsilon^{(j)}|\right)^2 \\
&= \sum_{i,j=1}^{r} (c_{s+t}^{-1} c_i c_j + \delta_{ij} c_j) \log|\varepsilon^{(i)}| \log|\varepsilon^{(j)}| \\
&= \sum_{\mu,\nu=1}^{r} x_\mu x_\nu \left(\sum_{i,j=1}^{r} (c_{s+t}^{-1} c_i c_j + \delta_{ij} c_j) \log|E_\mu^{(i)}| \log|E_\nu^{(j)}|\right) \\
&=: \sum_{\mu,\nu=1}^{r} q_{\mu\nu} x_\mu x_\nu .
\end{aligned}$$

This shows that (6.7) is indeed a quadratic form. It is positive definite since (6.7) is always non-negative and becomes zero only in case all conjugates of ε are of absolute value 1. But then ε is in $TU(R)$ because of (2.5), hence (6.7) vanishes only in case of $x_1 = \cdots = x_r = 0$.

The next step will be the computation of the determinant of the quadratic form. It is easily seen that the matrix equation

$$(q_{\mu\nu})_{1 \leqslant \mu,\nu \leqslant r} = (c_k \log|E_\mu^{(k)}|)_{1 \leqslant \mu,k \leqslant r} (d_{i,j})_{1 \leqslant i,j \leqslant r} (c_l \log|E_\nu^{(l)}|)_{1 \leqslant l,\nu \leqslant r}$$

is satisfied for

$$d_{ij} = c_{s+t}^{-1} + \delta_{ij} c_i^{-1} \quad (1 \leqslant i,j \leqslant r).$$

The evaluation of the corresponding determinants yields

$$\det(q_{ij}) = \mathrm{Reg}(U(R))^2 2^{-t} n \tag{6.9}$$

because of $\sum_{i=1}^{r} c_i = n - c_{s+t}$, $\prod_{i=1}^{s+t} c_i^{-1} = 2^{-t}$ and exercise 1.

In view of chapter 3 (3.34) it remains to give a lower bound for (6.7) in case $\varepsilon \in U(R) \setminus TU(R)$. This will be done by analytic methods. We set

$$x_j := \log |\varepsilon^{(j)}| \quad (1 \leqslant j \leqslant n), \tag{6.10}$$

and then minimize

$$\sum_{j=1}^{n} x_j^2 \tag{6.11a}$$

subject to suitable side conditions coming from the properties of ε of $U(R)$. Obviously, we can require

$$\sum_{j=1}^{n} x_j = 0 \tag{6.11b}$$

because of $|N(\varepsilon)| = 1$. But then a criterion which excludes the solution $x_1 = \cdots = x_n = 0$ is most important. The image of R under the mapping

$$\left.\begin{array}{l} \psi: R \to \mathbb{R}^n: \\ \omega \mapsto (\omega^{(1)}, \ldots, \omega^{(s)}, 2^{\frac{1}{2}} \operatorname{Re} \omega^{(s+1)}, 2^{\frac{1}{2}} \operatorname{Im} \omega^{(s+1)}, \ldots, 2^{\frac{1}{2}} \operatorname{Re} \omega^{(s+t)}, 2^{\frac{1}{2}} \operatorname{Im} \omega^{(s+t)})^t \end{array}\right\} \tag{6.12}$$

is a lattice $\psi(R)$ of mesh $|d(R)|$. For the lattice vectors $\psi(\omega)$ the usual Euclidean norm in $\mathbb{R}^n$ is

$$\|\psi(\omega)\| = \left(\sum_{j=1}^{n} |\omega^{(j)}|^2 \right)^{1/2} =: T_2(\omega)^{1/2}. \tag{6.13}$$

It is no problem to compute the successive minima of $\psi(R)$ with respect to $\| \quad \|$ by chapter 3 (3.36). Usually it suffices to calculate only M_1, M_2, M_3 which coincide with $\|\psi(\omega_1)\|^2$, $\|\psi(\omega_2)\|^2$, $\|\psi(\omega_3)\|^2$ for a reduced basis $\psi(\omega_1), \ldots, \psi(\omega_n)$ of $\psi(R)$ (compare chapter 3 (3.32)). And a reduced basis for R was already used in the preceding section. It is easily seen that

$$M_1 = n \quad \text{for } \psi(1). \tag{6.14}$$

Namely, every $\omega \in R$, $\omega \neq 0$, satisfies $|N(\omega)| \geqslant 1$, hence $T_2(\omega) \geqslant n$ by the inequality between arithmetic and geometric means. The same argument yields

$$\|\psi(\omega)\| = n^{\frac{1}{2}} \Leftrightarrow \omega \in TU(R). \tag{6.15}$$

This implies that a basis of R consisting of roots of unity $\omega_1, \ldots, \omega_n$ satisfies $\|\psi(\omega_i)\|^2 = M_i = n$ $(1 \leqslant i \leqslant n)$. And this happens in all cyclotomic fields. On the other hand, for $TU(R) = \{\pm 1\}$ we get $M_2 > n$ and for $n = s$ we even have $M_2 \geqslant (3/2)n$ [8] (note that $T_2(\omega) = \operatorname{Tr}(\omega^2) \in \mathbb{Z}$ in this case).

Remark (6.16)

For $\varepsilon \in U(R) \setminus TU(R)$ we always have $\|\psi(\varepsilon)\|^2 \geqslant M_2$. However, in case $(1 + 5^{\frac{1}{2}})/2 \in U(R)$, for example, $M_2 \leqslant 3(n/2)$ independently of the discriminant of R. Therefore, in case of $\mathbb{Q}(R)$ having proper subfields, higher successive minima usually must be taken into consideration to obtain a good lower bound for (6.7).

Theorem (6.17)

Let $M^* := \min\{T_2(\omega) | \omega \in U(R) \setminus TU(R)\} = T_2(\omega^*)$. *Then* $M^* > n$ *and* $\varepsilon \in U(R) \setminus TU(R)$ *satisfies*

$$\sum_{j=1}^{n} (\log|\varepsilon^{(j)}|)^2 \geqslant \frac{n}{4}\left(\log\left(\frac{M^*}{n} + \left(\frac{M^{*2}}{n^2} - 1\right)^{1/2}\right)\right)^2 =: M_0.$$

Proof

We set $x_j := \log|\varepsilon^{(j)}|$ $(1 \leqslant j \leqslant n)$ and minimize

$$f(\mathbf{x}) := \sum_{j=1}^{n} x_j^2$$

subject to $\sum_{j=1}^{n} x_j = 0$ and $\sum_{j=1}^{n} e^{2x_j} \geqslant M^*$. A vector $\mathbf{x} \in \mathbb{R}$ which satisfies both side conditions is called a *feasible* solution. Obviously, there are feasible solutions $\mathbf{x}$, for example those corresponding to units $\varepsilon \in R \setminus TU(R)$. Now (6.15) implies $M^* > n$. Hence, each feasible $\mathbf{x}$ must have positive and negative coordinates. Let $\mathbf{y} \in \mathbb{R}^n$ be feasible. Then each $\mathbf{x} \in \mathbb{R}^n$ with $f(\mathbf{x}) \leqslant f(\mathbf{y})$ necessarily satisfies $-f(\mathbf{y})^{\frac{1}{2}} \leqslant x_j \leqslant f(\mathbf{y})^{\frac{1}{2}}$ $(1 \leqslant j \leqslant n)$. Hence, the existence of a global minimum is guaranteed. If the minimum is attained, let us say for the vector $\mathbf{z}$, the second side condition must be *active*, i.e. $\sum_{j=1}^{n} e^{2z_j} = M^*$. Otherwise we could decrease the maximal coordinate of $\mathbf{z}$ by a very small constant δ and increase the minimal coordinate of $\mathbf{z}$ by δ to obtain a new feasible solution $\mathbf{z}_\delta$ with $f(\mathbf{z}_\delta) < f(\mathbf{z})$. Therefore we can apply Lagrange's multiplier method yielding a system of equations in the variables $x_1, \ldots, x_n, \lambda, \mu$:

$$2x_j + \lambda + 2\mu e^{2x_j} = 0 \quad (1 \leqslant j \leqslant n),$$

$$\sum_{j=1}^{n} x_j = 0, \quad \sum_{j=1}^{n} e^{2x_j} = M^*.$$

For fixed μ the function $g(x) := x + \mu e^{2x}$ with derivative $g'(x) = 1 + 2\mu e^{2x}$ is strictly increasing for $\mu > 0$. But then $g(x) = -\lambda/2$ cannot have different solutions for $x < 0$ and $x > 0$, a contradiction to our conclusion from above. Hence, μ must be negative and we get exactly two solutions $u > 0$, $v < 0$ of $g(x) = -\lambda/2$. This means that we must minimize $su^2 + (n-s)v^2$ for $u > 0$, $v < 0$ and $s \in \{1, \ldots, n-1\}$ subject to $su + (n-s)v = 0$ and $se^{2u} + (n-s)e^{2v} = M^*$. Since ε and ε^{-1} yield the same value $f(\mathbf{x})$ we can assume $s \geqslant n/2$. We substitute $v = su/(s-n)$ and $u = ((n-s)/2)\log y$ $(y \in \mathbb{R}^{>1})$ and obtain the equivalent problem: Minimize $\frac{1}{4}ns(n-s)(\log y)^2$ subject to $G(s,y) := sy^n - M^*y^s + n - s = 0$ for $(s,y) \in [n/2, n-1] \times (1, \infty)$. Because of $G_y(s,y) = \frac{1}{y}n\left(sy^n - \frac{1}{n}sM^*y^s\right)$ the function $G(s,y)$ has exactly one minimum for fixed s. Hence, $G(s,1) = n - M^* < 0$ yields exactly one solution $y := h(s)$ of $G(s,y) = 0$ for fixed

s. We shall prove that $F(s) := s(n-s)(\log y)^2 = s(n-s)(\log h(s))^2$ is strictly increasing in s. Namely,

$$F'(s) = (\log h(s))^2(n-2s) + s(n-s)2\frac{\log h(s)}{h(s)}h'(s)$$
$$= \log h(s)\left((n-2s)\log h(s) + \frac{2s(n-s)}{h(s)}\left(-\frac{G_s(s,y)}{G_y(s,y)}\right)\right)$$
$$= \log y\left((n-2s)\log y - \frac{2(n-s)(y^n - 1 - \log yM^*y^s)}{ny^n - M^*y^s}\right).$$

Because of $y > 1$, $G(s,y) = 0$ and the denominator in the last term being positive we obtain the following chain of equivalent inequalities

$$F'(s) > 0,$$
$$(n-2s)\log y\,(ny^n - M^*y^s) > 2(n-s)(y^n - 1 - \log yM^*y^s),$$
$$\log y(n(n-2s)y^n + nM^*y^s) > 2(n-s)(y^n - 1),$$
$$\log y^{n/2} > (y^{n-1})/(y^{n+1}).$$

Setting $z = y^n$ we need to prove $\frac{1}{2}\log z > (z-1)/(z+1)$ for $z > 1$. But this last inequality follows from $1/(2t) > 2/(1+t)^2$ $(t > 1)$ by integrating both sides from 1 to z. Thus we have shown that $(n/4)s(n-s)(\log h(s))^2$ is strongly increasing (even for $1 \leqslant s \leqslant n-1$ (!)). Because of our assumption $s \geqslant n/2$ the minimum is attained for $s = n/2$. But then $G(s,y) = 0$ implies

$$y^n - \frac{2}{n}M^*y^{n/2} + 1 = 0,$$

i.e.

$$y^{n/2} = \frac{M^*}{n} + \left(\frac{M^{*2}}{n^2} - 1\right)^{1/2},$$

and the theorem is proved. □

Remarks (6.18)

(i) M^* can be easily calculated by chapter 3 (3.15).

(ii) In case n is odd the estimate of (6.17) can be slightly improved by computing $y > 1$ from $G((n+1)/2, y) = 0$ and then $(n/4)F((n+1)/2)$ as a lower bound for $\sum_{j=1}^{n}(\log|\varepsilon^{(j)}|)^2$.

(iii) It seems somewhat puzzling that we can stipulate $s = n/2$ even though $F(s)$ is strictly increasing in $[1, n-1]$. The reason for this is that ε and ε^{-1} yield the same value $\sum_{j=1}^{n}(\log|\varepsilon^{(j)}|)^2$ but $\|\psi(\varepsilon)\|$ and $\|\psi(\varepsilon^{-1})\|$ can differ substantially.

Corollary (6.19)

The regulator Reg$(U(R))$ *of* R *satisfies the inequality*

$$\text{Reg}\,(U(R)) \geqslant (M_0'\gamma_r^{-r}2^r n^{-1})^{1/2}.$$

Proof
By (6.7), (6.9), chapter 3 (3.34), (6.17). □

Examples (6.20)

(i) Let $R = \mathbb{Z}[\rho]$, ρ a zero of $t^3 + t^2 - 2t - 1 = 0$. The conjugates of ρ are $\rho = \rho^{(1)} = 1.247$, $\rho^{(2)} = -0.445$, $\rho^{(3)} = -1.802$ and the discriminant of R is $d_R = 49$. A reduced basis is $\omega_1 = 1, \omega_2 = \rho, \omega_3 = \rho^2 - 2$ yielding the successive minima $M_1 = 3$, $M_2 = M_3 = 5$. These data yield a lower regulator bound $\mathrm{Reg}(U(R)) \geqslant 0.45$ which is very close to the real value $\mathrm{Reg}(U(R)) = 0.53$.

(ii) Let $R = \mathbb{Z}[\rho]$, ρ a zero of $t^4 - 2t^2 - 1 = 0$. A reduced basis of R is $\omega_1 = 1$, $\omega_2 = \rho$, $\omega_3 = \rho^3 - 2\rho$, $\omega_4 = \rho^2 - 1$ providing $M_1 = 4$, $M_2 = M_3 = 4(2)^{\frac{1}{2}}$, $M_4 = 8$. Hence, we obtain 0.48 as a lower regulator bound whereas $\mathrm{Reg}(U(R)) = 1.35$.

If $\mathbb{Q}(R)$ contains proper subfields then the estimate (6.19) for $\mathrm{Reg}(U(R))$ may be too weak. In that case we need to take into consideration higher successive minima of (6.7). Let $M_i^* \in \mathbb{R}^{>0}$ be minimal such that there are independent units $\varepsilon_1, \ldots, \varepsilon_i$ in $U(R)$ satisfying $\|\varepsilon_j\|^2 \leqslant M_i^*$ $(1 \leqslant j \leqslant i)$ for some natural number i. As in (6.17) we set
$M_{oi} := (n/4)(\log((M_i^*/n) + ((M_i^{*2}/n^2) - 1)^{1/2}))^2$ and obtain

Theorem (6.21)
For $1 \leqslant i \leqslant r$ *there holds the estimate* $\mathrm{Reg}(U(R)) \geqslant (M_{o1} \cdot \cdots \cdot M_{oi}^{r+1-i} \gamma_r^{-r} 2^t n^{-1})^{1/2}$.

The proof hinges essentially on chapter 3 (3.34) and is left as an easy exercise to the reader.

Though (6.21) yields the best lower bound for $\mathrm{Reg}(U(R))$ which we know so far we also present some other explicit bounds at the end of this section. From the results of [3], [4] we excerpt

$$\mathrm{Reg}(U(R)) \geqslant \left(\left(\frac{3(\log(|d(R)|/n^n))^2}{(n-1)n(n+1) - 6t} \right)^r \frac{2^t}{n\gamma_r^r} \right)^{1/2} \qquad (6.22)$$

in case $\mathbb{Q}(R)$ is primitive over $\mathbb{Q}$ and $|d(R)| > n^n$.

Moreover, for $t = 0$ and $n \leqslant 11$ the constant n^n in (6.22) can be replaced by $4^{\lfloor n/2 \rfloor}$. If R contains proper subrings, the units of those subrings must be taken into consideration (compare Satz XII of [3]). The results in [3], [4] were stated only for maximal orders R but the methods also apply to non-maximal orders.

Lower regulator bounds can also be obtained by means of Analytic Number Theory. The best known result is due to Zimmert [14]. Satz 3 of his paper states for arbitrary $\gamma > 0$ and a maximal order R of a field F of degree $n = s + 2t$:

$$\frac{\operatorname{Reg}(U(R))}{\# TU(R)} \geqslant \frac{(1+\gamma)(1+2\gamma)}{2} \Gamma(1+\gamma)^{s+t} \Gamma\left(\frac{3}{2}+\gamma\right)^t 2^{-s-t} \pi^{-t/2}$$
$$\times \exp\left((-1-\gamma)\left((s+t)\frac{\Gamma'}{\Gamma}\left(\frac{1+\gamma}{2}\right) + t\frac{\Gamma'}{\Gamma}\left(1+\frac{\gamma}{2}\right) + \frac{2}{\gamma} + \frac{1}{1+\gamma}\right)\right). \quad (6.23)$$

This estimate yields good results for $n \geqslant 6$ and small discriminants. Optimal values for γ are in the interval $(0, 1)$. Unfortunately Zimmert's result does not depend on specific data of the field F, e.g. its discriminant.

We close this section by presenting also an upper estimate for the regulator of a field F of degree $n = s + 2t$ and discriminant d_F. Using an idea of Landau Siegel [9] proved by analytic methods

$$\frac{\operatorname{Reg}_F}{\# TU(F)} < 2^{-s} 4(2\pi)^{-t} \left(\frac{be \log |d_F|}{n-1}\right)^{n-1} |d_F|^{\frac{1}{2}} \quad (6.24)$$

for $b = (1 + \log \pi/2 + (t/n) \log 2)^{-1}$.

Exercises

1. Let $\alpha, \beta_1, \ldots, \beta_r \in \mathbb{R}$ and $\prod_{i=1}^r \beta_i \neq 0$. Prove
$$\det((\alpha + \delta_{ij}\beta_i)_{1 \leqslant i,j \leqslant r}) = \left(\prod_{i=1}^r \beta_i\right)\left(\alpha \sum_{i=1}^r \beta_i^{-1} + 1\right).$$
2. Let R be totally real. Then for all $\varepsilon \in U(R) \backslash TU(R)$ we have
$$\sum_{j=1}^n (\log |\varepsilon^{(j)}|)^2 \geqslant n\left(\log \frac{1+5^{\frac{1}{2}}}{2}\right)^2.$$
(This result is obtained in [4] in a completely different way.)
3. Let ρ be a zero of $t^4 + t^3 - 3t^2 - t + 1 = 0$. Show that $\rho, \rho + 1, \rho - 1$ are a system of fundamental units in $\mathbb{Z}[\rho]$. (On the other hand, the $\mathbb{Q}$-rank of $\psi(1), \psi(\rho), \psi(\rho + 1), \psi(\rho - 1)$ is only two.)
4. Compute a lower regulator bound for $R = \mathbb{Z}[\rho]$, ρ a zero of $t^4 + 2t^2 + 2 = 0$, and compare it to $\operatorname{Reg}(U(R))$.
5. Show $\|\psi(\varepsilon^k)\| \leqslant \|\psi(\varepsilon^{k+1})\|$ for $\varepsilon \in U(R)$ and $k \in \mathbb{Z}^{\geqslant 0}$.
6. Prove (6.21) and apply it to example (6.20) (ii) for $i = 2$.

5.7. Computation of fundamental units

From the two previous sections we assume that we can determine as many units $\varepsilon \in U(R)$ as will be needed and that we know a lower regulator bound for $\operatorname{Reg}(U(R))$. The computation of fundamental units will then be carried out in three steps. In step I we produce r independent units $\eta_1, \ldots, \eta_r$ of R.

In step II we use additional units for a potential enlarging of the subgroup $U_\eta := TU(R) \times \langle\eta_1\rangle \times \cdots \times \langle\eta_r\rangle$ of $U(R)$. Finally, in step III we determine $U(R)$ from U_η. Because of step II the last step will usually be a verification of $U(R) = U_\eta$. In extensive calculations of fundamental units the groups U_η and $U(R)$ turned out to be different after step II only in about 3% of all cases.

Step I: *construction of r independent units*
We assume that we know already $0 \leqslant m < r$ independent units $\eta_i (1 \leqslant i \leqslant m; m \in \mathbb{Z}^{\geqslant 0})$, hence also $\mathbf{b}_i := L_2(\eta_i)$ and the corresponding orthogonal vectors $\mathbf{b}_i^*$ (compare (6.1), chapter 3 (3.24)). Each time a new unit $\eta \in U(R) \backslash TU(R)$ is found by (5.25) we set $\mathbf{b}_{m+1} = L_2(\eta)$ and compute $\mathbf{b}_{m+1}^*$. In case of $\mathbf{b}_{m+1}^* = \mathbf{0}$ we increase m by 1. In this way we proceed until $m = r$ is obtained. Then we easily calculate

$$\operatorname{Reg}(U_\eta(R)) = \prod_{i=1}^{r} \|\mathbf{b}_i^*\| \quad \text{for } U_\eta(R) = \langle TU(R), \eta_1, \ldots, \eta_r\rangle.$$

We note that it can be difficult to check whether the (floating point) vector $\mathbf{b}_{m+1}^*$ is zero. Because of $\prod_{i=1}^{r} \|\mathbf{b}_i^*\| \geqslant \operatorname{Reg}(U(R))$ (for which we know a positive lower bound) the $\|\mathbf{b}_i^*\|$ cannot be too small in general. If we must assume a linear dependence, however, we either search for another unit η or we proceed as in step II.

Step II: *enlarging of* $U_\eta(R)$
After the computation of a subgroup $U_\eta(R) = \langle TU(R), \eta_1, \ldots, \eta_r\rangle$ of $U(R)$ of finite index we try to enlarge this subgroup by additional units η_{r+1} in case the quotient of $\operatorname{Reg}(U_\eta(R))$ and of a lower bound for $\operatorname{Reg}(U(R))$ obtained by the methods of section 6 is still $\geqslant 2$. Applying chapter 3 (3.48) to $L_2(\eta_i)$ $(1 \leqslant i \leqslant r+1)$ we get integers $m_1, \ldots, m_{r+1}$ subject to $\sum_{i=1}^{r+1} |m_i| > 0$, $\sum_{i=1}^{r+1} r_i L_2(\eta_i) = \mathbf{0}$ and units $\tilde{\eta}_1, \ldots, \tilde{\eta}_r$ such that $U_{\tilde{\eta}}(R) = \langle TU(R), \tilde{\eta}_1, \ldots, \tilde{\eta}_r\rangle \ni \eta_i$ $(1 \leqslant i \leqslant r+1)$. If U_η is not enlarged for – let us say – five more units we assume $U_\eta(R) = U(R)$ and proceed to step III.

Step III: *computation of a system of fundamental units*
As an easy consequence of the fundamental theorem on finitely generated abelian groups we obtain:

Theorem (7.1)
Let $\eta_1, \ldots, \eta_k$ $(0 \leqslant k < r)$ *be part of a system of fundamental units of R. Then* $\eta_{k+1} \in U(R)$ *also belongs to that system, if and only if the equation*

$$\eta_{k+1} = \zeta \eta_1^{m_1} \cdot \cdots \cdot \eta_k^{m_k} \omega^m \quad (\zeta \in TU(R); m_i, m \in \mathbb{Z}, 1 \leqslant i \leqslant k; \omega \in R) \qquad (7.2)$$

is unsolvable for $|m| \geqslant 2$.

A proof is easily derived from the elementary divisor theorem and chapter

3 (2.9) (see exercise 1). We note that for $k=0$ the theorem gives a criterion, whether η_1 is a fundamental unit. The theorem will be suited for constructive purposes only if we can test the solvability of (7.2) in finitely many steps. If (7.2) is solvable, then ω is clearly a unit. Therefore we can assume $m>0$ without loss of generality. Next we can choose the m_i to be non-positive and greater than $-m$ by replacing the solution ω by

$$\omega \prod_{\substack{i=1\\ m_i>0}}^{k} \eta_i^{\lceil m_i/m \rceil} \prod_{\substack{i=1\\ m_i<0}}^{k} \eta_i^{\lfloor m_i/m \rfloor}.$$

Thus the solvability of (7.2) is closely related to the solvability of

$$|\omega^{(j)m}| = \left| \prod_{i=1}^{k} (\eta_i^{(j)})^{-m_i} \eta_{k+1}^{(j)} \right| \quad (1 \leqslant j \leqslant n)$$

$$(m \in \mathbb{N},\ m \geqslant 2;\ m_i \in \mathbb{Z}^{\leqslant 0}, -m < m_i;\ 1 \leqslant i \leqslant k). \tag{7.3}$$

Namely, if (7.3) is not solvable for $\omega \in R$ then (7.2) is insolvable, too. A solution of (7.3) does not necessarily imply that (7.2) is solvable, however ω^m might differ from the power product of the units on the right by a root of unity not contained in $TU(R)$. Of course, this can be easily checked. As a consequence the solvability of (7.2) can be checked in finitely many steps since m can be bounded by

$$m \leqslant q, \tag{7.4}$$

q the quotient of the regulator of the computed unit group U_η and the lower bound of $\mathrm{Reg}(U(R))$ derived in section 6.

We present three different methods for testing the solvability of (7.3) under the restriction (7.4) for $k=0,1,\ldots,r-1$. Their application either proves that the computed unit group U_η already coincides with $U(R)$ or it produces larger subgroups $U_{\tilde\eta}$ of $U(R)$, until the whole unit group is obtained. It is clear that the tests need to be done only for prime numbers $p \leqslant q$. Namely, a solution of (7.2) provides also a solution for each divisor $\tilde m$ of m.

Method 1. In case q, r are small we recommend a straightforward attack on the problem. Any potential solution $\omega \in R$ of (7.2) has a presentation

$$\omega = e_1\omega_1 + \cdots + e_n\omega_n \quad (e_i \in \mathbb{Z},\ 1 \leqslant i \leqslant n), \tag{7.5}$$

where $\omega_1,\ldots,\omega_n$ denote a $\mathbb{Z}$-basis of R. As in section 3 we compute the corresponding dual basis $\omega_1^*,\ldots,\omega_n^*$ defined by

$$\mathrm{Tr}(\omega_i\omega_j^*) = \delta_{ij} \quad (1 \leqslant i, j \leqslant n). \tag{7.6}$$

Then (7.2) yields

$$e_i = \mathrm{Tr}(\omega\omega_i^*) = \sum_{j=1}^{n} (\zeta\eta_1^{(j)-m_1} \cdot \ldots \cdot \eta_k^{(j)-m_k} \eta_{k+1}^{(j)})^{1/m} \omega_i^{*(j)} \quad (1 \leqslant i \leqslant n). \tag{7.7}$$

Here the right-hand side is to be calculated numerically for all possibilities

$\zeta \in TU(R)$, $0 \leqslant -m_i < m$ $(1 \leqslant i \leqslant k)$, m a prime number below q. If the (floating point) calculation yields $e_i \in \mathbb{Z}$ $(1 \leqslant i \leqslant n)$, then the corresponding ω is a solution of (7.2) which again can be checked by integral computations. Since we have to compute mth roots this procedure can usually be recommended only for totally real fields and $m \neq 2$. For totally real fields and $m = 2$, the insolvability of (7.2) can in general be decided upon considering the signs of $\eta_i^{(j)}$. Since $m_i \in \{0, 1\}$ in that case we can easily check whether the products $\prod_{i=1}^{k} \eta_i^{(j)m_i}$ can be all negative or all positive $(1 \leqslant j \leqslant n)$.

Method 2. In more complicated cases we suggest applying the following method. Instead of computing the coefficients e_i of a possible solution ω for all possibilities of m, m_i, ζ $(1 \leqslant i \leqslant k,\ \zeta \in TU(R))$ we give worst case upper bounds for the conjugates of ω. Obviously,

$$\begin{aligned} R_j := \min_{2 \leqslant m \leqslant q} \left(|\varepsilon_{k+1}^{(j)}|^{1/m} \prod_{i=1}^{k} \min(1, |\varepsilon_i^{(j)}|^{(m-1)/m}) \right) &\leqslant |\omega^{(j)}| \\ \leqslant \max_{2 \leqslant m \leqslant q} \left(|\varepsilon_{k+1}^{(j)}|^{1/m} \prod_{i=1}^{k} \max(1, |\varepsilon_i^{(j)}|^{(m-1)/m}) \right) & \\ =: S_j \quad (1 \leqslant j \leqslant n), & \end{aligned} \tag{7.8}$$

and we can apply (3.8).

The units obtained still need to be tested whether they indeed satisfy (7.3) and finally (7.2). But this again is easy. We take logarithms on both sides of (7.3) and obtain the following system of linear equations

$$mL_2(\omega) + \sum_{i=1}^{k} m_i L_2(\eta_i) = L_2(\eta_{k+1}) \tag{7.9}$$

in the unknowns $m, m_1, \ldots, m_k$. If (7.9) is unsolvable over $\mathbb{Z}$ for all obtained units ω, then (7.2) is unsolvable also and we can increase k, repeating this process until we have obtained r independent units. If (7.9) has a solution $(m, m_1, \ldots, m_k)^t \in \mathbb{Z}^{k+1}$ with $|m| \geqslant 2$, then we need to check whether

$$\eta_{k+1} \left(\prod_{i=1}^{k} \eta_i^{-m_i} \right) \omega^{-m} \in TU(R). \tag{7.10}$$

If this is the case, we replace $\eta_{k+1}, U_\eta, \mathrm{Reg}(U_\eta)$ by

$$\eta_{k+1} \leftarrow \omega, \quad U_\eta \leftarrow TU(R) \times \langle \eta_1 \rangle \times \cdots \times \langle \eta_r \rangle, \quad \mathrm{Reg}(U_\eta) \leftarrow \mathrm{Reg}(U_\eta)/|m|. \tag{7.11}$$

If $\mathrm{Reg}(U_\eta)$ does not increase any more we replace k by $k+1$.

This procedure has the additional advantage that we need not carry it out for each prime number $p \leqslant q$ separately but only for the worst bounds (7.8) obtained for $m = q$. Once we have checked that our units $\eta_1, \ldots, \eta_k$ are not powers themselves, i.e. $\eta_i \neq \zeta\omega^m$ $(\omega \in U(R),\ \zeta \in TU(R),\ 1 \leqslant m \leqslant q)$, we can even handle the tests for all k $(1 \leqslant k < r)$ simultaneously by noting that the worst bounds (7.8) result for $k = r - 1$. This is of course due to the fact that the method of section 3 for solving (7.8) is so efficient that larger bounds $R_j, S_j (1 \leqslant j \leqslant n)$ barely increase the computation time. In spite of that fact it is of course always recommendable to start with units $\eta_1, \ldots, \eta_r$ whose conjugates are close to 1 in absolute value. This can usually be obtained by applying reduction to $L_2(\eta_1), \ldots, L_2(\eta_r)$.

Method 3. Though we believe that the method just described is best suited for testing the solvability of (7.2) (mainly because of the fast procedure of section 3) we conclude this section presenting a p-adic method which is first of all of theoretical interest but also seems to be very promising from the computational viewpoint. However, to our knowledge it has not been tested numerically so far.

Let $p_1, \ldots, p_s$ be all odd rational prime numbers below q. (The case of 2 dividing $(U(R):U_\eta)$ can be easily treated with one of the first two methods.) Then $U(R)$ is generated by U_η and those units ε of R for which some power ε^h of the form $h = \prod_{i=1}^s p_i^{v_i} \leqslant q$ $(v_i \in \mathbb{Z}^{\geqslant 0}, 1 \leqslant i \leqslant s)$ belongs to U_η. Again we discuss the solvability of (7.2) starting with $k = 0$, i.e. we try to solve

$$\varepsilon\zeta = \omega^m \; (\zeta \in TU(R), \omega \in \mathrm{R}, m \geqslant 2) \tag{7.12a}$$

for $\varepsilon = \eta_1$. For the coefficients e_i of the presentation (7.9a) of ω we obtain the bounds (7.10) with $k = 0$. Then we determine a rational prime number

$$p > \max\{2\max\{T_i | 1 \leqslant i \leqslant n\}, |d(R)|\} \tag{7.12b}$$

such that the order h_p of the prime residue class group $(R/pR)^\times$ is not divisible by $p_1, \ldots, p_s$. Hence, there exist positive rational integers q_i $(1 \leqslant i \leqslant s)$ such that $p_i q_i \equiv 1 \bmod h_p$. If $\omega \in R$ solves (7.12a) for $m = p_i$ $(1 \leqslant i \leqslant s)$, then

$$(\varepsilon\zeta)^{q_i} = \omega^{p_i q_i} \equiv \omega \bmod pR. \tag{7.12c}$$

Because of (7.12b) we just need to compute the congruence solutions ω of (7.12c) with the coefficients e_v satisfying $-p/2 < e_v < p/2$ $(1 \leqslant v \leqslant n)$ for $1 \leqslant i \leqslant s$ and then check, whether one of them satisfies (7.12a). This has to be done for each p_i $(1 \leqslant i < s)$ and each possibility for $\zeta \in TU(R)$. If we obtain a solution of (7.12a) in this way, then $\varepsilon = \eta_1$ is replaced by ω thus enlarging U_η.

From now on we assume that (7.12a) has no solution. Then $\mathbb{Q}(R)(\eta_1^{1/p_j}) \supset \mathbb{Q}(R)$ and the polynomial $t^{p_j} - \eta_1 \in R[t]$ is irreducible $(1 \leqslant j \leqslant s)$. By Tschebotareff [11] there exists a prime ideal $\mathfrak{p}_j \nmid d(R)$ such that $t^{p_j} - \eta_1$ remains irreducible in $R/\mathfrak{p}_j[t]$ and therefore $p_j | (N(\mathfrak{p}_j) - 1)$. (Namely, for $p_j \nmid (N(\mathfrak{p}_j) - 1)$ there is $q_j \in \mathbb{N}$ subject to $p_j q_j \equiv 1 \bmod (N(\mathfrak{p}_j) - 1)$ and $\xi := \eta_1^{q_j}$ satisfies $\xi^{p_j} = \eta_1^{p_j q_j} \equiv \eta_1 \bmod \mathfrak{p}_j$ in contradiction to the irreducibility of $t^{p_j} - \eta_1$ in $R/\mathfrak{p}_j[t]$.) Since $(R/\mathfrak{p}_j)^\times$ is cyclic with $p_j | \#(R/\mathfrak{p}_j)^\times$ and η_1 is not a p_jth power in $(R/\mathfrak{p}_j)^\times$ the order of η_1 is divisible by p_j. Hence, for every unit η_i $(1 \leqslant i \leqslant r)$ there is precisely one rational integer v_i $(0 \leqslant v_i < p_j)$ such that $\eta_1^{v_i}\eta_i$ is congruent to a p_jth power modulo $\mathfrak{p}_j$. Since this was derived for arbitrary j $(1 \leqslant j \leqslant s)$ we can now apply the Chinese remainder theorem to obtain units $\tilde{\eta}_2, \ldots, \tilde{\eta}_r$ $(\tilde{\eta}_i = \eta_1^{v_i}\eta_i,\ 2 \leqslant i \leqslant r)$ such that $\langle\eta_1\rangle \times \cdots \times \langle\eta_r\rangle = \langle\eta_1\rangle \times \langle\tilde{\eta}_2\rangle \times \cdots \times \langle\tilde{\eta}_r\rangle$ and every element of $\langle\tilde{\eta}_2\rangle \times \cdots \times \langle\tilde{\eta}_r\rangle$ is a p_jth power modulo $\mathfrak{p}_j$ $(1 \leqslant j \leqslant r)$.

Therefore every $\omega \in U(R)$ satisfying $\omega^{p_j} \in U_\eta$ is the product of some root of unity, some power of η_1 and a p_jth power $\zeta^{p_j} \in \langle\tilde{\eta}_2\rangle \times \cdots \times \langle\tilde{\eta}_r\rangle =: U_\eta^{(2)}$. Repeating this process we finally end up with a unit group $U_\eta^{(r)}$ with one generator. But we already considered the task of computing p_jth roots of a single unit. Hence, we obtain a system of independent units inductively such that it already contains all p_jth powers of units of $U(R)$ $(1 \leqslant j \leqslant r)$, hence a system of fundamental units.

Exercises

1. Use the fundamental theorem on finitely generated abelian groups and chapter 3 (2.9) to prove theorem (7.1).
2. Let $R = \mathbb{Z}[\alpha]$ and α a root of $t^3 + t^2 - 2t - 1$. Compute a system of fundamental units of R starting from $\varepsilon_1 = 2\alpha^2 + 5\alpha + 2$, $\varepsilon_2 = 2\alpha^2 + 3\alpha + 1 \in U(R)$.
3. Let $F = \mathbb{Q}(\alpha)$, α a root of $t^4 + t^3 - 3t^2 - t + 1$. Compute a system of fundamental units for o_F.

5.8. Remarks on computerization

For computing a system of fundamental units of R we need the characteristic properties of R as input data. In general we assume the knowledge of an irreducible polynomial

$$f(t) = t^n + a_1 t^{n-1} + \cdots + a_n \in \mathbb{Z}[t], \tag{8.1}$$

represented by $a_1, \ldots, a_n \in \mathbb{Z}$, such that $F = \mathbb{Q}(R) = \mathbb{Q}(\rho)$ for a zero ρ of f. For the computations in R – especially the many norm computations of (5.23) – it is advantageous to compute the roots

$$\rho^{(1)} = \rho, \quad \rho^{(2)}, \ldots, \rho^{(n)} \tag{8.2}$$

of f up to machine precision by the IMSL – subroutine ZPOLR, available at most computer centers nowadays. We order the zeros in the usual way such that

$$\rho^{(1)}, \ldots, \rho^{(s)} \in \mathbb{R}, \quad \rho^{(s+1)}, \ldots, \rho^{(n)} \in \mathbb{C}\backslash\mathbb{R}, \tag{8.3}$$

and

$$\rho^{(s+t+i)} = \overline{\rho^{(s+i)}} \quad (1 \leqslant i \leqslant t), \quad n = s + 2t.$$

A basis of R usually belongs to the input data, in case of $R = o_F$ it can be computed by the methods of chapter 4 within the program. In any case we assume

$$R = \mathbb{Z}\omega_1 + \cdots + \mathbb{Z}\omega_n \tag{8.4a}$$

for

$$\omega_i = \frac{1}{m}\sum_{j=1}^{n} m_{ij}\rho^{j-1}, \tag{8.4b}$$

i.e. the knowledge of

$$m \in \mathbb{N}, \quad M = (m_{ij}) \in \mathbb{Z}^{n \times n}. \tag{8.4c}$$

The importance of having the basis reduced was already pointed out in section 5. The implementation of the methods of sections 3–7 for the computation of $U(R)$ should be no problem. We therefore consider only a few necessary subroutines and make some remarks on potential difficulties.

Addition of numbers of R is done componentwise. For multiplication, however, we need the multiplication constants γ_{ijk} of

$$\omega_i\omega_j = \sum_{k=1}^{n} \gamma_{ijk}\omega_k. \tag{8.5}$$

To obtain them we first calculate the coefficients $\beta_{i+j,k}$ of

$$\rho^{i+j} = \rho^i\rho^j = \sum_{k=1}^{n} \beta_{i+j,k}\rho^{k-1} \tag{8.6}$$

via reducing powers ρ^m $(m \geqslant n)$ by means of $f(\rho) = 0$. Namely,

$$\beta_{i+j,k} = \delta_{i+j,k-1} \quad (0 \leqslant i+j \leqslant n-1), \tag{8.7a}$$

and for $i+j \geqslant n$ we compute successively

$$\beta_{n,k} = -a_{n+1-k} \quad (1 \leqslant k \leqslant n), \tag{8.7b}$$

$$\left.\begin{aligned} \beta_{m+1,k} &= \beta_{m,k-1} + \beta_{mm}a_{n+1-k} \quad (2 \leqslant k \leqslant n), \\ \beta_{m+1,1} &= -\beta_{mm}a_n, \end{aligned}\right\} \tag{8.7c}$$

for $m = n, n+1, \ldots, 2n-3$. Then

$$\begin{aligned} \omega_i\omega_j &= \left(\frac{1}{m}\sum_{\mu=1}^{n} m_{i\mu}\rho^{\mu-1}\right)\left(\frac{1}{m}\sum_{\nu=1}^{n} m_{j\nu}\rho^{\nu-1}\right) \\ &= m^{-2}\sum_{\mu,\nu=1}^{n} m_{i\mu}m_{j\nu}\rho^{\mu-1}\rho^{\nu-1} \\ &= m^{-2}\sum_{k=1}^{n}\left(\sum_{\mu,\nu=1}^{n} m_{i\mu}m_{j\nu}\beta_{\mu+\nu-2,k}\right)\rho^{k-1} \end{aligned}$$

and we denote the coefficient of ρ^{k-1} by λ_{ijk}. From $(\omega_1, \ldots, \omega_n) = (1, \rho, \ldots, \rho^{n-1})M^t(1/m)$ we obtain $(1, \rho, \ldots, \rho^{n-1}) = (\omega_1, \ldots, \omega_n)(M^t)^{-1}m$ and therefore

$$(\gamma_{ij1}, \ldots, \gamma_{ijn})^t = (M^t)^{-1}\frac{1}{m}(\lambda_{ij1}, \ldots, \lambda_{ijn})^t \quad (1 \leqslant i, j \leqslant n). \tag{8.8}$$

Mainly for the norm calculations it is useful also to know the $\omega_i^{(j)}$ $(1 \leqslant i, j \leqslant n)$ as floating point numbers. But this can be easily done by (8.3) and (8.4b). Having stored the values of $\omega_i^{(j)}$ a norm computation of $\alpha \in R$ via $N(\alpha) = \prod_{j=1}^n \alpha^{(j)}$ then requires $O(n^2)$ multiplications instead of $O(n^3)$, if we use the regular representation matrix M_α of α and compute its determinant by pure integral calculations. We note that we do not need to use complex numbers but rather work with the real numbers

$$D(i,j) := \begin{cases} \omega_i^{(j)} & \text{for } 1 \leqslant j \leqslant s \\ \operatorname{Re} \omega_i^{(j)} & \text{for } s+1 \leqslant j \leqslant s+t \quad (1 \leqslant i \leqslant n). \\ \operatorname{Im} \omega_i^{(j)} & \text{for } s+t+1 \leqslant j \leqslant n. \end{cases} \tag{8.9}$$

Then the absolute norm of $\alpha = \sum_{i=1}^n \alpha_i \omega_i$ $(\alpha_i \in \mathbb{Z})$ becomes

$$|N(\alpha)| = \left| \prod_{j=1}^{s} \left(\sum_{i=1}^{n} \alpha_i D(i,j) \right) \right| \times \prod_{j=s+1}^{s+t} \left(\left(\sum_{i=1}^{n} \alpha_i D(i,j) \right)^2 + \left(\sum_{i=1}^{n} \alpha_i D(i,j+t) \right)^2 \right). \tag{8.10}$$

The result will be correct if the round-off errors are less than 0.5. An error estimate will be given below.

The processing of the algebraic integers of (5.23) requires to compute α/β for $\alpha, \beta \in R$, $\beta \neq 0$, whenever the quotient belongs to R. This can be done by solving the following system of linear equations in the unknowns $x_1, \ldots, x_n$:

$$\begin{aligned} \alpha_1\omega_1 + \cdots + \alpha_n\omega_n &= (\beta_1\omega_1 + \cdots + \beta_n\omega_n)(x_1\omega_1 + \cdots + x_n\omega_n) \\ &= \sum_{i,j=1}^{n} \beta_i x_j \omega_i \omega_j \\ &= \sum_{k=1}^{n} \omega_k \sum_{i,j=1}^{n} \beta_i \gamma_{ijk} x_j. \end{aligned} \tag{8.11}$$

At first we calculate a solution by floating point arithmetic. Such a solution always exists because of $\alpha/\beta \in \mathfrak{Q}(R)$. If we assume that the coordinates of the solution vector are integers, we can check our result using integer arithmetic by just calculating the right hand side of (8.11) again. Another possibility requiring only integral arithmetic is to apply chapter 3 (3.48).

For many computations – see sections 3, 4, 7 – a dual basis is useful. A dual basis $\omega_1^*, \ldots, \omega_n^*$ of R is obtained upon solving the matrix equation

$$(\omega_i^{(j)})_{1 \leqslant i,j \leqslant n} (\omega_j^{*(i)})_{1 \leqslant i,j \leqslant n} = I_n. \tag{8.12}$$

Concerning the occurrence of round-off errors we just discuss the norm computation since it is most important. We assume that the zeros of f are calculated up to machine precision and therefore affected with an error ε (which is less than 10^{-14} for a CDC Cyber 76, for example). Then the error for the $D(i, j)$ of (8.9) is roughly estimated by

$$\varepsilon \sum_{k=2}^{n} (k-1)|\rho^{(j)}|^{k-2} \quad (1 \leqslant j \leqslant n). \tag{8.13a}$$

For $x = \xi_1\omega_1 + \cdots + \xi_n\omega_n \in R$ the error in computing $x^{(j)}$ is at most

$$\delta_j := \varepsilon \sum_{i=2}^{n} |\xi_i| \sum_{k=2}^{n} (k-1)|\rho^{(j)}|^{k-2} \quad (1 \leqslant j \leqslant n). \tag{8.13b}$$

The absolute value of the norm of x (assumed to be derived from some transformation of the fundamental parallelotope) is bounded by B of (5.3). Hence, the total error in computing $N(x)$ is roughly bounded by

$$\delta := B \sum_{j=1}^{n} \frac{\delta_j}{|x^{(j)}|}. \tag{8.13c}$$

It is clear that the coefficients of x must be bounded to obtain $\delta < 0.5$. On the other hand, we cannot expect to compute a unit whose coefficients are huge by single precision arithmetic.

We note that the choice of a reduced basis of R usually helps to keep the coefficients of the fundamental units small.

6

The class group of algebraic number fields

6.1. Introduction

The fundamental differences between the arithmetic in $\mathbb{Q}$ and that in finite algebraic extensions F of $\mathbb{Q}$ are described by the unit group U_F and the class group Cl_F of F. While we saw in the last chapter that algebraic number fields usually contain units other than ± 1 (the only units of $\mathbb{Z}$) we now discuss the deviation of the ring of integers o_F of F from being a factorial ring. Again this is extremely important for solving non-linear Diophantine equations. As we already mentioned in chapter 4, section 5 Kummer believed that he had established a proof of Fermat's last theorem that

$$x^m + y^m = z^m \quad (x, y, z \in \mathbb{Z},\ xyz \neq 0) \tag{1.1}$$

is insolvable for any $m \in \mathbb{Z}^{\geq 3}$, until Dirichlet pointed out to him that his proof was based on the (wrong) premise that o_F is factorial for all cyclotomic fields. Kummer's struggle to save his 'proof' accelerated the beneficial development of algebraic number theory in the second half of the past century.

We will demonstrate the importance of the question whether o_F is factorial or not by a much simpler example. Let us consider the equation

$$2y^3 = x^2 + 5 \ (x, y \in \mathbb{Z}). \tag{1.2}$$

It can also be written as

$$2y^3 = (x + (-5)^{\frac{1}{2}})(x - (-5)^{\frac{1}{2}}), \tag{1.3}$$

i.e. we obtain an equation in $\mathbb{Z}[(-5)^{\frac{1}{2}}]$. Let us assume that $o_F = \mathbb{Z}[(-5)^{\frac{1}{2}}]$ is a factorial ring of $F = \mathbb{Q}((-5)^{\frac{1}{2}})$. Which consequences does this have for solving (1.3)?

Let $\alpha, \pi \in o_F$ and $\bar{\ }$ denote the usual complex conjugation. Obviously, $\pi | \alpha \Leftrightarrow \bar{\pi} | \bar{\alpha}$, and for $y \in \mathbb{Z}$ this means $\pi | y \Leftrightarrow \bar{\pi} | y$. Now assume that π is an irreducible element of o_F and that $\pi^k | y, \pi^{k+1} \nmid y$ for some $k \in \mathbb{N}$ implying $\pi^{3k} | \alpha\bar{\alpha}$ for a solution $y, \alpha = x + (-5)^{\frac{1}{2}}$ of (1.3). If $\pi = \bar{\pi}$ then k must be even and we can set $\tilde{y} \leftarrow y\pi^{-k}$, $\tilde{\alpha} \leftarrow \alpha\pi^{-3k/2}$ yielding another solution of (1.3) for which y is not divisible by π any more. For $\pi \neq \bar{\pi}$ there are integers $\mu, \nu \in \mathbb{Z}^{\geq 0}$ such that $\pi^{\mu} | \alpha, \bar{\pi}^{\nu} | \alpha$ and $\mu + \nu = 3k$ for a solution of (1.3). Hence we can replace y by $y\pi^{-k}\bar{\pi}^{-k}$ and α by $\alpha\pi^{-\mu}\bar{\pi}^{-\nu}$, obtaining another solution of (1.3) for which y is not divisible by π anymore. Hence, if (1.3) is solvable at all the assumption of o_F being factorial yields the solvability of $2 = a^2 + 5b^2$ which is clearly impossible.

Thus either (1.3) is not solvable or $\mathbb{Z}[(-5)^{\frac{1}{2}}]$ cannot be factorial. But (1.3) obviously has the solution $y = 3$, $x = \pm 7$ and we shall see later that this is indeed the only solution (compare exercise 2 of section 2). Thus o_F is certainly not a factorial ring. As a kind of a substitute for the decomposition of each element into a unique power product of irreducible elements times a unit we find that in o_F every non-zero ideal is a unique power product of prime ideals. This is the concept of Dedekind rings which we developed in chapter 4, section 5. As a special case the ring of integers o_F of an algebraic number field F is considered as a Dedekind ring in section 2.

The essential facts are among others that o_F is factorial if and only if every ideal of o_F is principal (see (1.13)). If not every ideal of o_F is principal then the equivalence relation on the set of all non-zero ideals of o_F defined by

$$\mathfrak{a} \sim \mathfrak{b} :\Leftrightarrow \alpha\mathfrak{a} = \beta\mathfrak{b} \tag{1.4}$$

for suitable $\alpha, \beta \in o_F \setminus \{0\}$ and $\mathfrak{a}, \mathfrak{b}$ non-zero ideals of o_F yields $h_F > 1$ equivalence classes. The number of equivalence classes is called the *class number* of F. The number h_F measures the deviation of o_F from being factorial (see also (1.13)). Moreover, the equivalence classes with respect to $\sim$ form a finite abelian multiplicative group Cl_F, where the multiplication is defined on the representatives. The computation of this group (respectively its order) – via suitable representatives of the ideal classes – is the main goal of this chapter.

As in the computation of the unit group Minkowski's number geometric ideas form the basis. Namely, they imply that those representatives can be chosen from a finite set of ideals $\mathfrak{a}$ of o_F characterized by the fact that the index $(o_F : \mathfrak{a})$ is less than a certain bound depending only on the field degree $n = s + 2t$ and the discriminant d_F of F – the so-called Minkowski bound. Moreover, that finite set can be effectively constructed from the prime ideals occurring in the (unique) prime ideal factorization of po_F where p runs through all rational prime numbers below the Minkowski bound. This theory is treated as part of the ideal theory of o_F in section 2.

The three remaining sections deal with the task of turning that effective procedure of determining Cl_F into an efficient one. Section 3 contains the fundamentals about ideal arithmetic in o_F. As $\mathbb{Z}$-modules of finite index in o_F the non-zero ideals of o_F also have a $\mathbb{Z}$-basis, a useful fact for most computational problems concerning ideals. However, non-zero ideals $\mathfrak{a}$ of o_F can also be presented in the form $\mathfrak{a} = ao_F + \alpha o_F$ $(a \in \mathbb{N}, \alpha \in o_F)$. This is obviously of advantage for storing ideals (only $n+1$ instead of n^2 – respectively $n(n+1)/2$ (compare chapter 3 (2.10)) – rational integers are needed for each ideal). Beyond that we develop some canonical representation of ideals by two elements which allows ideal multiplication via their generators:

$$\mathfrak{a}\mathfrak{b} = ab o_F + \alpha\beta o_F \quad \text{for } \mathfrak{a} = a o_F + \alpha o_F, \quad \mathfrak{b} = b o_F + \beta o_F. \tag{1.5}$$

Since we are dealing with the multiplicative structure of the non-zero ideals of o_F this is clearly a big advantage. It is also a great help in factoring $p o_F$, p a rational prime number, into prime ideals of o_F.

Section 4 discusses the task how to decide whether two given ideals are equivalent in the sense of (1.4). The well-known methods for this are indeed effective but not very efficient. We make instead use of an idea of K. Mahler which enables us to transfer the problem of determining whether $k \in \mathbb{N}$ is a product of n linear forms (tantamount to the norm of a certain ideal element) to the computation of certain values of a finite number of positive definite quadratic forms via chapter 3 (3.10). (Parts of that concept were already considered in chapter 5, section 3.) A complexity analysis shows that the number of arithmetical operations needed by our new method is only about the square root of the number of operations required by the old one.

Finally, in section 5 we develop an algorithm for the computation of Cl_F using the results of the earlier sections. The remaining difficulty is to build up Cl_F as a direct product of cyclic subgroups. Beginning with a finite number of generators our information is in general not complete, i.e. there can and often will exist equivalence relations between the generating ideals which are not implied by the ones we know. Testing for those relations becomes much easier by the methods of section 4 but still consumes a lot of computation time. Hence, those tests should be kept to a minimum. How to do this is the essential content of section 5. Once the class group has been established those tests become quite easy as is shown at the end of that section.

From the theory of Dedekind rings developed in chapter 4, section 5 we shortly recall the following important results. Dedekind rings R were characterized in chapter 4 (5.6) as entire rings

which are Noetherian and integrally closed and in which every non-zero prime ideal is maximal; (1.6)

in which the non-zero R-fractional ideals form a group, the so-called **ideal group** I_R of R; (1.7)

in which every ideal is a (unique) product of prime ideals. (1.8)

In this chapter we denote the subgroup of I_R consisting of all principal ideals by H_R. In case the *ideal class group* $\mathrm{Cl}_R := I_R/H_R$ is finite its order is said to be the *class number* h_R of R. In case of $R = o_F$ (see (2.1)) we also write Cl_F, h_F instead of Cl_R, h_R in general. The following useful properties of non-zero ideals $\mathfrak{a}, \mathfrak{b}$ of a Dedekind ring R are easy consequences of (1.8).

$\mathfrak{a} \subseteq \mathfrak{b}$ *if and only if R contains an ideal $\mathfrak{c}$ satisfying* $\mathfrak{a} = \mathfrak{b}\mathfrak{c}$ (1.9)
(*compare chapter* 4 (5.2)). *In that case we also write* $\mathfrak{b}|\mathfrak{a}$, *respectively* $\mathfrak{b}|\alpha$ *if* $\mathfrak{a}$ *is the principal ideal* αR.

$\gcd(\mathfrak{a}, \mathfrak{b}) = \mathfrak{a} + \mathfrak{b}$. (1.10)

$\mathfrak{a}$ *is the greatest common divisor of two principal ideals. The generating element of one of them can be chosen arbitrarily from* $\mathfrak{a}\backslash\{0\}$ (*compare chapter* 4 (5.39)d). (1.11)

There exist an element ω of R and an ideal $\mathfrak{c}$ of R such that $\mathfrak{b} + \mathfrak{c} = R$ *and* $\mathfrak{a}\mathfrak{c} = R\omega$. (1.12)

Proof of (1.12)
Let $\mathfrak{p}_1, \ldots, \mathfrak{p}_r$ be all prime ideals of R which divide either $\mathfrak{a}$ or $\mathfrak{b}$. Then $\mathfrak{a}$ has a presentation $\mathfrak{a} = \prod_{i=1}^r \mathfrak{p}_i^{a_i}$ ($a_i \in \mathbb{Z}^{\geq 0}$, $1 \leq i \leq r$), and we set

$$\mathfrak{d}_i := \prod_{\substack{j=1 \\ j \neq i}}^{r} \mathfrak{p}_j^{a_j+1} \quad (1 \leq i \leq r).$$

(In case $r = 1$ obviously $\mathfrak{d}_1 = R$.) Clearly, $R = \mathfrak{d}_1 + \cdots + \mathfrak{d}_r = \gcd(\mathfrak{d}_1, \ldots, \mathfrak{d}_r)$. Hence, there exist elements $\delta_i \in \mathfrak{d}_i$ ($1 \leq i \leq r$) satisfying $1 = \delta_1 + \cdots + \delta_r$. Also there exist elements $\beta_i \in \mathfrak{p}_i^{a_i} \backslash \mathfrak{p}_i^{a_i+1}$ ($1 \leq i \leq r$), because of (1.8) for example. We set $\omega := \sum_{i=1}^r \beta_i \delta_i$ and obtain $\mathfrak{p}_i^{a_i} \| R\omega$ because of $\mathfrak{p}_i^{a_i+1} | R\delta_j$ ($1 \leq j \leq r, j \neq i$) and $\mathfrak{p}_i \nmid R\delta_i$. Hence $\prod_{k=1}^r \mathfrak{p}_k^{a_k} | R\omega$ and $R\omega = \mathfrak{a}\mathfrak{c}$ for some ideal $\mathfrak{c}$ of R which is not divisible by any of the $\mathfrak{p}_i$ ($1 \leq i \leq r$). Since $\mathfrak{b}$ is a power product of $\mathfrak{p}_1, \ldots, \mathfrak{p}_r$ by assumption we obtain $\mathfrak{b} + \mathfrak{c} = R$ according to (1.10). □

The following theorem shows that Cl_R and h_R describe how much the structure of the Dedekind ring R deviates from a factorial ring.

Theorem (1.13)
The class number h_R of a Dedekind ring R is one if and only if R is factorial.

Proof
In case $h_R = 1$, every ideal of R is principal and R is therefore a unique factorization ring.

On the other hand, if R is a factorial Dedekind ring it suffices to show that every prime ideal of R is principal. Let $\mathfrak{p}$ be a non-zero prime ideal of

R. Because of (1.8) there exists an element β of $\mathfrak{p}\backslash\mathfrak{p}^2$, i.e. $R\beta \subseteq \mathfrak{p}$, $R\beta \not\subseteq \mathfrak{p}^2$. The ideal $R\beta$ has a unique presentation as a product of prime ideals, say $R\beta = \prod_{i=1}^{n} \mathfrak{p}_i^{m_i}$ ($\mathfrak{p}_i$ distinct prime ideals, $m_i \in \mathbb{N}$, $1 \leqslant i \leqslant n$). Reordering the $\mathfrak{p}_i$, if necessary, we obtain $\mathfrak{p}_1 \subseteq \mathfrak{p}$ and therefore $\mathfrak{p}_1 = \mathfrak{p}$ because of (1.6). Since $R\beta$ is not contained in $\mathfrak{p}^2$ we must have $m_1 = 1$. In the factorial ring R the element β has a presentation as a product of irreducible elements, say $\beta = \varepsilon \prod_{j=1}^{s} \pi_j^{k_j}$ (π_j distinct irreducible elements of $R, k_j \in \mathbb{N}$, $1 \leqslant j \leqslant s$, $s \in \mathbb{N}$, $\varepsilon \in U(R)$). But then $R\pi_i$ $(1 \leqslant i \leqslant s)$ is a prime ideal of R and from $\prod_{j=1}^{s} (R\pi_j)^{k_j} = R\beta = \mathfrak{p} \prod_{i=2}^{r} \mathfrak{p}_i^{m_i}$ we obtain $\mathfrak{p} = R\pi_j$ for a suitable index j $(1 \leqslant j \leqslant s)$ by (1.8).

□

6.2. The ring o_F of algebraic integers as a Dedekind ring

In this section we consider the maximal order of an algebraic number field and discuss the most important consequences which follow from the fact of it being a Dedekind ring. In the sequel F always denotes an algebraic number field of degree n and o_F the integral closure of $\mathbb{Z}$ in F. A subring of o_F which is also a free $\mathbb{Z}$-module of rank n (order of F) is denoted by R. From (1.6) it is clear that R can be a Dedekind ring only in case $R = o_F$. Again from chapter 4 we recall (chapter 4 (5.9)):

The maximal order o_F of an algebraic number field F is a Dedekind ring. (2.1)

We note that every (fractional) ideal of an order R of F is a free $\mathbb{Z}$-module of rank n. Hence, every order R is Noetherian and each non-zero prime ideal of R is a maximal ideal. But in case $R \subset o_F$ the ring R is not integrally closed and therefore I_R not a group (see also exercise 4). For any non-zero R-ideal $\mathfrak{a}$ the number of residue classes of $R/\mathfrak{a}$ equals the absolute value of the determinant of a transition matrix from a $\mathbb{Z}$-basis of R to a $\mathbb{Z}$-basis of $\mathfrak{a}$ (compare chapter 3 (2.9)) and is therefore finite. This plays an important role in proving that – in case $R = o_F$ – the group I_R/H_R is finite. The following definition is a special case of chapter 4 (5.44).

Definition (2.2)

Let R be an order of the algebraic number field F and $\mathfrak{a}$ a non-zero ideal of R. Then $\mathrm{N}(\mathfrak{a}) := \# R/\mathfrak{a}$ *is called the* **norm of the ideal** $\mathfrak{a}$.

As in chapter 4 the ideal norm is a natural generalization of the norm of an element.

Proposition (2.3)

For $\beta \in R\backslash\{0\}$ we have $\mathrm{N}(R\beta) = |\mathrm{N}(\beta)|$.

Proof
Let $M_\beta \in \mathbb{Z}^{n\times n}$ be the regular representation of β with respect to a $\mathbb{Z}$-basis $\omega_1,\ldots,\omega_n$ of R: $(\omega_1,\ldots,\omega_n)\beta = (\omega_1,\ldots,\omega_n)M_\beta$. Then chapter 3 (2.9) and chapter 5 (5.5) yield

$$\#(R/R\beta) = |\det M_\beta| = |\mathrm{N}(\beta)|. \qquad \square$$

In the most important case $R = o_F$ the ideal norm is shown to be a multiplicative function by the next two lemmata. (Compare also exercises 29–33 of chapter 4, section 5.)

Lemma (2.4)
Let $R = o_F$, $\mathfrak{a}$ a non-zero ideal of R and α, β be arbitrary elements of R. The congruence $\alpha x \equiv \beta \bmod \mathfrak{a}$ is solvable in R if and only if $\beta \equiv 0 \bmod \gcd(\mathfrak{a}, \alpha R)$. If it is solvable, then it has modulo $\mathfrak{a}$ exactly $\mathrm{N}(\gcd(\mathfrak{a}, \alpha R))$ different solutions. Any solution x is uniquely determined modulo $\mathfrak{a}(\gcd(\mathfrak{a}, \alpha R))^{-1}$.

Proof
Let $\mathfrak{d} := \gcd(\mathfrak{a}, \alpha R)$ and $\mathfrak{c} = \mathfrak{a}\mathfrak{d}^{-1}$. If the congruence is solvable, there is an $\tilde{\alpha} \in \mathfrak{a}$ such that $\alpha x - \tilde{\alpha} = \beta$ and $\mathfrak{d}$ divides $R\alpha x$, $R\tilde{\alpha}$ and therefore $R\beta$, i.e. $\beta \in \mathfrak{d}$. On the other hand, if β is in $\mathfrak{d} = \mathfrak{a} + \alpha R$, then there are $\tilde{\alpha} \in \mathfrak{a}$, $x \in R$ such that $\alpha x + \tilde{\alpha} = \beta$, i.e. $\alpha x \equiv \beta \bmod \mathfrak{a}$.

Now let x, y be both solutions of the congruence. This implies $\alpha(x-y) \in \mathfrak{a} = \mathfrak{c}\mathfrak{d}$, hence $R\alpha\, R(x-y) = \mathfrak{c}\mathfrak{d}\mathfrak{n}$ for some ideal $\mathfrak{n}$ of R. Comparing the prime ideal factorizations on both sides we obtain $R(x-y) \subseteq \mathfrak{c}$ because of $\mathfrak{d} = \gcd(\mathfrak{a}, R\alpha)$. Therefore any solution x of the congruence is uniquely determined modulo $\mathfrak{c}$. (For $\gamma \in \mathfrak{c}$ obviously $\alpha(x+\gamma) \equiv \alpha x \bmod \mathfrak{a}$ because of $\alpha\gamma \in \mathfrak{d}\mathfrak{c} = \mathfrak{a}$.) In the case of solvability of the congruence the number of modulo $\mathfrak{a}$ incongruent solutions is equal to the number of residue classes of $\mathfrak{c}/\mathfrak{a}$. So, it remains to construct a bijection between $\mathfrak{c}/\mathfrak{a}$ and $R/\mathfrak{d}$. For this purpose choose $\omega \in R$ satisfying $\gcd(R\omega, \mathfrak{d}\mathfrak{c}) = \mathfrak{c}$ by (1.12).

Then we map

$$\mathfrak{c}/\mathfrak{a} = \{\xi\omega + \mathfrak{a} \mid \xi \in R\} \text{ onto } R/\mathfrak{d} = \{\xi + \mathfrak{d} \mid \xi \in R\}$$

via

$$\xi\omega + \mathfrak{a} \mapsto \xi + \mathfrak{d}.$$

Clearly, $\omega \in \mathfrak{c}$ and therefore $\xi\omega \in \mathfrak{c}$ for every $\xi \in R$. On the other hand, let $\gamma \in \mathfrak{c}$, then $\xi\omega \equiv \gamma \bmod \mathfrak{a}$ is solvable in R because of $R\omega + \mathfrak{a} = \mathfrak{c}$. Therefore the map $\xi\omega + \mathfrak{a} \mapsto \xi + \mathfrak{d}$ is independent from the representative $\xi\omega$. Namely, for $\xi\omega + \mathfrak{a} = \tilde{\xi}\omega + \mathfrak{a}$ we obtain $(\xi - \tilde{\xi})\omega \in \mathfrak{a}$, hence $\xi - \tilde{\xi} \in \mathfrak{d}$ because of $R\omega + \mathfrak{a} = \mathfrak{c}$. The map is clearly bijective, hence $\#\mathfrak{c}/\mathfrak{a} = \mathrm{N}(\mathfrak{d})$. $\square$

Lemma (2.5)
Let $\mathfrak{a}, \mathfrak{b}$ be non-zero ideals of $R = o_F$. Then $\mathrm{N}(\mathfrak{a}\mathfrak{b}) = \mathrm{N}(\mathfrak{a})N(\mathfrak{b})$.

Proof
According to (1.12) we choose $\omega \in R$ satisfying $\mathfrak{ab} + R\omega = \mathfrak{a}$. For $r := \mathrm{N}(\mathfrak{a})$, $s := \mathrm{N}(\mathfrak{b})$ let $\xi_1, \ldots, \xi_r \in R$ and $\eta_1, \ldots, \eta_s \in R$ be a full set of representatives for $R/\mathfrak{a}$, $R/\mathfrak{b}$, respectively. We set $\tau_{ij} := \xi_i + \omega\eta_j$ $(1 \leqslant i \leqslant r, 1 \leqslant j \leqslant s)$ and show that they form a full set of representatives for $R/\mathfrak{ab}$.

In case of $\tau_{ij} \equiv \tau_{kl} \bmod \mathfrak{ab}$ $(1 \leqslant i, k \leqslant r,\ 1 \leqslant j, l \leqslant s)$ we obtain successively

$$\begin{aligned} \xi_i &\equiv \xi_k \bmod \mathfrak{a}, \quad \text{hence } i = k,\\ \omega\eta_j &\equiv \omega\eta_l \bmod \mathfrak{ab},\\ \eta_j &\equiv \eta_l \bmod \mathfrak{b} \text{ (because of (2.4)), i.e. } j = l. \end{aligned}$$

On the other hand, let $\alpha \in R$. Then $\alpha \equiv \xi_i \bmod \mathfrak{a}$ for some i $(1 \leqslant i \leqslant r)$, $\alpha - \xi_i \equiv \eta\omega \bmod \mathfrak{ab}$ is solvable, and therefore η uniquely determined modulo $\mathfrak{b}$ according to (2.4). Hence, $\eta = \eta_j$ for a suitable index j $(1 \leqslant j \leqslant s)$ and $\alpha \equiv \xi_i + \omega\eta_j \bmod \mathfrak{ab}$. □

The last lemma enables us to prove the finiteness of I_R/H_R in case $R = o_F$. In addition to o_F being a Dedekind ring, we need the finiteness of $o_F/\mathfrak{a}$ for non-zero ideals $\mathfrak{a}$ of o_F for the proof. We apply Minkowski's Convex Body Theorem chapter 3 (4.2), to the set C of chapter 3 (4.6). Namely, the *Minkowski mapping*

$$\varphi: o_F \to \mathbb{R}^n: x \mapsto (x^{(1)}, \ldots, x^{(s)}, \operatorname{Re} x^{(s+1)}, \operatorname{Im} x^{(s+1)}, \ldots, \operatorname{Re} x^{(s+t)}, \operatorname{Im} x^{(s+t)})^t \quad (2.6)$$

maps o_F onto an n-dimensional lattice with $d(\varphi(o_F)) = 2^{-t}|d_F|^{1/2}$. If we restrict φ to a non-zero ideal $\mathfrak{a}$ of o_F, we obviously obtain $d(\varphi(\mathfrak{a})) = 2^{-t}\mathrm{N}(\mathfrak{a})|d_F|^{1/2}$ for the sublattice $\varphi(\mathfrak{a})$ by chapter 3 (3.6). Then the set S of chapter 3 (4.5) (and therefore of course C of chapter 3 (4.6)) does not contain a vector $\mathbf{x} \in \mathbb{R}^n$, $\mathbf{x} \neq \mathbf{0}$, such that $\varphi^{-1}(\mathbf{x})$ is a non-zero element of o_F, $\mathfrak{a}$, respectively. But the set $C_{s,t}(\lambda)$ (see chapter 3 (4.8)) for $\lambda^n = (2^{n-s}n!(2/\pi)^t d(\Lambda))$ contains a non-zero $\mathbf{x} \in \mathbb{Z}^n$ because of its compactness and $V(C_{s,t}(\lambda)) = 2^n d(\Lambda)$ by chapter 3 (4.2). This $\mathbf{x}$ satisfies

$$\begin{aligned} |\mathrm{N}(\varphi^{-1}(\mathbf{x}))| &= \prod_{i=1}^{s} |x^{(i)}| \prod_{\substack{i=s+1\\ i-s\,\mathrm{odd}}}^{n-1} (x^{(i)2} + x^{(i+t)^2})\\ &\leqslant \left(\frac{\lambda}{n}\right)^n = \frac{n!}{n^n}\left(\frac{8}{\pi}\right)^t d(\Lambda). \end{aligned}$$

Thus we have proved the following lemma.

Lemma (2.7)

Let $\mathfrak{a}$ be a non-zero ideal of o_F. Then $\mathfrak{a}$ contains a non-zero element x satisfying

$$|\mathrm{N}(x)| \leqslant \frac{n!}{n^n}\left(\frac{4}{\pi}\right)^t \mathrm{N}(\mathfrak{a})|d_F|^{1/2}.$$

Corollary (2.8)

Every ideal class C of I_F/H_F contains an ideal $\mathfrak{a}$ of o_F satisfying

$$\mathrm{N}(\mathfrak{a}) \leqslant \frac{n!}{n^n}\left(\frac{4}{\pi}\right)^t |d_F|^{1/2} =: M_F$$

M_F is called **Minkowski's constant** *of F.*

Proof

Let C be any ideal class of I_F/H_F and $\mathfrak{c}$ an integral ideal of C^{-1}. We choose $x \in \mathfrak{c}$ subject to (2.7). By (1.9) there is an integral ideal $\mathfrak{a}$ (necessarily in C) satisfying $xo_F = \mathfrak{a}\mathfrak{c}$. Hence, we obtain

$$\mathrm{N}(\mathfrak{a}) = \frac{|\mathrm{N}(x)|}{\mathrm{N}(\mathfrak{c})} \leqslant \frac{n!}{n^n}\left(\frac{4}{\pi}\right)^t |d_F|^{1/2}.$$

□

We note that there are better bounds than (2.8) which were obtained by analytic methods (see [14] of chapter 5). Those should be used for class group computations in number fields of degree $n \geqslant 4$.

Corollary (2.9)

The class number h_F of F is finite.

Proof

Because of (2.8), o_F being a Dedekind ring, and the multiplicativity of the ideal norm it suffices to show that there are at most finitely many prime ideals in o_F whose norms are less than M_F. Let $\mathfrak{p}$ be a non-zero prime ideal of o_F. Since $o_F/\mathfrak{p}$ is an additive group of order $\mathrm{N}(\mathfrak{p})$ and $1 \notin \mathfrak{p}$, we get $\mathrm{N}(\mathfrak{p})(1+\mathfrak{p}) = \mathfrak{p}$, hence $\mathrm{N}(\mathfrak{p})$ is a natural number contained in $\mathfrak{p}$. Obviously, $\mathrm{N}(\mathfrak{p}) \geqslant 2$ and $\mathfrak{p}$ occurs in the prime ideal factorization of $\mathrm{N}(\mathfrak{p})o_F$. Therefore $\mathfrak{p}$ divides po_F for some rational prime number p and p is uniquely determined by $\mathfrak{p}$ (see exercise 5). But this implies $\mathrm{N}(\mathfrak{p}) | p^n$ by (2.5), hence $\mathrm{N}(\mathfrak{p}) = p^f$ for some $f \in \mathbb{N}$, $1 \leqslant f \leqslant n$. Again, by (2.5) there can be only finitely many prime ideals $\mathfrak{p}$ containing po_F. Since there are only finitely many prime numbers p with $p \leqslant M_F$ this concludes the proof. □

For later applications we note part of the results of the last proof separately:

Lemma (2.10)

Every non-zero ideal $\mathfrak{a}$ of o_F satisfies $\mathrm{N}(\mathfrak{a}) \in \mathfrak{a}$. If $\mathfrak{a}$ is a prime ideal, then $o_F/\mathfrak{a}$ is a field with p^f elements, where $f \in \mathbb{N}$, $1 \leqslant f \leqslant n$, and p is the unique prime number contained in $\mathfrak{a}$.

We note that there are algebraic number fields for which the class number is arbitrarily large. For example, if d runs through the squarefree natural numbers the class numbers h_F of the fields $F = \mathbb{Q}(-d)^{\frac{1}{2}}$ satisfy $\lim_{d \to \infty}(\log h_F / \log d^{\frac{1}{2}}) = 1$ [6]. On the other hand it is an open question, if infinitely many algebraic number fields have class number one. For real quadratic fields C.F. Gauss conjectured that $h_F = 1$ for infinitely many F, and extensive computer calculations seem to support this conjecture.

If the Minkowski constant of (2.8) is less than 2, the corresponding algebraic number field F necessarily has class number $h_F = 1$. The following examples illustrate this.

Examples (2.11)

(i) Let $F = \mathbb{Q}(d^{\frac{1}{2}})$, $d \equiv 2, 3 \bmod 4$, $d \in \mathbb{Z}$ square-free. Then $h_F = 1$, if $(2!/2^2)(4/\pi)^t(4|d|)^{1/2} < 2$. For $t = 0$ this is fulfilled by $d = 2, 3$, for $t = 1$ in case of $d = -1, -2$.

(ii) $F = \mathbb{Q}(e^{2\pi i/5})$ is totally complex of degree $n = 4 = 2t$. Because of chapter 2 (11.12) and chapter 4 (6.10) the discriminant is $d_F = 5^3$. Hence, $M_F = 15(5)^{\frac{1}{2}}/(2\pi^2) < 2$ and $h_F = 1$.

In order to determine h_F and I_F/H_F lemma (2.8) suggests to consider the factorization of all rational prime numbers $p \leqslant M_F$ into prime ideals of F. And we are about to develop the tools for this task.

We start with a slightly more general situation. Namely, we consider a relative finite algebraic extension G of an algebraic number field F. The rings of algebraic integers of F, G are denoted by R, S respectively. The degree $[G:F]$ is denoted by n, as usual. If $\mathfrak{p}$ is a prime ideal of R, then $S\mathfrak{p} := \{\sum_{i=1}^{m} s_i p_i \mid s_i \in S,\ p_i \in \mathfrak{p},\ m \in \mathbb{N}\}$ is obviously an ideal of S. Because of (2.1) S is a Dedekind ring and there exists a prime ideal factorization $S\mathfrak{p} = \mathfrak{P}_1^{e_1} \cdot \ldots \cdot \mathfrak{P}_r^{e_r}$ of $S\mathfrak{p}$ into a power product of prime ideals of S. Hence, we need to determine the prime ideals $\mathfrak{P}_i$, the exponents e_i and the natural number r.

Lemma (2.12)

Let $\mathfrak{p}$, $\mathfrak{P}$ be non-zero prime ideals of R, S respectively. Then the following conditions are equivalent

(i) $\mathfrak{P} \mid S\mathfrak{p}$,
(ii) $\mathfrak{P} \supseteq S\mathfrak{p}$,
(iii) $\mathfrak{P} \supseteq \mathfrak{p}$,
(iv) $\mathfrak{P} \cap R = \mathfrak{p}$,
(v) $\mathfrak{P} \cap F = \mathfrak{p}$.

Proof

The equivalence of (i) and (ii) is just (1.9).

(ii)$\Rightarrow$(iii) is obvious since $1 \in S$ and therefore $S\mathfrak{p} \supseteq \mathfrak{p}$.

(iii)$\Rightarrow$(iv): It is clear that $\mathfrak{P} \cap R \supseteq \mathfrak{p}$. On the other hand, $\mathfrak{P} \cap R$ is an ideal properly contained in R. Hence, $\mathfrak{P} \cap R = \mathfrak{p}$ because of the maximality of $\mathfrak{p}$.

(iv)$\Rightarrow$(v): $\mathfrak{P}\cap F=(\mathfrak{P}\cap S)\cap F=\mathfrak{P}\cap(S\cap F)=\mathfrak{P}\cap R=\mathfrak{p}$.

(v)$\Rightarrow$(ii): Clearly $\mathfrak{P}\supseteq\mathfrak{p}$ and therefore $\mathfrak{P}=S\mathfrak{P}\supseteq S\mathfrak{p}$. □

Definition (2.13)

If one of the conditions (2.12) (i)–(v) *is satisfied, we say* $\mathfrak{P}$ **lies over** $\mathfrak{p}$, *or* $\mathfrak{p}$ **lies under** $\mathfrak{P}$.

The following corollary is obvious.

Corollary (2.14)

Let $\mathfrak{p}$ *be a non-zero prime ideal of* R *and* $S\mathfrak{p}=\mathfrak{P}_1^{e_1}\cdot\ldots\cdot\mathfrak{P}_r^{e_r}$ *the prime ideal factorization of* $S\mathfrak{p}$ *in* S. *Then* $\mathfrak{P}_1,\ldots,\mathfrak{P}_r$ *are exactly the prime ideals of* S *lying over* $\mathfrak{p}$.

Definition (2.15)

Let $\mathfrak{p}$ *be a non-zero prime ideal of* R *and* $\mathfrak{P}$ *a prime ideal of* S *such that* $\mathfrak{P}^e|S\mathfrak{p}$, $\mathfrak{P}^{e+1}\nmid S\mathfrak{p}$. *Then the exponent* $e=e(\mathfrak{P}|\mathfrak{p})$ *is called the* **ramification index** *of* $\mathfrak{P}$ *over* $\mathfrak{p}$. *The prime ideal* $\mathfrak{p}$ *is called* **ramified** *in* G, *if there is a prime ideal* $\mathfrak{P}$ *of* S *lying over* $\mathfrak{p}$ *with* $e(\mathfrak{P}|\mathfrak{p})>1$, *otherwise* **unramified.**

For the determination of the number r of prime ideals of S lying over a non-zero prime ideal $\mathfrak{p}$ of R we shall need a further invariant which compares the residue class fields $R/\mathfrak{p}$ and $S/\mathfrak{P}$. Since $R/\mathfrak{p}$ is a priori not necessarily contained in $S/\mathfrak{P}$ we construct a natural embedding.

Lemma (2.16)

Let $\mathfrak{p}$ *be a non-zero prime ideal of* R *and* $\mathfrak{P}$ *a prime ideal of* S *lying over* $\mathfrak{p}$. *Then* $\kappa\colon R/\mathfrak{p}\to S/\mathfrak{P}\colon r+\mathfrak{p}\mapsto r+\mathfrak{P}$ *is a ring-monomorphism which embeds* $R/\mathfrak{p}$ *in* $S/\mathfrak{P}$. *In the sequel we can therefore identify* $R/\mathfrak{p}$ *and* $\kappa(R/\mathfrak{p})$.

Proof

Apply (2.12). □

Definition (2.17)

Let $\mathfrak{p}$ *be a non-zero prime ideal of* R *and* $\mathfrak{P}$ *a prime ideal of* S *lying over* $\mathfrak{p}$. *Then* $f=f(\mathfrak{P}|\mathfrak{p}):=[S/\mathfrak{P}:R/\mathfrak{p}]$ *is called the* **degree of inertia of** $\mathfrak{P}$ *over* $\mathfrak{p}$.

This definition clearly implies

Theorem (2.18)

Let $\mathfrak{P}$, $\mathfrak{p}$ *be as in* (2.17). *Then*

$$\mathrm{N}(\mathfrak{P})=\#S/\mathfrak{P}=\mathrm{N}(\mathfrak{p})^{f(\mathfrak{P}|\mathfrak{p})}.$$

This enables us to prove a first major result on the prime ideal factorization in S.

Theorem (2.19)
Let $\mathfrak{p}$ be a non-zero prime ideal of R and $S\mathfrak{p} = \mathfrak{P}_1^{e_1} \cdot \ldots \cdot \mathfrak{P}_r^{e_r}$ the corresponding prime ideal factorization in S. Then

$$n = \sum_{i=1}^{r} e_i f_i = \sum_{\mathfrak{P} \supseteq \mathfrak{p}} e(\mathfrak{P}|\mathfrak{p}) f(\mathfrak{P}|\mathfrak{p}). \tag{2.20}$$

Proof
Because of $N(S\mathfrak{p}) = N(\mathfrak{p})^{e_1 f_1 + \cdots + e_r f_r}$ it suffices to show $N(S\mathfrak{p}) = N(\mathfrak{p})^n$. If h_F is the class number of F, then $\mathfrak{p}^{h_F} = \lambda R$ for some $\lambda \in R$. Hence,

$$\begin{aligned} N(\mathfrak{p}S)^{h_F} &= N((\mathfrak{p}S)^{h_F}) = N(\mathfrak{p}^{h_F} S) = N(\lambda S) \\ &= |N_{G/\mathbb{Q}}(\lambda)| = |N_{F/\mathbb{Q}}(N_{G/F}(\lambda))| \\ &\quad \text{(by exercise 8 of chapter 2, section 3)} \\ &= |N_{F/\mathbb{Q}}(\lambda^n)| = |N_{F/\mathbb{Q}}(\lambda)|^n \\ &= N(\mathfrak{p}^{h_F})^n = N(\mathfrak{p})^{h_F n} \end{aligned}$$

which concludes the proof. □

Having established (2.20) we need a method for constructing all prime ideals $\mathfrak{P}$ lying over $\mathfrak{p}$. Before we attack this problem we shortly discuss the influence of automorphisms on ideal factorizations.

Lemma (2.21)
Let $\mathfrak{P}$, $\mathfrak{p}$ as in (2.17) and $\sigma \in \mathrm{Aut}(G/F)$. Then

(i) *$\sigma(\mathfrak{P})$ is a prime ideal of S,*
(ii) *$\sigma(\mathfrak{P}) \supseteq \mathfrak{p}$,*
(iii) *$f(\mathfrak{P}|\mathfrak{p}) = f(\sigma(\mathfrak{P})|\mathfrak{p})$,*
(iv) *$e(\mathfrak{P}|\mathfrak{p}) = e(\sigma(\mathfrak{P})|\mathfrak{p})$.*

Proof
(i) and (ii) are obvious. For (iii) we note that σ implies a vector space isomorphism

$$\hat{\sigma}: S/\mathfrak{P} \to \sigma(S)/\sigma(\mathfrak{P}) = S/\sigma(\mathfrak{P}),$$

where $S/\mathfrak{P}$, $S/\sigma(\mathfrak{P})$ are considered as $R/\mathfrak{p}$ vector spaces. Finally, to show (iv) we apply σ to the factorization $S\mathfrak{p} = \mathfrak{P}_1^{e_1} \cdot \ldots \cdot \mathfrak{P}_r^{e_r}$ to obtain $S\mathfrak{p} = \sigma(S\mathfrak{p}) = \sigma(\mathfrak{P}_1)^{e_1} \cdot \ldots \cdot \sigma(\mathfrak{P}_r)^{e_r}$. □

In particular, for a Galois extension G/F we get

Theorem (2.22)

Let G/F be a Galois extension and $\mathfrak{p}$ a non-zero prime ideal of F. Then the Galois group $H = \mathrm{Gal}(G/F)$ operates transitively on the prime ideals $\mathfrak{P}_i$ $(1 \leqslant i \leqslant r)$ which lie over $\mathfrak{p}$.

Proof

We assume that there are indices $1 \leqslant i < j \leqslant r$ such that $\{\sigma(\mathfrak{P}_i) \mid \sigma \in H\} \cap \{\sigma(\mathfrak{P}_j) \mid \sigma \in H\} = \emptyset$. By the Chinese remainder theorem, chapter 2 (2.17), we construct $s \in S$ subject to

$$s \equiv 1 \bmod \sigma(\mathfrak{P}_i) \text{ for all } \sigma \in H,$$
$$s \equiv 0 \bmod \sigma(\mathfrak{P}_j) \text{ for all } \sigma \in H.$$

For this element s we obtain

$$\mathrm{N}_{G/F}(s) = \prod_{\sigma \in H} \sigma^{-1}(s) \in \mathfrak{P}_j \cap R = \mathfrak{p},$$

and on the other hand

$$\mathrm{N}_{G/F}(s) = \prod_{\sigma \in H} \sigma^{-1}(s) \notin \mathfrak{P}_i \supseteq \mathfrak{p},$$

clearly a contradiction. Hence, the assumption was false. □

As an easy consequence of (2.21) and (2.22) we obtain

Corollary (2.23)

Let G, F, $\mathfrak{p}$ as in (2.22). Then the prime ideal factorization of $\mathfrak{p}$ in S is of the form $S\mathfrak{p} = (\mathfrak{P}_1 \cdot \ldots \cdot \mathfrak{P}_r)^e$, the ramification indices $e = e(\mathfrak{P} \mid \mathfrak{p})$, respectively the degrees of inertia $f = f(\mathfrak{P} \mid \mathfrak{p})$ coincide for all $\mathfrak{P}$ lying over $\mathfrak{p}$ implying $n = efr$.

The determination of the prime ideals $\mathfrak{P}$ of S lying over the non-zero prime ideal $\mathfrak{p}$ of R is always a little complicated. The main reason for this is that we usually know G only as $G = F(\rho)$ for some $\rho \in S$ and in general there is not even a basis of S over R. Hence, we have direct access only to the ideal theory of $R[\rho]$ and not of $S = \mathrm{Cl}(R, F)$. The transition from $R[\rho]$ to S is then possible via the so-called conductor:

Definition (2.24)

Let $\mathfrak{o}$ be any order of G. Then $\mathfrak{F} := \{x \in \mathfrak{o} \mid xS \subseteq \mathfrak{o}\}$ is called the **conductor** of $\mathfrak{o}$ *in S.*

It is easily seen that $\mathfrak{F}$ is an ideal of $\mathfrak{o}$ as well as of S. Also we note $\mathfrak{F} \supseteq mS$ for $m := (S:\mathfrak{o})$. Before we state the general result we need two preparatory lemmata.

Lemma (2.25)

Let $\mathfrak{o}$ be an order of S of conductor $\mathfrak{F}$. Then

$D_S := \{\mathfrak{a} \subseteq S \mid \mathfrak{a}$ a non-zero ideal of S such that $\mathfrak{a} + \mathfrak{F} = S\}$ and $D_{\mathfrak{o}} := \{\tilde{\mathfrak{a}} \subseteq \mathfrak{o} \mid \tilde{\mathfrak{a}}$ a non-zero ideal of $\mathfrak{o}$ such that $\tilde{\mathfrak{a}} + \mathfrak{F} = \mathfrak{o}\}$ are multiplicative monoids with cancellation law.

Proof
All $\mathfrak{a}, \mathfrak{b} \in D_S$ satisfy

$$\begin{aligned} S = SS = (\mathfrak{a} + \mathfrak{F})(\mathfrak{b} + \mathfrak{F}) &= \mathfrak{a}\mathfrak{b} + \mathfrak{F}(\mathfrak{a} + \mathfrak{b} + \mathfrak{F}) \\ &= \mathfrak{a}\mathfrak{b} + \mathfrak{F}S = \mathfrak{a}\mathfrak{b} + \mathfrak{F}, \end{aligned}$$

hence D_S is a monoid with unit element S. Analogously we show that $D_{\mathfrak{o}}$ is a monoid. In D_S common factors can be divided out because of $D_S \subseteq I_S$. The proof of this property for $D_{\mathfrak{o}}$ is contained in the proof of the next lemma. □

Lemma (2.26)
Let $\mathfrak{o}$ be an order of G of conductor $\mathfrak{F}$ and $D_{\mathfrak{o}}$, D_S as in (2.25).

(i) *$\kappa: D_{\mathfrak{o}} \to D_S: \tilde{\mathfrak{a}} \mapsto S\tilde{\mathfrak{a}}$ is an isomorphism with inverse $\kappa^{-1}: D_S \to D_{\mathfrak{o}}: \mathfrak{a} \mapsto \mathfrak{a} \cap \mathfrak{o}$.*
(ii) *For $\mathfrak{a} \in D_S$ we have $S/\mathfrak{a} \simeq \mathfrak{o}/\mathfrak{a} \cap \mathfrak{o}$.*
(iii) *Every ideal $\tilde{\mathfrak{a}}$ of $D_{\mathfrak{o}}$ has in $\mathfrak{o}$ a unique presentation as a product of maximal ideals.*

We note that (2.26) (iii) proves the statement of (2.25) about the cancellation law for $D_{\mathfrak{o}}$.

Proof
(i) For $\tilde{\mathfrak{a}} \in D_{\mathfrak{o}}$ we obtain

$$S \supseteq \kappa(\tilde{\mathfrak{a}}) + \mathfrak{F} = S\tilde{\mathfrak{a}} + \mathfrak{F} = S\tilde{\mathfrak{a}} + S\mathfrak{F} = S(\tilde{\mathfrak{a}} + \mathfrak{F}) = S\mathfrak{o} = S,$$

hence κ is a homomorphism of $D_{\mathfrak{o}}$ into D_S. (The homomorphy of κ is obvious.) For the injectivity of κ it suffices to prove $\tilde{\mathfrak{a}} = S\tilde{\mathfrak{a}} \cap \mathfrak{o}$ ($\tilde{\mathfrak{a}} \in D_{\mathfrak{o}}$) since then $S\tilde{\mathfrak{a}} = S\tilde{\mathfrak{b}}$ implies $\tilde{\mathfrak{a}} = \tilde{\mathfrak{b}}$. We conclude as follows:

$$\begin{aligned} \tilde{\mathfrak{a}} \subseteq S\tilde{\mathfrak{a}} \cap \mathfrak{o} &= S\tilde{\mathfrak{a}} \cap (\tilde{\mathfrak{a}} + \mathfrak{F}) \\ &= \tilde{\mathfrak{a}} + S\tilde{\mathfrak{a}} \cap \mathfrak{F} \\ &= \tilde{\mathfrak{a}} + S\tilde{\mathfrak{a}}\mathfrak{F} \quad (S\,\tilde{\mathfrak{a}} \text{ and } \mathfrak{F} \text{ are comaximal}) \\ &= \tilde{\mathfrak{a}} + \tilde{\mathfrak{a}}\mathfrak{F} \\ &= \tilde{\mathfrak{a}}. \end{aligned}$$

To prove the surjectivity of κ we choose an arbitrary $\mathfrak{a} \in D_S$. Then $S = \mathfrak{a} + \mathfrak{F}$ implies $\tilde{\mathfrak{a}} := \mathfrak{a} \cap \mathfrak{o} \in D_{\mathfrak{o}}$. It therefore remains to show $S(\mathfrak{a} \cap \mathfrak{o}) = \mathfrak{a}$:

$$\begin{aligned} \mathfrak{a} = \mathfrak{a}\mathfrak{o} &= \mathfrak{a}(\mathfrak{o} \cap \mathfrak{a} + \mathfrak{F}) \\ &= \mathfrak{a}(\mathfrak{o} \cap \mathfrak{a}) + \mathfrak{F}\mathfrak{a} \end{aligned}$$

$$= \mathfrak{a}(\mathfrak{o} \cap \mathfrak{a}) + \mathfrak{F}(\mathfrak{o} \cap \mathfrak{a})$$
$$\text{(because of } \mathfrak{F}\mathfrak{a} = \mathfrak{F} \cap \mathfrak{a} = \mathfrak{F} \cap (\mathfrak{o} \cap \mathfrak{a}) = \mathfrak{F}(\mathfrak{o} \cap \mathfrak{a}))$$
$$= S(\mathfrak{o} \cap \mathfrak{a})$$
$$= S\tilde{\mathfrak{a}}.$$

(ii) We consider the residue class mapping $\varphi\colon \mathfrak{o} \twoheadrightarrow S/\mathfrak{a}\colon r \mapsto r + \mathfrak{a}$. Because of the homomorphism theorem for rings we need to show: (a) $\operatorname{Ker}\varphi = \mathfrak{o} \cap \mathfrak{a}$, (b) φ is surjective. Part (a) is obvious. For (b) let r be any element of S. Because of $S = \mathfrak{a} + \mathfrak{F}$ there exist $\tilde{r} \in \mathfrak{a}$ and $f \in \mathfrak{F}$ such that $r = \tilde{r} + f$. Since f is in $\mathfrak{o}$ we obtain $\varphi(f) = r + \mathfrak{a}$.

(iii) Let $\tilde{\mathfrak{a}} \in D_{\mathfrak{o}}$ and $\kappa(\tilde{\mathfrak{a}}) = S\tilde{\mathfrak{a}} = \mathfrak{a} \in D_S$. Then $\mathfrak{a}$ has a unique presentation of prime ideals of S:

$$\mathfrak{a} = \mathfrak{P}_1^{e_1} \cdot \ldots \cdot \mathfrak{P}_r^{e_r}.$$

Since the $\mathfrak{P}_i$ $(1 \leqslant i \leqslant r)$ contain $\mathfrak{a}$ they must also belong to D_S. Therefore we obtain a factorization in $\mathfrak{o}$: $\tilde{\mathfrak{a}} = \kappa^{-1}(\mathfrak{P}_1)^{e_1} \cdot \ldots \cdot \kappa^{-1}(\mathfrak{P}_r)^{e_r}$. Because of (ii) all $\mathfrak{o}/(\mathfrak{P}_i \cap \mathfrak{o})$ are fields, hence $\kappa^{-1}(\mathfrak{P}_i) = \mathfrak{P}_i \cap \mathfrak{o}$ maximal ideals of $\mathfrak{o}$. The presentation of $\tilde{\mathfrak{a}}$ as product of maximal ideals is unique, since S is a Dedekind ring and κ an isomorphism. □

In the following principal theorem we obtain the prime ideals $\mathfrak{P}$ of S which lie over a non-zero prime ideal $\mathfrak{p}$ of R from the factors of the minimal polynomial of ρ $(G = F(\rho))$ modulo $\mathfrak{p}$.

Theorem (2.27)

(*Decomposition of prime ideals*). *Let G, F be algebraic number fields with maximal orders S, R respectively. Let $G = F(\rho)$, $\rho \in S$, and $f(t) \in R[t]$ the minimal polynomial of ρ. Finally, let $\mathfrak{F}$ denote the conductor of $R[\rho]$ in S. Then the decomposition of prime ideals $\mathfrak{p}$ of R satisfying $S\mathfrak{p} + \mathfrak{F} = S$ into prime ideals of S is the following: Let $\bar{f}(t) = \bar{f}_1(t)^{e_1} \cdot \ldots \cdot \bar{f}_r(t)^{e_r}$ be the decomposition of $f(t)$ into distinct monic irreducible polynomials over $R/\mathfrak{p}[t]$, and let $f_i(t) \in R[t]$ be distinct monic irreducible polynomials which are mapped onto $\bar{f}_i(t)$ under the residue class mapping $R[t] \to R/\mathfrak{p}[t]$ $(1 \leqslant i \leqslant r)$. Then $S\mathfrak{p}$ has a unique presentation $S\mathfrak{p} = \mathfrak{P}_1^{e_1} \cdot \ldots \cdot \mathfrak{P}_r^{e_r}$ as a power product of prime ideals in S, where*

$$\mathfrak{P}_i = \mathfrak{p}S + f_i(\rho)S,$$
$$e(\mathfrak{P}_i \mid \mathfrak{p}) = e_i,$$
$$f(\mathfrak{P}_i \mid \mathfrak{p}) = \deg(f_i) \quad (1 \leqslant i \leqslant r).$$

Proof

The underlying idea is to study the prime ideals of $R[\rho]$ which lie over a given non-zero prime ideal $\mathfrak{p}$ of R subject to $S\mathfrak{p} + \mathfrak{F} = S$. The transition from $R[\rho]$ to S is then via (2.26). For abbreviation we denote the images under the residue class mapping $R \to R/\mathfrak{p}$, respectively, $R[t] \to R[t]/\mathfrak{p}R[t]$, by $^-$, i.e.

$\sum_{i=0}^m a_i t^i \mapsto \sum_{i=0}^m \bar{a}_i t^i$. We start by showing

$$R[\rho]/\mathfrak{p}R[\rho] \simeq \bar{R}[t]/\bar{f}(t)\bar{R}[t]. \tag{2.28a}$$

This result is obtained by applying the ring homomorphism theorem to the mapping

$$\Phi\colon R[\rho] \to \bar{R}[t]/\bar{f}(t)\bar{R}[t]\colon h(\rho) \mapsto \bar{h}(t) + \bar{f}(t)\bar{R}[t],$$

where h is an arbitrary polynomial of $R[t]$. We have to show that Φ is a well-defined homomorphism, that $\ker\Phi = \mathfrak{p}R[\rho]$ and that Φ is surjective. The latter is, of course, obvious as well as the fact that Φ is a homomorphism. To show that Φ is well defined we take $h_1, h_2 \in R[t]$ such that $h_1(\rho) = h_2(\rho)$. This implies in $R[t]$: $h_1(t) - h_2(t) = q(t)f(t) + r(t)$ with $\deg(r) < \deg(f)$ since f was monic. The specialization $t \mapsto \rho$ yields $r(\rho) = 0$ and therefore $r = 0$ because of the irreducibility of f. Applying $^-$ to the equation above we obtain in $\bar{R}[t]$: $\bar{h}_1(t) - \bar{h}_2(t) = \bar{q}(t)\bar{f}(t)$, hence $\Phi(h_1(\rho)) = \Phi(h_2(\rho))$. To determine $\ker\Phi$ let $h(\rho) \in R[\rho]$ such that $\Phi(h(\rho)) = 0$. This implies $\bar{h}(t) = \bar{q}(t)\bar{f}(t)$ for some polynomial $\bar{q}(t) \in \bar{R}[t]$, hence $h(t) = q(t)f(t) + r(t)$ with $r(t) \in \mathfrak{p}R[t]$ in $R[t]$. The specialization $t \mapsto \rho$ then yields $h(\rho) = r(\rho) \in \mathfrak{p}R[\rho]$. On the other hand, for $h(\rho) \in \mathfrak{p}R[\rho]$ we have $h(t) \in \mathfrak{p}R[t]$ and therefore $\bar{h}(t) = \bar{0} \in \bar{R}[t]$.

Because of (2.28a) it suffices to determine the prime ideals of $\bar{R}[t]/\bar{f}(t)\bar{R}[t]$. But $\bar{R}[t]$ is a principal ideal ring and the only irreducible elements of $\bar{R}[t]$ dividing $\bar{f}(t)$ are $\bar{f}_1(t), \ldots, \bar{f}_r(t)$. Hence, $\bar{f}_i(t)\bar{R}[t]$ are the only prime ideals of $\bar{R}[t]$ lying over $\bar{f}(t)\bar{R}[t]$ implying that $\bar{f}_i(t)/\bar{f}(t)\bar{R}(t)$ generate exactly the non-zero prime ideals of $\bar{R}[t]/\bar{f}(t)\bar{R}[t]$ $(1 \leqslant i \leqslant r)$. Let us denote the isomorphism of (2.28a) by Φ. Then obviously $\Phi^{-1}(\bar{f}_i(t)/\bar{f}(t)\bar{R}[t])$ generate all non-zero prime ideals of $R[\rho]/\mathfrak{p}R[\rho]$; they are of the form $f_i(\rho)R[\rho]/\mathfrak{p}R[\rho]$.

Hence, we obtain that all prime ideals of $R[\rho]$ which contain $\mathfrak{p}R[\rho]$ are $f_i(\rho)R[\rho] + \mathfrak{p}R[\rho]$ $(1 \leqslant i \leqslant r)$. Now our premise $S\mathfrak{p} + \mathfrak{F} = S$ and (2.26) imply that all prime ideals of S lying over $S\mathfrak{p}$ are given by

$$\mathfrak{P}_i := S\mathfrak{p} + f_i(\rho)S \quad (1 \leqslant i \leqslant r). \tag{2.28b}$$

It remains to show

$$f(\mathfrak{P}_i|\mathfrak{p}) = \deg(f_i), \tag{2.28c}$$

$$e(\mathfrak{P}_i|\mathfrak{p}) = e_i \quad (1 \leqslant i \leqslant r). \tag{2.28d}$$

The proof of (2.28c) is as follows:

$$\begin{aligned}
\mathrm{N}(\mathfrak{p})^{f(\mathfrak{P}_i|\mathfrak{p})} &= |R/\mathfrak{p}|^{(S/\mathfrak{P}_i:R/\mathfrak{p})} = |S/\mathfrak{P}_i| \\
&= |R[\rho]/\mathfrak{P}_i \cap R[\rho]| \quad \text{(by (2.26))} \\
&= |(R[\rho]/\mathfrak{p}R[\rho])/((\mathfrak{P}_i \cap R[\rho])/\mathfrak{p}R[\rho])| \\
&= |(\bar{R}[t]/\bar{f}(t)\bar{R}[t])/(\bar{f}_i(t)\bar{R}[t]/\bar{f}(t)\bar{R}[t])| \quad \text{(by (2.28a))} \\
&= |\bar{R}[t]/\bar{f}_i(t)\bar{R}[t]| \\
&= |R/\mathfrak{p}|^{\deg(\bar{f}_i)}.
\end{aligned}$$

For the proof of (2.28d) we show $e_i \geqslant e(\mathfrak{P}_i|\mathfrak{p})$ at first. Namely, we have

$$\prod_{i=1}^{r} \mathfrak{P}_i^{e_i} = \prod_{i=1}^{r} (\mathfrak{p}S + f_i(\rho)S)^{e_i} \subseteq \mathfrak{p}S + f_1(\rho)^{e_1} \cdot \ldots \cdot f_r(\rho)^{e_r} S$$

$$\subseteq \mathfrak{p}S + \mathfrak{p}R[\rho]S \subseteq \mathfrak{p}S = \prod_{i=1}^{r} \mathfrak{P}_i^{e(\mathfrak{P}_i|\mathfrak{p})}$$

because of $f(t) - f_1(t)^{e_1} \cdot \ldots \cdot f_r(t)^{e_r} \in \mathfrak{p}R[t]$ and therefore $f_1(\rho)^{e_1} \cdot \ldots \cdot f_r(\rho)^{e_r} \in \mathfrak{p}R[\rho]$. On the other hand, (2.20) yields

$$n = \sum_{i=1}^{r} e(\mathfrak{P}_i|\mathfrak{p}) f(\mathfrak{P}_i|\mathfrak{p}) = \sum_{i=1}^{r} e(\mathfrak{P}_i|\mathfrak{p}) \deg(f_i)$$

$$\leqslant \sum_{i=1}^{r} e_i \deg(f_i) = \deg(f) = n.$$

Since all constants involved are positive this implies $e_i = e(\mathfrak{P}_i|\mathfrak{p})$ for all i $(1 \leqslant i \leqslant r)$. □

Before we give some applications of this theorem let us consider how we can check for a given non-zero prime ideal $\mathfrak{p}$ of R whether $S\mathfrak{p} + \mathfrak{F} = S$ is satisfied. A direct attack of this problem is not very promising because of the definition of $\mathfrak{F}$. However, in most cases the following lemma does the job.

Lemma (2.29)

Let R, S, $\mathfrak{p}$, $\mathfrak{F}$ be as in (2.27). Then sufficient conditions for $S\mathfrak{p} + \mathfrak{F} = S$ are:

(a) $\mathfrak{p} + Rm = R$ *for* $m = (S{:}R[\rho])$, *or*
(b) $\mathfrak{p}$ *does not divide* $d(f)R$, $d(f)$ *the discriminant of* $f(t)$.

Proof

Since the prime ideal $\mathfrak{p}$ is also maximal ($R/\mathfrak{p}$ is finite) it suffices to show that $m = (S{:}R[\rho])$ and $d(f)$ are in $\mathfrak{F}$. Namely, this at once implies $S = SR = S\mathfrak{p} + Sk \subseteq S\mathfrak{p} + S\mathfrak{F} \subseteq S$ for $k \in \{m, d(f)\}$. But m is in $\mathfrak{F}$ per definition. For $d(f)$ we conclude as follows. As previously shown $S \subseteq \frac{1}{d(f)} R[\rho]$ and this yields $d(f)S \subseteq R[\rho]$ and therefore $d(f) \in \mathfrak{F}$, too. □

As a consequence of this lemma all but a finite number of prime ideals satisfy the conditions of (2.27). From (2.27) we derive

Corollary (2.30)

Let $F, G, R, S, f, \mathfrak{p}$ as in (2.27). If $\mathfrak{p}$ does not divide $d(f)$, then $\mathfrak{p}$ is unramified in G.

Proof

Because of (2.27), (2.29) it suffices to show that $f(t) \in R(t)$ remains separable in $R/\mathfrak{p}[t]$. Let K be the splitting ring of f and $\mathfrak{q}$ a prime ideal of K lying over $\mathfrak{p}$. We denote the residue class mapping of $K[t]$ into $K/\mathfrak{q}[t]$ by Φ.

Then for $f(t)=\prod_{i=1}^n(t-\xi_i)$ in $K[t]$ we obtain $\Phi(f)=\prod_{i=1}^n(t-\Phi(\xi_i))$ in $K/\mathfrak{q}[t]$. If there were indices i, j $(1 \leqslant i < j \leqslant n)$ such that $\Phi(\xi_i)=\Phi(\xi_j)$, this would imply $\mathfrak{q} \mid (\xi_i - \xi_j)$ and therefore $\mathfrak{q} \mid d(f)$ in K. But then also $d(f) \in \mathfrak{q} \cap F = \mathfrak{p}$ contrary to the premise $\mathfrak{p} \nmid d(f)$. Hence, the $\Phi(\xi_i)$ $(1 \leqslant i \leqslant n)$ must be pairwise distinct and f therefore remains separable in $R/\mathfrak{p}\,[t]$. □

Example (2.31)

As a consequence of (2.28) we obtain the factorization of rational numbers p into prime ideals in quadratic fields $G=\mathbb{Q}(m^{\frac{1}{2}})$, $m \in \mathbb{Z}$ square-free. For the application of (2.27) we note that $F=\mathbb{Q}$, $R=\mathbb{Z}$, $f(t)=t^2-m$, and

$$S=\begin{cases} \mathbb{Z}[m^{\frac{1}{2}}] \text{ for } m \equiv 2,3 \bmod 4, \text{ therefore } \mathfrak{F}=S, \\ \mathbb{Z}\left[\dfrac{1+m^{\frac{1}{2}}}{2}\right] \text{ for } m \equiv 1 \bmod 4, \text{ therefore } \mathfrak{F}=2S. \end{cases}$$

We discuss several cases for p.

(a) Let p divide m. Then $S\mathfrak{p}+\mathfrak{F}=S$ is satisfied. The polynomial $f(t)=t^2-m$ splits in $\mathbb{Z}/p\mathbb{Z}[t]$ into t^2 yielding $pS=(m^{\frac{1}{2}}S+pS)^2$.

(b) Let m be odd and $p=2$. For $m \equiv 3 \bmod 4$ we obtain that $f(t)=t^2-m$ splits in $\mathbb{Z}/p\mathbb{Z}[t]$ into $(t+1)^2$ yielding $2S=((m^{\frac{1}{2}}+1)S+2S)^2$. However, for $m \equiv 1 \bmod 4$ we cannot directly apply (2.27). Therefore we choose a different polynomial $\tilde{f}(t)=t^2-t-(m-1)/4$ for generating G over $\mathbb{Q}$. Then we obtain $S=\mathbb{Z}[\tilde{\rho}]$ for the root $\tilde{\rho}=(1+m^{\frac{1}{2}})/2$ of $\tilde{f}$ again. Now $\tilde{f}(t)$ splits in $\mathbb{Z}/2\mathbb{Z}[t]$ into $t(t-1)$ for $m \equiv 1 \bmod 8$ and remains irreducible for $m \equiv 5 \bmod 8$, yielding

$$2S=\begin{cases} \left(\dfrac{1+m^{\frac{1}{2}}}{2}S+2S\right)\left(\dfrac{1-m^{\frac{1}{2}}}{2}S+2S\right) & \text{for } m \equiv 1 \bmod 8 \\ 2S & \text{for } m \equiv 5 \bmod 8 \end{cases}$$

by (2.27).

(c) Let p be odd and $p \nmid m$. Again (2.27) can be applied. $f(t)=t^2-m$ splits in $\mathbb{Z}/p\mathbb{Z}[t]$ into $(t-n)(t+n)$ for $m \equiv n^2 \bmod p$ and remains irreducible if m is not a square modulo p. Hence, (2.27) yields

$$pS=\begin{cases} ((n+m^{\frac{1}{2}})S+pS)((-n+m^{\frac{1}{2}})S+pS) & \text{for } m \equiv n^2 \bmod p \\ pS & \text{for } m \notin (\mathbb{Z}/p\mathbb{Z})^2 \end{cases}.$$

Using the Legendre symbol, the results (a)–(c) together yield the decomposition behavior of rational prime numbers p in quadratic number fields as follows:

Proposition (2.32)

Let $G=\mathbb{Q}(m^{\frac{1}{2}})$, $0 \neq m$ a square-free rational integer, and d_G the discriminant of G.

With respect to $S = \mathrm{Cl}(\mathbb{Z}, G)$ the prime number p is said to be

$$\left.\begin{cases} \textit{decomposed} \\ \textit{ramified} \\ \textit{undecomposed} \end{cases}\right\} \Leftrightarrow \left(\frac{d_G}{p}\right) = \begin{cases} 1 \Leftrightarrow d_G \in (\mathbb{Z}/p\mathbb{Z})^2 \\ 0 \Leftrightarrow p \mid d_G \\ -1 \Leftrightarrow d_G \notin (\mathbb{Z}/p\mathbb{Z})^2. \end{cases}$$

At the end of this section we consider the powers of rational prime numbers dividing the discriminant of an algebraic number field in dependence of the decomposition behavior of that prime number.

Lemma (2.33)

Let G, F be algebraic number fields with maximal orders S, R respectively. Let $[G:F] = n$ and $\mathfrak{p}$ a non-zero prime ideal of R which decomposes in S into $S\mathfrak{p} = \mathfrak{P}_1^{e_1} \cdot \ldots \cdot \mathfrak{P}_r^{e_r}$. Furthermore let β_{ik} $(1 \leqslant k \leqslant f(\mathfrak{P}_i/\mathfrak{p}), 1 \leqslant i \leqslant r)$ be a basis of $S/\mathfrak{P}_i$ over $R/\mathfrak{p}$ and $\alpha_{ij} \in (\mathfrak{P}_i^{j-1} \backslash \mathfrak{P}_i^j) \cap (\bigcap_{\nu=1,\nu \neq i}^{r} \mathfrak{P}_\nu^{e_\nu})$ $(1 \leqslant j \leqslant e_i)$. Then the elements $\alpha_{ij}\beta_{ik}$ $(1 \leqslant j \leqslant e_i, 1 \leqslant k \leqslant f_i = f(\mathfrak{P}_i | \mathfrak{p}); 1 \leqslant i \leqslant r)$ form a basis of $S/\mathfrak{p}S$ over $R/\mathfrak{p}$.

Proof

First we show $(S/\mathfrak{p}S : R/\mathfrak{p}) = n$. Namely, if h_F denotes the class number of F, then $\mathfrak{p}^{h_F} = \lambda R$ for suitable $\lambda \in R$ and this implies

$$\begin{aligned} \mathrm{N}(\mathfrak{p}S)^{h_F} &= \mathrm{N}(\lambda S) = \mathrm{N}_{F/\mathbb{Q}}(N_{G/F}(\lambda))| = |\mathrm{N}_{F/\mathbb{Q}}(\lambda^n)| \\ &= |\mathrm{N}_{F/\mathbb{Q}}(\lambda)|^n = \mathrm{N}(\mathfrak{p})^{h_F n}. \end{aligned}$$

For an ideal $\mathfrak{p}$ of R let $\Phi_{\mathfrak{p}S}: S \to S/\mathfrak{p}S$ denote the corresponding residue class mapping. A subset M of S is called *modulo* $\mathfrak{p}$ *independent*, if $\{\Phi_{\mathfrak{p}S}(m) | m \in M\}$ is linearly independent over $R/\mathfrak{p}$. For $i = 1, \ldots, r$ we choose subsets $M_i = \{\beta_{i1}, \ldots, \beta_{if_i}\}$ of S such that $\Phi_{\mathfrak{P}_i}(M_i)$ is a basis of $S/\mathfrak{P}_i$ over $R/\mathfrak{p}$, hence $|\Phi_{\mathfrak{P}_i}(M_i)| = f(\mathfrak{P}_i | \mathfrak{p}) = f_i$. Then we choose $\gamma_{ij} \in \mathfrak{P}_i^{j-1} \backslash \mathfrak{P}_i^j$ $(1 \leqslant j \leqslant e_i)$ and determine $\delta_{i\nu} \in S$ subject to

$$\begin{aligned} \delta_{i\nu} &\equiv -\gamma_{ij} \bmod \mathfrak{P}_\nu^{e_\nu} \quad (1 \leqslant \nu \leqslant r, \nu \neq i) \\ \delta_{i\nu} &\equiv 0 \bmod \mathfrak{P}_i^{e_i} \end{aligned}$$

by the Chinese remainder theorem, chapter 2 (2.17) $(1 \leqslant i \leqslant r)$. Then the $\alpha_{ij} = \delta_{ij} + \gamma_{ij}$ are in $(\mathfrak{P}_i^{j-1} \backslash \mathfrak{P}_i^j) \cap (\bigcap_{\nu=1,\nu \neq i}^{r} \mathfrak{P}_\nu^{e_\nu})$ $(1 \leqslant j \leqslant e_i; 1 \leqslant i \leqslant r)$ and it remains to show that the $\alpha_{ij}\beta_{ik}$ $(1 \leqslant j \leqslant e_i, 1 \leqslant k \leqslant f_i; 1 \leqslant i \leqslant r)$ are linearly independent elements of $S/\mathfrak{p}S$ over $R/\mathfrak{p}$. Any relation

$$\sum_{i=1}^{r} \sum_{j=1}^{e_i} \sum_{k=1}^{f_i} \gamma_{ijk}\alpha_{ij}\beta_{ik} \equiv 0 \bmod \mathfrak{p} \quad (\gamma_{ijk} \in R)$$

of course implies

$$\sum_{j=1}^{e_i} \sum_{k=1}^{f_i} \gamma_{ijk}\alpha_{ij}\beta_{ik} \equiv 0 \bmod \mathfrak{P}_i^{e_i} \quad (1 \leqslant i \leqslant r).$$

We fix the index i and carry out induction over j.

$j=1$: $\sum_{k=1}^{f_i} \gamma_{i1k}\alpha_{i1}\beta_{ik} \equiv 0 \bmod \mathfrak{P}_i$ implies $\sum_{k=1}^{f_i} \gamma_{i1k}\beta_{ik} \equiv 0 \bmod \mathfrak{P}_i$ because of the choice of the α_{1j}, hence $\gamma_{i1k} \equiv 0 \bmod \mathfrak{p}$ because of the choice of the β_{ik} $(1 \leqslant k \leqslant f_i)$.

$j-1 \Rightarrow j$: $\sum_{\kappa=1}^{j-1}\sum_{k=1}^{f_i} \gamma_{i\kappa k}\alpha_{i\kappa}\beta_{ik} + \sum_{k=1}^{f_i} \gamma_{ijk}\alpha_{ij}\beta_{ik} \equiv 0 \bmod \mathfrak{P}_i^j$ with the double sum being congruent 0 modulo $\mathfrak{p}$. Then $\sum_{k=1}^{f_i} \gamma_{ijk}\alpha_{ij}\beta_{ik} \equiv 0 \bmod \mathfrak{P}_i^j$ and $\sum_{k=1}^{f_i} \gamma_{ijk}\beta_{ik} \equiv 0 \bmod \mathfrak{P}_i$, $\gamma_{ijk} \equiv 0 \bmod \mathfrak{p}$ $(1 \leqslant k \leqslant f_i)$ as above. □

We apply (2.33) in the case $F = \mathbb{Q}$.

Lemma (2.34)
Let G be an algebraic number field of degree n and $S = \mathrm{Cl}(\mathbb{Z}, G)$. Let p be a rational prime number and $\alpha_1, \ldots, \alpha_n$ elements of S such that the canonical images $\bar{\alpha}_1, \ldots, \bar{\alpha}_n$ of $\alpha_1, \ldots, \alpha_n$ in S/pS form a basis of S/pS over $\mathbb{Z}/p\mathbb{Z}$. Then the index of the $\mathbb{Z}$-order $\mathfrak{o}$ generated by $\alpha_1, \ldots, \alpha_n$ in S is not divisible by p.

Proof
We assume $p|(S:\mathfrak{o})$. Then there is $\beta \in S \backslash \mathfrak{o}$ such that $p\beta \in \mathfrak{o}$ and $\overline{p\beta} = \bar{0}$. Hence, $p\beta = \sum_{i=1}^n m_i\alpha_i$, not all $m_i \equiv 0 \bmod p$, yielding a non-trivial presentation of $\bar{0}$ by $\bar{\alpha}_1, \ldots, \bar{\alpha}_n$ of S/pS, certainly a contradiction. □

An easy application of chapter 3 (2.9) yields

Corollary (2.35)
The discriminants $d(S)$, $d(\mathfrak{o})$ of the $\mathbb{Z}$-modules $S, \mathfrak{o}$ of (2.34) satisfy $d(\mathfrak{o}) = a^2 d(S)$ $(a \in \mathbb{N}, p \nmid a)$.

Lemma (2.36)
Let $\gamma_1, \ldots, \gamma_n$ be the elements $\alpha_{ij}\beta_{ik}$ of (2.33) in the case $F = \mathbb{Q}$, $\mathfrak{p} = p\mathbb{Z}$. Then $p^k | (\det(\gamma_i^{(j)})_{1 \leqslant i,j \leqslant n})^2$ for $k = \sum_{i=1}^r (e_i - 1)f_i = n - \sum_{i=1}^r f_i$.

Proof
Let γ_ν be of the form $\alpha_{ij}\beta_{ik}$ with $j \geqslant 2$. Then γ_ν is contained in every prime ideal $\mathfrak{P}_\mu$ of G $(1 \leqslant \mu \leqslant r)$. Let L be the normal extension of $G/\mathbb{Q}$ and $\mathfrak{q}$ be a prime ideal of L lying over $p\mathbb{Z}$. Then $\mathfrak{q} \cap S$ is a prime ideal of S containing $\mathfrak{p}$ and therefore γ_ν. Hence, γ_ν lies in all prime ideals $\mathfrak{q}$ of L which contain $p\mathbb{Z}$. If Φ is the extension of a $\mathbb{Q}$-isomorphism ϕ of G to L, then $\Phi(\gamma_\nu) \in \mathfrak{q}$ for a fixed prime ideal $\mathfrak{q}$ of L lying over $p\mathbb{Z}$ (namely, $\Phi^{-1}(\gamma_\nu) \in \Phi^{-1}(\mathfrak{q})$ which itself

is a prime ideal of L lying over $p\mathbb{Z}$ and therefore containing γ_ν). Thus we obtain

$$\mathrm{Tr}(\gamma_\nu\gamma_\mu) = \sum_{i=1}^{n} \Phi_i(\gamma_\nu)\Phi_i(\gamma_\mu) \in \mathfrak{q} \cap \mathbb{Z} = p\mathbb{Z}.$$

Therefore all rows ν of the determinant $\det(\mathrm{Tr}(\gamma_\nu\gamma_\mu)_{1 \leqslant \nu,\mu \leqslant n})$, which do not correspond to indices $(i1, ik)$ of $\alpha_{ij}\beta_{i\gamma}$ are divisible by p. □

Corollary (2.37)

$$p^{n - \sum_{i=1}^{r} f_i} \mid d(S).$$

As another corollary we again obtain (2.30).

Exercises

1. Let $\mathfrak{p}$ be a non-zero prime ideal in o_F. Prove that the least natural number contained in $\mathfrak{p}$ is a prime number. What about the reversion? If $\mathfrak{a}$ is a non-zero ideal of o_F for which the least natural number contained in $\mathfrak{a}$ is a prime number, is $\mathfrak{a}$ necessarily a prime ideal?
2. Using (2.31) and the product formula of the norm show that $h_F = 2$ for $F = \mathbb{Q}(-5)^{\frac{1}{2}}$. Then prove that $2y^3 = x^2 + 5$ has only the solutions $x = \pm 7$, $y = 3$ over $\mathbb{Z}$.
3. Let $f(t) = \sum_{i=0}^{n} a_i t^{n-i} \in \mathbb{Z}[t]$ be monic and irreducible. Let p be a rational prime number such that $p^r \mid a_n$, $p^{r+1} \nmid a_n$, $p^r \mid a_i$ $(1 \leqslant i \leqslant n-1)$ for some $r \in \mathbb{N}$. Let ρ be a zero of f and $F = \mathbb{Q}(\rho)$.

   (a) Prove that for $R = \mathrm{Cl}(\mathbb{Z}, F)$ the ideal $p^r R$ is the nth power of an integral ideal.
   (b) Determine $m \in \mathbb{N}$ maximal such that p^m divides the discriminant d_F of F.
   (Note: For $r = 1$ the polynomial f is an Eisenstein polynomial and therefore necessarily irreducible.)
4. Let $R = \mathbb{Z}[2i] = \mathbb{Z} + 2i\mathbb{Z}$, $i^2 = -1$. Give straightforward proofs of

   (a) R is not integrally closed;
   (b) I_R is not a group;
   (c) R is Noetherian,
   (d) every non-zero prime ideal of R is maximal.
5. Let R be a Dedekind ring containing $\mathbb{Z}$. Prove:

   (a) For any non-zero prime ideal $\mathfrak{p}$ of R there is exactly one prime number p subject to $\mathfrak{p} \supseteq Rp$.
   (b) For $U(R) \cap \mathbb{Z} = \{\pm 1\}$ the ring R contains infinitely many prime ideals.

6.3. Ideal calculus

To determine the ideal class group $\mathrm{Cl}_F := I_{o_F}/H_{o_F}$ of an algebraic number field F it suffices to know a representing element of each ideal class. The appropriate choice is an integral ideal $\mathfrak{a}$ of norm $\mathrm{N}(\mathfrak{a}) \leqslant M_F$, where M_F is the Minkowski constant for F. Each such ideal $\mathfrak{a}$ has a presentation

$\mathfrak{a} = \prod_{i=1}^{r} \mathfrak{p}_i^{e_i}$ ($\mathfrak{p}_i$ distinct prime ideals, $e_i \in \mathbb{N}$, $1 \leqslant i \leqslant r$, $r \in \mathbb{N}$). Because of the multiplicativity of the ideal norm each $\mathfrak{p}_i$ also satisfies

$$\mathrm{N}(\mathfrak{p}_i) \leqslant M_F \quad (1 \leqslant i \leqslant r).$$

Each prime ideal $\mathfrak{p}_i$ divides exactly one rational prime number $p = p(\mathfrak{p}_i)$ and $\mathrm{N}(\mathfrak{p}_i)$ is a power of p. Hence, it suffices to consider all rational prime numbers

$$p_1, \ldots, p_s \tag{3.1}$$

such that $p_1 = 2, p_2 = 3, p_3 = 5, \ldots$, and $p_s \leqslant M_F < p_{s+1}$. For each p_i $(1 \leqslant i \leqslant s)$ we need the prime ideal factorization of $p_i o_F$ in F:

$$p_i o_F = \prod_{j=1}^{r_i} \mathfrak{p}_{ij}^{e_{ij}} \quad (r_i \in \mathbb{N};\ e_{ij} \in \mathbb{N};\ 1 \leqslant j \leqslant r_i;\ 1 \leqslant i \leqslant s), \tag{3.2}$$

to obtain a full set of representing elements for Cl_F among those ideals

$$\mathfrak{a} = \prod_{i=1}^{s} \prod_{j=1}^{r_i} \mathfrak{p}_{ij}^{m_{ij}} \quad (m_{ij} \in \mathbb{Z}^{\geqslant 0};\ 1 \leqslant j \leqslant r_i;\ 1 \leqslant i \leqslant s) \tag{3.3}$$

satisfying

$$\mathrm{N}(\mathfrak{a}) \leqslant M_F. \tag{3.4}$$

This process will become efficient after some additional considerations with which we shall deal in section 5. One of the greatest difficulties in this context is to develop a method for testing whether two ideals are equivalent. This will be done in section 4. In this section we shall mainly be concerned with the task of developing methods permitting efficient operation with (integral) ideals.

It is clear that the abstract definition of an ideal is not useful for constructive purposes. However, we already know that the ideals $\mathfrak{a}$ of o_F can be presented in two ways:

via a $\mathbb{Z}$*-basis* $\mathfrak{a} = \mathbb{Z}\alpha_1 + \cdots + \mathbb{Z}\alpha_n$, (3.5a)

by two generating elements, one of which is arbitrary in $\mathfrak{a}$ *and usually chosen from* $\mathfrak{a} \cap \mathbb{N}$ (*for example,* $\mathrm{N}(\mathfrak{a})$ *will do*). (3.5b)

This was shown in the two preceding sections.

Clearly, the second presentation is less storage consuming. For $[F:\mathbb{Q}] = n$ the presentation of an integer of o_F via a fixed integral basis $\omega_1, \ldots, \omega_n$ requires n rational integers. Therefore a presentation of an integral ideal $\mathfrak{a}$ according to (3.5b) only takes $n + 1$ rational integers compared to n^2 (respectively $n(n+1)/2$ as shown below) using (3.5a). This clearly favors (3.5b). However, fundamental problems – for example, the computation of the norm of the ideal $\mathfrak{a}$ – can hardly be solved with this method of presentation. So we are bound to use both possibilities of representing $\mathfrak{a}$ by rational integers. In the sequel we shall consider in detail for which problems either one of those presentations should be used. We remark that the presentation of a fractional ideal $\mathfrak{a}$ just requires the storage of an additional positive integer b such that $b\mathfrak{a}$ is integral.

First we shall discuss ideal presentations of the form (3.5a) in case the integral ideal $\mathfrak{a}$ is given in the form $\mathfrak{a} = a o_F + \alpha o_F$. We easily derive a $\mathbb{Z}$-basis of $\mathfrak{a}$ as follows. Let M_a, M_α be the regular representation matrices of a, α, respectively. The composite matrix $M_{\mathfrak{a}} := (M_a \vdots M_\alpha) \in \mathbb{Z}^{n \times 2n}$ is reduced into its Hermite normal form $H(M_{\mathfrak{a}})$ for which the last n columns are all $\mathbf{0}$. By $HM_{\mathfrak{a}}$ we denote the $n \times n$ matrix consisting of the first n columns of $H(M_{\mathfrak{a}})$. We claim that $(\omega_1, \ldots, \omega_n)HM_{\mathfrak{a}}$ is a $\mathbb{Z}$-basis of $\mathfrak{a}$, hence $\mathfrak{a}$ can be represented by $n(n+1)/2$ rational integers. To prove this we note that $a\omega_1, \ldots, a\omega_n$, $\alpha\omega_1, \ldots, \alpha\omega_n$ is a system of generating elements of $\mathfrak{a}$ over $\mathbb{Z}$. But the transformation to the $2n$ elements consisting of the inner products of $(\omega_1, \ldots, \omega_n)$ with the columns of $H(M_{\mathfrak{a}})$ was by a unimodular (hence invertible) matrix. Similarly we obtain a $\mathbb{Z}$-basis for any ideal $\mathfrak{a}$ given by a finite number of generators $\alpha_1, \ldots, \alpha_k \in o_F$ ($k \in \mathbb{N}$).

Therefore we can consider integral ideals $\mathfrak{a}$ presented by a $\mathbb{Z}$-basis of the form

$$\alpha_i = \sum_{j=i}^{n} h_{ji}\omega_j \quad (1 \leqslant i \leqslant n;\ HM_{\mathfrak{a}} = (h_{ji})). \tag{3.6a}$$

Because of chapter 3 (2.9) we obtain for the norm of $\mathfrak{a}$

$$\mathrm{N}(\mathfrak{a}) = (o_F : \mathfrak{a}) = \det(HM_{\mathfrak{a}}) = \prod_{i=1}^{n} h_{ii}. \tag{3.6b}$$

An element $\beta = \sum_{j=1}^{n} b_j\omega_j$ of o_F is contained in $\mathfrak{a}$, if and only if there are $x_1, \ldots, x_n \in \mathbb{Z}$ satisfying

$$\sum_{j=1}^{n} b_j\omega_j = \beta = \sum_{i=1}^{n} x_i\alpha_i = \sum_{i=1}^{n} x_i \sum_{j=1}^{n} h_{ji}\omega_j = \sum_{j=1}^{n} \omega_j\left(\sum_{i=1}^{j} h_{ji}x_i\right).$$

Hence,

$$\beta \in \mathfrak{a} \Leftrightarrow x_j := \frac{1}{h_{jj}}\left(b_j - \sum_{i=1}^{j-1} h_{ji}x_i\right) \in \mathbb{Z} \quad (j = 1, 2, \ldots, n), \tag{3.6c}$$

where the terms on the right-hand side are computed recursively.

Another useful application of that presentation is the determination of the smallest natural number x contained in $\mathfrak{a}$. If we choose the integral basis $\omega_1, \ldots, \omega_n$ in such a way that $\omega_n = 1$, then we have $b_1 = b_2 = \cdots = b_{n-1} = 0$ in the presentation $x = b_1\omega_1 + \cdots + b_n\omega_n$ and (3.6c) implies that $x_1 = x_2 = \cdots = x_{n-1} = 0$ and $x = b_n = h_{nn}x_n$. Hence,

$$\min \mathfrak{a} \cap \mathbb{N} = h_{nn} \tag{3.7}$$

for this choice of an integral basis $\omega_1, \ldots, \omega_n$ of o_F.

We proceed to a 2-element presentation (3.5b) of an ideal $\mathfrak{a}$. Again we consider the problem to decide whether $\beta=\sum_{j=1}^{n} b_j\omega_j \in o_F$ is contained in $\mathfrak{a}$. For $\mathfrak{a}=ao_F+\alpha o_F$ ($a\in\mathbb{N}, \alpha=a_1\omega_1+\cdots+a_n\omega_n\in o_F$) we have $\beta\in\mathfrak{a}$, if and only if there exist elements $\xi=\sum_{j=1}^{n} x_j\omega_j$, $\eta=\sum_{j=1}^{n} y_j\omega_j$ of o_F satisfying

$$\beta=a\xi+\alpha\eta. \tag{3.8a}$$

We transform (3.8a) into a system of linear equations by setting

$$\omega_i\omega_j=\sum_{k=1}^{n} m_{ijk}\omega_k \quad (m_{ijk}\in\mathbb{Z}), \tag{3.8b}$$

$$b_{kj}:=\sum_{i=1}^{n} m_{ijk}a_i \quad (1\leqslant k,j\leqslant n;\ B:=(b_{kj})). \tag{3.8c}$$

Then (3.8a) is equivalent to

$$\mathbf{b}=(aI_n\ \ B)\begin{pmatrix}\mathbf{x}\\ \mathbf{y}\end{pmatrix}. \tag{3.9}$$

Therefore (3.8a) is solvable precisely if the system of linear congruences

$$\mathbf{b}\equiv B\mathbf{y} \bmod a \tag{3.10a}$$

has a solution $\mathbf{y}\in\mathbb{Z}^n$. For

$$a=\prod_{i=1}^{u} p_i^{k_i} \quad (u,k_i\in\mathbb{N},\ p_i \text{ distinct prime numbers},\ 1\leqslant i\leqslant u), \tag{3.10b}$$

the solvability of (3.10a) is tantamount to the one of

$$\mathbf{b}\equiv B\mathbf{y}^{(i)} \bmod p_i^{k_i} \quad (1\leqslant i\leqslant u). \tag{3.10c}$$

But the solutions of (3.10c) (if there are any) are easily calculated for each i ($1\leqslant i\leqslant u$) by a Gaussian type algorithm. If we indeed obtain a solution $\mathbf{y}^{(i)}$ for each i, then we compute $m_1,\ldots,m_u\in\mathbb{Z}$ subject to

$$1=\sum_{i=1}^{u} m_i(ap_i^{-k_i}), \tag{3.10d}$$

and thus, finally, a solution $\mathbf{y}$ of (3.10a) via

$$\mathbf{y}=\sum_{i=1}^{u} m_i(ap_i^{-k_i})\mathbf{y}^{(i)}. \tag{3.10e}$$

This method has the advantage that all numbers occurring during the calculations have an absolute value bounded by a. It is especially favorable for our considerations, since the natural numbers a which have to be discussed in connection with class group computations usually contain very few prime divisors (compare section 5).

As already mentioned in the introduction we are first of all interested in special 2-element presentations of ideals which allow to multiply the ideals in terms of their generators. Indeed, it turns out that we can present $\mathfrak{a}, \mathfrak{b}$ by (a, α), (b, β), respectively, such that $\mathfrak{ab}$ is presented by $(ab, \alpha\beta)$.

Let $\mathfrak{p}_1, \ldots, \mathfrak{p}_m$ be all prime ideals of o_F dividing a, α, b, β such that we have presentations $ao_F = \prod_{i=1}^{m} \mathfrak{p}_i^{v_i(a)}$, and so on. Then $\mathfrak{ab} = (ab, \alpha\beta)$ holds if we require

$$\begin{aligned} &\min(v_i(a) + v_i(b), v_i(a) + v_i(\beta), v_i(\alpha) + v_i(b), v_i(\alpha) + v_i(\beta)) \\ &\quad = \min(v_i(a) + v_i(b), v_i(\alpha) + v_i(\beta)) \quad (1 \leqslant i \leqslant m). \end{aligned} \tag{3.11}$$

In the sequel we denote by $\mathbb{P}$ the set of all prime numbers and by P a (possibly empty) subset of $\mathbb{P}$. Furthermore, let

$$P_F := \{\mathfrak{p} \in I_F \mid \mathfrak{p} \text{ a prime ideal, } \mathfrak{p} \text{ divides } po_F \text{ for some } p \in P\} \tag{3.12}$$

be a set of prime ideals of o_F, each prime ideal of P_F lying over some prime number p of P. If P is finite, then there is a natural number a ($a = 1$ for $P = \emptyset$) such that a is exactly divisible by the prime numbers $p \in P$. In that case we also write

$$P(a) \quad \text{instead of } P. \tag{3.13}$$

Definition (3.14)
Let $P \subseteq \mathbb{P}$ and $\mathfrak{a} \in I_F$. A couple $(a, \alpha) \in \mathbb{N} \times F^\times$ is called a **P-normal presentation** *of $\mathfrak{a}$, if the following four conditions are satisfied:*

(i) $\mathfrak{a} = ao_F + \alpha o_F$,
(ii) $\mathfrak{a} = \prod_{\mathfrak{p} \in P_F} \mathfrak{p}^{v_\mathfrak{p}(\mathfrak{a})} \quad (v_\mathfrak{p}(\mathfrak{a}) \in \mathbb{Z})$,
(iii) $ao_F = \prod_{\mathfrak{p} \in P_F} \mathfrak{p}^{v_\mathfrak{p}(a)} \quad (v_\mathfrak{p}(a) \in \mathbb{Z}^{\geqslant 0})$,
(iv) *No $\mathfrak{p} \in P_F$ occurs in the prime ideal presentation of $\alpha\mathfrak{a}^{-1}$.*

This definition seems to be somewhat artificial, but the connection to (3.11) will be immediate via the following criterion. Also we note that a P-normal presentation in the sense of (3.14) need not even exist. Its existence will be proved only under additional restrictions on P (see theorem (3.21)). For example, a P-normal presentation in case of $P = \emptyset$ implies $\mathfrak{a} = o_F$, $a = 1$ and α arbitrary in o_F. This also follows from

Lemma (3.15)
Let $P \subseteq \mathbb{P}$ and $\mathfrak{a} \in I_F$. Then $(a, \alpha) \in \mathbb{N} \times F^\times$ is a P-normal presentation of $\mathfrak{a}$, if and only if the following three conditions are satisfied:

(i) $\mathfrak{a} = \prod_{\mathfrak{p} \in P_F} \mathfrak{p}^{v_\mathfrak{p}(\mathfrak{a})} \quad (v_\mathfrak{p}(\mathfrak{a}) \in \mathbb{Z})$,
(ii) $ao_F = \prod_{\mathfrak{p} \in P_F} \mathfrak{p}^{v_\mathfrak{p}(a)} \quad (v_\mathfrak{p}(a) \in \mathbb{Z}^{\geqslant 0},\ v_\mathfrak{p}(a) \geqslant v_\mathfrak{p}(\mathfrak{a}))$
(iii) $\alpha o_F = \prod_{\mathfrak{p} \in P_F} \mathfrak{p}^{v_\mathfrak{p}(\mathfrak{a})} \prod_{\mathfrak{q} \in (\mathbb{P} \setminus P)_F} \mathfrak{q}^{v_\mathfrak{q}(\alpha)} \quad (v_\mathfrak{q}(\alpha) \in \mathbb{Z}^{\geqslant 0})$.

Proof
"$\Leftarrow$" Let $(a,\alpha)\in\mathbb{N}\times F^\times$ satisfy (i)–(iii) of the lemma. Clearly, $ao_F+\alpha o_F=\mathfrak{a}$, and (i)–(iii) of (3.14) are fulfilled. Furthermore, $\alpha\mathfrak{a}^{-1}=\prod_{\mathfrak{q}\in(\mathbb{P}\setminus P)_F}\mathfrak{q}^{v_\mathfrak{q}(\alpha)}$, hence also (iv) of (3.14) is satisfied.

"$\Rightarrow$" According to (3.14) we at once obtain (i), (ii) of the lemma. We note that (iv) of (3.14) implies

$$\alpha\mathfrak{a}^{-1}=\prod_{\mathfrak{q}\in(\mathbb{P}\setminus P)_F}\mathfrak{q}^{v_\mathfrak{q}(\alpha\mathfrak{a}^{-1})}\quad(v_\mathfrak{q}(\alpha\mathfrak{a}^{-1})\in\mathbb{Z}).$$

Because of

$$\mathfrak{a}=ao_F+\alpha o_F=ao_F+\alpha\mathfrak{a}^{-1}\mathfrak{a}=\prod_{\mathfrak{p}\in P_F}\mathfrak{p}^{v_\mathfrak{p}(a)}+\prod_{\mathfrak{q}\in(\mathbb{P}\setminus P)_F}\mathfrak{q}^{v_\mathfrak{q}(\alpha\mathfrak{a}^{-1})}\prod_{\mathfrak{p}\in P_F}\mathfrak{p}^{v_\mathfrak{p}(\mathfrak{a})}$$
$$=\prod_{\mathfrak{p}\in P_F}\mathfrak{p}^{\min(v_\mathfrak{p}(a),v_\mathfrak{p}(\mathfrak{a}))}\prod_{\mathfrak{q}\in(\mathbb{P}\setminus P)_F}\mathfrak{q}^{\min(0,v_\mathfrak{q}(\alpha\mathfrak{a}^{-1}))}$$

and (ii) of (3.15) already being established we must necessarily have $v_\mathfrak{q}(\alpha\mathfrak{a}^{-1})\geqslant 0$ for all $\mathfrak{q}\in(\mathbb{P}\setminus P)_F$. □

We note that this criterion is not suitable for practical computations since it requires full knowledge of the prime ideal factorization of $\mathfrak{a}$, ao_F, αo_F. However, it is a very useful tool for developing ideal arithmetics in the sequel.

Proposition (3.16)
Let $P\subseteq\mathbb{P}$ and $\mathfrak{a},\mathfrak{b}\in I_F$. If (a,α), (b,β) are P-normal presentations of $\mathfrak{a},\mathfrak{b}$, respectively, then $(ab,\alpha\beta)$ is a P-normal presentation of $\mathfrak{ab}$.

The proof is straightforward using (3.11) and (3.15) (see exercise 1).

A special case of multiplication is forming powers of an ideal.

Corollary (3.17)
Under the premises of (3.16) the couple (a^m,α^m) is a P-normal presentation of $\mathfrak{a}^m$ $(m\in\mathbb{N})$.

We note, however, that for any presentation $\mathfrak{a}=ao_F+\alpha o_F$ we obtain the powers $\mathfrak{a}^m$ in the form $\mathfrak{a}^m=a^m o_F+\alpha^m o_F$ (see exercise 2).

We proceed to compute the inverse of a normally presented integral ideal.

Proposition (3.18)
Let $P\subseteq\mathbb{P}$ and $\mathfrak{a}\in I_F$ be integral with P-normal presentation (a,α). Then each

$b \in \mathbb{N} \cap \alpha o_F$ is the product of two natural numbers c, d such that

$$c o_F = \prod_{\mathfrak{p} \in P_F} \mathfrak{p}^{v_{\mathfrak{p}}(c)}, \quad d o_F = \prod_{\mathfrak{q} \in (\mathbb{P} \setminus P)_F} \mathfrak{q}^{v_{\mathfrak{q}}(d)},$$

and $(1, d\alpha^{-1})$ is a P-normal presentation of $\mathfrak{a}^{-1}$.

Proof
For $b \in \mathbb{N} \cap \alpha o_F$ we set

$$c := \prod_{p \in P} p^{v_p(b)}, \quad d := \prod_{p \in \mathbb{P} \setminus P} p^{v_p(b)}$$

and the first statement of the proposition is immediate. For $\mathfrak{a} = \prod_{\mathfrak{p} \in P_F} \mathfrak{p}^{v_{\mathfrak{p}}(\mathfrak{a})}$ obviously $\mathfrak{a}^{-1} = \prod_{\mathfrak{p} \in P_F} \mathfrak{p}^{-v_{\mathfrak{p}}(\mathfrak{a})}$ with $-v_{\mathfrak{p}}(\mathfrak{a}) \in \mathbb{Z}^{\leqslant 0}$ since $\mathfrak{a}$ was integral. Therefore $1 \in o_F$ with $1 o_F = \prod_{\mathfrak{p} \in P_F} \mathfrak{p}^0$ satisfying (ii) of (3.15). Finally,

$$d\alpha^{-1} o_F = \prod_{\mathfrak{p} \in P_F} \mathfrak{p}^{-v_{\mathfrak{p}}(\mathfrak{a})} \prod_{\mathfrak{q} \in (\mathbb{P} \setminus P)_F} \mathfrak{q}^{-v_{\mathfrak{q}}(\alpha) + v_{\mathfrak{q}}(d)}$$

and the exponents $-v_{\mathfrak{q}}(\alpha) + v_{\mathfrak{q}}(d)$ are non-negative because of $d \in \alpha o_F$. Hence, the conditions of (3.15) are satisfied, and $(1, d\alpha^{-1})$ is a P-normal presentation of $\mathfrak{a}^{-1}$. □

If the natural number a in a P-normal presentation (a, α) of an integral ideal $\mathfrak{a}$ has specific properties, it yields the decomposition of $\mathfrak{a}$ into a product of two ideals.

Proposition (3.19)
Let $\mathfrak{a} \in I_F$ be an integral ideal with $P(a)$-normal presentation (a, α). Any factorization $a = bc$ ($b, c \in \mathbb{N}$, $\gcd(b, c) = 1$) then implies a factorization of $\mathfrak{a}$ via $\mathfrak{a} = \mathfrak{b}\mathfrak{c}$ for $\mathfrak{b} = b o_F + \alpha o_F$, $\mathfrak{c} = c o_F + \alpha o_F$, and (b, α) is a $P(b)$-normal presentation of $\mathfrak{b}$, (c, α) a $P(c)$-normal presentation of $\mathfrak{c}$.

Proof
The case $b = 1$ (or $c = 1$) is trivial. For $b > 1$, $c > 1$ we split the set $P(a)$ into $P(a) = P(b) \dot{\cup} P(c)$ and obtain

$$b o_F = \prod_{\mathfrak{p} \in P(b)_F} \mathfrak{p}^{v_{\mathfrak{p}}(b)}, \quad c o_F = \prod_{\mathfrak{p} \in P(c)_F} \mathfrak{p}^{v_{\mathfrak{p}}(c)}.$$

Clearly, (b, α) and (c, α) are normal presentations of $\mathfrak{b}$, $\mathfrak{c}$, respectively, since $\mathfrak{a}$ was integral. Also $\mathfrak{b}\mathfrak{c} = \mathfrak{b} \cap \mathfrak{c} \supseteq (bc, \alpha)$ and on the other hand, obviously, $\mathfrak{b}\mathfrak{c} = (bc, b\alpha, c\alpha, \alpha^2) \subseteq (bc, \alpha)$. □

The normal presentations also allow us to compute quotients of suitable integral ideals very easily. This will be used later on. Let $\mathfrak{a}, \tilde{\mathfrak{a}}$ be non-zero integral ideals of F with $P(a)$-normal presentations (a, α), $(a, \tilde{\alpha})$, respectively. Let $\min(\tilde{\alpha} o_F \cap \mathbb{N}) = cd$ such that $c = \prod_{p \in P(a)} p^{v_p(c)}$, $d = \prod_{p \in \mathbb{P} \setminus P(a)} p^{v_p(d)}$.

According to (3.18) we have

$(1, d\tilde{\alpha}^{-1})$ is a $P(a)$-normal presentation of $\tilde{\mathfrak{a}}^{-1}$, (3.20a)

hence by (3.16)

$(a, \alpha d\tilde{\alpha}^{-1})$ is a $P(a)$-normal presentation of $\mathfrak{a}\tilde{\mathfrak{a}}^{-1}$,

and, of course, (3.20b)

$$\mathfrak{a} \subseteq \tilde{\mathfrak{a}} \Leftrightarrow \mathfrak{a}\tilde{\mathfrak{a}}^{-1} \subseteq o_F \Leftrightarrow \alpha d\tilde{\alpha}^{-1} \in o_F. \tag{3.20c}$$

Before we consider the application of normal presentations to the task of factoring an ideal into prime ideals we certainly must prove the existence of such presentations under suitable conditions.

Theorem (3.21)

Let $a, b \in \mathbb{N}$, $\alpha \in o_F$, $\alpha \neq 0$, and $\mathfrak{a} = a o_F + b^{-1}\alpha o_F$ an ideal of F. Then $\mathfrak{a}$ has a $P(ab)$-normal presentation.

Proof

Let

$$a o_F = \prod_{\mathfrak{p} \in P(ab)_F} \mathfrak{p}^{v_\mathfrak{p}(a)}, \quad b o_F = \prod_{\mathfrak{p} \in P(ab)_F} \mathfrak{p}^{v_\mathfrak{p}(b)}, \quad \alpha o_F = \prod_{\mathfrak{p} \in P(ab)_F} \mathfrak{p}^{v_\mathfrak{p}(\alpha)} \prod_{\mathfrak{q} \in (P \setminus P(ab))_F} \mathfrak{q}^{v_\mathfrak{q}(\alpha)}.$$

Firstly, we show the existence of a $P(ab)$-normal presentation of the integral ideal

$$b\mathfrak{a} = ab o_F + \alpha o_F = \prod_{\mathfrak{p} \in P(ab)_F} \mathfrak{p}^{\min(v_\mathfrak{p}(a) + v_\mathfrak{p}(b), v_\mathfrak{p}(\alpha))}.$$

For each $\mathfrak{p} \in P(ab)_F$ choose

$$\beta_\mathfrak{p} \in \mathfrak{p}^{\min(v_\mathfrak{p}(a) + v_\mathfrak{p}(b), v_\mathfrak{p}(\alpha))} \setminus \mathfrak{p}^{\min(v_\mathfrak{p}(a) + v_\mathfrak{p}(b), v_\mathfrak{p}(\alpha)) + 1}.$$

Then by the Chinese remainder theorem, chapter 2 (2.17), we obtain $\beta \in o_F$ satisfying $\beta \equiv \beta_\mathfrak{p} \bmod \mathfrak{p}^{\min(v_\mathfrak{p}(a) + v_\mathfrak{p}(b), v_\mathfrak{p}(\alpha)) + 1}$ for all $\mathfrak{p} \in P(ab)_F$, hence

$$\beta o_F = \prod_{\mathfrak{p} \in P(ab)_F} \mathfrak{p}^{\min(v_\mathfrak{p}(a) + v_\mathfrak{p}(b), v_\mathfrak{p}(\alpha))} \prod_{\mathfrak{q} \in (P \setminus P(ab))_F} \mathfrak{q}^{v_\mathfrak{q}(\beta)},$$

($v_\mathfrak{q}(\beta) \in \mathbb{Z}^{\geq 0}$, $v_\mathfrak{q}(\beta) = 0$ for all but a finite number of $\mathfrak{q}$). Therefore (ab, β) is a $P(ab)$-normal presentation of $b\mathfrak{a}$.

In the second part of the proof we show that $P(ab)$-normal presentations (ab, β) of $ab o_F + \beta o_F$ and $(a, b^{-1}\beta)$ of $a o_F + b^{-1}\beta o_F$ ($a, b \in \mathbb{N}$, $\beta \in o_F$, $\beta \neq 0$) are in 1–1-correspondence. Using (3.15) we obtain the following chain of equivalences: $b^{-1}\mathfrak{a} := a o_F + b^{-1}\beta o_F$ has $P(ab)$-normal presentation $(a, b^{-1}\beta)$

$$\Leftrightarrow v_\mathfrak{p}(\beta) - v_\mathfrak{p}(b) = v_\mathfrak{p}(b^{-1}\mathfrak{a}) \leq v_\mathfrak{p}(a) \quad \forall \mathfrak{p} \in P(ab)_F$$

$$\Leftrightarrow v_\mathfrak{p}(\beta) = v_\mathfrak{p}(\mathfrak{a}) \leq v_\mathfrak{p}(a) + v_\mathfrak{p}(b) \quad \forall \mathfrak{p} \in P(ab)_F$$

$\Leftrightarrow \mathfrak{a} = ab o_F + \beta o_F$ has $P(ab)$-normal presentation (ab, β). □

We note that the underlying set P of prime numbers – $P(ab)$ in (3.21) – has to be finite for the application of the Chinese remainder theorem. Hence, P-normal presentations in the generality of definition (3.14) need not always exist.

Unfortunately the proof of (3.21) is not suitable for constructive purposes, either, since it makes use of prime ideal factorizations which are difficult to obtain in general. In the sequel we therefore develop other methods of determining normal presentations of ideals. A first step into this direction is the following criterion.

Lemma (3.22)

Let $a \in \mathbb{N}$, $\alpha \in o_F$, $\alpha \neq 0$, and $\mathfrak{a} = a o_F + \alpha o_F$. Then (a, α) is a $P(a)$-normal presentation of $\mathfrak{a}$, if and only if

$$\gcd\left(a, \frac{\min(\alpha o_F \cap \mathbb{N})}{\gcd(\min(\alpha o_F \cap \mathbb{N}), a)}\right) = 1. \tag{3.23}$$

Proof

We note that condition (3.23) means that for every prime number p which divides a and for which p^k divides $\min(\alpha o_F \cap \mathbb{N})$ also p^k divides a, i.e. the exponent of p has to be larger for a than for $\min(\alpha o_F \cap \mathbb{N})$.

Using (1.11) we obtain for $m := \min(\alpha o_F \cap \mathbb{N})$:

$$m o_F = \prod_{\mathfrak{p} \in P(a)_F} \mathfrak{p}^{v_\mathfrak{p}(m)} \prod_{\mathfrak{q} \in (\mathbb{P} \setminus P(a))_F} \mathfrak{q}^{v_\mathfrak{q}(m)}, \tag{3.24}$$

with

$$v_\mathfrak{p}(m) \geqslant v_\mathfrak{p}(\alpha), \quad v_\mathfrak{q}(m) \geqslant v_\mathfrak{q}(\alpha) \geqslant 0.$$

Hence, if (3.23) is satisfied, we have

$$0 = \min(v_\mathfrak{p}(a), v_\mathfrak{p}(m) - \min(v_\mathfrak{p}(m), v_\mathfrak{p}(a))) \quad \text{for all } \mathfrak{p} \in P(a)$$

and therefore $v_\mathfrak{p}(m) \leqslant v_\mathfrak{p}(a)$ because of $v_\mathfrak{p}(a) > 0$ for all $\mathfrak{p} \in P(a)_F$. This yields $v_\mathfrak{p}(\alpha) \leqslant v_\mathfrak{p}(a)$ for all $\mathfrak{p} \in P(a)_F$, and (a, α) is a $P(a)$-normal presentation of $\mathfrak{a}$.

On the other hand, let (a, α) be a $P(a)$-normal presentation of $\mathfrak{a}$ and let us again assume (3.24). For $\mathfrak{q} \in (\mathbb{P} \setminus P(a))_F$ clearly $v_\mathfrak{q}(a) = 0$, hence $\min(v_\mathfrak{q}(a), v_\mathfrak{q}(m) - \min(v_\mathfrak{q}(m), v_\mathfrak{q}(a))) = 0$. To establish (3.23) it therefore remains to show that $\min(v_\mathfrak{p}(m), v_\mathfrak{p}(a)) = v_\mathfrak{p}(m)$ for all $\mathfrak{p} \in P(a)_F$. For this purpose we factorize m into $m = bd$ such that

$$b o_F = \prod_{\mathfrak{p} \in P(a)_F} \mathfrak{p}^{v_\mathfrak{p}(m)}, \quad d o_F = \prod_{\mathfrak{q} \in (\mathbb{P} \setminus P(a))_F} \mathfrak{q}^{v_\mathfrak{q}(m)},$$

and $\gcd(b, d) = 1$. Then we obtain

$$a d o_F = \prod_{\mathfrak{p} \in P(a)_F} \mathfrak{p}^{v_\mathfrak{p}(a)} \prod_{\mathfrak{q} \in (\mathbb{P} \setminus P(a))_F} \mathfrak{q}^{v_\mathfrak{q}(m)} \subseteq \alpha o_F$$

because of $v_\mathfrak{p}(a) \geqslant v_\mathfrak{p}(\alpha)$, $v_\mathfrak{q}(m) \geqslant v_\mathfrak{q}(\alpha)$. But $m = \min(\alpha o_F \cap \mathbb{N})$ implies $m \mid ad$ (otherwise $ad = Q(ad, m)m + R(ad, m)$ and $R(ad, m)$ would be a smaller natural number than m in αo_F). This of course yields $b|a$ and therefore $v_{\mathfrak{p}}(m) \leqslant v_{\mathfrak{p}}(a)$ for all $\mathfrak{p} \in P(a)_F$. □

This criterion is very useful for a normalization procedure. Because of the second part of the proof of (3.21) it suffices to develop such a procedure for integral ideals $\mathfrak{a} = ao_F + \alpha o_F$ ($a \in \mathbb{N}, \alpha \in o_F, \alpha \neq 0$). We remark that the trivial cases $a = 0$ or $\alpha = 0$ have obvious normalizations:

$\mathfrak{a} = ao_F$ *has the* $P(a)$*-normal presentation* (a, a); (3.25a)

$\mathfrak{a} = \alpha o_F$ *has the* $P(|N(\alpha)|)$*-normal presentation* $(|N(\alpha)|, \alpha)$. (3.25b)

In the general case we know that $\mathfrak{a}$ has a $P(a)$-normal presentation, i.e. only the generator α must be changed appropriately. We already noted that the straightforward method of the proof of (3.21) is not to be recommended. Instead we use a probabilistic approach which turned out to be highly successful in actual calculations. To obtain an appropriate element α' from α such that (a, α') is a $P(a)$-normal presentation of $\mathfrak{a} = ao_F + \alpha o_F$ it suffices to consider elements

$$\alpha' = \alpha + a\xi \quad (\xi \in o_F). \tag{3.26a}$$

Hence, we just search for potential candidates ξ among those elements of o_F whose coordinates in the given integral basis are small, i.e. we choose bounds $S_i \in \mathbb{N}$ $(1 \leqslant i \leqslant n)$ such that the coefficients of ξ in

$$\xi = x_1\omega_1 + \cdots + x_n\omega_n \quad (x_i \in \mathbb{Z}) \tag{3.26b}$$

satisfy

$$|x_i| \leqslant S_i \quad (1 \leqslant i \leqslant n). \tag{3.26c}$$

This yields the following 'heuristic' algorithm.

Algorithm for the computation of a normal presentation of an ideal (3.27)

Input. An integral basis $\omega_1, \ldots, \omega_n$ of o_F, $\alpha \in o_F$, $a \in \mathbb{N}$, bounds S_i $(1 \leqslant i \leqslant n)$.

Output. Either a $P(a)$-normal presentation (a, α') of $\mathfrak{a} = ao_F + \alpha o_F$ or 'No normal presentation found'.

Step 1. (Initialization). Set $v_i \leftarrow -S_i - 1$ $(1 \leqslant i \leqslant n)$.

Step 2. (Change of v-coordinates). Set $i \leftarrow n$.

Step 3. (Increase v_i). Set $v_i \leftarrow v_i + 1$. For $v_i > S_i$ go to 5.

Step 4. (Construct α'). Set $\alpha' \leftarrow \alpha + a\sum_{i=1}^{n} v_i\omega_i$ and check, whether (3.23) holds for (a, α'). If this is the case, print solution (a, α') and terminate. Else go to 2.

Step 5. (Decrease i). For $i = 1$ print 'No normal presentation found' and terminate. Else set $v_i \leftarrow -S_i - 1$, $i \leftarrow i - 1$ and go to 3.

Remarks

The algorithm should be carried out only if (a, α) is not yet a $P(a)$-normal presentation. In case of $(1/a)\alpha \in o_F$ we obviously have the solution $\alpha' = a$ (see (3.25a)) and should therefore not use the algorithm. The bounds S_i should be less than a of course. Making use of the second part of the proof of (3.21) the algorithm can also be used to compute a normal presentation of a fractional ideal.

Generally the computation of 2-element respectively 2-element-normal presentations of ideals in connection with class group computations is much easier. Namely, for those prime numbers p not dividing the index $(o_F:\mathbb{Z}[\rho])$ theorem (2.27) yields a 2-element presentation for those prime ideals $\mathfrak{p}$ containing po_F in the form $\mathfrak{p} = po_F + \alpha o_F$, $\alpha = g(\rho)$, where $g(t) \in \mathbb{Z}[t]$ is a monic polynomial dividing the minimal polynomial of ρ in $p\mathbb{Z}[t]$.

Clearly, $(p, g(\rho))$ is a $P(p)$-normal presentation of $\mathfrak{p}$, if the ramification index $e = e_{\mathfrak{p}}$ of $\mathfrak{p}$ is greater than one. For $e = 1$ it is still a $P(p)$-normal presentation in case of $g(\rho) \notin \mathfrak{p}^2$. The latter can be easily tested. Finally, for $e = 1$ and $g(\rho) \in \mathfrak{p}^2$ a $P(p)$-normal presentation of $\mathfrak{p}$ is given by $(p, g(\rho) \pm p)$.

Prime numbers p subject to $p|(o_F:\mathbb{Z}[\rho])$ are somewhat more difficult to deal with. Similarly to (2.27) we obtain a factorization of po_F into ideals $\mathfrak{a}_i$ $(1 \leqslant i \leqslant k)$ of the form

$$\mathfrak{a}_i = po_F + f_i(\rho)o_F \quad (f_i(t) \in \mathbb{Z}[t] \text{ monic and non-constant}). \tag{3.28}$$

Then we need to test whether those ideals $\mathfrak{a}_i$ are prime ideals. Since the non-zero prime ideals of o_F are maximal the following proposition is immediate.

Proposition (3.29)

For any non-zero proper ideal $\mathfrak{a}$ of o_F we have

(i) *$\mathfrak{a}$ is a prime ideal, if and only if $\mathfrak{a} + xo_F = o_F$ for all $x \in o_F \backslash \mathfrak{a}$.*
(ii) *All $y, z \in o_F$ subject to $y - z \in \mathfrak{a}$ satisfy $yo_F + \mathfrak{a} = zo_F + \mathfrak{a}$.*
(iii) *$yo_F + \mathfrak{a} = -yo_F + \mathfrak{a}$ for all $y \in o_F$.*

Because of (3.29) (ii) it suffices to check all x of a complete residue system of $o_F/\mathfrak{a}$ in (*i*), whether $xo_F + \mathfrak{a} = o_F$, and (iii) further restricts the number of elements x to be tested. A complete residue system of $o_F/\mathfrak{a}$ is given by the elements of (compare (3.6a))

$$R := \left\{ \sum_{i=1}^{n} y_i \omega_i \,|\, y_i \in \mathbb{Z}, \quad -\frac{h_{ii}}{2} < y_i \leqslant \frac{h_{ii}}{2}, \quad 1 \leqslant i \leqslant n \right\}. \tag{3.30}$$

This follows from the construction of $HM_{\mathfrak{a}}$ and chapter 3 (2.9). Because of (3.29) (iii) we can restrict possible candidates to the set R^+ of those elements of R for which the non-vanishing coefficient of lowest index i is positive. If

$$y o_F + \mathfrak{a} = o_F \quad \text{for all } y \in R^+, \tag{3.31a}$$

respectively

$$1 \in y o_F + \mathfrak{a} \quad \text{for all } y \in R^+, \tag{3.31b}$$

then $\mathfrak{a}$ is a prime ideal according to (3.29). Otherwise we obtain an element $\tilde{y} \in R^+$ such that

$$o_F \supset \tilde{\mathfrak{a}} := \tilde{y} o_F + \mathfrak{a} \supset \mathfrak{a}. \tag{3.32}$$

If $\mathfrak{a}$ is represented by two elements a, α, however, we still need to construct a 2-element presentation for $\tilde{\mathfrak{a}}$. As a first generator for $\tilde{\mathfrak{a}}$ we choose $\tilde{a} := p$ since $p o_F \subset \mathfrak{a}$ yields $p = \min \mathfrak{a} \cap \mathbb{N} = \min \tilde{\mathfrak{a}} \cap \mathbb{N}$. A second generator $\tilde{\alpha}$ subject to $\tilde{\mathfrak{a}} = \tilde{a} o_F + \tilde{\alpha} o_F$ can be chosen from finitely many appropriate elements of o_F (see exercise 4). Extensive calculations showed that a suitable element $\tilde{\alpha}$ is very likely contained in the set

$$\left\{ \beta = \sum_{i=1}^{n} k_i \tilde{\alpha}_i \,|\, \beta \neq 0, \quad k_i \in \{-1, 0, 1\}, \quad 1 \leqslant i \leqslant n; \quad k_j = 1 \quad \text{for} \right. \tag{3.33}$$

$$\left. j = \min\{i \,|\, k_i \neq 0\} \right\},$$

where $\tilde{\alpha}_1, \ldots, \tilde{\alpha}_n$ is a $\mathbb{Z}$-basis of $\tilde{\mathfrak{a}}$ being calculated from the representation matrices of $a, \alpha, \tilde{y}$.

If $\tilde{\mathfrak{a}}$ is a prime ideal lying over $p o_F$, we obtain from its two generators $\tilde{a}, \tilde{\alpha}$ a $P(p)$-normal presentation as before for those prime ideals $\mathfrak{q}$ lying over $q o_F$ for $q \nmid (o_F : \mathbb{Z}[\rho])$. Using (3.20) it is then easy to determine the exact power m of $\tilde{\mathfrak{a}}$ dividing $p o_F$ and to reduce the task of factoring $p o_F$ into prime ideals to the factorization of $p \tilde{\mathfrak{a}}^{-m}$.

Finally, we must discuss how to multiply ideals $\mathfrak{a} = a o_F + \alpha o_F$, $\mathfrak{b} = b o_F + \beta o_F$ in $P(a)$-, $P(b)$-normal presentations, respectively, where $P(a) \neq P(b)$. Then (3.16) can be applied only after we have computed $P(ab)$-normal presentations for both ideals. The latter is done inductively by calculating $P(aq)$-normal presentations for $P(a)$-normally presented ideals $\mathfrak{a} = a o_F + \alpha o_F$ and q a prime number not dividing a. For the following construction we additionally require $v_{\mathfrak{p}}(a) \geqslant 1$ for all $\mathfrak{p} \in P(a)_F$. In the beginning, when a is a prime number, this is obviously satisfied.

If no prime ideal $\mathfrak{q} \in \mathbb{P}_F$ subject to $\mathfrak{q} \,|\, q o_F$ divides αo_F, i.e. $q \nmid \mathrm{N}(\alpha)$, then (aq, α) is already a $P(aq)$-normal presentation of $\mathfrak{a}$. This is because $\mathfrak{q}$ does not divide $\mathfrak{a}$. In case of $q \,|\, \mathrm{N}(\alpha)$ we order the set $P(q)_F = \{\mathfrak{q}_1, \ldots, \mathfrak{q}_k, \mathfrak{q}_{k+1}, \ldots, \mathfrak{q}_m\}$ of prime ideals lying over $q o_F$ such that $\alpha \in \mathfrak{q}_i$ $(1 \leqslant i \leqslant k)$, $\alpha \notin \mathfrak{q}_j$ $(k+1 \leqslant j \leqslant m)$. Using the $P(q)$-normal presentations $\mathfrak{q}_i = q o_F + \beta_i o_F$ $(1 \leqslant i \leqslant m)$ we compute

$$\tilde{\alpha} := \alpha + \beta_{k+1} \cdot \ldots \cdot \beta_m a^2 \tag{3.34}$$

and obtain $(aq, \tilde{\alpha})$ as $P(aq)$-normal presentation of $\mathfrak{a}$. Namely, for $\mathfrak{p} \in P(a)_F$ the exponents $v_\mathfrak{p}(\mathfrak{a})$ and $v_\mathfrak{p}(\alpha)$ are equal and the exponents $v_\mathfrak{p}(a)$ satisfy $v_\mathfrak{p}(a) \geq \max\{v_\mathfrak{p}(\alpha), 1\}$ according to our assumption, hence $\tilde{\alpha} \in \mathfrak{p}^{v_\mathfrak{p}(\mathfrak{a})} \backslash \mathfrak{p}^{v_\mathfrak{p}(\mathfrak{a})+1}$. For $1 \leq i \leq k$ we have $\tilde{\alpha} \notin \mathfrak{q}_i$ because of $\beta_{k+1} \cdot \ldots \cdot \beta_m a^2 \notin \mathfrak{q}_i$ and, finally, for $k+1 \leq j \leq m$ the relations $\alpha \notin \mathfrak{q}_i$, $\beta_{k+1} \cdot \ldots \cdot \beta_m a^2 \in \mathfrak{q}_i$ yield $\tilde{\alpha} \notin \mathfrak{q}_i$.

However, during class group computations it will usually be faster to test a few elements $k \in \{1, 2, \ldots, q-1\}$, whether $q \nmid \mathrm{N}(\alpha + ka^2)$, since in that case $(a, \alpha + ka^2)$ as well as $(aq, \alpha + ka^2)$ are $P(aq)$-normal presentations of $\mathfrak{a}$.

Exercises

1. Prove proposition (3.16).
2. Let $\mathfrak{a} = ao_F + \alpha o_F$ $(a \in \mathbb{N}, \alpha \in o_F)$ be an integral ideal. Show that $\mathfrak{a}^n = a^n o_F + \alpha^n o_F$.
3. Compute the prime ideal decomposition of po_F for those prime numbers dividing $d(f)$ for $f(t) = t^3 - m$ and $m \in \{3, 5, 6, 7\}$ by the method of this section.
4. Let $\mathfrak{a} = ao_F + \alpha o_F + \beta o_F$ be an integral ideal. Using the results of the next section determine a finite set of elements of o_F which is guaranteed to contain an element γ satisfying $\mathfrak{a} = ao_F + \gamma o_F$.

6.4. On solving norm equations II

As already noted at the beginning of the preceding section one of the main subtasks for the construction of the class group of an algebraic number field F is to decide whether two given ideals are equivalent. This is tantamount to the decision, whether a given ideal $\mathfrak{a}$ of F is a principal ideal. It is clear that it suffices to solve this taks for integral ideals since the presentations (3.5a), (3.5b) for fractional ideals contain $b \in \mathbb{N}$ such that $b\mathfrak{a} \subset o_F$.

Therefore $\mathfrak{a}$ always denotes a non-zero integral ideal of F in the sequel. The following criterion is quite obvious:

Lemma (4.1)

A non-zero ideal $\mathfrak{a}$ of o_F is a principal ideal, if and only if $\mathfrak{a}$ contains an element α such that $|\mathrm{N}(\alpha)| = \mathrm{N}(\mathfrak{a})$.

Proof

Let $\alpha \in \mathfrak{a}$. Then $\alpha o_F \subseteq \mathfrak{a}$ and there exists a non-zero integral ideal $\mathfrak{b}$ of o_F such that $\mathfrak{a}\mathfrak{b} = \alpha o_F$ by (1.9). Hence, $\mathrm{N}(\mathfrak{a}) = \mathrm{N}(\alpha o_F)$ is equivalent to $\mathrm{N}(\mathfrak{b}) = 1$. But the latter holds precisely for $\mathfrak{b} = o_F$ because of $\mathrm{N}(\mathfrak{b}) = (o_F : \mathfrak{b})$. □

For a given non-zero integral ideal $\mathfrak{a}$ its norm is easily calculated as $\det(HM_\mathfrak{a})$ from the presentation (3.5a) of $\mathfrak{a}$. Therefore the task of deciding whether $\mathfrak{a}$ is principal is essentially the task of deciding whether an element $\alpha \in \mathfrak{a} \subset o_F$ with $|\mathrm{N}(\alpha)| = \mathrm{N}(\mathfrak{a})$ exists. This task is in principle solved by the following theorem.

Theorem (4.2)

For given $a \in \mathbb{N}$ there are only finitely many non-associate elements α of o_F such that $|\mathrm{N}(\alpha)| = a$; those can be effectively computed.

Proof

The existence was already shown in chapter 5 (2.3). For computation let

$$L: o_F \backslash \{0\} \to \mathbb{R}^{s+t}: \xi \mapsto (c_1 \log|\xi^{(1)}|, \ldots, c_{s+t} \log|\xi^{(s+t)}|)^t,$$

where

$$c_j = \begin{cases} 1 & \text{for } 1 \leqslant j \leqslant s \\ 2 & \text{otherwise} \end{cases},$$

i.e. $L|_{U_F} = L_1$ of chapter 5 (6.1). It was shown in chapter 5, section 6 that for independent units $\eta_1, \ldots, \eta_r$ $(r = s + t - 1)$ of o_F the vectors $\mathbf{c} := (c_1, \ldots, c_{s+t})^t$, $L(\eta_1)^t, \ldots, L(\eta_r)^t$ form a basis of $\mathbb{R}^{s+t}$. Hence, for $\xi \in o_F$, $\xi \neq 0$, there are $x_1, \ldots, x_{s+t} \in \mathbb{R}$ such that

$$L(\xi) = \sum_{i=1}^{r} x_i L(\eta_i) + x_{s+t}\mathbf{c}. \tag{4.3a}$$

Adding up the coordinates of the vectors on both sides of (4.3a) we obtain

$$\begin{aligned} \log|\mathrm{N}(\xi)| &= \sum_{j=1}^{s+t} c_j \log|\xi^{(j)}| = \sum_{j=1}^{s+t} \left(\sum_{i=1}^{r} x_i c_j \log|\eta_i^{(j)}| + x_{s+t} c_j \right) \\ &= \sum_{i=1}^{r} x_i \sum_{j=1}^{s+t} c_j \log|\eta_i^{(j)}| + x_{s+t} \sum_{j=1}^{s+t} c_j \\ &= n x_{s+t}. \end{aligned} \tag{4.3b}$$

We set

$$m_i := \{x_i\}, \quad \lambda_i := x_i - m_i, \tag{4.3c}$$

such that $|\lambda_i| \leqslant \frac{1}{2}$ $(1 \leqslant i \leqslant r)$ and then

$$\tilde{\xi} := \xi \eta_1^{-m_1} \cdot \ldots \cdot \eta_r^{-m_r}, \tag{4.3d}$$

$\tilde{\xi}$ being associate to ξ, yielding

$$\log|\tilde{\xi}^{(j)}| = \sum_{i=1}^{r} \lambda_i \log|\eta_i^{(j)}| + \frac{\log|\mathrm{N}(\xi)|}{n} \quad (1 \leqslant j \leqslant s + t) \tag{4.3e}$$

for conjugates of $\tilde{\xi}$ because of (4.3b). Therefore the $\tilde{\xi}^{(j)}$ satisfy

$$|\tilde{\xi}^{(j)}| \leqslant \exp\left(\frac{1}{2} \sum_{i=1}^{r} |\log|\eta_i^{(j)}|| + \frac{\log|\mathrm{N}(\xi)|}{n} \right) =: S^{(j)}, \tag{4.3f}$$

$$|\tilde{\xi}^{(j)}| \geqslant \exp\left(-\frac{1}{2} \sum_{i=1}^{r} |\log|\eta_i^{(j)}|| + \frac{\log|\mathrm{N}(\xi)|}{n} \right) =: R^{(j)} \quad (1 \leqslant j \leqslant n). \tag{4.3g}$$

Hence, a complete set of non-associate solutions $\alpha \in o_F$ subject to $|\mathrm{N}(\alpha)| = a$ can be easily computed by chapter 5 (3.8), with $k = a$. □

Remarks
Using the ideas of (4.2) the amount of computations for solving $|\mathrm{N}(\alpha)| = a$ strongly depends on the bounds $R^{(j)}$, $S^{(j)}$ of (4.3f, g) ($1 \leqslant j \leqslant n$). Hence, the absolute values of the conjugates of $\eta_1, \ldots, \eta_r$ should be close to 1 which can often be achieved by applying reduction to $L(\eta_1), \ldots, L(\eta_r)$.

If we search for an element α of an ideal $\mathfrak{a}$ solving $|\mathrm{N}(\alpha)| = a$ we just need to replace the basis $\omega_1, \ldots, \omega_n$ of o_F by a $\mathbb{Z}$-basis $\alpha_1, \ldots, \alpha_n$ of $\mathfrak{a}$, i.e. we replace $R = o_F$ by $R = \mathfrak{a}$.

The advantage of solving norm equations by chapter 5 (3.8) rather than by the old standard method described in chapter 5 (3.5–6) is very well demonstrated by the following example from [1].

Example (4.4)
Let F be generated by a root ρ of the polynomial $f(t) = t^4 - 32t^3 + 154t^2 + 1632t - 44$. In o_F we want to solve $|\mathrm{N}(\alpha)| = k$ for $k \in \{2, 25, 100, 257, 295, 512\}$. In t_1 seconds of computation time a program using chapter 5 (3.8) could show that no solution exists, whereas the old enumeration strategy would have needed t_2 seconds:

k	2	100	257	512
t_1	0.362	0.851	1.545	2.541
t_2	7600	385 000	963 000	20 000 000

The new ellipsoid method found a solution for $k = 25,295$ in about 0.08 seconds in each case, but the box enumeration strategy did not succeed in determining one in a reasonable amount of time (28 seconds). All CPU-times refer to a CDC Cyber 76.

As already mentioned in chapter 5, section 3, we shall use the rest of this section to show how the new idea for solving norm equations originated from K. Mahler's paper [3]. For his concept of ceilings we need to introduce normalized valuations on the algebraic number field F. Clearly,

$$v_{\infty,j}(x) := |x^{(j)}| \quad (1 \leqslant j \leqslant n) \quad \text{with } v_{\infty,j} = v_{\infty,j+t} \quad (s+1 \leqslant j \leqslant s+t) \tag{4.5}$$

are archimedean valuations of F. Non-archimedean valuations of F are obtained via the prime ideals. Let $\mathfrak{p}$ be a non-zero prime ideal of F, and for each $x \in F^{\times}$

$$x o_F = \prod_{\mathfrak{p} \in \mathbb{P}_F} \mathfrak{p}^{\nu_{\mathfrak{p}}(x)} \quad (\nu_{\mathfrak{p}}(x) \in \mathbb{Z}) \tag{4.6}$$

be the prime ideal factorization of $x o_F$. Then we define $v = v_{\mathfrak{p}}$ for all $\mathfrak{p} \in \mathbb{P}_F$ via

$$v_{\mathfrak{p}}(x) := \mathrm{N}(\mathfrak{p})^{-\nu_{\mathfrak{p}}(x)}, \quad v_{\mathfrak{p}}(0) = 0. \tag{4.7}$$

It is easily seen that the $v_\mathfrak{p}$ are non-archimedian valuations of F. We leave this as exercise 2. They are *normalized valuations with respect to* $\mathfrak{p}\in\mathbb{P}_F$. The normalization effect is expressed in the *product formula*:

$$1=\prod_{v_\mathfrak{p}\in\mathfrak{N}} v_\mathfrak{p}(x)\prod_{j=1}^{n} v_{\infty,j}(x)\quad (x\in F^\times), \tag{4.8}$$

where we denoted by $\mathfrak{N}$ the set of all valuations $v_\mathfrak{p}$ of (4.7) for $\mathfrak{p}\in\mathbb{P}_F$. The proof of (4.8) is straightforward:

$$\begin{aligned}\prod_{v_\mathfrak{p}\in\mathfrak{N}} v_\mathfrak{p}(x)\prod_{j=1}^{n} v_{\infty,j}(x) &= \prod_{\mathfrak{p}\in\mathbb{P}_F} \mathrm{N}(\mathfrak{p})^{-v_\mathfrak{p}(x)}|\mathrm{N}(x)|\\ &= \prod_{\mathfrak{p}\in\mathbb{P}_F} \mathrm{N}(\mathfrak{p})^{-v_\mathfrak{p}(x)}\prod_{\mathfrak{p}\in\mathbb{P}_F} \mathrm{N}(\mathfrak{p})^{v_\mathfrak{p}(x)}\\ &\quad \text{(because of } |\mathrm{N}(x)|=\mathrm{N}(x\mathfrak{o}_F))\\ &=1.\end{aligned}$$

Having established the product formula for normalized valuations it is now easy to introduce ceilings in a similar way as K. Mahler did in his 1964 paper 'Inequalities for ideal bases in algebraic number fields' [3]. It is convenient to set $\mathfrak{B}:=\mathfrak{N}\cup\{v_{\infty,j}\,|\,1\leqslant j\leqslant n\}$.

Definition (4.9)

A **ceiling** *is a function* $\lambda:\mathfrak{B}\to\mathbb{R}^{>0}$ *with the properties*

(i) $\lambda(v_\mathfrak{p})=\mathrm{N}(\mathfrak{p})^{-h_\mathfrak{p}}\quad (h_\mathfrak{p}\in\mathbb{Z})$,
(ii) $\lambda(v_\mathfrak{p})=1$ *for almost all* $v_\mathfrak{p}\in\mathfrak{N}$,
(iii) $\prod_{v\in\mathfrak{B}}\lambda(v)=1$.

The connection between ideals and ceilings can be easily described. For an arbitrary ceiling λ we set

$$Q_\lambda:=\prod_{j=1}^{n}\lambda(v_{\infty,j}),\quad R_\lambda:=\prod_{v_\mathfrak{p}\in\mathfrak{N}}\lambda(v_\mathfrak{p}). \tag{4.10a}$$

Then the following properties are immediate:

$$Q_\lambda, R_\lambda\in\mathbb{R}^{>0},\quad Q_\lambda R_\lambda=1. \tag{4.10b}$$

If we define the ideal $\mathfrak{a}_\lambda$ via

$$\mathfrak{a}_\lambda:=\prod_{\mathfrak{p}\in\mathbb{P}_F}\mathfrak{p}^{h_\mathfrak{p}} \tag{4.10c}$$

($h_\mathfrak{p}\in\mathbb{Z}$, compare (4.9) (i)), then $\mathfrak{a}_\lambda$ is clearly in I_F and satisfies

$$\mathrm{N}(\mathfrak{a}_\lambda)=\prod_{\mathfrak{p}\in\mathbb{P}_F}\mathrm{N}(\mathfrak{p}^{h_\mathfrak{p}})=\frac{1}{R_\lambda}=Q_\lambda. \tag{4.10d}$$

On the other hand, for every ideal $\mathfrak{a}\in I_F$ we have $\mathfrak{a}=\prod_{\mathfrak{p}\in\mathbb{P}_F}\mathfrak{p}^{h_\mathfrak{p}}$ $(h_\mathfrak{p}\in\mathbb{Z})$

and obtain a (in general infinitely many) ceiling(s) λ as in (4.9), since we have the freedom of choosing the values $\lambda(v)$ for $v \in \mathfrak{V} \setminus \mathfrak{N}$ arbitrarily with (4.9) (iii) as only side condition. (But compare (4.12).)

Similar to principal ideals we can define principal ceilings:

Definition (4.11)
A ceiling λ is called a **principal ceiling**, *if there exist $x \in F^\times$ and $h \in \mathbb{N}$ such that $\lambda(v)^h = v(x)$ for all $v \in \mathfrak{V}$.*

Lemma (4.12)
Let F be an imaginary quadratic number field. Then every ceiling of F is principal.

Proof
Since F has exactly one archimedean valuation v_∞ the value $\lambda(v_\infty)$ of any ceiling λ is completely determined by $\lambda(v_\mathfrak{p})$ for all $\mathfrak{p} \in \mathbb{P}_F$ because of (4.9) (iii). Let h_F be the class number of F. Then for all $\mathfrak{p} \in \mathbb{P}_F$ we have $\mathfrak{p}^{h_F} = x_\mathfrak{p} o_F$ for a suitable $x_\mathfrak{p} \in F^\times$. Hence, we set $x := \prod_{\mathfrak{p} \in \mathbb{P}_F} x_\mathfrak{p}^{h_\mathfrak{p}}$ and obtain

$$\lambda(v_\mathfrak{p})^{h_F} = \mathrm{N}(\mathfrak{p})^{-h_F h_\mathfrak{p}} = v_\mathfrak{p}(x) \quad \text{for all } v_\mathfrak{p} \in \mathfrak{N}.$$

Especially, $\lambda(v_\infty)^{2h_F} = |x|^2$ by (4.9) (iii). □

Corollary (4.13)
Let F be an algebraic number field properly containing $\mathbb{Q}$ and F not an imaginary quadratic field. Then there exist non-principal ceilings for F.

Ceilings have the advantage that they allow us to transform the task of deciding whether an integral ideal $\mathfrak{a}$ is principal – i.e. a decision, whether the equation

$$|\mathrm{N}(\xi)| = \mathrm{N}(\mathfrak{a}) \tag{4.14}$$

is solvable for $\xi \in \mathfrak{a}$ – into a decision, whether a suitable positive definite quadratic form has minimum $n = [F:\mathbb{Q}]$. Namely, let λ be a ceiling of the algebraic number field F under consideration which corresponds to the ideal $\mathfrak{a}$ in the sense of (4.10c). For a (fixed) basis $\alpha_1, \ldots, \alpha_n$ of $\mathfrak{a}$ we set

$$\Phi_\lambda(\mathbf{x}) := \sum_{j=1}^{n} \lambda(v_{\infty,j})^{-2} v_{\infty,j}\left(\sum_{i=1}^{n} x_i \alpha_i\right)^2 \quad (\mathbf{x} \in \mathbb{Z}^n). \tag{4.15}$$

Clearly, Φ_λ is a positive definite quadratic form which of course depends on the parameters $\lambda(v_{\infty,j})$ $(1 \leqslant j \leqslant n)$. The inequality between arithmetic and geometric means yields

$$\left(\frac{\Phi_\lambda(\mathbf{x})}{n}\right)^n \geqslant \prod_{j=1}^{n} \frac{v_{\infty,j}\left(\sum_{i=1}^{n} x_i\alpha_i\right)^2}{\lambda(v_{\infty,j})^2} = \frac{\left|\mathrm{N}\left(\sum_{i=1}^{n} x_i\alpha_i\right)\right|^2}{Q_\lambda^2} \geqslant \frac{N(\mathfrak{a})^2}{Q_\lambda^2} = 1 \quad (4.16)$$

for $\mathbf{x} \neq \mathbf{0}$ because of (4.10d).

Theorem (4.17)

Let $\mathfrak{a}$ be an integral ideal of the algebraic number field F of degree n. Then $\mathfrak{a}$ is principal, if and only if there is a ceiling λ of F such that

$$\min\{\Phi_\lambda(\mathbf{x}) \mid \mathbf{x} \in \mathbb{Z}^n,\ \mathbf{x} \neq \mathbf{0}\} = n. \quad (4.18)$$

Proof

We note that equality in (4.16) is tantamount to $\lambda(v_{\infty,j}) = v_{\infty,j}(\alpha)$ for $\mathfrak{a} = \alpha o_F$, $\alpha = \sum_{i=1}^{n} x_i\alpha_i$ $(1 \leqslant j \leqslant n)$. □

The remaining difficulty in using (4.17) constructively is of course, that Φ_λ depends on $s+t$ positive real parameters $\lambda(v_{\infty,j})$ $(1 \leqslant j \leqslant s+t)$. Hence, we use a discretization method developed in [2] to obtain the result of 5 (3.8).

A complexity analysis of this method shows that the number of arithmetic operations used by the new method is at most the square root of the number of operations of the old one provided that the box to be covered is large enough.

Exercises

1. Show that the $v_\mathfrak{p}$ of (4.7) are non-archimedean valuations.
2. With respect to the use of chapter 3 (3.10–12) for solving chapter 5 (3.8e) develop an appropriate order for the vectors $\mathbf{r} \in \mathbb{Z}^n$ of 5 (3.8f–h). (Compare exercise 1 of chapter 5 section 3.)
3. Write a program for solving norm equations.

6.5. Computation of the class group

In the two preceding sections we developed the appropriate means to attack the problem of computing the ideal class group $\mathrm{Cl}_F = I_F/H_F$ of an algebraic number field $F = \mathbb{Q}(\rho)$. By computing we understand the determination of

structure constants $n_1, \ldots, n_u \in \mathbb{Z}^{\geqslant 2}$ $(u \in \mathbb{Z}^{\geqslant 0})$ such that $n_i | n_{i+1}$ $(1 \leqslant i < u)$ and Cl_F is isomorphic to the direct product of u cyclic groups C_{n_j} of order n_j $(1 \leqslant j \leqslant u)$, and of (5.1a)

ideals $\mathfrak{a}_1, \ldots, \mathfrak{a}_u$ of I_F such that $C_{n_j} = \langle \mathfrak{a}_j H_F \rangle$. (5.1b)

Then (5.1a, b) yield

$$\left.\begin{aligned} &\mathrm{Cl}_F = \prod_{i=1}^{u} \langle \mathfrak{a}_i H_F \rangle, \\ &\mathfrak{a}_i^{n_i} = a_i o_F \text{ for suitable } a_i \in F^\times \quad (1 \leqslant i \leqslant u). \end{aligned}\right\} \quad (5.2)$$

By (2.8) we know that the $\mathfrak{a}_i$ can be chosen integrally with norm below the Minkowski bound:

$$N(\mathfrak{a}_i) \leqslant M_F. \tag{5.3}$$

As was discussed already at the beginning of section 3 we generate a list P_F of all rational prime numbers p subject to the restriction $p \leqslant M_F$. By (2.27) – or in case of p dividing $(o_F:\mathbb{Z}[\rho])$ by the methods of section 3 – we compute the prime ideal factorization of po_F for each $p \in P_F$. This produces a second list $\mathbb{P}_F$ of all non-zero prime ideals $\mathfrak{p}$ of o_F dividing some $p \in P_F$. Obviously, we can eliminate those $\mathfrak{p}$ from $\mathbb{P}_F$ of which we know that they are principal ideals or that $N(\mathfrak{p}) > M_F$. But for practical reasons it is recommended to remove only those $\mathfrak{p}$ from $\mathbb{P}_F$ for which all prime ideals $\mathfrak{q}$ dividing the same rational prime number p have $N(\mathfrak{q}) > M_F$. Hence, we assume that we have a list $\mathbb{P}_F = \{\mathfrak{p}_1, \ldots, \mathfrak{p}_v\}$ $(v \in \mathbb{Z}^{\geqslant 0})$ of prime ideals such that

(i) all non-zero prime ideals $\mathfrak{p}$ of F with $N(\mathfrak{p}) \leqslant M_F$ are contained in $\mathbb{P}_F$ in case $\mathfrak{p} \neq po_F$ $(p \in P_F)$,
(ii) whenever $p \in P_F$ is divisible by a non-principal prime ideal $\mathfrak{p}$ with $N(\mathfrak{p}) \leqslant M_F$, then all prime ideals dividing po_F are contained in $\mathbb{P}_F$.

It remains to determine the ideals $\mathfrak{a}_i$ and their orders n_i of (5.1) from $\mathbb{P}_F$ $(1 \leqslant i \leqslant u)$ in an efficient way.

A straightforward attack of this problem – i.e. choosing $\mathfrak{p}_1$ from $\mathbb{P}_F$ computing its order $\mathrm{ord}(\mathfrak{p}_1 H_F)$ in Cl_F, doing the same for $\mathfrak{p}_2$ and then constructing the subgroup $\langle \mathfrak{p}_1 H_F, \mathfrak{p}_2 H_F \rangle$ of Cl_F and so on – is usually too time consuming even for quite small values of v. An exception is the trivial case of $\mathbb{P}_F = \emptyset$, whence $h_F = 1$ and $\mathrm{Cl}_F = \{H_F\}$. In the sequel we stipulate $\mathbb{P}_F \neq \emptyset$.

Firstly, we construct a *class-group-matrix* $C = (c_{ij}) \in \mathbb{Z}^{v \times v}$ which sums up our information about equivalence relations among the ideals of $\mathbb{P}_F$. The row index i refers to the ith prime ideal $\mathfrak{p}_i$ in $\mathbb{P}_F$ and the column index j to an element $\beta_j \in F^\times$ (usually an integer) for which the prime ideal decomposition of $\beta_j o_F$ is a power product of $\mathfrak{p}_1, \ldots, \mathfrak{p}_v$ the exponents just being the entries c_{ij} $(1 \leqslant i \leqslant v)$:

$$\beta_j o_F = \prod_{i=1}^{v} \mathfrak{p}_i^{c_{ij}} \quad (c_{ij} \in \mathbb{Z}). \tag{5.4}$$

Of course we shall choose $\beta_j = p_{t_j} (1 \leqslant j \leqslant k)$ where p_{t_j} denotes the jth prime number of P_F for which all prime ideal divisors are contained in $\mathbb{P}_F$. Clearly,

$$k < v. \tag{5.5}$$

Hence there arises the subtask of computing further elements $\beta_{k+1}, \ldots, \beta_v \in F^\times$

subject to (5.4). That will be considered separately following algorithm (5.18). Here we just note that we know how to construct elements of small absolute norm of o_F (and similarly of some $\mathfrak{p}_i (1 \leqslant i \leqslant v)$) from chapter 5, section 5. Usually they are still at hand from the computation of independent units of F and just need to be tested whether their norm is a power product of the $p_{i_j} (1 \leqslant j \leqslant k)$.

Having determined $\beta_1, \ldots, \beta_v$ and the corresponding exponents c_{ij} $(1 \leqslant i, j \leqslant v)$ we apply column reduction to C and compute its Hermite normal form $H(C)$. Let us denote the entries of $H(C)$ by $\tilde{c}_{ij}$. The operations carried out correspond to the construction of elements $\tilde{\beta}_1, \tilde{\beta}_2, \ldots$ of $F^\times$ which are power products of the β_j $(1 \leqslant j \leqslant v)$ with exponents in $\mathbb{Z}$. Clearly, $\tilde{\beta}_j$ is divisible at most by prime ideals $\mathfrak{p}_i$ with $i \geqslant j$. In case C is regular we know that $\mathfrak{p}_j^{\tilde{c}_{jj}}$ divides $\tilde{\beta}_j o_F$ with $\tilde{c}_{jj} > 0$.

If C is not regular then $H(C)$ contains $v - w$ zero columns for $w = \operatorname{rank}(C)$. Then we will produce new elements $\hat{\beta}_1, \ldots, \hat{\beta}_{v-w}$ of $F^\times$ of the type (5.4) and insert the corresponding exponents into the columns $w+1, \ldots, v$ of $H(C)$. Then we replace C by $H(C)$ and start again computing the Hermite normal form of the new matrix.

Thus we finally end up with a class-group-matrix $C = (c_{ij})$ which is regular and in Hermite normal form, i.e. a regular lower triangular matrix. Then $\det(C) = \prod_{i=1}^{v} c_{ii}$ is already an upper bound for h_F and C contains essential information about Cl_F. These remarks are summarized and extended by the following theorem.

Theorem (5.6)

Let I_K denote the set of all ideals of F which are power products of prime ideals of $\mathbb{P}_F = \{\mathfrak{p}_1, \ldots, \mathfrak{p}_v\}$. Let H_K be the set of all principal ideals contained in I_K. Let Λ be the sublattice of $\mathbb{Z}^v$ consisting of all $\mathbf{c} \in \mathbb{Z}^v$ such that $\prod_{i=1}^{v} \mathfrak{p}_i^{c_i} \in H_K$. Finally, let the class-group-matrix C for $\mathbb{P}_F$ be regular. Then we have

(i) $\mathrm{Cl}_F \simeq I_K / H_K$,
(ii) $I_K / H_K \simeq \mathbb{Z}^v / \Lambda$,
(iii) $h_F \mid |\det(C)|$.

Proof

(i) We consider the map

$$\sigma: I_F/H_F \to I_K/H_K: \mathfrak{a}H_F \mapsto \tilde{\mathfrak{a}}H_K,$$

where $\tilde{\mathfrak{a}} \in I_K$ is chosen such that $\mathfrak{a}H_F = \tilde{\mathfrak{a}}H_F$. This is possible since I_K contains representatives of all ideal classes of F as is guaranteed by the Minkowski bound. We must show that σ is a group isomorphism.

We start to prove that σ is well defined. Let $\mathfrak{a}, \mathfrak{b} \in I_F$ with $\mathfrak{a}H_F = \mathfrak{b}H_F$ and $\tilde{\mathfrak{a}}, \tilde{\mathfrak{b}} \in I_K$ such that $\sigma(\mathfrak{a}H_F) = \tilde{\mathfrak{a}}H_K$, $\sigma(\mathfrak{b}H_F) = \tilde{\mathfrak{b}}H_K$. This implies

$\tilde{\mathfrak{a}}H_F = \mathfrak{a}H_F = \mathfrak{b}H_F = \tilde{\mathfrak{b}}H_F$ and $\tilde{\mathfrak{a}}\tilde{\mathfrak{b}}^{-1} \in H_F$. But then $\tilde{\mathfrak{a}}\tilde{\mathfrak{b}}^{-1} \in H_F \cap I_K = H_K$ and $\tilde{\mathfrak{a}}H_K = \tilde{\mathfrak{b}}H_K$ follows.

Now, σ is clearly a surjective group homomorphism and it remains to show that it is also injective. Let $\mathfrak{a}, \mathfrak{b} \in I_F$ with $\sigma(\mathfrak{a}H_F) = \tilde{\mathfrak{a}}H_K = \tilde{\mathfrak{b}}H_K = \sigma(\mathfrak{b}H_F)$. We conclude that $\tilde{\mathfrak{a}}\tilde{\mathfrak{b}}^{-1} \in H_K \subseteq H_F$ implying $\mathfrak{a}H_F = \tilde{\mathfrak{a}}H_F = \tilde{\mathfrak{b}}H_F = \mathfrak{b}H_F$.

(ii) We consider the map

$$\tau: I_K \to \mathbb{Z}^v/\Lambda: \prod_{i=1}^{v} \mathfrak{p}_i^{c_i} \mapsto (c_1, \ldots, c_v)^t + \Lambda.$$

Obviously, τ is a group epimorphism with kernel H_K. Hence, the isomorphism theorem for groups yields $I_K/H_K \simeq \mathbb{Z}^v/\Lambda$.

(iii) The columns $\mathbf{c}_1, \ldots, \mathbf{c}_v$ of the class-group-matrix C generate a sublattice $\tilde{\Lambda}$ of Λ of equal rank. Hence, $\tilde{\Lambda}$ is of finite index in Λ and we obtain $h_F = (\mathbb{Z}^v : \Lambda) = (\mathbb{Z}^v : \tilde{\Lambda})/(\Lambda : \tilde{\Lambda})$ because of (i) and (ii). But $(\mathbb{Z}^v : \tilde{\Lambda})$ is just the absolute value of the determinant of the transition matrix of the standard basis of $\mathbb{Z}^v$ to the basis of $\tilde{\Lambda}$ consisting of the columns of C (chapter 3 (2.9)). We note that $(\mathbb{Z}^v : \tilde{\Lambda}) = \det(C)$ if C is in Hermite normal form. □

Remark (5.7)

For $|\det(C)| = 1$ we get $h_F = 1$ even without having to know any units of F.

In the sequel we therefore assume $C = (c_{ij})$ in Hermite normal form and $\det(C) > 1$. In this case the information contained in C will in general not be complete. Any column j $(1 \leqslant j \leqslant v)$ of C with $c_{jj} = 1$ is superfluous for further considerations since $c_{j\nu} = 0$ $(1 \leqslant \nu < j)$ and the ideal class of $\mathfrak{p}_j$ is representable by the prime ideals $\mathfrak{p}_\mu$ $(\mu > j)$ because of $\mathfrak{p}_j H_F = (\prod_{\mu=j+1}^{v} \mathfrak{p}_\mu^{-c_{\mu j}})H_F$. By deleting all such columns and the corresponding rows from C we obtain a matrix $RC = (b_{ij})$ of much smaller dimension, say w. Again RC is in Hermite normal form and $\det(RC) = \det(C)$. Let $\mathfrak{p}_{i_1}, \ldots, \mathfrak{p}_{i_w}$ be the prime ideals corresponding to the rows $1, \ldots, w$ of RC and β_{i_ν} $(1 \leqslant \nu \leqslant w)$ be the elements of $F^\times$ corresponding to the columns of RC. For simplicity's sake we again denote them by $\mathfrak{p}_1, \ldots, \mathfrak{p}_w$, $\beta_1, \ldots, \beta_w$ in the sequel. (This corresponds to a permutation of the rows and columns of C.) Before we continue to discuss the processing of RC a simple example will illustrate our reductions.

Example (5.8)

Let $F = \mathbb{Q}(\rho)$ for $\rho^2 - 243\rho - 1 = 0$. We compute $d_F = 59\,053$, $n = s = 2$, $M_F = 121.5$, and obtain 36 prime ideals in $\mathbb{P}_F$ from the prime numbers 109, 103, 83, 79, 73, 67, 61, 59, 53, 47, 43, 37, 29, 23, 19, 11, 7, 3.

For C in Hermite normal form we obtain

$$\begin{array}{c|ccc|ccc|}
 & \beta_1 & \cdots & \beta_{33} & \beta_{34} & \beta_{35} & \beta_{36} \\
\hline
\mathfrak{p}_1 & & & & & & \\
\vdots & & I_{33} & & & 0 & \\
\mathfrak{p}_{33} & & & & & & \\
\hline
\mathfrak{p}_{34} & * & \cdots & * & 5 & 0 & 0 \\
\mathfrak{p}_{35} & 0 & \cdots & 0 & 0 & 1 & 0 \\
\mathfrak{p}_{36} & * & \cdots & * & 1 & 1 & 5 \\
\hline
\end{array}$$

and thus for RC

$$\begin{array}{c|cc|}
 & \beta_1 & \beta_2 \\
\hline
\mathfrak{p}_1 & 5 & 0 \\
\mathfrak{p}_2 & 1 & 5 \\
\hline
\end{array}\ ,$$

where $\mathfrak{p}_1 | 7o_F$, $\mathfrak{p}_2 | 3o_F$. Hence, $h_F \in \{1, 5, 25\}$. Next we convince ourselves that $|\mathrm{N}(\xi)| = 3$ is not solvable in o_F. Therefore the smallest power of $\mathfrak{p}_2$ belonging to H_F is 5, $\mathfrak{p}_2^5 = ((245 + 59\,053^{\frac{1}{2}})/2)o_F$. But then $\mathfrak{p}_1^5$ (and thus $\mathfrak{p}_1$ itself) cannot be a principal ideal and we conclude $h_F = 25$, $\mathrm{Cl}_F = \langle \mathfrak{p}_1 H_F \rangle$.

Remark (5.9)

It may give a little surprise that the prime ideals dividing larger prime numbers correspond to smaller indices. This is due to the following heuristic argument. Since the norms of the β_j $(1 \leqslant j \leqslant v)$ are of small absolute value it is unlikely that for prime ideals $\mathfrak{p}_i$ with $\mathrm{N}(\mathfrak{p}_i)$ big all entries c_{ij} $(1 \leqslant j \leqslant v)$ of the ith row of the class-group-matrix have a common divisor greater than one. Hence, in the Hermite normal form $H(C)$ of C the diagonal elements in the upper left hand corner will usually be 1 which makes it a little easier to construct RC from $H(C)$. Also during the computation of $H(C)$ from C the elements of the rows of higher index tend to grow. So we will usually find the biggest diagonal entries of $H(C)$ in the lower right-hand corner. But those entries will then correspond to prime ideals of smaller norm, therefore such tests as will be needed for the solvability of norm equations will be less time consuming.

So far we have only applied column operations to the class-group-matrix. From now on we also use row operations. Then the columns of the truncated class-group-matrix RC do not correspond to prime ideals $\mathfrak{p}_1, \ldots, \mathfrak{p}_w$ any longer but to possibly fractional ideals $\mathfrak{a}_1, \ldots, \mathfrak{a}_w$ for which we keep stored the exponents of their presentations by the $\mathfrak{p}_j$:

$$\mathfrak{a}_i := \prod_{j=1}^{w} \mathfrak{p}_j^{m_{ij}} \quad (m_{ij} \in \mathbb{Z};\ \ 1 \leqslant i \leqslant w). \tag{5.10}$$

Our goal is to transform RC into a diagonal matrix $\mathrm{diag}(n_1, \ldots, n_u)$ $(u \leqslant w)$ such that the ideal $\mathfrak{a}_i$ corresponding to row i is of order n_i in Cl_F and

$\mathrm{Cl}_F = \prod_{i=1}^{u} \langle \mathfrak{a}_i H_F \rangle$. The information contained in RC will in general not be sufficient for this purpose, i.e. the Smith normal form of RC can look quite different from $\mathrm{diag}(n_1, \ldots, n_u)$. Thus we need to construct Cl_F from RC under possibly existing hidden additional side conditions. Instead of trying to exhibit all of those – which is obviously not very efficient – we use as additional information that we can decide whether a given ideal is principal or not. We note that $\mathfrak{p}_1, \ldots, \mathfrak{p}_w$ should be given in $P(p)$-normal presentation from now on. The matrix $\mathrm{diag}(n_1, \ldots, n_u)$ will be constructed from RC starting in the last row. Our essential information about $\mathfrak{a}_w := \mathfrak{p}_w$ is

$$\mathfrak{p}_w^{\,b_{ww}} = \beta_w o_F \in H_F. \tag{5.11a}$$

In the usual way (compare chapter 2 (5.2, 3)) we compute

$$\mathrm{ord}(\mathfrak{p}_w H_F) =: s_w \tag{5.11b}$$

by the methods of section 4. Clearly s_w is a natural number dividing b_{ww}. In case of $1 < s_w < b_{ww}$ we reduce the elements b_{wj} $(j < w)$ of the last row of RC modulo s_w and obtain

$$0 \leqslant \beta_{wj} < s_w \quad (1 \leqslant j < w). \tag{5.11c}$$

Then we set

$$b_{ww} \leftarrow s_w. \tag{5.11d}$$

For $s_w = 1$ we delete line w from RC, replace w by $w-1$ and carry out (5.11*b*–*d*) anew. In this way we either get $w = 0$ in which case we terminate with $h_F = 1$, $\mathrm{Cl}_F = \langle H_F \rangle$, or $s_w > 1$, $w \geqslant 1$, whence we proceed.

The remaining transformations of RC are done by an inductive procedure. We assume in the sequel that the lower triangular matrix $RC = (b_{ij}) \in (\mathbb{Z}^{\geqslant 0})^{w \times w}$ and the integral ideals $\mathfrak{a}_1, \ldots, \mathfrak{a}_w$ corresponding to its rows satisfy after ρ steps $(1 \leqslant \rho < w)$:

$$\mathfrak{a}_i = \mathfrak{p}_i \quad (1 \leqslant i \leqslant w - \rho) \tag{5.12a}$$

$$\langle \mathfrak{a}_{w-\rho+1} H_F, \ldots, \mathfrak{a}_w H_F \rangle = \prod_{j=0}^{\rho-1} \langle \mathfrak{a}_{w-j} H_F \rangle, \tag{5.12b}$$

$$s_j := \mathrm{ord}(\mathfrak{a}_j H_F) = b_{jj} \quad (w - \rho + 1 \leqslant j \leqslant w), \tag{5.12c}$$

$$s_{w-\rho+1} | s_{w-\rho+2} | \cdots | s_w, \tag{5.12d}$$

$$b_{ij} = 0 \quad (w - \rho + 1 \leqslant j \leqslant w, i > j). \tag{5.12e}$$

The computation of s_w in (5.11) is therefore tantamount to the step $\rho = 1$. In the sequel we assume $1 \leqslant \rho < w$.

In each succeeding step we then need to determine the minimal exponent m such that $\mathfrak{p}_{w-\rho}^m \in G_\rho := \prod_{j=0}^{\rho-1} \langle \mathfrak{a}_{w-j} H_F \rangle$. This new information is entered into column $w - \rho$ of RC and the elementary divisor theorem applied to the submatrix $(b_{ij})_{w-\rho \leqslant i,j \leqslant w}$ of RC. The result will be (5.12a–e) also for $\rho + 1$.

Hence, we search for a minimal $m \in \mathbb{N}$ such that

$$\mathfrak{p}_{w-\rho}^m \prod_{j=0}^{\rho-1} \mathfrak{a}_{w-j}^{h_j} \in H_F \tag{5.13}$$

for suitable $h_j \in \mathbb{Z}$ subject to $0 \leqslant h_j < s_{w-j}$, $0 \leqslant j < \rho$. The following properties of m are immediate:

$$m \mid b_{w-\rho,w-\rho}, \tag{5.14a}$$

$$\mathfrak{p}_{w-\rho},\quad \mathfrak{p}_{w-\rho}^m,\quad \mathfrak{p}_{w-\rho}^{b_{w-\rho,w-\rho}} \notin H_F, \tag{5.14b}$$

if there is an index j, $0 \leqslant j < \rho$, such that $b_{w-j,w-\rho} > 0$,

$$\mathfrak{p}_{w-\rho}^{b_{w-\rho,w-\rho}} \prod_{j=0}^{\rho-1} \mathfrak{a}_{w-j}^{b_{w-j,w-\rho}} \in H_F. \tag{5.14c}$$

As we already saw in example (5.8) the occurrence of an entry $b_{w-j,w-\rho} > 0$ for an index j with $0 \leqslant j < \rho$ can be of advantage. Indeed, for $b_{w-\rho,w-\rho} = mk$ according to (5.14a) we obtain from (5.13), (5.14c) the conditions

$$kh_j \equiv b_{w-j,w-\rho} \bmod s_{w-j} \quad (0 \leqslant j < \rho). \tag{5.15a}$$

If the entries $b_{w-j,w-\rho}$ do not vanish for all j $(0 \leqslant j < \rho)$, those conditions usually diminish the number of possibilities for m drastically. For example, if $b_{w-\rho,w-\rho}$ is a prime number and there is an index j $(0 \leqslant j < \rho)$ such that $b_{w-\rho,w-\rho} \mid s_{w-j}$, $b_{w-j,w-\rho} \not\equiv 0 \bmod b_{w-\rho,w-\rho}$ then m is different from 1 implying $m = b_{w-\rho,w-\rho}$, $h_\nu = b_{w-\nu,w-\rho}$ $(0 \leqslant \nu < \rho)$, as in example (5.8).

For all solutions of the system of congruences (5.15a) we need to test (5.13) with the methods of the preceding section. Since we usually expect $m = b_{w-\rho,w-\rho}$ we choose the order of the candidates for m to be tested again as in chapter 2 (5.2, 3). After those tests we know a solution of (5.13) and replace column $w-\rho$ of RC via

$$b_{w-\rho,w-\rho} \leftarrow m, \quad b_{w-j,w-\rho} \leftarrow h_j \quad (0 \leqslant j < \rho). \tag{5.15b}$$

Then we apply the elementary divisor theorem to the free $\mathbb{Z}$-modules $M = \mathbb{Z}^{\rho+1} = \bigoplus_{i=1}^{\rho+1} \mathbb{Z}\mathbf{e}_i$ and its submodule $T = \bigoplus_{i=1}^{\rho+1} \mathbb{Z}\mathbf{t}_i$ given by $\mathbf{t}_1 = (b_{w-\rho,w-\rho}, \ldots, b_{w,w-\rho})^t$, $\mathbf{t}_i = s_{w-\rho+i-1}\mathbf{e}_i$ $(2 \leqslant i \leqslant \rho+1)$. This means to transform the matrix A satisfying

$$(\mathbf{e}_1, \ldots, \mathbf{e}_{\rho+1})A = (\mathbf{t}_1, \ldots, \mathbf{t}_{\rho+1}) \tag{5.16a}$$

into its Smith normal form by unimodular row and column operations. Hence, we compute unimodular matrices U, $V \in \mathbb{Z}^{(\rho+1)\times(\rho+1)}$ (as well as U^{-1}) subject to

$$UAV = \operatorname{diag}(d_1, \ldots, d_{\rho+1}) \quad (d_{\rho+1}, d_i \in \mathbb{N},\ d_i \mid d_{i+1},\ 1 \leqslant i \leqslant \rho) \tag{5.16b}$$

as shown in chapter 3. The entries $\tilde{u}_{ij}$ $(1 \leqslant i,j \leqslant \rho+1)$ of U^{-1} are then used to replace $\mathfrak{p}_{w-\rho}, \mathfrak{a}_{w-\rho+1}, \ldots, \mathfrak{a}_w$ by $\mathfrak{b}_1, \ldots, \mathfrak{b}_{\rho+1}$ via

$$\mathfrak{b}_i := \mathfrak{p}_{w-\rho}^{\tilde{\mu}_{1,i}} \prod_{j=1}^{\rho} \mathfrak{a}_{w-\rho+j}^{\tilde{\mu}_{j+1,i}}. \tag{5.16c}$$

Following that we must update the matrix RC. We begin by setting

$$\mathfrak{a}_{w-j} \leftarrow \mathfrak{b}_{\rho+1-j}, \; b_{w-j,w-j} \leftarrow s_{w-j} \leftarrow d_{\rho+1-j} \quad (0 \leqslant j \leqslant \rho). \tag{5.17a}$$

The new (partial) columns $\mathfrak{c}_i := (b_{w-\rho,i}, \ldots, b_{w,i})^t$ $(1 \leqslant i \leqslant w-\rho)$ are obtained from the old ones upon multiplication by U:

$$\mathfrak{c}_i \leftarrow U\mathfrak{c}_i \quad (1 \leqslant i < w-\rho). \tag{5.17b}$$

Afterwards the new elements $b_{w-j,i}$ still need to be reduced modulo s_{w-j} so that they again satisfy

$$0 \leqslant b_{w-j,i} < s_{w-j} \quad (1 \leqslant i < w-\rho, \; 0 \leqslant j \leqslant \rho). \tag{5.17c}$$

The new column $w-\rho$ is to contain zeros below its diagonal entry:

$$b_{i,w-\rho} \leftarrow 0 \quad (w-\rho < i \leqslant w). \tag{5.17d}$$

For $s_{w-\rho} = 1$ we then eliminate row $w-\rho$ and column $w-\rho$ from RC:

$$b_{ij} \leftarrow b_{i+1,j} \quad (1 \leqslant j < w-\rho), \; b_{ii} \leftarrow b_{i+1,i+1} \quad \text{for } i = w-\rho, w-\rho+1, \ldots, w-1, \tag{5.17e}$$

and

$$w \leftarrow w-1, \quad \rho \leftarrow \rho-1. \tag{5.17f}$$

Now we have again the possibilities of $b_{w-\rho,w-\rho}$ being equal to or greater than one. For $b_{w-\rho,w-\rho} = 1$ we carry out (5.17e, f) anew. For $b_{w-\rho,w-\rho} > 1$ we increase ρ by one and proceed to the next step unless $w-\rho$ becomes zero in which case we terminate.

If the criterion of termination $w-\rho = 0$ is obtained, the last matrix RC yields the structure of the class group of F. To get (5.1a, b) we just need to replace u by w and n_i by s_i $(1 \leqslant i \leqslant u)$. The class number of F is given by

$$h_F = \prod_{i=1}^{w} s_i. \tag{5.18}$$

As stated in connection with (5.5) we still must discuss the subtask of constructing elements $\beta_j \in F^\times$ $(k+1 \leqslant j \leqslant v)$ subject to (5.4) which are needed to get a regular class-group-matrix C.

Clearly, $\beta_j \in F^\times$ satisfies (5.4) if and only if

$$|\mathrm{N}(\beta_j)| = \prod_{q \in P_F'} q^{m_q} \quad (m_q \in \mathbb{Z}), \tag{5.19}$$

where P_F' denotes the subset of P_F consisting of those prime numbers p of P_F the prime ideal divisors of which are in $\mathbb{P}_F$.

In an initial step we consider the elements of small (absolute) norm constructed for the determination of a system of independent (or fundamental) units of F. Then we compute $H(C)$ of the class-group-matrix C by factoring the elements considered initially. Either $H(C)$ is already regular, or we are in need of special elements $\beta \in F^\times$ divisible by certain prime ideals $\mathfrak{p}$ of $\mathbb{P}_F$ which can be read off from $H(C)$. Then we apply the process of constructing integers of F of small absolute norm of chapter 5, section 5, but with a basis $\pi_1, \ldots, \pi_n$ of $\mathfrak{p}$ instead of $\omega_1, \ldots, \omega_n$ of o_F. Thus all obtained elements are at least divisible by $\mathfrak{p}$ and we choose those which pass the norm test (5.19). The determination of a factorization (5.4) is then obtained by the methods of (4.8–10), respectively (3.20c).

Remark (5.20)

Since we cannot assume that the information about Cl_F contained in C is complete even if $H(C)$ becomes regular it is recommended to construct a few more elements $\beta_j \in F^\times$ subject to (5.4) and to check whether adding the corresponding c_{ij} of (5.4) as additional columns to C decreases $\det(H(C))$. If this is not the case for about five consecutive β_j's we would expect that the information obtained on Cl_F is 'nearly complete' and go on to compute RC, etc.

Since the development of the means for computing Cl_F and h_F took place throughout several sections (sections 1–5) of this chapter we now present a summarization of all necessary steps. The whole procedure is split into four major subtasks.

I. Input and preparations

The following data are INPUT:

N – degree of the field F under consideration;

$F(I)\,(1 \leqslant I \leqslant N)$ – the coefficients of a generating polynomial

$$f(t) = t^N + \sum_{I=1}^{N} F(I)t^{N-I} \quad \text{for } F, \text{ i.e. } F = \mathbb{Q}(\rho)$$

for a zero ρ of f.

From those all other pieces of F which are needed can be either computed by subroutines or by different programs and then also be added to the input. We especially need:

$Z \in \mathbb{C}^N$ – zeros of $f(t)$, i.e. $f(Z(I)) = 0 \quad (1 \leqslant I \leqslant N)$;

$S, T \in \mathbb{Z}^{\geqslant 0}$ – the number of real respectively complex conjugates of F, $N = S + 2T$;

$MD \in \mathbb{N}, M \in \mathbb{Z}^{N \times N}$ – coefficients of an integral basis $\omega_1, \ldots, \omega_N$ of F,

$$\text{i.e. } \omega_I = \frac{1}{MD} \sum_{J=1}^{N} M(I,J)\rho^{J-1} \quad (1 \leqslant I \leqslant N)$$

(compare chapter 5 (8.4b)); in general $\omega_1, \ldots, \omega_N$ should be reduced;

d_F – the discriminant of F;

$\Omega \in \mathbb{C}^{N \times N}$ – $\Omega(I, J)$ contains the Jth conjugate of ω_I ($1 \leqslant I, J \leqslant N$);

$\Gamma \in \mathbb{Z}^{N \times N \times N}$ – Γ contains the multiplication constants for

$$\omega_I \omega_J = \sum_{K=1}^{N} \Gamma(I, J, K) \omega_K \quad \text{(compare chapter 5 (8.5));}$$

$E \in \mathbb{Z}^{R \times N}$ – coefficients of a maximal set $\{\varepsilon_1, \ldots, \varepsilon_R\}$ of independent (or even fundamental) units of F, i.e.

$$\varepsilon_I = \sum_{J=1}^{N} E(I, J) \omega_J \quad (1 \leqslant I \leqslant R;\ R = S + T - 1);$$

usually we assume that the vectors $L_2(\varepsilon_I)$ ($1 \leqslant I \leqslant R$) are reduced so that the conjugates of ε_I will be of absolute value close to 1 (see chapter 5 (6.1) and the remark after chapter 6 (4.3)).

We note that Z, Ω can be chosen in $\mathbb{R}^N$, $\mathbb{R}^{N \times N}$, respectively, if we separate the real and imaginary parts of complex conjugates of ρ (compare chapter 5 (8.9)). To calculate the coefficients of an element of o_F given as a sum of powers of ρ in the integral basis we also need M^{-1}. We further stipulate the existence of subroutines for computing norms of elements and a dual basis with respect to a basis of an order or of an ideal of o_F.

II. Ideal factorization

The Minkowski bound M_F is computed from d_F, N, T ((2.8)). Then we construct a list P_F of all prime numbers $p \leqslant M_F$ (P_F can also be obtained as part of a list of sufficiently many prime numbers contained among the input data). From P_F we obtain a list $\mathbb{P}_F$ of prime ideals $\mathfrak{p}$ as described at the beginning of this section. We list those $\mathfrak{p}$ in their representation by two generators together with their norm, their ramification index $e_\mathfrak{p}$ and their degree of inertia $f_\mathfrak{p}$. (Though the first generator for $\mathfrak{p}$ will usually be the prime number p divisible by $\mathfrak{p}$ and $\mathrm{N}(\mathfrak{p})$ is therefore determined by p and $f_\mathfrak{p}$, the storing of $f_\mathfrak{p}$ is usually easier than to compute it anew when it will be needed in the program.)

We also establish a subroutine which computes a $\mathbb{Z}$-basis for an integral ideal given by two generators. This itself needs a subroutine for the computation of the Hermite normal form of a matrix. Finally, we use a subroutine for factoring polynomials in $\mathbb{Z}/p\mathbb{Z}[t]$.

III. Computation of the class-group-matrix C

The class group matrix C is computed as described in the current section. Besides the entries corresponding to prime numbers p of P_F we need subroutines for:

(i) determining elements of small norm in o_F or some integral ideal given by its $\mathbb{Z}$-basis;

(ii) checking whether some integer β of o_F whose norm is divisible by $\mathrm{N}(\mathfrak{a})$

is contained in that integral ideal $\mathfrak{a}$ (here $\mathfrak{a}$ is some power $\mathfrak{p}^k$ of a prime ideal $\mathfrak{p}$, and $\mathfrak{p}^k$ is obtained from $\mathfrak{p} = p o_F + \pi o_F$ as $\mathfrak{p}^k = p^k o_F + \pi^k o_F$, see exercise 2 of section 3 and of this section).

IV. Computation of Cl_F

This was discussed in detail in this section. From C in Hermite normal form we compute RC and obtain $h_F | \det(RC)$. Especially we need subroutines for the computation of the Smith normal form of a matrix and for solving norm equations (compare chapter 3, section 2 and chapter 6, section 4).

Let us finally remark that not all required subroutines are mentioned in detail. For example we need a subroutine for computing the Hermite normal form of a matrix, preferably also one using modular arithmetic. Also we did not go into detail about the construction of two generators of those ideals dividing $(o_F : \mathbb{Z}[\rho])$ since this was already done in section 3. Of course the principal ideal test of section 4 is needed as most important subroutine.

Once we know generators $\mathfrak{a}_1, \ldots, \mathfrak{a}_u$ of Cl_F subject to (5.1a, b) the usually difficult problem of deciding whether two given ideals $\mathfrak{a}, \mathfrak{b} \in I_F$ are equivalent can easily be solved. This will be discussed in the remainder of this section.

Let $\mathfrak{a}$ be a non-zero ideal of o_F with $\mathbb{Z}$-basis $\alpha_1, \ldots, \alpha_n$. Then each $\xi \in \mathfrak{a}$ has a representation

$$\xi = \sum_{i=1}^{n} x_i \alpha_i \quad (x_i \in \mathbb{Z}) \tag{5.21}$$

and

$$Q_{\mathfrak{a}}(\mathbf{x}) := \sum_{j=1}^{n} |\xi^{(j)}|^2 \tag{5.22}$$

is a positive definite quadratic form of determinant $\mathrm{N}(\mathfrak{a})^2 |d_F|$. Let $\tilde{\xi} = \sum_{i=1}^{n} \tilde{x}_i \alpha_i$ such that

$$Q_{\mathfrak{a}}(\tilde{\mathbf{x}}) = \min \{ Q_{\mathfrak{a}}(\mathbf{x}) \mid \mathbf{x} \in \mathbb{Z}^n, \ \mathbf{x} \neq \mathbf{0} \}. \tag{5.23}$$

By chapter 3 (3.34a), (3.35a, b) we get

$$Q_{\mathfrak{a}}(\tilde{\mathbf{x}})^n \leqslant \gamma_n^n \mathrm{N}(\mathfrak{a})^2 |d_F| \tag{5.24}$$

and the inequality between arithmetic and geometric means yields

$$\mathrm{N}(\tilde{\xi})^2 \leqslant \left(\frac{Q_{\mathfrak{a}}(\tilde{\mathbf{x}})}{n} \right)^n \tag{5.25}$$

implying

$$|\mathrm{N}(\tilde{\xi})/\mathrm{N}(\mathfrak{a})| \leqslant (\Gamma_n |d_F|)^{1/2} =: M_E \tag{5.26}$$

for

$$\Gamma_n = \gamma_n^n n^{-n}. \tag{5.27}$$

Hence, every integral ideal $\mathfrak{a}$ of I_F contains an element α such that $|\mathrm{N}(\alpha)/\mathrm{N}(\mathfrak{a})|$ is bounded by M_E, a constant only depending on F but not on $\mathfrak{a}$.

Analogously as in the beginning of this section we determine a list $\mathbb{P}_E$ of non-zero prime ideals $\mathfrak{p}_1, \ldots, \mathfrak{p}_E$ of o_F satisfying $|\mathrm{N}(\mathfrak{p}_i)| \leqslant M_E$ $(1 \leqslant i \leqslant E)$ and then a list of presentations

$$\mathfrak{p}_i H_F = \left(\prod_{j=1}^{u} \mathfrak{a}_j^{k_{ij}} \right) H_F \quad (k_{ij} \in \mathbb{Z}^{\geqslant 0}, 0 \leqslant k_{ij} < n_j, 1 \leqslant j \leqslant u, 1 \leqslant i \leqslant E) \quad (5.28)$$

together with non-zero elements $\pi_i, \tilde{\pi}_i$ of o_F such that

$$\pi_i \mathfrak{p}_i = \tilde{\pi}_i \prod_{j=1}^{u} \mathfrak{a}_j^{k_{ij}}. \quad (5.29)$$

For the $\mathfrak{a}_j$ $(1 \leqslant j \leqslant u)$ we assume (5.2). Then we can easily solve the following two problems for integral ideals $\mathfrak{a}, \mathfrak{b} \in I_F$:

Procedure for determining whether $\mathfrak{a}$ ***is principal*** (5.30)

Compute $\tilde{\xi} \in \mathfrak{a}$ subject to (5.23) by chapter 3 (3.15). Compute the prime ideal factorization

$$\tilde{\xi}\mathfrak{a}^{-1} = \prod_{i=1}^{E} \mathfrak{p}_i^{v_i} \quad (v_i \in \mathbb{Z}^{\geqslant 0}). \quad (5.31)$$

Then $\mathfrak{a}$ is principal if and only if

$$\sum_{i=1}^{E} v_i k_{ij} \equiv 0 \bmod n_j \quad (1 \leqslant j \leqslant u) \quad (5.32)$$

because of (5.28).

In case $\mathfrak{a}$ is principal there hold equations $\sum_{i=1}^{E} v_i k_{ij} = \mu_j n_j$ with non-negative rational integers μ_j $(1 \leqslant j \leqslant u)$. Hence we obtain

$$\left(\prod_{i=1}^{E} \pi_i^{v_i} \right) \tilde{\xi}\mathfrak{a}^{-1} = \prod_{i=1}^{E} \tilde{\pi}_i^{v_i} \prod_{j=1}^{u} \mathfrak{a}_j^{k_{ij} v_i} = \prod_{i=1}^{E} \tilde{\pi}_i^{v_i} \prod_{j=1}^{u} a_j^{\mu_j} o_F$$

by (5.29), (5.31), (5.32) and (5.2). But the latter implies

$$\mathfrak{a} = \tilde{\xi} \prod_{i=1}^{E} (\pi_i/\tilde{\pi}_i)^{v_i} \prod_{j=1}^{u} a_j^{-\mu_j} o_F \quad (5.33)$$

and establishes a generating element for $\mathfrak{a}$.

Procedure for deciding equivalence of two ideals $\mathfrak{a}, \mathfrak{b}$. (5.34)

We compute $\alpha \in o_F$, $\mathfrak{c}$ a non-zero ideal of o_F such that

$$\mathfrak{a}\mathfrak{b}^{-1} = \alpha^{-1}\mathfrak{c} \quad (5.35)$$

by the methods of section 3. Then the equivalence of $\mathfrak{a}$ and $\mathfrak{b}$ is tantamount to $\mathfrak{c}$ being a principal ideal. This is then tested by (5.30). If the outcome of the test is positive we also obtain $\beta \in o_F$ subject to $\mathfrak{c} = \beta o_F$ by (5.33). But then obviously

$$\alpha\mathfrak{a} = \beta\mathfrak{b}. \quad (5.36)$$

We note that (5.30), (5.34) can also be applied to fractional ideals $\mathfrak{a}, \mathfrak{b}$ since there are integers $a, b \in o_F$ such that $a\mathfrak{a}, b\mathfrak{b}$ are integral. This was already pointed out at the beginning of section 3.

Finally, it is of some interest by how much the constants M_F and M_E differ since that is a measure for the amount of computations which are necessary to establish (5.28). Since the Γ_n can be given explicitly only for $n \leqslant 8$ we present a list for M_E/M_F for those n.

List of M_E/M_F for $n \leqslant 8$ (5.37)

n	s	t	M_E/M_F
2	2	0	1.15
	0	1	0.91
3	3	0	1.22
	1	1	0.96
4	4	0	1.33
	2	1	1.05
	0	2	0.82
5	5	0	1.32
	3	1	1.03
	1	2	0.81
6	6	0	1.39
	4	1	1.09
	2	2	0.85
	0	3	0.67
7	7	0	1.44
	5	1	1.13
	3	2	0.89
	1	3	0.70
8	8	0	1.63
	6	1	1.28
	4	2	1.00
	2	3	0.79
	0	4	0.62

The list (5.37) seems to suggest that the bounds M_F and M_E don't differ by much for small field degrees and small absolute values of the discriminants. Thus the following result is somewhat surprising.

Proposition (5.38)

$M_E/M_F \to 0$ for $n \to \infty$.

Proof

Instead of Hermite's constants γ_n^n we use the upper bounds of chapter 3 (3.35b), and obtain

$$M_E/M_F \leqslant \left(n^n\left(\frac{2}{\pi}\right)^s\right)^{1/2} \Gamma\left(\frac{n}{2}+2\right)2^{-t}(n!)^{-1}. \tag{5.39a}$$

Then the application of Stirling's formula yields for $n \geqslant 50$:

$$\begin{aligned}\log(M_E/M_F) &\leqslant \tfrac{1}{2}(n\log n + s\log 2 - s\log\pi) - \left(\frac{n}{2}+2\right)\\ &\quad + \left(\frac{n}{2}+\frac{3}{2}\right)\log\left(\frac{n}{2}+2\right) + \tfrac{1}{2}\log(2\pi) - t\log 2\\ &\quad - \tfrac{1}{2}\log(2\pi) - (n+\tfrac{1}{2})\log n + n\\ &= \frac{n}{2}\left(\log\left(\frac{n}{2}+2\right) - \log n\right) + \frac{n}{2} + \frac{s}{2}(\log 2 - \log\pi) - t\log 2\\ &\quad + \tfrac{3}{2}\log\left(\frac{n}{2}+2\right) - \tfrac{1}{2}\log n - 2\\ &= \frac{n}{2}\left(\log\frac{n+4}{2n}+1\right) + \frac{s}{2}(\log 2 - \log\pi) - t\log 2 + \tfrac{1}{2}\log\frac{\left(\frac{n}{2}+2\right)^3}{n} - 2\\ &\leqslant 0.192n - 0.225s - 0.7t + \tfrac{1}{2}\log\left(\frac{n^2}{8}+3n+6+\frac{8}{n}\right) - 2\\ &< -0.033n - 0.25t + \log\frac{n}{2} - 2.\end{aligned} \tag{5.39b}$$

We note that $\log(M_E/M_F)$ is less than -1.38, -28.78 for $n = 100$, 1000, respectively. □

Thus for large n the list of prime ideals needed to establish ideal equivalence is much shorter than the one actually needed for the computation of the class group.

Exercises

1. Let $\mathfrak{a}$ be a fractional ideal with prime ideal decomposition $\mathfrak{a} = \prod_{j=1}^k \mathfrak{p}_j^{m_j}$ ($m_j \in \mathbb{Z}$) and assume $\mathrm{N}(\mathfrak{p}_j) = p^{f_j}$ ($1 \leqslant j \leqslant k$). Determine $m \in \mathbb{N}$ (as small as possible) such that $m\mathfrak{a}$ is an integral ideal.
2. Let $\mathfrak{a} = \mathbb{Z}\alpha_1 + \cdots + \mathbb{Z}\alpha_n$ be a non-zero prime ideal of o_F and $\omega_1, \ldots, \omega_n$ an integral basis such that $\alpha_i = \sum_{j=1}^n a_{ij}\omega_j$ ($1 \leqslant i \leqslant n$). Develop an algorithm which determines for $\beta \in F^\times$ the exact power of $\mathfrak{a}$ dividing βo_F.
3. Compute h_F and Cl_F for $F = \mathbb{Q}(m^{\frac{1}{2}})$, $m \in \{-21, -14, 79, 223\}$.
4. Compute h_F and Cl_F for $F = \mathbb{Q}(m^{\frac{1}{3}})$, $m \in \{3, 5, 6, 7\}$.

7

Recent developments

7.1. Introduction

Since the first printing of this book in 1989 algorithmic algebraic number theory has attracted rapidly increasing interest. This is documented, for example, by a regular meeting, ANTS (algebraic number theory symposium), every two years whose proceedings [1], [2] give a good survey about ongoing research. Also there are several computer algebra packages concentrating on number theoretical computations. At present the most prominent ones, which are available for free, are KANT [30], PARI [41] and SIMATH [47]. KANT comes with a data base for algebraic number fields, already containing more than a million fields of small degree. KANT is developed by the research group of the author at Berlin and will be presented in some detail in this chapter. We note that almost all of KANT and PARI is also contained in the MAGMA system [39].

In the sequel we shortly discuss the improvements which were obtained for the computation of the important invariants of algebraic number fields. On the other hand, in computational algebraic number theory the interest has gradually turned from absolute extensions to relative extensions of number fields and we will sketch the important developments in that area. If subfields exist, the information about the invariants of those subfields can be lifted and used in the field under consideration. This relative point of view permits computations in fields of much larger degrees and has important applications, for example to class field computations. We were able to compute integral bases of fields of degree beyond 1000 and to calculate Hilbert class fields of degree over 100 for base fields of degree 4.

As before, we place special emphasis on algorithms for fast calculations. Readers who are mainly interested in complexity aspects of the problems are referred to the survey article by H. W. Lenstra [36].

7.2. Galois groups

In the sequel

$$f(t) = t^n + a_1 t^{n-1} + \cdots + a_n \in \mathbb{Z}[t] \tag{2.1}$$

denotes an irreducible polynomial. Its Galois group $\Gamma := \mathrm{Gal}(f)$ is considered as a subgroup of $\mathfrak{S}_n$, the symmetric group of n letters as in chapter 2. We note that the assumption of f being irreducible is in this section only made because it greatly simplifies the presentation.

Usually, in a first step Γ is approximated from below. Decomposing the polynomial $f(t)$ into its prime factors modulo suitably chosen prime numbers tells us that Γ contains permutations of certain cycle types (compare chapter 2). This excludes those transitive subgroups of $\mathfrak{S}_n$ which are too small. In general one would expect that one of the minimal remaining transitive subgroups of $\mathfrak{S}_n$ coincides with Γ. We remark, however, that the non-occurrence of a special cycle type may also hinge on the fact that sufficiently many primes p were not yet considered.

In the second step Γ is then approximated from above which includes the actual verification. We can exclude large subgroups of $\mathfrak{S}_n$ as candidates by an attempt to construct the splitting field of f over $\mathbb{Q}$. We set $F_0 = \mathbb{Q}$, $f_0(t) = f(t) \in \mathbb{Q}[t]$, $i = 0$ and then inductively $F_{i+1} = F_i(\rho_{i+1})$ for a root ρ_{i+1} of $f_i(t)$, $f_{i+1}(t) = f_i(t)/(t-\rho_{i+1}) \in F_{i+1}[t]$. We note that $f_{i+1}(t)$ need not be irreducible in $F_{i+1}[t]$. Consequently, we would rather work with a zero ρ_{i+1} of its largest irreducible factor. As factorization methods over number fields have not been very efficient so far (at least not current implementations), this method is bound to fail in practice for $[F_{i+1}:\mathbb{Q}] > 200$. On the other hand, it can be easily applied to exhibit G_{168} as the Galois group of $t^7 - 7t + 3$, for example. Carrying out a few steps will usually produce a smaller upper bound for Γ.

Eventually, we arrive at a situation, in which we know transitive subgroups $G < H < \mathfrak{S}_n$ with Γ contained in H (with small index), and we need to decide whether Γ is contained in a conjugate subgroup of G. This decision is possible through the use of appropriate resolvent polynomials, so-called **indicator functions** (see chapter 2). The construction of appropriate indicator functions is still a cumbersome task. Up to $n = 7$ there are the tables of Stauduhar [49] (see the appendix: Numerical tables). For $8 \leqslant n \leqslant 11$ they have been recently calculated by Eichenlaub and Olivier [21] and for $n = 12$ by K. Geißler [57]. Difficulties in further developments are due mainly to the huge number of transitive subgroups in higher degrees. We note that one needs to store coset representatives and an indicator function for each pair of transitive subgroups (G, H) of $\mathfrak{S}_n$, where H is minimal with respect to properly containing G.

7.3. Integral basis

Let F be an algebraic number field of degree n over $\mathbb{Q}$. We assume that F is presented in the form

$$F = \mathbb{Q}(\rho) \cong \mathbb{Q}[t]/f(t)\,\mathbb{Q}[t], \tag{3.1}$$

with, say, $\rho = \rho^{(1)}$ a root of the polynomial $f(t)$ of (3.1). The ring of integers of F, denoted o_F, has a $\mathbb{Z}$-basis $\omega_1, \ldots, \omega_n$ (**integral basis** of F).

The following complexity result on computing integral bases was derived by Chistov [12] in 1989:

Theorem (3.2)

The computation of an integral basis of an algebraic number field is polynomial time equivalent to the calculation of the largest square factor of a given positive integer.

We note that there is no known 'good' algorithm for exhibiting that square factor.

The proof of Chistov's theorem is as follows. One direction is immediately clear if we consider the computation of an integral basis of $\mathbb{Q}(\sqrt{m})$. The other part is obtained from a careful analysis of the Round-2-Algorithm of Zassenhaus. In chapter 4 following Lemma (5.53) we remarked on this algorithm that it consumes more time and storage than the embedding algorithm presented in section 4.6 (the latter usually being referred to as Round-4). For extensions of small degree n this is not correct, however. We recall that o_F is computed by merging all so-called p-maximal overorders of the equation order of $f(t)$ for the prime numbers p whose square divides the discriminant of f. If n is less than, say, 10 and the index of the equation order in its p-maximal overorder is a small p-power, then the Round-2-Algorithm will be preferable. Its basics are already presented following the proof of Lemma 4.(5.53). For details we refer to [42] and [13]. Cohen's book contains the nice observation that for each prime number p the Dedekind test already yields the multiplicator ring of the arithmetic radical under consideration, hence the first overorder of the equation order constructed with Round-2.

At present, work concentrates on generalizations of these ideas to relative extensions of number fields. For example, a relative version of the Round-2-Algorithm was developed by C. Friedrichs [58].

7.4. Unit group and class group

Since both of these tasks are closely related, a joint treatment seems to be appropriate. We restrict ourselves to the maximal order o_F in the sequel. It should be noted, however, that almost all ideas presented can also be applied to non-maximal orders R if one excludes those prime ideals of R containing $(o_F : R)$.

All known methods for computing fundamental units are based on ideas already used in our proof of Dirichlet's Theorem in chapter 5.2. Roughly speaking, a sequence of elements of bounded norm in o_F is determined. Within the sequence, there must be elements of the same norm. Among these elements, there are pairs (α, β) such that $\frac{\alpha}{\beta} \in U_F$. Appropriately choosing r such sequences, units which generate a subgroup U of U_F of finite index are obtained. U_F is then computed by one of the procedures described below. The explicit construction of those sequences is based on the LLL-algorithm which allows us to carry out the original ideas (of the proof of Dirichlet's Theorem) in a computationally satisfactory way. This approach was initially carried out by Buchmann and Pethö [9], and later on generalized and refined by the author [43].

One of the weak parts of that method is the management of the calculated elements. In numerical examples, it is usually much faster to detect products of powers of several elements which are units than to obtain units from quotients of appropriate pairs of elements. A first approach which takes advantage of this observation was given by Zassenhaus and the author for quadratic fields in [44]. In this approach, the principal ideals generated by the considered elements are factored into a product of prime ideals and then arithmetic is performed on the exponent vectors. Clearly, any non-trivial linear combination of zero yields a unit.

The latter idea of doing linear algebra on exponent vectors also plays a key role in algorithms for class group computations (compare chapter 6). We choose a set $\mathcal{S}$ of prime ideals in a way which guarantees

$$Cl_F = \langle \mathfrak{p} H_F \mid \mathfrak{p} \in \mathcal{S} \rangle. \tag{4.1}$$

For example, it is sufficient to consider all prime ideals with norm below an appropriate bound, for example the Minkowski bound. Since Minkowski's bound is too large in general, we usually choose $\mathcal{S}$ much smaller. We either assume the correctness of the **Generalized Riemann Hypothesis** (GRH), which implies [4]

$$Cl_F = \langle \mathfrak{p} H_F \mid \mathrm{N}(\mathfrak{p}) \leqslant 12(\log |d_F|)^2 \rangle, \tag{4.2}$$

or simply let $\mathcal{S}$ consist of several (up to several hundred or even thousand, if appropriate from numerical experience) prime ideals such that

$$\mathcal{S} = \{\mathfrak{p} \mid \mathrm{N}(\mathfrak{p}) \leqslant B\} \tag{4.3}$$

for a suitable bound B. In the latter case, the prime ideals $\mathfrak{p}$ with $B < \mathrm{N}(\mathfrak{p}) \leqslant M_F$ are dealt with at a later stage. The set $\mathcal{S}$ is usually called a **factor basis**.

Then we search for elements of the group

$$U(\mathcal{S}) := \{x \in F^\times \mid v_\mathfrak{p}(x) = 0 \,\forall\, \mathfrak{p} \notin \mathcal{S}\}, \tag{4.4}$$

so-called $\mathcal{S}$**-units**. In order to apply methods from the geometry of numbers, we need to change the multiplicative structure of $U(\mathcal{S})$ to an additive one. This can be done by two different well-known procedures. Let $\mathcal{S} = \{\mathfrak{p}_1, \dots, \mathfrak{p}_v\}$ and $x \in U(\mathcal{S})$. We consider the two mappings

$$\varphi_1 : U(\mathcal{S}) \to \mathbb{Z}^v : x \mapsto (v_{\mathfrak{p}1}(x), \dots, v_{\mathfrak{p}_v}(x)) \tag{4.5}$$

and

$$\varphi_2 : U(\mathcal{S}) \to \mathbb{R}^r : x \mapsto (c_1 \log |x^{(1)}|, \dots, c_r \log |x^{(r)}|)$$

$$\text{with } c_j = \begin{cases} 1 & \text{for } j \leqslant r_1 \\ 2 & \text{otherwise.} \end{cases} \tag{4.6}$$

To collect the information of the finite and infinite primes, we combine both mappings, φ_1 and φ_2, to

$$\Phi : U(\mathcal{S}) \to \mathbb{Z}^v \times \mathbb{R}^r : x \mapsto (\varphi_1(x), \varphi_2(x)). \tag{4.7}$$

Obviously, $\Phi(U(\mathcal{S}))$ is a lattice in Euclidean $(v+r)$-space. In the special case of $\mathcal{S}$ satisfying (4.1), that lattice contains complete information about the unit group U_F and the class group Cl_F of F. In that case, it is not difficult to see from our results in chapters 5 and 6 that the determinant of that lattice is

$$d(\Phi(U(\mathcal{S}))) = h_F \operatorname{Reg}_F, \tag{4.8}$$

the product of the class number and of the regulator of F.

So far our investigations are of a theoretical nature in that we do not know h_F, Reg_F, nor $U(\mathcal{S})$. Even though $U(\mathcal{S})$ is finitely generated, no efficient methods for establishing a system of generators are known. Hence, these considerations are mainly lacking due to incomplete information about $U(\mathcal{S})$. This difficulty can be overcome in two ways, the first one assuming GRH.

1. Method I (Assuming GRH)

Computing the Euler product for the quotient of the Dedekind zeta function of F and the Riemann zeta function up to a sufficient precision

$$g(x) := \prod_{\substack{p \in \mathbb{P} \\ p \leqslant x}} (1 - p^{-1}) \prod_{\substack{\mathfrak{p} \in \mathbb{P}_F \\ \mathfrak{p} \ni p}} (1 - \mathrm{N}(\mathfrak{p})^{-1})^{-1}, \tag{4.9}$$

where $\mathbb{P}_F$ denotes the set of all prime ideal of o_F, the residue of the Dedekind zeta function at $s = 1$ and hence the product $h_F \operatorname{Reg}_F$ can be estimated. All we need is a constant $C \in \mathbb{R}^{>0}$ satisfying

$$C \leqslant h_F \operatorname{Reg}_F < 2C. \tag{4.10}$$

A lower bound for x which guarantees such a result can be derived analogous to Buchmann and Williams in [10].

With a fixed $C \in \mathbb{R}^{>0}$ satisfying (4.10), we compute elements α of $U(\mathcal{S})$ as well as their images $\Phi(\alpha)$ until the set of image vectors $\mathcal{M}$ contains $v+r$ linearly independent elements. This implies that the elements of $\mathcal{M}$ span a sublattice of $\Phi(U(\mathcal{S}))$ of finite index. By Hermite normal form techniques, we determine $v+r$ vectors in span($\mathcal{M}$) generating span($\mathcal{M}$) along with the absolute value D of the determinant of the matrix whose rows are these $v+r$ vectors. (We note that the class group matrix of chapter 6 is part of this matrix.) D is an integral multiple of $h_F \operatorname{Reg}_F$. If $D < 2C$, those vectors span $\Phi(U(\mathcal{S}))$ and, thus, h_F and Reg_F can be read off from the Hermite normal form of the matrix under consideration. If $D \geqslant 2C$, we need to add the images of further $\mathcal{S}$-units to enlarge the sublattice of $\Phi(U(\mathcal{S}))$ already constructed. Though the determination of $\mathcal{S}$-units was already discussed in chapter 6, this subject will be taken up again later.

2. Method II (Unconditional)

The additional costs for not assuming GRH are substantial. In that case our incomplete information on $U(\mathcal{S})$ requires either

the knowledge of a full system of independent units of F combined with a method for solving norm equations

or

the knowledge of fundamental units of F.

Both approaches are carried out in detail in chapters 5 and 6. Hence, we only describe several improvements found in the last few years.

For example, we need the potential of taking a p-th root of a **relation** ($\mathcal{S}$-unit)

$$\alpha o_F = \prod_{i=1}^{v} \mathfrak{p}_i^{m_i} \quad (m_i \in \mathbb{Z}), \tag{4.11}$$

in case it exists in F. If trivial checks (like $|N(\alpha)|$ has to be a p-th power) seem to indicate that some element $\alpha \prod_{k=0}^{r} E_k^{v_k}$ $(0 \leqslant v_k < p)$, where $E_0, \dots, E_r$ denote generators of U_F, is indeed a p-th power, the p-th root can be obtained by a simple generalization of the method for computing U_F from a subgroup U of finite index described in chapter 5.7. This was observed by J. von Schmettow in his thesis [45].

Also, we will explain in greater detail how to construct $\mathcal{S}$-units α. More precisely, an additional $\mathcal{S}$-unit α should be computable for which $\Phi(\alpha)$ is either linearly independent from the subspace of $\mathbb{R}^{v+r}$ generated by the Φ-images of all previously calculated $\mathcal{S}$-units or is at least not contained in the lattice spanned by them. Besides the methods of chapter 6 we present several new ideas for generating $\mathcal{S}$-units.

(i) Computing a power of an ideal $\mathbf{a}$ which is necessarily principal.
This procedure consists of constructing sequences $\{\alpha_n\}, \{\beta_n\}$ of elements of o_F and of integral ideals $\{\mathbf{a}_n\}, \{\mathbf{b}_n\}$ subject to the following conditions:

(i) $\mathbf{a}_1 = \mathbf{a}$,
(ii) α_n is the first basis element of a LLL-reduced basis of $\mathbf{a}\mathbf{a}_n, \mathbf{b}_n = \alpha_n(\mathbf{a}\mathbf{a}_n)^{-1}$,
(iii) β_n is the first basis element of a LLL-reduced basis of $\mathbf{b}_n, \mathbf{a}_{n+1} = \beta_n(\mathbf{b}_n)^{-1}$.

The elements and ideals constructed in this way have the following properties.

Lemma (4.12)

(i) *The norms of the ideals* $\mathbf{a}_n$ *and* $\mathbf{b}_n$ *are bounded.*
(ii) *The ideals* $\mathbf{a}_n \mathbf{a}^{-n}$ *and* $\mathbf{a}^{n+1}\mathbf{b}_n$ *are principal.*
(iii) *There exist* $n_1, n_2 \in \mathbb{Z}^{>0}, n_2 > n_1$, *satisfying*

$$\mathbf{a}^{n_2-n_1} = \prod_{i=n_1}^{n_2-1} \frac{\alpha_i}{\beta_i} o_F. \tag{4.13}$$

Although Zassenhaus had already considered reducing bases of a lattice and its reciprocal alternatively, the construction presented here was found by J. von Schmettow in his thesis [45].

(ii) Applying ideal reduction.
For quadratic fields, Hafner and McCurley obtained a subexponential algorithm for computing class groups by using ideal reduction [26]. A generalization of these concepts to algebraic number fields of higher degree was given by J. Buchmann [8].
We recall that F is equipped with the usual scalar product

$$\langle\ ,\ \rangle : F \times F \to \mathbb{R}^{\geq 0} : (x, y) \mapsto \sum_{j=1}^{n} x^{(j)}\overline{y^{(j)}}, \tag{4.14}$$

where $\bar{\ }$ denotes complex conjugation. Let $\mathbf{a}$ be an ideal of F which is not contained in mo_F for any $m \in \mathbb{Z}^{\geq 2}$. Set $m_{\mathbf{a}} := \min \mathbf{a} \cap \mathbb{Z}^{>0}$. Then $\mathbf{a}$ is called **reduced** if the only element $\beta \in \mathbf{a}$ satisfying $|\beta^{(j)}| < m_{\mathbf{a}}$ $(1 \leq j \leq n)$ is 0. For an arbitrary ideal $\mathbf{a}$ of F, we obtain a corresponding reduced ideal $\mathbf{a}' = \frac{a}{\alpha}\mathbf{a}$ in $\mathbf{a}H_F$ by computing $\alpha_{\mathbf{a}} \in \mathbf{a}$ subject to $\langle \alpha_{\mathbf{a}}, \alpha_{\mathbf{a}} \rangle = \min\{\langle \beta, \beta \rangle \mid \beta \in \mathbf{a}, \beta \neq 0\}$ and $a \in \mathbb{Z}^{\geq 1}$ minimal subject to $\frac{a}{\alpha}\mathbf{a} \subseteq o_F$. To generate relations, an exponent vector $(m_1, \ldots, m_v) \in \mathbb{Z}^v$ (or $(\mathbb{Z}^{\geq 0})^v$) is chosen randomly, and then the ideal $\mathbf{a} = \prod_{i=1}^{v} \mathbf{p}_i^{m_i}$ is reduced to an integral ideal $\mathbf{a}'$ in the ideal class of $\mathbf{a}$. If $\mathbf{a}'$ is again a product of

$\mathfrak{p}_1, \ldots, \mathfrak{p}_v$, then the prime ideal factorization of $\mathbf{a}\mathbf{a}'^{-1}$ is an $\mathscr{S}$-unit. In practice this procedure will be speeded up by choosing exponent vectors with small entries. Instead of computing a shortest element of $\mathbf{a}$ for the reduction process, we can check, for the first element β of a LLL-reduced basis, whether $\beta^{-1}\mathbf{a}$ splits over the factor basis. This second possibility is especially important for fields of degree $n > 10$.

The idea of computing a LLL-reduced basis, say $\beta_1, \ldots, \beta_n$, of an ideal $\mathbf{a}$ and subsequently checking whether $\beta_i^{-1}\mathbf{a}$ splits over the factor basis is also an excellent tool in the case that the initially chosen set $\mathscr{S}$ of prime ideals is not guaranteed to satisfy (4.1) but only (4.3). Then we compute Cl_F as before with high probability. For all prime ideal $\mathfrak{p}$ subject to $B < \mathrm{N}(\mathfrak{p}) < M_F$ we then try to factor $\beta_i^{-1}\mathfrak{p}$ for LLL-reduced basis elements β_i of $\mathfrak{p}$ over $\mathscr{S}$. If we succeed, we actually prove that the computed group is indeed Cl_F. We note that this method should be carried out with increasing value of $\mathrm{N}(\mathfrak{p})$, thus increasing $\mathscr{S}$ after each successful test. Those rare prime ideals which fail the test need to be treated separately. To such ideals we may apply a few computational tricks which are similar to those for handling exceptional primes in sieve factoring methods.

For complexity results we refer to the paper of Lenstra [36] and the recent thesis by Ch. Thiel [50] which contains the first rigorous complexity analysis of an algorithm for computing a system of fundamental units.

7.5. Examples and applications

The progress in the computation of invariants of algebraic number fields during the last seven years has been tremendous. Before 1989, hardly any computations of invariants of fields with degree beyond 10 were known. Now fields of degree 20 or more (units, class groups) are no longer exceptional, the major obstacle being Minkowski's unrealistically large bound. Assuming GRH, fields of higher degrees and not too large discriminants are certainly attackable [29].

For integral bases computations, the critical part is the factorization of the polynomial discriminant. Hence, it is no surprise that Böffgen [6] successfully treated trinomials of degree beyond 100. Recently, M. Daberkow [15] managed to compute an integral basis of a radical extension of degree $n = 1332$.

Implementations of the presented algorithms in the software package KANT [30] were used to compute extensive tables of integral bases, units, and class groups. We only mention that we obtained results for

- all quartic fields with absolute discriminant less than one million;

- all quintic fields with absolute discriminant less than 2×10^7 (totally real fields), respectively 5×10^6 (other signatures);
- miscellaneous fields of degree between 6 and 9.

In total, we computed data for more than one million algebraic number fields. They are available in the form of a data base system on a public server [51] which can be accessed from the KANT shell (compare section 7).

There is a wide area of potential applications for the developed methods and computed data. In the remainder of this section, we give a short overview on some of the most important ones.

A natural application is certainly to nonlinear diophantine problems. The calculation of the solutions of a nonlinear diophantine equation is carried out, for example, by factoring that equation in an appropriate algebraic number field. Important examples are norm equations [59], norm form equations, index form equations and discriminant form equations. Of special interest among norm form equations are Thue- and Thue-Mahler-equations. For example, it has been shown in [25] that the resolution of index form equations in a quartic number field F can be reduced to the computation of all solutions of a finite number of Thue equations. This can be considered as a major step forward since formerly known methods for the resolution of an index form equation require operations in the Galois closure of the original field, hence in an extension of degree up to 24 in this case, whereas the Thue equations can be solved in F. Algorithms for the resolution of Thue equations (and even Thue inequalities) were much improved by the ideas of Bilu [5]. Recently, K. Wildanger [60] developed efficient methods for solving unit equations and computing all integral points on Mordell curves $Y^2 = X^3 + k$. They are now sufficiently advanced and stable to handle field degrees n up to twenty. Implementations are available in KANT [30].

Another application which made news in the last few years is in the so-called **number field sieve** (NFS) for factoring large integers M. In 1989, Pollard came up with the idea to use a ring homomorphism $\varphi: \mathbb{Z}[\rho] \to \mathbb{Z}/M\mathbb{Z}: \rho \mapsto m$, for suitable $m \in \mathbb{Z}^{>1}$, $\rho \in \mathbb{C}$ such that ρ is a zero of a monic polynomial $f(t) \in \mathbb{Z}[t]$ of degree d satisfying $f(m) \equiv 0 \bmod M$. This has the advantage that the factor basis used in factorization methods, such as the **quadratic sieve**, is split into two parts, ideals of degree one and their norms. This reduction of the size of the factor basis makes NFS the factoring algorithm with the best expected running time. Major difficulties of that method are

- finding an appropriate polynomial $f(t)$ for arbitrarily large numbers; for numbers M of 120 digits, the degree d of $f(t)$ should be 4 and then each of the coefficients of $f(t)$ will be of size about 30 digits, which makes computations in the corresponding ring $\mathbb{Z}[\rho]$ extremely slow (Pollard initially treated only special M, such as Fermat numbers, for which $f(t)$ has small coefficients).

- calculating numbers $a+b\rho \in \mathbb{Z}[\rho]$ which are power products of prime ideals of the factor basis; because these elements belong to a non-full module, their number is finite; in the terminology of section 4, we would need a method for computing $\mathscr{S}$-units in a restricted module; no algorithm for solving this task is known.

These two obstacles are essentially the reason that factorizations of arbitrarily large numbers by the number field sieve are not yet faster than by the quadratic sieve and variants of the latter. (The better expected running time concerns numbers which are still larger than the ones factored now.) For a survey of this subject we recommend [37].

7.6. Relative extensions I: Kummer extensions

In this section we mainly consider special relative extensions, so-called Kummer extensions. These are finite abelian extensions E of an algebraic number field F, say of relative degree n, in case F contains the p-th roots of unity for every prime number p dividing n. For these extensions there is a rich theory in the literature and it is not surprising that many of the theoretical results can be used for developing powerful algorithms for actual computations. Especially, we discuss the calculation of generators of the ring o_E of integers of E as an o_F-module and of relative integral bases in case they exist. This includes the calculation of relative normal forms for modules/ideals. Also we sketch methods for the detection of subfields and for their embedding. The algorithms for these tasks are useful for various applications. The probably most important will be considered in the subsequent section: the arithmetic construction of Hilbert class fields.

The tasks for general Kummer extensions are easily reduced to those for Kummer extensions of prime degree. The following theorem is well known.

Theorem (6.1)

Let k be a field and n a positive integer such that k contains a primitive n-th root of unity. For any subgroup B of $k^\times$ containing $k^{\times n}$ we set $K_B := k(B^{1/n})$. This yields a natural bijection between the subgroups of $k^\times$ containing $k^{\times n}$ and the abelian extensions of k of exponent n.

Since these field extensions are abelian and since we are only interested in finite extensions we can restrict ourselves to the case that $n \in \mathbb{P}$ is a prime number and k contains a primitive p-th root of unity in the sequel.

We consider the following situation. Let F be an algebraic number field of degree m over $\mathbb{Q}$. Let p be a prime number such that F contains a primitive p-th root of unity. Then any abelian extension E of F of relative degree p is obtained in the form $E = F(\sqrt[p]{\mu})$ with an element μ of F which is not a p-th power of an element of F. In this section we discuss the computation of relative integral bases of E in case they exist.

Probably the first result in this direction from a computational point of view is due to J. Sommer [48] in 1907. He treats the simplest case $m = p = 2$, i.e. a relative quadratic extension E of a quadratic number field F. In case F has class number one he gives explicitly a basis of the free o_F-module o_E. If the class number of F is larger than one, he gives explicit generators of o_E as an o_F-module as well as a $\mathbb{Z}$ basis of o_E.

Almost 90 years later methods for the computation of a relative integral basis (a system of generators) for o_E as an o_F-module for $p = 2$ but arbitrary m were developed in [16]. We note that in case $p = 2$ every base field F contains a primitive p-th root of unity. Eventually, M. Daberkow generalized and refined these methods in his thesis [15] to obtain a system of generators for o_E over o_F for arbitrary prime numbers p. Those results form the core part of this section.

In the sequel we assume that E is generated over F by the p-th root of an element $\mu \in o_F$. Since we can always multiply (or divide) μ by p-th powers of elements of o_F we can stipulate without loss of generality, that μ is not contained in the p-th power of any prime ideal of o_F which lies over po_F:

$$v_{\mathfrak{p}}(\mu) \in \{0, \dots, p-1\} \tag{6.2}$$

for all prime ideals $\mathfrak{p}$ of o_F with $v_{\mathfrak{p}}(p) > 0$.

For a description of o_E as an o_F-module we make use of the **relative discriminant** $\mathfrak{d}_{E/F}$ of the extension E/F. (This is exactly the discriminant ideal $\mathfrak{D}(o_E/o_F)$ introduced in Definition 4 (5.46).) Especially we need to know all prime ideals dividing $\mathfrak{d}_{E/F}$. For Kummer extensions of prime degree this is no problem because of the next theorem [28].

Theorem (6.3)

The non-zero prime ideals $\mathfrak{q}$ of o_F decompose in E in the following way.
If p does not divide $v_{\mathfrak{q}}(\mu)$, then $\mathfrak{q}$ is totally ramified in E:

$$\mathfrak{q}o_E = \mathfrak{Q}^p.$$

If p divides $v_{\mathfrak{q}}(\mu)$ there are 2 possibilities:

(i) *For $p \notin \mathfrak{q}$ the decomposition of $\mathfrak{q}$ depends on the solvability of the congruence $x^p \equiv \mu \bmod \mathfrak{q}$ in o_F.*
If it is solvable, then $\mathfrak{q}$ splits completely: $\mathfrak{q}o_E = \mathfrak{Q}_1 \cdot \ldots \cdot \mathfrak{Q}_p$.
If it is not solvable, then $\mathfrak{q}$ remains inert: $\mathfrak{q}o_E = \mathfrak{Q}$.

(ii) *For $p \in \mathfrak{q}$ we set $e = v_{\mathfrak{q}}(p)/(p-1)$. There are 3 possibilities:*

(a) $\mathfrak{q}o_E = \mathfrak{Q}_1 \cdot \ldots \cdot \mathfrak{Q}_p$,
if the congruence $x^p \equiv \mu \bmod \mathfrak{q}^{ep+1}$ *is solvable;*

(b) $\mathfrak{q}o_E = \mathfrak{Q}$,
if the congruence $x^p \equiv \mu \bmod \mathfrak{q}^m$ *is solvable for* $m = ep$, *but not for* $m = ep+1$;

(c) $\mathfrak{q}o_E = \mathfrak{Q}^p$,
if the congruence $x^p \equiv \mu \bmod \mathfrak{q}^{ep}$ *is not solvable in* o_F.

We need to know the relative discriminant explicitly, hence we compute $k_\mathfrak{q} = v_\mathfrak{q}(\mathfrak{d}_{E/F})$ for all non-zero prime ideals $\mathfrak{q}$ of o_F. We note that $k_\mathfrak{q} = 0$ for all unramified prime ideals. For the ramified prime ideals we have $v_\mathfrak{q}(p) + v_\mathfrak{q}(\mu) > 0$, hence, the following lemma.

Lemma (6.4)

Let $\mathfrak{q}$ *be a non-zero prime ideal of* o_F *which ramifies in E. The exponent* $k_\mathfrak{q}$ *of* $\mathfrak{q}$ *in the prime ideal decomposition of* $\mathfrak{d}_{E/F}$ *has the value:*

(1) $k_\mathfrak{q} = p - 1 + p v_\mathfrak{q}(p)$ *in case* $v_\mathfrak{q}(\mu) > 0$;

(2) $k_\mathfrak{q} = (p-1)(ep - m + 1)$ *for* $v_\mathfrak{q}(\mu) = 0$ *and*
$m = \max\{1 \leqslant k < pe \mid \exists x \in o_F\colon x^p \equiv \mu \bmod \mathfrak{q}^k\}$.

Algorithms for the computation of $k_\mathfrak{q}$ can be found in [15].

Generators for o_E over o_F are obtained by local methods. We need to introduce a few more constants. From the preceding lemma we know that $p^p \mu^{p-1} \mathfrak{d}_{E/F}^{-1}$ is the square of an integral ideal which we denote by $\mathfrak{a}$.

Remark

We note that o_E has a relative integral basis over o_F precisely if $\mathfrak{a}$ is a principal ideal [3]. Methods for principal ideal testing were described in previous chapters.

We split $\mathfrak{a}$ into a product of two integral ideals Φ and $\tilde{\Phi}$, where $\tilde{\Phi}$ is the product of those prime ideals $\mathfrak{q}$ containing $\mathfrak{a}$ (counted with multiplicities) who satisfy $v_\mathfrak{q}(\mathfrak{d}_{E/F}) > 0$, i.e. Φ is not contained in any of those prime ideals $\mathfrak{q}$. For arbitrary non-zero prime ideals $\mathfrak{q}$ of o_F we set:

$$\kappa_\mathfrak{q} := v_\mathfrak{q}(\Phi)\frac{2}{p(p-1)} \in \mathbb{Z}^{\geqslant 0}, \quad e_\mathfrak{q} := v_\mathfrak{q}(p)/(p-1) \in \mathbb{Z}^{\geqslant 0},$$

and in case $v_\mathfrak{q}(\mathfrak{d}_{E/F}) > 0$ also

$$l(\mathfrak{q}) := \begin{cases} \max\{0 < k < pe_\mathfrak{q} \mid \exists x \in o_F\colon x^p \equiv \mu \bmod \mathfrak{q}^k\} & \text{for } v_\mathfrak{q}(\mu) = 0, v_\mathfrak{q}(p) > 0 \\ v_\mathfrak{q}(\mu) & \text{for } v_\mathfrak{q}(\mu) > 0, v_\mathfrak{q}(p) > 0. \\ v_\mathfrak{q}(\mu) - p\lfloor v_\mathfrak{q}(\mu)/p \rfloor & \text{for } v_\mathfrak{q}(\mu) > 0, v_\mathfrak{q}(p) = 0 \end{cases}$$

Then p and $l(\mathfrak{q})$ are coprime and there are rational integers $r = r(\mathfrak{q})$,

$s = s(\mathfrak{q}) \in \mathbb{Z}^{\geq 0}$ satisfying $rl(\mathfrak{q}) - sp = 1$. Next we compute $\gamma \in o_F$ with the following properties:

(1) $\gamma^p \equiv \mu \bmod \mathfrak{q}^{l(\mathfrak{q})}$ for all $\mathfrak{q} \in \mathbb{P}_F$ with $v_\mathfrak{q}(\mathfrak{d}_{E/F}) > 0$ and $v_\mathfrak{q}(\mu) = 0$,
(2) $\gamma^p \equiv \mu \bmod \mathfrak{q}^{e_\mathfrak{q} p}$ for all $\mathfrak{q} \in \mathbb{P}_F$ with $v_\mathfrak{q}(\mathfrak{d}_{E/F}) = 0$ and $v_\mathfrak{q}(p) > 0$,
(3) $\gamma \in \mathfrak{q}^{\frac{1}{p} v_\mathfrak{q}(\mu)}$ for all $\mathfrak{q} \in \mathbb{P}_F$ with $v_\mathfrak{q}(\mathfrak{d}_{E/F}) = 0$ and $v_\mathfrak{q}(\mu) > 0$.

These constants are now used to construct a (small) set Ω of o_F-generators of o_E. The set Ω is the union of a distinguished set Ω_1 containing generators for o_E with respect to the prime ideals not containing $\mathfrak{d}_{E/F}$ and sets $\Omega_\mathfrak{q}$ for each prime ideal $\mathfrak{q}$ containing the relative discriminant $\mathfrak{d}_{E/F}$. The first set Ω_1 contains $2p-1$ generators and each of the sets $\Omega_\mathfrak{q}$ contains $p-1$. We start with the set Ω_1. We choose a non-zero element δ in the ideal $\mathfrak{a} := \prod_{\mathfrak{q} \not\supseteq \mathfrak{d}_{E/F}} \mathfrak{q}^{\kappa_\mathfrak{q}}$ and obtain an integral ideal $\mathfrak{b} = \beta_1 o_F + \beta_2 o_F$ satisfying $\delta o_F = \mathfrak{b}\mathfrak{a}$. Setting

$$\rho_i := \beta_i \frac{\gamma - \sqrt[p]{\mu}}{\delta} \quad (i = 1, 2)$$

we get

$$\Omega_1 = \{1, \rho_1, \ldots, \rho_1^{p-1}, \rho_2, \ldots, \rho_2^{p-1}\} \subset o_E.$$

The other sets of generators $\Omega_\mathfrak{q}$ are of the form

$$\Omega_\mathfrak{q} = \{\rho_\mathfrak{q}, \ldots, \rho_\mathfrak{q}^{p-1}\}$$

for suitable integers $\rho_\mathfrak{q}$. They are constructed in the following way. For a fixed prime ideal $\mathfrak{q}$ containing the relative discriminant $\mathfrak{d}_{E/F}$ we choose:

(1) $\pi \in \mathfrak{q} \backslash \mathfrak{q}^2$ with corresponding integral ideal $\mathfrak{b} := \pi o_F \mathfrak{q}^{-1}$,
(2) $\beta_\mathfrak{q} \in \mathfrak{b}$ subject to $v_\mathfrak{q}(\beta_\mathfrak{q}) = 0$.

If $v_\mathfrak{q}(\mu)$ is zero, we obtain

$$\rho_\mathfrak{q} := \beta_\mathfrak{q}^{s(\mathfrak{q})} \frac{(\gamma - \sqrt[p]{\mu})^{r(\mathfrak{q})}}{\pi^{s(\mathfrak{q})}}.$$

However, if $v_\mathfrak{q}(\mu)$ is greater than zero, we set $k := s(\mathfrak{q})$ for $v_\mathfrak{q}(p) > 0$, $k := s(\mathfrak{q}) + r(\mathfrak{q})(v_\mathfrak{q}(\mu) - l(\mathfrak{q}))/p$ for $v_\mathfrak{q}(p) = 0$, respectively, and define

$$\rho_\mathfrak{q} := \left(\frac{\beta_\mathfrak{q}}{\pi}\right)^k \sqrt[p]{\mu}^{r(\mathfrak{q})}.$$

Theorem (6.5)

The set

$$\Omega := \{1, \rho_1, \ldots, \rho_1^{p-1}, \rho_2, \ldots, \rho_2^{p-1}\} \cup \{\rho_\mathfrak{q}, \ldots, \rho_\mathfrak{q}^{p-1} \mid \mathfrak{d}_{E/F} \subseteq \mathfrak{q}\}$$

is a system of o_F-generators for o_E.

The proof can be found in [15]. Since all elements are given explicitly and since the computation of the relative discriminant is straightforward we have indeed developed an algorithm for the computation of a set Ω of o_F-generators for o_E. Once we have calculated Ω we can apply a reduction procedure putting it into a normal form. This will be discussed in the sequel.

Remark

Choosing the elements π_q appropriately the reduction of new generators modulo the ones calculated before becomes very easy. Many of the new generators can be eliminated right away. In practice, we usually obtained a system of generators of at most $2p$ elements after this.

For a general reduction procedure we consider a generalization of the Hermite normal form concept for relative extensions of algebraic number fields which was developed in [7]. We note that another algorithm for this task is given by H. Cohen in [14]. Slightly more general than before we consider a finitely generated o_F-module M of full rank in E. The presented algorithm is based on the following theorem [40].

Theorem (6.6)

For the module M there are ideals $\mathfrak{a}_1, \ldots, \mathfrak{a}_n$ in o_F and algebraic numbers $\xi_1, \ldots, \xi_n$ in E such that

$$M = \mathfrak{a}_1 \xi_1 + \cdots + \mathfrak{a}_n \xi_n. \tag{6.7}$$

Algorithm (Computation of a relative normal form) (6.8)

Input: *An o_F-generating system $\eta_1, \ldots, \eta_k$ of M.*

Output: *Ideals $\mathfrak{a}_1, \ldots, \mathfrak{a}_n$ contained in o_F and algebraic numbers $\xi_1, \ldots, \xi_n$ in E satisfying* (6.7).

Init: *Find an F-basis $\omega_1, \ldots, \omega_n$ of E. Define $U_t := \omega_t F + \cdots + \omega_n F$, $U_{n+1} := \{0\}$ and $M_t := M \cap U_t$ for $1 \leqslant t \leqslant n$.*

Loop: *For $1 \leqslant t \leqslant n$ do:*

Step 1: *For $1 \leqslant i \leqslant k$ compute coefficients $\gamma_{i,j}^{(t)} \in F$ satisfying*

$$\eta_i = \sum_{j=1}^{t-1} \gamma_{i,j}^{(t)} \xi_j + \sum_{j=t}^{n} \gamma_{i,j}^{(t)} \omega_j.$$

Step 2: *Compute the ideal*

$$\mathfrak{a}_t := \gamma_{1,t}^{(t)} o_F + \cdots + \gamma_{k,t}^{(t)} o_F$$

and its inverse $\mathfrak{a}_t^{-1}$.

Step 3: *Determine $\mu_1, \mu_2 \in \mathfrak{a}_t$ and $\nu_1, \nu_2 \in \mathfrak{a}_t^{-1}$ such that*

$$\mu_1 \nu_1 + \mu_2 \nu_2 = 1.$$

Step 4: *Determine $h_i \in M_t$ and $u_i \in U_{t+1}$ such that $\mu_i \omega_t = h_i + u_i$ ($i = 1, 2$).*

Step 5: *Set $\xi_t := \omega_t - (\nu_1 u_1 + \nu_2 u_2) = \nu_1 h_1 + \nu_2 h_2$.*

End Loop:

Adjust: *For* $1 \leqslant t \leqslant n$ *calculate* $\alpha_t \in F \backslash \{0\}$ *such that* $\alpha_t \mathfrak{a}_t \subseteq o_F$ *and substitute* $\mathfrak{a}_t, \xi_t$ *by* $\alpha_t \mathfrak{a}_t, \alpha_t^{-1} \xi_t$.

A detailed discussion of the steps 1 to 5 is given in [7]. Similar to Hermite normal form algorithms, the computation of that representation suffers from large entries occurring in intermediate calculations. This can partially be avoided by adding a reduction procedure after step 5 (see [17]).

We still need to explain the calculation of a relative integral basis in case it exists. (A criterion for its existence was given in the remark following Lemma (6.4).) For this the normal form of Theorem (6.6) needs to be refined.

Corollary (6.9)

Under the premises of the preceding theorem there is an ideal $\mathfrak{a}$ *in* o_F *and there are algebraic numbers* $\eta_1, \ldots, \eta_n$ *satisfying*

$$M = o_F \eta_1 + \cdots + o_F \eta_{n-1} + \mathfrak{a} \eta_n.$$

We note that the proof given in [40] can be easily turned into a constructive one yielding an algorithm for actually computing $\mathfrak{a}$ and $\eta_1, \ldots, \eta_n$. Obviously, if $\mathfrak{a}$ is a principal ideal, we get a relative integral basis.

For an arbitrary pure extension $L = K(\sqrt[p]{\mu})$ we have to make the detour via the corresponding Kummer extension. This means that we calculate subsequently the extensions $F = K(\zeta_p)$ (ζ_p a primitive p-th root of unity), $E = F(\sqrt[p]{\mu})$ as well as their rings of integers and then obtain $o_L = L \cap o_E$ by means of linear algebra. We illustrate the whole method by an example.

Example

Let K be a pure extension of the rational number field generated by a root ρ of $f(t) = t^4 + 65$. We want to compute the ring of integers in $L = K(\sqrt[3]{\mu})$ for $\mu = 10$. We begin by computing $F = K(\zeta_3)$ of degree 2 over K. F is generated by a root δ of the polynomial

$$f(t) = t^8 + 4t^7 + 10t^6 + 16t^5 + 149t^4 + 276t^3 - 380t^2 - 516t + 4161.$$

An integral basis of F is given by $\omega_1, \ldots, \omega_8$ subject to

$$\begin{aligned}
\omega_i &= \delta^{i-1} \quad (1 \leqslant i \leqslant 6) \\
\omega_7 &= (222388 + 2108490\delta + 83920\delta^2 + 10490\delta^3 \\
&\quad + 139517\delta^4 + 3147\delta^5 + 1049\delta^6)/263299, \\
\omega_8 &= (197142 + 150164\delta + 130051\delta^2 + 181030\delta^3 \\
&\quad + 179522\delta^4 + 150802\delta^5 + 528\delta^6 + \delta^7)/263299.
\end{aligned}$$

Hence, the discriminant of F is 40035484569600000.

Next we consider the extension $E = F(\sqrt[3]{10})$ and construct a set Ω generating o_E as an o_F-module. We obtain the following nine elements:

$$\xi_i = \mu^{i-1} \quad (1 \leqslant i \leqslant 3),$$

$$\xi_4 = \frac{1}{3}((2\omega_2 + 2\omega_4 + \omega_5) + (\omega_2 + \omega_4 + \omega_5)\mu),$$

$$\xi_5 = \frac{1}{3}((2\omega_1 + 2\omega_3 + \omega_4 + \omega_5 + \omega_8) + (\omega_1 + \omega_2 + \omega_4 + 2\omega_6 + \omega_7 + \omega_8)\mu + (2\omega_2 + \omega_3 + \omega_4 + 2\omega_5 + \omega_6 + 2\omega_7 + \omega_8)\mu^2),$$

$$\xi_6 = \frac{1}{2}(\omega_1 + \omega_4 + \omega_5 + \omega_7)\mu,$$

$$\xi_7 = \frac{1}{2}(\omega_1 + \omega_3 + \omega_5)\mu^2,$$

$$\xi_8 = \frac{1}{5}(2\omega_1 + \omega_2 + 3\omega_3 + \omega_4 + 4\omega_5 + 3\omega_7)\mu,$$

$$\xi_9 = \frac{1}{5}(3\omega_2 + 3\omega_3 + 3\omega_4 + 3\omega_5 + 2\omega_7)\mu^2.$$

Finally, we intersect o_E with L and get an integral basis $\alpha_1, \ldots, \alpha_{12}$ of L:

$$\alpha_i = \rho^{i-1} \quad (1 \leqslant i \leqslant 4),$$

$$\alpha_i = \mu\rho^{i-5} \quad (5 \leqslant i \leqslant 7),$$

$$\alpha_8 = \frac{1}{10}(5 + 5\rho + 5\rho^2 + \rho^3)\mu,$$

$$\alpha_9 = \frac{1}{3}(1 + \mu + \mu^2),$$

$$\alpha_{10} = \frac{1}{3}(\rho + \rho\mu + \rho\mu^2),$$

$$\alpha_{11} = \frac{1}{30}((20 + 10\rho^2) + (20 + 10\rho^2)\mu + (5 + \rho^2)\mu^2),$$

$$\alpha_{12} = \frac{1}{30}((20\rho + 10\rho^3) + (15 + 5\rho + 15\rho^2 + \rho^3)\mu + (5\rho + \rho^3)\mu^2).$$

Knowing the integral basis of L the calculation of the field discriminant is immediate:

$$d_L = 28146547071811584000000000000.$$

Using this method integral bases of algebraic number fields E with $[E:\mathbb{Q}] > 1000$ were computed [15]. It was not surprising that the corresponding field discriminants had up to several thousand digits.

7.7. Relative extensions II: Hilbert class fields

The tools of the preceding section as well as powerful methods for computing unit groups and class groups in algebraic number fields (KANT, MAGMA, PARI) allow the purely arithmetic construction of class fields in case of moderate degree of the base field and small exponent of its (ray) class group. We exclusively concentrate on the computation of Hilbert class fields since this requires less machinery and therefore the presentation becomes simpler. We note that since the paper of Hasse [27] in 1964, in which he computes Hilbert class fields of imaginary quadratic fields of class numbers 3 and 5, the first progress about arithmetic class field computations was reported in [17] less than two years ago.

In the sequel we consider an algebraic number field k and the task of computing its Hilbert class field, i.e. the maximal abelian extension Γ of k which is unramified at all places. In case the order of the class group of k is a prime number p a solution is presented in a joint article with M. Daberkow [18]. Hence, it suffices to reduce the general task to the calculation of a sequence of class field extensions each of which is of prime order.

A *subgroup* H of the group I_k of all fractional ideals of k always consists of the ideals of I_k which are contained in the ideal classes of a subgroup $\bar{H}$ of the class group Cl_k of k. In general, we make no distinction between $\bar{H}$ and H in the sequel. As a first result we note the following lemma from [19] which is a consequence of the existence theorem of class field theory.

Lemma (7.1)

Let $Cl_k = H_1 \times \cdots \times H_n$ be the canonical decomposition of the class group Cl_k of k into subgroups H_i of prime power order. We denote the subgroup $H_1 \times \cdots \times H_{i-1} \times H_{i+1} \times \cdots \times H_n$ by $\tilde{H}_i$ for abbreviation $(1 \leqslant i \leqslant n)$. If Γ_i is the class field of k belonging to $\tilde{H}_i$ then the Hilbert class field of k is just $\Gamma = \Gamma_1 \cdot \ldots \cdot \Gamma_n$.

This reduces our task to the computation of class fields for those subgroups H of Cl_k for which the factor group Cl_k/H is cyclic of prime power order, say p^s. We consider the descending chain of subgroups $Cl_k = H_0 \supset H_1 \supset \cdots \supset H_s = H$ with $|H_{i-1}/H_i| = p$ and we denote by k_i the class field corresponding to H_i $(1 \leqslant i \leqslant s)$. The next lemma of [19] reduces our task to steps of relative degree p.

Lemma (7.2)

$N_{k_i/k}^{-1}(H_{i+1})$ is a subgroup of the class group of k_i and the corresponding class field of k_i is the class field of k with respect to H_{i+1}.

Example

Let k be the number field generated by a root ρ of the polynomial $f(t) = t^4 - 5t^2 + 196$. We compute the following invariants of k:

1. k is totally complex,
2. k has an integral basis $\omega_1 = 1$, $\omega_2 = \rho$, $\omega_3 = (\rho+\rho^2)/2$, $\omega_4 = (14+9\rho+\rho^3)/28$,
3. k has the discriminant $d_k = 576081$,
4. k has the regulator $R_k = 7.656$,
5. k has class number $h_k = 36$ and the class group $Cl_k = H_1 \times H_2 \times H_3$ with $H_i \simeq \mathbb{Z}/3\mathbb{Z}$ $(i = 1,2)$ and $H_3 \simeq \mathbb{Z}/4\mathbb{Z}$.

For the reduction to class fields of prime degree we set $\tilde{H}_1 = H_2 \times H_3$, $\tilde{H}_2 = H_1 \times H_3$, $\tilde{H}_3 = H_1 \times H_2$. Clearly, the class fields belonging to $\tilde{H}_i$ for $i = 1,2$ are already of prime degree $p = 3$. For the remaining subgroup $\tilde{H}_3$ we denote the unique subgroup of H_3 of order 2 by U_3. We determine the class field of k corresponding to the group $H_1 \times H_2 \times U_3$ in a first step and call it $\tilde{k}$. In a second step we calculate the class field of k for $\tilde{H}_3$ as class field of $\tilde{k}$ for the appropriate subgroup of index 2 in the class group of $\tilde{k}$ in accordance with the preceding lemma.

The construction of the Hilbert class field of k is therefore reduced to the computation of class fields corresponding to subgroups of prime index p in the class group of k, respectively extensions of k whose degree is a power of p. We note that all these fields are contained in the Hilbert class field and are therefore abelian over k and totally unramified at all places. This reduces the number of candidates, especially if k is totally real.

The arithmetic construction of abelian (totally unramified) p-extensions is well understood only in the case of Kummer extensions. The transfer from an arbitrary base field k to a field F containing the p-th roots of unity is via the next theorem from [35].

Theorem (7.3)

Let k, $H \subset Cl_k$ as above and F be a cyclic extension of k. We set $G = N_{F/k}^{-1}(H) \subseteq Cl_F$. If E is the class field of F corresponding to G, then the class field Γ_H of k corresponding to H is a subfield of E. In that case E is abelian over k, and E is the composite of F and Γ_H.

This puts us into a situation in which we can apply the methods of [18].

Example (cont.)

Let k be the totally complex quartic field of class number 36 discussed in the previous example. Using the reduction procedures described and the ideas from [18] the Hilbert class field Γ of k was computed with the KASH-system

[30] (see also section 7.9). We just give relative generators for Γ since for an absolute generation by a root of a polynomial of degree 144 the polynomial coefficients would require more space than the remainder of this article. We obtain $\Gamma = k(\alpha, \beta)$ for zeros

α of

$$\begin{aligned} t^9 &+ (-9+6\omega_4)\,t^8 - (119+12\omega_2+19\omega_4)\,t^7 + (902+82\omega_2-58\omega_3-380\omega_4)\,t^6 \\ &+ (3013+798\omega_2+213\omega_3+581\omega_4)\,t^5 \\ &- (19168+3580\omega_2-1453\omega_3-7746\omega_4)\,t^4 \\ &- (1702+2174\omega_2+5785\omega_3+12224\omega_4)\,t^3 \\ &+ (74948+31710\omega_2+5013\omega_3-21554\omega_4)\,t^2 \\ &- (126690+47540\omega_2-10700\omega_3-5180\omega_4)\,t \\ &+ (48475+9550\omega_2-10150\omega_3-19600\omega_4), \end{aligned}$$

β of

$$t^4 + (714+192\omega_2-192\omega_4)\,t^2 - (72-144\omega_4)\,t + (909-704\omega_2+704\omega_4).$$

7.8. Computation of subfields

The contents of this section treat arbitrary relative extensions, but the results are very useful for arithmetic class field computations via a detour over Kummer extensions as well. We consider the following task. Let $E = \mathbb{Q}(\rho)$ be a number field of degree mn for a root ρ of a monic irreducible polynomial $f(t) \in \mathbb{Z}[t]$ (generating polynomial for E). We want to determine all subfields F of E of fixed degree m over $\mathbb{Q}$. Each such subfield F is presented by a pair of polynomials (g, h) with $g(t) \in \mathbb{Z}[t]$ being a generating polynomial for F and $h(t) \in \mathbb{Q}[t]$ describes the embedding of F into E, i.e. $h(\rho) = \beta$ for a root β of $g(t)$. It is straightforward that there is a 1-1-correspondence between subfields of E and such pairs of polynomials $(g, h) \in \mathbb{Z}[t] \times \mathbb{Q}[t]$ satisfying $g(h(t)) \equiv 0 \bmod f(t)\,\mathbb{Z}[t]$.

Remark

Clearly, a subfield F of degree m exists only if there is an integer d_F of absolute value greater than one whose n-th power divides the discriminant d_E of E. Any $|d_F|$ satisfying this divisibility condition is a potential absolute discriminant of a subfield of degree m. If there are only a few candidates for d_F often a short search in available tables of number fields (essentially for $m \leqslant 6$) yields a short list of potential subfields and an embedding algorithm tests for all the fields F of that list, whether they are actually subfields of E.

In all other cases, however, the method developed in the sequel will be superior. The underlying idea is easily explained. Let Ω denote the set of all zeros $\rho_1, \ldots, \rho_{mn}$ of $f(t)$ in $\mathbb{C}$, say $\rho = \rho_1$. The Galois group G of $f(t)$ operates on

Ω by permuting the zeros. Any non-empty subset Δ of Ω with the property that for all $\pi \in G$ the image $\pi(\Delta)$ either coincides with Δ or is disjoint from Δ is called a block of G (in Ω). A set of blocks $\{\Delta_1, \ldots, \Delta_s\}$ is called a block system if it is invariant under G. Clearly, every block is contained in (precisely one) block system.

It is an immediate consequence of Galois theory that E contains a subfield F of degree m if and only if there is a block Δ of G of n elements, i.e. of blocksize n. Therefore we obtain a correspondence between subfields and block systems. Let us assume that $\{\Delta_1, \ldots, \Delta_m\}$ is a block system, that H is the subgroup of G fixing Δ_1, and that F is the subfield of E fixed by H. We set

$$\delta_i := \prod_{\gamma \in \Delta_i} \gamma \quad (1 \leqslant i \leqslant m).$$

Hence, $\delta_1 \in F$, and the δ_i are conjugates of δ_1. Therefore

$$g(t) := \prod_{i=1}^{m} (t - \delta_i)$$

is the characteristic polynomial of $\delta_1 \in F$. If $g(t)$ is irreducible, its root δ_1 generates F. Otherwise $g(t)$ is a power of an irreducible polynomial (of exponent greater than one), i.e. several δ_i coincide. In this case we can apply a simple Tschirnhaus transformation to ρ. Substituting $f(t)$ by $f(t-j)$ for a few rational integers j we are guaranteed to obtain a generator for F [20].

Hence, the computation of subfields is reduced to the calculation of appropriate blocks, respectively block systems. Of course, these considerations are purely theoretical so far, since we neither know the Galois group G nor its operation on Δ.

For practical applications Dixon [20] suggested making use of van der Waerden's criterion (see section 2.9.8). If a prime number p does not divide the discriminant of the polynomial $f(t)$ then $f(t)$ has a congruence factorization

$$f(t) \equiv \prod_{i=1}^{r} f_i(t) \bmod p\mathbb{Z}[t],$$

into a product of monic polynomials $f_i(t) \in \mathbb{Z}[t]$ which are irreducible and coprime modulo $p\mathbb{Z}[t]$. Then the Galois group G of $f(t)$ contains a permutation π of cycle type decomposition $\pi = (\pi_1) \ldots (\pi_r)$ with $n_i := |\pi_i| = deg(f_i)$ $(1 \leqslant i \leqslant r)$. We note that π generates a cyclic subgroup of G. We use this information to generate a set of possible candidates for blocks (block systems). These candidates are called potential blocks (potential block systems) in the sequel.

More precisely, we assume that we know $\pi \in G$ generating a cyclic subgroup of G. As above we denote the lengths of the cycles π_i of π by n_i $(1 \leqslant i \leqslant r)$. A subset Δ of Ω of n elements is called a potential block, if $\pi^j(\Delta) \cap \Delta \in \{\emptyset, \Delta\}$

for $(1 \leqslant j \leqslant |\langle\pi\rangle|)$. A set $\mathscr{S} = \{\Delta_1, \ldots, \Delta_m\}$ of potential blocks is called a potential block system, if $\Omega = \bigcup_{i=1}^{m} \Delta_i$, $\Delta_i \cap \Delta_j = \varnothing$ for $i \neq j$, and $\pi(\Delta_i) \in \mathscr{S}$ $(1 \leqslant i, j \leqslant m)$.

From [32] we excerpt two further important conditions on potential block systems which are used to decrease the number of candidates considerably. For $\Delta_i \in \mathscr{S}$ we call the smallest positive exponent k_i satisfying $\pi^{k_i}(\Delta_i) = \Delta_i$ the exponent of Δ_i (with respect to π).

1. If $\Delta_i \in \mathscr{S}$ contains an element $\rho_\mu \in \Omega$ subject to μ being contained in the cycle π_l of π, then Δ_i contains exactly n_l/k_i elements of Ω the indices of which belong to π_l.
2. If Δ_i, Δ_j for $i \neq j$ contain elements of Ω whose indices belong to the same cycle of π, then the exponents of Δ_i and Δ_j coincide and $\Delta_i = \pi^\nu(\Delta_j)$ for an integer ν satisfying $1 \leqslant \nu \leqslant k_i$.

With these remarks in mind the computation of a full set of candidates of potential block systems is essentially a combinatorial problem. Omitting details we present the following algorithm.

Algorithm (Computation of potential block systems) (8.1)

Input: *A generating polynomial $f(t) \in \mathbb{Z}[t]$ for E of degree mn and a permutation π of cycle type decomposition $\pi = (\pi_1) \ldots (\pi_r)$ with $|\pi_i| = n_i$ $(1 \leqslant i \leqslant r)$.*

Output: *A list of potential block systems corresponding to a subfield F of degree m.*

Step 1: *Set $k = 1$, $Z = \{(\pi_1), \ldots, (\pi_r)\}$, $u = r$.*

Step 2: *Determine all subsets B of $\{2, \ldots, u\}$ satisfying $nk - n_1 = \sum_{b \in B} n_b$ and $k \mid n_b \forall b \in B$. For each such subset B do:*

(i) *Remove the cycles belonging to B from Z.*

(ii) *If Z is then empty print the found potential block system. If Z is not empty, the algorithm needs to be applied recursively to the new set Z.*

Step 3: *For $k = n$ terminate. Else increase k to the next larger divisor of n and go to 2.*

The presentation of step 2 is kept somewhat vague, a precise version (see [32]) becomes highly technical and requires much more space.

Example

Let $m = n = 2$ and π be of cycle type decomposition $(\pi_1)(\pi_2)$ of lengths $n_i = 2$ $(i = 1, 2)$. For $k = 1$ we obtain the block system $(\rho_1, \rho_2)(\rho_3, \rho_4)$. For $k = 2$ there are two more potential block systems, namely $(\rho_1, \rho_3)(\rho_2, \rho_4)$ and $(\rho_1, \rho_4)(\rho_2, \rho_3)$. Next the found block systems are used to calculate candidates for subfields of degree m. The algorithm below generalizes Dixon's ideas in as much as exponents k larger than one can also be handled. The underlying theory can be found in [31].

Algorithm (Computation of subfields) (8.2)

Input: *A generating polynomial $f(t) \in \mathbb{Z}[t]$ for E of degree mn, a prime number p with potential block system $\{\Delta_1, \dots, \Delta_m\}$ and corresponding exponents $k_1, \dots, k_m$.*

Output: *A generating polynomial $g(t)$ for a potential subfield F of degree m.*

Step 1: *For $1 \leqslant i \leqslant m$ determine the cycles and corresponding polynomials which contain elements in Δ_i, factorize these polynomials in an extension of degree k_i and determine the zeros belonging to Δ_i.*

Step 3: *Factorize $f(t)$ in an extension of degree $k = lcm(k_1, \dots, k_m)$.*

Step 4: *Lift those factors to a sufficient precision by Hensel's method.*

Step 5: *Compute approximations δ_i of the product of the zeros belonging to block Δ_i.*

Step 6: *Compute $g(t) = \prod_{i=1}^{m} (t - \delta_i)$.*

The output consists of polynomials $g(t) \in \mathbb{Z}[t]$ which are potential generating polynomials for subfields F of E of degree m. For the embedding of subfields there are well-known methods basing on the factorization of polynomials over number fields and/or LLL-reduction. For larger field degrees they tend to become slow. Hence, we rather use our knowledge about potential block systems to exhibit a polynomial $h(t) \in \mathbb{Q}[t]$ of degree at most $mn - 1$ satisfying $h(\rho) = \beta$ for a root β of $g(t)$. We know that the roots $\rho_1, \dots, \rho_{mn}$ of $f(t)$ can be partitioned into blocks $\Delta_1, \dots, \Delta_m$ such that the roots β_i of $g(t)$ are just the products of all elements belonging to Δ_i $(1 \leqslant i \leqslant m)$. Also the denominators of the coefficients of $h(t)$ can be easily estimated from above. Hence, approximations of the corresponding zeros up to a sufficient precision by Hensel's lemma will actually yield $h(t)$ exactly. The following algorithm is explained in [31] in greater detail.

Algorithm (Embedding of potential subfields) (8.3)

Input: *A generating polynomial $f(t) \in \mathbb{Z}[t]$ for E of degree mn, and a polynomial $g(t)$ generated by the previous algorithm for a prime number p and a block system $\{\Delta_1, \dots, \Delta_m\}$.*

Output: *A polynomial $h(t) \in \mathbb{Q}[t]$ satisfying $g(h(t)) \equiv 0 \bmod f(t)$, if $g(t)$ generates a subfield F of E, or the result that $g(t)$ does not generate a subfield of E.*

Step 1: *Calculate $h_0(t) \in \mathbb{Z}[t]$ satisfying $h_0(\rho_j) \equiv \beta_i \bmod p$ for all $\rho_j \in \Delta_i$ $(1 \leqslant i \leqslant m)$.*

Step 2: *Lift $h_0(t)$ to a sufficient precision $h_k(t)$ modulo $p^{2^k}\mathbb{Z}[t]$ by Hensel's method.*

Step 3: *Retrieve from $h_k(t)$ a polynomial $h(t) \in \mathbb{Q}(t)$. If $f(t)$ divides $g(h(t))$ print $h(t)$, else print that $g(t)$ does not generate a subfield of E.*

We note that an appropriate polynomial $h(t)$ will not be found in Step 3, if the potential block system is not a block system.

We conclude this section with 2 examples. Tables of subfield computations are contained in [32].

Examples

(i) Let E be generated by a root ρ of $f(t) = t^6 + 108$. We choose $m = 3$, $n = 2$. For $p = 7$ we get $f(t) \equiv (t^3+2)(t^3+5) \bmod 7\mathbb{Z}[t]$. Hence, $\pi = (123)(456)$, and we need to search for potential blocks of size 2. There are no such blocks of exponent 1. For exponent 3 we can combine one zero of index belonging to $\pi_1 = (123)$ with any one whose index bleongs to $\pi_2 = (456)$. We obtain three potential block systems: $\{\{\rho_1, \rho_4\}, \{\rho_2, \rho_5\}, \{\rho_3, \rho_6\}\}$, $\{\{\rho_1, \rho_5\}, \{\rho_2, \rho_6\}, \{\rho_3, \rho_4\}\}$, and $\{\{\rho_1, \rho_6\}, \{\rho_2, \rho_4\}, \{\rho_3, \rho_5\}\}$.

For all three systems the corresponding polynomial $g(t)$ is $g(t) = t^3 - 108$. The 3 subfields of E of degree 3 can be distinguished only by the so-called embedding polynomial $h(t)$. We calculate

$$h(t) \in \{-t^2, (t^5 + 6t^2)/12, (-t^5 + 6t^2)/12\}.$$

This example is considered again in subsection 10.

(ii) The second example was given by D. Lazard. He asked for all subfields of the field F generated by a root of the following polynomial of degree 40:

$$\begin{aligned}
&x^{40} + 8x^{39} - 972x^{38} - 4438x^{37} + 368411x^{36} + 319263x^{35} - 70822744x^{34}\\
&+ 158288849x^{33} + 7272303005x^{32} - 33340410389x^{31} - 412637940242x^{30}\\
&+ 2860211279898x^{29} + 12408126952305x^{28} - 136057114420269x^{27}\\
&- 122973006802033x^{26} + 395430884437770x^{25} - 4414713412013052x^{24}\\
&- 7175245256294431 1x^{23} + 200736719645454859x^{22} + 745141821965932237x^{21}\\
&- 385600533813446057x^{20} - 2357344486297791023x^{19}\\
&+ 4210254326336514678 3x^{18} - 4186749151323359325 4x^{17}\\
&- 2652839349310553929 17x^{16} + 646725310260982947946x^{15}\\
&+ 769494645028196970860x^{14} - 4418058564958867582535x^{13}\\
&+ 1440918665088174301840x^{12} + 16999989172236574721320x^{11}\\
&- 22653473046721498313233x^{10} - 32675347429924388587607x^{9}\\
&+ 91700548881113811658426x^{8} - 27129684140284614054 18x^{7}\\
&- 173138564231129651232857x^{6} + 138524234196323700539396x^{5}\\
&+ 100184208695576521111933x^{4} - 187966742518186234660038x^{3}\\
&+ 78450078315486832226357x^{2} - 106295208031937099038 80x\\
&+ 70578012754570633858 93.
\end{aligned}$$

The preceding algorithm produced 4 subfields given by generating polynomials
$x^4+122111359x^3+441284550591715x^2-2868510407839796790x$
$+70578012754570633858 93,$

$x^5+280045x^4-30846661800x^3+796683011921109x^2$
$-4229952378356895951x-70578012754570633858 93,$

$x^8-36688x^7-353574160x^6-189512351885x^5+945725045119287x^4$
$-732150450978572042x^3+182289427558444750649x^2$
$-9248024243433337911 26x+70578012754570633858 93,$

$x^{20}+316x^{19}+20497x^{18}-1220315x^{17}-120719738x^{16}-858476459x^{15}$
$+107284763042x^{14}+2364093617871x^{13}-11436266583320x^{12}$
$-804111496163231x^{11}-5293728100770241x^{10}+80084901023882949x^9$
$+1181007213585185691x^8+6977271527223658 05x^7-6961773161603033237x^6$
$-38912831120937361 7964x^5+3004805032687410012 48x^4$
$+85228401842863729648 97x^3+24921107246172262623162x^2$
$+2258567284936914938 2110x+70578012754570633858 93.$

Afterwards a search for elements of small T_2-value in the corresponding subfields yields generating polynomials with much smaller coefficients:
$x^4+x^3-5x^2-4x+5,$
$x^5-x^4-4x^3+3x^2+3x-1,$
$x^8-x^7+x^6-4x^5-4x^4+17x^3-6x+1,$
$x^{20}-5x^{19}-4x^{18}+55x^{17}-38x^{16}-217x^{15}+284x^{14}+352x^{13}-679x^{12}-147x^{11}+$
$626x^{10}-72x^9-128x^8-176x^7+159x^6-3x^5+24x^4-22x^3-14x^2+4x+1.$
Unfortunately, a listing of the corresponding embedding polynomials would be lengthy.

In his thesis [61] J. Klüners developed several improvements of this algorithm as well as methods for the computation of the automorphism group of a normal extension.

7.9. KANT and the KANT shell KASH

KANT [30] is a software system for computations with algebraic numbers and algebraic number fields. It is developed by the author's research group at the Technical University Berlin. It is freely available at the internet address given at the end of this section.

For a proper use of KANT [30], the user needs to have some experience with programming in C and an understanding of the memory management in MAGMA [39]. Because of this disadvantage, we built a shell around the C–library KANT, which combines the functionality of KANT with a comfortable user interface based on GAP a software package for group theory [46].

The interpreter consists of several units, e.g. the KANT–package, system–dependent functions, an additional memory manager, an internal function library, etc.

In principle, a simple main–loop is performed:

READ Reading the command–line from the keyboard, out of a file or from another process.

EVAL Evaluating the input: the input is tokenized and a multiway tree for evaluation is created. By recursion, the root of the tree is evaluated to a single result (value).

PRINT Printing the result on the screen, into a file or sending it to another process.

Within the shell, the user can do arithmetical operations with integers, rationals, real and complex numbers (with arbitrary precision), matrices, polynomials or – after the definition of an order – with algebraic numbers, ideals, etc. Of course, all results can be assigned to variables for later use.

Furthermore the user can make use of two different kinds of functions, the 'internal functions' and the 'user functions'. The first are built-in functions of the internal function library, i.e. they are written in C, linked to KASH and cannot be changed. In contrast, the user can create personalized (user) functions: With the PASCAL-like programming language, the user can create loops, conditional branches, functions etc. and use all internal and user functions. In this environment, it is even possible to write sophisticated programs. All user functions and programs can be stored as (external) text files which build a user function library. Additionally, KASH possesses an interface to the public domain PVM-software which allows distributed computing and is very easy to handle.

Presently, there are more than 350 internal functions installed, 200 additional predefined user functions and comprehensive references are available. Because KASH grows weekly, updates will be made more often than for the KANT-library.

Accessible from KASH is an SQL-database for number fields [51]. The database is designed to give easy and fast access to more than one million number fields. Currently the following invariants are stored (if known) and can be used as keys in a selection:

- a generating polynomial together with its signature
- an integral basis, the field discriminant
- the unit group and the regulator
- the class group with structural information
- the Galois group

In accordance with PARI [41] we choose a special form of the generating polynomial $f(t) = t^n + a_1 t^{n-1} + \cdots + a_n$ as a unique key for the fields in the database (at least for number fields of low degree). As a generating element we take an algebraic interger ρ subject to the following conditions:

(1) $T_2(\rho) = \sum_{j=1}^{n} |\rho^{(j)}|^2$ is minimal,
(2) the index of the equation order $\mathbb{Z}[\rho]$ in the maximal order is minimal,
(3) $a_1 > 0$ is minimal,
(4) $|a_i|$ $(2 \leqslant i \leqslant n)$ are minimal.

Isomorphy can be tested with KASH. In a first step one can check some invariants and if all tests are successful there is the possibility to choose between several algorithms for proving the isomorphy. The underlying SQL-database (Postgres95) is public domain and available for every system supported by KASH.

As an example we will find all totally real cubic fields with discriminant less than 10,000 and common inessential discriminant divisors. A well known sufficient and necessary criterion is that exactly three different prime ideals divide 2, so the following program is straightforward:

```
kash> DbOpen();                            # open the
                                             database
true
kash> query :=                             # we are interested
                                             only
> "degree=3 and disc<10000
  and [number of real zeroes]=3";;
kash>                                      # in small  totally
                                             real cubic fields
kash> DbCountQuery(query);
382
kash> DbQuery(query);
true
kash> L := [];
kash> repeat
> o := DbNextOrder();
> if o<>false and
>      OrderIndex(o)mod 2 = 0 and          # 2 has to divide
                                             the index,
                                           # this is a fast
                                             criterion
>      Length(Factor(2*o))=3 then
                                           # exactly 3
                                             different primes
```

```
>     Add(L, o);
> fi;
> until o=false;
kash> Length(L);                    # we found 14 fields
14
```

7.10. Examples

In the sequel we give some examples of KASH.

Computation of the maximal order, the unit group and the class group
We start with the equation order $\mathbb{Z}[\rho]$ for $\rho^4 - 117 = 0$ and compute a set of fundamental units.
First we create the order $\mathbb{Z}[\sqrt[4]{117}]$ of degree 4 over $\mathbb{Z}$.

```
kash> o := Order(Z,4,117);
Generating polynomial: x^4 - 117
```

We compute the fundamental units in the equation order. Setting $\rho = \sqrt[4]{117}$ the first fundamental unit is $649 - 60\rho^2$.

```
kash> OrderUnitsFund(o);
[ [649, 0, -60, 0], [26618086, -8093388, 2460843,
  -748234] ]
```

To calculate the index of the unit group of the equation order in the unit group of the corresponding maximal order, we proceed as follows.

```
kash> O := OrderMaximal(o);
   F[1]
    |
   F[2]
  /
 /
Q
F [ 1]      Given by transformation matrix
F [ 2]      x^4 - 117
Generating polynomial: x^4 - 117
Discriminant: -316368
```

A transformation matrix from a basis of `o` to a basis of `O` is stored but not printed (it can be obtained with the command `OrderTransformation Matrix`).

```
kash> OrderUnitsFund(O);
[ [2, 0, -1, 0], [1, -1, 1, 0] ]
```

The units are represented in the basis of the maximal order. After determining a set of fundamental units, we get the index as the quotient of the two regulators.

```
kash> OrderReg(o)/OrderReg(O);
36
```

Finally, we calculate the class group structure:

```
kash> OrderClassGroup(O);
[ 4, [2, 2] ]
```

This means that the class group is of order 4 and is isomorphic to $C_2 \times C_2$.

Computing subfields

The following example demonstrates the computation of subfields: We start by creating the equation order $\mathbb{Z}[\rho]$ for $\rho^6 + 108 = 0$.

```
kash> o :=Order(Z,6,-108);
Generating polynomial: x^6+108
```

The computation of proper subfields of the quotient field $\mathbb{Q}(\rho)$ of o yields the following list of equation orders.

```
kash> L :=OrderSubfield(o);
[ Generating polynomial: x^3 - 108
    , Generating polynomial: x^3 - 108
    , Generating polynomial: x^3 - 108
    , Generating polynomial: x^2 + 108
     ]
```

There are 3 subfields of degree 3 which are isomorphic but not identical and one subfield of degree 2. Let ρ_1, ρ_2, ρ_3 denote the roots of $x^3 - 108$. `L[i]` denotes the i-th equation order in `L`.

```
kash> r1 :=Elt(L[1],[0,1,0]);
[0, 1, 0]
kash> r2 :=Elt(L[2],[0,1,0]);;
kash> r3 :=Elt(L[3],[0,1,0]);;
```

The elements look identical, but they are indeed different which is detected upon moving them into the order o.

```
kash> EltMove(r1,o)           # This produces the
                                element
                              # (6*rho^2-rho^5)/12
```

```
[0, 0, 6, 0, 0, -1] / 12
kash> EltMove(r2,o);          # This produces the
                                element -rho^2
[0, 0, -1, 0, 0, 0]
kash> EltMove(r3,o);          # This produces the
                                element
                              # (6*rho^2+rho^5)/12
[0, 0, 6, 0, 0, 1] / 12
```

Any element of L[i] can be lifted in an analogous way.

Solution of Thue-equations

Given an irreducible form $f \in \mathbb{Z}[X, Y]$ of degree $\geqslant 3$ and an integer a, we compute all $(x, y) \in \mathbb{Z}^2$ subject to $f(x, y) = a$.
Let $f(X, Y) := X^3 + X^2Y - 6XY^2 + 2Y^3$ and solve $f(x, y) = 2$. The corresponding number field F is created by a root of the irreducible polynomial $f(X, 1) \in \mathbb{Z}[X]$.

```
kash> t := Thue([1,1,-6,2]); # [1,1,-6,2] are the
                               coefficients of f.
X^3 + X^2 Y - 6 X Y^2 + 2 Y^3
kash> Solve(t,2);              # Compute a list of all
                                 solutions [x,y].
[ [ -724, -411 ], [ -4, -11 ], [ -3, 1 ],
[ -1, -1 ], [ 0, 1 ], [ 2, 1 ] ]
```

Additionally, we can solve Thue–equations up to sign on the right hand side, for example $X^7 + X^6Y - 6X^5Y^2 - 5X^4Y^3 + 8X^3Y^4 + 5X^2Y^5 - 2XY^6 - Y^7 = \pm 1$.

```
kash> t := Thue([1,1,-6,-5,8,5,-2,-1]);
X^7 + X^6 Y - 6 X^5 Y^2 - 5 X^4 Y^3 + 8 X^3 Y^4
  + 5 X^2 Y^5 - 2 X Y^6 - Y^7
kash> Solve(t,1,"abs");        # Compute a list of all
                                 solutions [x,y].
[ [ -2, -1 ], [ -1, -1 ], [ -1, 0 ], [ -1, 1 ],
  [ 0, -1 ], [ 0, 1 ], [ 1, -1 ], [ 1, 0 ], [ 1, 1 ],
  [ 2, 1 ] ]
```

Solution of norm equations

We consider the relative extension E/F for

$$F = \mathbb{Q}(\alpha), \alpha^2 - 2 = 0, \quad \text{and} \quad E = F(\beta), \beta^2 + 1 = 0.$$

For all $\theta = i\alpha + j$ $(-4 \leqslant i, j \leqslant 4)$ we want to know if there exists $\mu \in o_E$ with $\mathrm{N}_{E/F}(\mu)/\theta \in \mathrm{T}U_F$.

```
kash> F := Order(Z, 2, 2);;
kash> E := Order(F, 2, -1);;
kash> l1 :=[];                   # Create an empty list.
kash> zero := Elt(F, 0));;       # Create the zero
                                   element in F
kash> for i in [0..4] do
>         for j in [-4..4] do
>           z := Elt(F, [i,j]);
>           if z<>zero then
>             Add(l1, [z, OrderNormEquation(E, z)]);
>           fi;
>         od;
>       od;
```

The output consists of a list containing pairs $[\theta, \mu]$ if there is a solution μ and $[\theta,$ `false`$]$ otherwise (optional).

```
[ [ [0, -4], false ], [ [0, -3], false ],
  [ [0, -2], false ], [ [0, -1], false ],
  [ [0, 1], false ], [ [0, 2], false ], [ [0, 3],
  false ],
...
  [ [1, -1], false ], [ 1, [ [[0, -1],
  [0, 1]] / 2 ] ],
...
  [ [4, 3], false ], [ [4, 4], false ] ]
```

It took 17 seconds to solve these 44 norm equations.

Computation of Hilbert class fields

The following is an example for the computation of the Hilbert class field for $F := \mathbb{Q}(\rho)$ where $\rho^3 + \rho^2 - 42\rho - 107 = 0$.
We start by reading the equation order of F.

```
kash> f :=Poly(Zx,[1,1,-42,-107]);;    # f(x) =
                                         x^3+x^2-42x-1
kash> F :=Order(f);;                   # Create the
                                         equation order
                                       # of the
                                         polynomial f
kash> F :=OrderMaximal(F);;
```

```
kash> OrderUnitsFund(F);;
kash> OrderClassGroup(F, "euler");;   # option
                                        "euler" is
                                        necessary
                                      # for the
                                        function
                                      # OrderHilbert
                                        ClassField
kash> F;
Generating polynomial: x^3 + x^2 - 42*x - 107
Discriminant: 70313
Regulator: 21.20506
Units:
[3, 1, 0]     [9, 12, 2]
class number 2
class group structure C2
cyclic factors of the class group:
<5, [3, 0, 1]>
```

The discriminant is always the discriminant of the order under consideration. The cyclic factors are given in a 2 element normal representation.
We apply the user function `OrderHilbertClassField` to it.

```
kash> Y := OrderHilbertClassField (F);
Starting Class Field Computation
   Degree      : 3
   Signature   : [ 3, 0 ]
   Class Group : [ 2, [ 2 ]]
--------------------------------------
Checking cyclic group C2
--------------------------------------
Computing class field for cyclic subgroup C2
```

We obtain the following 4 elements $\alpha_1, \ldots, \alpha_4$, a power product of which yields a generating element.

```
List of Generators :
[ [1299, 255, -62], -1, [3, 1, 0], [9, 12, 2] ]
```

We compute a generating element $\mu = \alpha_1^{e_1} \cdot \ldots \cdot \alpha_4^{e_4}$ for $(e_1, \ldots, e_4) \in (\mathbb{Z}/2\mathbb{Z})^4$. Only unramified extensions $F(\sqrt[2]{\mu})$ of F are processed further.

```
Exponent Vector [ 1, 1, 0, 1 ] -->[79, 5, -2]
```

Hence, the only solution is $\mu = 79 + 5\rho - 2\rho^2$.

Since we obtain just one unramified extension, it has to be the class field. Of course, there is also a built in checking routine.

Availability

KASH is freely available via `ftp.math.tu-berlin.de` at `pub/algebra/Kant/Kash`. Further information can be obtained from: http://www.math.tu-berlin.de/algebra. It has been ported to the following architectures:

- HP 7000 (HP-UX 9.01),
- IBM RS 6000 (AIX 3.2.5),
- Intel 486 (Linux Kernel 2.0.13),
- Intel 486 (MS DOS 5.0, Windows95, OS/2),
- Silicon Graphics (IRIX 5.3),
- Sun SPARC (SunOS 4.1.3),
- Sun SPARC (SunOS 5.4),
- DEG Alpha (OSF1 V3.2).

APPENDIX: NUMERICAL TABLES

Permutation groups

1.1 Table of the primitive groups of degree $n \leqslant 12$

Column 1 contains the degree n and its number k with respect to the degree in the form n.k, i.e. 7.4 means the fourth primitive group of degree 7. Column 2 contains the order of the group, column 3 gives the notation or a description as used in chapter 2, section 9. In column 4 we list the transitivity t of the group followed by a letter p, if the group is t-fold primitive. Whenever the group is doubly primitive so that the subgroup fixing the last integer on which the group acts is a primitive group of degree $n-1$, this subgroup G_{n-1} occurs in column 5 by its degree and number. A "+" sign in the last column means that the group consists of even permutations. (Reference: Charles C. Sims, Computational methods in the study of permutation groups, in *Computational Problems in Abstract Algebra* (pp. 169–83), J. Leech (editor), Pergamon Press, Oxford and New York, 1970.)

degree no.	order	notation or description	t	G_{n-1}	+
2.1	2	$\mathfrak{S}_2$	2		
3.1	3	$\mathfrak{A}_3$			+
3.2	6	$\mathfrak{S}_3$	3	2.1	
4.1	12	$\mathfrak{A}_4$	$2p$	3.1	+
4.2	24	$\mathfrak{S}_4$	4	3.2	
5.1	5	C_5			+
5.2	10	D_{10}			+
5.3	20	$\mathrm{Hol}(C_5)$	2		
5.4	60	$\mathfrak{A}_5$	$3p$	4.1	+
5.5	120	$\mathfrak{S}_5$	5	4.2	
6.1	60	$PSL(2,5)$	$2p$	5.2	+
6.2	120	$PGL(2,5)$	3	5.3	
6.3	360	$\mathfrak{A}_6$	$4p$	5.4	+
6.4	720	$\mathfrak{S}_6$	6	5.5	
7.1	7	C_7			+
7.2	14	D_{14}			
7.3	21	$\mathrm{Hol}(C_7)\cap\mathfrak{A}_7$			+
7.4	42	$\mathrm{Hol}(C_7)$	2		
7.5	168	$PSL(3,2)$	2		+
7.6	2520	$\mathfrak{A}_7$	$5p$	6.3	+
7.7	5040	$\mathfrak{S}_7$	7	6.4	
8.1	56	$\{z\to az+b \mid 0\neq a\in\mathbb{F}_8,\ b\in\mathbb{F}_8\}$	$2p$	7.1	+
8.2	168	$\{z\to az^{2^k}+b \mid a,b\in\mathbb{F}_8, a\neq 0, k=0,1,2\}$	$2p$	7.3	+
8.3	168	$PSL(2,7)$	$2p$	7.3	+
8.4	336	$PGL(2,7)$	3	7.4	
8.5	1344	$\mathrm{Hol}(C_2\times C_2\times C_2)$	3	7.5	+
8.6	20 160	$\mathfrak{A}_8$	$6p$	7.6	+
8.7	40 320	$\mathfrak{S}_8$	8	7.7	
9.1	36	$\{z\to az+b \mid a\in\mathbb{F}_9^2, b\in\mathbb{F}_9\}$			
9.2	72	$\{z\to az^{3^k}+b \mid a\in\mathbb{F}_9^2, b\in\mathbb{F}_9, k=0,1\}$			
9.3	72	$\{z\to az+b \mid a,b\in\mathbb{F}_9, a\neq 0\}$	2		
9.4	72	$\langle 9.1, \{z\to\beta z^3 \mid \beta\in\mathbb{F}_9\backslash\mathbb{F}_9^2\}\rangle$	2		+
9.5	144	$\{z\to az^{3^k}+b \mid a,b\in\mathbb{F}_9, a\neq 0, k=0,1\}$	2		
9.6	216	$\mathrm{Hol}(C_3\times C_3)\cap\mathfrak{A}_9$	2		+
9.7	432	$\mathrm{Hol}(C_3\times C_3)$	2		
9.8	504	$PGL(2,8)$	$3p$	8.1	+
9.9	1512	$\mathrm{Hol}(9.8)$	$3p$	8.2	+
9.10	$\frac{1}{2}\cdot 9!$	$\mathfrak{A}_9$	$7p$	8.6	+
9.11	$9!$	$\mathfrak{S}_9$	9	8.7	
10.1	60	$\mathfrak{A}_5$			+
10.2	120	$\mathfrak{S}_5$			
10.3	360	$PSL(2,9)$	$2p$	9.1	+
10.4	720	$\mathfrak{S}_6$	$2p$	9.2	+
10.5	720	$PGL(2,9)$	3	9.3	
10.6	720	$\langle PSL(2,9), \{z\to\beta z^3 \mid \beta\in\mathbb{F}_9^2, z\in\mathbb{F}_9\cup\{\infty\}\}\rangle$	3	9.4	+
10.7	1440	$\mathrm{Hol}(PSL(2,9))$	3	9.5	
10.8	$\frac{1}{2}\cdot 10!$	$\mathfrak{A}_{10}$	$8p$	9.10	+
10.9	$10!$	$\mathfrak{S}_{10}$	10	9.11	
11.1	11	C_{11}			+

(Contd.)

degree no.	order	notation or description	t	G_{n-1}	+
11.2	22	D_{22}			
11.3	55	$\mathrm{Hol}(C_{11})\cap\mathfrak{A}_{11}$			+
11.4	110	$\mathrm{Hol}(C_{11})$	2		
11.5	660	$PSL(2,11)$	$2p$	10.1	+
11.6	7920	M_{11}	4	10.6	+
11.7	$\frac{1}{2}\cdot 11!$	$\mathfrak{A}_{11}$	$9p$	10.8	+
11.8	11!	$\mathfrak{S}_{11}$	11	10.9	
12.1	660	$PSL(2,11)$	$2p$	11.3	+
12.2	1320	$PGL(2,11)$	3	11.4	
12.3	7920	M_{11}	$3p$	11.5	+
12.4	95 040	M_{12}	5	11.6	+
12.5	$\frac{1}{2}\cdot 12!$	$\mathfrak{A}_{12}$	$10p$	11.7	+
12.6	12!	$\mathfrak{S}_{12}$	12	11.8	

1.2 Table of transitive permutation groups of degree $n \leqslant 7$ as Galois groups

As in Table 1.1 we list in columns 1–3 the degree, the order and the notation of the groups, respectively. Column 4 contains the corresponding indicator function which is used to determine for a given monic irreducible polynomial of $\mathbb{Z}[t]$ of that degree, whether its Galois group is contained in the group of that row (compare chapter 2, section 10). An example of a suitable polynomial $f(t)\in\mathbb{Z}[t]$ whose Galois group is exactly the one of column 3 is given in column 5. Finally, column 6 contains the discriminant of that polynomial.

For the computation of the Galois group of a given monic irreducible nth degree polynomial $f(t)\in\mathbb{Z}[t]$ ($3\leqslant n\leqslant 7$) using Table 1.2 we refer to chapter 2, section 10. For the choice of the appropriate indicator function Table 1.3 is useful. (References: L. Soicher & J. McKay, Computing Galois groups over the rationals, *J. Number Theory*, **20** (1985), 273–81. R.P. Stauduhar, The determination of Galois groups, *Math. Comp.*, **27** (1973), 981–96.)

1.3 Diagrams of transitive permutation groups of degree $4\leqslant n\leqslant 7$

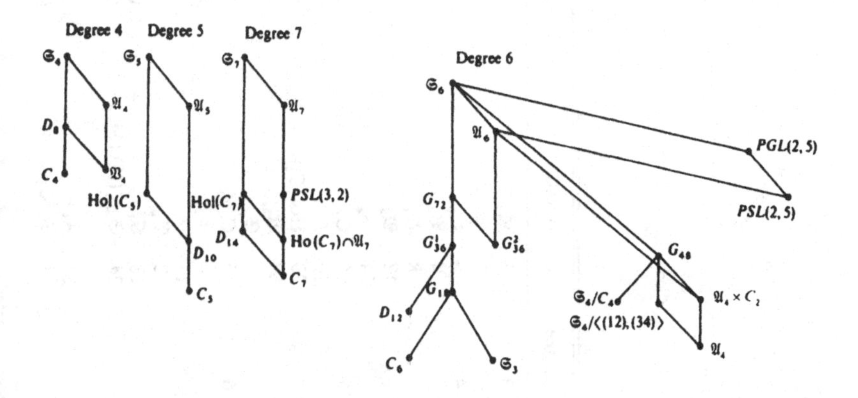

degree	order	notation	indicator function	polynomial of discriminant	
3	3	$\mathfrak{A}_3$	$d(f)$	t^3+t^2-2t-1	7^2
	6	$\mathfrak{S}_3$		t^3+2	-2^23^3
4	4	C_4	$x_1x_2^2+x_2x_3^2+x_3x_4^2+x_4x_1^2$	$t^4+t^3+t^2+t+1$	5^3
	4	$\mathfrak{B}_4$		t^4+1	2^8
	8	D_8	$x_1x_3+x_2x_4$	t^4-2	-2^{11}
	12	$\mathfrak{A}_4$	$d(f)$	$t^4+8t+12$	$2^{12}3^4$
	24	$\mathfrak{S}_4$		t^4+t+1	229
5	5	C_5	$x_1x_2^2+x_2x_3^2+x_3x_4^2+x_4x_5^2+x_5x_1^2$	$t^5+t^4-4t^3-3t^2+3t+1$	11^4
	10	D_{10}		$t^5-5t+12$	$2^{12}5^6$
	20	$\mathrm{Hol}(C_5)$	$(x_1x_2+x_2x_3+x_3x_4+x_4x_5+x_5x_1-x_1x_3-x_2x_5-x_5x_2-x_2x_4-x_4x_1)^2$	t^5+2	2^45^5
	60	$\mathfrak{A}_5$	$d(f)$	$t^5+20t+16$	$2^{16}5^6$
	120	$\mathfrak{S}_5$		t^5-t+1	$19\cdot151$
6	6	C_6	$x_1x_6^2+x_2x_4^2+x_3x_5^2+x_4x_2^2+x_5x_1^2+x_6x_2^2$	$t^6+t^5+t^4+t^3+t^2+t+1$	-7^5
	6	$\mathfrak{S}_3$	$x_1x_4+x_2x_6+x_3x_5$	t^6+108	$-2^{16}3^{21}$
	12	D_{12}	$x_1x_4+x_2x_5+x_3x_6$	t^6+2	$-2^{11}3^6$
	12	$\mathfrak{A}_4$		t^6-3t^2-1	2^63^8
	18	G_{18}	$(x_1-x_2)(x_2-x_3)(x_3-x_1)+(x_4-x_5)(x_5-x_6)(x_6-x_4)$	t^6+3t^3+3	-3^{11}
	24	$\mathfrak{S}_4/\langle(12),(34)\rangle$	$(x_1+x_2-x_3-x_4)(x_3+x_4-x_5-x_6)(x_5-x_6-x_1-x_6)$ $\times(x_1-x_2)(x_3-x_4)(x_5-x_6)$	t^6-3t^2+1	-2^63^8
	24	$\mathfrak{A}_4\times C_2$		t^6-4t^2-1	2^6229^2
	24	$\mathfrak{S}_4/C_4$	$(x_1+x_2-x_3-x_4)(x_3+x_4-x_5-x_6)(x_5+x_6-x_1-x_2)$	$t^6-3t^5+6t^4-7t^3+2t^2+t-4$	229^3

6	36	G_{36}^1		$t^6 + 2t^3 - 2$	$2^8 3^9$
	36	G_{36}^2	$(x_1 - x_2)(x_2 - x_3)(x_3 - x_1)(x_4 - x_5)(x_5 - x_6)(x_6 - x_4)$	$t^6 + 6t^4 + 2t^3 + 9t^2 + 6t - 4$	$2^{10} 3^6 5^4$
	48	G_{48}	$x_1x_2 + x_3x_4 + x_5x_6$	$t^6 + 2t^2 + 2$	$-2^{11} 5^2 7^2$
	60	$PSL(2,5)$		$t^6 + 10t^5 + 55t^4 + 140t^3$	$2^3 6 5^8$
				$+ 175t^2 + 170t + 25$	
	72	G_{72}	$x_1x_2x_3 + x_4x_5x_6$	$t^6 + 2t^4 + 2t^3 + t^2 + 2t + 2$	$-2^8 733$
	120	$PGL(2,5)$	$(x_1x_2 + x_3x_5 + x_4x_6)(x_1x_3 + x_4x_5 + x_2x_6)(x_3x_4 + x_1x_6 + x_2x_5)$	$t^6 + 10t^5 + 55t^4 + 140t^3$	$5^{20} 19^3$
			$\times (x_1x_5 + x_2x_4 + x_3x_6)(x_1x_4 + x_2x_3 + x_5x_6)$	$+ 175t^2 - 3019t + 25$	$5^{20} 19^3 151^3$
	360	$\mathfrak{A}_6$	$d(f)$	$t^6 + 24t - 20$	$2^{16} 3^6 5^6$
	720	$\mathfrak{S}_6$		$t^6 + t + 1$	$-101 \cdot 431$
7	7	C_7		$t^7 + t^6 - 12t^5 - 7t^4 + 28t^3$	$17^2 29^6$
				$+ 14t^2 - 9t + 1$	
	14	D_{14}	$x_1x_2 + x_2x_3 + x_3x_4 + x_4x_5 + x_5x_6 + x_6x_7 + x_7x_1$	$t^7 + 7t^3 + 7t^2 + 7t - 1$	$-3^6 7^9$
	21	$\mathrm{Hol}(C_7) \cap \mathfrak{A}_7$		$t^7 - 14t^5 + 56t^3 - 56t + 22$	$2^6 7^{10}$
	42	$\mathrm{Hol}(C_7)$	$x_1x_2x_4 + x_1x_2x_6 + x_1x_3x_4 + x_1x_3x_7 + x_1x_5x_6 + x_1x_5x_7 + x_2x_3x_5$	$t^7 + 2$	$-2^6 7^7$
			$+ x_2x_3x_7 + x_2x_4x_5 + x_2x_6x_7 + x_3x_4x_6 + x_3x_5x_6 + x_4x_5x_7 + x_4x_6x_7$		
	168	$PSL(3,2)$	$x_1x_2x_4 + x_1x_3x_7 + x_1x_5x_6 + x_2x_3x_5 + x_2x_6x_7 + x_3x_4x_6 + x_4x_5x_7$	$t^7 - 7t^3 + 14t^2 - 7t + 1$	$7^8 17^2$
	2520	$\mathfrak{A}_7$	$d(f)$	$t^7 + 7t^4 + 14t + 3$	$3^6 7^8$
	5040	$\mathfrak{S}_7$		$t^7 + t + 1$	$-11 \cdot 239 \cdot 331$

Fundamental units and class numbers

2.1 Fundamental units $\varepsilon > 1$ and class numbers h of real quadratic number fields $F = \mathbb{Q}(m^{\frac{1}{2}})$, $m < 300$

Column 1 contains the discriminant d_F of F which is m in case $m \equiv 1 \bmod 4$, $4m$ otherwise.

Column 2 gives the coefficients $a, b, c \in \mathbb{N}$ of the fundamental unit $\varepsilon = (a + bm^{\frac{1}{2}})/c > 1$ of F, c being listed only for $c > 1$.

Column 3 contains the class number h of F and column 4 the structure of the class group Cl_F, if h is greater than one and Cl_F not cyclic.

discriminant d_F $d_F = m$ or $d_F = 4m$	fundamental unit ε a, b, c of $\varepsilon = (a + bm^{\frac{1}{2}})/c$ c listed only if $c \neq 1$			class number h	type of class group if not cyclic
4·2	1	1		1	
4·3	2	1		1	
5	1	1	2	1	
4·6	5	2		1	
4·7	8	3		1	
4·10	3	1		2	
4·11	10	3		1	
13	3	1	2	1	
4·14	15	4		1	
4·15	4	1		2	
17	4	1		1	
4·19	170	39		1	
21	5	1	2	1	
4·22	197	42		1	
4·23	24	5		1	
4·26	5	1		2	
29	5	1	2	1	
4·30	11	2		2	
4·31	1520	273		1	
33	23	4		1	
4·34	35	6		2	
4·35	6	1		2	
37	6	1		1	
4·38	37	6		1	
4·39	25	4		2	
41	32	5		1	
4·42	13	2		2	
4·43	3482	531		1	
4·46	24 335	3588		1	
4·47	48	7		1	
4·51	50	7		2	
53	7	1	2	1	
4·55	89	12		2	
57	151	20		1	
4·58	99	13		2	
4·59	530	69		1	
61	39	5	2	1	
4·62	63	8		1	
65	8	1		2	
4·66	65	8		2	

(Contd.)

discriminant d_F $d_F = m$ or $d_F = 4m$	fundamental unit ε a, b, c of $\varepsilon = (a + bm^{\frac{1}{2}})/c$ c listed only if $c \neq 1$			class number h	type of class group if not cyclic
4·67	48 842	5967		1	
69	25	3	2	1	
4·70	251	30		2	
4·71	3480	413		1	
73	1068	125		1	
4·74	43	5		2	
77	9	1	2	1	
4·78	53	6		2	
4·79	80	9		3	
4·82	9	1		4	
4·83	82	9		1	
85	9	1	2	2	
4·86	10 405	1122		1	
4·87	28	3		2	
89	500	53		1	
4·91	1574	165		2	
93	29	3	2	1	
4·94	2 143 295	221 064		1	
4·95	39	4		2	
97	5604	569		1	
101	10	1		1	
4·102	101	10		2	
4·103	227 528	22 419		1	
105	41	4		2	
4·106	4005	389		2	
4·107	962	93		1	
109	261	25	2	1	
4·110	21	2		2	
4·111	295	28		2	
113	776	73		1	
4·114	1025	96		2	
4·115	1126	105		2	
4·118	306 917	28 254		1	
4·119	120	11		2	
4·122	11	1		2	
4·123	122	11		2	
4·127	4 730 624	419 775		1	
129	16 855	1484		1	
4·130	57	5		4	$C_2 \times C_2$
4·131	10 610	927		1	
133	173	15	2	1	
4·134	145 925	12 606		1	
137	1744	149		1	
4·138	47	4		2	
4·139	77 563 250	6 578 829		1	
141	95	8		1	
4·142	143	12		3	
4·143	12	1		2	
145	12	1		4	
4·146	145	12		2	
149	61	5	2	1	
4·151	1 728 148 040	140 634 693		1	
4·154	21 295	1716		2	
4·155	249	20		2	
157	213	17	2	1	

(Contd.)

discriminant d_F $d_F = m$ or $d_F = 4m$	fundamental unit ε a, b, c of $\varepsilon = (a + bm^{\frac{1}{2}})/c$ c listed only if $c \neq 1$			class number h	type of class group if not cyclic
4·158	7743	616		1	
4·159	1324	105		2	
161	11 775	928		1	
4·163	64 080 026	5 019 135		1	
165	13	1	2	2	
4·166	1 700 902 565	132 015 642		1	
4·167	168	13		1	
4·170	13	1		4	$C_2 \times C_2$
173	13	1	2	1	
4·174	1451	110		2	
177	62 423	4692		1	
4·178	1601	120		2	
4·179	4 190 210	313 191		1	
181	1305	97	2	1	
4·182	27	2		2	
4·183	487	36		2	
185	68	5		2	
4·186	7501	550		2	
4·187	1682	123		2	
4·190	52 021	3774		2	
4·191	8 994 000	650 783		1	
193	1 764 132	126 985		1	
4·194	195	14		2	
4·195	14	1		4	$C_2 \times C_2$
197	14	1		1	
4·199	16 266 196 520	1 153 080 099		1	
201	515 095	36 332		1	
4·202	3141	221		2	
4·203	57	4		2	
205	43	3	2	2	
4·206	59 535	4148		1	
209	46 551	3220		1	
4·210	29	2		4	$C_2 \times C_2$
4·211	278 354 373 650	19 162 705 353		1	
213	73	5	2	1	
4·214	695 359 189 925	47 533 775 646		1	
4·215	44	3		2	
217	3 844 063	260 952		1	
4·218	251	17		2	
4·219	74	5		4	
221	15	1	2	2	
4·222	149	10		2	
4·223	224	15		3	
4·226	15	1		8	
4·227	226	15		1	
229	15	1	2	3	
4·230	91	6		2	
4·231	76	5		4	$C_2 \times C_2$
233	23 156	1517		1	
4·235	46	3		6	
237	77	5	2	1	
4·238	11 663	756		2	
4·239	6 195 120	400 729		1	
241	71 011 068	4 574 225		1	
4·246	88 805	5662		2	

(Contd.)

discriminant d_F $d_F = m$ or $d_F = 4m$	fundamental unit ε a, b, c of $\varepsilon = (a + bm^{\frac{1}{2}})/c$ c listed only if $c \neq 1$			class number h	type of class group if not cyclic
4·247	85 292	5427		2	
249	8 553 815	542 076		1	
4·251	3 674 890	231 957		1	
253	1861	117	2	1	
4·254	255	16		3	
4·255	16	1		4	$C_2 \times C_2$
257	16	1		3	
4·258	257	16		2	
4·259	847 225	52 644		2	
4·262	104 980 517	6 485 718		1	
4·263	139 128	8579		1	
265	6072	373		2	
4·266	685	42		2	
4·267	2402	147		2	
269	82	5		1	
4·271	115 974 983 600	7 044 978 537		1	
273	727	44		2	
4·274	1407	85		4	
277	2613	157	2	1	
4·278	2501	150		1	
281	1 063 532	63 445		1	
4·282	2351	140		2	
4·283	138 274 082	8 219 541		1	
285	17	1	2	2	
4·286	561 835	33 222		2	
4·287	288	17		2	
4·290	17	1		4	$C_2 \times C_2$
4·291	290	17		4	
293	17	1	2	1	
4·295	2 024 999	117 900		2	
4·298	409 557	23 725		2	
4·299	415	24		2	

3.1–7.1. Tables of fundamental units and class numbers of algebraic number fields F of small absolute discriminant and degree $n = s + 2t$, $3 \leq n \leq 7$

Column 1 contains the discriminant of F.

Column 2 contains the coefficients $a(1), \ldots, a(n)$ of a generating polynomial $f(t) = t^n + \sum_{i=0}^{n-1} a(n-i)t^i$, i.e. $F = \mathbb{Q}(\rho)$ for a zero $\rho \in \mathbb{C}$ of $f(t)$.

Column 3 contains a pairwise reduced $\mathbb{Z}$-basis $\omega_1, \ldots, \omega_n$ of $o_F = \mathrm{Cl}(\mathbb{Z}, F)$. The ω_i are given as $\mathbb{Q}$-linear combinations of $1, \rho, \ldots, \rho^{n-1}$.

Column 4 contains the coefficients of the $r = s + t - 1$ fundamental units of F with respect to their presentation by $\omega_1, \ldots, \omega_n$, i.e. one row of entries $e(1), \ldots, e(n)$ represents the fundamental unit $\varepsilon = e(1)\omega_1 + \cdots + e(n)\omega_n$.

Column 5 contains the regulator of F rounded to two decimal places.

The class numbers of all those fields are one except for the complex cubic field of discriminant -283 which has $h_F = 2$. Therefore the class number h_F is not listed separately. For fourth degree fields the Galois groups are also listed in an additional column.

3.1 Table of fundamental units of totally real cubic fields with $d_F < 1000$

discriminant d_F	coeff. of gen. polyn. $a(i), i = 1, \ldots, 3$	int. basis	two fund. units $e(i), i = 1, \ldots, 3$	R_F
49	$1, -2, -1$	$1, \rho, -2 + \rho^2$	$0,1,0$ $1,1,0$	0.53
81	$0, -3, 1$	$1, \rho, -2 + \rho + \rho^2$	$0,1,0$ $-1,1,0$	0.85
148	$3, -1, -1$	$1, \rho, -2 + 2\rho + \rho^2$	$0,1,0$ $4,2,1$	1.66
169	$1, -4, 1$	$1, \rho, -3 + \rho + \rho^2$	$0,1,0$ $-1,1,0$	1.37
229	$0, -4, 1$	$1, \rho, -3 + \rho^2$	$0,1,0$ $2,1,0$	2.36
257	$3, -2, -1$	$1, \rho, -1 + 3\rho + \rho^2$	$0,1,0$ $-1,1,0$	1.97
316	$1, -4, -2$	$1, \rho, -3 + \rho^2$	$-2,1,1$ $-26,11,21$	3.91
321	$1, -4, -1$	$1, \rho, -3 + \rho + \rho^2$	$0,1,0$ $0, -1, 1$	2.57
361	$-2, -5, -1$	$1, \rho, -3 - 2\rho + \rho^2$	$0,1,0$ $1,1,0$	1.95
404	$1, -5, 1$	$1, \rho, -3 + \rho + \rho^2$	$0,1,0$ $2, -2, 1$	3.76
469	$4, -2, -1$	$1, \rho, -1 + 4\rho + \rho^2$	$0,1,0$ $10,3,4$	3.85
473	$0, -5, 1$	$1, \rho, -3 + \rho^2$	$0,1,0$ $-2,1,0$	2.84
564	$-2, -4, 2$	$1, \rho, -3 - 2\rho + \rho^2$	$-8,2,5$ $-3,1,0$	5.40
568	$1, -6, 2$	$1, \rho, -4 + 2\rho + \rho^2$	$1,0,2$ $1,1,-1$	6.09
621	$3, -3, -2$	$1, \rho, -2 + 3\rho + \rho^2$	$-1,1,0$ $9,2,-4$	5.40
697	$3, -4, -1$	$1, \rho, -3 + 3\rho + \rho^2$	$0,1,0$ $-1,1,0$	2.71
733	$-2, -6, -1$	$1, \rho, -4 - 2\rho + \rho^2$	$0,1,0$ $8, -4, 3$	5.31
756	$0, -6, 2$	$1, \rho, -4 + \rho^2$	$1,0,-1$ $17, -3, -9$	5.69
761	$1, -6, 1$	$1, \rho, -4 + \rho + \rho^2$	$0,1,0$ $3,1,0$	3.53
785	$-2, -5, 1$	$1, \rho, -3 - 2\rho + \rho^2$	$0,1,0$ $-1,1,1$	4.10
788	$1, -7, 3$	$1, \rho, -4 + 2\rho + \rho^2$	$2, -1, 0$ $9,2,-4$	5.99
837	$0, -6, 1$	$1, \rho, -4 + \rho^2$	$0,1,0$ $21,14,6$	6.80
892	$-2, -7, -2$	$1, \rho, -4 - 3\rho + \rho^2$	$35, -8, 9$ $-7,2,2$	8.32
940	$3, -4, -2$	$1, \rho, -3 + 3\rho + \rho^2$	$14,4,3$ $10,2,-5$	8.91
961	$-2, -9, 2$	$1, \rho, \frac{1}{2}(-6 - \rho + \rho^2)$	$11,6,4$ $5345, 1244, -3162$	12.18
985	$1, -6, -1$	$1, \rho, -4 + \rho + \rho^2$	$0,1,0$ $3,1,0$	3.72
993	$1, -6, -3$	$1, \rho, -4 + \rho^2$	$1,0,-1$ $-9,2,4$	5.55

3.2 Table of fundamental units of complex cubic fields with $|d_F| < 300$

discriminant d_F	coeff. of gen. polyn. $a(i), i = 1,\ldots,3$	int. basis	1 fund. unit $e(i), i = 1,\ldots,3$	R_F
-23	$0,-1,1$	$1,\rho,$ $-1+\rho^2$	$0,1,0$	0.28
-31	$0,1,1$	$1,\rho,$ ρ^2	$0,1,0$	0.38
-44	$1,-1,1$	$1,\rho,$ $-1+\rho+\rho^2$	$0,1,0$	0.61
-59	$0,2,1$	$1,\rho,$ $1+\rho^2$	$0,1,0$	0.79
-76	$3,1,1$	$1,\rho,$ $-1+2\rho+\rho^2$	$0,1,0$	1.02
-83	$1,1,2$	$1,\rho,$ $\rho+\rho^2$	$1,1,0$	1.04
-87	$-2,-1,-1$	$1,\rho,$ $-1-2\rho+\rho^2$	$0,1,0$	0.93
-104	$3,2,2$	$1,\rho,$ $2\rho+\rho^2$	$1,-1,1$	1.58
-107	$1,3,2$	$1,\rho,$ $1+\rho^2$	$1,1,0$	1.26
-108	$0,0,2$	$1,\rho,$ ρ^2	$1,-1,1$	1.35
-116	$1,0,2$	$1,\rho,$ $\rho+\rho^2$	$1,-2,1$	1.72
-135	$3,0,1$	$1,\rho,$ $-2+2\rho+\rho^2$	$0,1,0$	1.13
-139	$4,6,1$	$1,\rho,$ $1+2\rho+\rho^2$	$0,1,0$	1.66
-140	$0,2,2$	$1,\rho,$ $1+\rho^2$	$1,1,0$	1.47
-152	$4,3,2$	$1,\rho,$ $-2+2\rho+\rho^2$	$1,1,1$	2.13
-172	$-2,0,-2$	$1,\rho,$ $-2\rho+\rho^2$	$1,0,-1$	1.88
-175	$1,2,3$	$1,\rho,$ ρ^2	$1,1,0$	1.29
-199	$4,1,1$	$1,\rho,$ $-1+3\rho+\rho^2$	$0,1,0$	1.34
-200	$1,2,-2$	$1,\rho,$ ρ^2	$-1,1,1$	2.60
-204	$1,1,3$	$1,\rho,$ ρ^2	$-1,1,1$	2.35
-211	$3,1,2$	$1,\rho,$ $-1+2\rho+\rho^2$	$3,1,0$	2.24
-212	$1,4,2$	$1,\rho,$ $2+\rho+\rho^2$	$7,-2,4$	2.71
-216	$0,3,2$	$1,\rho,$ ρ^2	$1,1,-1$	3.02
-231	$1,0,-3$	$1,\rho,$ ρ^2	$2,2,1$	1.75
-239	$3,2,3$	$1,\rho,$ $2\rho+\rho^2$	$1,1,1$	2.10
-243	$3,3,4$	$1,\rho,$ $2\rho+\rho^2$	$1,0,-1$	2.52
-244	$-5,4,-2$	$1,\rho,$ $1-4\rho+\rho^2$	$5,-2,2$	3.30

(Contd.)

3.2 (Contd.)

discriminant d_F	coeff. of gen. polyn. $a(i), i = 1,\dots,3$	int. basis	1 fund. unit $e(i), i = 1,\dots,3$	R_F
-247	0,1,3	$1,\rho,$ ρ^2	1,1,0	1.54
-255	1,0,3	$1,\rho,$ $\rho+\rho^2$	2,1,0	1.99
-268	$-2,-2,-2$	$1,\rho,$ $-2-2\rho+\rho^2$	$3,-1,0$	2.52
-283	0,4,1	$1,\rho,$ $2+\rho^2$	0,1,0	1.40
-300	4,2,2	$1,\rho,$ $-1+3\rho+\rho^2$	6,2,1	3.15

4.1 Table of fundamental units of totally real quartic fields with $d_F < 5000$

discriminant d_F	coeff. of gen. polyn. $a(i)$, $i = 1,\dots,4$	int. basis	three fund. units $e(i)$, $i = 1,\dots,4$	R_F	Galois group
725	$1,-3,-1,1$	$1,\rho,$ $-1+\rho+\rho^2,$ $-1-2\rho+\rho^2+\rho^3$	0,1,0,0 $-1,1,0,0$ 1,1,0,0	0.83	D_8
1125	$1,-4,-4,1$	$1,\rho,$ $-2+\rho^2$ $-1-3\rho+\rho^3$	0,1,0,0 1,1,0,0 $-1,0,1,0$	1.17	C_4
1600	$-4,0,8,-1$	$1,\rho,$ $\frac{1}{2}(-1-2\rho+\rho^2),$ $\frac{1}{2}(3-\rho-3\rho^2+\rho^3)$	0,1,0,0 $0,1,0,-1$ $1,0,-1,0$	1.54	$\mathfrak{V}_4$
1957	$0,-4,1,1$	$1,\rho,$ $-2+\rho^2,$ $1-3\rho+\rho^3$	0,1,0,0 $1,-1,0,0$ $0,0,1,-1$	1.92	$\mathfrak{S}_4$
2000	$-4,1,6,1$	$1,\rho,$ $-1-2\rho+\rho^2,$ $2-3\rho^2+\rho^3$	0,1,0,0 1,1,0,0 $1,0,-1,0$	1.85	C_4
2048	$-4,2,4-1$	$1,\rho,$ $-1-2\rho+\rho^2,$ $2-3\rho^2+\rho^3$	0,1,0,0 $0,1,0,-1$ $1,0,-1,0$	2.44	C_4
2225	$5,4,-5,-1$	$1,\rho,$ $-2+2\rho+\rho^2,$ $\frac{1}{2}(-1+2\rho+4\rho^2+\rho^3)$	0,1,0,0 4,2,1,1 9,4,2,2	2.06	D_8
2304	$0,-4,0,1$	$1,\rho,$ $-2+\rho^2,$ $-3\rho+\rho^3$	0,1,0,0 0,0,1,1 $1,1,-1,-1$	2.66	$\mathfrak{V}_4$
2525	$6,8,-3,-1$	$1,\rho,$ $3\rho+\rho^2,$ $-1+2\rho+4\rho^2+\rho^3$	0,1,0,0 $1,1,-1,0$ $1,0,-1,0$	2.09	D_8
2624	$2,-3,-2,1$	$1,\rho,$ $-1+2\rho+\rho^2,$ $-1-2\rho+2\rho^2+\rho^3$	0,1,0,0 $1,-1,0,0$ 1,1,0,0	2.19	D_8
2777	$2,-3,-5,1$	$1,\rho,$ $-2+\rho+\rho^2,$ $-1-3\rho+\rho^2+\rho^3$	0,1,0,0 $2,-1,-1,2$ $1,0,0,-1$	3.04	$\mathfrak{S}_4$

(Contd.)

4.1 (Contd.)

discriminant d_F	coeff. of gen. polyn. $a(i)$, $i=1,\ldots,4$	int. basis	three fund. units $e(i)$, $i=1,\ldots,4$	R_F	Galois group
3600	$-4,-3,14,1$	$1,\rho,$ $\frac{1}{3}(-2-2\rho+\rho^2),$ $\frac{1}{3}(5-3\rho-3\rho^2+\rho^3)$	0,1,0,0 1,0, − 1,0 2,0,0,1	2.62	$\mathfrak{V}_4$
3981	$1,-4,-2,1$	$1,\rho,$ $-2+\rho+\rho^2,$ $-1-3\rho+\rho^2+\rho^3$	0,1,0,0 1,1,0,0 1,1, − 1,0	3.19	$\mathfrak{S}_4$
4205	$1,-5,1,1$	$1,\rho,$ $-2+2\rho+\rho^2,$ $1-5\rho+\rho^2+\rho^3$	0,1,0,0 1, − 1,0,0 1, − 1,1,0	2.82	D_8
4225	$-4,-3,14,-4$	$1,\rho,$ $\frac{1}{2}(-4-\rho+\rho^2),$ $\frac{1}{4}(4-4\rho-3\rho^2+\rho^3)$	1,1, − 1, − 1 2, − 1,0,3 3, − 1,0,3	3.19	$\mathfrak{V}_4$
4352	$0,-8,8,1$	$1,\rho,$ $-4+\rho+\rho^2,$ $-2-5\rho+2\rho^2+\rho^3$	0,1,0,0 1,0,0, − 1 1,0, − 1,0	4.18	D_8
4400	$-8,17,-4,-1$	$1,\rho,$ $1-4\rho+\rho^2,$ $-2+9\rho-6\rho^2+\rho^3$	0,1,0,0 1, − 1,1,0 1,0, − 1,0	3.29	D_8
4525	$5,2,-10,1$	$1,\rho,$ $-3+2\rho+\rho^2,$ $\frac{1}{3}(-5+\rho+4\rho^2+\rho^3)$	0,1,0,0 1, − 1,0,0 0,0,0,1	3.06	D_8
4752	$2,-3,-4,1$	$1,\rho,$ $-2+\rho+\rho^2,$ $-1-3\rho+\rho^2+\rho^3$	0,1,0,0 1,1,0,0 0, − 1,1,0	3.71	D_8
4913	$1,-6,-1,1$	$1,\rho,$ $-3+\rho+\rho^2,$ $\frac{1}{2}(3-6\rho+\rho^3)$	0,1,0,0 0,0,1,1 5,3,3,2	3.46	C_4

4.2 Table of fundamental units of quartic fields with two complex conjugates and $|d_F| < 300$

discriminant d_F	coeff. of gen. polyn. $a(i)$, $i=1,\ldots,4$	int. basis	two fund. units $e(i)$, $i=1,\ldots,4$	R_F	Galois group
− 275	$1,0,-2,-1$	$1,\rho,$ $\rho^2,$ ρ^3	0,1,0,0 1,1,0,0	0.37	D_8
− 283	$0,-2,1,1$	$1,\rho,$ $\rho^2,$ ρ^3	1,1,0,0 1,− 1,0,0	0.38	$\mathfrak{S}_4$
− 331	$0,-2,3,-1$	$1,\rho,$ $\rho^2,$ ρ^3	0,1,0,0 1,− 1,0,0	0.43	$\mathfrak{S}_4$
− 400	$0,-1,0,-1$	$1,\rho,$ $\rho^2,$ ρ^3	0,1,0,0 1, − 1,0,0	0.51	D_8
− 448	$2,1,2,1$	$1,\rho,$ $\rho^2,$ ρ^3	0,1,0,0 1,1,0,0	0.56	D_8
− 475	$1,-2,2,-1$	$1,\rho,$ $\rho^2,$ ρ^3	0,1,0,0 1,− 1,0,0	0.58	D_8

(Contd.)

4.2 (Contd.)

discriminant d_F	coeff. of gen. polyn. $a(i)$, $i = 1,\ldots,4$	int. basis	two fund. units $e(i)$, $i = 1,\ldots,4$	R_F	Galois group
−491	1, −1, −3, −1	$1, \rho, \rho^2, \rho^3$	0,1,0,0 1,1,0,0	0.63	$\mathfrak{S}_4$
−507	1, −1,1,1	$1, \rho, \rho^2, \rho^3$	0,1,0,0 1,1,0,0	0.65	D_8
−563	1, −1,1, −1	$1, \rho, \rho^2, \rho^3$	0,1,0,0 1, −1,0,0	0.70	$\mathfrak{S}_4$
−643	1,0,2,1	$1, \rho, \rho^2, \rho^3$	0,1,0,0 1,1,0,0	0.72	$\mathfrak{S}_4$
−688	0,0,2, −1	$1, \rho, \rho + \rho^2, 1 + \rho^3$	0,1,0,0 1,0,1,1	1.00	$\mathfrak{S}_4$
−731	0, −2,1, −1	$1, \rho, \rho^2, \rho^3$	0,1,0,0 1, −1,0,0	0.87	$\mathfrak{S}_4$
−751	1, −1,2, −1	$1, \rho, \rho^2, \rho^3$	0,1,0,0 2,1,0,0	1.07	$\mathfrak{S}_4$
−775	1,0,3, −1	$1, \rho, \rho + \rho^2, \frac{1}{2}(1 + 2\rho^2 + \rho^3)$	0,1,0,0 1,1,0,1	0.87	D_8
−848	0, −1,2,1	$1, \rho, \rho^2, \rho^3$	0,1,0,0 1,1,0,0	0.99	$\mathfrak{S}_4$
−976	0, −3,2, −1	$1, \rho, \rho^2, \rho^3$	0,1,0,0 1, −1,0,0	0.99	$\mathfrak{S}_4$
−1024	0, −2,0, −1	$1, \rho, -1 + \rho^2, -2\rho + \rho^3$	0,1,0,0 1,1,1,1	1.35	D_8
−1099	0, −4,3, −1	$1, \rho, \rho^2, \rho^3$	0,1,0,0 1, −1,0,0	1.01	$\mathfrak{S}_4$
−1107	2,0,1, −1	$1, \rho, 2\rho + \rho^2, 1 + 2\rho^2 + \rho^3$	0,1,0,0 1,1, −1,0	1.29	$\mathfrak{S}_4$
−1156	1, −2,1,1	$1, \rho, -1 + \rho + \rho^2, -2\rho + \rho^2 + \rho^3$	0,1,0,0 1,1,1,1	1.53	D_8
−1192	1,2, −1, −1	$1, \rho, \rho^2, -1 + 2\rho + \rho^2 + \rho^3$	0,1,0,0 3,2,1,1	1.57	$\mathfrak{S}_4$
−1255	0, −1,3, −1	$1, \rho, -1 + \rho + \rho^2, 1 - \rho + \rho^3$	0,1,0,0 1, −1,1,2	1.58	$\mathfrak{S}_4$

4.3 Table of fundamental units of totally complex quartic fields with $d_F < 600$

discriminant d_F	coeff. of gen. polyn. $a(i)$, $i = 1,\ldots,4$	int. basis	one fund. unit $e(i)$, $i = 1,\ldots,4$	R_F	Galois group
117	0,2,3,1	$1,\rho,$ $1+\rho^2,$ $2+2\rho+\rho^3$	0,1,−1,1	0.54	D_8
125	1,1,1,1**	$1,\rho,$ $\rho^2,$ ρ^3	1,1,0,0	0.96	C_4
144	0,−1,0,1	$1,\rho,$ $\rho^2,$ ρ^3	1,−1,0,0	1.32	C_4
189	2,0,−1,1	$1,\rho,$ $\rho^2,$ ρ^3	0,1,0,0	0.86	D_8
225	1,2,−1,1	$1,\rho,$ $1+\rho+\rho^2,$ $\frac{1}{2}(1+2\rho+2\rho^2+\rho^3)$	0,1,0,0	0.96	$\mathfrak{V}_4$
229	0,0,1,1	$1,\rho,$ $\rho^2,$ ρ^3	0,1,0,0	0.34	$\mathfrak{S}_4$
256	0,2,4,2	$1,\rho,$ $\rho^2,$ $\frac{1}{3}(1+\rho^2+\rho^3)$	0,1,0,1	1.76	$\mathfrak{V}_4$
257	0,1,1,1	$1,\rho,$ $\rho^2,$ ρ^3	0,1,0,0	0.44	$\mathfrak{S}_4$
272	0,1,2,1	$1,\rho$ $\rho^2,$ ρ^3	0,1,0,0	0.73	D_8
320	2,0,−2,1	$1,\rho,$ $\rho^2,$ ρ^3	0,1,0,0	1.06	D_8
333	5,7,3,3	$1,\rho,$ $2\rho+\rho^2,$ $1+2\rho+3\rho^2+\rho^3$	1,0,−1,0	1.46	D_8
392	2,6,−2,1	$1,\rho,$ $\frac{1}{2}(3+2\rho+\rho^2),$ $\frac{1}{4}(-3+5\rho+\rho^2+\rho^3)$	0,0,1,1	0.63	D_8
400	0,3,0,1	$1,\rho$ $\rho^2,$ ρ^3	0,1,0,0	0.96	$\mathfrak{V}_4$
432	−4,3,2,1	$1,\rho,$ $-2\rho+\rho^2,$ $2\rho-3\rho^2+\rho^3$	0,1,0,0	1.66	D_8
441	0,5,0,1	$1,\rho,$ $\frac{1}{2}(3+\rho+\rho^2),$ $\frac{1}{2}(1+4\rho+\rho^3)$	0,1,0,0	1.57	$\mathfrak{V}_4$
512	0,2,0,2	$1,\rho,$ $1+\rho^2,$ $\rho+\rho^3$	0,1,−1,0	1.53	C_4
513	4,9,16,13	$1,\rho,$ $\frac{1}{2}(1+\rho+\rho^2),$ $\frac{1}{2}(5+4\rho+2\rho^2+\rho^3)$	1,0,−1,0	1.96	D_8
549	6,10,3,1	$1,\rho,$ $-1+2\rho+\rho^2,$ $3\rho+4\rho^2+\rho^3$	0,1,0,0	2.11	D_8

(Contd.)

4.3 (Contd.)

discriminant d_F	coeff. of gen. polyn. $a(i)$, $i = 1, \ldots, 4$	int. basis	one fund. unit $e(i)$, $i = 1, \ldots, 4$	R_F	Galois group
576	$0, -2, 0, 4$	$1, \rho,$ $\frac{1}{2}\rho^2,$ $\frac{1}{2}\rho^3$	1,1,1,0	2.29	$\mathfrak{D}_4$
576	0,2,0,4	$1, \rho,$ $\frac{1}{2}\rho^2,$ $\frac{1}{2}\rho^3$	0,1,1,1	1.76	$\mathfrak{D}_4$
592	0,2,2,1	$1, \rho,$ $\rho^2,$ ρ^3	0,1,0,0,	0.92	$\mathfrak{S}_4$

5.1 Table of fundamental units of totally real quintic fields with $d_F < 102\,000$

discriminant d_F	coeff. of gen. polyn. $a(i), i = 1, \ldots, 5$	int. basis	four fund. units $e(i)$, $i = 1, \ldots, 5$	R_F
14 641	$1, -4, -3, 3, 1$	$1, \rho,$ $-2 + \rho^2,$ $-3\rho + \rho^3,$ $1 - 2\rho - 3\rho^2 + \rho^3 + \rho^4$	$0,1,0,0,0$ $1,-1,0,0,0$ $1,1,0,0,0$ $1,0,-1,0,0$	1.64
24 217	$0, -5, -1, 3, 1$	$1, \rho,$ $-2 + \rho^2,$ $-1 - 4\rho + \rho^3,$ $2 - 5\rho^2 + \rho^4$	$0,1,0,0,0$ $1,-1,0,0,0$ $1,1,0,0,0$ $1,-1,1,0,0$	2.40
36 497	$1, -5, -3, 2, 1$	$1, \rho,$ $-2 + \rho^2,$ $-2 - 4\rho + \rho^2 + \rho^3,$ $1 + 2\rho - 5\rho^2 + \rho^4$	$0,1,0,0,0$ $1,1,0,0,0$ $1,2,1,-4,-2$ $3,4,2,-5,-3$	3.55
38 569	$1, -5, -1, 4, -1$	$1, \rho,$ $-2 + \rho + \rho^2,$ $-3\rho + \rho^2 + \rho^3,$ $3 - 2\rho - 5\rho^2 + \rho^3 + \rho^4$	$0,1,0,0,0$ $1,-1,0,0,0$ $1,1,0,0,0$ $1,0,1,0,0$	3.16
65 657	$1, -5, -2, 5, -1$	$1, \rho,$ $-2 + \rho + \rho^2,$ $-1 - 3\rho + \rho^2 + \rho^3,$ $2 - 2\rho - 4\rho^2 + \rho^3 + \rho^4$	$0,1,0,0,0$ $1,-1,0,0,0$ $1,1,-1,0,0$ $1,0,1,0,0$	5.50
70 601	$1, -5, -2, 3, 1$	$1, \rho,$ $-2 + \rho + \rho^2,$ $-1 - 4\rho + \rho^2 + \rho^3,$ $2 - 2\rho - 5\rho^2 + \rho^3 + \rho^4$	$0,1,0,0,0$ $1,-1,0,0,0$ $1,1,0,0,0$ $0,-1,1,0,1$	4.61
81 509	$2, -4, -6, 4, 1$	$1, \rho,$ $-2 + \rho + \rho^2,$ $-1 - 3\rho + \rho^2 + \rho^3,$ $1 - 5\rho - 3\rho^2 + 2\rho^3 + \rho^4$	$0,1,0,0,0$ $6,3,0,-2,-3$ $2,1,-1,-1,-1$ $1,0,-1,0,0$	7.63
81 589	$2, -4, -8, 0, 1$	$1, \rho,$ $-2 + \rho^2,$ $-4\rho + \rho^3,$ $-4\rho - 4\rho^2 + \rho^3 + \rho^4$	$0,1,0,0,0$ $1,2,-2,0,3$ $3,1,-2,1,3$ $2,0,-1,0,1$	7.61
89 417	$2, -4, -7, 2, 1$	$1, \rho,$ $-2 + \rho + \rho^2,$ $1 - 4\rho + \rho^3,$ $1 - 3\rho - 4\rho^2 + \rho^3 + \rho^4$	$0,1,0,0,0$ $1,1,-1,0,0$ $2,1,0,0,0$ $1,0,1,0,0$	6.74

(Contd.)

5.1 (Contd.)

discriminant d_F	coeff. of gen. polyn. $a(i)$, $i=1,\ldots,5$	int. basis	four fund. units $e(i)$, $i=1,\ldots,5$	R_F
101 833	2, − 5, − 5,1,1	$1,\rho,$ $-2+\rho+\rho^2,$ $-3-4\rho+2\rho^2+\rho^3,$ $1+\rho-6\rho^2+\rho^3+\rho^4$	0,1,0,0,0 1,1,0,0,0 130, − 228,110, − 27,13 1,1,0, − 1,0	6.33

5.2 Table of fundamental units of quintic fields with two complex conjugates and $|d_F| < 10\,000$

discriminant d_F	coeff. of gen. polyn. $a(i)$, $i=1,\ldots,5$	int. basis	three fund. units $e(i)$, $i=1,\ldots,5$	R_F
− 4511	0, − 2, − 1,0,1	$1,\rho,$ $-1+\rho^2,$ $-2\rho+\rho^3,$ $-\rho-2\rho^2+\rho^4$	0,1,0,0,0 1, − 1,0,0,0 1,1,0,0,0	0.63
− 4903	1, − 3, − 1,2, − 1	$1,\rho,$ $-1+\rho+\rho^2,$ $1-2\rho+\rho^3,$ $2-\rho-3\rho^2+\rho^3+\rho^4$	0,1,0,0,0 0,0,1,0,0 1,1,0,0,0	0.67
− 5519	0, − 3, − 1,1,1	$1,\rho,$ $-1+\rho^2,$ $-2\rho+\rho^3,$ $1-\rho-3\rho^2+\rho^4$	0,1,0,0,0 1, − 1,0,0,0 1,1,0,0,0	0.73
− 5783	0, − 2,3,4,1	$1,\rho,$ $-1+\rho^2,$ $1-\rho+\rho^3,$ $3+3\rho-2\rho^2+\rho^4$	0,1,0,0,0 1,1,0,0,0 0,1,1, − 1,2	0.76
− 7031	0, − 1, − 1, − 1,1	$1,\rho,$ $-1+\rho^2,$ $-\rho+\rho^3,$ $-1-\rho-\rho^2+\rho^4$	0,1,0,0,0 1, − 1,0,0,0 1,1,0,0,0	0.89
− 7367	0, − 4, − 1,2,1	$1,\rho,$ $-2+\rho^2,$ $-3\rho+\rho^3,$ $2-\rho-4\rho^2+\rho^4$	0,1,0,0,0 1, − 1,0,0,0 1,1,0,0,0	0.90
− 7463	0, − 4, − 1,4,1	$1,\rho,$ $-2+\rho^2,$ $-1-2\rho+\rho^3,$ $2-\rho-3\rho^2+\rho^4$	0,1,0,0,0 1, − 1,1,0,0 1,1,0,0,0	0.93
− 8519	1,0,1, − 1, − 1	$1,\rho,$ $-1+\rho^2,$ $\rho^2+\rho^3,$ $-1+\rho+\rho^3+\rho^4$	0,1,0,0,0 1, − 1,0,0,0 1,1,0,0,0	1.00
− 8647	0, − 3, − 2,2,1	$1,\rho$ $-1-\rho+\rho^2,$ $-1-2\rho+\rho^3,$ $1-\rho-3\rho^2+\rho^4$	0,1,0,0,0 1, − 1,0,0,0 1,1,0,0,0	1.03
− 9439	0, − 3, − 2,4,1	$1,\rho,$ $-1+\rho^2,$ $-1-\rho+\rho^3,$ $2-\rho-2\rho^2+\rho^4$	0,1,0,0,0 1, − 1,0,0,0 1,0, − 1,0,0	1.21

5.2 (Contd.)

discriminant d_F	coeff. of gen. polyn. $a(i)$, $i = 1,\ldots,5$	int. basis	three fund. units $e(i)$, $i = 1,\ldots,5$	R_F
-9759	$1,-3,-2,1,-1$	$1,\rho,$ $-2+\rho^2,$ $-2\rho+\rho^3,$ $1-2\rho-3\rho^2+\rho^3+\rho^4$	0,1,0,0,0 1,1,0,0,0 1,0,2,1,1	1.24

5.3 Table of fundamental units of quintic fields with four complex conjugates and $d_F < 4000$

discriminant d_F	coeff. of gen. polyn. $a(i)$, $i = 1,\ldots,5$	int. basis	two fund. units $e(i)$, $i = 1,\ldots,5$	R_F
1609	$0,-3,0,2,1$	$1,\rho,$ $-1+\rho^2,$ $-2\rho+\rho^3,$ $\rho-2\rho^2+\rho^4$	0,0,0,1,0 0,0,1,1,0	0.27
1649	$0,-3,-1,3,1$	$1,\rho,$ $-1+\rho^2,$ $-1-\rho+\rho^3,$ $1-2\rho^2+\rho^4$	1,1,0,0,0 $1,-1,0,0,0$	0.27
1777	$0,-2,-1,2,1$	$1,\rho,$ $-1+\rho^2,$ $-1-\rho+\rho^3,$ $1-\rho-\rho^2+\rho^4$	$1,0,-1,0,1$ 1,1,0,0,0	0.29
2209	$0,-1,2,-2,1$	$1,\rho,$ $-1+\rho+\rho^2,$ $1-\rho+\rho^2+\rho^3,$ $-2+2\rho-\rho^2+\rho^4$	0,1,0,0,0 $1,-1,0,0,0$	0.35
2297	0,1,1,1,1	$1,\rho,$ $\rho^2,$ $1+\rho+\rho^3,$ $\rho+\rho^4$	0,1,0,0,0 1,1,0,0,0	0.36
2617	$0,-2,3,-2,1$	$1,\rho,$ $-1+\rho+\rho^2,$ $1-2\rho+\rho^2+\rho^3,$ $-1+2\rho-3\rho^2+\rho^4$	0,1,0,0,0 $1,0,-1,0,0$	0.39
2665	$0,1,0,-2,1$	$1,\rho,$ $\rho^2,$ $1+\rho+\rho^2+\rho^3,$ $-1+\rho+2\rho^2+\rho^3+\rho^4$	0,1,0,0,0 $1,-1,0,0,0$	0.40
2869	$0,-2,0,1,1$	$1,\rho,$ $-1+\rho^2,$ $-\rho+\rho^3,$ $1-2\rho^2+\rho^4$	1,1,0,0,0 $1,-1,0,0,0$	0.43
3017	$0,-1,0,0,1$	$1,\rho,$ $-1+\rho^2,$ $-\rho+\rho^3,$ $-\rho^2+\rho^4$	0,1,0,0,0 $1,-1,0,0,0$	0.44
3089	$0,-1,0,2,1$	$1,\rho,$ $\rho^2,$ $-\rho^2+\rho^3,$ $-\rho^2+\rho^4$	0,1,0,0,0 1,1,0,0,0	0.49

6.1 Table of fundamental units of totally real sextic fields with $d_F < 1\,229\,000$

discriminant d_F	coeff. of gen. polyn. $a(i)$, $i = 1, \ldots, 6$	int. basis	five fund. units $e(i)$, $i = 1, \ldots, 6$	R_F
300 125	1, −7, −2,7,2, −1	$1, \rho,$	0,1,0,0,0,0	3.28
		$-2+\rho+\rho^2,$	1, −1,0,0,0,0	
		$-1-5\rho+\rho^2+\rho^3,$	1,1,0,0,0,0	
		$2-\rho-6\rho^2+\rho^3+\rho^4,$	0,1,1,1,0,0	
		$-4+3\rho+6\rho^2-7\rho^3+\rho^5$	1,0, −1, −1,0,0	
371 293	1, −5, −4,6,3, −1	$1, \rho,$	0,1,0,0,0,0	3.78
		$-2+\rho^2,$	1, −1,0,0,0,0	
		$-3\rho+\rho^3,$	1,1,0,0,0,0	
		$2-4\rho^2+\rho^4,$	1, −1,1,0,0,0	
		$1+3\rho-3\rho^2-4\rho^3+\rho^4+\rho^5$	1,0, −1,0,0,0	
434 581	0, −8,0,12, −7,1	$1, \rho,$	0,1,0,0,0,0	4.19
		$-3+\rho^2,$	1, −1,0,0,0,0	
		$3-5\rho-\rho^2+\rho^3,$	1,0,0,1,0,0	
		$3-2\rho-6\rho^2+\rho^4,$	2,1,0,0,0,0	
		$-3+7\rho-\rho^2-7\rho^3+\rho^5$	1,1,0, −1,0,0	
453 789	1, −6, −6, −8,8,1	$1, \rho,$	0,1,0,0,0,0	4.40
		$-2+\rho^2,$	1,1,0,0,0,0	
		$-3\rho+\rho^3,$	0,0,1,0,0,0	
		$2+3\rho-4\rho^2-\rho^3+\rho^4,$	1, −1,1,0,0,0	
		$2+5\rho-\rho^2-5\rho^3+\rho^5$	1,0, −1,0,0,0	
485 125	2, −4, −8,2,5,1	$1, \rho,$	0,1,0,0,0,0	4.53
		$-2+\rho^2,$	−1,1,0,0,0,0	
		$-2-3\rho+\rho^2+\rho^3,$	1,1,0,0,0,0	
		$2+\rho-4\rho^2+\rho^4,$	1, −1,1,0,0,0	
		$3+5\rho-4\rho^2-5\rho^3+\rho^4+\rho^5$	0,1,1,0,0,0	
592 661	−5,2,18, −11, −19,1	$1, \rho,$	0,1,0,0,0,0	5.23
		$-3-\rho+\rho^2,$	1,1,0,0,0,0	
		$2-2\rho-2\rho^2+\rho^3,$	1,0, −1,1,0,0	
		$2+7\rho-2\rho^2-3\rho^3+\rho^4,$	1,0, −1,0,0,0	
		$6\rho+8\rho^2-3\rho^3-3\rho^4+\rho^5$	1,0,1,0,0,0	

(Contd.)

6.1 (Contd.)

discriminant d_F	coeff. of gen. polyn. $a(i)$, $i=1,\dots,6$	int. basis	five fund. units $e(i)$, $i=1,\dots,6$	R_F
703 493	1, −7, −2,14, −5, −1	$1,\rho,$	0,1,0,0,0,0	5.71
		$-2+\rho+\rho^2,$	1, −1,0,0,0,0	
		$-3\rho+\rho^2+\rho^3,$	0,1, −1,0,0,0	
		$2-4\rho-3\rho^2+2\rho^3+\rho^4,$	1,0, −1,0,0,0	
		$-1+5\rho-5\rho^2-4\rho^3+2\rho^4+\rho^5$	0,0,1,0,0,0	
722 000	1, −6, −7,4,5,1	$1,\rho,$	0,1,0,0,0,0	6.41
		$-2+\rho^2,$	1, −1,0,0,0,0	
		$-1-4\rho+\rho^3,$	1,1,0,0,0,0	
		$2-\rho-5\rho^2+\rho^4,$	1,1,0, −1,0,0	
		$3+6\rho-2\rho^2-6\rho^3+\rho^5$	0,1,1,0,0,0	
810 448	3, −2, −9,0,5,1	$1,\rho,$	0,1,0,0,0,0	6.89
		$-2+\rho+\rho^2,$	1, −1,0,0,0,0	
		$-1-3\rho+\rho^2+\rho^3,$	1,1,0,0,0,0	
		$-3\rho-2\rho^2+2\rho^3+\rho^4,$	0,1,0, −1,0,0	
		$2+4\rho-6\rho^2-4\rho^3+2\rho^4+\rho^5$	0,0,1,0,0,0	
820 125	0, −9,4,9, −3, −1	$1,\rho,$	0,1,0,0,0,0	6.28
		$-3+\rho+\rho^2,$	1, −1,0,0,0,0	
		$-1-6\rho+\rho^2+\rho^3,$	23, −23,7,28,17, −50	
		$3+4\rho-8\rho^2+\rho^4,$	−16,13, −5, −9, −6,19	
		$\frac{1}{19}(11-36\rho-39\rho^2+2\rho^3+7\rho^4+\rho^5)$	1,1,0,0,0, −1	
905 177	1, −7, −9,7,9, −1	$1,\rho,$	0,1,0,0,0,0	6.91
		$-2+\rho^2,$	1, −1,0,0,0,0	
		$1-4\rho-\rho^2+\rho^3,$	1,1,0,0,0,0	
		$3+\rho-5\rho^2-\rho^3+\rho^4,$	1,1, −1,0,0,0	
		$-2+7\rho+3\rho^2-6\rho^3-\rho^4+\rho^5$	1,1,1,0,0,0	
966 125	3, −3, −10,3,8, −1	$1,\rho,$	0,1,0,0,0,0	7.43
		$-2+\rho+\rho^2,$	1, −1,0,0,0,0	
		$-2-2\rho+2\rho^2+\rho^3$	1,1,0,0,0,0	
		$-5\rho-\rho^2+3\rho^3+\rho^4,$	0,1, −1,0,0,0	
		$-1+5\rho-3\rho^2-4\rho^3+2\rho^4+\rho^5$	1,0, −1,0,0,0	

980 125	0, − 9,9,4, − 3, − 1	$1, \rho,$	0,1,0,0,0,0	7.12
		$-3 + \rho + \rho^2,$	1, − 1,0,0,0,0	
		$-2 - 5\rho + 2\rho^2 + \rho^3,$	4,3,1,1,2,1	
		$1 + 3\rho - 7\rho^2 + \rho^3 + \rho^4,$	0,0,0,1,0,0	
		$-1 - 3\rho + 10\rho^2 - 8\rho^3 + \rho^5$	8,17,10, − 36,22,30	
1 075 648	6,8, − 8, − 13,6,1	$1, \rho,$	0,1,0,0,0,0	7.70
		$-1 + 2\rho + \rho^2,$	1, − 1,0,0,0,0	
		$-2 + 3\rho^2 + \rho^3,$	1,1, − 1,0,0,0	
		$-1 - 4\rho + 2\rho^2 + 4\rho^3 + \rho^4,$	0,1, − 1,0,0,0	
		$1 - 5\rho - 5\rho^2 + 5\rho^3 + 5\rho^4 + \rho^5$	1,0, − 1,0,0,0	
1 081 856	0, − 6,2,7, − 2, − 1	$1, \rho,$	0,1,0,0,0,0	7.76
		$-2 + \rho^2,$	1, − 1,0,0,0,0	
		$-1 - 3\rho + \rho^2 + \rho^3,$	1,1,0,0,0,0	
		$2 - 2\rho - 4\rho^2 + \rho^3 + \rho^4,$	1, − 1,1,0,0,0	
		$5\rho - 2\rho^2 - 5\rho^3 + \rho^4 + \rho^5$	1,0, − 1,0,0,0	
1 134 389	1, − 6, − 7,5,6,1	$1, \rho,$	0,1,0,0,0,0	7.82
		$-2 + \rho^2,$	1, − 1,0,0,0,0	
		$-1 - 4\rho + \rho^2 + \rho^3,$	1,1,0,0,0,0	
		$4 + 3\rho - 5\rho^2 - \rho^3 + \rho^4,$	1,1,1,0,0,0	
		$3 + 7\rho - 2\rho^2 - 6\rho^3 + \rho^5$	1, − 1,1,0,0,0	
1 202 933	1, − 6, − 2,6,0, − 1	$1, \rho,$	0,1,0,0,0,0	8.74
		$-2 + \rho + \rho^2,$	1, − 1,0,0,0,0	
		$-1 - 4\rho + \rho^2 + \rho^3,$	1,1,0,0,0,0	
		$2 - \rho - 5\rho^2 + \rho^3 + \rho^4,$	1,0,1,0,0,0	
		$6\rho - 2\rho^2 - 6\rho^3 + \rho^4 + \rho^5$	0,1,1,0,0,0	

6.2 *Fundamental units of the sextic field with two complex conjugates and minimum discriminant*

d_F	coeff. of gen. polyn. $a(i)$, $i = 1,\ldots,6$	int. basis	four fund. units $e(i)$, $i = 1,\ldots,6$	R_F
−92 779	1, −2, −3, −1,2,1	$1,\rho,$ $-1+\rho^2,$ $-2\rho+\rho^3,$ $-2\rho^2+\rho^4,$ $2+\rho-\rho^2-3\rho^3+\rho^5$	0,1,0,0,0,0 1, −1,0,0,0,0 1,1,0,0,0,0 0,1, −1,0,0,0	1.26

6.3 *Fundamental units of the sextic field with four complex conjugates and minimum discriminant*

d_F	coeff. of gen. polyn. $a(i)$, $i = 1,\ldots,6$	int. basis	three fund. units $e(i)$, $i = 1,\ldots,6$	R_F
28 037	2,0, −3,0,2, −1	$1,\rho,$ $\rho+\rho^2,$ $-1+\rho+2\rho^2+\rho^3,$ $-\rho+\rho^2+2\rho^3+\rho^4,$ $1-2\rho-2\rho^2+2\rho^3+3\rho^4+\rho^5$	1,0, −1,0,0,0 1, −1,0,0,0,0 1,1,0,0,0,0	0.48

6.4 *Fundamental units of totally complex sextic fields with $|d_F| < 23\,100$*

discriminant d_F	coeff. of gen. polyn. $a(i)$, $i = 1,\ldots,6$	int. basis	two fund. units $e(i)$, $i = 1,\ldots,6$	R_F
−9747	0,1,1, −2, −1,1	$1,\rho,$ $\rho^2,$ $\rho+\rho^3,$ $-1+\rho+2\rho^2+\rho^4,$ $-1-2\rho+\rho^2+\rho^3+\rho^5$	0,0,0,0,0,1 1,1, −1,1,2,3	0.60
−10 051	1,2,2,2,2,1	$1,\rho,$ $\rho^2,$ $\rho+\rho^3,$ $\rho^2+\rho^4,$ $\rho^3+\rho^5$	0,1,0,0,0,0 1,1,0,0,0,0	0.21
−10 571	2,2,1,2,2,1	$1,\rho,$ $\rho^2,$ $\rho^2+\rho^3,$ $1+\rho^2+\rho^3+\rho^4,$ $1+2\rho+\rho^4+\rho^5$	0,0, −1,0,1,0 0,0, −1,0,1,1	0.21
−10 816	2,0, −2, −1,0,1	$1,\rho,\rho^2,$ ρ^3,ρ^4,ρ^5	0,1,0,0,0,0 1,1,0,0,0,0	0.43
−11 691	1, −1,0,0, −1,1	$1,\rho,$ $\rho+\rho^2,$ $\rho^2+\rho^3,$ $\rho^3+\rho^4,$ $-1+\rho^2+\rho^4+\rho^5$	0,1,0,0,0,0 1, −1,0,0,0,0	0.69

(Contd.)

6.4 (Contd.)

discriminant d_F	coeff. of gen. polyn. $a(i)$, $i = 1, \ldots, 6$	int. basis	two fund. units $e(i)$, $i = 1, \ldots, 6$	R_F
− 12 167	3,5,5,5,3,1	$1, \rho,$ $1 + \rho + \rho^2,$ $\rho + \rho^2 + \rho^3,$ $1 + \rho + 2\rho^2 + 2\rho^3 + \rho^4,$ $2\rho + 2\rho^2 + 3\rho^3 + 2\rho^4 + \rho^5$	0,1,0,− 1,0,1 0,0,0,0,0,1	0.24
− 14 283	1,1,2,1,0,1	$1, \rho,$ $\rho^2,$ $1 + \rho^3,$ $\rho + \rho^4,$ $\rho + \rho^2 + \rho^3 + \rho^4 + \rho^5$	0,1,0,0,0,0 1,1,0,0,0,0	0.80
− 14 731	1,0,− 1,− 1,0,1	$1, \rho,$ $\rho^2,$ $\rho^3,$ $\rho^3 + \rho^4,$ $-1 - \rho + \rho^3 + \rho^4 + \rho^5$	0,1,0,0,0,0 1,1,0,0,0,0	0.28
− 16 551	2,2,3,3,1,1	$1, \rho,$ $\rho + \rho^2,$ $1 + \rho^2 + \rho^3,$ $2 + 2\rho + 2\rho^2 + 2\rho^3 + \rho^4,$ $-1 + \rho + \rho^2 + \rho^4 + \rho^5$	0,1,0,0,0,0 1,1,0,0,0,0	0.93
− 16 807	1,1,1,1,1,1	$1, \rho,$ $\rho^2,$ $\rho^3,$ $\rho^4,$ ρ^5	1,1,0,0,0,0 1,− 1,1,0,0,0,	2.10
− 18 515	0,2,1,2,0,1	$1, \rho,$ $1 + \rho^2,$ $\rho + \rho^3,$ $1 + \rho + \rho^2 + \rho^4,$ $\rho + \rho^2 + 2\rho^3 + \rho^5$	0,1,0,0,0,0 0,0,1,0,0,0	0.33
− 19 683	0,0,1,0,0,1	$1, \rho,$ $\rho^2,$ $\rho^3,$ $\rho^4,$ ρ^5	1,1,0,0,0,0 1,0,1,0,0,0	3.40
− 20 627	1,1,2,2,1,1	$1, \rho,$ $\rho^2,$ $1 + \rho^3,$ $\rho + \rho^4,$ $\rho + \rho^2 + \rho^4 + \rho^5$	0,1,0,0,0,0 1,0,1,0,0,0	0.39
− 21 168	1,− 2,− 1,4,− 3,1	$1, \rho,$ $-1 + \rho + \rho^2,$ $-1 + 2\rho^2 + \rho^3,$ $2 - 2\rho + 2\rho^3 + \rho^4,$ $-2 + 4\rho - \rho^2 - 2\rho^3 + \rho^4 + \rho^5$	0,1,0,0,0,0 1,− 1,0,0,0,0	1.12

(Contd.)

6.4 (Contd.)

discriminant d_F	coeff. of gen. polyn. $a(i)$, $i = 1,\ldots,6$	int. basis	two fund. units $e(i)$, $i = 1,\ldots,6$	R_F
−21 296	1,2,3,2,1,1	$1,\rho,$ $\rho^2,$ $1+\rho+\rho^3,$ $\rho+\rho^2+\rho^4,$ $\rho^2+\rho^3+\rho^5,$	−1, −1,0,1,1,1 0,1,0,0,0,0	0.37
−22 291	1,0,1,1,0,1	$1,\rho,$ $\rho^2,$ $1+\rho^3,$ $1+\rho+\rho^3+\rho^4,$ $\rho+\rho^2+\rho^4+\rho^5$	0,1,0,0,0,0 1,1,0,0,0,0	0.37
−22 592	0, −1,0,2,2,1	$1,\rho,$ $\rho^2,$ $-\rho^2+\rho^3,$ $1+\rho-\rho^3+\rho^4,$ $2+2\rho-\rho^3+\rho^5$	0,1,0,0,0,0 1,1,0,0,0,0	0.75
−22 707	1,4,4,5,3,1	$1,\rho,$ $1+\rho^2,$ $2\rho+\rho^3,$ $2+\rho+3\rho^2+\rho^4,$ $1+2\rho+\rho^2+3\rho^3+\rho^5$	0,1,0,0,0,0 0,0,1,0,0,0	1.26
−22 747	1,0,2,1, −1,1	$1,\rho,$ $\rho^2,$ $1+\rho^2+\rho^3,$ $-1+\rho+\rho^4,$ $-1+\rho+\rho^2+\rho^4$ $+\rho^5$	1, −1,1,0,0,0 1,1,0,0,0,0	0.39
−23 031	0,1,1,1,2,1	$1,\rho,$ $\rho^2,$ $\rho^3,$ $\rho+\rho^4,$ $1+\rho^2+\rho^5$	0,1,0,0,0,0 1,0,1,0,0,0	1.19

7.1 Fundamental units of the totally real seventh degree field with minimum discriminant

d_F	coeff. of gen. polyn. $a(i)$, $i = 1,\ldots,7$	int. basis	six fund. units $e(i)$, $i = 1,\ldots,7$	R_F
20 134 393	1, −6, −5,8,5, −2, −1	$1,\rho,\rho^2,$ $\rho^3,\rho^4,\rho^5,\rho^6$	−2, −4,4,5, −1, −1,0 −2,1,8, −4, −6,1,1 −3,2,15, −4, −12,1,2 0,1,0,0,0,0,0 −2,0,1,0,0,0,0 2,2, −8, −1,6,0, −1	14.45

8 Integral bases

We present two examples for the computation of the maximal order of an equation order by the embedding algorithm of chapter 4, section 6.

1. For $f(t) = t^{11} + 101t^{10} + 4151t^9 + 87851t^8 + 976826t^7 + 4621826t^6 - 5948674t^5 - 113111674t^4 - 12236299t^3 + 1119536201t^2 - 1660753125t - 332150625$

we obtain the reduced discriminant

$$d_r(f) = 810256533911915751014400000 = 2^{22} \times 3^{12} \times 5^4 \times 29^4 \times 82231$$

and the discriminant

$$d(f) = 2^{130} \times 3^{12} \times 5^{12} \times 29^{18} \times 82231^6$$

The algorithm produces:
$p = 2$
idempotents $-1421478951492431/256\xi^{10} + 86970691550133/32\xi^9 + 959425090967179/128\xi^8 - 140484061157699/32\xi^7 - 654028913747701/128\xi^6 - 139900389254233/32\xi^5 - 3264518827489/4\xi^4 - 234005131717649/32\xi^3 - 650184017173519/256\xi^2 + 33744914112585/4\xi - 333711335100551/128$

$1421478951492431/256\xi^{10} - 86970691550133/32\xi^9 - 959425090967179/128\xi^8 + 140484061157699/32\xi^7 + 654028913747701/128\xi^6 + 139900389254233/32\xi^5 + 3264518827489/4\xi^4 + 234005131717649/32\xi^3 + 650184017173519/256\xi^2 - 33744914112585/4\xi + 333711335100679/128$

the factorization

$t^7 + 851162717265t^6 + 14329665348285t^5 + 2322157331885t^4 + 10822901210111t^3 + 3196794082595t^2 + 2100643537999t + 15204983818911,$

$t^4 + 16741023327252t^3 + 2982968628006t^2 + 8480582660020t + 809919201793,$

further idempotents

$390730948745463/128\xi^6 - 1974053779852231/64\xi^5 - 746745761669899/128\xi^4 - 191811737241957/32\xi^3 - 1004454580604047/128\xi^2 - 179581618641623/64\xi - 433530700975229/128,$

$-390730948745463/128\xi^6 + 197405377985231/64\xi^5 + 746745761669899/128\xi^4 + 191811737241957/32\xi^3 + 1004454580604047/128\xi^2 + 179581618641623/64\xi + 433530700975357/128,$

a further factorization

$t^6 + 5196405170734t^5 + 14148831447667t^4 + 16109228567028t^3 + 8507932353751t^2 + 11870227502526t + 10540983492437$
$t + 13246943590947,$

the 2-minimal basis

$\omega_{11} = 1/2048\xi^{10} + 1/1024\xi^9 + 1/2048\xi^8 + 1/256\xi^7 + 1/1024\xi^6 + 7/512\xi^5 + 21/1024\xi^4 - 11/256\xi^3 - 99/2048\xi^2 + 25/1024\xi + 53/2048$

$\omega_{10} = 1/512\xi^9 + 1/512\xi^8 - 3/256\xi^5 - 3/256\xi^4 + 1/64\xi^3 + 1/64\xi^2 - 3/512\xi - 3/512$

$\omega_9 = 1/256\xi^8 - 3/128\xi^4 + 1/32\xi^2 - 3/256$

$\omega_8 = 1/128\xi^7 + 1/128\xi^6 + 1/128\xi^5 + 1/128\xi^4 - 5/128\xi^3 - 5/128\xi^2 + 3/128\xi + 3/128$

$\omega_7 = 1/64\xi^6 + 1/64\xi^4 - 5/64\xi^2 + 3/64$

$\omega_6 = 1/32\xi^5 + 1/32\xi^4 + 1/16\xi^3 + 1/16\xi^2 - 3/32\xi - 3/32$

$\omega_5 = 1/16\xi^4 + 1/8\xi^2 - 3/16$

$\omega_4 = 1/8\xi^3 + 1/8\xi^2 - 1/8\xi - 1/8$

$\omega_3 = 1/4\xi^2 - 1/4$

$\omega_2 = 1/2\xi + 1/2$

$\omega_1 = 1.$

$p = 3$
factors mod 3

$t^6 + t^5 + 2t^3 + t^2 + 2t + 1$
$t^3 + t^2 + t + 2$
$t,$

factorization mod 3^{24}

$t^6 + 120783431803t^5 + 8973604878t^4 + 61904146670t^3 + 31825819645t^2 + 183824600885t + 30301271638$

$t^3 + 156274299151t^2 + 97643734609t + 173505680846$
$t^2 + 5371805628t + 22263657813,$

3-minimal basis

$\omega_{11} = 1/729\xi^{10} + 101/729\xi^9 - 223/729\xi^8 - 358/729\xi^7 - 34/729\xi^6 - 34/729\xi^5 - 34/729\xi^4 - 34/729\xi^3 - 34/729\xi^2 - 34/729\xi$

$\omega_{10} = \xi^9$

$\omega_9 = \xi^8$

$\omega_8 = \xi^7$

$\omega_7 = \xi^6$

$\omega_6 = \xi^5$

$\omega_5 = \xi^4$

$\omega_4 = \xi^3$

$\omega_3 = \xi^2$

$\omega_2 = \xi$

$\omega_1 = 1$

$p = 5$

factors mod 5

$t + 4$
$t + 1$
t ,

factorization mod 5^8

$t^4 + 187926t^3 + 272826t^2 + 260801t + 308101$
$t^5 + 336550t^4 + 176650t^3 + 80725t^2 + 165425t + 246226$
$t^2 + 256875t + 161875,$

idempotents

$-52091/5\xi^4 - 884924/5\xi^3 - 871611/5\xi^2 + 16121/5\xi + 278999/5$

$52091/5\xi^4 + 884924/5\xi^3 + 871611/5\xi^2 - 16121/5\xi - 278994/5,$

factorization

$t^4 + 309039t^3 + 157846t^2 + 157544t + 347441$
$t + 27511,$

idempotents

$-4543197/25\xi - 49587$
$4543197/25\xi + 49588,$

factorization

$t + 725$
$t + 256150,$

5-minimal basis

$\omega_{11} = 1/25\xi^{10} + 1/25\xi^9 + 1/25\xi^8 + 1/25\xi^7 + 1/25\xi^6 + 1/25\xi^5 + 1/25\xi^4 + 1/25\xi^3 + 1/25\xi^2 + 1/25\xi$

$\omega_{10} = 1/5\xi^9 + 1/5\xi^7 + 1/5\xi^5 + 1/5\xi^3 + 1/5\xi$

$\omega_9 = \xi^8$

$\omega_8 = \xi^7$

$\omega_7 = \xi^6$

$\omega_6 = \xi^5$

$\omega_5 = \xi^4$

$\omega_4 = \xi^3$

$\omega_3 = \xi^2$

$\omega_2 = \xi$

$\omega_1 = 1.$

p = 29

factors mod 29

$$t^4 + 27t^3 + 9t^2 + 10t + 24$$
$$t + 28$$
$$t + 12,$$

factorization mod 29^8

$$t^4 + 506054657381t^3 + 52694279576t^2 + 495928586992t + 216188927047$$
$$t^3 + 179037728472t^2 + 480036204243t + 194325614858$$
$$t^4 + 270603218852t^3 + 311400732346t^2 + 383680458976t + 294504268343,$$

idempotents

$$170950383710294/841\xi^2 - 124930383118041/841\xi + 116845930821609/841$$
$$-170950383710294/841\xi^2 + 124930383118041/841\xi - 116845930820768/841,$$

factorization

$$t^2 + 235370557884t + 112494502447$$
$$t + 443913583549,$$

idempotents

$$3441901651958/29\xi - 5081864234391/29, -3441901651958/29\xi + 5081864234420/29,$$

factorization

$$t + 300710030106$$
$$t + 434906940739$$

29-minimal basis

$$\omega_{11} = 1/24389\xi^{10} + 2/24389\xi^9 + 9/24389\xi^8 + 279/24389\xi^7 + 325/24389\xi^6 - 11199/24389\xi^5 - 11647/24389\xi^4 + 9277/24389\xi^3 + 1911/24389\xi^2 + 1329/24389\xi + 9713/24389$$

$$\omega_{10} = 1/841\xi^9 + 10/841\xi^7 - 11/841\xi^6 + 90/841\xi^5 - 1/29\xi^4 + 379/841\xi^3 - 158/841\xi^2 - 153/841\xi - 129/841$$

$$\omega_9 = 1/841\xi^8 - 9/841\xi^7 + 4/841\xi^6 + 330/841\xi^5 - 357/841\xi^4 - 383/841\xi^3 + 259/841\xi^2 + 5/841\xi + 150/841$$

$$\omega_8 = 1/29\xi^7 + 10/29\xi^5 + 10/29\xi^4 - 11/29\xi^3 - 3/29\xi^2 + 11/29\xi + 11/29$$

$$\omega_7 = 1/29\xi^6 + 9/29\xi^5 + 4/29\xi^4 - 12/29\xi^3 - 3/29\xi^2 - 1/29\xi + 2/29$$

$\omega_6 = \xi^5$

$\omega_5 = \xi^4$

$\omega_4 = \xi^3$

$\omega_3 = \xi^2$

$\omega_2 = \xi$

$\omega_1 = 1$

Integral basis ($\xi = t/f$)

$$\omega_{11} = 1/910314547200\xi^{10} - 85607/455157273600\xi^9 + 1352801/910314547200\xi^8 - 921683/113789318400\xi^7 - 84877487/455157273600\xi^6 + 507753959/227578636800\xi^5 + 7760693413/455157273600\xi^4 - 741690983/113789318400\xi^3 - 19749911299/910314547200\xi^2 + 40692408193/455157273600\xi - 4064571/49948672$$

$\omega_{10} = 1/2152960\xi^9 + 1/430592\xi^8 - 1/33640\xi^7 + 7/26912\xi^6 - 223/37120\xi^5 + 2533/215296\xi^4 - 8863/269120\xi^3 + 3963/53824\xi^2 + 115901/2152960\xi - 43283/430592$

$\omega_9 = 1/215296\xi^8 + 5/53824\xi^7 + 1/53824\xi^6 + 155/53824\xi^5 + 387/107648\xi^4 - 1437/53824\xi^3 + 5089/53824\xi^2 + 10173/53824\xi - 56719/215296$

$\omega_8 = 1/3712\xi^7 + 1/3712\xi^6 - 39/3712\xi^5 - 15/3712\xi^4 - 197/3712\xi^3 + 139/3712\xi^2 + 619/3712\xi - 509/3712$

$\omega_7 = 1/1856\xi^6 - 5/464\xi^5 + 33/1856\xi^4 + 13/232\xi^3 + 171/1856\xi^2 - 109/464\xi + 147/1856$

$\omega_6 = 1/32\xi^5 + 1/32\xi^4 + 1/16\xi^3 + 1/16\xi^2 - 3/32\xi - 3/32$

$\omega_5 = 1/16\xi^4 + 1/8\xi^2 - 3/16$

$\omega_4 = 1/8\xi^3 + 1/8\xi^2 - 1/8\xi - 1/8$

$\omega_3 = 1/4\xi^2 - 1/4$

$\omega_2 = 1/2\xi + 1/2$

$\omega_1 = 1$

The index of the equation order in its maximal order is

$$2^{56} \times 3^6 \times 5^3 \times 29^9.$$

2. For $f(t) = t^{55} - 3080t + 3024$ we obtain the reduced discriminant
$d_r(f) = 9147600$
$= 2^4 \times 3^3 \times 5^2 \times 7 \times 11^2$ and the discriminant
$d(f) = 2^{216} \times 3^{162} \times 5^{56} \times 7^{54} \times 11^{56}$.

The algorithm produces:

$p = 2$

idempotents

$11/4\xi^{54} + 111/2\xi^{53} + 3\xi^{52} + 14\xi^{51} - 20\xi^{50} - 8\xi^{49} + 48\xi^{48} - 32\xi^{47} - 64\xi^{46} + 128\xi^{45} - 235/4\xi^{36} - 69\xi^{35} + 73\xi^{34} - 48\xi^{33} + 8\xi^{32} - 96\xi^{31} + 32\xi^{30} + 64\xi^{28} + 53/2\xi^{18} - 101\xi^{17} + 84\xi^{16} - 8\xi^{15} - 88\xi^{14} + 48\xi^{13}$

$- 11/4\xi^{54} - 111/2\xi^{53} - 3\xi^{52} - 14\xi^{51} + 20\xi^{50} + 8\xi^{49} - 48\xi^{48} + 32\xi^{47} + 64\xi^{46} + 128\xi^{45} + 235/4\xi^{36} + 69\xi^{35} - 73\xi^{34} + 48\xi^{33} - 8\xi^{32} + 96\xi^{31} - 32\xi^{30} - 64\xi^{28} - 53/2\xi^{18} + 101\xi^{17} - 84\xi^{16} + 8\xi^{15} + 88\xi^{14} - 48\xi^{13} + 1$

factorization mod 2^8

$$t^{36} + 60t^{35} + 148t^{34} + 64t^{33} + 160t^{32} + 128t^{31} + 128t^{30} + 66t^{18} + 188t^{17} + 16t^{16} + 96t^{15} + 32t^{14} + 192t^{13} + 4$$

$$t^{19} + 196t^{18} + 124t^{17} + 96t^{16} + 48t^{15} + 64t^{14} + 64t^{13} + 190t + 52,$$

idempotents

$$97/2\xi^{18} - 17\xi^{17} + 4\xi^{16} - 104\xi^{15} - 120\xi^{14} - 16\xi^{13} + 128\xi^{12}$$

$$-97/2\xi^{18} + 17\xi^{17} - 4\xi^{16} + 104\xi^{15} + 120\xi^{14} + 16\xi^{13} + 128\xi^{12} + 1,$$

factorization

$$t^{18} + 158t^{17} + 8t^{16} + 48t^{15} + 16t^{14} + 224t^{13} + 190$$
$$t + 38$$

Integral basis

$$\omega_{55} = 1/8\xi^{54}$$
$$\omega_{54} = 1/4\xi^{53}$$
$$\omega_{53} = 1/4\xi^{52}$$
$$\vdots$$
$$\omega_{37} = 1/4\xi^{36}$$
$$\omega_{36} = 1/2\xi^{35}$$
$$\omega_{35} = 1/2\xi^{34}$$
$$\vdots$$
$$\omega_{19} = 1/2\xi^{18}$$

$$\omega_{18} = \zeta^{17}$$
$$\omega_{17} = \zeta^{16}$$
$$\vdots$$
$$\omega_2 = \zeta$$
$$\omega_1 = 1$$

The index of the equation order in its maximal order is 2^{57}

Both examples were computed by R. Böffgen of Saarbrücken, who implemented an earlier version of the embedding algorithm on a Siemens 7560 computer. The CUP times were 73 and 1192 seconds. The first polynomial has Galois group M_{11} the second $\mathfrak{A}_{55}$. The polynomials were found by B.H. Matzat of Karlsruhe.

(References: R. Böffgen, Der Algorithmus von Ford/Zassenhaus zur Berechnung von Ganzheitsbasen in Polynomalgebren, *Ann. Univ. Saraviensis, Ser. Math.*, **1**, 3 (1987), 60–129; B. H. Matzat, Konstruktion von Zahl-und Funktionenkörpern mit vorgegebener Galoisgruppe, *J. Reine Angew. Math.* **399** (1984), 179–220).

Algorithms

REFERENCES

Chapter 1

1 D.E. Knuth, *The Art of Computer Programming*, vol. 2, 2nd edn, Addison-Wesley, 1981.

2 S. Lang, *Algebra*, Addison-Wesley, 1971.

Chapter 2

1 J.R. Bastida, *Field extensions and Galois theory, Encycl. of Math. 22*, Addison-Wesley, 1984.

2 D. Cantor & H. Zassenhaus, A new algorithm for factoring polynomials over finite fields, *Math. Comp.*, **36** (1981), 587-92.

3 N. Tschebotareff, Die Bestimmung der Dichtigkeit einer Menge von Primzahlen, welche zu einer gegebenen Substitutionsklasse gehören, *Math. Ann.*, **95** (1926), 191-228.

4 E. Galois, *Oeuvres*, V–X and 1–61, Paris, 1897.

5 C.F. Gauss, *Werke, vol.* 1 (*Disquisitiones Arithmeticae*), *vol.* 2 (*Sectio Octava*), Göttingen, 1876.

6 D. Hilbert, *Gesammelte Abhandlungen, Band 2*, 393–400, Springer Verlag, Berlin, 1933.

7 F. Klein, *Vorlesungen über das Ikosaeder*, Teubner Verlag, Leipzig, 1884.

8 D.E. Knuth, *The Art of Computer Programming*, vol. 2, sec. edn, Addison-Wesley 1981.

9 R. Land, Computation of Pólya polynomials of primitive permutation groups, *Math. Comp.*, **36** (1981), 267–78.

10 S. Lang, *Algebra*, Addison-Wesley, 1971.

11 R. Lidl & H. Niederreiter, *Finite Fields, Encycl. of Math. 20*, Addison-Wesley, 1983.

12 H. Niederreiter, Quasi-Monte Carlo methods and pseudo-random numbers, *Bull. of the AMS*, **84** (1978), 957–1041.

13 Št. Schwarz, Contribution à la réductibilité des polynômes dans la théorie des congruences, *Česka Společniet Nauk Prague, Trida Mathematička Prirodovedecka Vestnik* (1939), 1–7.

14 Št. Schwarz, Sur le nombre des racines et des facteurs irréductibles d'une congruence donnée, *Časopis Pro Pestovani Matematiky A Fysiky, Prague V*, **69** (1940), 128–45.

15 Št. Schwarz, On the reducibility of polynomials over a finite field, *Quart. J. Math. Oxford Ser.* (2), **7** (1956), 110–24.

16 W. Trinks, Ein Beispiel eines Zahlkörpers mit der Galoisgruppe PSL(3, F_2) über **Q**, Manuscript, Univ. Karlsruhe, West-Germany, 1968.

17 H. Wielandt, *Finite Permutation Groups*, Academic Press, New York, London, 1964.

18 H. Zassenhaus, *The Theory of Groups*, sec. edn, Chelsea, 1958.

Chapter 3

1 A. Bachem & R. Kannan, *Lattices and the Basis Reduction Algorithm*, Report 84, 6, Mathematisches Institut, Universität zu Köln, 1984.

2 H.F. Blichfeldt, A new principle in the geometry of numbers, with some applications, *Transactions Amer. Math. Soc.*, **15** (1914), 227–35.

3 J.W.S. Cassels, *An Introduction to the Geometry of Numbers*, 2nd edn, Springer Verlag, Berlin, Heidelberg, New York, 1971.

4 U. Fincke & M. Pohst, Improved methods for calculating vectors of short length in a lattice, including a complexity analysis, *Math. Comp.*, **44** (1985), 463–71.

5 G. Havas & L. Sterling, Integer matrices and abelian groups; p. 431–56 in *Symbolic and Algebraic Computation, Lecture Notes in Computer Science, 72*, Springer Verlag, Berlin, Heidelberg, New York, 1979.

6 R. Kannan & A. Bachem, Polynomial algorithms for computing the Smith and Hermite normal forms of an integer matrix, *Siam J. Comput.*, **8** (1979), 499–507.

7 D.E. Knuth, *The Art of Computer Programming, vol. 2*, sec. edn, Addison–Wesley, 1981.

8 S. Lang, *Algebra*, Addison–Wesley, 1971.

9 C.G. Lekkerkerker, *Geometry of Numbers*, Wolters–Noordhoff Publishing, Groningen, and North-Holland Publishing Company, Amsterdam, 1969.

10 A.K. Lenstra, H.W. Lenstra, Jr., & L. Lovász, Factoring polynomials with rational coefficients, *Math. Ann.*, **261** (1982), 515–34.

11 M. Mignotte, An inequality about factors of polynomials, *Math. Comp.*, **28** (1974), 1153–7.

12 D.R. Musser, Multivariate polynomial factorization, *JACM*, **22**, no. 2 (1975), 291–308.

13 M. Pohst, On computing isomorphisms of equation orders, *Math. Comp.* **48** (1987), 309–14.

14 J. Renus, *Gitterreduktionsverfahren mit Anwendungen auf lineare diophantische Gleichungssysteme*, Diplomarbeit, Universität Düsseldorf, 1986.

Chapter 4

1 G.E. Collins, The calculation of multivariate polynomial resultants, *JACM*, **18** (1971), 515–32.

2 E. Noether, Abstrakter Aufbau der Idealtheorie in algebraischen Zahl - und Funktionenkörpern, *Math. Ann.*, **96** (1927), 26–61.

3 M. Pohst, On the computation of number fields of small discriminants including the minimum discriminants of sixth degree fields, *J. Number Theory*, **14** (1982), 99–117.

4 M. Pohst, P. Weiler & H. Zassenhaus, On effective computation of fundamental units II, *Math. Comp.*, **38** (1982), 293–329.

5 H. Zassenhaus, On an embedding algorithm of an equation order into its maximal order for algebraic function fields. To appear in *Monatshefte für Mathematik*.

Chapter 5

1 U. Fincke, Ein Ellipsoidverfahren zur Lösung von Normgleichungen in algebraischen Zahlkörpern, Thesis, Düsseldorf 1984.

2 K. Mahler, Inequalities for ideal bases in algebraic number fields, *J. Austral. Math. Soc.*, **4** (1964), 425–47.

3 M. Pohst, Regulatorabschätzungen für total reelle algebraische Zahlkörper, *J. Number Theory*, **9** (1977), 459–92.

4 M. Pohst, Eine Regulatorabschätzung, *Abh. Math. Sem. Univ. Hamburg*, **47** (1978), 221–31.

5 M. Pohst & H. Zassenhaus, On effective computation of fundamental units I, *Math. Comp.*, **38** (1982), 275–91.

6 M. Pohst, P. Weiler & H. Zassenhaus, On effective computation of fundamental units II, *Math. Comp.*, **38** (1982), 293–329.

7. R. Remak, Über die Abschätzung des absoluten Betrages des Regulators eines algebraischen Zahlkörpers nach unten, *J. Reine Angew. Math.*, **167** (1932), 360–78.

8 C.L. Siegel, The trace of totally positive and real algebraic integers, *Annals of Math.*, **46** (1945), 302–12.

9 C.L. Siegel, Abschätzung von Einheiten, *Nachr. Akad. Wiss. Göttingen Math. Phys. Kl.* (1969), 71–86.

10 B.M. Trager, Algebraic Factoring and Rational Function Integration, *Proc. of the 1976* ACM *Symposium on Symbolic and Algebraic Computation* ('*SYMSAC 76*') pp. 219–26.

11 N. Tschebotareff, Die Bestimmung der Dichtigkeit einer Menge von Primzahlen, welche zu einer gegebenen Substitutionsklasse gehören, *Math. Ann.*, **95** (1926), 191–228.

12 H. Zassenhaus, On Hensel Factorization I, *J. Number Theory*, **1** (1969), 291–311.

13 H. Zassenhaus, On the units of orders, *J. Algebra*, **20** (1972), 368–95.

14 R. Zimmert, Ideale kleiner Norm in Idealklassen und eine Regulatorabschätzung, *Invent. Math.*, **62** (1981), 367–80.

Additional references

15 J. Buchmann, On the computation of units and class numbers by a generalization of Lagrange's algorithm, *J. Number Theory*, **26** (1987), 8–30.

16 J. Buchmann, Generalized continued fractions and number theoretic computations, *Bericht Nr. 269 der math.-stat. Sektion in der Forschungsgesellschaft Joanneum*, Graz, 1986.

17 B.N. Delone & D.K. Fadev, The theory of irrationalities of the third degree, *Amer. Math. Soc. Transl. of Math. Monographs* **10**, 1964.

18 V. Ennola & R. Turunen, On totally real cubic fields, *Math. Comp.*, **44** (1985), 495–519.

19 E.L. Ince, *Cycles of reduced ideals in quadratic fields*, reissued by Cambridge University Press, London 1968.

20 H.W. Lenstra Jr., On the calculation of class numbers and regulators of quadratic fields, *Lond. Math. Soc. Lect. Note Ser.* **56** (1982), 123–50.

21 D. Shanks, The infrastructure of real quadratic fields and its applications, *Proc. 1972 Numb. Th. Conf.*, Boulder (1972), 217–24.

22 R.P. Steiner, On the units in algebraic number fields, *Proc. 6th Manitoba Conf., Num. Math.* (1976), 415–35.

23 H.C. Williams, Continued fractions and number theoretic computations, *Rocky Mountain J. Math.*, **15** (1985), 621–55.

24 H.G. Zimmer, Computational problems, methods and results in algebraic number theory, *Springer Lect. Notes in Math.* **262** (1972).

Chapter 6

1 U. Fincke, Ein Ellipsoidverfahren zur Lösung von Normgleichungen in algebraischen Zahlkörpern, Thesis, Düsseldorf 1984.

2 U. Fincke & M. Pohst, A procedure for determining algebraic integers of given norm, *Proc. Eurosam 83, Springer Lecture Notes in Computer Science* **162** (1983), 194–202.

3 K. Mahler, Inequalities for ideal bases in algebraic number fields, *J. Austral. Math. Soc.*, **4** (1964), 425–47.

4 D.A. Marcus, *Number Fields*, Universitext, Springer Verlag, New York, Heidelberg, Berlin, 1977.

5 M. Pohst & H. Zassenhaus, Über die Berechnung von Klassenzahlen und Klassengruppen algebraischer Zahlkörper, *J. Reine Angew. Math.*, **361** (1985), 50–72.

6 C.L. Siegel, Über die Klassenzahl quadratischer Zahlkörper, *Acta Arithmetica*, **1** (1935), 83–6.

Chapter 7

1 L.M. Adleman & M.-D. Huang (Eds.), *Algorithmic Number Theory Symposium*, Proceedings ANTS I, Cornell University, 1994.

2 H. Cohen (Ed.), *Algorithmic Number Theory, Proc. Sec. Int. Symp., ANTS II, Bordeaux 1996*, Lect. Notes in Comp. Sc. 1122, Springer Verlag, 1996.

3 E. Artin, *Questions de base minimale dans la théorie des nombres algébriques*, The collected papers of Emil Artin, Addison–Wesley 1965, pp. 229–31.

4 E. Bach, Explicit bounds for primality testing and related problems, *Math. Comp.*, **55** (1990), 355–80.

5 Y. Bilu & G. Hanrot, *Solving Thue Equations of High Degree*, Mathematica Gottingensis, Schriftenreihe des Sonderforschungbereichs Geometrie und Analysis; Heft 11, Göttingen, 1995.

6 R. Böffgen, *Der Algorithmus von Ford/-Zassenhaus zur Berechnung von Ganzheitsbasen in Polynomalgebren*, Ann. Univ. Saraviensis, Ser. Math., Vol. 1, No. 3, 1987.

7 W. Bosma & M. Pohst, Computations with finitely generated modules over Dedekind rings. *Proc. ISSAC'91* (1991), 151–6.

8 J. Buchmann & S. Düllmann, A probabilistic class group and regulator algorithm and its implementation, in A. Pethö, M.E. Pohst, H.C. Williams, H.G. Zimmer (eds.), *Computational Number Theory*, Walter de Gruyter (1991), 53–72.

9 J. Buchmann & A. Pethö, On the computation of independent units in number fields by Dirichlet's method, *Math. Comp.*, **52** (1989), 149–59.

10 J. Buchmann & H.C. Williams, On the computation of the class number of an algebraic number field, *Math. Comp.*, **53** (1989), 679–88.

11 N. Čebotarev, Die Bestimmung der Dichtigkeit einer Menge von Primzahlen, welche zu einer gegebenen Substitutionsklasse gehören, *Math. Ann.*, **95** (1926), 191–228.

12 A.L. Chistov, The complexity of constructing the ring of integers of a global field, *Soviet Math. Dokl.*, **39** (1989), 597–600.
13 H. Cohen, *A Course in Computational Algebraic Number Theory*, Graduate Texts in Mathematics 138, Springer Verlag, 1993.
14 H. Cohen, Hermite and Smith normal form algorithms over Dedekind domains, to appear in *Math. Comp.*
15 M. Daberkow, Über die Bestimmung der ganzen Elemente in Radikalerweiterungen algebraischer Zahlkörper, Thesis, Berlin, 1995.
16 M. Daberkow & M. Pohst, On integral bases in relative quadratic extensions, *Math. Comp.*, **65** (1996), 319–29.
17 M. Daberkow & M. Pohst, Computations with relative extensions of number fields with an application to the construction of Hilbert class fields, *Proc. ISSAC'95*, ACM Press, 1995, pp. 68–76.
18 M. Daberkow & M. Pohst, *On Computing Hilbert Class Fields of Prime Degree*, pp. 67–74 in *Algorithmic Number Theory, Proc. Sec. Int. Symp., ANTS II, Bordeaux 1996*, H. Cohen (ed.), Lect. Notes in Comp. Sc. 1122, Springer Verlag, 1996.
19 M. Daberkow & M. Pohst, On the computation of Hilbert class fields, submitted for publication.
20 J. Dixon, Computing subfields in algebraic number fields, *J. Austral. Math. Soc.* (Series A), **49** (1990), 434–48.
21 Y. Eichenlaub & M. Oliver, Computation of Galois groups for polynomials with degree up to eleven, to appear.
22 C. Fieker, A. Jurk & M. Pohst, On solving relative norm equations in algebraic number fields; to appear in *Math. Comp.*
23 U. Fincke & M. Pohst, *A Procedure for Determining Algebraic Integers of Given Norm*, *Proceedings EUROCAL 83*, Springer LNCS Series No. 162, Springer Verlag, 1983.
24 D.J. Ford, The construction of maximal orders over a Dedekind domain, *J. Symb. Comp.*, **4** (1987), 69–75.
25 I. Gaál, A. Pethö & M. Pohst, Simultaneous representation of integers by a pair of ternary quadratic forms – with an application to index form equations in quartic number fields, *J. Number Theory*, **57** (1996), 90–104.
26 J.L. Hafner & K.S. McCurley, A rigorous subexponential algorithm for the computation of class groups, *J. Amer. Math. Soc.*, **4** (1989), 837–850.
27 H. Hasse, Uber den Klassenkörper zum quadratischen Zahlkörper mit der Diskriminante-47, *Acta Arithmetica*, **9** (1964), 419–34.
28 E. Hecke, *Lectures on the Theory of Algebraic Numbers*, Springer Verlag, 1981.
29 F. Heß, Zur Klassengruppenberechnung in algebraischen Zahlkörpern, Diploma Thesis, Technical University Berlin, 1996.
30 M. Daberkow, C. Fieker, J. Klüners, M. Pohst, K. Roegner, M. Schörnig & K. Wildanger, KANT V4, to appear in *J. Symb. Comp.*
31 J. Klüners, Über die Berechnung von Teilkörpern algebraischer Zahlkörper, Diploma Thesis, TU–Berlin, 1995.
32 J. Klüners & M. Pohst, On computing subfields, to appear in *J. Symb. Comp.*
33 S. Landau, Factoring polynomials over algebraic number fields, *SIAM J. of Comp.*, **14** (1985), 184–95.
34 S. Landau & G.L. Miller, Solvability by radicals is in polynomial time, *J. Comp. System Sci.*, **30** (1985), 179–208.
35 S. Lang, *Algebraic number theory*, Grad. Texts in Math. 110, Springer Verlag, 1986.
36 H.W. Lenstra Jr., Algorithms in algebraic number theory, *Bull. Amer. Math. Soc.*, **26** (1992), 211–44.
37 A.K. Lenstra & H.W. Lenstra Jr., *The development of the number field sieve*, Lecture Notes in Mathematics 1554, Springer Verlag, 1993.
38 A.K. Lenstra, H.W. Lenstra Jr. & L. Lovasz, Factoring polynomials with rational coefficients, *Math. Ann.*, **261** (1982), 515–34.
39 J. Cannon, Magma, to appear in *JSC*.
40 O.T. O'Meara, *Introduction to Quadratic Forms*, Springer Verlag, 1963.
41 D. Bernardi, C. Batut, H. Cohen & M. Olivier, *User's guide to PARI-GP*, version 1.39, Publ. Université Bordeaux I, 1991.
42 M. Pohst, *Computational Algebraic Number Theory*, Birkhäuser Verlag, 1993.
43 M. Pohst, On computing fundamental units, *J. Number Theory*, **47** (1994), 93–105.
44 M. Pohst & H. Zassenhaus, On unit computation in real quadratic fields, in *EU-ROSAM, '79, Marseille, 1979*,

pp. 95–107, Lecture Notes in Computer Science 72, Springer Verlag, 1979.

45 J. von Schmettow, Beiträge zur Klassengruppenberechnung, Dissertation, Düsseldorf, 1991.

46 M. Schönert, *et al.*: *GAP – Groups, Algorithms and Programming, version 3 release 3*; RWTH Aachen, Lehrstuhl D für Mathematik; Templergraben 64, 52062 Aachen, F.R.G.; 1993.

47 C. Hollinger & P. Serg, SIMATH – a computer algebra system, in *Computational Number Theory*, eds. A. Pethö, M.E. Pohst, H.C. Williams & H.G. Zimmer, de Gruyter Verlag, 1991, pp. 331–42.

48 J. Sommer, *Einführung in Zahlentheorie*, Teubner Verlag, Leipzig, 1907.

49 R.P. Stauduhar, The determination of Galois groups, *Math. Comp.*, **27** (1973), 981–96.

50 Ch. Thiel, On the complexity of some problems in algorithmic algebraic number theory, Dissertation, Saarbrücken, 1995.

51 A. Weber & M. Daberkow, *A database for number fields*, in J. Calmet & C. Limongelli (Eds.), *Design and Implementation of Symbolic Computation Systems*, pp. 320–30, Springer LNCS 1128, 1996.

52 H. Zassenhaus, The group of an equation, *Nachr. Akad. Wiss. Göttingen, Math.-Phys. Kl. II* (1967), 147–66.

53 H. Zassenhaus, Ein Algorithmus zur Berechnung einer Minimalbasis über gegebener Ordnung, in *Funktionalanalysis*, Birkhäuser, 1967, pp. 90–103.

54 H. Zassenhaus, On the group of an equation, in *Computers in Algebra and Number Theory, Proceedings of a Symposium in Applied Mathematics, 1970*, pp. 69–88, 1971.

55 H. Zassenhause, On the van der Waerden criterion for the group of an equation, in *EU-ROSAM, '79, Marseille, 1979*, pp. 95–107, Lecture Notes in Computer Science, Vol. 72, Springer Verlag, 1979.

56 H. Zassenhaus, *R. Land's Verfeinerung des D. Ford'schen Ordmax-Algorithmus*, Preprint, Saarbrücken, 1984.

Additional references

57 K. Geißler, Zur Berechnung von Galoisgruppen, Diplomarbeit, TU Berlin, 1997.

58 C. Friedrichs, Berechnung relativer Ganzheitsbasen mit dem Round-2-Algorithmus, Diplomarbeit, TU Berlin, 1997.

59 C. Ficker, Über relative Normgleichungen in algebraischen Zahlkörpern, Thesis, TU Berlin, 1997.

60 K. Wildanger, Über das hösen von Einheiten- und Indexform-gleichungen in algebraischen Zahlkörpern mit einer Anwendung auf die Bestimmung aller ganzen Punkte einer Mordellschen Kurve, Thesis, TU Berlin, 1997.

61 J. Klüners, Über die Berechnung von Automorphismen und Teilkörpern algebraischer Zahlkörper, Thesis, TU Berlin 1997.

INDEX

Printed in the United States
By Bookmasters